Photonic Band Gap Materials

NATO ASI Series

Advanced Science Institutes Series

A Series presenting the results of activities sponsored by the NATO Science Committee, which aims at the dissemination of advanced scientific and technological knowledge, with a view to strengthening links between scientific communities.

The Series is published by an international board of publishers in conjunction with the NATO Scientific Affairs Division

A Life Sciences	Plenum Publishing Corporation
B Physics	London and New York
C Mathematical and Physical Sciences	Kluwer Academic Publishers
D Behavioural and Social Sciences	Dordrecht, Boston and London
E Applied Sciences	
F Computer and Systems Sciences	Springer-Verlag
G Ecological Sciences	Berlin, Heidelberg, New York, London,
H Cell Biology	Paris and Tokyo
I Global Environmental Change	

PARTNERSHIP SUB-SERIES

1. Disarmament Technologies	Kluwer Academic Publishers
2. Environment	Springer-Verlag / Kluwer Academic Publishers
3. High Technology	Kluwer Academic Publishers
4. Science and Technology Policy	Kluwer Academic Publishers
5. Computer Networking	Kluwer Academic Publishers

The Partnership Sub-Series incorporates activities undertaken in collaboration with NATO's Cooperation Partners, the countries of the CIS and Central and Eastern Europe, in Priority Areas of concern to those countries.

NATO-PCO-DATA BASE

The electronic index to the NATO ASI Series provides full bibliographical references (with keywords and/or abstracts) to more than 50000 contributions from international scientists published in all sections of the NATO ASI Series.
Access to the NATO-PCO-DATA BASE is possible in two ways:

– via online FILE 128 (NATO-PCO-DATA BASE) hosted by ESRIN,
Via Galileo Galilei, I-00044 Frascati, Italy.

– via CD-ROM "NATO-PCO-DATA BASE" with user-friendly retrieval software in English, French and German (© WTV GmbH and DATAWARE Technologies Inc. 1989).

The CD-ROM can be ordered through any member of the Board of Publishers or through NATO-PCO, Overijse, Belgium.

Series E: Applied Sciences - Vol. 315

Photonic Band Gap Materials

edited by

Costas M. Soukoulis

Ames Laboratory and
Department of Physics and Astronomy,
Iowa State University,
Ames, Iowa, U.S.A.

Kluwer Academic Publishers

Dordrecht / Boston / London

Published in cooperation with NATO Scientific Affairs Division

Proceedings of the NATO Advanced Study Institute on
Photonic Band Gap Materials
Elounda, Crete, Greece
June 18–30, 1995

A C.I.P. Catalogue record for this book is available from the Library of Congress

ISBN 0-7923-3991-6

Published by Kluwer Academic Publishers,
P.O. Box 17, 3300 AA Dordrecht, The Netherlands.

Kluwer Academic Publishers incorporates the publishing programmes of
D. Reidel, Martinus Nijhoff, Dr W. Junk and MTP Press.

Sold and distributed in the U.S.A. and Canada
by Kluwer Academic Publishers,
101 Philip Drive, Norwell, MA 02061, U.S.A.

In all other countries, sold and distributed
by Kluwer Academic Publishers Group,
P.O. Box 322, 3300 AH Dordrecht, The Netherlands.

Printed on acid-free paper

TABLE OF CONTENTS

PHOTONIC BAND GAPS: 1D AND 2D STRUCTURES

PHOTONIC BAND GAPS AND LOCALIZATION

viii

PREFACE

This volume contains the papers presented at the NATO Advanced Study Institute (ASI) *Photonic Band Gap Materials* held at the Porto Elounda Mare Hotel in Elounda, Crete, June 18-30, 1995.

Photonic band gap crystals offer unique ways to tailor light and the propagation of electromagnetic waves (EM). In analogy to electrons in a crystal, EM waves propagating in a structure with a periodically-modulated dielectric constant are organized into photonic bands which are separated by gaps where propagating states are forbidden. There have been proposals for novel applications of these photonic band gap (PBG) crystals, with operating frequencies ranging from microwave to the optical regime, that include zero-threshold lasers, low-loss resonators and cavities, and efficient microwave antennas. Spontaneous emission is suppressed for photons in the photonic band gap, offering novel approaches to manipulating the EM field and creating high-efficiency light-emitting structures. Innovative ways to manipulate light can have profound influence on science and technology.

The objectives of this NATO-ASI were (i) to assess the state-of-the-art in experimental and theoretical studies of structures exhibiting photonic band gaps, (ii) discuss how such structures can be fabricated to improve technologies in different areas of physics and engineering, and (iii) to identify problems and set goals for further research. This was accomplished by the excellent job done by the lecturers and invited talks, which have paid special attention to the tutorial aspects of their contributions. The location of the NATO-ASI was a perfect and idyllic setting which allowed the participants to develop scientific interactions and friendships.

All of these objectives were met in the ASI and three areas within the field of photonic band gap structures were identified as the most promising and hope to receive considerable attention within the next few years. The first area of effort is in materials fabrication. This involves the creation of high quality, low loss, periodic dielectric structures either in 3D or 2D, especially on the optical scale. A variety of photonic crystals have been made by micromachining silicon wafers which have been oriented along (110), some with lattice constants as small as 500 micrometers. These are the smallest photonic crystals yet fabricated and are ready to be used in microwave applications. However, optical photonic crystals may be best fabricated from self-organizing materials and techniques, than with conventional techniques used in the semiconductor industry. The second area of consideration is in applications, and spin offs which may have technological and economic importance. Several contributors presented possible applications of PBGs in microwave and optical regimes. Microwave mirrors, directional antennas, resonators (especially for the 2 GHz region), filters, waveguides, Y splitters, and resonant microcavities were discussed. It was also pointed out that 2D photonic crystals might yield much richer structures and designs for optical applications. Finally, the third area is the studies of fundamentally new physical phenomena in condensed matter physics and quantum optics associated with localization and photonic band gaps.

This book, which compiles the lectures presented at the *Photonic Band Gap Materials* NATO ASI meeting, presents an excellent review of the recent developments of this rapidly expanding field. The lectures cover theoretical, experimental, and the application aspects of the PBGs. The collection of papers is roughly balanced between theory and experiment. It contains chapters that present latest research results appropriate of an advanced research workshop, as well as ones that review a particular field, with the goal of providing the reader with a sufficient overview and extensive references for a more detailed study. The book is divided into five chapters representing the various topics discussed at the ASI. Chapter I gives a historical overview of the PBG field and provides a review of the new experimental structures. Chapter II provides a detailed review of the transfer matrix techniques, applied to either dielectric or metallic structures. Experiments on metallic structures are presented in this chapter. Chapter III provides an up-to-date review of the potential applications of the PBGs, with specific examples in both the millimeter and optical regime. Chapter IV covers the experimental and theoretical developments of 1D and 2D PBG structures. Finally, Chapter V provides an excellent review of the localization aspects of the PBG field. Fundamentally new physical phenomena in condensed matter physic and quantum optics associated with localization and PBGs are discussed in this last chapter. We hope this book will not only prove interesting and stimulating to researchers active in the PBG field, but also serve as a useful reference to non-specialists, because of the introductory lectures.

The advanced study institute was made possible through the generous support of the NATO Scientific Affairs Division, Brussels, Belgium and the Ames Laboratory, operated by the U.S. Department of Energy by Iowa State University under Contract No. W-7405-Eng-82. I would like to thank the organizing committee, E. N. Economou, S. John, and J. Pendry for their valuable help on the organization of the program and the workshop. I would like to express my appreciation to Rebecca Shivvers and Susan Elsner, who prepared the conference materials and edited the manuscripts for this book. Finally, I wish to express my deepest appreciation to all the participants for making this a lively and enjoyable conference.

C. M. Soukoulis
Physics and Astronomy Department
Ames Laboratory
Iowa State University
Ames, Iowa 50011 U.S.A.

AN INTRODUCTION TO PHOTONIC CRYSTALS

J. D. Joannopoulos
Department of Physics
Massachusetts Institute of Technology
Cambridge, Masssachusetts 02139-4307

1. Introduction

During the past five years, there has emerged a new class of materials called photonic band gap materials or, more simply, photonic crystals. The underlying concept behind these materials stems from early notions by Yablonovitch [1] and John [2]. In a nutshell, the basic idea is to design materials so they can effect the properties of photons in much the same way ordinary solids or crystals effect the properties of electrons. Now, the properties of electrons are governed by Schroedinger's equation

$$\left\{ -\frac{\nabla^2}{2} + V(r) \right\} \psi(r) = E\psi(r) \tag{1}$$

and properties of photons by Maxwell's equations, which can be cast in a form very reminiscent of the Schroedinger equation,

$$\left\{ \nabla \times \frac{1}{\varepsilon(r)} \nabla \times \right\} H(r) = \omega^2 H(r). \tag{2}$$

Equations (1) and (2) are linear eigenvalue problems whose solutions are determined entirely by the properties of the potential, $V(r)$, or dielectric function, $\varepsilon(r)$, respectively. Therefore, if one were to construct a crystal consisting of a periodic array of macroscopic uniform dielectric "atoms,"

1

C. M. Soukoulis (ed.), Photonic Band Gap Materials, 1–21.
© *1996 Kluwer Academic Publishers. Printed in the Netherlands.*

then, as in the case of electrons, the photons could be described in terms of a band structure. And if one can have a band structure, one might be able to have a complete photonic band gap. A photonic band gap is a range of frequencies for which photons are forbidden to travel through the crystal in *any* direction of propagation. Forbidden, that is, unless there is a defect in the otherwise perfect photonic crystal. A defect can act like an internal surface within which a photon can be confined or localized and this capability gives one a new "dimension" in ones ability to "mold" or control the properties of light. Therein lies the true exciting potential of photonic crystals.

Actually, the first photonic crystal ever conceived is the traditional multilayer film, which has been studied since the 19th century. This is shown in the top panel of Figure 1. In this case, a photonic band gap only exists for one direction of propagation. However, if one were to include periodicity in two-, and eventually three-dimensions as shown in the other panels, one could create a photonic crystal which, in principle at least, could have a complete photonic band gap. In most of what follows in this tutorial, we will concentrate on two-dimensional photonic crystals for simplicity. All of the concepts and ideas that we will develop will be applicable to three-dimensions as well. We will return to three-dimensions, in the last section, when we discuss some interesting new developments in the design of photonic crystals for use in the optical regime.

2. The Photonic Band Gap

One appealing aspect of Maxwell's equations, is that, for all practical purposes, they can be solved exactly. With linear materials there are no

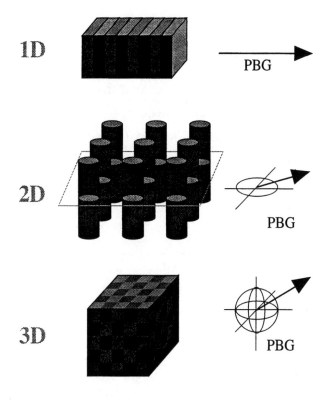

Figure 1. Periodic dielectric material in 1, 2, and 3 dimensions.

interactions between photons so that one is left with a fairly standard single-particle problem. This means that theoretical computations can be very helpful and very useful, working side-by-side with experimental efforts.

In Figure 2 we illustrate a comparison between theory and experiment for the dispersion relations of photons in a square lattice of alumina rods (ε=8.9) along the [10] direction by Robertson et al. [3]. For the measurements, 7 x 25 rods of diameter 0.74 mm were arranged in a square array, as indicated in the insets, with a lattice constant of 1.87 mm.

4

Coherent microwave transient spectroscopy measurements were then performed to measure the frequency and wavevector of the propagating photon. Because of the presence of a mirror symmetry plane, as shown in the insets, the photons decouple into transverse magnetic (TM) and transverse electric (TE) modes. The comparison between experiment and theory is excellent for both the TM and TE modes. We note also that for the TM modes there is an indication of a large photonic band gap between the first and second bands. To determine whether a complete photonic

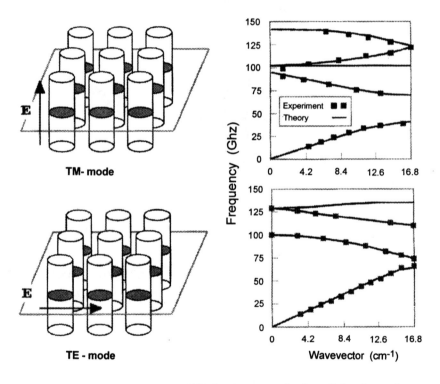

Figure 2. Band structure along [10] for the square lattice of alumina rods. Comparison of experiment and theory from Robertson et al. [3].

band gap exists, one needs to explore all possible directions of propagation. In Figure 3 we show the results of such an exercise for the high-symmetry directions of the Brillouin zone. A complete band gap does indeed exist between the first and the second TM bands. There is, however, no corresponding band gap for the TE modes. It should be possible to explain such a significant fact.

If we examine the displacement field pattern associated with the lowest TM band, we find that it is strongly concentrated in the dielectric regions. This is in sharp contrast to the field pattern associated with the

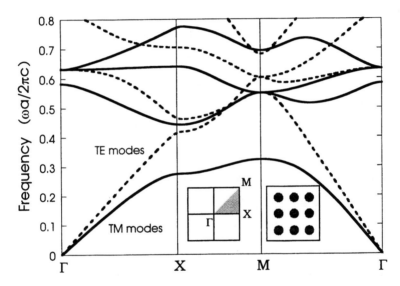

Figure 3. Photonic band structure for a square lattice
of dielectric (ε=8.9) rods with $r = 0.2a$.

second TM band which has most of its energy in the air regions. We can quantify these statements by calculating the fraction, f, of electrical energy inside the dielectric regions. For the modes at the X-point, for example, we obtain f=0.8 and f=0.3 for bands 1 and 2, respectively. The first band has most of its power in the dielectric regions and has a low frequency; the second has most of its power in the air region, and has a much higher frequency.

The fractions f for the TE modes do not contrast as strongly. At the X-point, for example, we find $f=0.2$ and $f=0.1$ for the first and second bands, respectively. In this case, both modes have significant amplitude in the air regions, raising their frequencies. They have no other choice; the field lines must be continuous so they are forced to penetrate the air regions. This is the origin of the small values of f and explains the absence of a band gap for the TE modes. Note that the vector nature of the photon field is central to this argument. The scalar D_z field of the TM modes can be localized within the rods, but the continuous field lines of the TE modes are compelled to penetrate the air regions to connect neighboring rods. As a result, consecutive TE modes do not exhibit markedly different f factors, and band gaps do not appear.

Although we will not discuss it any further here, it is interesting to note that one finds exactly the opposite behavior for TE and TM modes in the case of a crystal with a *connected* dielectric lattice. The interested reader is referred to Meade et al. [4] for an in-depth discussion of this and other aspects of the nature of the photonic band gap. We shall only state the general rule of thumb: *TM band gaps are favored in a lattice of isolated high-ε regions, and TE gaps are favored in a connected lattice.*

We conclude this section with a rather general observation. It turns out to be quite typical that the bands, above and below a band gap, can be distinguished by where the power lies—in the high-ε regions, or in the low-ε (usually air) regions. For this reason it is convenient to refer to the band *above* a photonic band gap as the "air band" and the band *below* a gap as the "dielectric band." The situation is analogous to the electronic band structure of semiconductors, in which the "conduction band" and the "valence band" surround the fundamental gap.

3. Imperfections

The key observation in the previous section was that the periodicity of the crystal induced a gap into its band structure. No electromagnetic modes are allowed in the gap. But if this is indeed the case, what happens when we send a photon (with frequency in the band gap) onto the face of the crystal from the outside? No real wave number exists for any mode at that frequency. Instead, the wave number is complex. In this case the modes are evanescent, decaying exponentially into the photonic crystal. Although these evanescent modes are genuine solutions of the eigenvalue problem, they do *not* satisfy the translational symmetry boundary conditions of the crystal. There is no way to excite them in a perfect crystal of infinite extent. An imperfection or defect in an otherwise perfect crystal, however, could, in principle, sustain such a mode. Examples of such defects are illustrated in Figure 4.

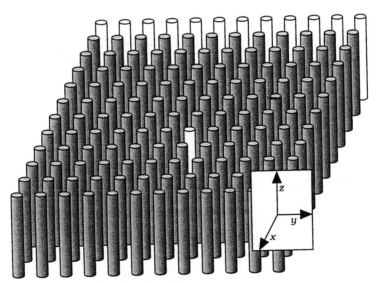

Figure 4. Schematic illustration of possible types of defects, shown in white, in a two-dimensional photonic crystal.

3.1. THE SURFACE

We can characterize a given surface of a photonic crystal by its inclination and termination. Surface inclination specifies the angles between the surface normal and the crystal axes. Surface termination specifies exactly where the surface cuts across the unit cell. With photonic crystals, unlike traditional electronic crystals, one has great liberty as to where to perform this cut.

Surface modes occur when there are electromagnetic modes near the surface which are not allowed to extend into the crystal at that frequency because of a photonic band gap. But we must consider the modes' behavior not just as a function of frequency, but also as a function of the wave vector, k, along the surface. A surface mode must have the appropriate combination (ω_S, k) which is disallowed in the crystal, not just the appropriate ω_S.

To determine where these regions exist, we pick a specific (ω_S, k) and ask if there is any wavevector perpendicular to the surface, $k_\perp$, for which

$$\omega_s = \omega_n(k, k_\perp) \quad , \tag{3}$$

where the $\omega_n(k)$ are the allowed modes of the bulk crystal. If we can, then there is at least one extended state in the crystal with the combination $\omega_S(k)$. If we tried to set up a surface mode with those parameters, it would not be localized; it would leak into the crystal.

This process of searching all possible $k_\perp$ for each k is aptly called "projecting the bulk band structure onto the surface Brillouin zone." We take all the information of the full crystal band structure and extract the information relevant to the surface. A true surface mode must be evanescent both inside the crystal and outside, in the air region, so we must project the band structure of both the photonic crystal and the air region.

In Figure 5 we show the projected bands for the (10) surface of the square lattice of dielectric rods. Each section of the plot is labeled with two letters; the first tells whether the states are extended or decaying in the air region, the second letter tells the same for the crystal region. The union of EE and DE regions represents the projected bands of the crystal, and the union of EE and ED regions the light cone. The termination of this

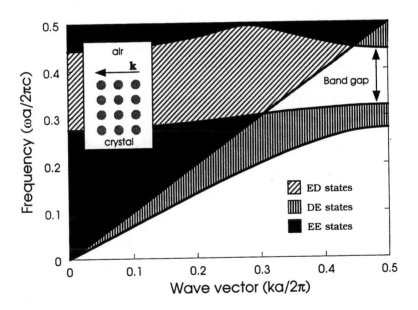

Figure 5. Projected band structure of the (10) surface of the square lattice of dielectric rods. The shading denotes regions in which light is transmitted (EE), internally reflected (DE), and externally reflected (ED). For this termination there are no surface photon modes.

surface is indicated in the inset. Note there are no surface modes in the allowed band gap region for this case. There is, however, a band of bona fide surface states when the surface is terminated by cutting the dielectric rods in half. This is shown in Figure 6.

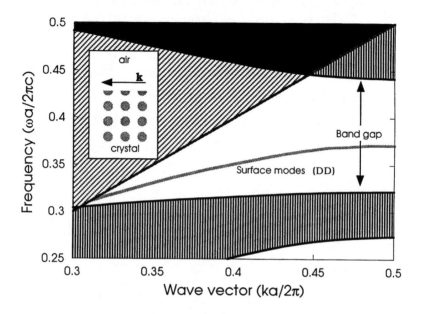

Figure 6. Projected band structure of the (10) surface of the square lattice of rods at a termination for which the outermost rods are cut in half. Same convention as in Figure 5. Note the presence of a band of surface states.

Actually, if we were to continue to slice away at the rods, dielectric would be removed and the surface band would rise in frequency sweeping across the gap until it vanished into the air band. Indeed, by slicing from one termination to an identical termination one layer below, we have removed one bulk unit cell and have decreased the total number of states in the dielectric band by one (at each **k**-point). This suggests a general claim: *for a photonic crystal with a band gap, and a surface of given inclination, one can always find some termination which will allow* *surface states.* This can have important consequences in allowing one to tune a photonic crystal to one's needs. A detailed discussion of these ideas can be found in Meade et al. [5].

3.2 THE CAVITY

We have seen that surface modes are evanescent waves trapped outside the crystal. But, we can also create imperfections that may trap light within the crystal. One class of imperfections of this type involves changing the dielectric medium in some local region of the crystal, deep within its bulk. As a simple example, consider making a change to the isolated rod shown in Figure 4 by modifying its dielectric constant, modifying its size, or simply removing it from the crystal. The effect of creating a vacancy in the square lattice of rods is illustrated in Figure 7. A defect state does indeed appear in the photonic band gap leading to a strongly localized

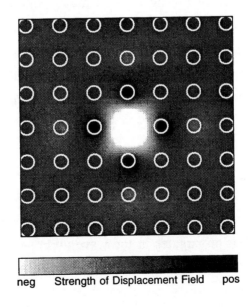

neg   Strength of Displacement Field   pos

Figure 7. Displacement field (normal to the page) for a vacancy in the square lattice of dielectric rods. Note the very localized nature of the photon.

state as shown in the figure. By removing a rod from the lattice, we create a *cavity* which is effectively surrounded by reflecting walls. If the cavity has the proper size to support a mode in the band gap, then light cannot escape, and we can pin the mode to the defect.

If the defect involves removal of dielectric (an "air defect" as in the case of the vacancy) then the cavity mode evolves from the dielectric band and can be made to sweep across the gap by adjusting the amount of dielectric removed. Similarly, if the defect involves the addition of extra dielectric material (a "dielectric-defect") then the cavity mode drops from the air band. In both cases, the defect state can be tuned to lie anywhere in the gap. Further information about air- and dielectric-defects can be found in Refs. [6-8].

This flexibility in tuning defects makes photonic crystals a very attractive medium for the design of novel types of filters, couplers, laser microcavities, etc. [9,10]. One intriguing new design of a laser microcavity involves a high Q, single-mode, bridge configuration as discussed by Villeneuve et al. [11] in these proceedings.

3.3 THE WAVEGUIDE

We can use point-like defects to trap light, as we have just seen. By using line defects, we can also *guide* light from one location to another. The basic idea is to carve a waveguide out of an otherwise-perfect photonic crystal. Light that propagates in the waveguide with a frequency within the band gap of the crystal is confined to, and can be directed along, the waveguide. This is a truly novel mechanism for the guiding of light. Traditionally, visible light is guided within dielectric waveguides such as fiber-optic cables, which rely exclusively on total internal reflection. However, if a fiber-optic takes a tight curve, the angle of incidence is too large for total internal reflection to occur, so light escapes at the corners and is lost. Photonic crystals, on the other hand, continue to confine light even around tight corners.

To illustrate these ideas, we turn again to the square lattice of dielectric rods as a simple example. In Figure 8 we plot the projected bands along the direction of propagation for a waveguide formed by removing one vertical row of rods, as shown in the inset. The shaded regions correspond to states that can propagate through the crystal. The band of states within the gap region corresponds to guided modes, which can travel freely within the narrow waveguide channel. The nature of a guided mode near mid-gap is illustrated in Figure 9.

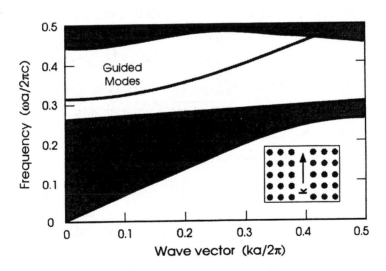

Figure 8. Projected bands for a waveguide in a square lattice of dielectric rods.
The waveguide is formed by removing one row of rods as shown in the inset.
Note the band of guided modes in the gap.

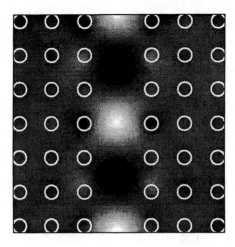

Figure 9. Displacement field of light propagating down a waveguide carved out of a square lattice of dielectric rods. Same convention as Figure 7.

Once light is induced to travel along the waveguide, it really has no where else to go. Since the frequency of the guided mode lies within the photonic band gap, the mode is forbidden to escape into the crystal. The primary source of loss can only be reflection back out the waveguide entrance. This suggests that we may use a photonic crystal to guide light around tight corners. This is shown in Figure 10. Even though the radius of curvature of the bend is less than the wavelength of the light, nearly all the light that goes in one end comes out the other!

Finally, an intriguing aspect of photonic crystal waveguides is they provide the only means possible to guide visible light, tractably, and efficiently, through narrow channels of *air*.

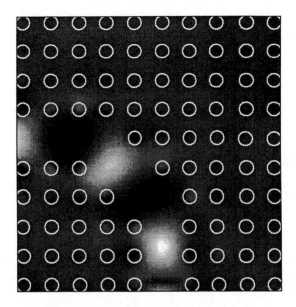

Figure 10. Displacement field of light traveling around a sharp bend.
Same convention as Figure 7.

4. A 3D Photonic Crystal for Submicron Fabrication

Ho et al. [12] were the first theorists to correctly predict that a particular three-dimensional photonic crystal would have a complete gap. Their crystal was a diamond lattice of spheres. They found that a complete band gap exists whether one embeds dielectric spheres in air, or air spheres in a dielectric medium, as long as one chooses the sphere radius appropriately. Such a structure is not easily fabricated, however. A different structure which has proven easier to manufacture in the laboratory consists of a dielectric medium which has been drilled along three of the axes of the diamond structure. It has been named *Yablonovite* after its discoverer E.

Yablonovitch [1,13]. It has the distinction of being the *first* three-dimensional photonic crystal with a complete band gap to be fabricated.

Other three-dimensional photonic crystal designs that offer complete photonic band gaps are found in references [14-17]. Of these, the Ho et al. structure [16] is the smallest three-dimensional photonic crystal with a complete gap to be manufactured to date. Özbay et al. [18] have used a clever technique of stacking thin micromachined (110) silicon wafers to fabricate these photonic crystals for wavelengths approaching 600 microns.

The ultimate goal, of course, is to design and fabricate photonic crystals for use at 1.5 microns, which is the canonical wavelength of light used in the optoelectronics industry. This, however, requires photonic crystal geometries with feature sizes that are at submicron length scales. A new class of photonic crystals, designed specifically to be amenable for fabrication at submicron length scales, has recently been introduced by Fan et al. [19]. One embodiment of this type of photonic crystal is shown in Figure 11. The crystal is designed to be built in a layered fashion, using two dielectric materials (e.g., Si and SiO_2), with a series of holes etched at normal incidence through the top surface after growth is completed. In order to create a crystal with a larger dielectric contrast, one of the two dielectric materials can be chosen so it can be removed at the end by selective etching. The sequence of "growth" steps illustrated in Figure 12 can help the reader visualize the basic elements that make up the crystal structure.

One begins by depositing a layer of Si of thickness d on a substrate of choice and by etching grooves into the Si layers as shown in part (a). The

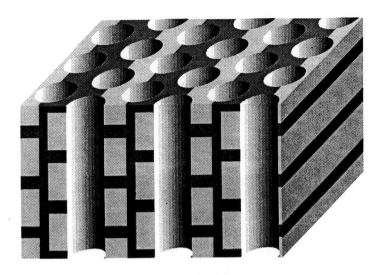

Figure 11. 3D photonic crystal for submicron fabrication.
Dark gray and light gray regions correspond to Si and SiO$_2$,
respectively. The long vertical columns are filled with air.

grooves run normal to the page and are separated by a distance a; they have a depth d and a width w. The grooves are then filled with SiO$_2$. The next step consists in growing another Si layer of height h on top of the previous layer, as shown in part (b), and etching long grooves of depth d and width w into this layer, as shown in part (c). We note these grooves actually extend into the first layer and are translated by a distance $a/2$ with respect to the first layer. After filling the grooves with SiO$_2$, another Si layer of height h is deposited on the top surface and long parallel groves are etched. The grooves are translated again by a distance $a/2$ with respect to the previous layer, as shown in part (d). From this point on, the structure repeats itself every two layers. Once this process is completed, an array of long cylindrical holes is etched into the top surface

18

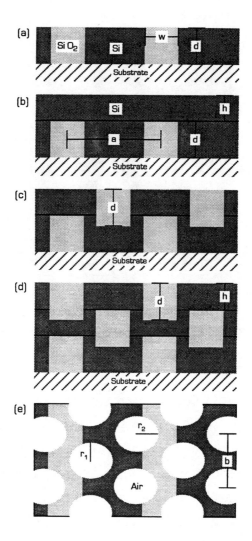

Figure 12. Schematic growth sequence for the structure in Figure 11.
(a)-(d) are cross-sectional views, (e) is a plan view.

of the structure, at normal incidence. In general, the cross section of the holes can be either circular or elliptical with parameters r_1 and r_2, as shown in part (e).

The design of this structure has many degrees of freedom which can be used to optimize the size of the gap. Using a dielectric constant for Si

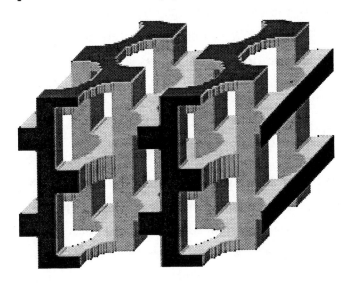

Figure 13. The photonic crystal of Figure 11 with the SiO$_2$ removed. The structure consists of a dielectric skeleton of Si embedded in air.

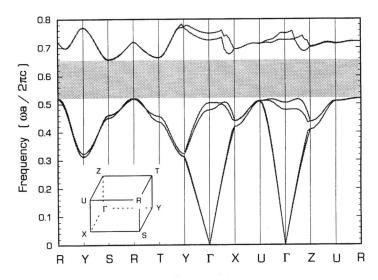

Figure 14. Band structure for the 3D photonic crystal of Figure 13. The shaded region identifies a 23% band gap.

of 12.096 at 1.53 μm and 2.084 for silica, gives an optimized photonic band gap of about 14%. A very significant improvement can be made by simply removing the oxide. The resulting photonic crystal structure is illustrated in Figure 13. With optimized parameters $w=0.36a$, $d=0.51a$, and $r_1=r_2=0.24a$, one finds a 23% gap as shown in the band structure of Figure 14.

5. Conclusion

Of course all of these results are purely theoretical predictions at this time. The proof of the pudding will come when it is demonstrated that such photonic crystals can indeed be fabricated in a tractable fashion at submicron lengthscales. Although more easily manufacturable one- and two-dimensional photonic crystals have already been fabricated at submicron lengthscales, it is the three-dimensional photonic crystal that will ultimately provide the most complete control possible for the localization and propagation of light.

6. Acknowledgments

The author gratefully acknowledges collaborations and contributions of K. Brommer, S. Fan, I. Kurland, R. Meade, A. Rappe, and P. Villeneuve to all aspects of this work.

7. References

1. Yablonovitch, E. (1987) *Phys. Rev. Lett.* **58**, 2059.
2. John, S. (1987) *Phys. Rev. Lett.* **58**, 2486.
3. Robertson, W., Arjavalingam, G., Meade, R., Brommer, K., Rappe, A., and Joannopoulos, J. (1992) *Phys. Rev. Lett.* **68**, 2023.

4. Meade, R., Brommer, K., Rappe, A., and Joannopoulos, J. (1993) *J.Opt.Soc.Am.* **10**, 328.

5. Meade, R., Brommer, K., Rappe, A., and Joannopoulos, J. (1991) *Phys. Rev. B***44**, 10961.

6. Meade, R., Brommer, K., Rappe, A., and Joannopoulos, J. (1991) *Phys. Rev. B***44**, 13772.

7. Yablonovitch, E., Gmitter, T., Meade, R., Rappe, A., Brommer, K., and Joannopoulos, J. (1991) *Phys. Rev. Lett.* **67**, 3380.

8. Fan, S., Winn, J., Devenyi, A., Chen, J., Meade, R., and Joannopoulos, J. (1995) *Opt. Soc. Am* ., in press.

9. Slusher, R. (1993) *Optics and Photonics News* **Feb.**, 8.

10. Meade, R., Devenyi, A., Joannopoulos, J., Alerhand, O., Smith, D., and Kash, K. (1994) *J. Appl. Phys.* **75**, 4753.

11. Villeneuve, P., Fan, S., Chen, J., Joannopoulos, J., Lim, K., Petrich, G., Kolodziejski, L., and Reif, R., in these proceedings.

12. Ho, K., Chan, C., and Soukoulis, C. (1990) *Phys. Rev. Lett.* **65**, 3152.

13. Yablonovitch, E., Gmitter, T., and Leung, K. (1991) *Phys. Rev. Lett.* **67**, 2295.

14. Chan, C., Ho, K., and Soukoulis, C. (1991) *Europhys. Lett.* **16**, 563.

15. Sözüer, H., and Haus, J. (1993) J. Optic. Soc. Am. B **10**, 296;

16. Ho, K., Chan, C., Soukoulis, C., Biswas, R., and Sigalas, M. (1994) *Solid State Commun.* **89**, 413.

17. Sözüer, H., and Dowling, J. (1994) *J. Mod. Opt.* **41**, 231.

18. Özbay, E., Michel, E., Tuttle, G., Sigalas, M., Biswas, R., and Ho, K. (1994) *Appl. Phys. Lett.* **64**, 2059; and in these procedings.

19. Fan, S., Villeneuve, P., Meade, R., and Joannopoulos, J. (1994) *Appl. Phys. Lett.* **65**, 1466.

PHOTONIC BAND GAP MATERIALS

R. BISWAS, C.T. CHAN, M. SIGALAS,
C.M. SOUKOULIS, and K.M. HO
*Ames Laboratory, Microelectronics Research Center, and
Department of Physics and Astronomy, Iowa State
University, Ames, IA 50011, U.S.A.*

Abstract

A new class of layer-by-layer structures that have full three-dimensional photonic band gaps is described. Each layer has a particularly simple arrangement of one-dimensional rods. These new structures have gaps of appreciable width. The dependence of the gap on the dielectric contrast, filling ratio, cross sectional geometry of the rods, and the structure of the unit cell is systematically studied. The gap is relatively insensitive to the exact geometrical properties of the rods in each layer. Such layer-by-layer structures have already been fabricated over a wide range of microwave and millimeter wave frequencies. Other structures with three-dimensional photonic bandgaps are surveyed.

1. Introduction and History

Photonic band gap crystals offer unique ways to tailor light and the propagation of electromagnetic waves. In analogy to electrons in a crystal, electromagnetic waves propagating in a structure with a periodically modulated dielectric constant are organized into photonic bands which are separated by gaps where propagating states are forbidden. Photonic band gap crystals are periodic dielectric structures which possess a frequency gap in which electromagnetic waves are forbidden, analogous to the electronic band gap in a semiconductor. There have already been proposals[1,2] for novel applications of these photonic band gap crystals, with operating frequencies ranging from the microwave to the optical regime, that include zero-threshold lasers, low-loss resonators and cavities, and efficient microwave antennas. Spontaneous emission is suppressed for photons in the photonic band gap, offering novel approaches to manipulating the electromagnetic field and creating high-

C. M. Soukoulis (ed.), Photonic Band Gap Materials, 23–40.
© 1996 *Kluwer Academic Publishers. Printed in the Netherlands.*

efficiency light-emitting structures. Innovative ways to manipulate light can have profound influence on science and technology.

Most of the above applications rely on the availability of structures with full three-dimensional photonic band gaps. The first structure that exhibited a full three-dimensional band was predicted by Ho, Chan, and Soukoulis to be the diamond structure[3] from theoretical calculations of the propagation of electromagnetic waves in periodic dielectric media. They predicted that the diamond structure possesses a full-three dimensional fundamental photonic band gap over the entire surface of the Brillouin zone. This was also predicted by several calculations by other groups[4,5]. The vectorial nature of the electromagnetic field was essential for the calculation of the photonic bands, whereas scalar wave calculations may not produce correct results. The photonic band gap exists for refractive index contrasts greater than 2 and a wide range of filling ratios for a lattice of spheres at the site of the diamond lattice. The basic elements for obtaining a photonic band gap structure are the connectivity, dielectric contrast, and the filling ratio.

Subsequently, an experimental 'three-cylinder' structure[6] with the symmetry of the diamond structure, was fabricated by Yablonovitch et al. [6] at microwave length scales by drilling sets of three holes at each triangular lattice point on the surface of a high dielectric material (stycast). This structure had a full 3-dimensional photonic band gap (PBG) between 10 and 13 GHz, and was an important proof-of-concept demonstration of photonic band gaps. The theoretical gap/midgap ratio was about 19% for this structure (in n=3.6 material) for a structure with about 85% air- somewhat smaller than the highest that could be obtained with a diamond structure with air spheres (29%) (Table 1). An important direction in subsequent work was to design and fabricate structures with three-dimensional photonic band gaps at smaller length scales, eventually directed towards optical and infrared frequencies. It appeared difficult to extend the 3-hole drilling process to smaller length scales, and alternative structures easier to fabricate were needed.

The fcc structure does not have a fundamental photonic band gap between the lowest bands (bands 2 and 3) as predicted earlier with scalar wave calculations. Full vector-wave band calculations[3,4,5] for air spheres in an fcc lattice revealed a crossing of the second and third bands for p-polarized waves along the surface of the Brillouin zone between U and X. The second and third bands are degenerate at the W point of the zone. These features lead to a 'pseudogap' or a region of low densities of states between the second and third bands. It is worth noting that the fcc structure does have a true higher photonic band gap between the eighth and ninth bands as discussed later in section 4 and found in recent calculations.

TABLE 1. 3-d photonic band gaps for structures with different lattice symmetry with the different dielectric elements noted. All gap/midgap ratios are for the refractive index contrast of 3.6 unless noted

Symmetry	element	material of element	bands	gap/midgap n=3.6
fcc	spheres	air	2 - 3	pseudogap
fcc	spheres	air	8 - 9	0.08
diamond	spheres	dielectric spheres	2 - 3	0.16
		air spheres	2 - 3	0.29
diamond	3-cylinder	air cylinders	2 - 3	0.19
		dielectric cylinders	2 - 3	0.13
FCT 4-layer	rectangular bar	no overlap air/diel	2 - 3	0.19
	overlapping	air bars	2 - 3	0.22
	cylinders -touching	dielectric	2 - 3	0.18
	cylin+glue sphere	dielectric	2 - 3	0.21
	cylinders	air-circular	2 - 3	0.25
	cylinders	air-elliptical	2 - 3	0.27
FCT 2-layer	bar/cylinders		2 - 3	no gap
distorted FCT	bars	dielectric	2 - 3	0.13 (n=3.41)
double-etch 70.5°	overetched bars	dielectric	2 - 3	0.16 (n=3.41)
hexagonal 6-layer	rectangular bars	dielectric	4 - 5	0.02
distorted-diamond	3-rectangular bars (planar) +cylinder	dielectric	2 - 3	0.25
" "	3-rectangular bars	dielectric	2 - 3	no gap
diamond	3+1 cylinders	dielectric	2 - 3	0.30
	" "	air	2 - 3	0.28
simple cubic	overlapping	air spheres	5 - 6	0.07
	cylinders or bars	dielectric	2 - 3	~0.03
rhombohedral A-7	cylinders	dielectric	2 - 3	0.31
" "		air	2 - 3	0.29

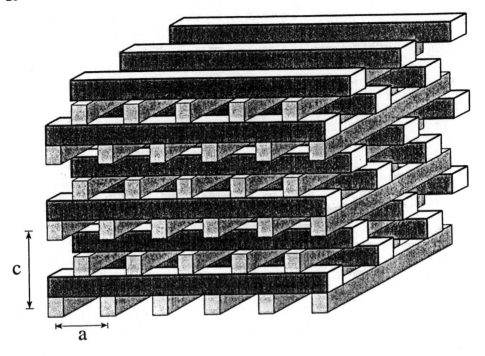

Figure 1. The new layer-by-layer structure producing full three-dimensional photonic band gaps. The structure is constructed by an orderly stacking of dielectric rods, with a simple one-dimensional pattern of rods in each layer. Although rods of rectangular cross-section are shown here, the rods may also have cylindrical or elliptical cross sections.

The basic direction pursued by our Iowa State group was to design a structure with a full three-dimensional PBG that could be easily fabricated experimentally. The diamond structure was constructed[7] by connecting the sites of the diamond structure with dielectric cylinders. This structure yielded gaps as large as 30% for dielectric rods or 28% for air rods[7]. The direction we pursued was to start with the diamond structure but reduce the dimensionality of the building blocks of this structure that would still yield a 3-D photonic band gap. The diamond structure constructed by interconnecting cylinders was distorted by flattening all the [111] planes of diamond to a hexagonal mesh and then connecting cylinders in the [111] direction between the hexagonal nodes in different layers. This retains the fourfold connectivity similar to diamond, although angles between cylinders are 120° and 90°. The ABC stacking was maintained and this structure was found to have extremely large gap/midgap ratios of 25 to 30% (Table 1), for either rectangular or cylindrical bars, respectively. These structures were part of the A-7 family of structures that were later studied. Fabrication of the two-dimensional hexagonal pattern in the [111] layers may still be feasible with a complex mask, but it would be experimentally very difficult to fabricate the [111] cylinders perpendicular to the layers. Eliminating the [111] cylinders and simply stacking hexagonal dielectric lattices does not maintain the photonic band gap.

The search for simplifying the structure and reducing the dimensionality of the structural building blocks continued. The Iowa State group has designed[8] a novel three-dimensional layer-by-layer structure that has a full three-dimensional photonic band gap over a wide range of structural parameters. The new structure (Fig. 1) consists of layers of one-dimensional rods with a stacking sequence that repeats every fourth layer with a repeat distance of c. Within each layer the rods are arranged in a simple one-dimensional pattern and are separated by a distance a, a significant simplification from the two-dimensional grid found earlier. The rods in the next layer are rotated by an angle θ with respect to rods in the previous layer. θ has the value of 90° but, in general, could vary from 90° to 60° but still have a full three-dimensional photonic band gap. The rods in the second neighbor plane are shifted by half the spacing a relative to rods in the first plane in a direction perpendicular to the rods. The rods in every alternate layer are parallel (Fig. 1). This structure has the symmetry of a face centered tetragonal (fct) lattice. For the special case of $c/a = \sqrt{2}$, the lattice can be derived from a fcc unit cell with a basis of two rods. This layered structure can be derived from a diamond structure by replacing the 110 chains of atoms with the rods.

2.   Layer-by-Layer Photonic Band Gap Structures

This new three-dimensional layer-by-layer structure has a full three-dimensional photonic band gap over a robust and wide range of structural parameters. Several variations of the structure shown in Fig. 1 exist. (1) Although Fig. 1 displays rods of rectangular cross section, the cross section of the rods can be circular, elliptical, or rectangular with different aspect ratios and are not crucial to determine the photonic band gap. (2)The rods can either be dielectric in a background of low dielectric material (air). Alternatively the structure can be constructed out of air holes (or low dielectric material) in a block of high dielectric material. (3) An important property of the structure in Fig. 1 (with rectangular cross section rods) is that it is self-dual, i.e., the structure may be described by touching air rods of thickness $(a-r_1)$ in an background dielectric rather than the dielectric rods of thickness r_1 in the air background. This dual character is beneficial for the calculations. (4) The rods need not be touching as shown in Fig. 1, but can be overlapping. In fact, the multiply-connected geometry of overlapping rods produces larger photonic band gaps.

We optimize the photonic band gap as a function of the structural parameters. These structural parameters consist of a) the c/a ratio, b) the filling ratio of the structure that is controlled by varying the width of the rods in Fig. 1, c) the dielectric contrast between the two materials, and d) the degree of overlap between the dielectric rods in the case of the overlapping rods.

This structure was first fabricated[9] in the microwave regime by stacking alumina cylinders and demonstrated a full three-dimensional photonic band gap at microwave frequencies (12-14 GHz). A similar structure was also fabricated with alumina rods that had a band gap between 18 and 24 GHz.

Since these initial structures had cylindrical rods, we first display the calculations for rods of cylindrical cross section. For touching cylindrical rods of radius r and center-to-center separation a in each layer the filling ratio is $f = \pi r/2a$. The radius controls the c/a ratio since $c = 8r$. Photonic band structure calculations were performed as a function of filling ratios f (by varying a) which also changes the c/a ratio. This structure has a full three-dimensional band gap between the lowest bands (bands 2 and 3). Figure 2 displays the magnitude of this gap for dielectric materials with a refractive index contrast n=3.6 similar to GaAs and air. The optimum gap/midgap ratio of 18% occurs for a filling ratio of about 0.22, and a c/d ratio close to 1.27. For the case of alumina rods with n=3.1, the maximum gap/midgap ratio is about 12.7% optimized at a filling ratio close to 0.25.

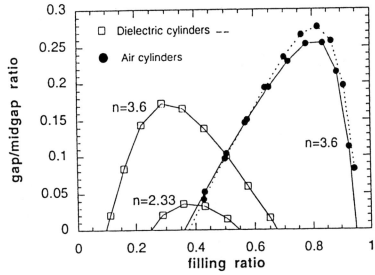

Figure 2. Calculated photonic band gap as a function of filling ratio for the new 3D layer-by-layer structure for the case of touching dielectric cylinders (squares) with refractive indices of 3.6 and 2.33. The photonic band gaps (circles) for overlapping air cylinders in background dielectric of refractive index 3.6 are shown. The solid line is for cylindrical holes whereas the dotted line is for elliptical cross section and aspect ratio of 1.25, for which the largest band gap was found. As in all figures, the photonic bandgap is the ratio of the gap to the midgap frequency.

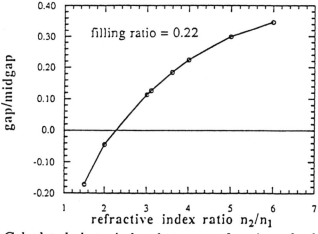

Figure 3. Calculated photonic band gap as a function of refractive index contrast for touching dielectric cylinders with a c/a ratio of 1.13. This c/a ratio is for the experimentally fabricated structure with alumina rods.

It is notable (Fig. 2) the inverse structure with cylindrical air cylinders in a dielectric material has a substantially higher band gap of about 25% for cylindrical air rods and 27% for air rods of elliptical cross section (Table 1). These structures with 80-85% air have highly overlapping air rods resulting in a sparse multiply-connected geometry. Generally, multiply-connected geometries have larger photonic band gaps. The preferred approach to fabricating such small-scale multiply-connected structures with air rods is to use ion-beam etching of a dielectric material, as has been demonstrated by Yablonovitch et al. for the 3-cylinder structure[6].

The experimentally fabricated microwave model of touching alumina cylinders had a c/a ratio of 1.13 and a filling ratio close to 0.22. For this structure it is instructive to plot the gap/midgap which increases monotonically as a function of increasing refractive index contrast (Fig. 3). A minimum refractive index contrast of about 2.2 is needed to obtain a photonic band gap. An added complication in the experimentally fabricated microwave structure is the presence of glue holding together each intersection between the dielectric rods. We performed calculations where glue spheres of the same dielectric material were added to the intersections of rods in each layer. We found (Fig. 4) that for glue spheres of up to 10% of the dielectric rod volume, the band gaps were actually enhanced over the case of touching dielectric cylinders. In the alumina case (n=3.1) the gap/midgap ratio increased from 12.7% to about 15.7% for glue spheres of about 10% the dielectric volume. The glue spheres increase the filling ratio and result in a multiply-connected geometry, which generally increases the photonic band gap.

The calculated photonic bands along the [110] direction compare very well with the measurements for the microwave model structure (Fig. 5). In particular, the top of the valence band occurs at K (.75,.75,0) at 11.7 GHz comparing very well with the experimental measurement of 11.9 GHz, for EM waves polarized perpendicular to the cylinders and an upper edge of 13.7 GHz comparing well with 14.0 GHz experimentally. As evident in Fig. 5, the valence band edge in the [110] direction is sensitive to the polarization of the electromagnetic (EM) waves and a lower valence band edge is found both experimentally and theoretically for EM waves polarized in the plane of the rods. The overall agreement between calculated and measured band structure is excellent over the entire Brillouin zone (Fig. 5). The band gap is much larger for propagation perpendicular to the layers (along the stacking direction) with calculated band edges at X' of 11.1 and 16.8 GHz compared to 10.7 and 17.4 GHz (Table 2). The minimum of the conduction band occurs at L (.5,.5,.5), and the bands are quite flat over the surface of the Brillouin zone.

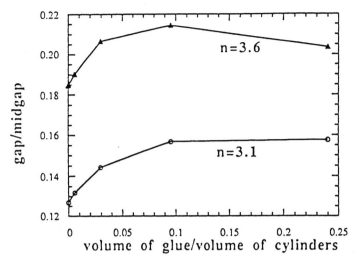

Figure 4. Calculated change in photonic band gap when small glue spheres are introduced at the intersections of the touching cylinders to bind the layers together. For simplicity the glue sphere has the same refractive index as the dielectric cylinder. Results are plotted as a function of the ratio of the glue to the dielectric cylinder volume, for two different refractive indices, n.

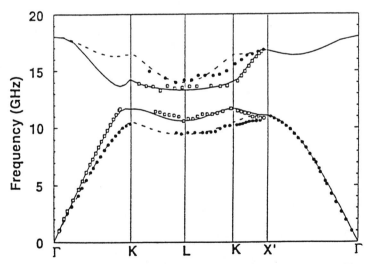

Figure 5. Frequency versus wave-vector dispersion along the 110 direction of the Brillouin zone. The dashed and solid curves are calculations for propagation with the E field parallel and perpendicular to the rods. The circles and squares are transmission measurements performed for the E field parallel and perpendicular to the rod axes.

TABLE 2. Experimental and calculated band edge frequencies for propagation along the stacking direction in the layer-by-layer structure

Material	refract ive index n	measur- ed midgap GHz	lower edge expt GHz	lower edge calc GHz	upper edge expt GHz	upper edge calc GHz
Alumina	3.1	12.6	10.7	11.1	17.4	16.8
Alumina	3.1	20.1	17.1	16.9	24.0	24.95
Silicon	3.41	98.5	81.0	81.4	122.	119.4
Silicon	3.41	325.	265	276	390	388
Silicon	3.41	439	360	378	524	518

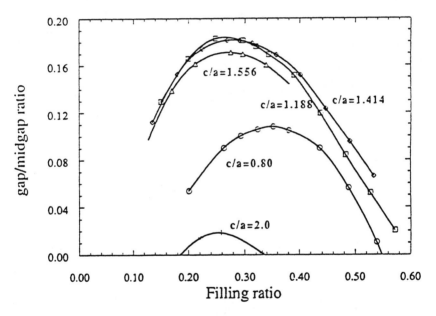

Figure 6. Photonic band gap for touching rectangular rods (as in Figure 1) as a function of the filling ratio. The curves are for different c/a ratios. In each curve the filling ratio was varied by changing the width of the rods and keeping c and a constant. Dielectric rods with refractive index 3.6 are used in all cases.

2.1.   RECTANGULAR ROD STRUCTURES

We have fabricated[10-12] the layer-by-layer structure with rectangular rods of silicon by micromachining silicon [110] wafers, using anisotropic etching properties of silicon and an orderly stacking procedure. The calculated photonic band gaps for different c/a ratios (Fig. 6) demonstrate that c/a values around 1.2-1.6 are optimum, with optimum filling ratios between 25 and 30%. In these calculations the c/a ratio is held constant as the width of the bars (or filling ratio) is varied. The micromachined silicon PBG structure was fabricated with a midgap at 94 GHz very close to the optimized value. Increasing the filling ratio decreases both the size of the gap and the midgap frequency. The maximum in the valence band lies along the 110 direction very close to K (.75,.75,0), whereas the minimum of the conduction band is at L(.5,.5,.5).

We have also calculated the bandgaps for the inverse structure with air rods (Fig. 7). When the rods do not overlap, the air and dielectric materials are symmetric and yield the same maximum gap/midgap of 18% for air fraction of 0.7-the same result as for the dielectric rods. When the air rods are allowed to overlap, the gap/midgap ratio increases to ~22% (Fig. 7) (when the overlap is about 30% of the rod height), similar to the increased gap in the multiply-connected cylindrical structures. It does seem that overlapping rectangular air rods cannot produce as large a gap as in an overlapping cylindrical structure.

The structure with rectangular Si-rods has been fabricated for three different length scales producing midgap frequencies of 95, 140 and 450 GHz using progressively thinner silicon wafers. In all three cases, the band edge frequencies are in excellent agreement with the calculated values. The structure with midgap at 94 GHz has also been fabricated by laser machining alumina wafers, illustrating the usefulness of our layer-by-layer structure.

2.2.   DOUBLE-ETCHED STRUCTURE

A further variation of this structure[12] is when the rods in the next layer are rotated by an angle with respect to rods in the previous layer, where θ has values different from 90°. θ is 70.53° for double etched Si(110) wafers where the one-dimensional rod pattern is etched on both sides of the wafer and hence θ is the angle between [111] directions. In general, θ can vary from 90° to 60° but still have a full three-dimensional photonic band gap. This structure is obtained by shearing the originally proposed layer-by-layer structure (with θ of 90°) to produce a different angle θ between rods in successive layers.

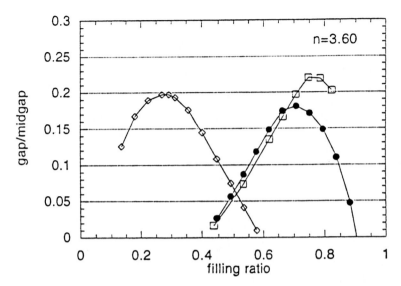

Figure 7. Calculated photonic band gap as a function of filling ratio for touching rectangular air rods (circles) and overlapping rectangular air rods (squares). Also shown for comparison is the result for touching rectangular dielectric rods (diamonds). A refractive index contrast of n=3.60 is used in all cases.

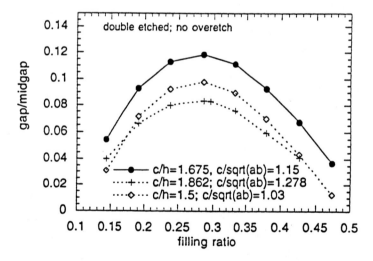

Figure 8. Calculated photonic band gap for touching double etched rectangular rods as a function of filling ratio. The angle between the rods is 70.53°. h is the distance between the rods in each layer. The c/h ratio is kept fixed for each curve. a and b are unit cell dimensions in the layers.

The calculated photonic band gap for the double etched structure (Fig. 8) is about 12% for filling ratios close to 30%. To achieve the best gap of about 12% it is very important to optimize the c/h ratio to a value close to 1.675 (or c/√ab close to 1.15) (Fig. 8), whereas the gap/midgap ratio decreases significantly at both larger and smaller c/h values. The gaps shown in Fig. 8 refer to the minimum gap in the Brillouin zone.

Reducing the angle θ to 70.53° does reduce the gap/midgap ratio to about 12%, which is still sizable enough for applications. The gap/midgap ratio can be increased by overetching this structure[12].

3. Search for Photonic Band Gap Structures

There have been extensive studies of two-dimensional photonic band gap structures[13,14] where the EM waves can have a gap in a two-dimensional plane but propagate along the third direction. Since these will be described in detail in separate articles in this volume, we do not discuss this area further here .

The basic idea in producing photonic band gaps is the midgap frequency ω_{mg} has a wavevector k_{mg} equal to the magnitude of the average wavevector at the first Brillouin zone surface k_{BZ}. k_{BZ} is dictated by the structure being proportional to the inverse of the lattice constant. The midgap frequency is $\omega_{mg} = c_{av} k_{mg}$, where c_{av} is the effective speed of electromagnetic waves in the media. Whether a particular structure has a gap or not depends on the overlap between the lowest frequency valence band and the higher frequency conduction band along the surface of the zone. In the diamond structure, the smallest wavevector (L) to the BZ surface corresponds to the bottom of the conduction band whereas the longest wavevector to the zone boundary (W or K) provides the bottom of the conduction band. The geometry of the diamond or the fct structure is just right that these two band frequencies at these zone boundary points do not overlap. In most structures the two frequencies at these two points would overlap and no photonic band gap would exist.

We illustrate this band overlap concept by surveying a number of three-dimensional structures with photonic band gaps. We revisit the fcc structure which may be realized as a close packed array of low-dielectric (refractive index n_1) spheres in a high-dielectric background material (refractive index n_2). We have calculated the photonic band structure by the plane wave expansion method and then calculated the photonic densities of states by the well-established tetrahedron method. The density of states (Fig. 9) shows i) the well-known pseudogap between the second and third bands and ii) a *full* three-dimensional band gap between the eighth and ninth band. These results have also found in recent

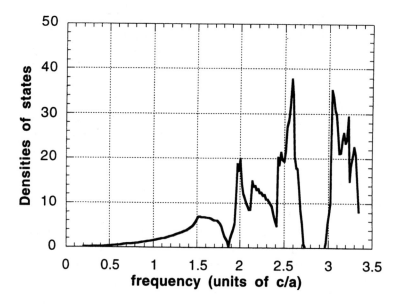

Figure 9. Densities of states for close packed low dielectric spheres (filling ratio 0.74) in a background high dielectric material, with refractive indices n_1 and n_2, respectively. The refractive index contrast of $n_2/n_1 =$ 4.46 is used with $n_1 = 1.59$. There is a full three-dimensional photonic band gap between bands 8 and 9, whereas a 'pseudogap' or region with a low densities of states appears between bands 2 and 3. Frequencies are in the units where c is the speed of light and a is the cubic lattice constant.

calculations[15,16]. For a contrast ratio of $n_2/n_1 = 4.46$ with low dielectric of 1.59 (e.g. polystyrene) we obtain the sizable gap/midgap ratio of 8%. Searching for such a higher gap may be very interesting experimentally.

As found in earlier calculations[15], the simple cubic structure composed of air spheres at the simple cubic lattice sites has a small higher gap of ~7% between the 5-6 bands. However, when the simple cubic lattice is connected by cylinders of bars, it does exhibit a small but fundamental band between bands 2 and 3 of ~3% (Table 1). Such small gaps may be difficult to realize experimentally.

We examined another structure produced by shearing our layer-layer structure to produce angles of 60° between bars in subsequent layers and stacking these bars to maintain four-fold connectivity. This structure has a spiral axis. A periodicity of 6 layers in the stacking direction was used. This structure however resulted in only an almost vanishing bandgap of about ≤2% (Table 1)

An interesting class of structures is the A7-family of structures[17]. These structures have rhombohedral symmetry and can be generated by connecting lattice points of the A7 structure by cylinders. The A7 class of structures can be described with two structural parameters - an internal displacement u and a shear angle α-that can be varied to optimize the gap. For special values of the parameters, the structure reduces to simple cubic, diamond, and the Yablonovitch 3-cylinder structure. Gaps as large as 30% are found[17] in the A7 class of structures for well optimized values of the structural parameters and fabrication of these structures would be most interesting.

An alternative layer-by-layer structure has been recently proposed by Fan et al.[18]. This consists of a layered structure of two dielectric materials in which a series of air columns is drilled into the top surface. The structural parameters have been optimized to yield 3-dimensional photonic gap/midgap ratios of 14% using Si, SiO_2 and air, to 23% using Si and air (i.e. the SiO_2 layers are replaced by air).

4.   Conclusions

In conclusion, we find that the photonic band gap depends crucially on i) the local connectivity or geometry, ii) the refractive index contrast and iii) the filling ratio of a structure. Multiply-connected geometries produce larger gaps than simply connected structures.

We have designed a new layer-by-layer structure that has a full three-dimensional photonic band gap. Each layer consists of a set of one-dimensional pattern of parallel dielectric rods. The rods were rotated by 90° between neighboring layers and shifted by half the distance a between second neighbor layers. This stacking procedure led to a unit cell of four layers. This structure has been fabricated by stacking alumina rods producing full 3-dimensional photonic band gaps between 12 and 14 GHz. The structure has been fabricated by micromachining silicon wafers and stacking the wafers in an orderly fashion producing millimeter wave photonic band gap structures at progressively smaller length scales. We have achieved these photonic band gap structures with midgap frequencies of 100 and 500 GHz. A number of applications of the microwave and millimeter wave PBG crystals may be realized with the structures we have already fabricated. This layer-by-layer structure is very promising for the extension of photonic band crystals into the infrared and optical regimes an area that will surely lead to new areas in basic physics together with novel applications.

We acknowledge very productive collaborations with E. Ozbay and G. Tuttle. The Ames Laboratory is operated for the U.S. Department of Energy by Iowa State University under contract W-7405-Eng-82. This work is supported by the Director of Energy Research, Office of Basic Energy Sciences and Advanced Energy Projects, and the Center of Advanced Technology Development, Iowa State University with funding from the U.S. Department of Commerce.

REFERENCES

1. For a recent review, see C.M. Soukoulis, ed., *Photonic bandgaps and Localization* (Plenum, New York, 1993).
2. Articles in Development and applications of materials exhibiting photonic band gaps, *J. Opt. Soc. Am. B* **10** (1993), a special feature edited by C.M. Bowden, J.P. Dowling, and H.O. Everitt.
3. K.M. Ho, C.T. Chan, and C.M. Soukoulis, Existence of a photonic band gap in periodic dielectric structures, *Phys. Rev. Lett..* **65**, 3152 (1990).
4. K.M. Leung and Y.F. Liu, Full vector wave calculation of photonic band structures in fcc dielectric media, *Phys. Rev. Lett.* **65**, 2646 (1990).
5. Z. Zhang and S. Satpathy, Electromagnetic propagation in periodic structure: Bloch wave solution of Maxwell's equation, *Phys. Rev. Lett..* **65**, 2650 (1990).
6. E. Yablonovitch, T.J. Gmitter and K.M. Leung, Photonic band structure: the face-centered-cubic case employing nonspherical atoms, *Phys. Rev. Lett..* **67**, 2295 (1991).
7. C.T. Chan, K.M. Ho and C.M. Soukoulis, Photonic band gaps in experimentally realizable periodic dielectric structures, *Europhys. Lett.* **16**, 563 (1991).
8. K.M. Ho, C.T. Chan, C.M. Soukoulis, R. Biswas, and M. Sigalas, Photonic band gaps in three dimensions: New layer-by-layer periodic structures, *Solid State Comm.* **89**, 413 (1994).
9. E. Ozbay, A. Abeyta, G. Tuttle, M. Tringides, R. Biswas, M. Sigalas, C.M. Soukoulis, C.T. Chan and K.M. Ho, Measurement of three-dimensional photonic band gap in new crystal structure made of dielectric rods, *Phys. Rev. B* **50**, 1945 (1994).
10. E. Ozbay, G. Tuttle, R. Biswas, M. Sigalas, and K.M. Ho, Micromachined millimeter wave photonic band gap crystals, *Appl. Phys. Lett..* **64**, 2059 (1994).
11. E. Ozbay, E. Michel, G. Tuttle, R. Biswas, K.M. Ho, J. Bostak, and D.M. Bloom, Terahertz spectroscopy of three-dimensional photonic band gap crystals, *Optics Lett.*, **19**, 1155 (1994).
12. E. Ozbay, G. Tuttle, R. Biswas, K.M. Ho, J. Bostak, and D.M. Bloom, Double-etch geometry for millimeter-wave photonic band gap crystals, *Appl. Phys. Lett..* **65**, 1617 (1994).
13. P.L. Gourley, J.R. Wendt, G.A. Vawter, T.M. Brennan, and B.E. Hammons, Optical properties of two-dimensional photonic lattices fabricated as honeycomb nanostructures in compound semiconductors, *Appl. Phys. Lett..* **64**, 687 (1994).

14. R.D. Meade, K.D. Brommer, A.M. Rappe, and J.D. Joannopoulos, Existence of a photonic band gap in two dimensions, *Appl. Phys. Lett..* **61**, 495 (1992).
15. H.S. Sozuer, J.W. Haus, and R. Inguva, Photonic bands: Convergence problems with the plane wave method, *Phys. Rev. B* **45**, 13962 (1992).
16. T. Suzuki and P. Yu, Emission power of an electric dipole in the photonic band structure of the fcc lattice, *J. Opt. Soc. of Am. B* **12**, 571 (1995).
17. C.T. Chan, S. Datta, K.M. Ho and C.M. Soukoulis, The A-7 structure: A family of photonic crystals, *Phys. Rev. B* **49**, 1988 (1994).
18. S. Fan, P. R. Villeneuve, R.D. Meade, and J. D. Joannopoulos, Design of 3D photonic crystals at submicron lengthscales, *Appl. Phys. Lett..* **65**, 1466 (1994).

MICROMACHINED PHOTONIC BAND GAP CRYSTALS: FROM MICROWAVE TO THE FAR-INFRARED

EKMEL ÖZBAY
Department of Physics
Bilkent University
Bilkent, Ankara 06533
TURKEY

1. Introduction

In May 1992, I accepted a position at Iowa State University (ISU) and started working on photonic band gap materials with Gary Tuttle, assistant professor of electrical engineering. By that time, the classical papers in the field had already been published [1, 2]. Concepts like conduction and valence bands, acceptor and donor defects [3], had been introduced and widely used. There were series of articles in *Physical Review Letters*, about physical mechanisms around these materials [4,5]. People were excited about possible "photonic" applications and experiments, where the spontaneous emission would be reduced to nil at optical frequencies. To my surprise, all of the experimental work was limited by the original work done by Eli Yablonovitch [2] which was only performed at 15 GHz!! There had been some work by the IBM group to test two-dimensional alumina rods which showed a band gap around 70 GHz, but only for one of the polarizations[6]. I could not find any other experimental work which showed a full band gap in all directions. Besides, there were no published efforts to push the frequency performance to a frequency higher than the original 15 GHz.

The real reason for the lack of further experimental work was the lack of suitable geometry. The initial fcc structure suggested by Yablonovitch et al. [7] was simple, but did not have a full band gap [1]. The diamond structure suggested by Ho et al.[1] had a full band gap, but at that time no one was able to develop an easy technique to produce a full diamond structure. Yablonovitch was able to convert the diamond symmetry to a new structure[2,8], which was made by drilling a dielectric block from three different angles. But the awkward geometry of the structure resulted in more admiration to the machine shop

41

C. M. Soukoulis (ed.), Photonic Band Gap Materials, 41–61.
© 1996 *Kluwer Academic Publishers. Printed in the Netherlands.*

technician, who must have spent weeks finishing a single structure. Although Yablonovitch suggested using reactive ion beam etching (RIBE) techniques to leapfrog the frequency performance directly to the optical regime[9], the initial results which came directly from the same research group were not encouraging. So, by mid-1992 it was clear a new geometry for building a full band gap structure at smaller scales was needed. In this tutorial, I will try to summarize our efforts to push the frequency limit to the far-infrared regime.

2. Layer-by-layer design: Microwave measurements

In Iowa State University, I worked with Kai-Ming Ho, Costas M. Soukoulis, and Che-Ting Chan (who had suggested the original diamond structure) as well as Rana Biswas, and Michael Sigalas. These theorists were well aware of the lack of a suitable geometry, and were in an intense search for a geometry suitable for down-scaling. A few months later they came up with a new geometry which seemed plausible. Figure 1 shows the structure which we decided to call "layer-by-layer structure" since it needs to be built by stacking the layers.

Of concern was not the layer. To obtain dielectric parallel rods was at least possible by conventional microfabrication methods. But the stacking part of the geometry did not have a clear solution at that time. However, we believed the new structure was superior to the existing structures in terms of down-scaling, thus we decided to concentrate our efforts on this new structure.

Although the simulations predicted a full band gap over a certain frequency range, we still needed to confirm this with experiment. For such a purpose, a large-scale model of the new structure would be ideal as it would be much easier to test at microwave frequencies. Therefore, we decided to build our large-scale structure out of cylindrical alumina rods (with a refractive index of 3.1) which were commercially available. The alumina rods were 0.318 ± 0.013 cm in diameter and 15.24 cm long and a center-to-center separation of 1.123 cm was chosen to obtain a dielectric filling ratio of ~0.26.

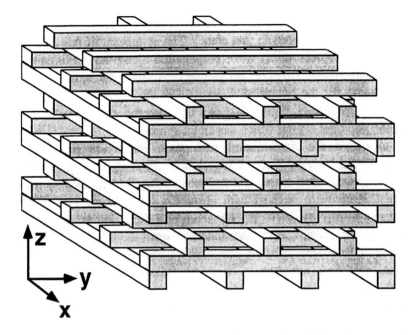

Figure 1. A schematic depicting the basic structure of the new photonic band gap crystal. The structure repeats every four layers in the stacking direction.

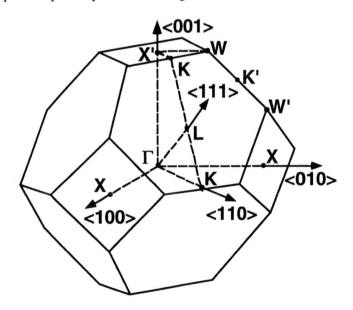

Figure 2. First Brillouin zone of the photonic gap structure which carries an fct symmetry. <001> direction corresponds to propagation along the stacking direction. <011> direction corresponds to propagation along the rod axes.

We define the plane of each rod layer as the x-y plane with the rod axes defining the (110) directions in this plane. The symmetry of this crystal is such that the electromagnetic (EM) wave propagation along the z axis is degenerate for both polarizations. However for propagation along the rods (110 direction), the propagation for the polarization vector **e** in the x-y plane is non-degenerate with **e** along the z axis. Figure 2 shows the corresponding first Brillouin zone (BZ) of the photonic crystal which has an fct symmetry.

We measured the transmission and phase dispersion properties of the structure using a Hewlett-Packard 8510A network analyzer[10]. Standard gain horn antennas were used to transmit and receive the EM radiation. Surroundings of the test setup were covered with absorbers to build an anechoic chamber resulting in a sensitivity of 70 dB. The setup was calibrated for phase and power measurements of the EM wave transmitted between the antennas. The structure was placed in the beampath of the EM wave transmitted from the source antenna. Additional absorbers are used to surround the side faces of the structure to minimize the leakage due to surface states. Two separate pairs of antennas were used to cover the 6-20 GHz measurement range. A pair of monopole antennas were used to obtain the phase dispersion at lower frequencies (1-6 GHz). Propagation characteristics along the z-axis were obtained by facing the top surface (x-y plane) of the structure to the source antenna. Propagation along (110) direction was obtained by facing the side surfaces of the structure to the source antenna. Different polarizations were achieved by rotating the antennas 90 degrees. To map out different regions of the first BZ, the incidence angle was changed by rotating the photonic gap crystal.

Transmission measurements were performed to determine the photonic gaps of the crystal for propagation along major axes. For propagation along the rods with the polarization vector **e** along the z-axis, the lower edge of the photonic gap occurred at 11.9 GHz, while the upper edge was at 14.0 GHz. This is very close to the calculated bandgap edges of 11.7 and 13.7 GHz. [11]. The experimental gap frequencies along several directions were in excellent agreement with the calculated gap values. We also mapped the first BZ and found the structure had a full band gap between 11.7 and 13.5 GHz.

The results we obtained from the large-scale model were very encouraging and now we knew we had a structure with a full band gap, we began searching for methods to fabricate such structures at smaller length scales.

3. Silicon Micromachined Photonic Crystals: 100 GHz Structure

In order to fabricate the structure at smaller scales, we began to look into semiconductor etching techniques. Semiconductors have a typical refractive index of 3, which yields a high enough contrast ratio with the air to achieve full photonic gap performance. During a discussion with Doug Robinson of ISU, he suggested using the anisotropic etching properties of silicon (110) surface. After reading a few articles[12,13] on silicon (110) oriented wafers, we determined this would be an ideal system for our new structure.

The new method utilizes the anisotropic etching of silicon by aqueous potassium hydroxide (KOH), which etches the {110} planes of silicon very rapidly while leaving the {111} planes relatively untouched. Thus, using (110)-oriented silicon, it was possible to etch arrays of parallel rods into wafers, and the patterned wafers can then be stacked in the correct manner to make the photonic bandgap crystal[14].

The (110) silicon wafers used in our work were each three inches in diameter and 390 µm thick. We chose relatively high-resistivity wafers (180 Ω-cm) in order to minimize absorption losses in the silicon. The fabrication was carried out in batches of twenty wafers at a time. The first step in the process was the growth of silicon dioxide on the front and back surfaces of the wafers. The oxide served as a mask during the anisotropic etching step, and we found that a 2.0 µm thick oxide was sufficient to provide the necessary protection. After oxidation, the front-surface oxide was patterned by conventional photolithography and buffered hydrofluoric acid etching. The pattern consists of 23 parallel rods, each 340 µm wide and separated by 935 µm wide gaps. These rod dimensions and the wafer thickness determine the center of the forbidden photonic gap—calculated to be 94 GHz in this case. The rods were aligned parallel to the {111} plane of the silicon, as defined by the major flat of the wafer. The stripes were 3.0 cm long so the 23 stripes form a square 3.0 cm x 3.0 cm pattern. A 1.15 cm wide border around the stripe array is protected by photoresist, with the outer regions of the wafer left exposed so a square wafer will be left after etching. Within the border region are four small rectangular openings that will serve as guide holes for the stacking process. A layer of photoresist protects the back surface oxide during patterning of the front surface.

After the oxide layer has been patterned, the wafers are dipped into an aqueous KOH etching solution. A typical etch performed in a 20% KOH solution at a temperature of 85°C takes approximately four hours to etch entirely through the wafer. Once the etch is completed, the remaining oxide is removed from both surfaces.

The individual wafers were stacked to form the photonic crystal using a holder having pins that align to the guide holes etched into the wafers. The guide holes are not symmetric with respect to the stripe patterns, but instead are offset slightly in the direction perpendicular to the rods. The separation between the edge of the rod pattern and the guide holes on one side is larger than the corresponding separation on the other side by one-half the rod-to-rod spacing. Thus, if one wafer is rotated by 180° with respect to a second wafer while keeping the guide holes aligned, the rods of the first wafer will be translated by one-half the rod spacing with respect to the rods of the second wafer. The photonic crystal is then easily assembled by stacking the patterned silicon wafers one-by-one on the holder, with a rotation of the holder by 90° before each wafer is put in place. This results in a stacking sequence that repeats every four layers, as shown in Fig. 1.

Once the fabrication was completed, the transmission properties of the crystal were measured with a W-band (75-110 GHz) measurement arrangement. A Ku-band frequency synthesizer was used to generate a signal that was first amplified and then multiplied in frequency by six times to reach the W band. The high-frequency signal was radiated by a standard-gain horn antenna (aperture size of 1.17 cm x 1.45 cm) with the photonic crystal in the path of the beam, and the transmitted radiation was collected by a second horn antenna. The amplitude of the received signal was measured using a harmonic mixer and a network analyzer. The setup was calibrated for power measurements of the EM wave transmitted from the source antenna. In order to improve the sensitivity of the test set up, EM absorbing pads were stacked around the test set to create a pseudo-anechoic environment. The resulting sensitivity was about 60 dB throughout the W-band.

Figure 4(a) shows the transmission measurement for propagation along the z-axis (which is degenerate for both polarizations along the rods) of a crystal which has 28 stacked wafers (7 unit cells). The drop-off in transmitted power, corresponding to the valence band edge for the crystal, starts at 81 GHz, which matches the calculated value exactly. The upper band edge for this orientation, predicted to be at 120 GHz, was beyond the upper measuring limit of the W-band test set and thus does not appear in Fig. 3(a). In the forbidden band, the average attenuation was 50 dB, a limit which was most likely due to leakage of the EM power around the sample, either through surface states of the crystal or through other leakage paths around the crystal.

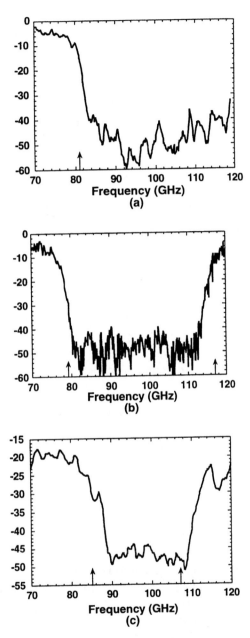

Figure 3. EM wave transmission through the micromachined crystal in which the wavevector of the incident EM waves is (a) normal to the wafer surfaces, (b) 35° from the normal to the wafer surfaces, and (c) parallel to the wafer surfaces. The arrows indicate calculated band edge frequencies.

The upper band edge was detected when the crystal was rotated so that the EM wave was incident at an angle of 35° with respect to the wafer surface normal (while the polarization vector **e** kept parallel to the wafer surfaces). Then the complete band gap fell within our measurement range, and both edges of the band gap were well-defined, as shown in Fig. 3(b). The valence band edge occurred at 76 GHz, while the conduction band edge was at 116 GHz, values in good agreement with the calculated values of 79 and 117 GHz.

For EM propagation in the stacking direction, we also measured the bandgap attenuation as a function of the number of stacked unit cells. The result was an average of 16 dB attenuation per unit cell, close to theoretical expectations of 17 dB per unit cell. This result is especially important for device considerations of these structures since three unit cells (for a total of 12 wafers) will yield a photonic band gap with 45-50 dB attenuation which is large enough for most applications.

Figure 3(c) shows the transmission characteristics for the case when the EM waves are incident on the side surface of a 40 wafer stack, with the polarization vector **e** along the z-axis. The lower edge of the gap was near 87 GHz, while the upper edge was near 110 GHz. Again, these values are in reasonably good agreement with the calculated band edges of 85 and 107 GHz. The average attenuation within the band gap was around 45 dB which was again mostly due to the leakage of the EM power around the crystal.

4. Extension of the Micromachining Method to the Far-Infrared

Although the structure used in this work was designed to have a photonic band gap near 100 GHz, the frequency range of this technology can be changed by simply scaling all the dimensions of the structure. The 100 GHz work was done by readily available wafers with a thickness of 390 μm. But, to extend this performance to sub-millimeter waves or the far-infrared regime, we had to find a company which could supply us with thin silicon (110) wafers. Virginia Semiconductor agreed to supply 100 μm thick wafers, and we designed a 500 GHz crystal using our theoretical simulation tools.

In order to build the 500 GHz crystal, we used fabrication techniques similar to the one used to build 100 GHz crystals. Fabrication again consisted of defining stripes that were parallel to (111) planes and subsequently etching the wafers in an KOH etch solution. The aforementioned anisotropic etching properties of aqueous KOH etch solutions resulted in parallel rods with rectangular cross sections, similar to one of the single layers described in Fig. 1.

The fabricated stripes were 2.0 cm long and 50 microns wide, separated by 185 micron wide gaps. A total of 86 parallel stripes were fabricated on each wafer, resulting in a square 2.0 x 2.0 cm pattern. The individual etched wafers were then stacked to form the photonic crystal using a holder with pins that aligned the guide holes that were etched through the wafers. The (110) silicon wafers used in this work were each two inches in diameter and 100 ± 5 µm thick. Relatively high-resistivity wafers (>100 Ω-cm) were chosen to minimize absorption losses in the silicon. During the course of fabrication, we were cautious in handling the thin silicon wafers which were susceptible to breakage. A layer of oxide, used to protect the back surface of the silicon wafer during the KOH etch, was intentionally kept after the etch. This gave a robust support to the relatively long stripes, preventing individual stripes from bending and bonding to the adjacent stripes[15].

Testing was performed by using a terahertz free space spectroscopy set-up shown in Fig. 4 [16,17]. EM pulses were generated and detected by an all-electronic approach designed and built around nonlinear transmission line technology along with magnetic dipole (slot) antennas. Two phaselocked microwave synthesizers with an offset frequency of 7 Hz were used to drive the generator and detector circuits. A spectrum analyzer was used to directly compute the Fourier spectra of the detected signals. The dynamic range of the system was around 30 dB for frequencies up to 550 GHz. We carried out the free space spectroscopic measurements of the photonic crystal by placing the structure on the beampath of the radiated signal. By comparing the detected signals with and without the crystal in place, we obtained the phase and magnitude transmission properties of the structure as a function of frequency. Since the pulses were periodic, the frequency domain information was limited to the harmonics of the input signal frequency (6.4 GHz). The setup used high resistivity silicon lenses to collimate and focus the radiation generated and detected by the antennas. The resulting test beam was highly collimated where 95% of the output radiation remained within a 10° radiation cone.

We used a structure consisting of 16 stacked silicon wafers (corresponding to 4 unit cells) for transmission measurements. Characteristics along the stacking direction were obtained by placing the structure on the beam path, so the transient radiation propagated in a plane perpendicular to the top surface of the structure.

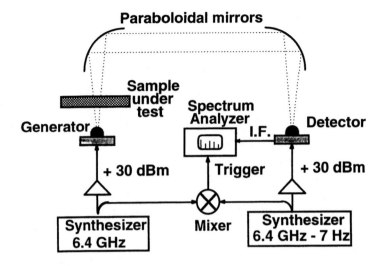

Figure 4. Diagram of the all electronic THz free space spectroscopy system.

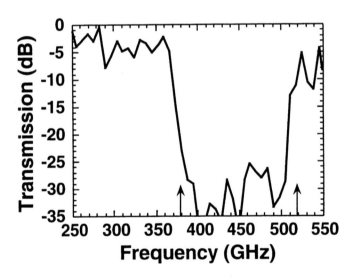

Figure 5. EM wave transmission through the micromachined crystal in which the wavevector of the incident radiation is normal to the wafer surfaces. The arrows indicate calculated band edge frequencies.

For the propagation along the stacking direction, the transmission characteristics were essentially independent of the polarization of the incident radiation as the structure had a four-fold axis of symmetry along this direction. The only polarization dependence came from the orientation of the first layer rods with respect to the polarization of the incident radiation. Although this resulted in an expected band gap edge difference between two polarizations, the difference (~3 GHz) was smaller than the minimum frequency resolution of the experimental set-up. We placed the structure about 1 cm away from the generator chip, so most of the incident radiation remained perpendicular to the surface. Figure 5 shows the transmission characteristics of the propagation along the stacking direction. The lower edge of the photonic gap is at 370 GHz, while the upper edge is at 520 GHz. This is very close to the calculated bandgap edges of 378 and 518 GHz. The average measured attenuation within the band gap was around 30 dB, limited by the dynamic range of the experimental setup. Our calculations for the band gap frequencies predicted the attenuation to be around 65 dB.

The transmission characteristics of the structure in other crystal directions could not be measured due to experimental limitations. In order to measure the transmission from the side surface of the crystal, we needed a crystal that was at least 2 cm thick. This corresponded to a structure with 200 wafers, too expensive to build. Although it was possible to make transmission measurements (along other crystal directions) by rotating the crystal, the refraction at the surface of the crystal resulted in a misalignment of the detected signal by steering the beam away from the original path. This misalignment resulted in uncalibrated measurements when we rotated the crystal from its original orientation. This prevented us from measuring the full band gap of the structure in all propagation directions. Although we were able to measure the properties of the photonic band gap only along the stacking direction, our theoretical calculations predicted a full three-dimensional band gap from 425 GHz (which occurs when the sample is rotated 55 degrees) to 518 GHz (which occurs along the stacking direction).

By using special silicon thinning methods and double etching[18] the wafers on both surfaces, we expect the frequency range of this fabrication technology could be extended to build structures with photonic bandgaps as high as 3 THz. We also expect the same three-dimensional dielectric design to be useful at even smaller scales for the eventual goal of fabricating photonic band gap structures at infrared and optical frequencies.

5. Defects

The new layer-by-layer structure exhibits a sizable and robust photonic band gap over a wide range of structural parameters. The aforementioned performance also puts the new structure in the frequency range where a number of millimeter and sub-millimeter wave applications have been proposed, including efficient millimeter wave antennas, filters, sources, and waveguides [19-22]. However, most of these applications are based on the presence of defect or cavity modes which are obtained by locally disturbing the periodicity of the photonic crystal [3,23,24]. The frequencies of these modes lie within the forbidden band gap of the pure crystal, and the associated fields are localized around the defect.

It was clear we needed to investigate the existence of such cavity structures built around the layer-by-layer PBG crystal. In this investigation of defect structures we followed the approach of Yablonovitch *et al.* [3] wherein we either removed portions of the dielectric lattice or inserted extra dielectric materials into the air gaps. The basis for our defect study was again the large-scale model used for microwave measurements which exhibited a full three-dimensional bandgap between 11.7 and 13.5 GHz.

We first investigated the properties of defect structures formed by removing a portion of a single rod, as illustrated in Fig. 6(a). The length of the missing section is d, and to facilitate possible comparison to similar PBG crystals with different lattice constants, we define the defect volume ratio d/a where a is the center-to-center separation between rods of the PBG crystal. We measured the transmission and phase dispersion properties of structures with defects using a Hewlett-Packard 8510A network analyzer with standard gain horn antennas to transmit and receive the EM radiation.

Figure 6(b) shows the transmission characteristics of propagation along the z-axis (the stacking direction) for a structure with 4 unit cells (16 layers) along the z-axis. The defect was placed at the 8th layer of the structure and had a d/a ratio of 1. The electric field polarization vector of the incident EM wave e was parallel to the rods of the defect layer. The lower edge of the band gap along this propagation direction starts at 10.7 GHz, while the upper edge is around 17.4 GHz. The defect transmission was centered at 12.85 GHz, and the peak transmission was 30 dB below the incident signal. Using an expanded frequency scale, we measured the Q-factor (quality factor, defined as the center frequency divided by the peak's full-width at half-maximum) of the peak to be greater than 1,000.

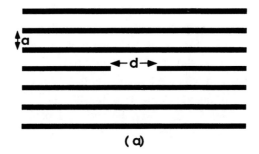

(a)

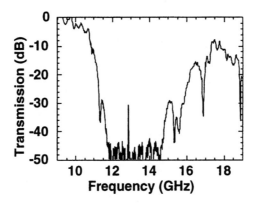

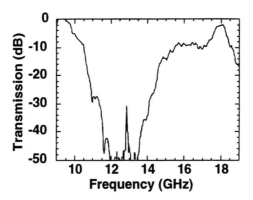

Figure 6. (a) Missing material defect structure is obtained by removing a part of the alumina rod. The removed part has a width of *d*, while the center-to-center separation between two adjacent rods is *a*. (b) Transmission characteristics (along the stacking direction) obtained from a defect structure with a *d/a* ratio of 1. (c) Transmission characteristics of the same defect structure when it is rotated 35 degrees. The defect frequency remains constant at different angles.

We have measured the defect transmission at different incidence angles by rotating the sample with respect to the antennas while keeping *e* parallel to the rods of the defect layer. Figure 6(c) shows the transmission properties of the same defect structure when the sample was rotated 35°. Although the upper and lower edges of the gap changed significantly due to different propagation directions, the defect frequency remained the same (12.85 GHz). We also found the transmission via the defect was polarization dependent — no defect mode was present when *e* was perpendicular to the rods of the defect layer.

We used the transfer-matrix method (TMM), recently introduced by Pendry and MacKinnon[25], to calculate the EM transmission through a photonic crystal with defects[26,27]. Since the TMM method requires periodicity in the directions parallel to the interfaces, we examined the case of a periodic defect, made by removing every other rod from a single layer (see Fig. 7(a)). The experiments are done using a structure consisting of 4 unit cells (16 layers), where the 8th layer is chosen as the defect layer. Figure 7(b) compares the predicted theoretical transmission with the experimental results. As can be seen from the plot, the agreement between theory and experiment is especially good for the defect frequency. The measured defect frequency is 12.62 GHz, while theory predicts 12.58 GHz. Again, we observed no defect modes when *e* is perpendicular to the defect rods. This is in accordance with theoretical results which show the peak transmission for the waves with *e* parallel to the removed rods is three orders of magnitude higher than the peak transmission obtained from the other polarization. Figure 7(c) shows the characteristics of the cavity mode on an expanded frequency scale. Although the measured Q-factor (750) of the cavity mode is very close to the theoretical Q-factor (800), the measured peak transmission of −17 dB is far below the predicted value of −4 dB. The discrepancy is probably due to the finite length of the rods of the test structure, while our theoretical calculations assumed a crystal made of infinitely long rods. Calculations show the defect frequency and Q-factor are not very sensitive to separation between the defects. In particular, the defect frequency increases from 12.58 to 12.68 to 12.39 GHz as separation between the removed rods increases from 2*a* to 3*a* to 4*a*. Also, the Q-factor increases rapidly for increasing thickness of the PBG crystal. For a structure with 8 unit cells (32 layers), where the defect layer structure shown in Fig. 7(a) is chosen as the 16th layer, calculations predict a defect frequency at 12.61 GHz with a Q-factor greater than 1.4×10^6. This suggests the possibility of building cavity structures with very high Q-factors from this geometry.

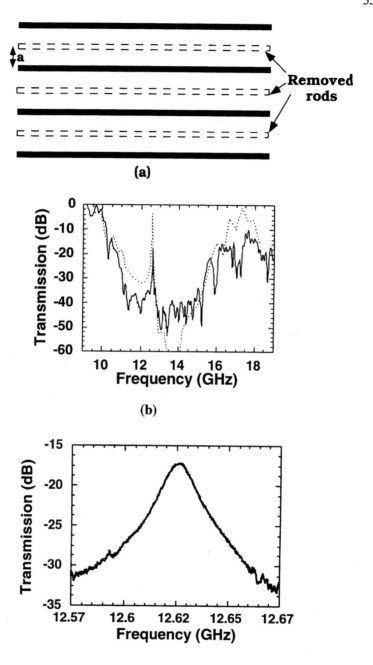

Figure 7. (a) Every other rod of a single layer is removed to generate the defect layer. (b)Comparison of the theoretical (dashed line) and experimental (solid line) transmission characteristics of the defect structure. (c) Expanded frequency scale for the defect mode which has a Q-factor of 750.

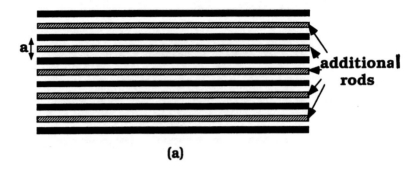

(a)

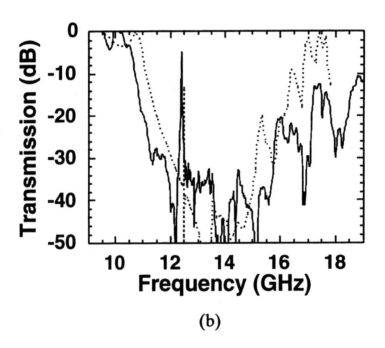

(b)

Figure 8. (a) The defect layer is built by additional rods are placed at the midpoints of adjacent rods. (b) Comparison of the theoretical (dashed line) and experimental (solid line) transmission characteristics of the defect structure.

Finally, we used TMM to study a periodic additive defect. Figure 8(a) depicts the periodic defect layer, in which an extra dielectric rod was inserted halfway between each lattice rod in one layer of the PBG crystal. This defect layer was the 5th layer of a structure consisting of 3 unit cells (12 layers total). Theoretical and experimental spectra for transmission in the z-direction are compared in Fig. 8(b). The experimental (12.4 GHz) and theoretical (12.5 GHz) defect frequencies are in good agreement along with experimental and theoretical peak transmissions, which are both ~10 dB lower than the incident signal. However, theory predicts large Q-values (>10,000) while the measured Q-values are approximately 2,000. Again, the difference is likely due to the finite size of the experimental PBG structure.

6. Extension of the Defects to the Millimeter-Wave Regime

In our initial investigations of defects in millimeter-wave structures, we looked at defects incorporated into our earlier silicon micromachined structures. Although silicon micromachining has proven to be very successful for building three-dimensional photonic crystals, we found that defects incorporated into our structures had relatively poor qualities. In particular, peak transmission of the defect mode was low (-20 dB below the incident signal) and the Q-factor (quality factor, defined as the center frequency divided by the peak's full-width at half-maximum) of the resonance was also low (~ 100). We attribute these effects to the relatively low resistivity of our silicon wafers (30-180 ohm-cm). Our calculations show the transmission at the top of the defect peak and the Q-factor decrease drastically as one increases the conductivity of the Si wafers. As an alternative to using even higher resistivity silicon, we investigated the use of alumina as the dielectric material for defect structures. Alumina with a typical resistivity six orders of magnitude larger than the resistivity of silicon would be a good dielectric material to build high-Q cavity structures. To build alumina-based millimeter wave photonic crystals, we patterned the wafers by means of laser machining.

The square-shaped alumina wafers (Kyocera 96% alumina: resistivity $> 10^{10}$ Ω-cm, refractive index = 2.86 @ 100 GHz) used in this work were each 5.1 cm x 5.1 cm, and 400 µm thick. The pattern consisted of 24 parallel stripes, each 460 µm wide and separated by 840 µm wide gaps for a filling ratio of 0.36. These stripe dimensions and the wafer thickness determine the center of the forbidden photonic gap—calculated to be 94.7 GHz in this case. The stripes are 3.1 cm long so the 24 stripes form a square 3.1 cm x 3.1 cm pattern. The stripes are cut using a high power carbon dioxide laser[28].

58

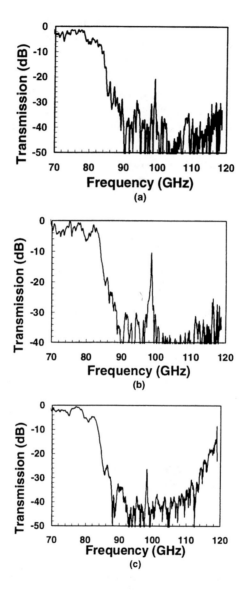

Figure 9. (a) Transmission characteristics (along the stacking direction) obtained from a defect structure with a *d/a* ratio of 2. (b) Transmission characteristics (along the stacking direction) obtained from a defect structure with a *d/a* ratio of 24. (c) Transmission characteristics of the same defect structure(*d/a* ratio=24) when it is rotated 25°. The defect frequency remains constant at different angles.

The laser-machined wafers were stacked to form the photonic crystal, using a holder with pins aligning the guide holes machined through the alumina wafers. Using a W-band network analyzer, the transmission characteristics of the crystal were measured. The transmission characteristics of the structure along the several directions are all in good agreement with our theoretical calculations which predicted a full band gap from 89.6 to 99.8 GHz.

In addition to testing the transmission properties of the new crystal, we also fabricated defect structures by removing material from the crystal. The defect structure is similar to the one depicted in Figure 6 (a) where a portion of one of the alumina stripes is removed. Figure 9(a) shows the transmission characteristics of propagation along the z-axis (the stacking direction) for a structure with 4 unit cells (16 layers) along the z-axis. The defect was placed at the 8th layer of the structure and had a d/a ratio of 2. The electric field polarization vector of the incident EM wave e was parallel to the rods of the defect layer. The defect transmission was centered at 99.2 GHz, and the peak transmission was 20 dB below the incident signal. Using an expanded frequency scale, we measured the Q-factor of the peak to be 1,030. Figure 9(b) shows the transmission properties of a similar defect with a d/a ratio of 24 which corresponds to a single missing rod for our structure. The defect frequency is centered at 98.6 GHz, and has a peak transmission of 10 dB below the incident signal.

We also measured the defect transmission at different incidence angles by rotating the sample with respect to the antennas while keeping e parallel to the rods of the defect layer. Figure 9(c) shows the transmission properties of the same defect structure (d/a=24) when the sample is rotated 25°. Although the upper and lower edges of the gap changed significantly due to different propagation directions, the defect frequency remained the same (98.6 GHz). To our knowledge, these results correspond to the first demonstration of a three-dimensional photonic band gap defect structure at millimeter wavelengths.

7. Conclusion

In this tutorial article, I tried to summarize our efforts to push the frequency performance of the new layer-by-layer photonic band gap structure. We were able to demonstrate a 500 GHz crystal, along with defects around 100 GHz. Besides further down-scaling, we also expect the new structures to find use in millimeter wave sub-millimeter wave applications.

60

Acknowledgments

I would like to acknowledge G. Tuttle, R. Biswas, M.M. Sigalas, C.T. Chan, C. M. Soukoulis, and K. M. Ho for sharing their expertise with me during the course of the photonic band gap research described in this tutorial. Ames Laboratory is operated for the U.S. Department of Energy by Iowa State University under contract No. W-7405-Eng-82. This work is supported by the Director for Energy Research, office of Basic Energy Sciences and Advanced Energy Projects, the Center for Advanced Technology Development at ISU, and the Scalable Computing Laboratory which is funded by ISU and Ames Laboratory.

8. References

1. K. M. Ho, C. T. Chan, and C. M. Soukoulis, Phys. Rev. Lett. **65**, 3152 (1990).
2. E. E. Yablonovitch, T. J. Gmitter, and K. M. Leung, Phys. Rev. Lett. **67**, 2295 (1991).
3. E. Yablonovitch et al., Phys. Rev. Lett. **67**, 3380 (1991).
4. S. John, and J. Wang, Phys. Rev. Lett. **64**, 2418 (1990).
5. J. Martorell and N. M. Lawandy, Phys. Rev. Lett. **65**, 1877 (1990)
6. W. M. Robertson, G. Arjavalingam, R. D. Meade, K. D. Brommer, A. M. Rappe, and J. D. Joannopoulos, Phys. Rev. Lett. **68**, 2023 (1992).
7. E. Yablonovitch, and T. J. Gmitter, Phys Rev. Lett. **63**, 1950 (1989)
8. C. T. Chan, K. M. Ho, and C. M. Soukoulis, Europhysics Lett. **16**, 563 (1991).
9. E. Yablonovitch, in *Photonic Band Gaps and Localization*, C. M. Soukoulis, Ed. (Plenum, New York, 1993) p. 207.
10. E. Ozbay, A. Abeyta, G. Tuttle, M. Tringides, R. Biswas, C. Soukoulis, C. T. Chan, and K. M. Ho, Phys. Rev. **B50**, 1945 (1994).
11. K. M. Ho, C. T. Chan, C. M. Soukoulis, R. Biswas, and M. Sigalas, Solid State Comm. **89**, 413 (1994).
12. E. Bassous, in *Symposium on Electrochemical Technology in Electronics*, L. T. Romankiw, T. Osaka, Eds. (Electrochemical Society, Penington, 1987), p. 619.
13. D. L. Kendall and G. R. d. Guel, in *Micromachining and Micropackaging of Transducers*, C. D. Fung, P. W. Cheung, Eds. (Elsevier Science Publishers, Amsterdam, 1985), p. 107.
14. E. Ozbay, E. Michel, G. Tuttle, M. Sigalas, R. Biswas, and K. M. Ho, Appl. Phys. Lett., **64** 2059 (1994).
15. E. Ozbay, E. Michel, G. Tuttle, R. Biswas, K. M. Ho, J. Bostak, and D. M. Bloom, Optics Lett. **19**, 1155 (1994).
16. D. W. Van Der Weide, J. S. Bostak, B. A. Auld, and D. M. Bloom, Appl. Phys. Lett. **62**, 22 (1993).
17. J. S. Bostak, D. W. Van Der Weide, I. Aoki, B. A. Auld, and D. M. Bloom, in proceedings of *UltraFast Electronic and Optoelectronics'93*, ed. J. Shah, and U. Mishra, p. 112, (1993).
18. E. Ozbay, G. Tuttle, M. Sigalas, R. Biswas, K. M. Ho, J. Bostak, and D. M. Bloom, Appl. Phys. Lett. **65**, 1617 (1994).
19. See the articles in the J. Opt. Soc. Am. B **10** (1993), a special feature edited by C.M. Bowden, J.P. Dowling, and H.O. Everitt.

20. E. R. Brown, C. D. Parker, and E. Yablonovitch, J. Opt. Soc. B **10**, 404 (1993).
21. E.R. Brown, C.D. Parker, and O.B. McMahon, Appl. Phys. Lett. **64**, 3345 (1994).
22. C.J. Maggiore, A.M. Clogston, G. Spalek, W.C. Sailor, and F.M. Mueller, Appl. Phys. Lett. **64**, 1451 (1994).
23. S. L. McCall, P. M. Platzman, R. Dalichaouch, D. Smith, and S. Schultz, Phys. Rev. Lett. **67**, 2017 (1991).
24. K. M. Leung, J. Opt. Soc. B **10**, p. 303 (1993).
25. J. B. Pendry and A. MacKinnon, Phys. Rev. Lett. **69**, 2722 (1992).
26. M. M. Sigalas, C. M. Soukoulis, E. N. Economou, C. T. Chan and K. M. Ho, Phys. Rev. **B48**, 14121 (1993).
27. M. M. Sigalas, C. M. Soukoulis, C. T. Chan and K. M. Ho, Phys. Rev. **B49**, 11080 (1994).
28. The laser machining was performed by Accumet Engineering, Hudson, MA.

FABRICATION OF THREE-DIMENSIONAL PHOTONIC BAND GAP MATERIAL BY DEEP X-RAY LITHOGRAPHY

G. FEIERTAG[1], W. EHRFELD[1], H. FREIMUTH[1],
G. KIRIAKIDIS[2], H. LEHR[1], T. PEDERSEN[2,3], M. SCHMIDT[1],
C. SOUKOULIS[2,4] AND R. WEIEL[1]

[1]IMM Institut für Mikrotechnik GmbH, Carl-Zeiss-Straße 18-20,
D-55129 Mainz, Germany
[2]Foundation for Research and Technology Hellas (FORTH),
P.O. Box 1527, Vassilika Vouton, Heraklion, Crete, Greece
[3]Electrolux EPD HOT Tech-Centre, Sjaellandsgade 2, DK-7000
Fredericia, Denmark
[4]Ames Laboratory and Department of Physics and Astronomy,
Iowa State University, Ames, IA 50011, USA

Abstract. Yablonovitch et al. [1] constructed the "three cylinder" dielectric structure which exhibits a photonic band gap in the whole Brillouin zone. They implemented the diamond structure proposed by Ho et al. [2,3]. This structure can be fabricated mechanically by drilling three sets of holes 35° off vertical into the top of a solid dielectric. However, extremely small dimensions, which are required for high frequencies, cannot be achieved. This problem can be solved by using deep X-ray lithography, which allows the fabrication of structures that can be used up to the infrared range. Three irradiations were performed in which the tilted arrangement of mask and resist was rotated each time by 120°. PMMA resist layers with a thickness of 500 microns were irradiated at the DCI storage ring in Orsay, France. The lattice constants of the structures were 227 and 114 microns corresponding to midgap frequencies of 0.75 and 1.5 THz, respectively.

Since the dielectric constant of the PMMA is not high enough for the formation of a photonic band gap, a moulding step must be applied. The holes in the resist structure were filled with a solution of polyvinylsilazane in tetrahydrofuran. After the evaporation of the solvent, the samples were pyrolyzed at 1100°C under N_2 atmosphere. The resist decomposes into CO_2, CH_4, CO, and H_2O, whereas polyvinylsilazane is transformed into a SiCN ceramic. A lattice of ceramic rods corresponding to the holes in the resist structure remained.

C. M. Soukoulis (ed.), Photonic Band Gap Materials, 63–69.
© *1996 Kluwer Academic Publishers. Printed in the Netherlands.*

1. Introduction

It has been known for many years that the periodic arrangement of atoms in crystals leads to the formation of the energy bands of the electrons. A few years ago it was realised that the three-dimensional periodic arrangement of dielectric material results in energy bands for photons [4,5]. The interesting properties and possible applications have inspired researchers to look for dielectric structures which open up a photonic band gap [2,3,6]. Yablonovitch et al. [1] proposed a face-centered cubic structure which can be fabricated by drilling three sets of holes 35° off vertical into the top of a solid dielectric. Mechanical drilling and ion etching have their limitations when processing small structures with sufficient depth. We introduced deep X-ray lithography into the fabrication of photonic band gap material to overcome these limitations.

In this paper we describe the process which was used to produce photonic band gap material with lattice constants of 114 and 227 microns. The three-dimensional patterning is done by x-ray lithography with synchrotron radiation. To make high ε ceramic structures, a moulding step was applied.

2. Deep X-ray Lithography

The process of deep X-ray lithography and the following steps of the LIGA process are described in detail in [7] and [8]. A pattern is transferred from a mask into a thick layer of a radiation sensitive polymer (resist). For most applications the direction of the beam is perpendicular to the resist surface and the mask plane. When synchrotron radiation is used, the edges of the structured resist are extremely steep and precise.

To fabricate photonic band gap materials multiple irradiations, where mask and resist are tilted with respect to the beam, are necessary. The structure proposed by Yablonovitch et al. [1] can be fabricated by three tilted irradiations ("three cylinder" structure). Mask and resist must be tilted by 35° with respect to the beam. Between the irradiations the tilted arrangement of mask and resist must be rotated each time by 120°. For the "three cylinder" structure, the formation of a photonic band gap is only possible if the ratio of the refractive indices is larger than 2.1 [3]. Radiation sensitive polymers having a large enough refractive index in the 1 THz frequency range are not available yet. Instead, we considered adding a high refractive index ceramic powder in a resist material. Unfortunately, most ceramics have high X-ray absorption coefficients making the pattern transfer in a thick resist layer filled with ceramic powder by X-ray lithography difficult.

As a consequence we have chosen a fabrication process which comprises two main steps. The deep X-ray lithography is followed by a moulding step, where the holes in the resist are filled by a high refractive index ceramic material or a metal.

2.1. DESIGN OF THE MASK

A deep X-ray lithography mask with triangular arrays of holes in a 12 micron thick Au absorber was made. The mask comprises areas of 3.5 by 3 mm^2 for different lattice constants and filling ratios. Two lattice constants, 114 and 227 microns and seven different filling ratios for each were designed.

To calculate the volume of the crystal, which is produced by deep X-ray lithography, a Monte Carlo integration was performed. The calculation takes into account that the absorber has a thickness of 12 micron and that only points are irradiated that are within the holes at the bottom and the top of the absorber. The filling ratios obtained by the Monte Carlo integration are shown in Table 1.

TABLE 1. Diameters of the holes in the absorber of the mask and filling ratios of lattice to remaining resist.

Lattice Constant 227 micron		Lattice Constant 114 micron	
Diameter [micron]	Filling Ratio	Diameter [micron]	Filling Ratio
25.9	0.039	15.7	0.033
36.3	0.068	20.7	0.079
46.7	0.148	25.8	0.140
57.2	0.222	30.9	0.213
67.7	0.306	36.1	0.296
78.2	0.396	41.4	0.385
88.6	0.488	46.6	0.477

2.2. IRRADIATION

The irradiations were carried out at the synchrotron radiation source DCI in Orsay, France. We used an X-ray scanner built by JENOPTIK GmbH with a special module for tilted and rotated irradiations.

Mask and resist were fixed in a cassette. The distance of the mask to the surface of the resist was set to 100 microns. The thickness of the resist was 500 microns, in order to achieve a little more than two lattice constants for the 227 microns structure. Due to the low divergency of synchrotron radiation, the vertical width of the radiation fan is only some mm. To irradiate a larger area the mask and resist cassette must be scanned. We used polymethylmethacrylate (PMMA) as the resist material. PMMA can be dissolved in the developer if a dose of more than 4 KJ/cm^3 is deposited. For single normal exposures, we chose the irradiation time and the spectral distribution of the radiation in a way that 5 KJ/cm^3 are deposited at the bottom and less than 20 KJ/cm^3 at the top of the resist. When a dose of more than about 20 KJ/cm^3 is reached, PMMA is transformed into a powder whose volume is larger than the volume of the PMMA. To avoid damage to the microstructures, we usually keep the dose below 20 KJ/cm³ for single exposures. The channels irradiated for the photonic band gap structures intersect, so the dose in the intersection points is three times the dose of a single irradiation. Within a reasonable irradiation time we could not reach a dose of less than 40 KJ/cm³ in the intersection points at the top of the resist. We were afraid of cracks due to the increase in volume in these regions however, the PMMA structures came out perfectly well.

Figure 1 is a picture of a 500-micron thick PMMA plate with channels forming a band gap structure. The picture was taken with an optical microscope so the channels can be seen in the transparent PMMA. Figure 2 shows the intersection of three channels near the top of a PMMA plate. In the next process step, these PMMA structures are used as moulds to make the ceramic band gap material.

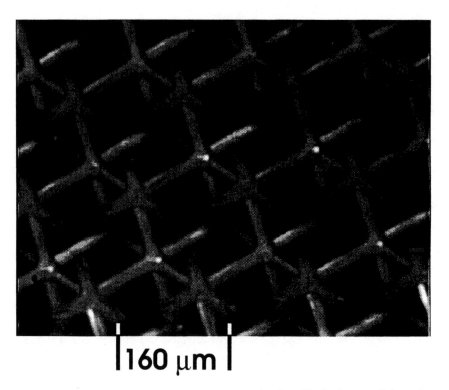

160 µm

Figure 1.   Channels in a 500-micron thick PMMA resist plate. The lattice out of channels can be seen inside the transparent PMMA.

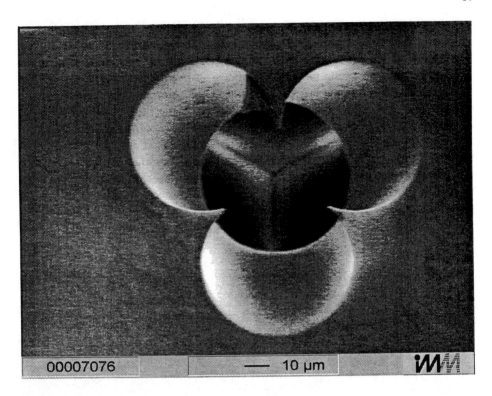

00007076   ——— 10 µm

Figure 2.   Intersection of three holes in PMMA resist.

3. Ceramic Structures

In order to transform the photonic band gap structure into a material with a sufficiently high dielectric constant, a moulding step had to be applied. We decided to use a preceramic polymer (polyvinylsilazane) which can be transformed into a siliconcarbonitride ceramic by a pyrolysis process. Since a solution of polyvinylsilazane in tetrahydrofurane (THF) wets the surface of PMMA, a complete filling of the microchannels can be achieved by simply putting a drop of the solution onto the PMMA structures. After the evaporation of the THF, a solid polymer remains. All process steps must be carried out in an inert atmosphere because the preceramic polymer is sensitive to air and humidity. Prior to the pyrolysis the polysilazane must be crosslinked so that it becomes unmeltable. This is simply done by exposure to air. The complicated geometric structure of photonic band gap material precludes the separation of the PMMA form the preceramic polymer. Therefore the PMMA must be removed during the firing as a lost mould. The pyrolysis must be carried out very carefully to minimise crack formation in the ceramic. The delicacy of this process can be understood if one realises that the transformation of polyvinylsilazane into siliconcarbonitide ceramic is accompanied by a volume shrinkage of about 50%. The samples were pyrolysed at T=1100°C under nitrogen atmosphere. At least partially crack-free ceramics can be obtained if the heating rate does not exceed 5 K/h. A detailed description of the pyrolysis process is given in [9]. A scanning electron microscope picture of a ceramic photonic band gap material is shown in Figure 3.

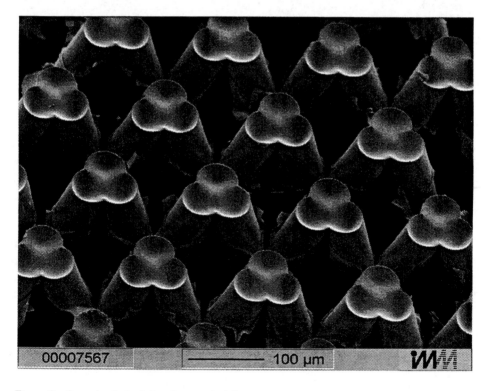

Figure 3. Ceramic photonic band gap material.

4. Summary

It has been demonstrated that deep X-ray lithography is suitable to produce photonic band gap *structures* with small lattice constants which cannot be fabricated by using other methods. The moulding step required to transform photonic band gap *structures* into ceramic photonic band gap *material* has also been shown.

Further work will include shifting the lattice constants of the photonic band gap materials into a range of less than 10 μm. Improvements to the moulding step to obtain larger pieces of photonic band gap material which are suitable for transmission measurements need to be carried out. Precursor materials other than polyvinylsilazane will be also tested.

5. Acknowledgements

This work was performed within the framework of the Human Capital and Mobility research network *Microfabrication with synchrotron radiation* supported by the European Commission under contract No. ERB CHRX CT93 0394.

C.S. works at the Ames Laboratory which is operated by the U.S. Department of Energy by Iowa State University under contract No. W-7405-Eng-82. His work was supported by the Director for Energy Research, Office of Basic Energy Sciences.

We want to thank Alain Labéque and Stefan Megtert from the French LIGA group for their support at the DCI synchrotron radiation source.

References

1. E. Yablonovitch, T.J. Gmitter and K.M. Leung, Phys. Rev. Lett. **67**, 2295 (1991).
2. K. M. Ho, C. T. Chan and C. M. Soukoulis, Phys. Rev. Lett. **65**, 3152 (1990).
3. C. T. Chan, K. M. Ho and C. M. Soukoulis; Europhys. Lett., **16** (6), 563-568 (1991).
4. E. Yablonovitch, Phys. Rev. Lett. **58**, 2059 (1987)
5. S. John, Rev. Lett. **58**, 2486 (1987)
6. E. Yablonovitch and T.J. Gmitter, Phys. Rev. Lett. **63**, 1950 (1989)
7. W. Ehrfeld. and D. Münchmeyer; Nuc. Inst. and Meth. in Phys. Res. **A 303**, 523 (1991).
8. W. Ehrfeld and H. Lehr; Radiat. Phys. Chem. **45**, No. 3, pp. 349 (1995).
9. H. Freimuth, V. Hessel, H. Kölle, W. Ehrfeld, T. Vaahs, M. Brück, submitted to: J. Am. Ceram. Soc.

BLOCH WAVE OPTICS IN PHOTONIC CRYSTALS: PHYSICS AND APPLICATIONS

P.ST.J. RUSSELL and T.A. BIRKS
Optoelectronics Research Centre,
University of Southampton,
Southampton SO17 1BJ,
United Kingdom

1. Introduction

We aim in this chapter to provide an introduction to the rich tapestry of physical phenomena involving wave propagation in wavelength-scale periodic structures, and to explore briefly their significance in present day research and technology.

The first clue to the presence of a periodic structure is often the reflection of waves at certain specific wavelengths and angles of incidence. The natural world is full of visually attractive examples of this - moth and butterfly wings [1,2], snake and fish scales [3], bird feathers, and some gem stones [4-6] show bright flashes or "rainbows" of colour upon rotation in sunlight. More and more examples of synthetic self-organised periodic structures are also emerging: certain co-polymer mixtures can form with regular arrays of cavities [7]; suspensions of mono-dispersed particles in liquids can crystallise into three-dimensional lattices [8]; iridoviridae self-organise into crystalline lattices [9]; and multiply periodic intensity patterns produced by interfering laser beams can be used to force atoms into regular arrays [10].

Periodic structures have of course been studied in many different areas of physics and technology, examples being phonons in atomic lattices, acousto-optical diffraction in solids and liquids [11], X-ray, electron and neutron diffraction in crystals [12], distributed feedback lasers [13,14], wavelength filtering in optical communications, spectrometry, transmission electron microscopy, multilayer coatings, phased-array microwave antennae, holograms [15] and — last but not least — electronic band structure [16]. Common to the physics underlying all these devices and phenomena are *stop-bands*: ranges of angle and frequency where waves are blocked from travelling through the periodic structure. The practical consequence is that light incident from an isotropic external medium is either reflected, or tunnels through to the other side; it cannot exist freely in the periodic medium itself. On the k-k and ω-k dispersion diagrams the stop-bands appear as *anti-crossing* points, and are also known variously as band gaps, momentum-gaps and regions of evanescence. In only a few cases, however, do they become so strong and numerous that they coalesce to cover all of wavevector space, forbidding

71

© 1996 *Kluwer Academic Publishers. Printed in the Netherlands.*

propagation in all directions within limited bands of frequency — the *photonic band gaps* [17].

Prior to the present interest in photonic band gaps, research had been carried out on photonic band structure and propagation in *strongly modulated* singly and multiply periodic planar waveguides [18-22]. Although the aim of that work was not to achieve a full photonic band gap, it generated a number of fabrication, characterisation and conceptual techniques helpful in working out the wider implications of photonic band structure in optoelectronics. The central theme was the use of Bloch waves (the normal modes of electromagnetic propagation in periodic media) and the development of an optics based on them [20]. Group velocity (which could be derived directly from the wavevector diagram) played a central role, leading to an understanding of phenomena such as negative and multiple refraction and Bloch wave interference [21]. The concepts of Bloch wave optics led to the development of some unique devices [19], and the approach also permitted accurate and detailed explanations for the complex and often beautiful phenomena that were seen in the periodic waveguides [21, 22,24].

In section 2, a brief illustrative introduction to wavevector diagrams and their uses is given. Simple approximate solutions for the scalar dispersion relation in the two-dimensional square and hexagonal cases are derived in section 3. These are used to illustrate a Hamiltonian optics description of the behaviour of Bloch waves in inhomogeneous photonic crystals (section 4), results which are relevant to the modelling of the fields at defects. Bloch wave interference and scattering are treated in sections 5 and 6, and nonlinear effects discussed in section 7. A brief foray into applications occupies section 8, and conclusions drawn in section 9.

2. Wavevector Diagrams and Their Uses

We shall be making extensive use of plots of allowed wavevectors in two dimensions at fixed optical frequency - the *wavevector diagram*. This is very valuable in discussions of the propagation of light in photonic crystals. For an isotropic medium, the diagram is a circle of constant radius. For an anisotropic birefringent medium it is an ellipse with semi-axes of length $2\omega n_{ord}/c$ and $2\omega n_{ext}/c$ where n_{ord} and n_{ext} are the ordinary and extraordinary refractive indices. The group velocity corresponding with a point on the curves in wavevector space is given by:

$$\mathbf{v}_g = \nabla_\delta \omega(\delta) = \frac{\partial \omega}{\partial \delta} = \frac{d\omega}{d\vartheta}\frac{\partial \vartheta}{\partial \delta} = \frac{c}{2n_o}\frac{\partial \vartheta}{\partial \delta} \tag{1}$$

and points normal to the curves on the wavevector diagram, in the direction of increasing frequency. An illustrative example of how the wavevector diagram assists in working out ray directions and boundary conditions is given in Figure 1. A ray incident from air on a parallel slab of birefringent crystal is refracted in two different ways depending on the state of electric field polarisation. In both cases the component of wavevector parallel to the slab boundaries (the minimum distance joining the origin of wavevector space to the construction line) is conserved. For polarisation in the isotropic plane, Snell's law is obeyed as usual. For

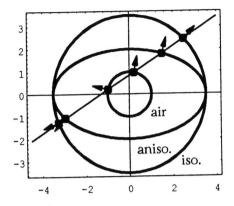 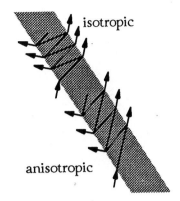

Figure 1. Example of the use of the wavevector diagram to predict the ray paths (group velocities) of multiple reflections within a parallel slab of birefringent crystal . The wavevector component along the boundaries is conserved, yielding the construction line that intersects with the circles and ellipse in wavevector space. The normals at these points of intersection give the ray directions and hence the zig-zag paths for isotropic and anisotropic states of polarization.

polarisation in the anisotropic plane, the refracted and reflected ray paths in the crystal, while unusual and difficult to predict, are readily obtained using the wavevector diagram. Another beautiful example of the non-coincidence of group and phase velocity is in ultrasonic pulse propagation in elastically anisotropic materials such as crystalline silicon, as discussed in a recent article [23].

3. "Toy models" for dispersion relation in two cases

To illustrate the discussion throughout this chapter, we first derive two approximate solutions for the Bloch wave dispersion relation, one for a two-dimensional (2-D) square crystal, and the other for a 2-D hexagonal crystal. Analytical expressions are accessible through reducing drastically the number of "partial" plane waves in the Bloch wave expansion to three for the hexagonal and four for the square case. Despite the inaccuracies introduced, little of the underlying physics is lost in this approximation, and although we do not have space to show this, it is not difficult to include the case where the electric field is polarised in the plane of the lattice wavevectors.

In each case the structure is described by a relative dielectric constant ϵ:

$$\epsilon / \epsilon_0 = 1 + M \sum_{j=1}^{N} \cos(\mathbf{K}_j \cdot \mathbf{r}) \tag{2}$$

where $\mathbf{K}_j$ is the lattice vector of the j-th set of planes with spacing $2\pi / |\mathbf{K}_j|$, M is the amplitude of the dielectric constant modulation around its average value ϵ_0, and $N = 3$ for the hexagonal and 2 for the square cases. The average wavevector in the structure is $k_0 = \omega n_0 / c$, where n_0 is the average index. In the two-dimensional case, n_0 is given by the square root of the *optical path area* of a unit cell, divided by the square root of its real area.

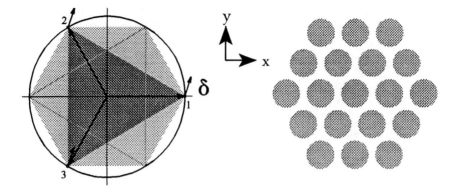

Figure 2. First tileable Brillouin zone (hexagonal lightly shaded region on the left) in wavevector space of the structure on the right. For the calculation it is sufficient to use a triangular sub-zone (darkly shaded region). The average index circle is drawn for the special case when it intersects the three numbered symmetry (S) points. The locus of points traced out by δ (identical for each S-point) yields the dispersion surfaces (see section 3.1).

As already mentioned, the wavevector diagram for an isotropic medium is a circle of constant radius $\omega n_o/c$: the *average index circle*. In a photonic crystal with a hexagonal or square microstructure, important symmetry points occur at the vertices of the regular polygon formed by concatenated lattice vectors (the Brillouin zone). At one particular optical frequency the vertices of this polygon lie on the average index circle; at the points of intersection, the Bragg condition is satisfied simultaneously at all sets of lattice planes. We shall call the vertices S-points; in the hexagonal case (Figure 2) there are six, and in the square case (Figure 3) four. As M increases, the loci of allowed wavevectors (the "dispersion curves") in the vicinity of the S-points becomes complicated, stop-bands and other features appearing.

Each point on the dispersion curves within the first Brillouin zone (i.e., the zone straddling the average index circle) is associated with a unique travelling Bloch wave; the featureless regions in between are populated with evanescent waves, which are excited only if boundaries or structural defects occur in the otherwise regular photonic crystal. The full wavevector diagram is obtained by tiling the first Brillouin zone to cover all of wavevector space. Each

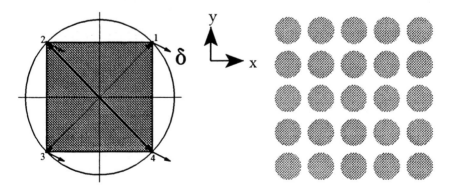

Figure 3. First Brillouin zone (square shaded region on the left) in wavevector space of the square crystal on the right. The average index circle is drawn for the special case when it intersects the four numbered symmetry (S) points. The locus of points traced out by δ (identical for each S-point) yields the dispersion surfaces (see section 3.2).

wavevector in the first Brillouin zone thus has an infinite number of associated wavevectors, one in each higher order Brillouin zone, linked by Bloch's theorem.

Thus, one approach to obtaining the Bloch wave function is to expand the fields in terms of an infinite sum of *partial* plane waves, one for each Brillouin zone. This sum is then truncated appropriately, and the resulting eigenvalue problem solved using standard techniques. We now make a drastic reduction in the number of partial waves (to four for the square and three for the hexagonal cases) in order to obtain illustrative analytical expressions for the dispersion relations.

3.1 HEXAGONAL CASE

The minimum number of partial waves required in this case is three, their wavevectors being expanded around the three S-points:

$$\mathbf{k}_i = \boldsymbol{\delta} + \hat{\mathbf{t}}_i K/\sqrt{3}, \quad K = |\mathbf{K}_j| \tag{3}$$

where $\mathbf{K}_j$ is the lattice vector from (2), $\boldsymbol{\delta}$ is the deviation of the Bloch wavevectors from their values when the multiple Bragg condition is satisfied, and

$$\hat{\mathbf{t}}_1 = \hat{\mathbf{x}}, \quad \hat{\mathbf{t}}_2 = (-\hat{\mathbf{x}} + \hat{\mathbf{y}}\sqrt{3})/2, \quad \hat{\mathbf{t}}_3 = (-\hat{\mathbf{x}} - \hat{\mathbf{y}}\sqrt{3})/2 \tag{4}$$

are unit vectors in the three S-directions. Substituting an Ansatz consisting of three plane waves with amplitudes V_1, V_2 and V_3 and wavevectors (3) into Maxwell's equations, and neglecting higher order terms in $\boldsymbol{\delta}$, leads to a 3×3 matrix whose determinant must equal zero for solutions:

$$\begin{pmatrix} \vartheta/2 - \boldsymbol{\delta}\cdot\hat{\mathbf{t}}_1 & \kappa & \kappa \\ \kappa & \vartheta/2 - \boldsymbol{\delta}\cdot\hat{\mathbf{t}}_2 & \kappa \\ \kappa & \kappa & \vartheta/2 - \boldsymbol{\delta}\cdot\hat{\mathbf{t}}_3 \end{pmatrix} \begin{pmatrix} V_1 \\ V_2 \\ V_3 \end{pmatrix} = 0 \tag{5}$$

where the coupling constant κ and the dephasing constant ϑ are defined by:

$$\kappa = k_o M/4, \quad \vartheta = 2(k_o - K/\sqrt{3}). \tag{6}$$

The dispersion relation turns out to have the form:

$$3\delta_y^2 = (\delta_x + \vartheta)^2 + \frac{2\kappa^2(4\kappa + 3\vartheta)}{2\delta_x - \vartheta}. \tag{7}$$

3.2 SQUARE CASE

The wavevectors of the four Bloch partial waves in this case are expanded around the four S-points:

$$\mathbf{k}_i = \boldsymbol{\delta} + \hat{\mathbf{t}}_i K/\sqrt{2}, \tag{8}$$

where $\boldsymbol{\delta}$ is the deviation of the Bloch wavevectors from their values when the multiple Bragg condition is satisfied and

$$\hat{t}_1 = (\pm\hat{x} + \hat{y})/\sqrt{2}, \quad \hat{t}_3 = (\mp\hat{x} - \hat{y})/\sqrt{2} \tag{9}$$

are unit vectors in the four S-directions; the upper subscript indices are taken with the upper signs. Putting an ansatz consisting of four plane waves with amplitudes V_1, V_2, V_3 and V_4 and wavevectors (8) into Maxwell's equations, and neglecting higher order terms in δ, leads to a 4×4 matrix whose determinant must equal zero for solutions. Its form is identical to (5), with one extra row for the fourth amplitude. The dephasing constant ϑ and coupling constant κ are defined for the square case by:

$$\kappa = k_o M/4, \quad \vartheta = 2k_o - K\sqrt{2}. \tag{10}$$

The resulting dispersion relation is:

$$\delta_y^2 = \delta_x^2 + (\vartheta/\sqrt{2})^2 \pm 2(\vartheta/\sqrt{2})\sqrt{\delta_x^2 + 2\kappa^2}. \tag{11}$$

3.3 SOLUTIONS FOR THE WAVEVECTOR DIAGRAMS

Some illustrative plots of the resulting dispersion curves (i.e., the locii of δ at fixed frequency plotted in 2-D wavevector space) around each S-point are given in Figures 4 and 5.

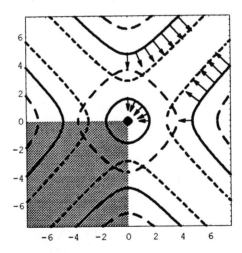

Figure 4. Wavevector diagram about the S_1 point for a square crystal with $\kappa=1$, $\vartheta=2.5$ (long-dashed curves), 5 (full curves) and 7.5 (short-dashed curves). The vertical and horizontal axes are δ_y and δ_x respectively. For $\vartheta=5$, the direction and relative magnitude of the group velocity is indicated in one quadrant using arrows of variable length. The shaded region is part of the first Brillouin zone (see Figure 3).

In the square case (Figure 4) for $\kappa=1$ and $\vartheta=7.5$, the central region approximates to a circle with equation:

$$\delta_x^2 + \delta_y^2 = \vartheta^2/4 - 4\kappa^2. \tag{12}$$

This is valid so long as terms of third order and higher in δ can be neglected. Note that (12) predicts that the circle vanishes for $(\vartheta/\kappa)^2 < 16$. The implications of (12) are intriguing; it states that the Bloch waves behave in many respects like plane waves in an isotropic medium, except that the velocity of light is reduced to a smaller value, and the dispersion with frequency is nonlinear. All the usual classical processes such as diffraction, interference and scattering will be present, except that Bloch waves on the outer stop-band branches not described by (12) will be excited at abrupt interfaces or aperiodicities. The direction of the group velocity at different points on the curves is marked by means of arrows whose length scales with the magnitude. For $\vartheta < 0$ they point in the opposite direction, the curves themselves having precisely the same shape (within the approximations of the model) as for $\vartheta > 0$.

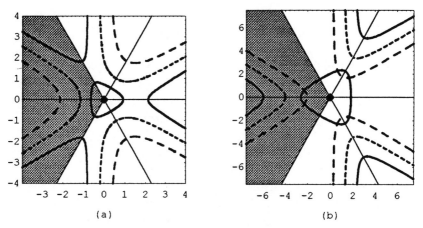

(a)   (b)

Figure 5. Wavevector diagrams about the S_1 point for a hexagonal crystal with (a) $\kappa = 1$, $\vartheta = -1.6$ (full curves), -0.5 (short dashed) and 0.6; and (b) $\vartheta = 4.5$ (full curves), 2.5 (short dashed) and 0.5. The vertical and horizontal axes are δ_y and δ_x respectively. The shaded region is part of the first hexagonal Brillouin zone (see Figure 2).

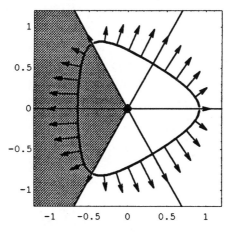

Figure 6. Magnified central rounded triangle from Figure 4 for $\kappa = 1$, $\vartheta = -1.6$, with the group velocity arrows drawn in. Notice the smaller velocities at the points of maximum curvature.

For the hexagonal case, the chosen parameters are $\kappa = 1$, $\vartheta = 0.6$, -0.5 and -1.6 (Figure 5a) and $\kappa = 1$, $\vartheta = 0.5, 2.5$ and 4.5 (Figure 5b). The central "rounded triangle" is redrawn for $\vartheta = -1.6$ in Figure 6, along with the group velocity arrows. Note the inverse correlation between curvature and group velocity. As the triangle becomes smaller, eventually disappearing, the curvature becomes tighter and the group velocity tends to zero.

An important consequence of this slowing down of light is that it has more time to interact with matter - dipoles, nonlinearities, scattering centres. Apparently weak perturbations in refractive index can result in strong scattering, and non-linear effects such as Brillouin and Raman scattering, the Pockels effect, and the optical Kerr effect, are enhanced. We shall explore some of these points in sections 6 and 7.

4. Geometrical Optics of Bloch Waves in Inhomogeneous Media

It is well-known that light can spiral around in a waveguide consisting of a cylindrically symmetric bell-shaped refractive index distribution. Because the transverse photon momentum is reduced by the presence of a large axial component of momentum, a weak potential well of higher refractive index is all that is needed to trap the light, as for example in the Ge-doped core of a silica optical fibre [25]. Trapping of light is much easier in photonic crystals, since in the vicinity of the band edge the photon momentum can be very small. Indeed, many important potential applications of photonic band gap materials rely on the use of spatial inhomogeneities to provide intra-band trapped states [26,27]. In addition the effects of inhomogeneities, in the band-windows where propagation is allowed, are not yet well understood. In this section we address the propagation in inhomogeneous periodic media, developing a Hamiltonian optics approach and using it to treat in particular the motion of the Bloch wave rays (given by the group velocity) around circularly symmetrical defects in which the properties vary slowly over many lattice periods.

Hamiltonian optics has been elegantly summarised by a number of authors, including Arnaud in his 1976 book *Beam and Fiber Optics* [28]. It can be applied where the dispersion relation in the homogeneous structure is known, and where, in the inhomogeneous real structure, parameters like average index vary slowly in space. It is essentially an analytical method for stepping through a non-uniform structure, matching phase velocities normal to the gradient of the inhomogeneity at each step, and propagating along the local group velocity to the next point. This process is described by solutions of Hamilton's equations, which take the general form:

$$\frac{d\mathbf{x}}{d\sigma} = \nabla_k H, \qquad \frac{d\mathbf{k}}{d\sigma} = -\nabla H \tag{13}$$

where $\mathbf{x} = \{x, y, z, -t\}$ is the four-vector for space-time, $\mathbf{k} = \{k_x, k_y, k_z, \omega\}$ the generalised wavevector, σ an arbitrary parameter (see below), and $H(\mathbf{x}, \mathbf{k})$ the Hamiltonian, which may be expressed directly from the dispersion relation for the waves. Note that in general $\mathbf{k}$ depends on position. Equation (13) can be re-cast in a Newtonian way [24,28]:

$$\frac{d^2\mathbf{x}}{d\sigma^2} = [\nabla_k\nabla_k H] : (-\nabla H) = [1/\mathbf{m}^*] : \mathbf{F} \tag{14}$$

in which the reciprocal effective mass tensor depends, in general, on position in real and reciprocal space. The Hamiltonian itself may be written in a number of equivalent ways, subject to the requirement that a phase front is given by the equation $H = 0$. In obtaining solutions to (13), it is important to distinguish total from partial differentiation. For the dispersion relation in (12), H can be chosen in the form:

$$H \equiv \left(\omega n_o/c - K/\sqrt{2}\right)^2 - \delta_x^2 - \delta_y^2 - 4\kappa^2 = 0. \tag{15}$$

For a central force field H depends only on distance r from an origin, and recognising that δ in (15) acts as the wavevector, (14) may be re-expressed in the form:

$$\frac{d^2\mathbf{r}}{d\sigma^2} = 2\frac{\partial H}{\partial r}\hat{\mathbf{r}} \tag{16}$$

where $\mathbf{r}$ is the position vector in two dimensions; if necessary the parameter σ can be related to real time t by the time component of the four-vector equations (13):

$$c(t-t_o) = -\int_{\sigma=\sigma_o}^{\sigma}\left(2\omega n_o/c - K\sqrt{2}\right)n_o\,d\sigma \tag{17}$$

where n_o and K depend in general on position (and hence on σ) as given by solutions of (16). The exact meaning of σ is not important if only the ray paths are sought; if, however, position as a function of time is required - such as when calculating the free spectral range of a cavity - (17) must be used. Working in cylindrical polar coordinates, we now follow the standard approach to motion in a central force field. Since the θ component of the left-hand-side of (16) must be zero, angular momentum h is conserved:

$$h = r^2\dot{\theta} = \text{constant} \tag{18}$$

where the dot denotes differentiation with respect to σ. Making the standard substitution $u = 1/r$, (16) may be reduced to the form:

$$\frac{d^2u}{d\theta^2} + u = \frac{2}{h^2}\frac{\partial H}{\partial u}. \tag{19}$$

To illustrate a particular case, we choose a square structure in which the modulation amplitude M, and hence the coupling constant κ, are radially dependent, K and n_o being kept constant:

$$\kappa^2 = (\kappa_o^2 + \kappa_1^2) - \kappa_1^2 u/u_1. \tag{20}$$

This leads to solutions which imitate the elliptical and hyperbolic orbits of a particle in a gravitational field:

$$u(\theta) = u_o(q + m\cos(\theta+\psi)) \tag{21}$$

where

$$q = 8\kappa_1^2/(h^2 u_o u_1), \qquad h = -2\delta_{oy}/u_o,$$

$$m = \sqrt{(1-q)^2 + (\delta_{ox}/\delta_{oy})^2}, \qquad \psi = \arccos[(1-q)/m]$$

(22)

are the definitions of the various parameters. The initial values of u, δ and θ are u_o, $\delta_o = (\delta_{ox}, \delta_{oy})$ and 0π respectively. Equation (21) yields the ray paths $r(\theta)$, which may be translated into $r(\sigma)$ via $d\theta/d\sigma = hu^2$, and then into $\mathbf{r}(t)$ via (17). A series of typical ray trajectories are plotted in Figure 7, for attractive ($u_1 > 0$) and repulsive ($u_1 < 0$) central force fields. In each plot, several paths are given for different directions of the initial ray at the point $(1,0)$ and the centre of the force field is at $(0,0)$.

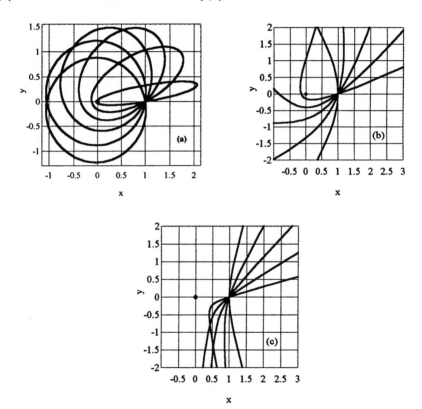

Figure 7. Trajectories of rays around a force field centred at $x=0$, $y=0$ in a square crystal in three cases: (a) attractive force (κ increasing with radius) at a frequency such that the light cannot escape; (b) attractive force at frequency where light can escape; (c) repulsive force. The parameter values are $\vartheta/4\kappa_o = -2.5$, $u_o = \kappa_o$ and $\kappa_1 = 0.2\kappa_o$, with (a) $u_1 = 0.004\kappa_o$, (b) $u_1 = 0.01\kappa_o$ and (c) $u_1 = -0.01\kappa_o$. In each case, the orbits for several different initial directions are given.

We now look in more detail at the case of a circular ray path, which occurs for initial conditions $\delta_{ox} = 0$ and $q = 1$. We shall derive the radius and the time taken for one cycle of revolution. From (21) for $q = 1$, recognising that $\delta_{oy} = \delta_o$ may be expressed using (15), the inverse radius is:

$$u_o = u_1 \left((\vartheta^2/4 - 4\kappa_o^2)/2\kappa_1^2 \right) \qquad (23)$$

and the time taken for one revolution, obtained from (17) using $d\theta/d\sigma = -2\delta_o u_o$, is:

$$T = \frac{\pi \vartheta n_o}{c u_o \sqrt{\vartheta^2/4 - 4\kappa_o^2}} = \frac{2\pi r_o}{2\delta_o c/\vartheta n_o} = \frac{2\pi r_o}{v_g} \qquad (24)$$

where $r_o = 1/u_o$. As indicated in (24), T can be expressed as the circumference of the circular path divided by the group velocity, exactly as intuition would suggest. The classical nature of the Hamiltonian approach does not predict quantization of these closed orbits; it merely gives us the particle trajectories (governed by the group velocity). In order to establish the optical frequencies at which discrete resonances appear, one must return to considering the underlying field amplitudes (governed by phase velocity). It is, however, possible to predict the frequency spacing between adjacent resonant modes - the free spectral range; this is simply $1/T$.

To put some "flesh" on these formulae, we consider a square structure fabricated along the lines of the techniques reported in [21], where coupling constants of 120/mm at an optical wavelength of 633 nm were achieved in etched Ta_2O_5 waveguides on borosilicate glass. For such a waveguide supporting TM polarised (the fields are then quasi-scalar) modes of effective index 1.7, a structure with $\Lambda = 260$ nm will have a Bragg wavelength of 624 nm. For $\vartheta/\kappa_o = 4.01$, i.e., operating at 633 nm, the rays will describe a circular trajectory of radius 100 μm with a period of 50 psec if the perturbation strength is $\kappa_1 \sqrt{r_1} = 3.8$ /mm$^{1/2}$. The free spectral range of the associated resonances will be 20 GHz. In a uniform medium of refractive index 1.7, light would take 3.6 psec to travel around a loop of the same dimensions, with a free spectral range of 0.28 THz. This illustrates both the substantial reductions in optical velocity that occur near a band edge, and the associated increase in the optical density of states.

5. Interference of Bloch Waves

Given their complicated dispersion, it is perhaps not surprising that Bloch waves interfere in a unique and complex manner, and that the characteristics of this interference provide clues to their behaviour when coupled together or scattered. In previous experimental work, singly and doubly periodic planar waveguides were studied [21,22]. Here we treat the two dimensional multiply periodic case. To start our discussion we take first a superposition of two different Bloch waves:

$$E(\mathbf{r}, t) = a_1(\mathbf{r}) \exp[-j(\delta_1 \cdot \mathbf{r} - \omega t)] + a_2(\mathbf{r}) \exp[-j(\delta_2 \cdot \mathbf{r} - \omega t)] \qquad (25)$$

where a_1 and a_2 are time-independent complex periodic functions of position whose structure mimics the lattice. For travelling Bloch waves the wavevectors δ are real-valued. If observing scattering from this pattern, coarse fringes would be seen with the form:

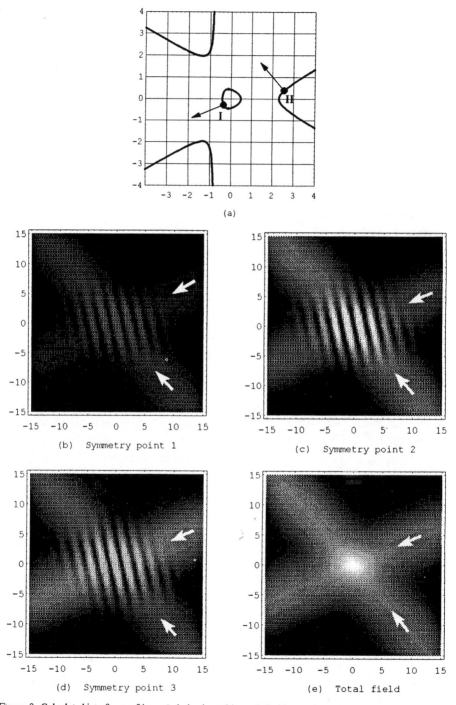

Figure 8. Calculated interference fringes (relative intensities scaled with gray-level) for the two orthogonal Bloch wave beams associated with the points on the wavevector diagram in (a), plotted for $\kappa = 1$ and $\vartheta = -1.6$. In (b), (c) and (d) the fringes created by the groups of partial waves at the 1st, 2nd and 3rd S-points are illustrated. In (e), the complete intensity pattern of the superimposed Bloch beams is given; note the complete absence of interference.

$$I(\mathbf{r}) = <|a_1(\mathbf{r})|^2 + |a_2(\mathbf{r})|^2> + <a_1(\mathbf{r})a_2^*(\mathbf{r}) + a_1^*(\mathbf{r})a_2(\mathbf{r})> \cos[(\delta_2 - \delta_1)\cdot\mathbf{r}] \quad (26)$$

where the averaging is over a unit cell and the variation in the argument of the cosine is assumed to be slow. Each of the two terms added together in the amplitude of the cosine turn out to be real, which is why no phase term appears in the argument of the cosine. The orientation of the fringes is given by $\delta_1 - \delta_2$, which bears no obvious relation to the directions of propagation as given by the group velocities. This has a number of bizarre consequences when a comparison with conventional plane wave interference is made. For example (defining α_1 and α_2 as the angles between the group velocities and $\delta_1 - \delta_2$), interference fringes can appear even when $\alpha_1 > 0$, $\alpha_2 < 0$; for two plane waves this fringe orientation would be impossible.

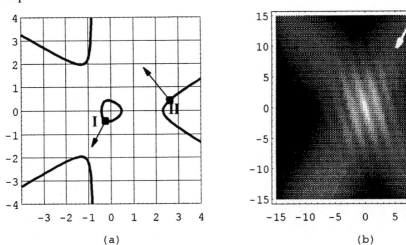

Figure 9. Complete fringe pattern (b) for the two non-orthogonal Bloch wave beams associated with the points on the wavevector diagram in (a). Note the presence, as expected, of visible interference fringes.

The fields forming the fringe patterns may be decomposed into *groups* of partial plane waves sharing approximately the same wavevector, gathered around each of the S-points (see e.g. Figures 1 and 2). Each group produces its own set of sub-fringes. If all these fringe sets are spatially in phase, a strong visible fringe pattern will be produced; if however the dips in intensity from one group are filled in by peaks in intensity from the other groups, the result will be a more uniform intensity distribution (orthogonal Bloch waves, with zero overlap integral in (26), produce a perfectly uniform intensity pattern, with no evidence of interference). This means that if a boundary with an isotropic medium is introduced at a point where the intensity from one group is high, a plane wave may appear in the isotropic medium, travelling in the general direction of the partial plane waves of that group. If a boundary cuts across regions where the intensity oscillates between different groups, then beamlets are produced, their directions corresponding to the groups that have the maximum local intensity; the result can often be striking. This process gives rise to the Pendellösung fringes in x-ray diffraction [29], and lies behind the operation of volume transmission gratings [15]; we have previously classified it as *exchange interference* [21], since each sub-group of plane waves interferes

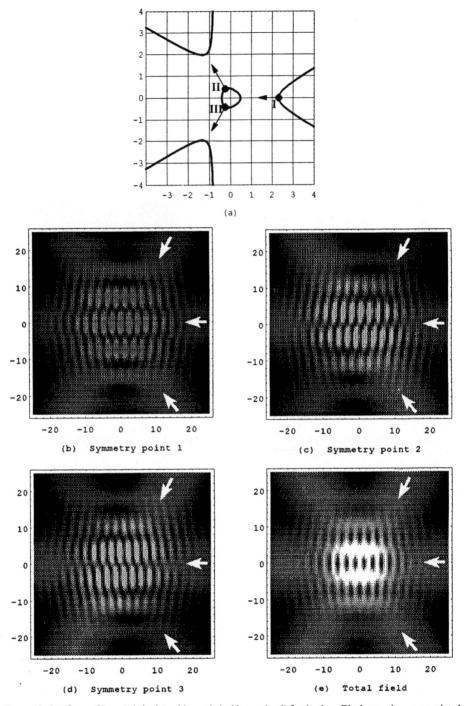

Figure 10. Interference fringes (relative intensities scaled with gray-level) for the three Bloch wave beams associated with the points on the wavevector diagram in (a). In (b), (c) and (d) the fringes created by the groups of partial waves at the 1st, 2nd and 3rd S-points are illustrated. In (e), the complete intensity pattern of the three superimposed Bloch beams is given.

constructively in different regions of space.

To illustrate these arguments, we now look at some specific examples in which the fringe intensity patterns of superimposed Bloch beams, and of their partial wave groups, are plotted for a two-dimensional hexagonal crystal. The points on the wavevector diagram, and the group velocity arrows, are also illustrated. First of all, an orthogonal pair of equal amplitude Bloch waves is selected (Figure 8). The interference patterns from the groups of partial waves at the S_1, S_2 and S_3 points are plotted successively in Figure 8b, 8c and 8d. Note the strong visible fringes. When, however, all three patterns are combined, the fringes vanish, leaving a uniform intensity that is higher in the middle simply because the Gaussian beam profiles overlap there. In a second example (Figure 9), a non-orthogonal Bloch wave pair is chosen, and this time the combined intensity field contains fringes. In the third example, three non-orthogonal Bloch waves are taken. The interference patterns here are more complex (Figure 10). Once again, the patterns from each sub-group, and from the combined field, are given. The S_1 subgroup produces a much weaker intensity pattern, as expected since the group velocities of the Bloch waves point predominantly in the $-x$ direction.

6. Coupling and Scattering of Bloch Waves

In this section we derive the coupled mode equations describing the interaction of two different Bloch waves in a two-dimensional periodic structure with a periodic distortion in properties - a *superlattice*. This analysis will provide a useful basis from which to discuss more generally the coupling and scattering of Bloch waves. The analysis is an extension of one presented previously for singly periodic structures [30,31]. The relative dielectric constant of the structure is taken in the form:

$$\epsilon(\mathbf{r}) = \epsilon_0 + \epsilon_p(\mathbf{r}) + \left\{ \epsilon_s = \epsilon_{so} \cos \mathbf{K}_s \cdot \mathbf{r} \right\} \tag{27}$$

where the subscript p denotes the undistorted "primary crystal," and s the weak "secondary grating," which together form the superlattice. The time-independent field ansatz is taken in the form of a superposition of two Bloch waves:

$$\begin{aligned}
\mathbf{E}(x,z) &= \hat{\mathbf{y}} \sum_{i=1}^{2} V_i(x,z) B_i(x,z) \\
&= \hat{\mathbf{y}} \sum_{i=1}^{2} V_i(x,z) a_i(x,z) \exp(-j\,\delta_i \cdot \mathbf{r}), \quad \delta_i \cdot \hat{\mathbf{y}} = 0
\end{aligned} \tag{28}$$

where the V_i are slowly-varying amplitudes and the B_i describe Bloch waves of the undistorted crystal; when the V_i are constant, no coupling is present and (28) is a solution of Maxwell's equations for the primary crystal. Substituting (28) into Maxwell's equations for the structure (27), neglecting second order derivatives of V_i, multiplying by $(B_1^* + B_2^*)$ and averaging over a unit cell of the primary crystal, we obtain the coupled Bloch wave equations:

$$\frac{\partial V_i}{\partial p_i} + j\,\kappa_{sij}\, V_j \exp(-j\,\vartheta_{ij} \cdot \mathbf{r}) = 0, \tag{29}$$

where the coupling constant and dephasing vector are given by:

$$\kappa_{sij} = \frac{\omega \epsilon_{so} I_{ij}}{4\epsilon_o(\hat{\mathbf{p}}_i \cdot \mathbf{v}_{gi}) I_{ii}}, \qquad \vartheta_{ij} = (\delta_j - \delta_i) \pm \mathbf{K}_s, \tag{30}$$

the sign is chosen to minimise the magnitude of the dephasing vector ϑ_{ij}, and

$$I_{ij} = \iint_{\text{cell}} a_i^* a_j \, dA. \tag{31}$$

The spatial coordinate p_i has a unit vector $\hat{\mathbf{p}}_i$ that can be chosen to point in any convenient direction not perpendicular to the group velocity $\mathbf{v}_{gi}$. Commonly, both p_1 and p_2 are chosen to lie parallel to one cartesian coordinate; this is useful if the secondary grating region has a straight flat boundary. If, however, two-dimensional coupling is to be treated, a good choice of $\hat{\mathbf{p}}_i$ points parallel to $\mathbf{v}_{gj}$, i.e., parallel to the group velocity of the *other* wave. It is straightforward to show that power is conserved by solutions of (29):

$$\frac{\hat{\mathbf{p}}_1 \cdot \mathbf{v}_{g1}}{I_{11}} \frac{\partial |V_1|^2}{\partial p_1} + \frac{\hat{\mathbf{p}}_2 \cdot \mathbf{v}_{g2}}{I_{22}} \frac{\partial |V_2|^2}{\partial p_2} = 0. \tag{32}$$

The first striking thing is that the coupling constant in (30) is inversely proportional to the group velocity. This means that, close to a band edge, the coupling constant becomes very large. Otherwise expressed: the Bloch waves become highly susceptible to scattering at perturbations which would not significantly affect a plane wave in an isotropic medium. A good example of this is the circle enclosing the Bragg point in the wavevector diagram of the square photonic crystal discussed in section 3.2. In the same waveguide, operating at $\vartheta/\kappa_o = 4.01$ with an average index of 1.7, the secondary grating period needed for phase-matched coupling of two Bloch waves travelling at right angles to one another is

$$\Lambda_s = \frac{2\pi}{\delta_o \sqrt{2}} = \frac{\pi}{2\sqrt{2} \, \kappa_o \sqrt{(\vartheta/4\kappa_o)^2 - 1}} \tag{33}$$

which works out at 0.26 mm for a primary grating coupling constant of 120/mm. The group velocity is 0.07 c/n_o and the coupling constant, for $\hat{\mathbf{p}}_1 = \hat{\mathbf{p}}_2$ bisecting the angle between the two group velocities, is

$$\kappa_{sij} = \kappa_{sji} = \frac{\pi \epsilon_{so}}{2\sqrt{2} n_o \lambda \sqrt{1 - (4\kappa_o/\vartheta)^2}} \tag{34}$$

which works out at $9.3 \times 10^3 \, \epsilon_{so}$/mm at 633 nm (note that the overlap integral $I_{12} \approx 1$ for these two waves). A secondary grating with an amplitude $\epsilon_{so} = 0.0001$ will couple the Bloch waves together at a rate of 0.93/mm, i.e., 100% coupling will be achieved in 1.7 mm. The effective modulation depth of the primary grating is $M \sim 0.028$, which implies that the amplitude of the dielectric constant modulation is 0.08 - some 800 times stronger than for the secondary grating.

The second striking thing is that the coupling constant depends on the overlap integral of the field microstructure of the Bloch waves. As we have seen already, the absence of real interference fringes between two different but co-propagating Bloch waves may be taken as evidence that they are orthogonal, i.e., that their overlap integral is zero. This means, for example, that a superlattice designed to couple together the two Bloch waves whose interference patterns are depicted in Figure 8 will not function. There is thus an intimate relationship between the presence of real interference fringes and the feasibility of coupling.

This analysis may be generalised to the case of superlattices whose secondary grating has multiple periodicities. For example, if the secondary "lattice" matches the hexagonal interference pattern created by interference of three Bloch waves (see, e.g., Figure 10), then coupling between these three waves can be achieved. The ramifications of this extension to the possibilities of optical superlattices (for obvious reasons not feasible in electronic superlattices made by MBE) remain to be explored in detail.

7. Nonlinear Scattering

The strong scattering that is possible between Bloch waves because of their very low momentum near a band-edge also means that nonlinear effects such as Brillouin and Raman scattering can be enhanced. There are several different but complementary ways of viewing this. In the first, one argues that the group velocity is low hence the light hangs about much longer; in the second, that the density of states is increased, so that by Fermi's Golden Rule the transition rate is higher; and in the third, that the low momentum of the Bloch photons means that they are easily scattered by weak (possibly nonlinear) perturbations. The first view leads naturally to the idea of an *effective optical path length* governed by the group velocity:

$$L_{eff} = L n_o \frac{c/n_o}{v_g} = L_{av} \frac{c/n_o}{v_g}. \tag{35}$$

Since near a band edge L_{eff} can be substantially larger than the "physical" path length based on the average index, the effectiveness of a whole range of nonlinear, electro-optic and acousto-optic effects will be enhanced. For example, an electro-optic modulator that needs to be 5 mm long for a desired performance could be reduced in length to 0.5 mm if the group velocity could be reduced by the same factor.

By way of a more detailed example we consider stimulated Brillouin scattering (SBS). In a photonic crystal consisting of holes drilled in high index materials such as Si or GaAs, the presence of significant Brillouin (and Raman) scattering will depend on the optical fields "seeing" the material, not the air. This means that only Bloch waves on the low frequency side of the stop-bands (when the field intensity peaks in the high index regions [32]) will produce a strong SBS signal. We also assume that acoustic waves whose wavelength covers many lattice periods in the photonic crystal see an "average" Young's modulus and density, and hence have a "sensible" dispersion relation. Following the formalism in [33], the Brillouin gain for interaction between two optical waves and an acoustic wave takes the form:

$$G_B = \frac{\omega \Omega_A (\partial \epsilon_r / \partial s)^2}{2 \alpha_A n_o^4 v_g^2 v_{gA} E} \tag{36}$$

in units of m/W. The acoustic frequency and power attenuation rate are $\Omega_A/2\pi$ and α_A, E is Young's modulus, v_{gA} and v_g the acoustic and optical group velocity magnitudes, and the partial derivative of ϵ_r with respect to strain s describes electrostriction. Comparing this expression with the gain in an isotropic medium of average index n_o, and assuming that all the parameters except E, the optical group velocities and the acoustic frequency remain unchanged (since both density and stiffness are reduced, the acoustic velocity might not change much), the enhancement in Brillouin gain is:

$$\frac{G_B}{G_{Bav}} = \frac{\Omega_A v_{gav}^2 E_{av}}{\Omega_{Aav} v_g^2 E} . \tag{37}$$

If we imagine two counter-propagating Bloch waves that interfere to produce fringes of spacing $10\lambda/n_o$, with a reduction in group velocity of 10× and a reduction in E of 2×, an enhancement of around 20× in Brillouin gain is expected. This is not dramatic, mainly because in the photonic crystal the Stokes frequency shift is reduced and hence the proportion of each pump photon that is converted into an acoustic phonon is smaller, resulting in less power in the acoustic wave and smaller SBS gain.

8. Various Applications and Related Issues

One benefit of comparing the behaviour of charge waves in electronic crystals and light waves in photonic crystals is that both the similarities and the differences can lead to interesting possibilities in both fields. For example, in semiconductor crystals the atoms or molecules forming the repeating unit are identical, whereas in a photonic crystal the unit cell can take almost any shape, since it contains a very large number of atoms (in a Si/air structure at a wavelength of 1 μm each unit cell has $\sim 10^9$ atoms). In addition, because the period in a photonic crystal is itself nothing special (unrelated to the size of an atom), it can be varied at will, along with all the other parameters typical of a periodic structure (average index, index contrast, crystal orientation and so on). Thus, two-dimensional photonic crystals can be imagined in which the crystal planes bend. An example is a crystal in which the "atoms" have sizes, shapes and positions given by the interference pattern generated by four cylindrical waves - see Figure 11. Electron beam lithography makes the creation of such complex patterns feasible, indeed no more difficult to produce than a perfectly regular periodic structure. Such patterns could be used for complex wavefront conversion within wavelength division multiplexing components, and for the design of microcavities in which spontaneous emission is controlled.

In solid state physics, the traditional emphasis has been on fully trapped states. In optics, because of the need to get light in and out of a device and use it, the emphasis has been on

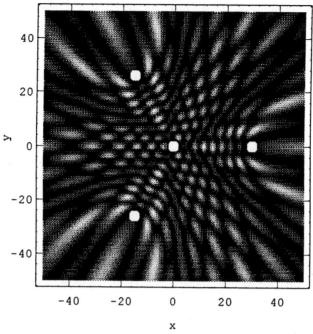

Figure 11. The fringe pattern generated by four cylindrical waves emanating from the origin and three points at 120° apart, equidistant from the origin (the white dots). If combined with very high modulation depths of refractive index, such holographically conceived patterns may exhibit rich and unusual behaviour.

semi-trapped states - ones that are coupled strongly to the outside world. The recent development of the quantum cascade semiconductor laser is the result of thinking about semi-trapped states for electrons [34].

The successful manipulation of light is a key ingredient in nearly all optoelectronic components. Photonic crystals, through the strong spatial and temporal dispersion exhibited by the Bloch waves near the band edges, are prime candidates for enhancing the interaction between light and matter and thus improving the performance of a whole range of optoelectronic components (lasers, modulators, wavefront convertors). The traditional reservation often expressed about the use of periodic structures for controlling light is that the bandwidth is narrow. While this is true of structures where the step in refractive index is small, it is much less a problem in photonic band gap materials, because the number of lattice periods needed for complete reflection is often in single figures, widening the bandwidth of reflection.

It is clear that the fabrication of extended photonic crystals at optical wavelengths stretches the capability of current nanotechnology, and indeed has not yet been successful. Fabrication of periodically etched high index films is less challenging, and moreover fits more naturally with many current devices produced in waveguide form, including lasers, amplifiers, modulators, couplers, filters and nonlinear elements. It may turn out that the first major uses of photonic crystals will be in thin films; however, before this can be achieved, the problems of maintaining waveguiding in deeply etched layers must be solved. Some recent work addresses this issue [35], demonstrating that solutions already exist.

Major future problems include getting light into a photonic crystal from the external world, and avoiding strong unwanted scattering near the band edges where the photon momentum is very low. An obvious solution to the first is to interpose an adiabatic transition region within which Hamiltonian optics is valid. The second may prove more difficult to manage, since it is likely to place tight tolerances on the fabrication accuracy of each and every unit cell.

9. Conclusions

The novelty, timeliness and promise of photonic crystals lies not just in the production of a full photonic band gap, or in the control of the photonic density of states and hence the spontaneous emission rate in active structures, but also in the exploitation of the strong spectral and spatial dispersion that exist near a band edge, and the very rapid spatial transformations in the optical fields this makes possible, to develop new active and passive devices. For example, there are good reasons for thinking it possible to reduce the length of waveguide components such as electro-optic modulators and directional couplers by orders of magnitude. This would allow much greater packing densities on optical chips, leading to large scale integration of optical functions. The Bloch wave optics outlined in this chapter may be a useful basis upon which to design and build this next generation of components in quantum optoelectronics and optical communications.

Acknowledgments

We gratefully acknowledge the assistance of Sarah Rutt in uncovering examples of periodic structures in the natural world, and the on-going support and interest of John Rarity, John Roberts and Terry Shepherd of DRA Malvern. The Optoelectronics Research Centre is an Interdisciplinary Research Centre of the U.K. Engineering and Physical Sciences Research Council.

References

1. H. Ghiradella, "Light and color on the wing: structural colors in butterflies and moths," *Appl. Opt.* **30** (3492-3500) 1991
2. H. F. Nijhout, "The developmental physiology of color patterns in lepidoptera," *Adv. Insect Physiology* **19** (181-247) 1985
3. P.J. Herring, "Reflective systems in aquatic animals," *Comp. Biochem. Physiol.* **109A** (513-546) 1994
4. N. Horiuchi, "New synthetic opal made of plastics," *The Australian Gemmologist* **14** (213-218) February 1992
5. P.J. Darragh, A.J. Gaskin and J.V. Saunders, "Opals," *Scientific American* **234** (84-95) 1976
6. J.P. Gauthier, "Natural and synthetic opals: TEM structural study," *Acta Cryst.* **A40** (C248) 1984
7. G. Widawski, M. Rawiso and B. François, "Self-organised honeycomb morphology of star-polymer polystyrene films," *Nature* **369** (387-389) 1994
8. P.N. Pusey, W.C.K. Wick, S.M. Ilett and P. Bartlett, "Phase behaviour and structure of colloidal suspensions," *J. Phys.-Condensed Matter* **6** No 23A (A29-A36) 1994
9. G. Devauchelle, D.B. Stoltz and F. Darcy-Tripier, "Comparative ultrastructure of iridoviridae," *Current Topics*

in Microbiology and Immunology, **116** (1-21) 1985

10. M.M. Burns, J.-M. Fournier and J.A. Golovchenko, "Optical matter: Crystallization and binding in intense optical fields," *Science* **249** (749-754) 1990

11. A. Korpel (Editor), "Selected Papers on Acoustooptics," SPIE Optical Engineering Press, volume **MS16**, Bellingham, Washington (1990).

12. J.M. Cowley, "Diffraction Physics," second revised edition, North-Holland (1981).

13. F.K. Kneubühl, "Theories of Distributed Feedback Lasers," Harwood Academic Publishers, Switzerland (1993).

14. S.L. McCall and P.M. Platzman, "An optimised π/2 distributed feedback laser," *IEEE J. Quant. Electr.* QE-**21** (1899-1904) 1985.

15. L. Solymar and D.J. Cooke, "Volume Holography and Volume Gratings," Academic Press, London (1981).

16. See, e.g., J.S. Blakemore, "Solid State Physics," Cambridge University Press, Cambridge, 1991; N.W. Ashcroft and N.D. Mermin, "Solid State Physics," Harcourt Brace College Publishers, Fort Worth (1976); H. Haken, "Quantum Field Theory of Solids," North-Holland, Amsterdam (1988); C. Kittel, "Quantum Theory of Solids," John Wiley & Sons, New York (1987).

17. E. Yablonovitch, "Photonic band gap structures," *J. Opt. Soc. Am.* 10 (283-295) 1993; J.D. Joannopoulos, R.D. Meade and J.N. Winn, "Photonic Crystals," Princeton University Press (1995); for an introduction to the use of wavevector diagrams see P.St.J. Russell, "Photonic band gaps," *Physics World* 5 (37-42) 1992.

18. P.St.J. Russell and R. Ulrich, "Elementary and coupled waves in periodic planar waveguides," *2nd European Conference on Integrated Optics*, Florence, IEE Conf. Publication **227** (88-91) October 1983.

19. P.St.J. Russell, "Novel thick-grating beam-squeezing device in Ta_2O_5 corrugated planar waveguide," *Electr. Lett.* **20** (72-73) 1984.

20. P.St.J. Russell, "Optics of Floquet-Bloch waves in dielectric gratings," *Appl. Phys.* **B39** (231-246) 1986.

21. P.St.J. Russell, "Interference of integrated Floquet-Bloch waves," *Phys. Rev. A* **33** (3232-3242) 1986.

22. R. Zengerle, "Light propagation in single and doubly periodic planar waveguides," *J. Mod. Opt.* **34** (1589-1617) 1987.

23. J.P. Wolfe, "Acoustic wavefronts in crystalline solids," *Physics Today* **48** (34-40) 1995.

24. P.St.J. Russell, T.A. Birks and F.D. Lloyd-Lucas, "Photonic Bloch Waves and Photonic Band Gaps," in *Confined Electrons and Photons: New Physics and Applications*, E. Burstein & C. Weisbuch (editors), (585-633), Plenum Press, 1995

25. A.W. Snyder and J.D. Love, "Optical Waveguide Theory," Chapman and Hall, London (1983).

26. E. Yablonovitch, T.J. Gmitter, R.D. Meade, A.M. Rappe, K.D. Brommer and J.D. Joannopolous, "Donor and acceptor modes in photonic band structure," *Phys. Rev. Lett.* **67** (3380-3383) 1991.

27. D.R. Smith, R. Dalichaouch, N. Kroll, S. Schultz, S.L. McCall and P.M. Platzman, "Photonic band structure and defects in one and two dimensions," *J. Opt. Soc. Am.* **10** (314-321) 1993.

28. J. A. Arnaud, "Beam and Fiber Optics," (Academic Press, New York, San Francisco, London, 1976)

29. Z.G. Pinsker, "Dynamical Scattering of X-rays in Crystals," Springer-Verlag, Berlin (1978).

30. P.St.J. Russell, "Bragg resonance of light in optical superlattices," *Phys. Rev. Lett.* **56** (596-599) 1986.

31. P.St.J. Russell, "Optical superlattices for modulation and deflection of light," *J. Appl. Phys.* **59** (3344-3355) 1986.

32. P.St.J. Russell, "Bloch wave analysis of dispersion and pulse propagation in pure distributed feedback structures", *J. Modern Optics*, **38** (1599-1619) 1991; Erratum: J. Capmany and P.St.J. Russell, *J. Mod. Optics* 1994.

33. P.St.J. Russell, D. Culverhouse and F. Farahi, "Theory of forward stimulated Brillouin scattering in dual-mode single-core fibres," *IEEE J. Quant. Electr.*, **27** (836-842) 1991.

34. J. Faist, F. Capasso, D.L. Sivco, A.L. Hutchinson, C. Sirtori and A.Y. Cho, "Quantum cascade laser - a new optical source in the mid infrared," *Infrared Physics & Technology* (1350-1395) **36** 1995.

35. D.M. Atkin, P.St.J. Russell, T.A. Birks and P.J. Roberts, "Photonic band structure of a periodically etched high-index waveguide layer: stationary resonant modes with a high Q-factor," proceedings of CLEO'95, Baltimore, U.S.A., paper CWF53, (223) 1995; full article in preparation (contact authors for preprint).

OPTICAL MEASUREMENTS OF PHOTONIC BAND STRUCTURE IN COLLOIDAL CRYSTALS

İ. İNANC TARHAN AND GEORGE H. WATSON
Department of Physics and Astronomy
University of Delaware
Newark, DE 19716, USA

Abstract. Polystyrene colloidal crystals form three-dimensional periodic dielectric structures which can be used for photonic band structure measurements in the visible regime. Kossel lines obtained from these crystals reveal the underlying photonic band structure of the lattice in a qualitative way. Also, Kossel lines are useful for locating symmetry points of the lattice for exact orientation of the crystals. From transmission measurements the photonic band structure of an fcc crystal has been obtained along the directions between the L-point and the W-point. A modified Mach-Zehnder interferometer has also been developed for accurately measuring relative phase shifts of light propagating in photonic crystals to determine the dispersion resulting from photonic band structure near the band edges.

1. Introduction

For the last eight years, the analogy between the propagation of electromagnetic waves in periodic dielectric structures, dubbed photonic band gap (PBG) or photonic crystals, and electron waves in atomic crystals has motivated the exploration of many new possibilities in this field [1–5]. Periodic modulations of the dielectric constant in PBG crystals, with spatial periods on the order of the wavelength of interest, cause electromagnetic waves incident along a particular direction to be diffracted for a certain range of frequencies, forming stop bands. When these stop bands are wide enough and overlap for both polarization states along all crystal directions, the material possesses a complete PBG. In such a crystal, propagation of electromagnetic waves within a certain frequency range is forbidden irrespective

C. M. Soukoulis (ed.), Photonic Band Gap Materials, 93–106.

of their direction of propagation. This leads to several interesting effects in quantum optics; one particular example is the suppression of spontaneous emission of light, where an excited atom embedded in a PBG crystal will not be able to make a transition to a lower energy state as readily if the frequency of the emitted photon lies within the bandgap; hence, increasing the lifetime of the excited state [1]. Of great interest to those searching for photon localization is the suggestion that the addition of some disorder to this otherwise ordered structure would require a reinterpretation of the Ioffe-Regel criterion for Anderson localization of photons, and that localized states would exist in the vicinity of the gap [2].

Although PBG crystals will have novel applications in the optical regime, the technological challenge of fabricating a three-dimensional PBG crystal exhibiting a complete band gap in the visible range has been difficult to surmount. Several important parameters must be controlled to build a structure with a complete PBG: lattice type, filling fraction, and sufficient contrast between the low and high dielectric regions. Structures exhibiting full photonic band gaps in the microwave [6, 7], millimeter [8], and submillimeter [9] regimes have already been fabricated; a submicron lengthscale PBG crystal has been proposed [10] but not yet realized experimentally.

To carry out photonic band structure studies in the optical regime, polystyrene colloidal crystals with lattice spacings comparable to the wavelength of light have been grown [11, 12]. These colloidal crystals consist of charged monodisperse polystyrene microspheres suspended in water, yielding a relative index of refraction of 1.20, which organize into face-centered-cubic (fcc) lattices under suitable conditions. Although such a crystal is not expected to exhibit large gaps because of the relatively low index contrast, it does permit study of photonic band structure, as reported here. The lattice parameter of these self-organizing structures can be easily tailored, providing flexibility in designing photonic crystals for use in specific spectral regions. With sufficient microsphere density, these crystals may yield a pseudogap, a depression of the density of propagating photonic states, suitable for photon localization studies. In addition, it is also possible to find that nonlinear optical properties are also modified as strongly as the linear properties considered to date.

2. Preparation of Colloidal Crystals

To prepare colloidal crystals, a stock solution of polystyrene microspheres was diluted to the desired volume fraction with deionized water and tumbled with ion exchange resin to remove stray ions from the suspension. The microspheres have a permanent net negative surface charge counterbalanced by positively charged counterions. Stray ions are also present in the suspension, decreasing the Debye screening length. Once the stray ions are removed with ion-exchange resin,[1] the microspheres interact with a short range repulsive Coulomb force, in addition to a long-range attractive Van der Waals force [12]. The colloidal suspension subsequently undergoes a phase transition from a disordered state to an fcc lattice (for a range of volume fractions and microsphere diameters) with the densest planes (111) parallel to the glass windows of the sample cell.

After the colloidal suspension is transferred into a thin cell [13], Bragg diffractions are immediately observed. Usually the surface of the crystal displays a diffusely mosaic looking structure, a sign of polycrystallinity. If such a crystal is undisturbed, occasionally high-quality single crystalline regions form over time (on the order of a month). Usually the sample remains polycrystalline, with the crystallites oriented with (111) planes parallel to the optical window of the cell but otherwise unoriented.

To obtain single colloidal crystals large enough to permit our measurements, the crystals were annealed. A shear flow is created in the crystal by mechanically rocking a sample cell about the axis normal to the faces of the cell [11]. In fcc crystals (111) planes are expected to slide over each other under the influence of the shear flow and achieve the correct relative orientation for fcc stacking. These samples were mechanically shear annealed with a typical rocking motion with peak-to-peak amplitute of 60 degrees at a frequency range of $\sim$ 0-50 Hz. After sufficiently large crystals formed in the cell (typically less than one day), their quality was confirmed by examining the Kossel line patterns [14, 15], which were also used to orient the single crystals.

3. Kossel Line Patterns

Illumination of a single colloidal crystal by a monochromatic laser beam can exhibit an interesting phenomenon: various dark rings and arcs (conic cross-sections) superimposed on a diffusely lit background form both in transmission and reflection geometries. These patterns are called Kossel lines [11,12,14–16] and are formed by attenuated bands of diffuse light which satisfy the Bragg condition along specific angular cones, as depicted in Fig. 1. Diffuse light originates by scattering from the impurities in thick crystals such as deviations in the polystyrene sphere diameter, dislocations, and vacancies in the lattice; for thin or dilute samples, a diverging source of light can be generated with an external diffuser [11].

Figure 2 depicts simulated Kossel lines [17] for an fcc crystal with a lattice parameter of 486 nm. In each case, the incident laser beam passes

[1]Ion exchange resin enclosed in nylon mesh was also placed in the cell to inhibit the destruction of the crystal by additional free ions leached into the suspension over time.

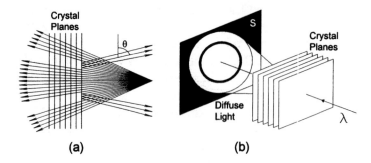

Figure 1. Formation of Kossel lines. (a) Parallel lines represent crystal planes in a photonic crystal and the probe beam is incident perpendicular to these planes. Some light is diffused and propagates through the crystal in all directions, as indicated by the arrows. For a given wavelength, light traveling only along a certain direction (in this case with an angle about θ from the crystal planes) can satisfy the Bragg condition. Such Bragg diffraction leads to a conical band of attenuation that forms the Kossel line pattern on intersection with a screen, as depicted in (b).

through the center of each frame; each circle or arc represents a band of attenuation arising from Bragg diffraction from a specific set of planes. Backscattered Bragg diffraction of the direct beam occurs at the L-point for 752 nm light and the X-point at 650 nm, where L and X are the symmetry points on the first Brillouin zone (FBZ) corresponding to [111] and [002] directions in the fcc crystal. The Bragg condition is satisfied by (111) and (002) planes, respectively, for these wavelengths, thus the beam is diffracted backwards and a circular dark spot appears in transmission at the center of each frame. For the transmission experiments discussed below (see Fig. 4), such Bragg diffraction of the direct beam leads to a strong attenuation feature in the spectrum. Simultaneous satisfaction of the Bragg condition (zone edge) by two sets of crystal planes can be observed where two circles touch at the center of the frame (*e.g.*, at K and U-points at 614 nm). At the W-point at 583 nm, the Bragg condition is satisfied simultaneously by three sets of crystal planes as seen by the intersection of three circles at the center of the frame in Fig. 3.

Kossel lines are useful for establishing single crystallinity and determining the lattice type, and are essential for orienting crystals where the symmetry directions are not known *a priori*. Even for structures fabricated with known symmetry directions, exact orientation may be possible *only* by inspection of Kossel lines,[2] since the index of refraction of a photonic crystal along a particular direction is not well known, owing to the anomalous

[2]One can also possibly rotate the crystal and observe Bragg spots, although this may be much less convenient.

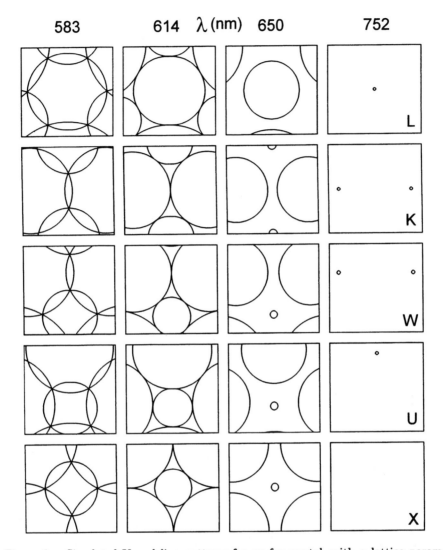

Figure 2. Simulated Kossel line patterns for an fcc crystal, with a lattice parameter of 486 nm, along several symmetry directions. Each row shows patterns obtained along the same direction for different wavelengths. Each frame represents a 90° field of view projected onto a plane.

index of refraction at the band edges [18, 19]. For a crystal with a complete PBG where no light is transmitted, the Kossel lines formed in reflection can still be used for exact alignment.

By varying the crystal orientation and illumination wavelength, Kossel

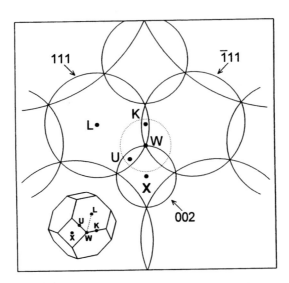

Figure 3. Simulated Kossel line pattern obtained along the [012] direction, corresponding to the W-point, for λ=583 nm. In our experiment, the wavelength of the dye laser is adjusted so that intersection of the 111, $\bar{1}$11 and 002 Kossel lines is obtained. Initially, the crystal was oriented so the incoming beam was along the [111] direction (L-point). Then the crystal was rotated from the L-point to the W-point until the center of the intersection region of the three Kossel lines blocked the transmitted beam. The area inside the broken circle indicates our experimentally observable area at this orientation. The symmetry points of first Brillouin zone for the fcc crystal are identified in the inset; the dashed line indicates the section of the FBZ scanned in Fig. 4.

line patterns can also provide a visual mapping of the photonic band structure. A complete PBG would be signified by the presence of Kossel lines along each direction for a given wavelength (center of each frame along a column of the matrix in Fig. 2). For example, at 583 nm this is only satisfied by the W-point; the Kossel pattern about the L-point indicates the absence of a band gap at this wavelength. Thus, the fcc lattice does not produce readily overlapping stop bands. A complete PBG is indicated when the widths of the Kossel line patterns are sufficient to generate overlap for all orientations.

4. Transmission Measurements

To measure photonic band structure in the visible regime, a 5.0% volume fraction colloidal crystal was prepared with 0.135 μm mean diameter polystyrene microspheres, monodisperse to within 4.2%.[3] The thickness of

[3] As reported by Duke Scientific Corporation, 2463 Faber Place, Palo Alto, CA 94303. Bulk sample volume fractions were determined by drying and weighing, as needed.

the crystal studied here was around 450 μm which resulted in about $\sim$1700 layers of (111) planes.

Since the only crystal direction known *a priori* is the [111] direction (*L*-point), the remaining orientational information was obtained via Kossel line analysis. The major criterion in the adjustment of the lattice spacing of the sample was to obtain *K*, *U*, and *W*-points (see Fig. 3) on the fcc FBZ (*i.e.* corresponding Kossel lines) within the wavelength range of the dye laser (R6G $\sim$ 570-640 nm). After aligning the crystals, a UV/VIS/NIR spectrophotometer with a collimated beam was used to obtain transmission spectra over a much wider range than possible with the dye laser used for observing Kossel lines. Transmission spectra were obtained for both orthogonal components of the plane-polarized beam along directions between the *L* and *W*-points. The transmitted beam and diffuse light were collected with a large aperture lens.

In Fig. 4 transmission measurements show optical stop bands opening between two and four orders of magnitude deep. The dramatic variation of the stop bands as a function of crystal angle is evident. In moving from the *L*-point to the *W*-point, the center wavelength of the stop band shifts to a shorter wavelength by $\sim$130 nm. This shift is not surprising, since the *L*-point is closest to the center of the FBZ while the *W*-point is the furthest from the center. Were the FBZ of the lattice more spherical than that of the fcc crystal, the frequency shift would be smaller. A convergence of two bands at the *W*-point is evident and the stop band at the *W*-point (0.145 eV) is much wider than at the *L*-point (0.021 eV), an unexpected result [20]. The bandwidths of the gaps are presented in Fig. 5, extracted from the fullwidths $\Delta\lambda$ of the normalized transmission bands in Fig. 4 at two orders of magnitude below unity transmission. Furthermore, on examination of the different polarization states in the crystal, the degeneracy expected at the *W*-point for a spherically symmetric basis (as with microspheres) [20] is not found for reasons yet uncertain. Additional theoretical work on photonic band structure of fcc polystyrene colloidal crystals would be helpful in elucidating the reasons for these discrepancies.

The decline in transmission around 500 nm, seen in Fig. 4(a), arises possibly from increase of the scattering cross section at shorter wavelengths, increased absorption of polystyrene, and a second-order Bragg minimum which is expected around 400 nm. The minimum of the second order gap is expected to be located at one-half of the first order gap minimum but may be shifted owing to anomalous dispersion.

100

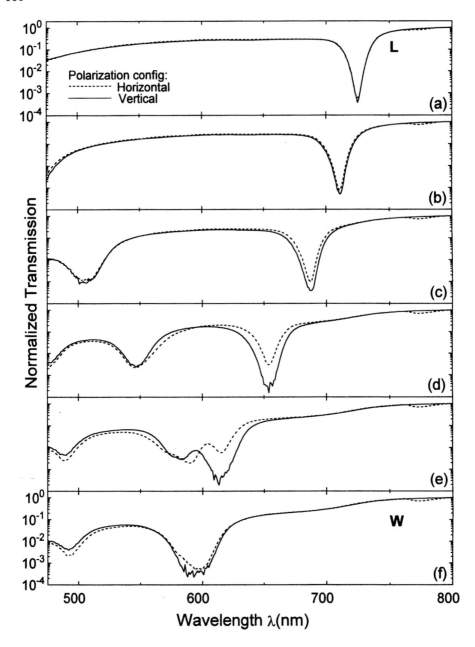

Figure 4. Normalized transmission spectra taken along the *L–W* directions for an fcc lattice formed by ordered 0.135 μm diameter polystyrene microspheres with a lattice constant of 469 nm. (a) *L*-point, (b) *L*+16°, (c) *L*+26°, (d) *L*+36°, (e) *L*+47°, and (f) *W*-point (*L*+51.1°), where the angles refer to the sample cell orientation relative to the incident light.

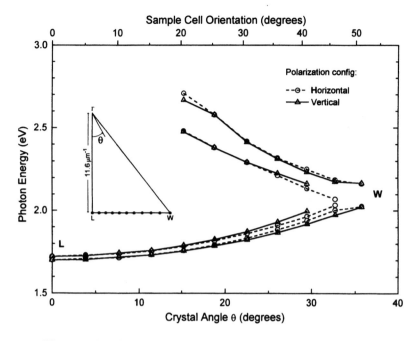

Figure 5. Photonic band structure, obtained from the transmission spectra in Fig. 4, for an fcc crystal of 0.135 μm diameter polystyrene microspheres with a lattice constant of 469 nm. Dashed (solid) lines are guides to the eye indicating horizontally (vertically) polarized incoming probe beam. Sample cell orientation on the top axis has been converted to crystal angle on the bottom axis by application of Snell's law. The inset shows a cross-section of part of the first Brillouin zone for this crystal, indicating the points on the zone edge (crystal orientation) where transmission spectra were obtained.

5. Interferometric Measurements

To observe the photonic dispersion relation directly at the band edges, a photonic crystal was grown from a colloidal suspension of polystyrene microspheres, having a diameter of 0.110 μm and monodisperse to within 4.3%. The transmission spectrum of the fcc crystal along the [111] direction exhibited a stop band centered around 599 nm with a bandwidth of 7 nm at 1% transmisson. Assuming 1.34 for the average index of refraction of the crystal, the distance between (111) planes was calculated from Bragg's law to be 0.224 μm, which corresponds to a lattice spacing of 0.387 μm and a volume fraction of 4.8%. Dynamical diffraction effects in these crystals are small, resulting in a correction factor that would lower the computed lattice parameter by about 1% [21–23].

The dispersion relation of the crystal along the [111] direction was obtained with a modified Mach-Zehnder interferometer as shown in Fig. 6 [18]. This version of the Mach-Zehnder interferometer was chosen because of its

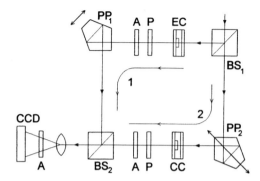

Figure 6. Modified Mach-Zehnder interferometer. The fringe pattern is positioned in front of the CCD camera by adjusting pentaprism PP_1. The optical paths are equalized by adjusting pentaprism PP_2 along a line perpendicular to PP_1. The relative intensity is optimized for maximum fringe visibility around the band edge by rotating the polarizer in the first arm. EC: Empty sample cell, CC: Crystal sample cell, BS's: Beam Splitter Cubes, A's: Analyzers, P's: Polarizers.

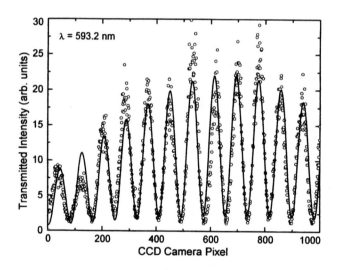

Figure 7. Typical interferometer fringe pattern with optimized fringe visibility. The solid curve is the resulting nonlinear least squares fit of Eq. 1.

orthogonality of adjustments [24]. The optical path lengths and dispersion were equalized in each arm as accurately as possible with the sample absent. The intensity in the first arm was adjusted relative to the second arm to increase the overall fringe visibility as the wavelength was tuned to the band edges. The fringe patterns were captured with a CCD camera for each step in wavelength, as shown in Fig. 7. A nonlinear least squares fit of the following equation to each fringe pattern was used to extract the desired

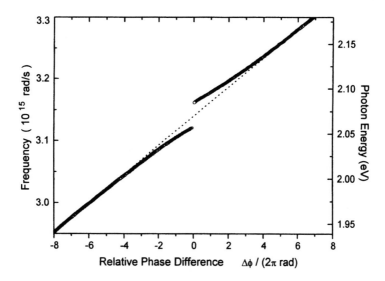

Figure 8. Wavelength dependence of relative phase of the fringe patterns. The broken region corresponds to the [111] stop band; the low transmitted intensity prohibits formation of a measureable fringe pattern.

phase information:

$$I(x) = I_0 \exp\left[-\left(\frac{x-\mu}{w}\right)^2\right]\{1 + \mathcal{V}\cos[2\pi f x + \Delta\phi]\}. \qquad (1)$$

Here I_0 is the intensity and μ and w are the center position and width of the gaussian envelope. The visibility, spatial frequency, and phase of the fringe pattern are $\mathcal{V}$, f, and $\Delta\phi$, respectively.

The absolute phase difference between the two interferometer arms can be considered in two parts, $\Delta\Phi(\lambda) = \phi_0(\lambda_0) + \Delta\phi(\lambda)$, where ϕ_0 is the initial phase difference arising from the insertion of the crystal at the initial wavelength, λ_0, and $\Delta\phi$ is the incremental phase difference (from Eq. 1) between the phase of the fringe pattern at a particular wavelength, λ, and λ_0. The relative phase shift across the tuning curve of the Rhodamine 6G dye laser is shown in Fig. 8.

The absolute phase difference between paths 1 and 2 is also given by

$$\Delta\Phi(\lambda) = 2\pi\left[n_c(\lambda) - 1\right]L/\lambda, \qquad (2)$$

where n_c is the index of refraction and L is the thickness of the crystal, determined to be 240 μm for this sample. Thus, the incremental change in the index of refraction of the crystal, $\Delta n_c(\lambda)$, can be related to the

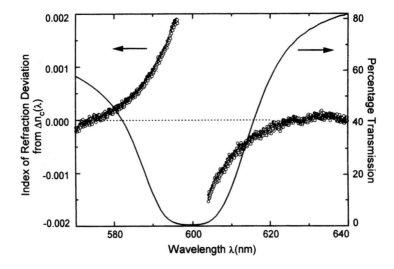

Figure 9. The deviation of the index of refraction near the stop band. The free photon dispersion has been linearly approximated and subtracted from the residual index of refraction obtained from Fig. 8. The transmission spectrum of the fcc crystal along the [111] direction is shown as the solid curve. The interferometrically-scanned region lies well inside the stop band.

incremental phase difference by,

$$\Delta n_c(\lambda) = n_c(\lambda) - n_c(\lambda_0) = \frac{\Delta\phi(\lambda)}{2\pi}\frac{\lambda}{L}. \tag{3}$$

The deviation from the free photon case around the band edges can be extracted by obtaining a residual from $\Delta n_c(\lambda)$ by linear approximation and subtraction. The resulting deviation $\Delta n \sim \pm 0.002$ at the band edges is clearly evident in Fig. 9, and is expected to grow as the center of the stop band is approached. The scanned region lies well inside the stop band, shown as percentage transmission in Fig. 9. The dramatic decrease of transmitted intensity makes measurement of fringe patterns difficult as the stop band is entered; hence, a gap appears in the relative phase and refractive index reported.

6. Conclusions

Observation of Kossel line patterns from photonic crystals provides a visual representation of the photonic band structure of the underlying lattice, opening up the possibility of another tool for investigating PBG crystals. Moreover, given a new candidate (lattice) for a PBG structure, its Kossel line patterns can provide a quick check for overlapping stop bands. Most importantly, Kossel lines also provide an exact experimental method for locating the symmetry directions of a single crystalline sample. In this study, the symmetry directions of a 3-D photonic crystal have been experimentally obtained with Kossel line analysis and the transmission measurements along these directions show optical stop bands opening up 2–3 orders of magnitude deep. As expected, the photonic band structure did not yield a complete gap along the $L–W$ points; however, the expected degeneracy at the W-point was not observed. The feasibility of measuring photonic band structure in colloidal crystals in the optical regime was also demonstrated by using a modified Mach-Zehnder interferometer.

Photonic crystals formed from polystyrene microspheres provide a novel system for examining photonic band structure in the optical regime. Colloidal crystals form readily by self-organization, requiring no lithography or advanced artifical structuring techniques. We are able to grow single crystals of good optical quality that are several mm across and thousands of lattice parameters thick.

The ability to readily tune the stop band to the wavelength of interest is an essential feature of using colloidal crystals for photonic band structure measurements, since the lattice parameter is controlled primarily via the volume fraction, rather than ball size as with ordinary close-packed structures. Microsphere size and relative dielectric strength of the microsphere dictate primarily the width of the stop band. Because of the limited tuning curve of the dye laser used to date, a relatively small microsphere diameter was selected to generate a narrow stop band to facilitate our initial studies. By expanding the effective range of the interferometer, complete mapping of photonic band structure and associated anomalous refractive index should be possible.

Wider gaps can be generated with balls closer to a Mie resonance. Generation of a pseudo-gap sufficient to study localization effects will necessitate generating W-point, L-point, and other stop bands that are wide enough for significant overlap. Our work, focused to date on growing and characterizing high quality single crystals, will be extended to larger microsphere diameter and volume fraction to satisfy as much as possible the criterion of overlapping stop bands. Following growth of a crystal with a suitable pseudogap, evidence for localized states will be sought around the impurity modes or by introduction of controlled disorder. An intriguing possibility is substitutional doping of the colloidal crystal with scatterers of the same size but having different dielectric strength, thereby preserving the crystalline order but introducing optical disorder.

7. Acknowledgments

We would like to thank Martin Zinkin for his initial efforts on this project and S. Weber, Ch. Schetelich and V. Greist [17] for providing us with their program for simulating Kossel line patterns. This work was supported by the National Science Foundation under grant DMR-9113618.

References

1. E. Yablonovitch, Phys. Rev. Lett., **58**, 2059 (1987).
2. S. John, Phys. Rev. Lett. **58**, 2486 (1987).
3. *Photonic Band Gaps and Localization*, edited by C. M. Soukoulis, (Plenum Press, New York, 1993).
4. C. M. Bowden, J. P. Dowling, and H. O. Everitt, J. Opt. Soc. Am. B **10**, special review issue (1993).
5. G. Kurizki and J. W. Haus, J. Mod. Opt. **41**, special review issue (1994).
6. E. Yablonovitch, T. J. Gmitter, and K. M. Leung, Phys. Rev. Lett. **67**, 2295 (1991).
7. E. Özbay, A. Abeyta, G. Tuttle, M. Tringides, R. Biswas, C. T. Chan, C. M. Soukoulis, and K. M. Ho, Phys. Rev. B **50**, 1945 (1994).
8. E. Özbay, E. Michel, G. Tuttle, R. Biswas, M. Sigalas, and K. M. Ho, Appl. Phys. Lett. **64**, 2059 (1994).
9. E. Özbay, E. Michel, G. Tuttle, R. Biswas, K. M. Ho, J. Bostak, and D.M. Bloom, Opt. Lett. **19**, 1155 (1994).
10. S. Fan, P. R. Villeneuve, R. D. Meade, and J. D. Joannopoulos, Appl. Phys. Lett. **65**, 1466 (1994).
11. N.A. Clark, A. J. Hurd, and B. J. Ackerson, Nature **281**, 57 (1979).
12. P. Pieranski, Contemp. Phys. **24**, 25 (1983).
13. A.J. Hurd, N. A. Clark, R. C. Mockler, and W. J. O'Sullivan, Phys. Rev. A **26**, 2869 (1982).
14. P. Pieranski, E. Dubois-Violette, F. Rothen, and L. Strzelecki, J. Physique **42**, 53 (1981).
15. T. Yoshiyama, I. Sogami and N. Ise, Phys. Rev. Lett. **53**, 2153 (1984).
16. T. Yoshiyama and I. Sogami, Phys. Rev. Lett. **56**, 1609 (1986).
17. S. Weber, Ch. Schetelich, and V. Greist, Cryst. Res. Technol. **29**, 727 (1994).
18. İ. İ. Tarhan, M. P. Zinkin, and G. H. Watson, Opt. Lett. **20**, 1571 (1995).
19. J. P. Dowling and C. M. Bowden, J. Mod. Opt. **41**, 345 (1994).
20. K. M. Leung and Y. F. Liu, Phys. Rev. Lett. **65**, 2646 (1990).
21. P. A. Rundquist, P. Photinos, S. Jagannathan, and S. A. Asher, J. Chem. Phys. **91**, 4932 (1989).
22. Y. Monovoukas, G. G. Fuller, and A. P. Gast, J. Chem. Phys. **93**, 8294 (1990).
23. W. L. Vos, et al., this volume.
24. P. Hariharan, Appl. Optics **8**, 1925 (1969).

INFLUENCE OF OPTICAL BAND STRUCTURES ON THE DIFFRACTION OF PHOTONIC COLLOIDAL CRYSTALS

WILLEM L. VOS[1], RUDOLF SPRIK[1], AD LAGENDIJK[1,2] AND GERARD H. WEGDAM[1]

[1] *van der Waals-Zeeman Instituut,*
Universiteit van Amsterdam,
1018 XE Amsterdam, The Netherlands

AND

ALFONS VAN BLAADEREN[2,3] AND ARNOUT IMHOF[3]
[2] *FOM Institute for Atomic and Molecular Physics*
1098 SJ Amsterdam, The Netherlands
[3] *van 't Hoff Laboratorium, Universiteit van Utrecht,*
3584 CH Utrecht, The Netherlands

Abstract. We have performed optical diffraction studies on colloidal crystals with large refractive index mismatches up to 1.45 and polarizibilities per volume as large as 0.6. These conditions push colloidal crystals into the regime where strong coupling between photonic crystals and the light field occurs. It is found that the photonic band structures result in apparent Bragg spacings that strongly depend on the wavelength of light. The dynamical diffraction theory that correctly describes weak photonic effects encountered in X-ray diffraction, also breaks down. Two simple models are presented that give a much better description of the diffraction of photonic crystals.

1. Introduction

Photonic crystals are 3-dimensional (3D) periodic composites of different dielectric materials, with lattice parameters of the order of the wavelength of light[1]-[7]. Light that travels through such structures experiences a pe-

C. M. Soukoulis (ed.), Photonic Band Gap Materials, 107–118.
© *1996 Kluwer Academic Publishers. Printed in the Netherlands.*

riodic variation of the refractive index, analogous to the periodic potential energy of an electron in an atomic crystal. Therefore, the dispersion curves of light become organized in bands in a Brillouin zone in reciprocal space called photonic band structures. Variation of the refractive index causes a splitting of the bands at the edges of the Brillouin zone, that are called stop gaps. No waves can propagate for energies within these gaps. The same principle underlies the functioning of dielectric mirrors [8], which can thus be regarded as 1D photonic crystals. The stop gaps widen with increasing modulation of the refractive index. Ultimately, the stopgaps in all directions will overlap and give rise to a complete photonic band gap [1, 2] for refractive index ratios larger than 1.9, a polarizibility α per 'atomic' volume v (times 4π) larger than 0.5, and suitable crystal structures - e.g. the diamond structure [9]. In this case, no wave with an energy within the gap can propagate through the crystal in *any* direction; the crystal acts as a 3D mirror. A fascinating situation arises when an excited atom or molecule with an excitation energy within the band gap is placed inside a photonic crystal. Spontaneous emission is no longer possible, and the atom is trapped in its excited state. This is a spectacular phenomenon in quantum electrodynamics, where it was believed for a long time that spontaneous emission is an intrinsic property of an atom [10]. Although this situation has not yet been observed experimentally, exciting prospects such as threshhold-less lasers and diodes have already been discussed [10]. In another analogy with electronics, John [2] has predicted that some disorder in the crystal will result in localized states inside the band gap, similar to excitons in a semiconductor band gap.

The term photonic band gap was introduced by Yablonovitch [1] but experimental observations have only been made in the microwave regime down to wavelengths of about $\lambda \sim 500$ μm [6]. It is clearly a tremendous challenge to scale the fabrication techniques of photonic structures down to reach the optical part of the spectrum ($400 < \lambda < 800$ nm). The group of Scherer (Caltech) has managed to drill holes of several hundred nm in slabs of GaAs [7]. A disadvantage of this method, however, is that crystals of only a few unit cells thickness can be made, which is insufficient for the creation of a complete optical band gap. In addition, this fabrication technique is very difficult.

A different approach is to make photonic crystals with colloidal suspensions [11]. These are, e.g., solid particles with radii r between roughly 1 and 1000 nm that are suspended in liquids or gases [12]. It is possible to synthesize particles with dimensions uniform to within a few percent, which can spontaneously nucleate crystals at sufficiently high density or low enough salt concentration [13]. If the lattice parameter a is of the order of optical wavelengths, crystal reflections can be visually observed as a beautiful iridescence [14]. Optical diffraction has been applied to identify various crystal structures in systems with low refractive index contrasts or low densities ($4\pi\alpha/v < 0.05$) [13], [15]-[17]. Optical photonic experiments have already been done on similar weakly photonic crystals by Asher *et al* [18], Martorell and Lawandy [19], and Herbert and Malcuit [20]. These demonstrate the use of colloidal crystals to applications in photonics.

In this paper, we present a study of colloidal crystals with refractive index ratios up to 1.45 and 'photonic strengths' ($4\pi\alpha/v < 0.05$) up to 0.6. The experiments are a significant advance towards the use of these systems in the strongly photonic regime. Diffraction experiments have been done to investigate both the optical band structures and the crystal structure. With increasing photonic strength, the band structures result in apparent Bragg spacings that strongly depend on the wavelength. Even the dynamical diffraction theory [21, 22], that is well-known in X-ray diffraction (where $4\pi\alpha/v \sim 10^{-4}$) fails to describe the observations. We present two simple models that give physically more reliable results for the lattice spacings of strongly scattering photonic crystals: firstly by combining Bragg's law with linear bands using an averaged refractive index, and secondly using periodically stratified dielectric media [8]. Earlier accounts of this work have appeared in Ref. [23, 24].

2. Theory

In a diffraction experiment, one observes crystal reflections if the incoming and outgoing wave vectors $\mathbf{k}_{in}$ and $\mathbf{k}_{out}$ yield a difference vector equal to a reciprocal lattice vector $\mathbf{G}_{hkl}$: $\mathbf{k}_{out} - \mathbf{k}_{in} = \mathbf{G}_{hkl}$, with hkl the Miller indices [25]. The length of the wave vector inside the crystal is $|\mathbf{k}| = (2\pi n)/\lambda$, with n the refractive index of the crystal, λ the wavelength of light in vacuum. The length of $\mathbf{G}_{hkl}$ is: $|\mathbf{G}_{hkl}| = 2\pi/d_{hkl}$, with d_{hkl} the lattice spacing of the hkl crystal planes. For weak photonic strengths, the dispersion relation between the energy E and the wave vectors is linear: $E = c|\mathbf{k}|$, and we obtain the well-known Bragg law: $d_{Bragg} = \lambda/(2n_{med}\sin\theta)$. Here, 2θ is the diffraction angle subtended between the incoming and outgoing wave vectors $\mathbf{k}_{in}$ and $\mathbf{k}_{out}$, n_{med} is the refractive index of the medium in which the particles are suspended (see Fig. 1). With increasing photonic strengths, the dispersion curves become non-linear, and stopgaps appear at the edges of the Brillouin zone. The diffraction condition is modified compared to the Bragg case and fulfilled at a lower energy (Fig. 1). In the limit of weak photonic strengths, improvements on the Bragg law can be calculated exactly with the dynamical diffraction theory [21, 22], that is well known for X-ray diffraction ($4\pi\alpha/v \sim 10^{-4}$). The relation between the lattice spacing d_{dyn}

and the diffraction angle is for specular reflections:

$$d_{\text{dyn}} = \frac{\lambda}{2n_{\text{med}}\sin\theta}\left(1 + \frac{\Psi}{2\sin^2\theta}\right)^{-1}, \tag{1}$$

with

$$\Psi = 3\phi\frac{m^2 - 1}{m^2 + 2}, \tag{2}$$

where ϕ is the volume fraction of colloidal spheres, m is the ratio of the refractive index of the spheres and the medium: $m = n_{\text{sph}}/n_{\text{med}}$, and Ψ is a photonic strength parameter that measures the interaction strength of the light and photonic crystal. Equation 1 can be heuristically derived - albeit with a different Ψ parameter - by considering Bragg diffraction as an interference between two light rays that are reflected from two crystalline layers (such as depicted in most elementary texts, e.g. [25]) and correcting for the extra optical path length that the second light ray experiences while twice traversing the upper layer. The parameter Ψ can be rewritten as:

$$\Psi = 4\pi\frac{\alpha}{v}. \tag{3}$$

Now it becomes clearer that Ψ is a good choice for a photonic strength parameter, because it can be physically interpreted as the ratio between the "optical" volume and the actual volume of the scattering particles. The effects of photonic strength become more apparent for smaller θ or larger d_{hkl}/λ, which corresponds to higher bands in Fig. 1. Therefore, it is very useful to do a diffraction experiment as a function of wavelength on a photonic crystal: in the short wavelength limit one will mostly probe the effects of photonic band structures (through $\mathbf{k}_{\text{in}}$ and $\mathbf{k}_{\text{out}}$), whereas in the long wavelength limit will mostly characterize the reciprocal lattice vectors and hence the crystal structure. Finally, we observe that an increase of Ψ for a given crystal structure and constant lattice parameter results in an increase of the diffraction angle 2θ.

3. Experimental

Crystals were grown from silica spheres ($n_{\text{sph}} = 1.45$) with radii between 108 and 525 nm, suspended in dimethylformamide (DMF), ethanol, and water ($n_{\text{med}} = 1.43$, 1.36, 1.33 respectively). For details of the synthesis and characterization of the spheres, see Refs. [26]-[28]. The samples crystallized after sedimentation under gravity in 400 μm thick and 4 mm wide glass capillaries (Vitro Dynamics). In addition, one charge stabilized system was studied, that started to crystallize immediately after loading and sealing capillaries. Crystals in air ($n_{\text{med}} = 1.00$) were made from sedimented crystals by letting the suspending liquid evaporate slowly in several weeks.

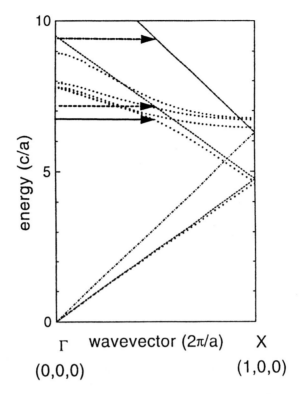

Figure 1. Bandstructures calculated with the plane-wave method [3, 4, 5] for the 100 (X) direction of an f.c.c. crystal with 74 vol% spheres (dots). For a complementary theoretical method, see Ref. [35]. Spheres with index $n_{sph} = 1.45$ and medium with $n_{med} = 1.0$ were taken, corresponding to $\Psi = 0.6$. At $E = 6.7$, the first folding of the Brillouin zone occurs. [25]. For clarity, only lower bands have been plotted. The incoming wavevectors $\mathbf{k}_{in}//\mathbf{100}$ taking part in the diffraction by a reciprocal lattice vector $\mathbf{G}_{hkl}$ with $hkl = 111$ have been indicated. The drawn arrow at $E = 6.7$ is $\mathbf{k}_{in}$ corresponding to the full photonic bandstructure. The dashed arrow at $E = 7.2$ is $\mathbf{k}_{in}$ corresponding to the linear bands with a slope that is inversely proportional to the average refractive index n_{cryst} at $|\mathbf{k}| = \mathbf{0}$ (dashed lines). The dashed-dotted arrow at $E = 9.4$ is $\mathbf{k}_{in}$ corresponding to the Bragg law: linear bands with a slope that is inversely proportional to the refractive index of the medium n_{med} (dashed-dotted lines).

The capillaries were mounted in a cilindrical bath containing glycerine ($n = 1.47$), that closely matches the refractive index of the capillaries. This bath was mounted on a stage that could be fully rotated (ω-circle), see Fig. 2. Because colloidal crystals often order with the close packed planes parallel to the walls of the cell (see e.g. [29, 30]), upon rotating the samples, the diffracted intensity of these crystal planes becomes very strong and clearly visible on a screen if it coincides with the specularly reflected laser beam. Incident monochromatic light with wavelengths λ between 458 and 785 nm was collimated, modulated, and polarized. Scattered radiation was

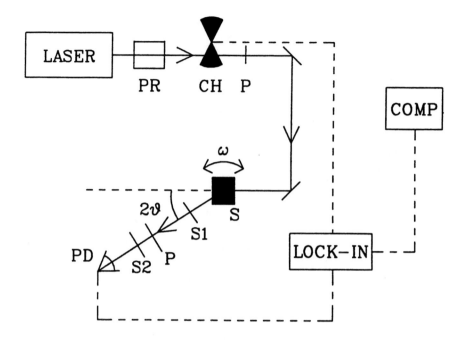

Figure 2. Schematic drawing of the light scattering setup. Light emanates from a laser, is polarization rotated (PR), modulated by a chopper (CH), and polarized (P). It is scattered in the sample (S) that is mounted in an index matching bath (not shown) that sits on a rotation table (ω-circle). Scattered light at an independent angle 2θ is collimated through slits S1 and S2, polarization analyzed (P), and collected with a photo diode (PD). The signal from the diode is discriminated with a lock-in amplifier and stored in a computer (comp).

collimated, polarization analyzed, and detected with a photo diode, all of which were mounted on an independently rotating stage (2θ-circle). In all experiments V-V polarization was used. The measured values of 2θ were corrected for refraction at the bath-capillary-crystal interfaces and offset errors were eliminated by scanning both positive and negative 2θ angles.

4. Results

Representative diffraction patterns of close packed lattice planes at $\lambda = 633$ nm for samples with increasing photonic parameters, but comparable lattice parameters are shown in Fig. 3. Crystals with $\Psi = 0.017$ reveal sharp diffraction lines near $2\theta = 75^o$ which corresponds to a lattice spacing $d_{hkl} = 380$ nm. For $\Psi = 0.115$, the diffraction peak is shifted to a larger scattering angle of 80°. This is caused by an increase of Ψ and also by a decrease of the total refractive index of the system, whereas $d_{\text{hkl}} = 370$ nm. For $\Psi = 0.60$, the diffraction angle has moved to 97°. This is caused by the increase of Ψ, in addition by the decrease of the total refractive index of the system, and a slightly smaller $d_{hkl} = 350$ nm. The diffracted signal appears to consist of several fringe-like features. Whereas we understand the position of the band as a whole, we have no explanation for the fringes but can exclude several possibilities; These are not caused by interference in the walls of the capillary (300 μm thick), which would give a fringe period of 0.04^o. Also, at the index contrast involved ($m = 1.45$), the Mie scattering of the spheres does not produce sharp and closely spaced resonances. Moreover, reducing Ψ of such a sample to 0.017 by reintroducing liquid DMF in the capillary results again in a single sharp diffraction peak near $2\theta = 75^o$, identical to the lower diffraction pattern in Fig. 3. Therefore, it is concluded that samples with increasing Ψ produce diffraction peaks at higher diffraction angle, which is so far consistent with the dynamical theory. We note that the volume fractions of the spheres in the crystals that are calculated from the lattice spacings and the known radii of the spheres, agree with estimates obtained from the volume fraction of the starting suspension and the ratio of the volumes of the crystal formed and the total sample.

From the measured diffraction angle 2θ, the Bragg spacing d_{Bragg} and the dynamic spacing d_{dyn} between the close packed lattice planes are calculated and plotted in Fig. 4. For $\Psi=0.017$, Fig. 4a reveals that d_{Bragg} decreases slightly as a function of wavelength. The values of d_{dyn} are lower and nearly independent of wavelength, the difference with d_{Bragg} decreasing to long wavelengths as expected. A similar result was obtained for a dilute charged system with $\Psi = 0.005$. In contrast, the results for $\Psi = 0.115$ show a strong increase of d_{Bragg} of 50% in going from 785 to 458 nm (Fig. 4b). This means that d_{Bragg} has become an unphysical estimate of the lattice spacing. A more reliable result is obtained with the dynamic theory, because d_{dyn} is independent of wavelength. At the longest wavelengths, d_{Bragg} converges to d_{dyn} as expected from Eq. (1). This confirms that long wavelength extrapolations may be used to estimate the real lattice spacings of photonic crystals. Similar results were obtained for a sample with $\Psi = 0.067$. For $\Psi = 0.60$, Fig. 4c reveals a giant increase of d_{Bragg} with decreasing wavelength. For this strongly scattering sample, it appears that the dynamic theory is also not appropriate anymore, because d_{dyn} clearly decreases with decreasing wavelength. Furthermore, the values of d_{dyn} would imply that the colloidal spheres are strongly overlapping. With increasing wavelength, both apparent lattice spacings converge to a value expected for touching close packed spheres (344 nm). We can exclude the possibility that the dispersive effects are caused by a smearing of the reciprocal lattice points due

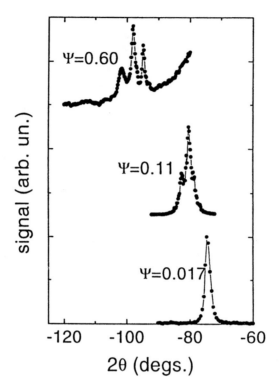

Figure 3. Diffraction patterns of colloidal crystals. The diffraction patterns have been normalized and offset for clarity. From bottom to top: spheres with $r = 211$ nm in DMF ($\Psi = 0.017$), in water ($\Psi = 0.11$), and in air ($\Psi = 0.6$).

to mixed hexagonal stacking, because smearing does not happen for close packed lattice planes [13, 31].

5. Discussion

We have used two simple models that go beyond dynamical diffraction theory to describe the relation between the diffraction angle, the wavelength, and the crystal structure and lattice parameter. The first model consists of approximating the apparent spacings d_{Bragg} and d_{dyn} of every hkl reflection with the apparent spacings of pairs of layers of dielectric materials. This is similar to the conventional approach to stratified dielectric systems such as dielectric mirrors [8]. One layer of each pair is assigned the refractive index n_{sph} of the colloidal spheres and the other that of the medium (n_{med}). The thickness d of each pair of layers is fixed to the expected lattice spacing d_{hkl}. The relative thickness of the layers is determined by the volume fraction of spheres ϕ, which is related to their radius r, and the lattice spacing d,

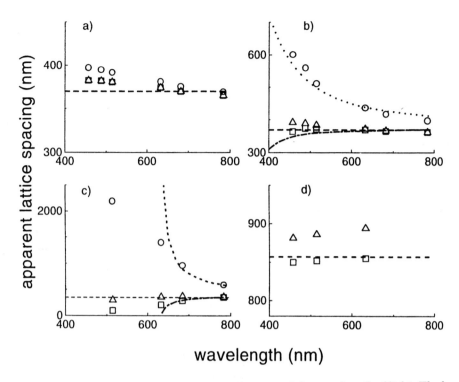

Figure 4. Close packed crystal spacings as a function of the wavelength of light. The long dashed line indicates the expected spacing of the spheres, the circles are Bragg spacings, the squares are spacings calculated with the dynamic theory, and the triangles are Bragg spacings using the average index of the crystal n_{cryst}. The short dashed curves are the Bragg spacings and the dashed-dotted curves the dynamic spacings of the periodically stratified media model. a) spheres with $r = 211$ nm in DMF ($\Psi = 0.017$), b) in H_2O ($\Psi = 0.11$), c) in air ($\Psi = 0.6$), d) spheres with $r = 525$ nm in DMF ($\Psi = 0.017$).

assuming hexagonal dense packing: $\phi = (16\pi)/3 \left(r/(d\sqrt{3}) \right)^3$. From the calculated reflection angle, the *apparent* spacings d_{Bragg} and d_{dyn} are found. In Fig. 4b, it is seen that the results of this parameter-free calculation closely track the experimental data for $\Psi = 0.11$. In Fig. 4c ($\Psi = 0.60$), the results are seen to agree with the experiments at longer wavelengths. This confirms that the observed diffraction features of Fig. 3 are caused by close packed layers of colloidal spheres. Thus, this simple model provides a reliable estimate of the crystal spacings in crystals with moderate photonic strengths, where the dynamic theory collapses. The deviations at short wavelengths or high energy bands can be explained as follows: in this case, waves are coupled that originate from many different possible hkl reciprocal lattice vectors. The stratified layer model however, only takes into account the vectors with Miller indices $m\,(hkl)$, with m an integer, which clearly gives a worse description for higher bands. Limitations of this model are that the shape of the calculated diffraction pattern differs from the experimental curves, and the width of the diffraction line is strongly overestimated. In fact, the width of the diffraction peak of a single crystal reflects the width of the stopbands. On the other hand however, the width of the diffraction peaks is also influenced by the size of the crystal grains [22], and we believe that in our polycrystalline samples this is the main determining factor.

In the second model, the dielectric constants of the constituent materials are combined to an average n_{cryst} by applying the Maxwell-Garnett theory (see e.g. [32]) to the crystalline configuration of colloidal spheres in the surrounding medium [33]. This produces an average refractive index n_{cryst} that agrees well with the exact solution in the limit $|\mathbf{k}| = 0$ (Fig. 1), but at the same time washes out all crystalline features as well. Therefore, point scatterers are assumed to be present on the positions of the colloidal spheres to account for the diffracted light. The lattice spacings are then calculated with the Bragg law, using n_{cryst} instead of the index of the medium n_{med}. Fig. 4c reveals that this model yields lattice spacings that agree much better with the expected values than both the Bragg or dynamic theories. Indeed Fig. 1 shows that for a dielectric crystal corresponding to Fig. 4c, the diffraction condition is met for 10 % higher energy than in the exact solution. This is in much better agreement with the exact results than the Bragg result.

In order to probe relatively higher energies in the bandstructures (see Fig. 1), we have also done experiments on close packed crystals ($\phi = 74$ vol%) of much larger spheres ($r = 525$ nm). This corresponds to d_{hkl}/λ up to 1.9, compared to 0.8 for the crystals described above. We find that for $\Psi = 0.017$, the dynamic estimate of the lattice spacing is in excellent agreement with the value based on the 2nd and 3rd harmonic of the close packed reflection (Fig. 4d), on independent experiments with confocal microscopy, and on the size of the spheres [34]. The Bragg law combined with the averaged refractive index n_{cryst} gives results that are about 3 % higher. The deviations between these two estimates are contrasted to their excellent agreement in Fig. 4a, and are likely caused by a greater difference between the linear bands using n_{cryst} (effective medium) and the exact bands (Fig. 1). This also indicates that a photonic parameter should include the relevant energy scale. Furthermore, these results illustrate why deviations from the dynamic diffraction theory have not been observed in [15]-[17]: first of all, in those experiments Ψ was less than about 0.05 and secondly, the experiments were done on crystals with much smaller lattice spacings, corresponding to $(d_{\text{hkl}}/\lambda) < 0.4$. Finally we note that the parameter Ψ should be improved to take into account excluded volume effects: it has the wrong limit when the volume fraction of high-index material goes to 1, in which case photonic effects are expected to go to zero again. Thus, it seems more reasonable that a photonic parameter scales in such a way that it vanishes in the limit of pure components, $\phi = 0$ or $\phi = 1$. Guidance may be provided by the suggestion of John that photonic band gaps are optimal for $\phi = 1/(2m)$ [11].

6. Conclusions

We have studied colloidal crystals with large refractive index mismatches and strong photonic strengths. We have found that the photonic band structures result in strong dispersion of the apparent Bragg and dynamic lattice spacings. Therefore, diffraction is a useful new method for studying photonic band structures. On the other hand, structural information on the strongly scattering colloidal crystals can be extracted with the aid of simple models. The large index of refraction ratios call for extensions of diffraction theory beyond the well-founded theories known from X-ray diffraction. It is a challenge to calculate the diffraction properties directly from the full band structure.

7. Acknowledgments

We thank Henk Lekkerkerker, Nynke Verhaegh, and David Weitz (Exxon) for useful discussions, and Boris Nieuwenhuis for preparing Fig. 2. This work was part of the research program of the Dutch Organization for Fundamental Research on Matter (FOM) and was made possible by financial support of the 'Nederlandse Organisatie voor Wetenschappelijk Onderzoek' (NWO).

References

1. E. Yablonovitch, Phys. Rev. Lett. **58**, 2058 (1987).
2. S. John, Phys. Rev. Lett. **58**, 2486 (1987).
3. E. Yablonovitch, T. J. Gmitter, and K. M. Leung, Phys.Rev. Lett. **67**, 2295 (1991).
4. "Photonic Band Gaps and Localization," Ed. by C. M. Soukoulis (Plenum, New York, 1993).
5. "Development and Applications of Materials Exhibiting Photonic Band Gaps," Ed. by C. M. Bowden, J. P. Dowling, and H. O. Everitt (J. Opt. Soc. Am. B **10**, 280 (1993)).
6. E. Özbay, E. Michel, G. Tuttle, R. Biswas, K. M. Ho, J. Bostak, and D. M. Bloom, Opt. Lett. **19**, 1155 (1994).
7. See other papers in these Proceedings.
8. M. Born and E. Wolf, "Principles of Optics" (Pergamon, Oxford, 1980).
9. K. M. Ho, C. T. Chan, and C. M. Soukoulis, Phys. Rev. Lett. **65**, 3152 (1990).
10. E. Yablonovitch, see Ref. [4], p. 207.
11. S. John, see Ref. [4], p. 1.

12. R. J. Hunter, "Foundations of Colloid Science," (Clarendon, Oxford, 1992).
13. P. N. Pusey, in "Liquids, Freezing, and the Glass Transition," Ed. by D. Levesque, J.-P. Hansen, and J. Zinn-Justin (Elsevier, Amsterdam, 1990).
14. P. N. Pusey and W. van Megen, Nature (London) **320**, 340 (1986).
15. T. Yoshiyama, I. Sogami, and N. Ise, Phys. Rev. Lett. **53**, 2153 (1984).
16. P. A. Rundquist, P. Photinos, S. Jagannathan, and S. A. Asher, J. Chem. Phys. **91**, 4932 (1989).
17. Y. Monovoukas, G. G. Fuller, and A. P. Gast, J. Chem. Phys. **93**, 8294 (1990).
18. P. L. Flaugh, S. E. O'Donnell, and S. A. Asher, Appl. Spectrosc. **38**, 847 (1984).
19. J. Martorell and N. M. Lawandy, Phys. Rev. Lett. **66**, 887 (1991).
20. C. J. Herbert and M. S. Malcuit, Opt. Lett. **18**, 1783 (1992).
21. W. H. Zachariasen, "Theory of X-ray Diffraction in Crystals," (Wiley, New York, 1945).
22. R. W. James, "The Optical Principles of the Diffraction of X-rays," (Bell, London, 1954).
23. W. L. Vos, R. Sprik, G. H. Wegdam, A. Lagendijk, A. Imhof, and A. van Blaaderen, in "Postdeadline digest, 1994 European Quantum Electronics Conference" (IEEE, Piscataway, 1994).
24. W. L. Vos, R. Sprik, A. Imhof, A. van Blaaderen, A. Lagendijk, and G. H. Wegdam, Phys. Rev. Lett. (submitted).
25. N. Ashcroft and D. Mermin, "Solid State Physics" (Holt, Rinehart, and Winston, New York, 1976).
26. A. van Blaaderen and A. P. M. Kentgens, J. Non-Cryst. Solids **149**, 161 (1992).
27. A. Imhof, A. van Blaaderen, G. Maret, J. Mellema, and J. K. G. Dhont, J. Chem. Phys. **100**, 2170 (1994).
28. A. van Blaaderen and A. Vrij, Langmuir **8**, 2921 (1992).
29. N. A. Clark, A. J. Hurd, and B. J. Ackerson, Nature (London) **281**, 57 (1979).
30. C. A. Murray, W. O. Sprenger, and R. A. Wenk, Phys. Rev. B **42**, 688 (1990).
31. P. N. Pusey, W. van Megen, P. Bartlett, B. J. Ackerson, J. G. Rarity, and S. M. Underwood, Phys. Rev. Lett. **63**, 2753 (1989).
32. See e.g. C. F. Bohren and D. R. Huffman, "Absorption and Scattering of Light by Small Particles," (Wiley, New York, 1983).
33. S. Datta, C. T. Chan, K. M. Ho, and C. M. Soukoulis, Phys. Rev. B **48**, 14936 (1993).
34. A. van Blaaderen and P. Wiltzius, Science, **270**, 1177 (1995).
35. J. B. Pendry, J. Mod. Optics **41**, 209 (1994).

FROM MICROMASER TO MICROLASER

O. BENSON, G. RAITHEL; H. WALTHER

Sektion Physik der Universität München and
Max-Planck-Institut für Quantenoptik
85748 Garching, Fed. Rep. of Germany

1. Overview

The simplest and most fundamental system for studying radiation-matter coupling is a single two-level atom interacting with a single mode of an electromagnetic field in a cavity. It received a great deal of attention shortly after the maser was invented, but at that time, the problem was of purely academic interest since the matrix elements describing the radiation-atom interaction are so small. The field of a single photon is not sufficient to lead to an atom field evolution time shorter than the other characteristic times of the system, such as the excited state lifetime, the time of flight of the atom through the cavity, and the cavity mode damping time. It was therefore not possible to test experimentally the fundamental theories of radiation-matter interaction, which predict, among other effects,

(a) a modification of the spontaneous emission rate of a single atom in a resonant cavity,
(b) oscillatory energy exchange between a single atom and the cavity mode, and
(c) the disappearance and quantum revival of Rabi nutation induced in a single atom by a resonant field.

The situation has drastically changed in the last few years with the introduction of frequency-tunable lasers, which can excite large populations of highly excited atomic states characterized by a high principal quantum number n of the valence electron. These states are generally called Rydberg states, since their energy levels can be described by the simple Rydberg formula. Such excited atoms are very suitable for observing quantum effects in radiation-atom coupling for three reasons. First, the states are very strongly coupled to the radiation field (the induced transition rates between neighbouring levels scale as n^4); second, transitions are in the millimetre wave region so that low-order mode cavities can be made large enough to allow rather long interaction times, and finally, Rydberg states have relatively long lifetimes with respect to spontaneous decay [1,2].

119

C. M. Soukoulis (ed.), Photonic Band Gap Materials, 119–141.

The strong coupling of Rydberg states to radiation resonant with transitions to neighbouring levels can be understood in terms of the correspondence principle: with increasing n the classical evolution frequency of the highly excited electron becomes identical with the transition frequency to the neighbouring level. Therefore, the atom corresponds to a large dipole oscillating with the resonance frequency; the dipole moment is very large since the atomic radius scales as n^2.

In order to understand the modification of the spontaneous emission rate in an external cavity, must remember that in quantum electrodynamics this rate is determined by the density of modes of the electromagnetic field at the atomic transition frequency ω_o, which depends on the square of the frequency in free space. If the atom is not in free space but in a resonant cavity, the continuum of modes is changed into a spectrum of discrete modes of which one may be in resonance with the atom. The spontaneous decay rate of the atom in the cavity γ_c will then be enhanced in relation to that in free space γ_f by a factor given by the ratio of the corresponding mode densities:

$$\gamma_c / \gamma_f = \rho_c(\omega_o) / \rho_f(\omega_o) = 2\pi Q / V_c \omega_o^3 = Q\lambda_o^3 / 4\pi^2 V_c,$$

where V_c is the volume of the cavity and Q is the quality factor of the cavity which expresses the sharpness of the mode. For low-order cavities in the microwave region, $V_c \approx \lambda_o^3$, which means that the spontaneous emission rate is increased by roughly a factor of Q. However, if the cavity is detuned, the decay rate will decrease. In this case, the atom cannot emit a photon, since the cavity is not able to accept it, and therefore the energy stays with the atom.

Recently, a number of experiments have been made with Rydberg atoms to demonstrate this enhancement and inhibition of spontaneous decay in external cavities or cavity-like structures. These experiments will be briefly reviewed in section 2.

More subtle effects due to the change of the mode density can also be expected: radiation corrections such as the Lamb shift and the anomalous magnetic dipole moment of the electron are modified with respect to the free space value, although changes are of the same order as present experimental accuracy. Roughly speaking, one can say that such effects are determined by a change of virtual transitions and not by real transitions as in the case of spontaneous decay. A review on this topic will not be given here, for details see Ref. [2]. For general reviews on cavity quantum electrodynamics see Refs. [3-5].

Section 3 describes the one-atom maser representing the idealized case of a two-level atom interacting with a single mode of a radiation field. The theory of this system was treated by Jaynes and Cummings [6] many years ago, and we shall concentrate on the dynamics of the atom-field interaction predicted by this theory. Some of the features are explicitly a consequence of the quantum nature of the electromagnetic field: the statistical and discrete nature of the photon field leads to new dynamic characteristics such as collapse and revivals in the Rabi oscillation. Furthermore, the photon statistics, the observation of quantum jumps, and the observation of atomic interference in the micromaser will be discussed.

In this review the experiments with optical cavities, in which dissipation of the energy is important, will not be discussed. However, a survey on the work performed with microcavities will be given.

2. Experimental Demonstration of the Modification of the Spontaneous Transition
Rate in Confined Space

The modification of the spontaneous decay rate of an excited atom in a cavity was noted for the first time years ago by Purcell [7]. To change the decay rate of an atom, in principle no resonator must be present; any conducting surface near the radiator affects the mode density and, therefore, the spontaneous radiation rate. Parallel-conducting planes can somewhat alter the emission rate but can only reduce the rate by a factor of 2 owing to the existence of TEM modes, which are independent of the separation.

In order to demonstrate experimentally the modification of the spontaneous decay rate, it is not necessary to go to single-atom densities in both cases. The experiments where the spontaneous emission is inhibited can also be performed with large atom numbers. However, in the opposite case, when the increase of the spontaneous rate is observed, a large number of excited atoms may disturb the experiment by induced transitions. The first experimental work on inhibited spontaneous emission was done by Drexhage (see Ref. [8]). The fluorescence of a thin dye film near a mirror was investigated. A reduction of the fluorescence decay by up to 25 % results from the standing wave pattern near the mirror. In recent years similar experiments were conducted by de Martini, et al. [9]. Inhibited spontaneous emission was also observed by Gabrielse and Dehmelt [10]. In these experiments with a single electron stored in a Penning trap, they observed that cyclotron orbits show lifetimes which are up to 10 times larger than those calculated for free space. The electrodes of the trap form a cavity which decouples the cyclotron motion from the vacuum radiation field leading to the longer lifetime.

Experiments with Rydberg atoms on inhibition of spontaneous emission have been performed by Hulet, et al. [11] and by Jhe, et al. [12] In the latter experiment, a 3.4 μm transition was suppressed. The first observation of enhanced atomic spontaneous emission in a resonant cavity was published by Goy, et al. [13]. This experiment was performed with Rydberg atoms of Na excited in the 23s state in a niobium superconducting cavity resonant at 340 GHz. Cavity tuning-dependent shortening of the lifetime was observed. The cooling of the cavity had the advantage of totally suppressing the black-body field. The latter precaution is not necessary if optical transitions are observed. The first experiment of this type was performed by Feld and collaborators [14]. They succeeded in demonstrating the enhancement of spontaneous transitions even in higher order optical cavities by using a concentric mirror arrangement where the transversal modes are degenerate.

In modern semiconductor devices, both electronic and optical properties can be tailored with a high degree of precision. Therefore, electron-hole systems producing recombination radiation analogous to radiating atoms can be localized in cavity-like structures, e.g. in quantum wells. Thus, optical microcavities of half or full wavelength size are obtained. Both suppression and enhancement of spontaneous emission in semi-conductor microcavities were demonstrated in experiments by Yamamoto and collaborators [15]. Similar structures have been used by Yablonovitch, et al. [16-18] exhibiting a photonic band gap with spontaneous emission forbidden in certain frequency regions. Microcavity structures have also been used to generate laser radiation; these experiments will be discussed in section 9 of this paper

3. Review of the One-Atom Maser

The most fundamental system to study the generation process of radiation in lasers and masers is to drive a single mode cavity by a single atom. This system, at first glance, seems to be another example of a Gedanken experiment but such a one-atom maser [19] really exists and can, in addition, be used to study the basic principles of radiation-atom interaction. The main features of the setup are:

(1)   It is the first maser which sustains oscillations with much less than one atom on the average in the cavity.
(2)   The maser allows one to study the dynamics of the energy exchange between an atom and a single mode of the cavity field as treated in the Jaynes-Cummings model [6].
(3)   The setup allows one to study in detail the conditions necessary to obtain non-classical radiation, especially radiation with sub-Poissonian photon statistics in a maser system directly.
(4)   It is possible to study a variety of phenomena of a quantum field including non-local aspects of the quantum measurement process.

What are the tools that make this device work? It is the enormous progress in constructing superconducting cavities with high quality factors together with the laser preparation of highly excited atoms – Rydberg atoms – that have made the realisation of such a one-atom maser possible [19]. Rydberg atoms are obtained when one of the outermost electrons of an atom is excited into a level close to the ionization limit. The main quantum number of the electron is then typically of the order of $60 - 70$. Those atoms have quite remarkable properties [1,2] which make them ideal for the maser experiments. The probability of induced transitions between neighbouring states of a Rydberg atom scales as n^4, here n denotes the principle quantum number. Consequently, a few photons are enough to saturate the transition between adjacent levels. Moreover, the spontaneous lifetime of a highly excited state is very large. We obtain a maser by injecting these Rydberg atoms into a superconducting cavity with a high quality factor. The injection rate is such that on the average there is much less than one atom present inside the resonator.

The experimental setup of the one-atom maser is shown in Fig.1. A highly colli-mated beam of rubidium atoms is used to pump the maser. Before entering the super-conducting cavity, the atoms are excited into the upper maser level $63p_{3/2}$ by the fre-quency-doubled light of a cw ring dye laser. The velocity of the Rydberg atoms and

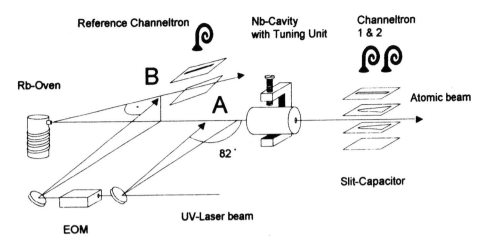

Figure 1. Scheme of the one-atom maser. In order to suppress blackbody-induced transitions to neighbouring states, the Rydberg atoms are excited inside a liquid-Helium-cooled environment.

thus their interaction time t_{int} with the cavity field was preselected by exciting a particular velocity subgroup with the laser. For this purpose, the laser beam irradiated the atomic beam at an angle of approximately 82°. As a consequence, the UV laser light (linewidth ≈ 2MHz) is blue-shifted by 50-200 MHz by the Doppler effect, depending on the velocity of the atoms [31].

The superconducting niobium maser cavity is cooled down to a temperature of 0.1 K by means of a ^{3}He / ^{4}He dilution refrigerator. At this temperature the number of thermal photons in the cavity is about 1×10^{-4} at a frequency of 21.5 GHz. The quality factor of the cavity can be as high as 3×10^{10} corresponding to a photon storage time of about 0.2 s. Two maser transitions from the $63p_{3/2}$ level to the $61d_{3/2}$ and to the $61d_{5/2}$ level are studied.

The Rydberg atoms in the upper and lower maser levels are detected by two separate field ionisation detectors. The field strength is adjusted to ensure that in the first detector the atoms in the upper level are ionised, but not those in the lower level; the lower level atoms are then ionised in a second field. To demonstrate maser operation, the cavity frequency is scanned across the respective transition frequency and the flux of atoms in the excited state is recorded simultaneously. Figure 2 shows the result for $63p_{3/2} - 61d_{3/2}$. Transitions from the initially prepared state to the $61d_{3/2}$ level (21.50658 GHz) are detected by a reduction of the electron count rate.

In the case of measurements at a cavity temperature of 0.5 K, shown in Fig.2, a reduction of the $63p_{3/2}$ signal can be clearly seen for atomic fluxes as small as 1750 atoms/s. An increase in flux causes power broadening and a small shift. Over the range from 1750 to 28000 atoms/s the field ionisation signal at resonance is independent of

124

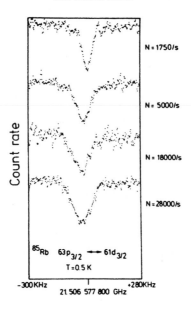

Figure 2. Maser operation of the one-atom maser manifests itself in a decrease of atoms in the excited state when the cavity is tuned through the resonance. The flux of excited atoms N governs the pump intensity. Power broadening of the resonance line demonstrates the multiple exchange of a photon between the cavity field and the atom passing through the resonator.

the particle flux which indicates the transition is saturated. This and the observed power broadening show a multiple exchange of photons exists between Rydberg atoms and the cavity field. For an average transit time of the Rydberg atoms through the cavity of 50 μs and a flux of 1750 atoms/s, we estimate approximately 0.09 Rydberg atoms are in the cavity on the average. According to Poissonian statistics this implies that more than 90 % of the events are due to single atoms. This clearly demonstrates that single atoms are able to maintain a continuous oscillation of the cavity with a field corresponding to a mean number of photons between unity and several hundred.

Among the studies performed with the one-atom maser are the measurements of the dynamics of the photon exchange between a single atom and a cavity mode [20,21]. Before we discuss some experiments with the one-atom maser the theory will be briefly reviewed.

4. Theory of the One-Atom Maser

The simplest form of interaction between a two-level atom and a single quantized mode of the electromagnetic field is described by the Jaynes-Cummings Hamiltonian [4,6].

$$H = \frac{1}{2}\hbar\omega_0\,\sigma_z + \hbar\omega\,a^\dagger a + \hbar\,(ga^\dagger\sigma_- + \text{adj.}).$$
$$= H_o + V$$

where

$$H_o = \frac{1}{2}\hbar\omega_0\,\sigma_z + \hbar\omega\,a^\dagger a,$$

and

$$V = \hbar\,(ga^\dagger\,\sigma_- + \text{adj.}).$$

Here, ω_0 is the atomic transition frequency, ω is the field frequency, a and $a^\dagger$ are the photon annihilation and creation operators of the field mode, with $[a, a^\dagger] = 1$, σ_z, and σ_- and σ_+ are atomic pseudo-spin operators, with $[\sigma_+,\sigma_-] = \sigma_z$, and

$$g = \frac{pE_\omega}{2\hbar}\sin KZ$$

is the electric dipole matrix element at the location Z of the atom, where E_ω is the "electric field per photon", $E_\omega = (\hbar\omega/\varepsilon_0\,V)^{1/2}$.

The Jaynes-Cummings model plays a central role in quantum optics owing to several reasons: it gives the simplest description of Rabi-flopping in a quantised field and the simplest illustration of spontaneous emission. Furthermore, it can be solved exactly and thus describes the true quantum dynamics observed with the one-atom maser such as collapse and revivals of the atomic inversion. The model describes the situation realised in the one-atom maser and allows a detailed investigation of the complexities of the atom-field dynamics in the simplest of all situations.

In the following, a few results following from the Jaynes-Cummings model are reviewed as relevant for the discussions in this paper.

The eigenstates of the Jaynes-Cummings Hamiltonian are

$$E_{1n} = \hbar\,[-\frac{1}{2}\omega_o + (n+1)\,\omega + \frac{1}{2}\,(\Omega_n + \delta)]$$
$$E_{2n} = \hbar\,[\frac{1}{2}\omega_o + n\,\omega - \frac{1}{2}\,(\Omega_n + \delta)],$$

where $\delta = \omega_0 - \omega$ is the atom-field frequency detuning. Ω_n is the generalized n-photon Rabi flopping frequency

$$\Omega_n = \sqrt{4g^2(n+1) + \delta^2}$$

The corresponding eigenvalues are the one of the dressed atom:

$$| 1,n \rangle = \sin \theta_n | a,n \rangle + \cos \theta_n | b,n + 1 \rangle$$

and

$$| 2,n \rangle = \cos \theta_n | a,n \rangle - \sin \theta_n | b,n + 1 \rangle ,$$

where the states $| a \rangle$ and $| b \rangle$ are the upper and lower atomic states, respectively, and $| n \rangle$ are the number states of the field mode with $a^\dagger a | n \rangle = n | n \rangle$. The angle θ is defined by means of the relations

$$\cos \theta_n = \frac{\Omega_n - \delta}{\sqrt{(\Omega_n - \delta)^2 + 4g^2(n+1)}} ,$$

$$\sin \theta_n = \frac{2g\sqrt{n+1}}{\sqrt{(\Omega_n - \delta)^2 + 4g^2(n+1)}} .$$

Note, in particular, that

$$\cos 2\theta_n = - \delta / \Omega_n$$

and

$$\sin 2\theta_n = 2g\sqrt{n+1} / \Omega_n .$$

In the vacuum field $n = 0$ and on resonance $\omega_o = \omega$ the dressed states are separated by the frequency $\Omega_o = 2g$ generally called vacuum Rabi splitting.

One of the interesting phenomena described by the Jaynes-Cummings model is the dynamical behaviour. When we assume the atom is initially in the upper state $| a \rangle$ and the field in the number state $| n \rangle$ it follows for the probability of an atom to be in the upper state

$$|C_{an}(t)|^2 = \cos^2(g\sqrt{n+1}\, t).$$

This shows the upper state population oscillates periodically at the Rabi flopping frequency, similar to the case of classical fields. If the field is initially described by the photon statistics p_n, the above results must be generalised to:

$$|C_a(t)|^2 = \sum_n p_n \cos^2(g\sqrt{n+1}\, t).$$

It has been shown that in the case where the field mode is initially in a coherent state, $|C_a(t)|^2$ undergoes a collapse followed by a series of revivals [20]. The collapse is due to the destructive interference of quantum Rabi floppings at different frequencies. A similar phenomenon may also occur with a classical field, however, the revivals are a purely quantum mechanical effect which originates in the discreteness of the quantum field. Collapse and revivals have been observed in a micromaser experiment [21]. In this experiment the interaction time of the atoms with the cavity was varied and the probability was investigated that the atoms leave the cavity in the excited state. As will be shown below, the photon statistics in the maser cavity changes when the interaction time is varied, therefore the photon statistics p_n is not a pure distribution. Nevertheless the revival shows up as was also confirmed in a computer simulation of the results on the basis of the Jaynes-Cummings model [21].

In the following we would like to summarise the results of the quantum-theory of the one-atom maser. Since the atom-field interaction takes place in a closed single mode cavity, there is no spontaneous emission rate into free space modes. Owing to the extremely high quality factors achieved in the superconducting cavities the photon lifetime is extremely long compared to the transit time of the atoms through the resonator. This means the cavity damping can be practically ignored while an atom interacts with the field. Since the atomic flux is kept small to have at most one atom present inside the cavity at a time. Hence, the cavity is empty most of the time. Therefore cavity damping can be neglected during the rare instances when an atom interacts with the cavity mode.

The one-atom maser theory is therefore based on the following strategy [22]: while an atom is in the cavity, the coupled atom-field system is described by the Jaynes-Cummings Hamiltonian and for the interval between atoms the evolution of the field density matrix is described by a master equation considering damping and in addition the mean number of thermal photons in the cavity. Besides this microscopic theory there is also a macroscopic theory based on the quantum theory of the laser [23].

The resulting probability distribution of the photons depends characteristically on the pump rate and on the interaction time t_{int} of the atoms with the cavity field. One obtains the following result for the probability of finding n photons in the maser cavity P(n) in steady state:

$$P(n) = P_0 \left(\frac{n_b}{n_b + 1} \right)^n \prod_{m=1}^{n} \left(1 + \frac{N}{n_b \gamma} \frac{\sin^2 (g \sqrt{m t_{int}})}{m} \right),$$

with N being the atomic pump rate, n_b the thermal photon number, and γ the cavity decay rate. P_0 is determined by the normalization condition $\Sigma \, P(n) = 1$. One can now evaluate the mean photon number $\langle n \rangle$ and the field variance in the form of the Q_f-parameter [22]: $Q_f = (\langle n^2 \rangle - \langle n \rangle^2 - \langle n \rangle) / \langle n \rangle$.

Figure 3 shows the mean photon number as a function of the interaction time of the atoms with the cavity. The photon number is scaled to N_{ex}; here N_{ex} is the average number of atoms that enter the cavity during the cavity decay time $N_{ex} = N/\gamma$. The maser thresholds occurs at $\Theta = 1$ and regions of sub-Poissonian photon statistics are between $\Theta = \pi$ and 2π as well as between 3π and 4π. The sub-Poissonian character leads to a negative Q_f. A large negative value is obtained close before 2π [6,22].

To get a more intuitive insight into this effect we recall the velocity selection preselects the velocity of the atoms. Hence, the interaction time is well defined, which leads to conditions usually not achievable in standard masers. This has a very impor-

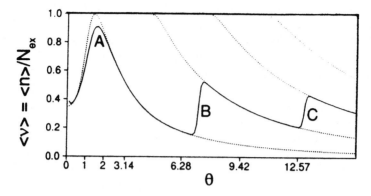

Figure 3. Mean value of the photon number $\nu = n/N_{ex}$ versus the pump parameter Θ, where the value of Θ is changed via N_{ex}. The solid line represents the micromaser solution for $\Omega = 36$ kHz, $t_{int} = 35$ μs, and temperature $T = 0.15$ K. The dotted lines are semiclassical steady-state solutions corresponding to fixed stable gain=loss equilibrium photon numbers. The crossing points between a line $\Theta = $ const. and the dotted lines correspond to the values where minima in the Fokker-Planck potential $V(\nu)$ occur. For details see text.

tant consequence when the intensity of the maser field grows as more and more atoms transfer their excitation energy to the field: even in the absence of dissipation this increase in photon number is halted when the increasing Rabi frequency leads to a situation where the atoms reabsorb the photon and leave the cavity in the upper state. For any photon number, this situation can be achieved by appropriately adjusting the velocity of the atoms. Then, the number distribution of the photons in the cavity is sub-Poissonian.

Unfortunately, the measurement of the nonclassical photon statistics in the cavity is not straightforward. The measurement process of the field involves the coupling to a measuring device, whereby losses lead inevitably to a destruction of the nonclassical properties. The ultimate technique to obtain information about the field employs the Rydberg atoms. For this purpose, the population and the statistics of the atoms in the upper and lower maser levels are probed when they leave the cavity. Accordingly, the atoms play a double role: (i) they pump the cavity and (ii) they are used for the diagnostics. These two roles interfere with one another because the detection of the atom in a known final state leads to a quantum mechanical reduction of the photon state inside the resonator. Frequent detection is accompanied by quasipermanent state reduction which can prevent the cavity field from relaxing to the steady state that would be reached if the atoms were left unobserved. Nevertheless, the steady-state properties determine the statistics of the clicks recorded by the atom detectors.

The theoretical treatment of the one-atom maser produces predictions about both the photon field and the emerging atoms. Only the latter can be tested experimentally, and the success of such tests feeds our confidence in the predictions about the quantized radiation field inside the cavity. The pump atoms are statistially independent in the standard one-atom-maser experiments, so their arrival times are subject to a Poissonian statistics; we shall, therefore, restrict the discussion to this standard Poissonian situation.

Inasmuch as the atoms arrive at random, they are recorded at equally random times, thus the only reproducible data are statistical. Consequently, one is led to studying the statistics of the detector clicks. Numerical simulations investigating the effect of repeated atomic measurements on the evolution of the cavity field have been performed by Meystre [24] as well as Meystre and Wright [25]. The relation between the counting statistics of the detected atoms emerging from the resonator and the photon-number statistics of the field inside the cavity has been studied analytically by Rempe and Walther [26] as well as by Paul and Richter [27]. In Ref. [26] the results are also compared with numerical simulations showing good agreement; experimental results are reported by Rempe, et al. [28]

In a recent paper a general method for the computation of various statistical properties of the click distribution were presented [29]. This method does not resort to numerical simulations. Naturally, the efficiencies of the detectors – far from the ideal 100 %, unfortunately – are taken into account. A central tool used in those calulations is a nonlinear master equation that governs the dynamics of the photon field in periods between detector clicks. The nonlinearity arises from the necessity to distinguish between the notions of observation and detection. When the detectors are active, all emerging atoms are observed but only a fraction is actually detected. Most of the time the experimenter is observing but does not detect.

In another approach on the basis of the concept of the counting statistics of light beams, the atomic counting probability, the waiting-time distribution, and the "two-atom correlation" function for a Poissonian atomic beam exciting the micro-maser cavity are calculated also. In an analytic treatment it is shown how the waiting-time distribution converges into the atom correlation function for vanishing detection efficiency [30].

Under steady-state conditions, as mentioned above, the photon number and the photon statistics of the maser field are essentially determined by the dimensionless parameter Θ. The quantity $\langle v \rangle = \langle n \rangle / N_{ex}$ shows the following generic behavior (see Fig. 3): It suddenly increases at the threshold value $\Theta = 1$ and reaches a maximum for $\Theta \approx 2$. As Θ further increases, $\langle v \rangle$ decreases and reaches a minimum at $\Theta \approx 2\pi$, and then abruptly increases to a second maximum. This general type of behavior recurs roughly at integer multiples of 2π, but becomes less pronounced with increasing Θ. The reason for the periodic maxima of $\langle v \rangle$ is that for $\Theta \approx 2\pi$ and multiples thereof the pump atoms perform an almost integer number of full Rabi flopping cycles, and start to flip over at a slightly larger value of Θ, this leading to enhanced photon emission. The maser threshold at $\Theta = 1$ shows the characteristics of a continuous phase transition, whereas the subsequent maxima in $\langle v \rangle$ can be interpreted as first-order phase transitions. In the intervals between the phase-transition points, the photon statistics is mostly sub-Poissonian. The field is super-Poissonian for all phase transitions [6,22] the large photon number fluctuations above $\Theta = 2\pi$ and multiples thereof being caused by the presence of two maxima in the photon number distribution P(n). They result from the fact that atoms in the upper maser level may or may not tip over to the lower level [6,22].

The phenomenon of the two coexisting neighbouring maxima in P(n) was also studied in a semi-heuristic Fokker-Planck (FP) approach [22]. There, the photon number distribution P(n) is replaced by a probability function $P(v,\tau)$ with continuous variables $\tau = t/\tau_{cav}$ and $v(n) = n/N_{ex}$, the latter replacing the photon number n. The steady-state solution obtained for $P(v,\tau), \tau \gg 1$, can be constructed by means of an effective potential $V(v)$ showing minima at positions where maxima of $P(v,\tau), \tau \gg 1$, are found. Close to $\Theta = 2\pi$ and multiples thereof, the effective potential $V(v)$ exhibits two equally attractive minima located at stable gain-loss equilibrium points of maser operation. The mechanism at the phase transitions mentioned is always the same: a minimum of $V(v)$ loses its global character when Θ is increased, and is replaced in this role by the next one. This reasoning is a variation of the Landau theory of first-order phase transitions, with $\sqrt{v}$ being the order parameter. This analogy actually leads to the notion that in the limit $N_{ex} \rightarrow \infty$ the change of the micromaser field around integer multiples of $\Theta = 2\pi$ can be interpreted as first-order phase transitions. In this region long field evolution time constants τ_{field} are expected. This phenomenon was experimentally demonstrated, as well as related phenomena, such as spontaneous quantum jumps between equally attractive minima of $V(v)$, and bistability and hysteresis [31]. Some of those phenomena are also predicted in the two-photon micromaser [32].

If there are no thermal photons in the cavity – a condition achievable by cooling the resonator to temperatures below 100 mK – very interesting features such as trapping states show up [33]. The investigation of the trapping states is discussed in detail in a recent review [34]. In the following we would like to review two experiments, one on the measurement of the photon statistics of the one-atom-maser and another one on the observation of quantum jumps and bistability in the maser field at the first order phase transitions points.

5. The Photon Statistics of the One-Atom Maser Field

As discussed above in the experiment on the photon statistics of the one-atom maser field [28] the number N of atoms in the lower maser level is counted for a fixed time interval T roughly equal to the storage time τ_{cav} of the photons in the cavity. By repeating this measurement many times the probability distribution of finding N atoms in the lower level is obtained. The normalized variance $Q_a = [\langle N^2 \rangle - \langle N \rangle^2 - \langle N \rangle] / \langle N \rangle$ is evaluated and is used to characterise the deviation from Poissonian statistics. A negative (positive) Q_a value indicates sub-Poissonian (super-Poissonian) statistics, while $Q_a = 0$ corresponds to a Poisson distribution with $\langle N^2 \rangle - \langle N \rangle^2 = \langle N \rangle$. The atomic Q_a can be related to the normalised variance Q_f of the photon number (for details see Refs. [26,28,29]). The experiment [26] shows that the photon field shows a variance $\langle n^2 \rangle - \langle n \rangle^2 = 0.3 \langle n \rangle$, which is 70 % below the shot noise level; this result agrees with the prediction of the theory [22,23].

We emphasise that the reason for the sub-Poissonian atomic statistics is the following: A changing flux of atoms changes the Rabi-frequency via the stored photon number in the cavity. By adjusting the interaction time, the phase of the Rabi-nutation cycle can be chosen so the probability for the atoms leaving the cavity in the upper maser level increases when the flux and therefore the photon number in the cavity is raised. We observe sub-Poissonian atomic statistics in the case where the number of atoms in the lower state is decreasing with increasing flux and photon number in the cavity. This feedback mechanism is also demonstrated when the anticorrelation of atoms leaving the cavity in the lower state is investigated. Measurements of this 'antibunching' phenomena have also been performed (see Ref. [34] for a detailed review and also Ref. [27]).

The fact that anticorrelation is observed shows that the atoms in the lower state are more equally spaced than expected for a Poissonian distribution. It means when two atoms enter the cavity close to each other the second one performs a transition to the lower state with a reduced probability.

The experiments demonstrate the sub-Poissonian photon statistics of the one-atom maser field. In addition, the maser experiment leads to an atomic beam with atoms in the lower maser level showing number fluctuations which are up to 40 % below those of a Poissonian distribution found in usual atomic beams. This is interesting, because atoms in the lower level have emitted a photon to compensate for cavity losses inevitably present. Although this is a purely dissipative process giving rise to fluctuations, nevertheless, the atoms still obey sub-Poissonian statistics.

6. Quantum Jumps of the Micromaser Field

The setup used for these measurements is the same as described above. As before, ^{85}Rb atoms were used to pump the maser. They are excited from the $5S_{1/2}$, $F = 3$ ground state to 63 $P_{3/2}$, $m_J = \pm 1/2$ states by linearly polarized light of a frequency-doubled c.w. ring dye laser. The polarisation of the laser light is linear and parallel to the likewise linearly polarised maser field, and therefore only $\Delta m_J = 0$ transitions are excited. Superconducting niobium cavities resonant with the transition to the 61 $D_{3/2}$, $m_J = \pm 1/2$ states were used. The cavity Q-values ranged from $4 \cdot 10^9$ to $8 \cdot 10^9$.

As before, information on the maser field and interaction of the atoms in the cavity can be obtained solely by state-selective field ionization of the atoms in the upper or lower maser level after they have passed through the cavity. The field ionization detector was recently modified, so there is now a detection efficiency of $\eta = (35 \pm 5)$ %. For different t_{int} the atomic inversion has been measured as a function of the pump rate; the coupling constant is about g = 40 rd/s.

Depending on the parameter range, essentially three regimes of the field evolution time constant τ_{field} can be distinguished [31]. We restrict the discussion here only to the results for intermediate time constants. The maser was operated under steady-state conditions close to the second first-order phase transition (C in Fig. 3). The interaction time was $t_{int} = 47$ μs and the cavity decay time $\tau_{cav} = 60$ ms. The value of N_{ex} necessary to reach the second first-order phase transition was $N_{ex} \approx 200$. For these parameters, the two maxima in P(n) are manifested in spontaneous jumps of the maser field between the two maxima with a time constant of ≈ 5s. This fact and the relatively large

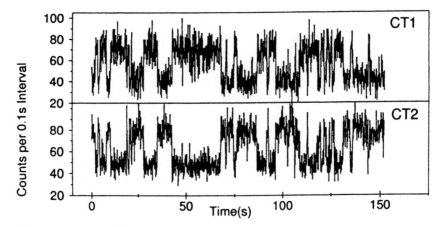

Figure 4. Quantum jumps between two equally stable operation points of the maser field. The channeltron counts are plotted versus time (CT1 = upper state, CT2 = lower state).

pump rate lead to the clearly observable field jumps shown in Fig. 4. Due to the large cavity field decay time, the average number of atoms in the cavity was still as low as 0.17. The two discrete values for the counting rates correspond to the metastable operating points of the maser, which correspond to ≈ 70 and ≈ 140 photons. In the Fokker-Planck description, the two values correspond to two equally attractive minima in the Fokker-Planck potential $V(\upsilon)$. If one considers, for instance, the counting rate of lower-state atoms (CT2 in Fig. 4), the lower (higher) plateaus correspond to time intervals in the low (high) field metastable operating point. If the actual photon number distribution is averaged over a time interval containing many spontaneous field jumps, the steady-state result P(n) of the micromaser theory is recovered.

In a parameter range where switching occurs much faster than in the case shown in Fig. 4, the individual jumps cannot be resolved any more owing to the reduced N_{ex} necessary in this case. Therefore, a different procedure must be chosen for the investigation. These experiments will not be discussed here; details are described in Ref. [31].

7. Atomic Interference in the One-Atom-Maser

In the following, we report on the observation of the maser resonance under conditions that atomic interference phenomena in the cavity get observable. Since a nonclassical field is generated in the maser cavity, we are able to investigate for the first time atomic interference phenomena under the influence of non-classical radiation. Interferences occur since a coherent superposition of dressed states is produced by mixing the dressed states at the entrance and exit hole of the cavity. Inside the cavity the dressed states develop differently in time giving rise to Ramsey-type interferences when the maser cavity is tuned through resonance.

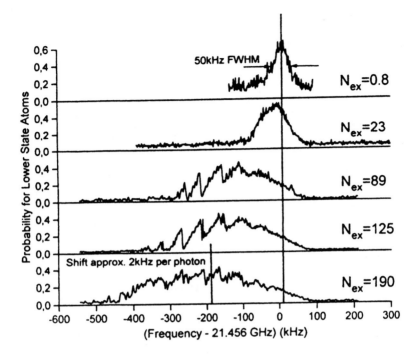

Figure 5. Maser resonance 63 $P_{3/2}$ - 61 $D_{5/2}$. The FWHM linewidth (50 kHz) of the upper plot sets an upper limit of $\approx$ 5 mV/cm to residual stray electric fields in the center of the cavity. The lower reonance lines are taken for the indicated large values of N_{ex}. The plots show that the center of the maser line shifts by about 2 kHz per photon. For $N_{ex} \geq$ the lines display periodic structures which are discussed in the text.

Figure 5 shows that for large values of N_{ex} ($N_{ex} > 89$) sharp, periodic structures appear. Those typically consist of a smooth red wing and a vertical step on the blue side. The clarity of the pattern rapidly reduces when N_{ex} increases to 190 or beyond. We will see later that those structures must be interpreted as interferences. Concerning the shift of the whole structure it is known from previous maser experiments there are small static electric fields in the entrance and exit holes of the cavity. These fields can, in fact, cause considerable line shifts. The N_{ex}-dependence of the line shift and other experimental facts result from a complicated interplay between the position dependent stray electric fields close to the holes and the sinusoidal envelope of the resonant cavity mode.

The usual formalism for the description of the coupling of an atom to the radiation field is the dressed atom approach introduced above, which is leading to a splitting of the coupled atom-field states depending on the vacuum Rabi-flopping frequency Ω, the photon number n, and the detuning δ. We face a special situation at the entrance and the exit hole of the cavity. There we have a position dependent variation of the cavity field, as a consequence Ω is position dependent. An additional variation results from the fact there are electric fields in the entrance and exit holes which may be caused either by the microstructure of the surface of the superconducting material or by contact potentials of Rb deposits. These electric fields lead to a position dependent atom-field detuning δ.

The Jaynes-Cummings-Hamiltonian only couples pairs of dressed states. Therefore, it is sufficient to consider the dynamics within such a pair of dressed states. In our case, prior to the atom-field interaction the system is in one of the two dressed states. For parameters corresponding to the periodic substructures in Fig. 5, the dressed states are mixed only at the beginning of the atom-field interaction and at the end. The mixing at the beginning creates a coherent superposition of the dressed states. Then, the system develops adiabatically, whereby the two dressed states accumulate a differential dynamical phase Φ which strongly depends on the cavity frequency. The mixing of the dressed states at the entrance and exit holes of the cavity, in combination with the intermediate adiabatic evolution, generates a situation similar to a Ramsey two-field interaction. In order to interpret the periodic substructures as a result of the variation of Φ with the cavity frequency, the phase Φ must be calculated from the atomic dynamics in the maser field.

The quantitative calculation can be performed in the following way. First, the variation of the static electric field in the cavity must be estimated. This is done by numerically solving the Laplace equation with the boundaries of the cavity and assuming a particular field strength in the cavity holes. Second, for different interaction times, photon numbers and cavity frequencies, the dynamics of the atom-field wavefunction is calculated by numerical integration of the Jaynes-Cummings model. This integration must be performed including the local variation of Ω inside the cavity owing to the mode structure of the microwave field. Furthermore, the variation of the detuning δ resulting from the static electric fields in the cavity holes must be considered. In order to use the micromaser model, we extract the probabilities that an atom emits a photon in a field of n - 1 photons prior to the interaction. In the second step of the calculation following the maser theory described in section 4, we obtain the photon number distribution P(n) of the cavity field under steady state conditions.

A theoretical result of $\langle n \rangle / N_{ex}$-obtained in this way is shown in Fig. 6. The uppermost plot shows the maser resonance line expected without any static electric field. With increasing DC field strength in the cavity holes the structure changes, the curve for 309 mV/cm coming very close to the ones displayed in Fig. 5 for N_{ex}-= 89 and 125. We stress that the field values indicated in Fig. 6 correspond to the maximum field strength in the cavity holes. The field value in the central part of the cavity is roughly 100 times smaller, and therefore without significance in low-flux maser experiments. Fig. 6 also shows the qualitative structure of the maser line is the same for all fields larger than about 200 mV/cm. Thus, the conditions required to find the periodic substructures experimentally are not very stringent.

The calculations reveal that on the vertical steps in the micromaser lines the photon number distribution has two well separate maxima. This situation is comparable to the conditions discussed in section 6, where maser field metastability is observed. Therefore, the maser field should exhibit metastability under the present conditions as well. Under steady-state operation of the maser on steep steps of the maser line, we observed spontaneous jumps of the maser field between metastable field states.

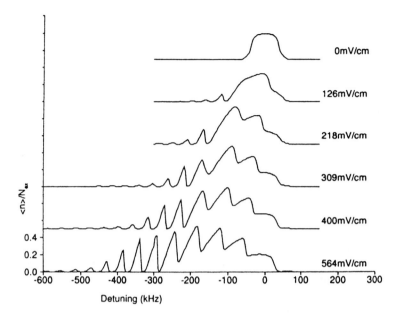

Figure 6. Theoretical maser lines for the indicated values of the static electric field strength in the cavity holes. In the calculation we use $N_{ex} = 100$ and $\Omega = 45$ krad/s. The interaction time is $t_{int} = 35$ μs, and the RMS deviation of the interaction time is 1 μs.

The calculations also show that on the smooth wings of the more pronounced spikes, the photon number distribution P(n) of the maser field is strongly sub-Poissonian. This leads us to the conclusion that we observe Ramsey-type interference induced by a nonclassical radiation field. The sub-Poissonian character of P(n) results from the fact that on the smooth wings of the spikes locally the photon gain reduces when the photon number is increased. This feedback mechanism stabilizes the photon number. The model is consistent with former results in the low-pump regime (this is not demonstrated in this section).

8. Linewidth and Phase Diffusion of the One-Atom Maser

In the following we would like to discuss another special feature of the one-atom maser, the spectrum. This spectrum is determined by the decay of the expectation value of the electric field [34]

$$\langle E(t)\rangle \sim \sum_{n=0}^{\infty}(n+1)^{1/2}\,\rho_{n,n+1}(t).$$

Hence, the micromaser spectrum is different from the other effects discussed so far since it involves the off-diagonal elements $\rho_{n,n+1}$ of the radiation density matrix rather than, e.g., the photon statistics, that is, the diagonal elements $\rho_{n,n}$; and it requires their time dependence rather than their steady state values. It could be shown that the linewidth D of the maser is given by

$$D = 4r\sin^2\left[\frac{gt_{int}}{4\sqrt{\langle n\rangle}}\right] + \frac{\gamma(2n_b+1)}{4\langle n\rangle}.$$

For small $g\tau/4\sqrt{\langle n\rangle}$ the sine function can be expanded; this leads to the familiar Schawlow-Townes linewidth

$$D = \frac{\alpha+\gamma(2n_b+1)}{4\langle n\rangle}$$

where

$$\alpha = \gamma(\sqrt{N_{ex}}\,gt_{int})^2.$$

The complicated pattern of the micromaser linewidth results, in part, from the complicated dependence of $\langle n\rangle$ on the pump parameter, which enters in the denominator. We emphasize that the one-atom maser linewidth goes beyond the standard Schawlow-Townes linewidth. The sine function suggests in the limit of large pumping parameters an oscillatory behaviour of the linewidth confirmed in an exact numerical treatment. (For details see also Refs. [35] and [36]).

It was pointed out that the maser linewidth can be measured when two phase coupled microwave fields are used; one before the atoms enter the microwave cavity and one after corresponding, in principle, to a modified Ramsey setup. The first micromaser field creates a superposition of the two maser states being then probed by the second one. The first field is only applied for an initial period in order to "seed" a phase in the micromaser cavity, in the second field the phase diffusion is probed. This technique will provide the basis for future experiments to measure the linewidth. It was tested by computer simulations and analytical calculations which showed that coherent pumping of the maser leads to new and interesting phenomena which cannot be discussed here. The first time coherent injection for the one atom maser was proposed in the paper by Krause, et al. [37] and discussed later in connection with the measurement of the phase diffusion linewidth [38]. Also the phase dynamics of the maser field in steady-state operation was discussed by Wagner, et al. [39] in the latter case only one microwave field is used for probing after the micromaser cavity. The latter scheme was further pursued in connection with the entanglement of states by Wagner, et al. [40]. The entangled state of the atom-field system occurs since a factorization in field and atom part is not possible. This fact leads to applications of the micromaser for Einstein-Podolsky-Rosen type of experiments. At this point we should also mention that the micromaser can also be used to test complementarity. For details refer to Refs. [41-43].

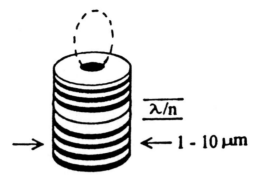

Figure 7. Micro-Fabry Perot cavity fabricated using molecular beam epitaxy to grow dielectric layered mirrors; the laser-active region is the one in the middle of the region indicated at the side by λ/n (see Ref. 44 for more details).

9. Microlasers

With respect to technical applications, it is interesting to investigate whether similar phenomena can also be observed in cavities for optical radiation. The simplest approach to fabricating an optical microcavity is to shrink the spacing between the mirrors of a Fabry-Perot resonator to λ/n (here n stands for the refractive index) while reducing the lateral dimensions to the same range, as shown in Fig. 7. This structure provides a single dominant longitudinal field mode that radiates into a narrow range of angles around the cavity axis as indicated by the dotted line.

The first optical microcavity experiments used dye molecules between high-reflectivity dielectric mirrors in the Fabry-Perot configuration [45]. Because spontaneous emission is a major source of energy loss, speed limitations, and noise in lasers, the capability to control spontaneous emission is expected to improve laser performance. If the fraction of spontaneous emission coupled into the lasing mode is made close to one, the "thresholdless" laser [45,46] is obtained in which the light output increases nearly linearly with the pump power instead of exhibiting a sharp turn-on at the pump threshold.

Semiconductor microcavities provide high-Q Fabry-Perot cavities for both basic studies and potential applications. Molecular beam epitaxy or organometallic chemical vapor deposition techniques are used to deposit high-reflectivity mirrors consisting of alternating quarter-wavelength layers of lattice-matched semiconductors. For example, 15 to 20 pairs of quarter-wave layers of $Al_{0.2}Ga_{0.8}As$ and AlAs result in a reflectivity greater than 99 % and Q values greater than 500. The optically active layer in such a microcavity is typically a GaAs quantum well located at the midplane of the cavity, where the field strength is a maximum. Figure 7 shows an example of the very small diameter microcavities that can be achieved by photolithographic patterning and etching of these layered materials. For detailed review see Refs. [44] and [46].

Whispering gallery modes are a second successful route of achieving high-Q microcavities. Total internal reflection along a curved boundary between two materials with different optical indices results in high-Q modes propagating within a half-wavelength of the boundary. Semiconductor whispering gallery mode resonators have been demonstrated in disk and cylindrical geometries. For a review, see Ref. [44].

It is remarkable these microcavities have an extremely low threshold that means their efficiency will also be high. Low cost, high density light source arrays and photonic circuits are possible because of the small size and low power consumption of such resonators. One will be able to produce entire wafers containing millions of microlasers with a multipole arrangement instead of cleaving each individual semiconductor laser as at present. This improvement will lead to higher yields and a lower cost per element. Another advantage of surface emitting microcavity sources is the efficient optical coupling of their stable symmetric mode patterns into optical fibers or waveguides. Surely these lasers will cause a revolution in optical computing and in optical communication in the future [44,46]!

9. References

1. Haroche, S. and Raimond, J.M. (1985) Radiative properties of Rydberg states in resonant cavities, in D. Bates and B. Bederson (eds.), *Advances in Atomic and Molecular Physics*, Academic Press, New York, Vol. 20, pp. 350-411, and Gallas, J. A., Leuchs, G., Walther, H. and Figger, H. (1985) Rydberg atoms: High resolution spectroscopy and radiation interaction – Rydberg molecules, in D. Bates and B. Bederson (eds.), *Advances in Atomic and Molecular Physics*, Academic Press, New York, Vol. 20, pp. 413-466.

2. Hinds, E. (1994) Cavity quantum electrodynamics, in P. Beerman (ed.), *Supplement 2 of Advances in Atomic, Molecular and Optical Physics*, Academic Press, New York, and Meschede, D. (1992) Radiating atoms in confined space: From spontaneous emission to micromasers, *Phys. Rep.* **211**, 201-250.

3. Haroche, S. (1992) Cavity quantum electrodynamics, in J. Dalibard, J. M. Raimond and J. Zinn-Justin (eds.), *Fundamental Systems in Quantum Optics*, Proceedings of Les Houches Summer School , Session LIII., North-Holland, pp. 767-940.

4. Meystre, P. (1992) Cavity quantum optics and the quantum measurement process, in E. Wolf (ed.), *Progress in Optics*, North Holland, Amsterdam, Vol. 30, pp. 261-355.

5. Raithel, G., Wagner, Ch., Walther, H., Scully, M., and Narducci, L. (1994) The micromaser: A proving ground for quantum physics, in P. Beerman (ed.), *Cavity Quantum Electrodynamics*, Supplement 2 of Advances in Atomic, Molecular, and Optical Physics, Academic Press, New York, pp. 57-121.

6. Jaynes, E. T. and Cummings, F. W. (1963) Comparison of quantum and semiclassical radiation theories with application to the beam maser, *Proc. IEEE*. **51**, 89-109.

7. Purcell, E. M. (1946) Spontaneous emission probabilities at radio frequencies, *Phys. Rev.* **69**, 681.

8. Drexhage, K. H. (1974) Interaction of light with monomolecular dye lasers, in E. Wolf (ed.), *Progress in Optics*, North-Holland, Amsterdam, Vol. 12, pp. XVII 165-229.

9. De Martini, F., Innocenti, G., Jacobovitz, G., and Mantolini, P. (1987) Anomalous spontaneous emission time in a microscopic optical cavity, *Phys. Rev. Lett.* **59**, 2955-2958.

10. Gabrielse, G. and Dehmelt, H. (1985) Observation of inhibited spontaneous emission, *Phys. Rev. Lett.* **55**, 67-70.

11. Hulet, R. G., Hilfer, E., and Kleppner, D. (1985) Inhibited spontaneous emission by a Rydberg atom, *Phys. Rev. Lett.* **55**, 2137-2140.

12. Jhe, W., Anderson, A., Hinds, E. A., Meschede, D., Moi, L., and Haroche, S. (1987) Suppression of spontaneous decay at optical frequencies: Test of vacuum-field anisotropy in confined space, *Phys. Rev. Lett.* **58**, 666-669.

13. Goy, P., Raimond, J. M., Gross, M., and Haroche, S. (1983) Observation of cavity-enhanced single-atom spontaneous emission, *Phys. Rev. Lett.* **50**, 1903-1906.

14. Heinzen, D. J., Childs, J. L., and Feld, M. S. (1987) Enhanced and inhibited visible spontaneous emission by atoms in a confocal resonator, *Phys. Rev. Lett.* **58**, 1320-1323.

15. Yamamoto, Y. S., Machita, K., Ikeda, K., and Björk, G. (1991) Controlled spontaneous emission in microcavity semiconductor lasers, in Y. Yamamoto (ed.), *Coherence Amplification and Quantum Effects in Semiconductor Lasers*, Wiley, New York.

16. Yablonovitch, E. (1987) Inhibited spontaneous emission in solid-state physics and electronics, *Phys. Rev. Lett.* **58**, 2059-2062.

17. Yablonovitch, E,. Gmitter, T. J., and Bhat, R. (1988) Inhibited and enhanced spontaneous emission from optically thin AlGaAs/GaAs double heterostructure, *Phys. Rev. Lett* **61**, 2546-2549.

18. Yablonovitch, E. and Gmitter, T. J. (1989) Photonic band structure: the face-centered-cubic case, *Phys. Rev. Lett.* **63**, 1950-1953.

19. Meschede, D., Walther, H., and Müller, G. (1985) The one-atom maser, *Phys. Rev.Lett.* **54**, 551-554.

20. See, for example: Eberly, J. H., Narozhny, N. B., and Sanchez-Mondragon, J. J. (1980) Periodic spontaneous collapse and revival in a simple quantum model, *Phys. Rev. Lett.* **44**, 1323-1326 and references therein.

21. Rempe, G., Walther, H., and Klein, N. (1987) Observation of quantum collapse and revival in a one-atom maser, *Phys. Rev. Lett.* **58**, 353-356.

140

22. Filipowicz, P., Javanainen, J., and Meystre, P. (1986) The microscopic maser, *Opt. Comm.* **58**, 327-330; (1986) Theory of a microscopic maser, *Phys. Rev. A* **34**, 3077-3087; (1986) Quantum and semiclassical steady states of a kicked cavity mode, *J. Opt. Soc. Am.* **B 3**, 906-910.

23. Lugiato, L., Scully, M. O., and Walther, H. (1987) Connection between microscopic and macroscopic maser theory, *Phys. Rev. A* **36**, 740-743.

24. Meystre, P. (1987) Repeated quantum measurements on a single-harmonic oscillator, *Opt. Lett.* **12**, 669-671; Meystre, P. (1988) Generation and detection of subpoissonian fields in micromasers, in P. Tombesi and E.R. Pike (eds.), *Squeezed and Nonclassical Light*, Plenum, New York, pp. 115-127.

25. Meystre, P. and Wright, E. M. (1988) Measurement induced dynamics of a micromaser, *Phys. Rev. A* **37**, 2524-2529.

26. Rempe, G. and Walther, H. (1990) Sub-Poissonian atomic statistics in a micromaser, *Phys. Rev. A* **42**, 1650-1655.

27. Paul, H. and Richter, T. (1991) Bunching and antibunching of de-excited atoms leaving a micromaser, *Opt. Comm.* **85**, 508-519.

28. Rempe, G., Schmidt-Kaler, F., and Walther, H. (1990) Observation of sub-Poissonian photon statistics in a micromaser, *Phys. Rev. Lett.* **64**, 2783-2786.

29. Briegel, H. J., Englert, B. G., Sterpi, N., and Walther, H. (1994) One-atom maser: Statistics of detector clicks, *Phys. Rev. A* **49**, 2962-2984.

30. Wagner, C., Schenzle, A., and Walther, H. (1994) Atomic waiting-times and correlation functions, *Opt. Comm.* **107**, 318-326.

31. Benson, O., Raithel, G., and Walther, H. (1994) Quantum jumps of the micromaser field – dynamic behavior close to phase transition points, *Phys. Rev. Lett.* **72**, 3506-3509.

32. Davidovich, L., Raimond, J. M., Brune, M., and Haroche, S. (1987) Quantum theory of the two-photon micromaser, *Phys. Rev. A* **36**, 3771-3786; Brune, M., Raimond, J. M., Goy, P., Davidovich, L., and Haroche, S. (1987) Realisation of a two-photon maser oscillator, *Phys. Rev. Lett.* **59**, 1899-1902.

33. Meystre, P., Rempe, G., and Walther, H. (1988) Very-low temperature behaviour of a micromaser, *Opt. Lett.* **13**, 1078-1080.

34. Raithel, G., Wagner, Ch., Walther, H., Narducci, L. M., and Scully, M. O. (1994) The micromaser: A proving ground for quantum physics, in P. Berman (ed.), *Advances in Atomic, Molecular, and Optical Physics,* Supplement 2, Academic Press, New York, pp. 57-121.

35. Scully, M. O., Walther, H., Agarwal, G. S., Quang, T., and Schleich, W. (1991) Micromaser spectrum, *Phys. Rev. A* **44**, 5992-5996.

36. Quang, T., Agarwal, G. S., Bergou, J., Scully, M. O., Walther, H., Vogel, K., and Schleich, W. P. (1993) Calculation of the micromaser spectrum I. Green's-function approach and approximate analytical techniques, *Phys. Rev. A* **48**, 803-812; Vogel, K., Schleich, W. P., Scully, M. O., and Walther, H. (1993) Calculation of the micromaser spectrum II. Eigenvalue approach, *Phys. Rev. A* **48**, 813-817.

37. Krause, J., Scully, M. O., and Walther, H. (1986) Quantum theory of the micromaser: Symmetry breaking via off-diagonal atomic injection, *Phys. Rev. A* **34**, 2032-2037.

38. Brecha, R. J., Peters, A., Wagner, C., and Walther, H. (1992) Micromaser and separated-oscillatory-field measurements, *Phys. Rev. A* **46**, 567-577.

39. Wagner, C., Brecha, R. J., Schenzle, A., and Walther, H. (1992) Phase diffusion and continuous quantum measurements in the micromaser, *Phys. Rev. A* **46**, R5350.

40. Wagner, C., Brecha, R. J., Schenzle, A., and Walther, H. (1993) Phase diffusion, entangled states, and quantum measurements in the micromaser, *Phys. Rev. A* **47**, 5068-5079.

41. Scully, M. O. and Walther, H. (1989) Quantum optical test of observation and complementarity in quantum mechanics, *Phys. Rev. A* **39**, 5229-5236.

42. Scully, M. O., Englert, B.-G., and Walther, H. (1991) Quantum optical tests of complementarity, *Nature* **351**, 111-116.

43. Englert, B.-G., Walther, H., and Scully, M. O. (1992) Quantum optical Ramsey fringes and complementarity, *Appl. Phys. B* **54**, 366-368.

44. Yamamoto, Y. and Slusher, R. S. (1993) Optical processes in microcavities, *Physics Today* **46**, 66-73.
45. De Martini, F., Jacobovitz, G. R. (1988) Anomalous spontaneous-stimulated decay phase transition and zero-threshold laser action in a microscopic cavity, *Phys. Rev. Lett.* **60**, 1711-1714.
46. Yokoyama, H. (1992) Physics and device applications of optical microcavities, *Science* **256**, 66-70.

ELASTIC WAVES IN PERIODIC COMPOSITE MATERIALS

M. KAFESAKI[1], E. N. ECONOMOU[1], AND M. M. SIGALAS[2]

[1] Research Center of Crete, FORTH, P.O. Box 1527, 71110 Heraklio, Crete, Greece, and Department of Physics, University of Crete.

[2] Ames Laboratory and Department of Physics and Astronomy Iowa State University, Ames IA 50011.

1. Introduction

There is a growing interest in recent years for the propagation of acoustic (AC) and elastic (EL) waves in random and periodic composite materials [1-15]. The interest among solid state physicists is mainly connected to the question of existence or not of spectral gaps in periodic systems or localized states in disordered systems in analogy with what happens to the electron wave propagation.

Apart from the many applications of the AC and EL waves (in geophysics, medicine, oil exploration, etc.), their advantages for experimental study as well as their rich physics, are the other two poles of attraction for the solid state physicists.

The rich physics stems from: *i*) the full vector character of EL waves with the different propagation velocity of the longitudinal and the transverse component; these two components, while travel independently in a homogeneous medium, are mixed together when the wave passes through inhomogeneities, *ii*) the term proportional to the mass density variation ($\nabla\rho$) appearing in the AC and EL wave equation which may cause the appearance of novel behavior, *iii*) the richness in parameters of the AC and EL wave propagation (on this we will return in more detail later on).

The advantages for experimental study - common for any type of classical waves (CW) - have to do with the absence of the complicated non linear interactions (such as electron-electron or electron-phonon interactions which appear in the study of the electrons) as well as from the fact that CW frequency can been controlled accurately and easily.

C. M. Soukoulis (ed.), Photonic Band Gap Materials, 143–164.

There is, however, the problem of the absorption and the problem of the difficulty to construct composites sustaining localized eigenstates. The solution of the latter was greatly facilitated by a suggestion by John and Rangarajan [16] and by Economou and Zdetsis [17]. They connected the spectral gaps of a periodic system with the mobility gaps (bands of localized eigenstates) of a random system by pointing out that when we gradually randomize a periodic system possessing gaps, tails of localized eigenstates start to appear inside the gap, changing it gradually to a band of localized eigenstates of the resulting random system. This connection is not surprising because both gap and localized states are due to the same mechanisms, i.e., multiple scattering and the destructive interference of the waves.

Following the above suggestion, attempts were focused in the examination of CW wave propagation in periodic systems. The study of the periodic systems is greatly facilitated due to the strong symmetry of the periodicity and to the computational experience from the study of the electrons in crystalline materials.

These periodic systems are binary composites consisting of scatterers, in most of the cases spheres or cylinders, embedded in a host material (matrix).

The two components of the composite can be either both connected to form a continuous network (network topology) or one of the two (the scatterers) can be completely surrounded by the other (cermet topology). The study of the classical wave propagation in periodic systems concerns mainly the examination of the possible existence of spectral gaps (frequency regions where no propagation of the wave exists) and the determination of the optimal conditions for the appearance of these gaps. Although the appearance of a gap in the electron wave propagation is something a priory guaranteed, the same is not true for CW (such as ACW and ELW). CW correspond to electron waves with energy higher than the maximum value of the potential as can be seen from the comparison of scalar classical wave equation with Schrödinger's equation. In such energy region, gaps appear under rather extreme conditions, as shown for both simple scalar waves (SSW) [5,15-19] and EM waves [5,15,18,20-23] which were the first studied experimentally [24]. In the EMW case it was found that gap appears with greater difficulty than in the scalar case due to their vector character [2]. On the basis of this result, elastic waves are expected to resist even more than EMW the opening of gaps due to their full vector character, as a result of which each of the longitudinal and the transverse component must develop gaps overlapping with each other in order for a full band gap to exist [5, 6]. The different propagation velocity of the longitudinal and the transverse component makes this overlapping more difficult and thus the situation more problematic.

However, the many parameters of the problem, and in particular the density contrast, offer the possibility for the gap to be opened. Indeed, computational and experimental data [3-6,15] concerning AC and EL wave propagation in periodic systems consisting mainly from spheres embedded in a host (3D) or from cylinders embedded in a host (2D) showed gaps under proper conditions. These conditions concern mainly the density and velocity contrast of the components of the composite, the volume fraction of one of the two components, the lattice structure and the topology; they are realized numerically in a Be or Si or SiO_2 matrix with embedded Au or Pb spheres placed periodically in an *fcc* lattice [6]. For a 2D square lattice, full band gaps have been found in Au cylinders in the Be host [4]. For 2D hexagonal lattices , Mo, Al_2O_3, Fe, and steel rods embedded in a Lucite host gave relatively wide gaps [18]. Recent numerical studies of the elastic wave band structure of 2D and 3D composite materials such as C or glass fibers or Pb spheres in epoxy host gave extremely large complete band gaps in those systems [7, 8]. From the experimental point of view, Kinra and Ker [9] have measured the phase velocity of longitudinal waves through 3D periodic polymer matrix composites as a function of frequency. Their results are consistent with the existing numerical results.

Recently, considerable progress in understanding the above results and in predicting favorable (for gap) material combinations were achieved by considering a plane AC or EL wave scattered by a single scatterer and by connecting the resonances of the single scattering cross-section with the gap and the other characteristics of the band structure [25]. The main idea was the qualitative extension of the linear combination of atomic orbitals (LCAO) method, well known from the study of the electrons, in the AC and EL wave case with the resonance to play the role of the eigenstates. The simplicity of the single scattering problem gave the possibility to understand physically some of the open problems of the study of the periodic systems. Also, single scattering results predicted new systems with extremely wide gaps [8].

In this paper we present first a review of previous results in combination with some recent results concerning the dependence of the AC and EL wave spectral gaps on various parameters of the problem. We sum up what we already know for the AC and EL wave propagation in periodic composites focusing on some unanswered questions. Since the single scattering may give the key for these questions, we will attempt to interpret some of the above results by using single scattering analysis. It is shown that single scattering can predict the basic characteristics of the band structure giving, also, some optimal conditions for the appearance of the gaps. In what follows, we present first the basic equations governing AC and EL wave propagation as well as a summary of our methods of calculation.

2. Equations and Methods of Calculation

The general equation of motion for an elastic solid can be written as :

$$\rho\frac{\partial^2 u^i}{\partial t^2} = \frac{\partial \sigma_{ik}}{\partial x_k}, \tag{1}$$

where u^i is the i-th component of the displacement vector, $\mathbf{u(r)}$, and $\rho(\mathbf{r})$ the mass density; σ_{ij} is *stress* tensor [26] which is related with the *strain* tensor

$$u_{ij} = \frac{1}{2}\left(\frac{\partial u^i}{\partial x_j} + \frac{\partial u^j}{\partial x_i}\right) \tag{2}$$

by:

$$\sigma_{ij} = C_{ijkl}u_{kl} \tag{3}$$

C_{ijkl} is the elasticity tensor.

For a *locally isotropic* medium the stress-strain relation (3) becomes :

$$\sigma_{ij} = 2\mu u_{ij} + \lambda u_{\ell\ell}\delta_{ij}, \tag{4}$$

where λ and μ are the so called Lamé coefficients of the medium [26]. The elastic wave equation (1) in this case can be written as:

$$\frac{\partial^2 u^i}{\partial t^2} = \frac{1}{\rho}\left\{\frac{\partial}{\partial x_i}\left(\lambda\frac{\partial u^\ell}{\partial x_\ell}\right) + \frac{\partial}{\partial x_\ell}\left[\mu\left(\frac{\partial u^i}{\partial x_\ell} + \frac{\partial u^\ell}{\partial x_i}\right)\right]\right\}. \tag{5}$$

For homogeneous media (λ, μ = constant) Eq. (5) gives rise to uncoupled longitudinal and transverse waves with velocities $c_l = \sqrt{(\lambda + 2\mu)/\rho}$ and $c_t = \sqrt{\mu/\rho}$, respectively.

For liquids, $\mu = 0$ (only longitudinal waves exist), and by introducing the pressure, $p = -\lambda\nabla\mathbf{u}$, Eq.(5) can be rewritten as

$$\frac{\partial^2 p}{\partial t^2} = \lambda\nabla\left(\frac{1}{\rho}\nabla p\right) \tag{6}$$

which for ρ = constant is reduced to a scalar Helmholtz equation.

In all the above formulae, repeated indices denote summation. Also, the indices are referred to a general coordinate system with the exception of (2) and consequently (5) which are valid only for cartesian coordinates (for the expression of u_{ij} in cylindrical or spherical coordinates see ref. 26 page 3).

For *periodic media* the displacement vector **u** which satisfies Bloch's theorem can be written as $\mathbf{u} = \mathbf{u_k(r)}e^{i\mathbf{kr}}$, where $\mathbf{u_k(r)}$ is a periodic function with the same periodicity as $\lambda(\mathbf{r})$, $\mu(\mathbf{r})$ and $\rho^{-1}(\mathbf{r})$ (see Eq.(5)). All these periodic functions can been expanded into a Fourier series (plane waves) with corresponding coefficients $\mathbf{u_{k+G}}$, $\lambda_{\mathbf{G}}$, $\mu_{\mathbf{G}}$ and $\rho_{\mathbf{G}}^{-1}$ respectively. **G** is a vector of the reciprocal lattice.

Approximating the infinite sums in the Fourier series by finite sums containing N terms, Eq. (5) and Eq. (6)) reduce to a $3N \times 3N$ ($N \times N$) arithmetically solvable matrix eigenvalue equation for the 3N (N) unknown coefficients $\mathbf{u_{k+G}}$ [5].

The above procedure is known as the PW (plane waves) method. The number N is increased until the desired convergence is achieved. In the calculations which we will present in this work, we kept N=400 **G** vectors to achieve convergence better than 5%.

The study of the *single scattering* is based on the calculation of the scattering cross-section for either a longitudinal or a transverse plane wave scattered by a homogeneous scatterer embedded in a homogeneous host.

The main step for the calculation of the cross-section is the calculation of the wave, **u**, inside and outside the scatterer. These waves are obtained from the general solution of the elastic wave equation in each region [25, 27] with the application of proper boundary conditions on the surface of the scatterer. The boundary conditions concern the continuity of the displacement vector **u** as well as the continuity of the stress vector, **P** (with components $P_i = \sigma_{ij}n_j$, where **n** is the outgoing unit vector normal to the surface of the scatterer) [25,28-30].

The scattering cross-section for an incident plane wave propagating in the **m**-direction is given by [25, 31]

$$\sigma = \int \frac{< n_i Re(\sigma_{ij}^{sc}) Re(\dot{u}_j^{sc}) >}{< m_i Re(\sigma_{ij}^{inc}) Re(\dot{u}_j^{inc}) >} r^2 d\Omega \quad for \ r \to \infty. \tag{7}$$

The angular brackets denote time average, the dot denotes time derivative, and the superscripts *sc* and *inc* refer to the scattered and the incident (which can be either longitudinal or transverse) fields, respectively. The quantities m_i are the components of the **m** unit vector; the symbol "*Re*" denotes the real part and the repeated indices indicate summation.

The single scattering calculations presented in this work are restricted to the case of spherical scatterers. In this case the spherical coordinate system is the most convenient. The displacements and the stresses of Eq. (7) are expanded into spherical waves giving the cross-section in the form of an infinite sum of partial cross-sections, each from the contribution of each partial spherical wave.

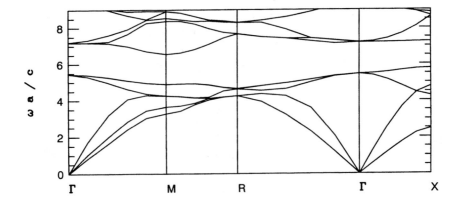

Figure 1. Elastic wave band structure for a *sc* periodic composite consisting of steel spheres in epoxy matrix. The volume fraction of the spheres is 0.268; ω is the frequency, a is the lattice constant and c the transverse wave velocity in the epoxy (c_{to}).

In all of our calculations, these infinite sums have been approximated by finite sums (by using a truncation criterion) containing, at most, 15 terms. For small frequencies, three or four terms were usually plenty to give satisfactory accuracy. The relative truncation error in all cases was less than 10^{-4}.

3. Study of the Periodic Systems

As mentioned earlier an extensive study of the acoustic and elastic wave propagation through periodic composites has been done concerning the dependence of the gap on various parameters of the composite [3-6,15]. We will present some of the main results of these studies which show the dependence of the gap on parameters such as the topology, the solidity of the scatterers and host, the density and velocity contrast between the two materials, the volume fraction, the lattice structure, and the shape of the scatterers.

Let us first discuss the role of the *topology* in the appearance of a gap. In Fig. 1 we show the band structure of a *sc* periodic composite consisting of steel spheres ($\rho = 7.8\,g/cm^3$, $c_l = 5.94\,km/s$, $c_t = 3.22\,km/s$) embedded in epoxy ($\rho = 1.180\,g/cm^3$, $c_l = 2.540\,km/s$, $c_t = 1.160\,km/s$). The volume fraction occupied by the spheres is 0.268, i.e., the spheres are unconnected (cermet topology). There is a complete band gap between the 6th and 7th band. The midgap frequency is $\omega_g a/c_{to} \approx 6.3$ with "a" being the lattice constant and c_{to} the transverse wave velocity in epoxy. The upper edge of the gap is at the M point while the lower edge is at the Γ point (indirect gap). The three lowest lying branches below the gap are the so-called acoustic branches. They are characterized by the fact that at the Γ point (long wave length limit) they tend linearly to zero. For high symmetry directions the two lowest lying acoustic branches are degenerate and are pure transverse waves while the third is pure longitudinal wave. For an arbitrary direction of k there is small admixture of the other polarization (compare the ΓX with the ΓM direction). The other, almost flat branches, corresponding to sharp peaks in the density-of-states, are the so-called optical branches and, in most of the cases, they have both longitudinal and transverse character. We note the above characteristics (acoustic and optical branches) are common characteristics in the elastic wave band structures of periodic composites with one scatterer in each primitive shell.

In Fig. 2 the steel spheres of Fig. 1 have been replaced by steel tetrahedral rods connecting nearest neighbors. This arrangement produces a network topology. We can see here that the wide gap of Fig. 1 has disappeared.

Figures 1 and 2, as well as, a variety of other computational data [4, 15] indicate that the preferable topology for the gap in both acoustic and elastic wave propagation is the cermet topology in contrast to the electromagnetic waves which seem to prefer the network. The most unexpected point in these results is that elastic waves are closer to the AC rather than to the EM waves, although one would expect the opposite (due to the vector character of both EL and EM waves). The isotropic scattering present in both the AC and EL case and absent in the EM case does not seem to explain this behavior because in the "hard" (with high c_t/c_l) solids, the role of this component is negligible as we will show later on. In the following, we will restrict ourselves to the structures with cermet topology. The subscript "o" (out) will denote the matrix material and the subscript "i" (in) will denote the scatterers.

Next, we discuss the effects of non constant *solidity* (ratio of the transverse to the longitudinal wave velocity, c_t/c_l) for scatterers and host. Although the role of this parameter seems to be not decisive for the gap, the existing results [5] seem to suggest that gap is slightly favored from the largest possible ratio c_t/c_l $(= 1/\sqrt{2})$ for both materials. This is not surprising because in high solidity materials due to the similar velocity of the longitudinal with the transverse mode, the stronger mixing of these modes occurs.

A very important parameter for the appearance of a gap, which is absent in the EM as well as in the simple scalar case, is the *density contrast*, ρ_i/ρ_o, between scatterers and host. This ratio gives the possibility for the gap in the ELW propagation to appear in many cases more easily than in the

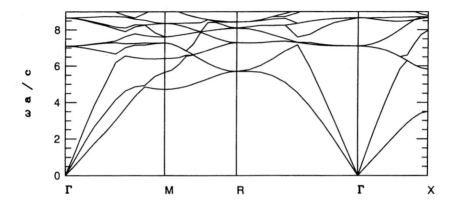

Figure 2. Elastic wave band structure for an *sc* periodic composite consisting of steel tetrahedral rods connecting nearest neighbors and embedded in epoxy. The volume fraction of the rods is 0.268; ω is the frequency, a is the lattice constant, and c the transverse wave velocity in the epoxy (c_{to}).

EMW propagation inspite of the full vector character of the elastic waves. What has been found [3-5] is that, for solids, the gap is favored from the largest ratio ρ_i/ρ_o, i.e., high density inclusions in a low density matrix, for liquids the opposite condition is required; thus, low density inclusions in a high density matrix is the ideal combination for a wide gap. To this strange and unexpected difference between liquids and solids we will return later on for a possible explanation.

Another parameter which affects the gap in a unclear way is the *velocity contrast* between scatterers and host. Here, we consider first the effects of velocity contrast in the absence of density contrast. In Fig. 3 we show the gap over midgap vs $r_c^2 = c_o^2/c_i^2$ for a 2D periodic composite consisting of liquid circular rods embedded in a liquid host in a square (B) or hexagonal (C) arrangement. The density of the rods are the same with the density of the host and the volume fraction occupied by the rods is 0.25. From Fig. 3 a critical value of r_c (r_c^{cr}) can be seen above where the first appearance of a gap occurs. As the ratio r_c increases from this critical value, the gap gets wider approaching a saturation in higher values of r_c. A gap has also been found for values of r_c less than one. In this case though, the critical velocity contrast ($1/r_c^{cr}$) for the first appearance of the gap is higher than that of the previous case, i.e., gap appears with more difficulty [4, 5]. The above tendencies have also been observed for spherical scatterers as well as for the full elastic case (ELW in solids) [5]. Considering the dependence of the critical value r_c^{cr} on the concentration of the scatterers and the lattice structure, it has been found [3-5] that r_c^{cr} has no strong dependence on the

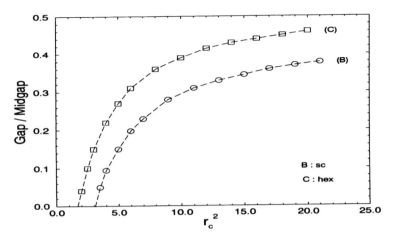

Figure 3. The width of the gap over the midgap frequency vs the velocity contrast $r_c^2 = c_o^2/c_i^2$ for a 2D periodic composite consisting of liquid circular columns embedded in a liquid in a square (B) or hexagonal (C) arrangement. Volume fraction of the columns $= 0.25$ and $\rho_i/\rho_o = 1$.

lattice structure while it remains nearly constant for a range of filling ratios around an optimum one, which depends on many factors; this optimum filling ratio is as low as 10% for elastic waves in composites such as Au in SiO_2 [6] and as high as 55% for W spheres in epoxy [8].

The widening of the gap as the velocity contrast increases is not unexpected because in the absence of density contrast, the impedance mismatch between the two materials (which gives the strong scattering and consequently the gap) is exclusively due to the velocity contrast.

Although the role of the velocity contrast in the absence of density contrast is rather clear, the same is not true if density contrast is present as well. According to the above discussion, one would expect a material combination of high density combined with high velocity contrast will be the ideal for the appearance of a gap. However, *wider* gaps were found [7, 8] in cases where r_c is far from its values considered as optimum according to the above. In Fig. 4 we show the band structure for an *fcc* periodic composite consisting of Ag spherical scatterers ($\rho = 10.635\,g/cm^3$, $c_l = 3.789\,km/s$, $c_t = 1.950\,km/s$) embedded in epoxy. The volume fraction of the scatterers is 0.35. We can see here an extremely wide gap with midgap frequency $\omega_g a/c_{to} \approx 10$ which appears between 6th and 7th band. Note also, the multitude of very flat bands corresponding to very sharp peaks in the elastic wave density-of-states. The Ag in epoxy case is a case of a wide gap, although the velocity contrast r_c is not that different from one. Some other results on this point are listed in Table 1 where with $max(\Delta\omega/\omega_g)$ we denote the width of the gap over the midgap frequency. These results concern spherical scatterers in fcc structures.

TABLE 1.

sphere/matrix	ρ_i/ρ_o	c_{lo}/c_{li}	c_{to}/c_{ti}	$max(\Delta\omega/\omega_g)$
W/Epoxy	15.85	0.49	0.41	0.750
Ni/Epoxy	7.60	0.43	0.36	0.550
Fe/Epoxy	6.66	0.42	0.35	0.500
Cu/Epoxy	7.60	0.54	0.50	0.500
Steel/Epoxy	6.61	0.43	0.36	0.500
Ag/Epoxy	8.77	0.67	0.59	0.500
Pb/Epoxy	9.62	1.18	1.35	0.500
Au/Si	8.36	2.63	4.35	0.055
Pb/Si	4.88	4.17	6.20	0.033

In the last two cases of Table 1 (Au/Si, Pb/Si), the gap appears between the higher acoustic branch and the lower optical [6], i.e., between the third and fourth branch as in all cases with r_c much larger than one which we have examined. The other cases in the table follow the Ag in epoxy system of Fig. 4 (i.e., the gap appears between the 6th and 7th band). What distinguishes the last two cases from the others is that in the last cases the high values of r_c give Lamé coefficients of the matrix higher than those of the scatterers, while the density contrast contributes in the opposite direction.

The above results indicate that in the presence of strong density contrast the velocity contrast, r_c, not only ceases as the dominant parameter for the gap, but it may affect its appearance in the opposite way than in the absence of density contrast. Later we will try to make this point clearer and to give a physical interpretation for the complicated case of the coexistence of density and velocity contrast.

In the following, we will discuss the role for the gap of the *volume fraction*, the *lattice structure*, and the *scatterer shape*.

In Fig. 5 we show the width of the gap over the midgap frequency vs volume fraction of the scatterers for a system consisting of gold (Au : $\rho = 19.49\,g/cm^3$, $c_l = 3.36\,km/s$, $c_t = 1.24\,km/s$) scatterers arranged in an fcc structure within a silicon (Si : $\rho = 2.33\,g/cm^3$, $c_l = 8.94\,km/s$, $c_t = 5.34\,km/s$) matrix. The scatterers are either spheres (A) or cubes (B) or cylinders with height over diameter being 0.92 (C) or 0.5 (D).

In Fig. 6 we show the gap over midgap vs volume fraction for Ag spheres in an epoxy matrix composite in fcc (solid line), bcc (dotted line), and sc (dashed line) structure. The dotted-long dashed line corresponds to an sc structure in which the Ag spheres have been replaced by Ag cubes.

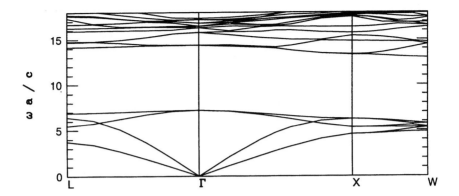

Figure 4. Elastic wave band structure for an *fcc* periodic composite consisting of Ag spheres in an epoxy matrix. The volume fraction of the spheres is 0.35; ω, a and c are the frequency, the lattice constant, and the transverse wave velocity in the epoxy (c_{to}), respectively.

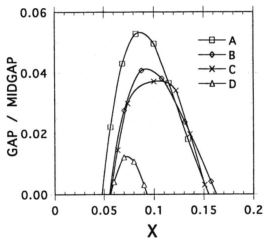

Figure 5. The width of the gap over the midgap frequency vs filling ratio, x, for an *fcc* periodic composite consisting of gold scatterers in a silicon matrix. The scatterers are spheres (A), cubes (B), or cylinders with height over diameter 0.92 (C), or 0.5 (D).

As can be seen in Fig. 5, spherical scatterers exhibit wider gaps than cylindrical or cubic. In most of the cases of our existing results, the more isotropic scatterer produces the wider gap [6]. There are some exceptions concerning mainly the *sc* structure in cases with r_c less than unity, where the cubic scatterers give a wider gap than the spheres (see Fig. 6). Comparing the results for the *sc* structure with scatterers, either spheres or cubes (see Fig.6), one can see, also, that cubes produce gaps in a wider range of

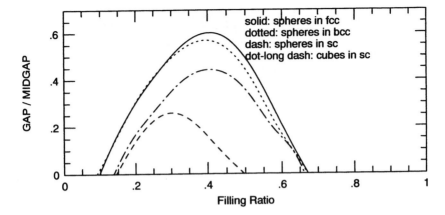

Figure 6. The width of the gap over the midgap frequency vs filling ratio for a periodic composite consisting of Ag spheres in epoxy matrix in a *fcc* (solid line), *bcc* (dotted line) and *sc* (dashed line) structure. The dotted-long dashed line corresponds to Ag cubes in epoxy matrix and *sc* structure.

filling ratios than the spheres. This is consistent with the preference of the cermet topology. The transition from the cermet to the network topology for the cubes in the *sc* structure occurs at filling ratio 1, thus, giving the chance for the gap to preserve itself for a wide range of filling ratios.

Considering the role of the lattice structure in the 3D case, *fcc*, *bcc* and diamond lattices seem to give better results than the *sc* (see Fig. 6), while the differences among them, especially in the cases with r_c larger than unity, are not dramatic. In the 2D case, hexagonal lattices have proven the most favorable for the gap (see Fig. 3).

Finally, we will consider the role of the volume fraction of the scatterers. As can be seen in Fig. 6 the optimum for the gap filling ratio for all structures except *sc*, is almost 0.4, while for *sc* it is around 0.3. In Fig.5, the wider gap appears for filling ratios less than 0.1. In all cases which we examined, the optimum for the gap filling ratio was between 0.05 and 0.5 (the EM waves gap is favored for values of filling ratio closer to 0.5 [2]). Usually the higher the number of branches below the midgap frequency, the larger the optimum filling ratio. For the *sc* structure, the wider gap usually appears in lower filling ratios than for the other structures, while the range of filling ratios in which gaps exist is narrower than that of the others. This last remark can be explained also by noting that the transition from the cermet to the network topology for the *sc* structure and spherical scatterers occurs at a lower concentration than for the other structures.

Considering the position of the gap vs filling ratio, it was found [25] that in the range in which a gap already exists, its position is almost independent from this parameter. The same almost independence is followed also by the flat bands of the periodic system (more detailed results on this point will be presented later). This relative independence is an indication that the multiple scattering does not influence appreciably this aspect of the band structure.

4. Single Scattering Study

The attempts to understand the above results and to find a simple way to predict optimum conditions for gap creation were focused on the examination of the single scattering. The relative insensitivity of certain important features of the band structure results on both the volume fraction of the scatterers and the lattice structure, as well as the flatness of certain bands, provide strong evidence for the dominant role of the single scattering in determining the positions of the flat bands and the gap(s). Furthermore, the conceptual and calculational simplicity of the single scattering allows a physical picture to emerge and an understanding to be achieved.

In the single scattering results which are presented in this work we have restricted ourselves to the case of spherical scatterer. Considering a plane wave scattered by a single spherical inclusion embedded in a homogeneous host (matrix), we calculated the total and the partial scattering cross-sections. We compared these cross-sections with certain features of the corresponding band structure results.

The main idea, as mentioned in the introduction, is to check the possible connection between the resonances of the cross-section and the flat bands or other characteristics of the band structure, to examine the possible extension of the LCAO method, in the AC and EL wave case with the resonances to play the role of the eigenstates. More specifically, the idea is to express the periodic field pattern $u(r)$ as a linear combination, $\sum_{\ell,a} c_{\ell,a} u_a(r - R_\ell)$, of the single resonance states $u_a(r - R_\ell)$, where a refers to the various resonances from a single scatterer and R_ℓ are the lattice vectors determining the position of the center of each scatterer in the periodic system. Two important differences from the corresponding electronic method should be stressed : i) Resonance states are not localized (i.e., decay too slowly, as $1/r$ as $r \to \infty$), and this may lead to divergences. ii) Besides the resonance states in the present problem, there is another channel of propagation employing at least partially, if not mainly, the host material. Both of these complications arise because in the classical wave case, as opposed to the electronic case, the host medium supports propagating solutions for every value of the frequency. Proper inclusion of the host propagation channel must face the formal problem of orthogonality to the resonance states and

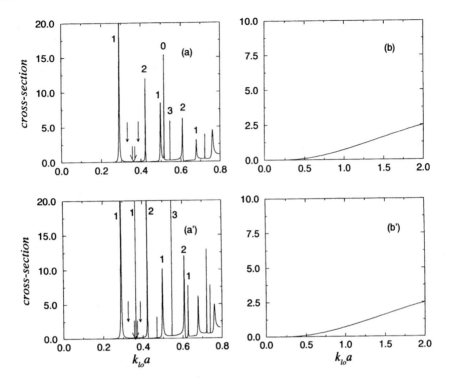

Figure 7. Total dimensionless cross-section, $\sigma/\pi a^2$, vs $k_{lo}a$ for longitudinal (upper panels) and transverse (lower panels) incident wave; $c_t/c_l = 1/\sqrt{2}$ for both sphere and matrix, $\rho_i/\rho_o = 1$ and $c_{lo}/c_{li} = c_{to}/c_{ti} = 8.66$ for (a) and (a$'$) and $1/8.66$ for (b) and (b$'$). a is the radius of the sphere and $k_{lo} = \omega/c_{lo}$ the longitudinal wave number in the matrix. The number next or above each resonance denotes the spherical harmonic responsible for its appearance. The arrows indicate the positions of the flat bands and the double arrow the position of the gap in the corresponding fcc periodic composite with volume fraction of the spheres 0.144.

may possibly cancel any divergence due to the long range nature of the resonances. In the present work, we shall not deal with the mathematical formulation of the linear combination of resonance modes (LCRM). We restrict ourselves to provide evidence supporting the proposition that an LCRM may work.

To establish a connection between the resonances and the flat bands, one must examine if there are some correspondences between them as well as if they depend in the same way on the parameters of the problem (it should be noted that by the term "flat bands," we denote peaks in the density-of-states rather than exact flatness).

Below, we discuss the relative dependence of resonances and flat bands

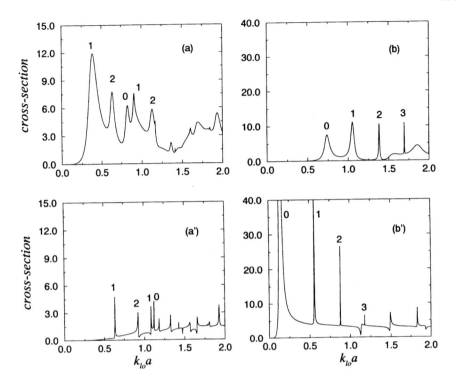

Figure 8. Total dimensionless cross-section, $\sigma/\pi a^2$, vs $k_{lo}a$ for $c_{lo}/c_{li} = 4$, $\rho_i/\rho_o = 10$ (a), (b) and 0.1 (a'), (b'). The left panels (a and a') correspond to a solid case with $c_t/c_l = 1/\sqrt{2}$ for both sphere and matrix and the right panels (b and b') to a fluid case ($c_t = 0$ for both materials). a : sphere radius; $k_{lo} = \omega/c_{lo}$. The incident wave is longitudinal and the number next or above each resonance denotes the corresponding spherical harmonic.

on parameters like velocity or density contrast of the two materials, mentioning, also, some of the main characteristics of the cross-section and attempting to answer some of the open questions from the band structure study.

In Fig. 7 we show the single scattering cross-section in a case of "hard" solids (with $c_t/c_l \approx 1/\sqrt{2}$) with the density of the scatterers to be the same as the density of the host and $c_{lo}/c_{li} = 8.66$ (a), (a') and $1/8.66$ (b), (b'). The upper panels (a, b) correspond to longitudinal incident waves while the lower (a', b') to transverse waves.

In the left panels of Fig. 7 one can see very sharp resonances in the cross-section. Each is due to the contribution of a partial spherical wave (mode) denoted by a number next or above the resonance; note that in the transverse incident case the pure longitudinal, $n = 0$, (spherically symmetric) mode does not exist. All the resonances of the longitudinal incident wave case are in the same position as the resonances of the transverse incident case. Both are associated with the same mode as indicated by the numbers. However, in the transverse incident wave case (see 7a$'$), there are additional resonances not present in the longitudinal case, which usually are very narrow. These narrow peaks correspond to long life times, i.e., low radiation field which means the scattered field is weak as compared with the field inside the sphere [25]. In all of the resonances, the higher amount of the scattered energy is of the form of transverse waves. The above features are common characteristics of the cross-sections in the case of solids with a high c_t/c_l ratio. The first resonance is usually due to the $n = 1$ oscillation, while the contribution of the $n = 0$ mode is negligible (see Fig. 7a). In the right panels of Fig.7 the cross-sections are very low and featureless.

The conclusion from Fig. 7 (comparison of Fig. 7a (7a$'$) with Fig. 7b (7b$'$)) and from other computational data on this point [25] is the role of the velocity contrast, c_{lo}/c_{li}, in the absence of density contrast is to form sharp, *closely placed resonances*. As the ratio c_{lo}/c_{li} decreases, these resonances move to higher frequencies and become lower and more well-separated, while the small background cross-section between them starts to grow. In the $c_{lo}/c_{li} = 1$ limit, the resonances as well as the background disappear (due to the absence of any contrast) and in the opposite direction, $c_{lo} < c_{li}$ (not completely examined), a very low and smooth cross-section is formed which remains almost unaffected after certain values of the contrast.

The existence of strong resonances in the low velocity scatterer case is not surprising because low velocity scatterers in the classical case correspond to deep potential wells in the electronic case. As is well known, deep potential wells in the electronic problem exhibit strong peaks in the scattering cross-section. The lowering of the resonances as the c_{lo}/c_{li} decreases is also expected due to the decreasing of the contrast between the two materials.

In order to examine the role of the velocity contrast in the presence of density contrast, we repeated the above calculations in the presence of a small density contrast. What was observed is that the role of velocity contrast (at least in the most well examined region of $c_{lo}/c_{li} \geq 1$) remains almost the same as what has been noticed above. In the $c_{lo}/c_{li} = 1$ limit, the scattering does not vanish in this case.

In summary, one can say that the *role of the velocity contrast, c_o/c_i, is to form* (by its increase) sharp, closely placed resonances, to move them to lower frequencies and to reduce the background scattering.

For the matrix-spherical inclusion combination shown in Figs. 7a and 7a$'$, the corresponding fcc periodic problem has been studied with volume fraction of the spheres 0.144 [5]. A relatively narrow gap has been found with midgap frequency $\omega_g a/c_{lo} \approx 0.36$ (a : sphere radius) which is denoted by a double arrow in Figs. 7a and 7a$'$. This band structure was found also to exhibit flat bands (corresponding to rather sharp peaks in the density of states) which in Figs. 7a and 7a$'$ are denoted by single arrows. We see the position of the gap is between the first and second resonance of Fig. 7a and between first and third resonance of Fig. 7a$'$. The width of the gap is much smaller than the frequency distance between these resonances. The first flat band is above the first resonance of Figs. 7a and 7a$'$ and the second below the second resonance of 7a and the third of 7a$'$. The second resonance of 7a$'$ is not connected with the characteristics of the band structure due to its extremely weak strength. For this reason, we ignore such very sharp resonances in many points of this discussion.

By lowering the density contrast from 8.66 to 5.48, flat bands, the gap, and the resonances moved in such a way as to retain their relative positions. [5, 25].

The existence of the gap between the first and second resonance in this range of filling ratios is something observed in all of the cases in which band structure results have been compared with the corresponding single scattering data. The flat bands in all these cases have been found to appear very close to the resonances, while the flatter the band the better the coincidence which has been observed (due to level repulsion and hybridization, an exact coincidence is not expected).

Extensive study of the position of the gap and the flat bands vs volume fraction [25] showed the first appearance of the gap (in the very low filling ratio limit) occurs very close to the first resonance. As the concentration increases, the gap moves away from this resonance approaching the midresonance position. These results - similar to what has been observed for scalar waves [35] - are consistent with the main idea of LCRM. The appearance of the gap very close to a resonance in the low concentration limit can be attributed to the fact that in these low concentrations the propagation of the wave takes place mainly through the host material and the scatterers only obstruct its propagation. As the concentration increases, the resonances of the neighboring spheres are mixed together and are broadened into bands giving the wave another, preferable in this range, channel of propagation. Thus, the resonances create flat bands and consequently, the gap (if any) moves to the region between them.

We will discuss next the role of the density contrast and its puzzling opposite effect on fluids and solids periodic media. In Fig. 8 we show the total scattering cross-section for longitudinal incident wave and for materials with $c_o/c_i = 4$ and $\rho_i/\rho_o = 10$ (upper panels) and 0.1 (lower panels). The left panels correspond to the "hard" solid case (with $c_t/c_l \approx 1/\sqrt{2}$), while the right panels to the fluid case (with $c_t \approx 0$). What distinguishes the solid from the fluid case, as Fig. 8 shows, is the role of the $n = 0$ mode. In the solid case the $n = 0$ mode has an insignificant contribution to the cross-section, while in the liquid case both the first resonance and the background (cross-section between the resonances) are due to $n = 0$ contribution.

What can be seen also from Fig. 8 is that in the solid case the resonances become significantly lower and narrower as the ratio ρ_i/ρ_o decreases. On the other hand, in the fluid case there is the opposite trend due again to the dominant role of the $n = 0$ mode. As ρ_i/ρ_o decreases, both the resonances and the background become significantly higher (note the change of the vertical scale). The most dramatic enhancement occurs for the first resonance (this resonance has been studied analytically [25] and in the low ρ_i/ρ_o limit its frequency is given by $\omega_o = (c_i/a)\sqrt{3\rho_i/\rho_o}$).

The dependence of the cross-section on the density ratio is consistent with the dependence of the gap on this ratio. The stronger single scattering corresponds to the wider gap. For the liquid case, the growth of the cross-section as the density ratio ρ_i/ρ_o decreases is due to the enormous increase of the $n = 0$ scattering which dominates the liquid case cross-section and is almost absent in solids (in solids, due to the high c_t, the transverse wave scattering dominates the cross-section). Thus, this enhanced $n = 0$ scattering must be responsible for the appearance of the gap for the liquids in the low ρ_i/ρ_o limit.

Corresponding band structure results for the specific cases of Fig. 8 do not exist. In some other cases of materials in which high density was combined with high velocity contrast, in which both band structure and single scattering results existed, the band structure results gave wide gaps [5]. The position of these gaps was usually between the first and second resonance of the corresponding single scattering cross-section. There were also flat bands very close in position with these resonances [25].

As discussed earlier, the surprising result on this point is that wider than the above gaps were found in cases with high density contrast, but velocity contrast was not much different from one (see Table 1).

In Fig. 9 we show the cross-section for Ag sphere in epoxy matrix (Fig. 9a) and for longitudinal incident wave. The Ag epoxy case (as show in Fig. 4) has an extremely wide gap, although the velocity contrast is not that large. In the second panel of Fig. 9, (b), the Ag sphere has been replaced by a rigid sphere ($\rho \to \infty$, $\mu \to \infty$, $\lambda \to \infty$, $c_l \to 0$, $c_t \to 0$), and in the third panel (c) the cross-section calculated by subtracting from the Ag sphere scattering amplitudes the rigid sphere scattering amplitudes are presented. The arrows in Fig.9c and the half arrows with the symbol "G" inside denote the flat bands and the gap, respectively, of the Ag in epoxy fcc periodic system with volume fraction of the spheres 0.35 (see Fig. 4).

Ag in epoxy

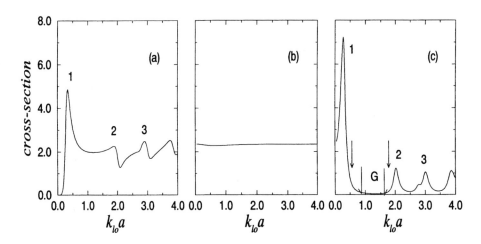

Figure 9. Total dimensionless cross-sections, $\sigma/\pi a^2$, vs $k_{lo}a$ for an Ag sphere in epoxy matrix (a) and a rigid sphere in epoxy matrix (b). The third panel (c) represents the cross-section calculated by subtracting from the Ag sphere scattering amplitude the rigid sphere scattering amplitude. The incident wave is longitudinal, a is the radius of the sphere, and $k_{lo} = \omega/c_{lo}$ the longitudinal wave number in the epoxy. The arrows indicate the positions of the flat bands and the half arrows (with the symbol G in between) the position of the gap in the Ag spheres in epoxy fcc periodic composite with volume fraction of the spheres 0.35. The number above each resonance indicates the mode responsible for its appearance.

There are some broad resonances in the Ag in epoxy cross-section well separated by a region of non negligible scattering. The scattering in this region is almost exclusively due to the contribution of the rigid sphere (for this analysis see ref. [32-34]) as can be seen in panel (c) where this contribution has been subtracted and the resonances emerge clearly.

This analysis clearly indicates that in the spectral region between the resonances, the wave can hardly penetrate the sphere (note that for a rigid sphere, the field inside and at the surface is zero). Calculations of the total elastic energy vs the distance from the center of the sphere in this spectral region verified this statement [25]. The same calculation showed that at the broad peaks (like those of Fig. 9) the higher amount of the energy is concentrated inside the sphere, although a significant leakage outside exists. The narrower the peak, the smaller the leakage.

Similar analysis, through single scattering of the other material combinations of Table 1 which exhibit extremely wide gaps, showed similar to Ag in epoxy cross-sections (with the broad and well separated resonances and the high background between them; the latter is associated with a rigid sphere) [8, 25].

It was noticed that the contribution of the background in these cases is not unexpected. From both density and velocity, and also from the Lamé coefficients contrast (the Lamé coefficients of the matrix are lower in these cases than those of the scatterer), one can see these composites are closer to the rigid sphere in epoxy composite than any system with very high or very low velocity contrast (c_o/c_i).

The above results strongly suggest that in order to obtain a wide gap in a composite periodic structure, a combination of inclusion-host is required such that the single inclusion cross-section exhibits *strong* and *well separated resonances* with a *considerable background in between*, attributed to either a rigid or maybe soft sphere (on this we will comment later on).

This basic result can be understood by considering two limiting channels of propagation in a composite. The first is mainly through the host material. The second is by hopping from scatterer to scatterer through the overlapping resonance modes (LCRM) (in most cases both channels seem to be operational).

On the basis of this argument, in the spectral region between first and second resonance of Fig. 9a, the wave can use neither the spheres for its propagation (since no resonances are nearby) nor the host (because of the high cross-section). Thus, it will be natural for a gap to open in that region. The fact that the peaks are well-separated (due to the absence of high velocity contrast) permits the appearance of a wide gap (note that the width of the gap is usually considerably less than the distance between the two resonances). On the other side, the considerable width of the resonances indicates strong leakage of the field outside each sphere and consequently a strong overlap and easy propagation through the resonance modes.

According to the above picture, one expects similar results in the cases for which well-separated strong resonances sandwich a high background due to a soft sphere contribution. This occurs because for a soft sphere $(\rho, \mu, \lambda, c_l, c_t \to 0)$ the field inside and at the surface is zero as in a rigid sphere. This situation may arise in the case of low density, low velocity scatterers in a high density, high velocity matrix, although the results of Fig. 8 indicate the resonances in this case may be neither strong nor well separated (for solid composites).

We will close this discussion with few comments concerning the role of the solidity, c_t/c_l. Most of our single scattering results seem to be much more sensitive to the solidity of the matrix than to the solidity of the scatterer. A solid sphere in a solid (liquid) host combination seems to give similar cross-sections with a liquid sphere in the same solid (liquid) host (with the velocity and density contrast between the two materials remaining the same). On the basis of this result, we expect for the characteristics of the band structure a similar insensitivity on the solidity of the scatterers.

The present work was supported by the EU grants ERBCHRX-CT93-0136, -0331, -0332, and MAS2-CT92-0019. Ames Laboratory is operated for the U. S. Department of Energy by Iowa State University under contract No. W-7405-ENG-82.

References

[1] Sheng, P. (ed.) (1990) *Scattering and Localization of Classical Waves in Random Media*, World Scientific, Singapore.

[2] Soukoulis, C.M. (ed.) (1993) *Photonic Band Gaps and Localization* , Plenum Press, New York (proceedings of the NATO ARW).

[3] Sigalas, M.M. and Economou, E.N. (1992) *J. Sound Vibration* **158**, 377.

[4] Sigalas, M.M. and Economou, E.N. (1993) *Solid State Commun.* **86**, 141.

[5] Economou, E.N. and Sigalas, M.M. (1993) in Soukoulis, C.M. (ed.), *Photonic Band Gaps and Localization*, Plenum Press, New York, pp. 317-338.

[6] Economou, E.N. and Sigalas, M.M. (1994) *J. Acoust. Soc. Am.* **95**, 1734.

[7] Vasseur, J.O., Djafari-Rohani, B., Dobrzynski, L., Kushwaha, M.S., and Halevi, P. (1994) *J. Phys.: Condens. Matter* **6**, 8759.

[8] Kafesaki, M., Sigalas, M.M. and Economou, E.N. (in press) *Sol. State. Comm.*

[9] Kinra, V.K. and Ker, E.L. (1983) *Int. J. Solids Structures* **19**, 393.

[10] Dowling, J.P. (1992) *J. Acoust. Soc. Am.* **91**, 2539.

[11] Ye, L., Cody, G., Zhou, M., Sheng, P., and Norris, A.N. (1992) *Phys. Rev. Lett.* **69**, 3080.

[12] Kushwaha, M.S., Halevi, P., Dobrzynski, L., and Djafari-Rouhani, B. (1993) *Phys. Rev. Lett.* **71**, 2022.

[13] Kushwaha, M.S. and Halevi, P. (1994) *Appl. Phys. Lett.* **64**, 1085.

[14] Kushwaha, M.S., Halevi, P., Martinez, G., Dobrzynski, L., and Djafari-Rouhani, B. (1994) *Phys. Rev. B* **49**, 2313.

[15] Economou, E.N. and Sigalas, M.M. (1993) *Phys. Rev. B* **48**, 13434.

[16] John, S. and Rangarajan, R. (1988) *Phys. Rev. B* **38**, 10101.

[17] Economou, E.N. and Zdetsis, A. (1989) *Phys. Rev. B* **40**, 1334.

[18] Sigalas, M.M., Economou, E.N., and Kafesaki, M. (1994) *Phys. Rev. B* **50**, 3393.

164

[19] Satpathy, S., Zhang, Z., and Salehpour, M.R. (1990) em *Phys. Rev. Lett.* **64**, 1239.

[20] Ho, K.M., Chan, C.T., and Soukoulis, C.M. (1990) *Phys. Rev. Lett.* **65**, 3152.

[21] Leung, K.M. and Liu, Y.F. (1990) *Phys. Rev. Lett.* **65**, 2646.

[22] Zhang, Z. and Satpathy, S. (1990) *Phys. Rev. Lett.* **65**, 2650.

[23] Joannopoulos, J., Meade, R., Winn, J. (1995) *Photonic Crystals*, Princeton University Press.

[24] Yablonovitch, E. and Gmitter, T.J. (1989) *Phys. Rev. Lett.* **63**, 1950.

[25] Kafesaki, M. and Economou, E.N. (in press) *Phys. Rev. B.*

[26] Landau, L.D. and Lifshitz, E.M. (1959) *Theory of Elasticity*, Pergamon, London.

[27] Straton, J.A. (1941) *Electromagnetic Theory*, McGraw-Hill, New York, pp. 392.

[28] Ying, C.F. and Truell, R. (1956) *J. Appl. Phys.* **27**, 1086.

[29] Einspruch, N., Witterhold, E., and Truell, R. (1960) *J. Appl. Phys.* **31**, 806.

[30] Johnson, G. and Truell, R. (1965) *J. Appl. Phys.* **36**, 3467.

[31] Gubernatis, J.E., Domany, E., and Krumhansl, J.A. (1977) *J. Appl. Phys.* **48** 2804.

[32] Brill, D. and Gaunaurd, G. (1987) *J. Acoust. Soc. Am.* **81**, 1.

[33] Flax, L., Gaunaurd, G. and Überall, H. (1981) in Mason, W.P. (ed.), *Physical Acoustics*, Academic, New York, vol. XV, pp. 191-294.

[34] Gaunaurd, G. and Überall, H. (1987) *J. Appl. Phys.* **50**, 4642.

[35] Datta, S., Chan, C.T., Ho, K.M., Soukoulis, C.M., and Economou, E.N. (1993) in Soukoulis, C.M. (ed), *Photonic Band Gaps and Localization*, Plenum, New York, pp. 289-297.

3-D METALLIC PHOTONIC BANDGAP STRUCTURES

D. F. SIEVENPIPER, M. E. SICKMILLER,
and E. YABLONOVITCH,
Electrical Engineering Department
University of California, Los Angeles
Los Angeles, CA 90095-1594

ABSTRACT

We have investigated the electromagnetic properties of a 3-dimensional wire mesh in a geometry resembling covalently bonded diamond. The frequency and wave vector dispersion show forbidden bands at those frequencies v_0, corresponding to the lattice spacing, just as dielectric photonic crystals do. But, they have a new forbidden band which commences at zero frequency and extends, in our geometry, to approximately one-half v_0, acting as a type of plasma cut-off frequency. Wire mesh photonic crystals appear to support a longitudinal plane wave, as well as 2 transverse plane waves. We identify an important new regime for microwave photonic crystals, an effective medium limit, in which electromagnetic waves penetrate deeply into the wire mesh through the aid of an impurity band.

In the past few years the concepts of solid state physics have been brought to bear on [1] electromagnetism and Maxwell's equations. Artificial 3-dimensional dielectric structures [2-5] have been discovered, which are the photonic analog of a semiconductor [6,7]. In these "Photonic Crystals," a photonic bandgap arises in which electromagnetic modes, spontaneous emission, and zero-point fluctuations are all absent. A dielectric photonic crystal containing a local defect forms an optical 3-d micro-cavity [8]. Since metals are usually quite lossy at optical frequencies, most photonic crystal work has focused on dielectric structures. Nevertheless, metallic 3-d structures can be valuable at microwave frequencies. Two-dimensional metallic structures are already being investigated [9] and theoretical work has recently begun [10-12] on 3-d metallic photonic crystals. We will show in this paper that 3-d wire mesh electromagnetic structures have very different physics from the dielectric photonic crystals which were studied previously.

In choosing a particular 3-d wire mesh geometry, we were influenced by an extension of the intuitively natural 2-dimensional hexagonal wire mesh geometry into

C. M. Soukoulis (ed.), Photonic Band Gap Materials, 165–171.

the third dimension. The electronic material analog of a hexagonal wire mesh is graphite structure, and the corresponding extension into three dimensions is diamond structure. Indeed, it has already been shown for 3-d dielectric photonic crystals, that diamond geometry [2] has an unusually intimate link with Maxwell's equations. Therefore, the wire mesh structure we have chosen consists of a diamond lattice, in which the "atoms" are geometrical points in space, and the valence bonds connecting these points are copper wires, 1 cm long. A photograph of our wire mesh structure is shown in Figure 1, together with a detailed view of the copper wire strips which snap together to make diamond geometry.

Our experiments consisted of microwave transmission measurements through the 3-d wire mesh structure, as a function of incidence angle and frequency. As expected, narrow forbidden gaps are observed centered at frequencies, v_o, corresponding to the spatial periodicity of the wires. Nonetheless, the most dominant electromagnetic feature of these wire mesh structures is not related to spatial periodicity, but happens to be a deep, sharp, and wide 3-dimensional stop band that extends from zero frequency up to a finite frequency, v_p. We observe in our geometry that v_p is approximately one-half of v_o. We regard v_p as being analogous to a type of plasma frequency associated with the motion of electrons in the continuous, interconnected, wire network, constituting our web-like structure. Alternately, it may be considered as a 3-d cut-off frequency for wavelengths too large to fit between the rows of wires. A transmission spectrum showing the steep edge at $v_p = 6.5 \text{GHz}$ is displayed in Figure 4(a), (To the left of the steep edge is a transmission peak produced by a defect mode.)

We attempted to measure the full electromagnetic dispersion in our diamond wire mesh photonic crystal. This required the identification of frequencies and wave-vectors for allowed modes. Mode frequencies were identified with step function changes in the transmission spectrum of external plane waves through the structure, while the corresponding internal wave vectors were determined by momentum matching at the surface of the photonic crystal. The projection of the external wave vector onto the surface of the photonic crystal fixed the internal $<k_x, k_y, 0>$, while k_z was

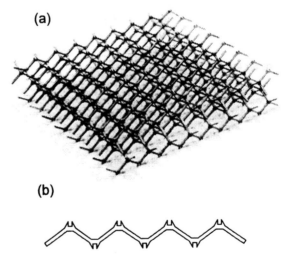

(a)

(b)

Figure 1. (a) A perspective photograph of our diamond geometry, 3-D, wire mesh photonic crystal. The individual wires play the role of "valence bonds" connecting the "atoms" which are merely geometrical points in space. The wires have a 1.25mm square cross-section, and are 1 cm long. (b) A detailed view of the copper wire strips which snap together to make diamond structure.

determined by physical reasoning as follows: For the lowest frequency step function cut-off, ν_p , the z-component was set at $k_z = 0$. For higher forbidden gaps due to periodicity, k_z was set so that $\vec{K}$ would fall at the surface of the Brillouin zone.

In the wire mesh photonic crystal shown in Figure 1, the top surface is a <1,0,0> cubic plane. Normal incidence puts the wave vector in the X-direction of reciprocal space, as shown in the Brillouin zone inset of Figure 2. Repeated spectra were taken, as the sample was rotated away from normal incidence, so the wave vector moved in the X-U-L direction. The procedure was repeated in the X-W-K direction for both s- and p- polarization. With the wave vector in the X-W-K direction, its component in the $k_z = 0$ plane, in the vicinity of the Γ-point, would be in the Γ-X direction, (one of the orthogonal X-points). Thus, an X-W-K scan also covered the Γ-X line. Likewise, an X-U-L scan covered the Γ-K line in the Γ-vicinity, as well. All of the various reciprocal space segments were then stitched together in Figure 2(a) for s-polarization and Figure 2(b) for p-polarization.

In an experiment such as this,

2(a)

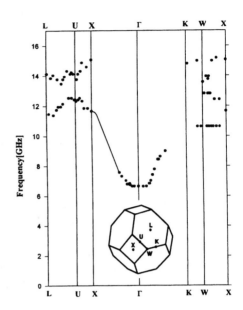

2(b)

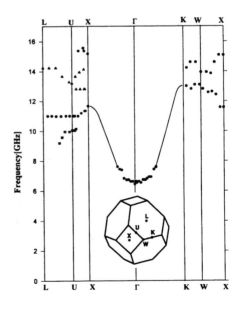

Figure 2. (a) Frequency and wave vector representing allowed plane waves inside our diamond wire mesh photonic crystal; s-polarization incident on a <1,0,0> face. (b) Dispersion relation for p-polarization. The triangular dots represent a mode which vanishes at normal incidence and may be a longitudinal plane wave. The square dots may be an experimental artifact.

there is no guarantee that every internal mode will be observed. Coupling to certain modes could be forbidden in our geometry, and some internal plane waves might simply couple inefficiently to external plane waves due to mode matching problems at the incident interface. In Figure 2(a), the modes detected at the K-point are certainly incomplete, since the K-point differs by a reciprocal lattice vector from the U-point, where additional modes are detected. The gap due to periodicity, at the surface of the Brillouin zone, falls between 12GHz and 13GHz, but is not clear that it is an absolute gap in all directions of space. The dominant feature is the low frequency gap from 0 to 6.5GHz associated with the motion of electrons in the continuous, interconnected, wire network. For incident p-polarization, in Figure 2(b), modes indicated by triangular dots (along X-U-L) had the property that they vanished at normal incidence. Therefore, these might be internal, longitudinally polarized plane waves similar to surface plasmons, but running along the internal wires of the structure. This implies that wire mesh photonic crystals support a longitudinal plane wave, as well as 2 transverse plane waves.

Our band structure results for metallic wire mesh photonic crystals may be summarized as follows: (a) Periodicity produces a stop band just as in dielectric photonic crystals. (In spite of some speculation that metallic photonic crystals would possess allowed rather than forbidden bands.) (b) Wire mesh photonic crystals have a new forbidden band which commences at zero frequency, and extends to approximately half the normal bandgap frequency. (c) Longitudinal modes are allowed in metallic photonic crystals. Therefore, the internal electromagnetic plane waves have three possible polarizations, rather than just two.

Let us now discuss the significance of these wire mesh photonic crystal properties. The most interesting point is the plasma edge, or cut-off frequency v_p. In the vicinity of v_p the mesh spacing is similar to the vacuum wavelength. Given that microwave-lengths are rather large, combined with the need for multiple mesh periods to form a 3-d crystal, these considerations would appear to demand fairly large structures. It would therefore be preferable to operate the wire mesh photonic crystal at mesh spacing much smaller than a vacuum wavelength. Under these conditions, the electromagnetic radiation generally penetrates only one or two unit cells, effectively interacting with only a 2-dimensional surface layer of the wire mesh. Such 2-dimensionally periodic metallic structures are already well-known [13-16] in electromagnetic engineering, where they are called "Frequency Selective Surfaces." To access a distinctive physical regime and a potentially useful one, we must design photonic crystal structures where electromagnetic waves penetrate a wire mesh, whose spacing $a \ll \lambda$ the vacuum wavelength. This condition defines a *new and different regime* of photonic crystals.

In the new regime $a \ll \lambda$, the crystal structures are actually effective media, described by a frequency-dependent dielectric constant, $\varepsilon(\omega)$, rather than the full 3-d wave vector dependence, $\varepsilon(\vec{k})$, of a true photonic crystal. Nevertheless, such effective media can have many unusual properties, including particularly a negative, predominantly real, dielectric constant at microwave frequencies, which would otherwise be difficult to find among solid metals. Since the local geometry of such an effective medium is probably not as important as it would be in a dielectric photonic crystal, a diamond wire mesh, a cubic [12] wire mesh, or perhaps a random wire mesh may perform similarly. The basic approach toward creating a penetrable medium in which a $<<\lambda$, is to introduce an impurity band within the low frequency forbidden gap, below v_p. In other words, a band structure within a band structure, or a super-lattice.

A single air gap in the metallic mesh is enough to dramatically change the transmission spectrum. Figure 4(a) shows a local mode within the bandgap created by a small cut in one wire. The frequency of this mode depends on the capacitance of the defect, with higher capacitance producing a lower frequency resonance. If a second defect is introduced nearby, it will affect the frequency of the first mode. Figure 3 shows the resonance frequency of two capacitive defects as one is tuned past the other. This familiar avoided crossing is seen in many systems of coupled resonators.

In Figure 4 of this paper, we present results on the initial stages of a transition from isolated defect modes to impurity band formation, by capacitive defects in the wire mesh, i.e., selectively cut wires. (Independently, Pendry [17] has introduced inductive loading of the individual wires to create such a medium.) In Figures 4(a) to 4(e) the defect count goes from 1 to 6. Remarkably, the electromagnetic spectrum of each defect array geometry is identical to the electronic energy levels of an analogous chemical molecule. For example, the defect array geometries correspond to a diatomic molecule, a triangular cyclopropenyl ion, up to benzene, 6 defects in a ring. Arrows in Figure 4 point to the electromagnetic defect modes; paired arrows representing doubly degenerate frequencies associated with the defect array symmetry. For the case of six defects in a ring, the six electromagnetic frequencies are grouped in degeneracies of 1:2:2:1 as can be found for benzene energy levels in most elementary [18] chemistry books.

To further reduce the frequency of the evolving impurity band, and to better satisfy the condition a $<<\lambda$, the gaps in the wires may be filled by high ε ceramic pellets or by inductive loading of the wires as suggested by Pendry et al. [17]. We believe that the effective medium limit, a $<<\lambda$, of a 3-d

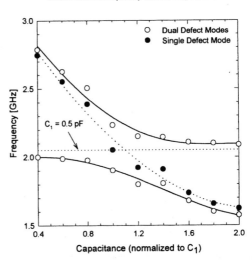

Metallic Diamond Structure with Capacitive Defects on Adjacent Bonds along [110]

Defect Mode Frequency versus Capacitance

Figure 3. Resonant frequencies of two capacitive defects in the metallic wire mesh photonic crystal. The capacitance of one defect is kept constant while the other is varied by insertion of small capacitors. The resonances display an avoided crossing effect as is common for many systems of coupled oscillators.

170

wire mesh photonic crystal is of both physical and practical interest. Having light weight, and easily modified properties, they are likely to play a more important role in microwave and millimeter wave technology. Those possible applications include: antenna structures. passive filter "windows," or frequency selective radomes, quasi-optical amplifying structures, and stealthy surfaces.

This work is supported by the Army Research Office and the National Science Foundation.

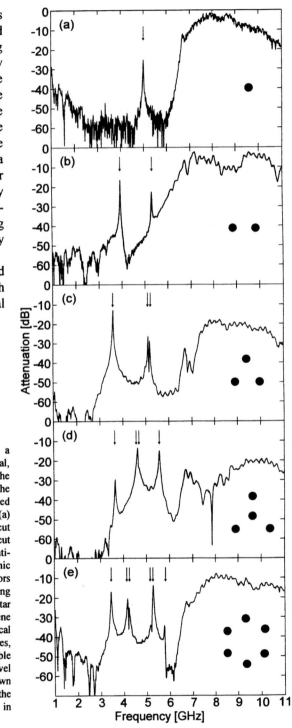

Figure 4. The evolution from a single defect in the photonic crystal, toward an impurity band within the 0 to 6.5GHz forbidden gap. The defects were all spatially located within the same <1,1,1> plane. (a) A defect consisting of a single cut wire. (b) Two neighboring cut wires, producing "bonding and anti-bonding modes" as in a diatomic molecule. (c) Three neighbors resembling a triangular ring molecule. (d) Four in a star configuration. (e) Six in a benzene ring configuration. The vertical arrows point to the allowed modes, paired arrows representing double degeneracy. The frequency level structure corresponds to that known for the energy levels of the corresponding molecules in chemistry.

REFERENCES:

1. See, for example, the articles in the special issue of J. Opt. Soc. Am. B10, (February, 1993).
2. K. M. Ho, C. T. Chan, and C. M. Soukoulis, Phys. Rev. Lett. 65, 3152 (1990).
3. E. Yablonovitch, T. J. Gmitter, and K. M. Leung, Phys. Rev. Lett. 67, 2295 (1991).
4. H. S. Sozuer, J. W. Haus, and R. Inguva, Phys. Rev. B45, 13962 (1992).
5. S. Fan, P. R. Villeneuve, R. D. Meade, and J. D. Joannopoulos, Appl. Phys. Lett. 65, 1466, (1994).
6. E. Yablonovitch, Phys. Rev. Lett. 58, 2059 (1987).
7. S. John, Phys. Rev. Lett. 58, 2486 (1987).
8. E. Yablonovitch, T. J. Gmitter, R. D. Meade, A. M. Rappe, K. D. Brommer, and J. D. Joannopoulos, Phys. Rev. Lett. 67, 3380 (1991).
9. D. R. Smith, S. Schultz, N. Kroll, M. M. Sigalas, K. M. Ho, and C. M. Soukoulis, Appl. Phys. Lett. 65, 645 (1994).
10. T. Suzuki, and P. K. L. Yu, J. Opt. Soc. Am. B12, 583 (1995).
11. A. R. McGurn, and A. A. Maradudin, Phys. Rev. B48, 17576 (1993).
12. M. M. Sigalas, C. T. Chan, K. M. Ho, and C. M. Soukoulis, Phys. Rev. B52, 11744, (1995).
13. J. P. Montgomery, IEEE Trans. Antennas & Prop., AP-23, 70 (1975).
14. C. H. Tsao and R. Mittra, IEEE Trans. Antennas & Prop., AP-32, 478 (1984).
15. G. Zarrillo and K. Aguiar, IEEE Trans. Antennas & Prop., AP-35, 1406 (1987).
16. S. Singh and D. R. Wilton, IEEE Trans. Antennas & Prop., AP-39, 190 (1991).
17. J. B. Pendry et al., to be published.
18. *Basic Principles of Chemistry*, H. B. Gray and G. P. Haight Jr., (W. A. Benjamin, New York, 1967), see page 317.

PHOTONIC BAND GAP STRUCTURES: STUDIES OF THE TRANSMISSION COEFFICIENT

M. SIGALAS, C. M. SOUKOULIS, C. T. CHAN, AND K. M. HO

Ames Laboratory and Department of Physics and Astronomy, Iowa State University, Ames, IA 50011

1. Introduction

Recently, there has been growing interest in the development of Photonic Band Gap (PBG) materials [1-21]. These are periodic dielectric materials exhibiting frequency regions where electromagnetic (EM) waves cannot propagate. The reason for the interest on PBG materials arises from the possible applications of these materials in several scientific and technical areas such as filters, optical switches, cavities, design of more efficient lasers, etc.[1, 2]. Most of the research effort has been concentrated in the development of two-dimensional (2D) and three-dimensional (3D) PBG materials consisting of positive and frequency independent dielectrics [1-18] because, in this case, one can neglect the possible problems related to the absorption [15, 19]. However, there is more recent work on PBG materials constructed from metals [20, 21] which suggests that these metallic structures may be very useful in the low frequency regions. In these regions, the metals become almost perfect reflectors.

In this chapter, we study 2D (section 2) and 3D (section 3) dielectric and metallic PBG structures. To find the possible advantages of metallic PBG materials over the dielectric materials, we compare the results for metals with the corresponding results for PBG materials constructed with positive and frequency independent dielectric constants. Also, by changing the scale of the structures, we find the frequency regions where metallic PBG materials can be used practically. For dielectric PBG materials, we study the effect of the absorption (section 4) and the effect of the disorder (section 5).

We use the transfer-matrix method (TMM), recently introduced by Pendry and MacKinnon [14], to calculate the EM transmission through

C. M. Soukoulis (ed.), Photonic Band Gap Materials, 173–202.
© *1996 Kluwer Academic Publishers. Printed in the Netherlands.*

the PBG materials. This method is presented in another chapter of this book. In the TMM, the total volume of the system is divided into small cells and the fields in each cell are coupled to those in the neighboring cells. Then, the transfer matrix is defined by relating the incident fields on one side of the PBG structure with the outgoing fields on the other side. Using the TMM, the band structure of an infinite periodic system can be calculated. The main advantage of this method is the calculation of the transmission and reflection coefficients for EM waves of various frequencies incident on a finite thickness slab of the PBG material. In this case, the material is assumed to be periodic in the directions parallel to the inter- faces. The TMM has previously been applied in studies of defects in 2D PBG structures [14], of PBG materials in which the dielectric constants are complex and frequency dependent[19], of 3D layer-by-layer PBG mate- rials [17], and of 2D metallic PBG structures [21]. In all these examples, the agreement between theoretical predictions and experimental measurements was very good.

For the metal, we use the following frequency dependent dielectric con- stant:

$$\epsilon(\nu) = 1 - \frac{\nu_p^2}{\nu(\nu - i\gamma)} \qquad (1)$$

where $\nu_p = 3600$ THz, $\gamma = 340$ THz are the plasma frequency and the absorp- tion coefficient, respectively. From the previous equation, it turns out that the conductivity is[22] $\sigma = \nu_p^2 \gamma / 2(\gamma^2 + \nu^2)$. For frequencies smaller than 100 THz, σ can be practically assumed independent of frequency and equal to $0.22 \times 10^5 (\Omega cm)^{-1}$ which is very close with the σ of Ti. However, the con- clusions are similar for other metals. The skin depth is $\delta = c(\mu\nu\sigma)^{-1/2}/2\pi$, where c is the velocity of light and μ is the magnetic permeability which is one in our case [22]. Thus, for $\nu = 100$ and 10 THz, the skin depth is 0.035 and 0.11μm, respectively.

2. 2D Photonic Band Gap Structures

2.1. DIELECTRIC PBG

We study a system consisting of dielectric cylinders with radius r=0.48 cm, and dielectric constant ϵ=9, forming a square lattice with lattice constant a=1.27 cm. This corresponds to a filling ratio of f=0.449. Figure 1 shows the transmission coefficient of waves propagating along the (1,0) direction and for E parallel and perpendicular to the axes of the cylinders (s- and p-polarized waves); there are 8 layers of cylinders along the (1,0) direction. For s-polarized waves (dotted line in Fig. 1), the transmission coefficient is considerably reduced around three frequencies (at 5, 10, and 15 GHz), corresponding to the three photonic band gaps. For p-polarized waves (solid

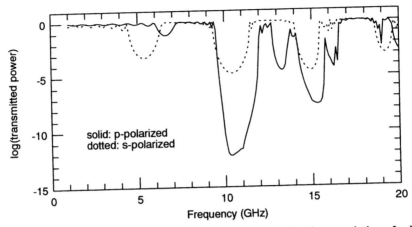

Figure 1. The transmitted power vs frequency for a square lattic e consisting of cylinders with f=0.449 surrounded by air for s- and p-polarized waves (dotted and solid lines, respectively) propagating along the (1,0) direction; there 8 layers of cylinders along this direction. The dielectric constant of the cylinders is 9.

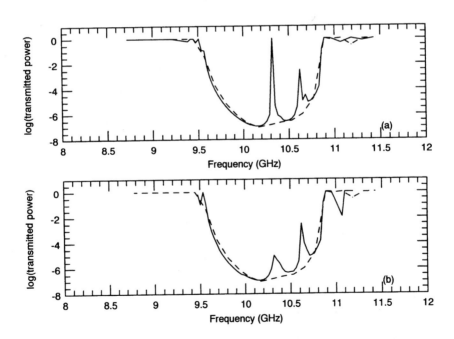

Figure 2. The transmitted power of s-polarized waves vs frequency for a square lattice consisting of 50 (5×10) cylinders with f=0.449 and ϵ=9 surrounded by air; the solid and dotted lines correspond to the perfect lattice and the defect lattice in which (a) two second- or (b) third-nearest neighbor cylinders have been removed.

line in Fig. 1), there is a photonic band gap around 11 GHz with a width of 2 GHz. This gap coincides with the first gap of the s-polarized waves along the (1,0) direction. There are also gaps around 13 and 15 GHz.

In order to study defect structures, we use a system similar with the previous one consisting of 10 layers of cylinders along the (1,0) direction. A supercell of 5 cylinders has been used along the (0,1) direction with periodic boundary conditions along the edges of the supercell. The EM waves propagate along the (1,0) direction with E parallel to the axes of the cylinders. The Green's function method has been used for the calculation of the transmission coefficient [15, 23, 24]. By removing two second nearest neighbor cylinders which are located close to the center of the system, two sharp peaks appear (see solid line in Fig. 2a). The first peak is located close to the upper edge of the gap while the second is in the middle of the gap. These results are in excellent agreement with the experimental results of the San Diego group [12, 15]. In the case of the third nearest neighbor defects (Fig. 2b), we also have two peaks which are located at nearly the same frequencies as in the previous case but the second peak in the middle of the gap is much smaller. It seems that the impurity state, which corresponds to that peak, is more delocalized in the second nearest neighbors defect case as a result of the interaction between the two defect states.

2.2. METALLIC PBG

In this section, we study 2D systems consisting of infinitely long metallic cylinders parallel to the z-axis, embedded in air and forming a square lattice with lattice constant a. The system has a finite thickness, L, along the y-axis and is infinite along the x-axis (similar systems with dielectric cylinders have been studied in Refs. 7, 9, and 12). The k-vector of the incident EM waves forms an angle, θ, with the y-axis and the E-field is either parallel or perpendicular (s- or p-polarization) to the z-axis. Each unit cell is divided into 20×20 cells resulting in a convergence of better than 3%.

Solid lines in Fig. 3 shows the transmission and absorption of s-polarized waves for a system consisting of metallic cylinders with filling ratio f=0.1, a=1.27 μm, and L=2a. For waves propagating along the (1,0) direction ($\theta = 0°$), there are two gaps in the transmission separated by a sharp peak located at around 125 THz. The lowest gap extends from zero up to a cut-off frequency, ν_c=125 THz and is very sharp, since there is a -23 dB per unit cell drop in the transmission at 50 THz. This cut-off frequency is an important feature of s-polarized waves propagating in 2D metallic PBG which distinguishes it from the corresponding case of 2D dielectric PBG. The second weak gap (from around 125 THz to 200 THz) is less sharp with -9 dB per unit cell drop in the transmission at 160 THz. The absorption (Fig. 3b) increases almost linearly with the frequency, except for a sharp peak at 125 THz. Note there is a peak in the transmission at around the same frequency. This is the result of the fact that the wave can actually propagate through the entire system while, for frequencies inside the gaps, the wave is actually reflected. Thus, one expects the absorption to be smaller in the frequencies where the gaps appear. Increasing the incident angle, the second gap tends to disappear while the first gap survives and it actually becomes wider; in particular, the cut-off frequency, ν_c is 140 and 150 THz for $\theta = 30°$ and $40°$, respectively. Also, the absorption becomes smaller (especially at low frequencies) as the incident angle increases.

The behavior of the transmission is different for the p-polarization (see dotted lines in Fig. 3). For $\theta = 0°$, there is only a small drop in the transmission at around 115 THz; this drop tends to disappear as the incident angle increases while other sharper drops appear in higher frequencies (see Fig. 3). In other words, the transmission profiles for the p-polarized EM waves propagating in metallic PBG structures are similar to the profiles of EM waves propagating in PBG materials consisting of positive and frequency independent dielectric constants. Absorption increases almost linearly as a function of frequency, and, in general, it increases as the incident angle increases. In order to explain the previous results, we make the plausible assumption that the cylinders form infinitely long metallic plates parallel to the x,y-plane with small thickness separated by d. For s-polarized waves, the previous assumption leads to a simple waveguide model with dimension d. In this case, the allowed propagating modes are given by the relation:

$$\nu_m = \frac{mc}{2d\sqrt{\epsilon_o}} \tag{2}$$

where m=1,2,3,..., c is the velocity of light in the air and ϵ_o is the dielectric constant for the material between the metallic plates. For ϵ_o=1 and assuming that d is roughly equal to the lattice constant, a=1.27 μm, the lowest allowed mode is $\nu_1 = 118$ THz which is in good agreement with the cut-off frequency of 125 THz for the s-polarization (solid lines in Fig. 3). Also, the previous formula (Eq. 2) suggests that the cut-off frequency depends on the dielectric constant of the surrounding medium (ν_c depends on $\epsilon_o^{1/2}$) which is also in agreement with the TMM results. For the p-polarization, the waveguide model does not apply and therefore no cut-off frequency is obtained in agreement with the TMM results (Fig. 2). Note that the results are qualitatively similar to those obtained from the electromagnetic theory of gratings [26]. However, in the case of gratings, one is dealing with one layer of perfectly arranged metallic cylinders while in the present case, we are studying cases with more than one layer as well as cases in which defects have been introduced in an initially periodic system.

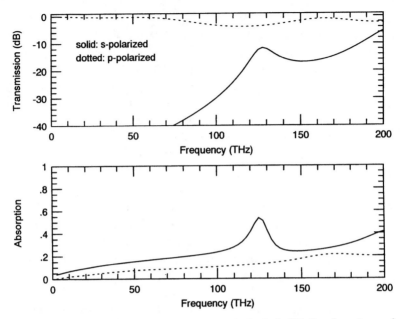

Figure 3. Transmission and absorption for s-polarized (solid lines) and p-polarized (dotted lines) EM waves propagating in a 2D square lattice consists of metallic cylinders with filling ratio f=0.1 embedded in air; the lattice constant is a=1.27 μm and the thickness is L= 2a; $\theta= 0°$.

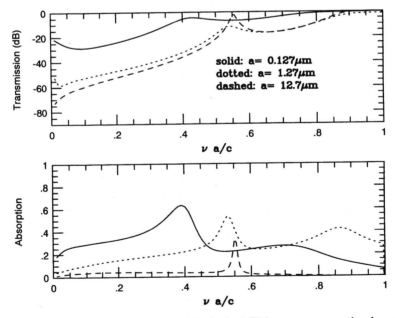

Figure 4. Transmission and absorption of s-polarized EM waves propagating in a system similar with the one described in Fig. 1. Solid, dotted, and dashed lines correspond to a= 0.127, 1.27, 12.7 μm, respectively; $\theta= 0°$.

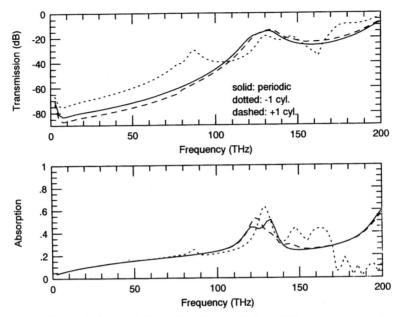

Figure 5. Transmission and absorption for s-polarized EM waves propagating in a 2D square lattice consisting of metallic cylinders with filling ratio f=0.1 embedded in air; the lattice constant is a=1.27 μm, the thickness is L= 3a, and θ= 0°. A supercell of width 3a has been used with periodic boundary conditions at the edges of the supercell. Results for the cases where one cylinder is added (dashed), one cylinder is removed (dotted), and the perfect lattice (solid) are shown.

Figure 4 shows the transmission and the absorption of s-polarized waves as a function of the dimensionless frequency $\nu a/c$ for L= 2a, f= 0.1, and $\theta = 0°$. In contrast to the dielectric PBG materials, the results change as we scale the dimensions of the structure. The dimensionless cut-off frequency $\nu_c a/c$ is 0.42, 0.54, 0.55 for a=0.127, 1.27, 12.7 μm, respectively. This indicates a convergence of the dimensionless cut-off frequency to a constant value, so, the cut off frequency in the microwave and millimeter regions can be predicted. Comparison of the transmission for a=1.27 and 12.7 μm (dotted and dash lines in Fig. 4) shows that the peak in the transmission at the cut off frequency becomes sharper as the lattice constant increases. The absorption (Fig. 4) becomes smaller as the lattice constant increases; in particular, the absorption at $\nu a/c$= 0.2 is 0.31, 0.13 and 0.04 for a=0.127, 1.27, and 12.7 μm, respectively. For a constant value of $\nu a/c$, the ratio of the skin depth over the lattice constant, a, is inversely proportional to the square root of a (this is correct for frequencies less than 100 THz). This means that by increasing the lattice constant, the part of the wave which penetrates into the metal (this part is responsible for the absorption) becomes smaller. So, the absorption must become smaller as the lattice constant increases and eventually, it will become negligible in the microwave and millimeter wave regions. Also, due to the fact that the penetration of the wave into the metal becomes smaller by increasing the lattice constant, the dimension, d, of the wave guide model (Eq. 2) decreases and the dimensionless cut-off frequencies increase in accordance with the TMM results.

In the following part of this section, we study how the transmission and absorption of s-polarized waves propagating in the periodic 2D systems, described previously, change by the introduction of defects. A supercell with a size of 3a along the x-axis has been used and periodic boundary conditions are imposed at the edges of the supercell along the x-axis; the system is finite along the y-axis with thickness L= 3a; a=1.27 μm, f=0.1 and k is along the y-axis. Two kinds of defects are introduced. First, by adding one cylinder in the system (dotted lines in Fig. 5); in this case the transmission and absorption change slightly from the corresponding values of the periodic case (compare solid and dotted lines in Fig. 5). Second, by changing the radius, r_d, of one of the cylinders; the ratio r_d/r (where r is the radius in the periodic case) gives the amount of discrepancy from the periodicity. For $r_d/r=0$ (this correspond to a case where one of the cylinders is removed; dash lines in Fig. 5) there is a peak in the transmission due to the defect at $\nu=$ 87 THz. The quality factor is defined as $Q = \nu_d/(\nu_+ - \nu_-)$; ν_d is the frequency where the peak appears and ν_+, ν_- are the frequencies above and below ν_d where the transmission is 3 dB smaller than at ν_d. For $r_d/r=0$, by increasing the lattice constant a, the Q and the transmission at the top of the peak increase, while the dimensionless frequency at the peak does not change. In particular, for a= 6.35 and 12.7 μm, the transmission at the top of the peak is -26 and -20 dB, respectively. Q is 39 and 34, respectively, while the dimensionless frequency of the defect is 0.38 in both cases. Finally, we find that Q decreases sharply and ν_d increases as the r_d/r increases. In particular, for a=1.28 μm and f=0.1, Q=2 and $\nu_d=$ 91 THz when $r_d/r=0.05$, while for $r_d/r=0.1$ there is only a small shoulder instead of a peak at 100 THz.

3. 3D Photonic Band Gap Structures

3.1. DIELECTRIC PBG

A layer-by-layer PBG crystal has been designed and fabricated [16, 17, 25]. The structure is assembled by stacking layers consisting of parallel rods with a center-to-center separation of a. The rods are rotated by 90° in each successive layer. Starting at any reference layer, the rods of every second neighboring layer are parallel to the reference layer, but shifted by a distance 0.5a perpendicular to the rod axes. This results in a stacking sequence that repeats every four layers. This lattice has face-centered-tetragonal (fct) lattice symmetry with a basis of two rods. The photonic band gap is not

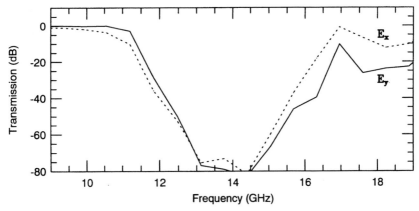

Figure 6. The transmission characteristics of EM waves propagating in a periodic layer-by-layer structure with k perpendicular to the plane of the rods (x-y plane). The systems consists of 16 layers of rods.

sensitive to the cross-section shape of the rods. This structure has a photonic band gap when both the filling ratio and the dielectric contrast meet certain requirements.

In the present case, we use circular rods with dielectric constant 9.61 and 0.318 cm in diameter; a=1.123 cm and the filling ratio of around 0.26. We define the plane of each layer as the x-y plane. Using the TMM, we calculate the transmission (Fig. 6) of a structure with four conventional unit cells along the z-axis (16 stacked layers). We use one unit cell along the x- and y-axis with periodic boundary conditions (the system is infinite along the x- and y-axes). Each conventional unit cell is divided into $7 \times 7 \times 8$ cells in order to ensure good convergence. For both polarizations, there is a gap between 11.2 and 16.9 GHz. This is in very good agreement with the experimental measurments as well as with the results of the plane wave (PW) method[25]. There are differences between the two polarizations, especially at high frequencies. It is worth noting that the difference between the two polarizations is due to the surface of the system (the rods on the surface are parallel to the z-axis). Band structure calculations of a corresponding infinite system do not show any difference between the two polarizations. The transmission at the middle of the gap is around -80 dB. Calculations for thinner structures show the drop of the transmission is about -19 dB per unit cell; by using just two unit cells (8 layers of rods), we can have a PBG with tansmission of about -38 dB which is enough for most of the possible applications.

Defects are created by removing or adding rods into the structure. We first investigate the properties of defect structure made by removing every other rod from the 8th layer. The system is otherwise similar with the one described in Fig. 7. A peak in the transmission appears at 12.58 GHz. The

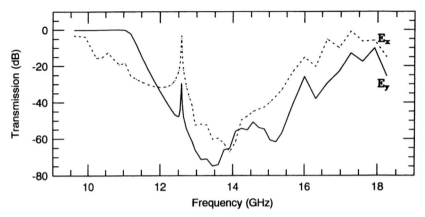

Figure 7. The transmission characteristics of EM waves propagating in a defect layer-by-layer structure constructed by removing every other rod from the 8th layer; the system consists of 16 layers of rods. k is perpendicular to the plane of the rods (x,y-plane).

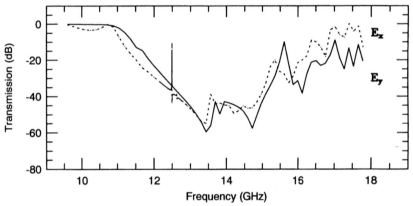

Figure 8. The transmission characteristics of EM waves propagating in a defect layer-by-layer structure constructed by adding rods in the midpoints of adjacent rods at the 5th layer; the system consists of 12 layers of rods. k is perpendicular to the plane of the rods (x,y-plane).

peak is sharper for the polarization where E is parallel to the axes of the removed rods (E_x in Fig. 7); in particular, Q=800 and the transmission at the peak is -4 dB for the wave with E//x while Q=900 and the transmission at the peak is -29.7 dB for the wave with E//y. The defect frequency and Q-factor are not very sensitive to separation between the defects. In particular, the defect frequency changes from 12.58 to 12.68 to 12.39 GHz as the separation between the removed rods increases from 2a to 3a to 4a. Also, the Q-factor increases rapidly by increasing the thickness of the PBG crystal. For a defect structure with 8 unit cells (32 layers) made by removing every other rod from the 16th layer, calculations predict a defect frequency at 12.61 GHz with a Q-factor more than 1.4×10^6. This suggests the possibility of building cavity structures with very high Q-factors from this geometry. We can also adjust the defect frequency by changing the radius of the rods. In a 4-unit cell system (16 layers), we assume that every other rod in the 8th layer has radius r_d instead of r. A defect peak inside the gap appears when r_d/r is around 0.65. By decreasing r_d/r, the defects frequency increases and for $r_d/r=0$ (this is the case where every other rod has been removed; see Fig. 7) the defect frequency is 12.58 GHz as discussed previously.

We also investigated defects obtained by adding pieces of rods or whole rods into the empty space of the PBG crystal (Fig. 8). In particular, an extra dielectric rod was inserted halfway between each lattice rod in the 5th layer of a structure that consisted of a 3-unit cell (12 layers of rods). For both polarizations, a sharp defect peak apears at 12.5 GHz with a Q-factor around 6×10^4. The transmission at the peak is slightly higher for the case where E is parallel to the added rods; it is -11 dB for E parallel to the added rods and -14 dB for E perpendicular to the added rods.

More theoretical and experimental results related to the layer-by-layer PBG structure are given in the paper by Ozbay et. al. in another chapter of this book.

3.2. METALLIC PBG

We study first 3D systems consisting of isolated metallic scatterers embedded in air (cermet topology[27, 28]). Figure 9 shows the transmission and absorption of EM waves propagating in simple cubic (s.c.) lattice consisting of metallic spheres with filling ratio f=0.03. The system is infinite along the x and y directions, while its thickness along the z-axis is L=4a and the incident waves with $\vec{k}$ along the z-axis. The results for both polarizations are the same, due to the symmetry of the lattice. For the present as well for all the following cases, each unit cell is divided into $10 \times 10 \times 10$ cells, which gives a convergence of better than 5% for the periodic cases, and better than 15% for the defect cases. There are two drops in the transmission (Fig. 9); the first around $\nu a/c=0.45$ and the second (and sharpest one) around 0.85. The wavevector, $\vec{k}$, parallel to the z-axis corresponds to the $\Gamma - X$ direction in the k-space; in this case, we expect the first gap to appear at the edge of the zone (in the X point) for $\nu a/c$ about 0.5 which is slightly higher than the frequency where the first drop in the transmission appears in this direction (Fig. 9). Due to the small filling ratio, there is no full band gap since the gaps in different directions do not overlap. We find similar results for fcc, bcc, and diamond structures with isolated metallic spheres or cubes. For the cases where the metal forms isolated scatterers,

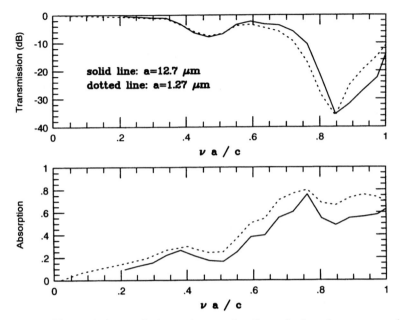

Figure 9. Transmission and absorption vs the dimensionless frequency $\nu a/c$ for EM waves propagating in a 3D s.c. lattice consisting of metallic spheres with f=0.03; L= 4a and $\theta= 0°$. Solid and dotted lines correspond to a= 1.27 and 12.7 μm, respectively.

the results are similar to those of the dielectric PBG materials. The present results for the isolated metallic scatterers are in agreement with the results of a recent work [29] in which monolayers consisting of metallic spheres with radius between 10-100 nm were studied.

From Fig. 9, we can also determine what happens as we scale the dimensions of the system and assume that the filling ratio of the system remains the same. By comparing the results for two lattice constants a= 1.27 and 12.7 μm (dotted and solid lines in Fig. 9), we find the transmission is almost the same for both cases as long as $\nu a/c$ is less than about 0.55. For higher frequencies, the transmission of the a=12.7 μm case is slightly higher than the one for the a=1.27 μm case. The absorption, however, is always smaller for the a=12.7 μm case. By increasing the lattice constant, the frequency decreases and the absolute value of the dielectric constant of the metal (see Eq. 1) becomes larger; this means the metal reflects more (and absorbs less) of the power. For this reason, the absorption decreases as the lattice constant increases. Also, for any lattice constant, there is a drop in the absorption at frequencies where there is a drop in the transmission as a result of the fact that only the few first layers of the material contribute to the absorption.

We now turn to the more interesting case of structures in which the metal forms a network; we shall refer to this case as a network topology

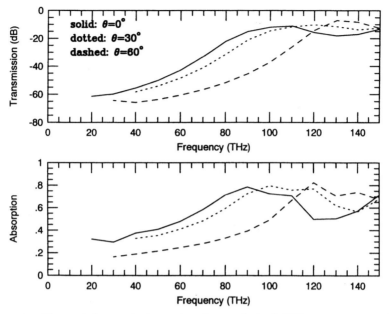

Figure 10. Transmission and absorption for s-polarized EM waves propagating in a 3D s.c. lattice consisting of metallic tetragonal rods connecting nearest neighbors with f=0.03 and L= 4a; a= 1.27 μm. Solid, dotted, and dashed lines correspond to θ= 0°, 30°, and 60°, respectively.

[27, 28]. Figure 10 shows the transmission and absorption for a system consisting of metallic tetragonal rods connecting nearest neighbor in a s.c. lattice (see Ref. [30] for more details about this structure). The system is infinite along the x and y directions while its thickness along the z-axis is L=4a; the lattice constant is a=1.27 μm and the filling ratio is f=0.03. For incident waves along the z-axis (in this case, the results are identical for both polarizations), there is a sharp drop in the transmission from zero up to a cut-off frequency, ν_c=105 THz; the drop at ν= 30 THz is 15 dB per unit cell. There is also a smaller drop (4.5 dB per unit cell) of the transmission between 110 THz and 160 THz. By increasing the incident angle, the cut-off frequency moves to higher frequencies (ν_c is 120 and 130 THz for θ = 30° and 60°, respectively). Therefore, the first (and sharpest) gap survives for any incident angle but the second (and smaller) disappears by increasing the incident angle. As in the isolated metallic scatterers case, the absorption increases as the frequency increases and it reaches a maximum in the frequencies where the transmission reaches a maximum. But, in general, the absorption is higher for the case of the network topology (compare Fig. 9 and 10). Since the skin depth for frequencies around 100 THz is much smaller than the thickness of the metal, we can assume the absorption is proportional to the surface area of the metal. But, for the

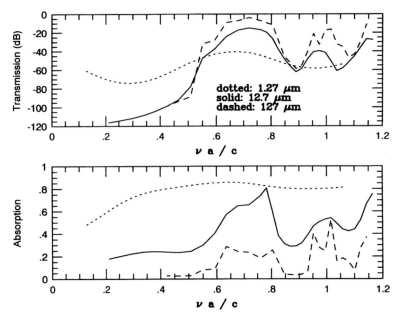

Figure 11. Transmission and absorption vs the dimensionless frequency $\nu a/c$ for EM waves propagating in a diamond lattice consisting of metallic cylinders connecting nearest neighbors with f=0.04, L= 4a and θ= 0°. Solid, dotted, and dashed lines correspond to a= 1.27, 12.7, and 127 μm, respectively.

same filling ratio, the surface of the metal is higher in the network topology, thus, the absorption will be higher for the network topology. As in the case of s-polarized EM waves in a 2D system, we can use the same waveguide model to predict the cut-off frequency. In contrast to the 2D case, the waveguide model for the 3D case is bounded for both polarizations, so, we expect the same cut-off frequency for both of them. The predicted cut-off frequency (Eq. 2) is 118 THz, for d= 1.27 μm which is very close to the cut-off frequency found by the TMM (ν_c=105 THz).

Figure 11 shows the transmission and absorption as a function of the dimensionless frequency $\nu a/c$, for metallic circular rods connecting nearest neighbors in a diamond structure (more details about this structure can be found in Ref. 4). The filling ratio is f= 0.04, the thickness of the system is L= 4a and the incident angle is $\theta = 0°$. For a=1.27 μm there is a peak of the transmission at around $\nu a/c$= 0.7 which separates two relatively small gaps. By increasing the lattice constant the transmission at the peak increases sharply; in particular, the transmission at $\nu a/c$= 0.7 is -40 dB, -16 dB, -4 dB for a= 1.27, 12.7, 127 μm. A peak of the transmission inside the second (higher) gap appears as the lattice constant increases; but, this gap does not survive as the incident angle increases. In contrast, the first gap survives for any incident angle. The absorption has the same behavior as in the s.c. case

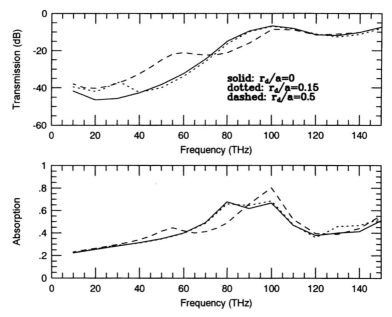

Figure 12. Transmission and absorption for EM waves propagating in a 3D s.c. lattice consisting of metallic tetragonal rods connecting nearest neighbors with f=0.03, L= 3a and $\theta = 0°$. A supercell of width 2a has been used with periodic boundary conditions at the edges of the supercell. Part of the metal, which is included in a sphere with a center in one of the crossing points of the rods at the second layer and radius r_d- has been removed. Solid, dotted, and dashed lines correspond to $r_d/a = 0$, 0.15 and 0.5, respectively.

with network topology, but, it is, in general, higher for the diamond case (compare solid lines in Figs 10, 11) This is due to the higher surface area of the metal in the diamond case. However, as the lattice constant increases, the absorption becomes smaller; the absorption at $\nu a/c = 0.5$ is 0.83, 0.24, 0.03 for a= 1.27, 12.7, 127 μm, respectively; so, we expect the absorption would be negligible in the microwave and millimeter regions. The cut-off frequency for the diamond case is different from the s.c. case ($\nu_c a/c = 0.44$, 0.7 for s.c. and diamond, respectively). Using a similar waveguide model as in the s.c. case and assuming that the size of the waveguide is given by the distance between the point $(1/4, 3/4, 3/4)a$ and the z=0 plane (which is 3a/4), we find the predicted cut-off frequency is 0.67 which is very close to the cut-off frequency of 0.60 calculated from the TMM.

In general, 3D metallic PBG structures with network topology exhibit a cut-off frequency for both polarizations below which there are no propagating modes. This feature puts them in the same category with the s-polarized waves propagating in 2D metallic PBG. But, the features of 3D metallic PBG structures with isolated metallic scatterers are similar to the features of the dielectric PBG as well as to the features of p-polarized waves in 2D metallic PBG. For all these cases there is no cut-off frequency.

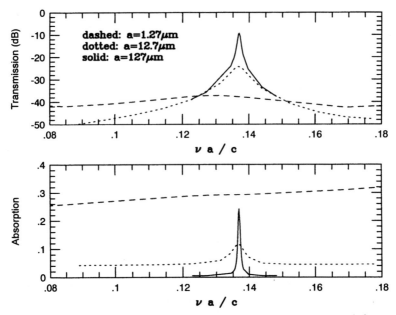

Figure 13. Transmission and absorption for EM waves propagating in a defect-structure similar with the one described in Fig. 8 with $r_d/a= 0.15$. Solid, dotted, and dashed lines correspond to a= 1.27, 12.7, and 127 μm, respectively.

We also study the effect of the introduction of defects in a periodic lattice. We use a s.c. lattice consisting of metallic tetragonal rods connecting nearest neighbors (network topology) with f= 0.03, a=1.27 μm, and the surrounding medium is air. A supercell has been used with width 2a along the x- and y-axis and periodic boundary conditions are imposed at the edges of the supercell; the system is finite along the z-axis with thickness L= 3a. A defect is introduced by removing part of the metal, which is included in a sphere of radius r_d centered at one of the crossing points of the rods in the second layer. Figure 12 shows the transmission and absorption for such a defect structure and incident waves with $\vec{k}$ parallel to the z-axis and a= 1.27 μm. Once again the results are identical for both polarizations due to the symetry of the structure. For r_d/a=0.15, a small peak in the transmission appears at around 31 THz; the quality factor is very small (Q= 3) and the transmission at the top of the peak is also small (-37.2 dB). Apart from the frequency region around the defect, the transmissions of the defect and the periodic structures are almost the same (compare dotted and solid lines in Fig. 12). For r_d/a=0.5 (dashed line in Fig. 12), there is a peak in the transmission at higher frequency (60 THz) which is even wider and with higher transmission at the top of the peak. Thus, one can adjust the frequency of the defect inside the gap by just changing the volume of the removed metal; the higher the amount of the removed metal, the higher the frequency where the defect peak appears. Studies in dielectric PBG materials [5] have shown that a defect band emerges from the lower edge of the gap and approaches the upper edge of the gap as the volume of the removed dielectric material increases. The behavior of the defect band is similar in the present case, despite the fact that the lower edge of the gap is actually at zero frequency. The absorption for $r_d/a=0.15$ is almost identical with the absorption of the periodic case except for a small maximum at the frequency where the defect peak appears (hardly noticed by comparing dotted and solid lines in Fig. 12). However, the differences in the absorption between the $r_d/a=0.5$ and the periodic cases (compare dashed and solid lines in Fig. 12) are more obvious with a more prominent peak of the absorption in the $r_d/a=0.5$ case at the frequency where defect appears (60 THz). Since the light is trapped around the defect region, one actually expects the absorption will be higher in this case.

Figure 13 shows transmission and absorption as functions of the dimensionless frequency $\nu a/c$ for a defect structure (similar to the one described in the previous paragraph) with $r_d/a=0.15$. As we mentioned earlier, the results change as we scale the dimensions of the structure due to the frequency dependence of the dielectric constant (Eq. 1). By increasing the lattice constant, the transmission at the top of the defect peak, T_d, and the Q factor increase by orders of magnitudes; in particular, $T_d=$ -37, -24, and -9 dB while Q= 3, 29, and 137 for a= 1.27, 12.7, and 127 μm. However, the dimensionless frequency of the defect, $\nu_d a/c$, increases slightly, reaching a constant value at high lattice constants ($\nu_d a/c=$ 0.1312, 0.1367, and 0.1369 for a= 1.27, 12.7, and 127 μm). The absorption at $\nu_d a/c$, on the other hand, exhibits a more peculiar behavior. In general (see Fig. 13), there is a peak in the absorption exactly at $\nu_d a/c$ (although, this peak is hardly noticeable for the a=1.27 μm) which becomes sharper as the lattice constant increases. As we mentioned earlier for the periodic cases, by increasing the lattice constant, the overall absorption decreases. Similarly in the present case for frequencies well above or below the defect frequency, the overall absorption (we call it background absorption) decreases as the lattice constant increases. At the defect frequency, however, the wave becomes more localized as the lattice constant increases, because the metal becomes a better reflector, hence, the transmission and absorption peaks become sharper.

4. Absorption

In the optical wavelength region, the dielectric constant of most of the materials has an appreciable imaginary part. In the following section, we examine how the gaps are affected by the absorption. We choose GaAs as

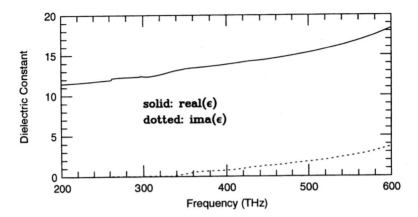

Figure 14. The experimental values of the real (solid line) and imaginary (dashed line) part of the dielectric constant of GaAs.

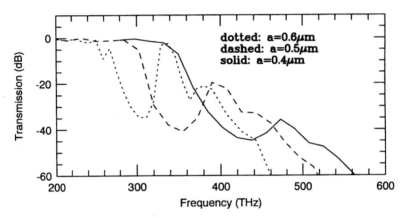

Figure 15. The transmission vs frequency for EM waves propagating along the (001) direction in a "3-cylinder" structure consisted of GaAs with f=0.221 and lattice constant a=0.4 μm (solid line), 0.5 μm (dashed line), and 0.6 μm (dotted line). The slab thickness is 4a.

a PBG material, since it is a possible candidate material for the development of PBG "crystals" in the optical wavelength region. However, the same conclusions can be reached for other absorbing materials. Recently, Yablonovitch [31] tried to develop a PBG "crystal" consisting of an array of microscopic crisscrossing holes, 0.5 μm in diameter and 0.7 μm apart, drilled into GaAs; this crystal has the "3-cylinder" structure [4, 5]. Theoretically, the first gap is expected to be between 200 and 270 THz, which is in the infrared region. As we can see from Fig. 14, the imaginary part of the dielectric constant is negligible in that region; but, as the frequency approaches the optical region, the imaginary part of the dielectric constant increases and becomes appreciable above 340 THz.

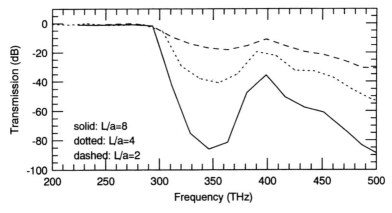

Figure 16. The transmission vs frequency for EM waves propagating along the (001) direction in a "3-cylinder" structure consisted of GaAs with f=0.221 and lattice constant a=0.5 μm The slab thickness is 2a (dashed line), 4a (dotted line), and 8a (solid line).

Using the experimental values of the GaAs dielectric constant [32], we calculated the transmission vs the frequency for three different lattice constants, a= 0.4, 0.5, and 0.6 μm (Fig. 15). In all cases, the structure is the "3-cylinder" structure with air holes in GaAs; the filling ratio of GaAs is 0.221 and the thickness of the slab is 4a. For a=0.6 μm (dotted line in Fig. 15) the first gap appears between 270 and 330 THz; since the imaginary part of ϵ is negligible in that region, the gap is not affected by the absorption. But above 340 THz, there is a continuous drop of the transmission as a result of the increase of the imaginary part of ϵ. For a=0.5 μm (dashed line in Fig. 15), the first gap starts to become affected by the absorption. The transmission at the upper edge of the gap is -20 dB, while for a=0.6 μm it is -2 dB. The effect of the absorption is even stronger for a=0.4 μm (solid line in Fig. 15); in this case, the transmission at the upper edge of the gap is -36 dB, while it is -44 dB in the center of the gap; thus, the drop due to the absorption becomes comparable with the drop of the transmission due to the underlying structure. For even smaller lattice constants, there is no recovery of the transmission at the upper edge of the gap and the transmission continues dropping. Thus, one must be very careful in designing photonic band gap materials in regions where ϵ is a complex number.

In nonabsorbing cases, it is commonly accepted that the examined PBG "crystal" must be as thick as possible because the transmission inside the gap is thickness dependent. So, for the thicker "crystal" the transmission will be lower and the gap more well developed. In Fig. 16, the transmission vs the frequency is given for a PBG "crystal" with a "3-cylinder" structure consisting of holes in GaAs with filling ratio f=0.221, a=0.5 μm, and three different thicknesses, ℓ=2a, 4a, 8a. At 300 THz, which is close to the lower

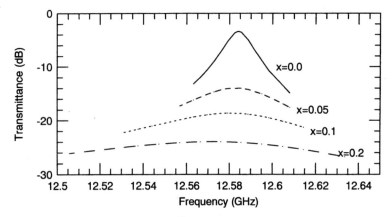

Figure 17. The transmission of EM waves propagating in a system similar with the one described in Fig. 7. The dielectric constant of the rods is $\epsilon = 9.61 + ix$. E is parallel to the axes of the removed cylinders.

edge of the gap, the transmission is -5 dB for $\ell = $ 2a, 4a, but it is -25 dB for $\ell = $8a. At 320 THz, which is well inside the gap but still not affected by the absorption, the thickness dependence of the transmission is more obvious; in particular, the transmission is -10, -30, and -60 dB for ℓ=2, 4, and 8a, respectively. These results support the previous statement for nonabsorbing PBG "crystals." But, above the upper edge of the gap (at 400 THz), the transmission is also thickness dependent because of the absorption; it is -10, -20, and -35 dB for ℓ=2, 4, and 8a, respectively. Assuming an experiment in which the background noise is -40 dB, the transmission at the upper edge of the gap cannot experimentally be distinguished from the noise; in contrast, for ℓ=2, and 4a, the transmission at the lower edge of the gap can be well measured and distinguished from the noise. Thus, for absorbing PBG materials, the thicker slabs are not the best candidates for giving gaps. In contrast with the nonabsorbing cases, thicker slabs give better photonic band gaps.

In order to study the effect of the absorption on the peak transmission due to defects, we calculated the transmission of a layer-by-layer structure with a defect. The structure is similar with the one described in Fig. 17, in which we removed every second rod from the 8th layer (the system contains 16 layers of rods). We assume the dielectric constant is given by: $\epsilon = 9.61 + ix$ (see Fig. 17). Increasing the imaginary part of the dielectric constant, the Q, as well as the transmission at the peak, decrease. In particular, Q= 800, 262, 163, and 90 for x= 0, 0.05, 0.1, and 0.2, respectively, while the transmission at the peak is -3.4, -14.0, -18.7, and -24.0. The introduction of the absorption makes the peak wider and the transmission on the peak smaller.

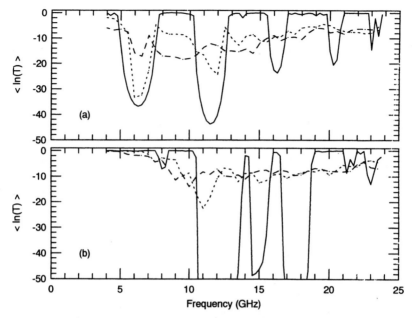

Figure 18. $< ln(T) >$ vs the frequency for 2D disordered media with d_r=0, 0.2 and 0.4 (solid, dotted and dashed lines, respectively) and s- and p-polarized waves (a and b panels). The filling ratio of the cylinders is around 0.29; the dielectric constants are 10 and 1 in the cylinders and in the surrounding medium, respectively; the lattice constant is a=1.28 cm and the thickness of the system is L=25 a.

5. Disorder

5.1. 2D DIELECTRIC PBG

As we pointed out in the introduction for 2D systems, all the states will become localized even with the smallest amount of disorder and the localization length, ℓ, is given by [23, 24]:

$$\ell = -\frac{2L}{< ln(T) >} \qquad (3)$$

where the $< ln(T) >$ is the logarithmic average of the transmission over the different configurations. In all the following cases, the thickness of the system is L and the lattice constant of the corresponding periodic system is a=1.28 cm (the results can be scaled in any lattice constant); the average lattice is a square with cylinders of circular cross-section and filling ratio, f, and the waves propagate along the y-axis while the axis of the cylinders is along the z-axis. A supercell which include three cylinders along the x-axis is used with periodic boundary conditions at the edges of the supercell; the results are almost the same for a supercell with four cylinders along the x-axis which, according to the scaling hypothesis for localization [23, 24], show the waves are localized. Due to computer time constraints, $< ln(T) >$ is calculated from an average over 10 configurations which gives less than a 20% error in the value of $< ln(T) >$. Finally, each unit cell is divided into 10×10 points; calculations with higher divisions per unit cell have less than a 5% difference with the present results.

Figure 18a shows the $< ln(T) >$ vs the frequency for a system consisting of cylinders with dielectric constant $\epsilon_i = 10$ and f= 0.28 embedded in air; the thickness is L=25a, the E-field of the wave is parallel with the axis of the cylinders (s-polarized). For the periodic case (solid line in Fig. 18a), there are two sharp drops in the transmission which correspond to the first two gaps and some smaller drops which correspond to the higher gaps. With the introduction of the first type of disorder where the radius of the cylinders changes randomly, the higher gaps disappear even with a small amount of disorder ($d_r = 0.2$, dotted line in Fig. 18a), while the first two gaps survive; especially the first one survives for d_r as high as 0.4. For the p-polarized waves (Fig. 18b), there are three large drops in the transmission of the periodic case (see solid line in Fig. 18b), but they do not correspond to full gaps since, as we have mentioned in the previous section, there are no full gaps for this polarization and for $\epsilon=10$. For this reason, these drops disappear easily with the introduction of the disorder; for $d_r =0.2$ (dotted line in Fig. 18b), only the first drop survives, although much smaller, while for $d_r =0.4$ (dashed line in Fig. 18b), there are no drops.

There are some common characteristics of the average transmission for both polarizations. For relatively high frequencies (ω higher than about 20 GHz in our case, or, in general, ω higher than about c/a), $< ln(T) >$ is almost independent of the frequency; we shall refer to this value as $< ln(T) >_s$, the saturated value of $< ln(T) >$. Also, this value is almost independent on the amount of the disorder within the accuracy of our statistical average. In particular, for p-polarized waves, $< ln(T) >_s$ is about -6 (Fig. 18b) for both $d_r = 0.2$ and 0.4 while for the s-polarized waves, $< ln(T) >_s$ is about -6.5 (Fig. 18a). Similar conclusions have been found for one-dimensional random bilayer systems[33] and they have been explained as follows. At high frequencies (small wavelengths), the phase coherence between the scattering at different interfaces is lost and the dominant factor is the scattering from each interface. But, the transmission and reflection coefficients at a sharp interface are frequency independent, therefore the $< ln(T) >$ and ℓ are also frequency independent at high frequencies [33]. In the present 2D case, for high frequencies, one expects the geometric limit to be reached and the dominant scattering factor to be from the interface, so, $< ln(T) >$ will be frequency independent. On the other hand, at low frequencies (high wavelengths), the wave is not affected by the inhomogeneity of the system, but it essentially "sees" a uniform medium with an effective dielectric constant, so we expect $< ln(T) >$ tends to zero as the frequency approaches zero and it is independent of the thickness of the system and the amount of the disorder. This is clearly shown for the p-polarization in Fig. 18b; $< ln(T) >$ is almost the same as in the perfect case for frequencies less than 6.5 and 5 GHz for the $d_r = 0.2$ and 0.4 cases, respectively. For the s-polarization, this happens for frequencies smaller than 4 GHz which is the lowest frequency in our calculations.

At intermediate frequencies, the drops of the $< ln(T) >$ are actually reminiscent of the corresponding gaps of the periodic case. Our calculations show the wider the gap of the periodic case, the larger the amount of disorder needed to close it. At frequencies inside the gaps, the $< ln(T) >$ and the localization length increase by increasing the disorder as a result of the fact that the density of states (DOS) become higher due to the creation of localized states inside the gap [33-37].

5.2. 3D DIELECTRIC PBG

In this section, we study PBG materials with a diamond lattice symmetry. The k-vector of the EM waves is along the (001) direction and the thickness of the system along that direction is 20 a (a is the lattice constant of the diamond lattice which is 1.27 cm in the following cases). Along the (100) and (010) directions, we use a supercell which contains M conventional cubic cells; periodic boundary conditions are assumed along the edges of the supercell. The disorder is introduced by moving the lattice points from their ideal positions in a diamond lattice, so the x, y, and z components of the position of the i-th lattice point in the random system differ from those of the periodic case by $\gamma_x a$, $\gamma_y a$, and $\gamma_z a$, respectively; γ_x, γ_y, and γ_z are random variables uniformly distributed over the interval $[-d, d]$. Due to computer time constraints, $< ln(T) >$ is calculated from an average over 8 configurations. Finally, each conventional unit cell is divided into $7 \times 7 \times 7$ points; calculations with higher divisions per unit cell having less than a 5% difference with the present results.

We first study a diamond lattice consisting of air spheres in a material with dielectric constant 10. The filling ratio of the air spheres is 0.77 and d=0.1. It is interesting to point out that each component of the vector which connects nearest neighbors is 0.25 a. Figure 19 shows the localization length, ℓ, as a function of the frequency for two different widths M=2 and 3. Although the results are not very convergent relative to the width M (except for the frequency 12 GHz), it is clearly seen that the gap (indicated by the vertical dashed lines) has almost disappeared. In its place, a region of localized states has appeared with a localization length of around 12a. Also note that the wave is delocalized at 15 GHz (at least for the particular

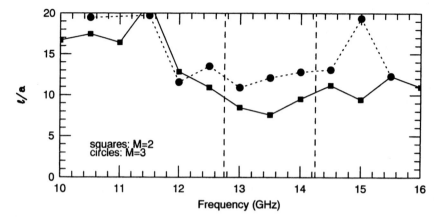

Figure 19. The localization length vs frequency for a PBG with diamond lattice symmetry consisting of air spheres with filling ratio f=0.77 in a material with dielectric constant 10. The lattice constant is a=1.28 cm and the strength of the disorder is d=0.1 (see text for more details).

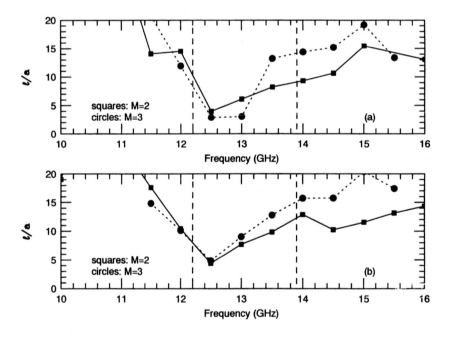

Figure 20. The localization length vs frequency for a PBG with diamond lattice symmetry consisting of rods connecting nearest neighbors surounded by air. The dielectric constant of the rods is 10 and their filling ratio is 0.25. The lattice constant is a=1.28 cm and the strength of the disorder is d=0.15 (a) and 0.2 (b) (see text for more details).

thickness that we study here), indicating that localized states appear in the region where the gap of the periodic case used to be.

We also study structures consisting of cylindrical rods connecting nearest neighbors in a diamond lattice. The dielectric constant of the rods is 10, the filling ratio is 0.25, and the surrounding medium is air. Each lattice point has the same set of nearest neighbors points in the periodic as well as in the disordered cases. For d=0.15 (Fig. 20a), the convergence relative to the width M is good for frequencies between 12 to 13 GHz, while it is poor for frequencies higher than 13.5 GHz. This indicates the wave has a large localization length (larger than the thickness of our system) or it is delocalized for frequencies higher than 13.5 GHz. However, there is a minimum in the localization length between 12.5 to 13 GHz which is a reminiscent of the gap of the periodic case. For d=0.2 (Fig. 20b), the convergence is good in a wider frequency region (between 11.5 and 13.5 GHz). There is also a minimum in the localization length at around 12.5 GHz, which shows the gap partially survived even for such a high amount of disorder (note that for d=0.25, the maximum displacement due to the disorder could be as high as the nearest neighbor distance). This is a surprising result which we attributed to the connectivity of the structure. However, further studies are needed in order to prove this statement.

We now focus on the frequency region between 11.5 and 12.5 GHz and M=3 (dotted lines in Figs 20a,b). For both d=0.15 and d=0.2, the localization length as a function of the frequency is given almost accurately by a line. The slop of this line gets bigger as the disorder increases, while there is a frequency (at 12.2 GHz which is also the lower edge of the gap) where the localization is almost the same for both d=0.15 and d=0.2. This means the states become more localized for frequencies smaller than 12.2 GHz.

6. Conclusions

The transfer matrix method has been used successfully for the calculation of the properties of PBG materials with dielectric and metallic components. The effects of the absorption and disorder are also examined.

For 2D systems consisting of metallic cylinders, there is considerable difference between the two polarizations. For p-polarized waves, the results are qualitatively similar to the dielectric PBG systems. Propagating modes are interrupted by band gaps appearing close to the edges of the Brillouin zone. On the other hand, for s-polarized waves, there is a cut off frequency ν_c. There are no propagating modes for frequencies between zero and ν_c, so the transmission has a very sharp drop in this frequency range. Above ν_c, there is the usual behavior of bands interrupted by gaps.

For 3D metallic PBG structures, the results are very sensitive on the topology of the structure. Systems with isolated metallic scatterers (cermet topology) exhibit similar behavior to the dielectric PBG materials. But, for metallic scatterers forming a continuous network (network topology), there are no propagating modes for frequencies smaller than a cut-off frequency for both polarizations and for any incident angle. Note that for dielectric PBG materials, there is no cut-off frequency for both types of the topology. We have shown this behavior, in both 2D and 3D cases, can be explained using a simple waveguide model where the ν_c is predicted with good accuracy. This cut-off frequency is well below the plasma frequency and is related to the structure of the system.

In all the periodic cases studied, the absorption can be largely neglected for metallic PBG structures with lattice constants, a, less than about 100 μm which correspond to frequencies below about 1 THz. Therefore, for frequencies less than about 1 THz, wide stop-band filters constructed from periodic metallic PBG materials can be used as alternatives to similar filters constructed from dielectric PBG.

By breaking the connections in the 3D metallic networks, defect states appear below the cut off frequency, resulting in a peak in the transmission. The smaller the volume of the removed metal, the smaller the frequency where the defect peak appears. As we explain in the next paragraph, this is a very interesting feature of the metallic PBG which, in connection with the fact that the filling ratio of the metal can be less than 0.01, can be used in the construction of narrow band-pass filters smaller in size than those constructed from dielectric PBG. By increasing the lattice constant, the Q factor and the transmission at the defect peak increase by order of magnitudes, while the dimensionless defect frequency remains almost constant. The absorption at the frequency where the defect peak appears increases as the lattice constant increases, an effect which may create problems in some of the possible applications.

An important advantage of metallic PBG structures is they could be smaller in size and lighter than the corresponding dielectric PBG materials. We can see this in the following example. A narrow band-pass filter operating at 323 GHz can be constructed using a metallic PBG with a defect structure similar with the one described in Fig. 9; this metallic structure would have the lattice constant, a=127 μm (see solid lines in Fig. 9) and the filling ratio of the metal less than 0.03. For a similar band-pass filter constructed from a dielectric PBG material, one needs a band gap at around 323 GHz. Assuming the midgap frequency of a dielectric PBG is given by $\nu a/c \equiv 1$, the lattice constant of the dielectric PBG would be around 929 μm and the filling ratio of the dielectric material would be higher than about 0.1. The dielectric PBG would be more than seven times larger and heavier than the corresponding metallic PBG material. Our calculations show that similar metallic PBG can be constructed with filling ratios of the metal less than 0.01 making them even more attractive for applications where the size or weight is a consideration. However, one must keep in mind that in the metallic case, there is also a considerable amount of absorption (see solid lines in Fig. 9) which could be completely neglected in the dielectric case. By decreasing γ in Eq. (1) (or, equivalently, by increasing the conductivity of the metal, σ), the absorption becomes smaller. This means the problem due to the absorption can be avoided by using a superconductor instead of a metal.

We have studied how the absorption affects the structural gaps and the possible difficulties of their experimental investigation. We have found that for absorbing cases, the transmission becomes thickness dependent for every frequency. For non-absorbing cases, the transmission is basically thickness independent, except for frequencies inside the structural gaps. This thickness dependence increases as the frequency increases. As a consequence, for a very thick absorbing system, the transmission in the upper edge of a structural gap could becomes so small that it can be impossible to find the recovery of the transmission at the upper edge of the gap since, experimentally, there is always a lower bound for a transmission measurment (noise level). Thus, for absorbing PBG materials, the thicker slab is not necessarily better in contrast with what is commonly accepted for non-absorbing PBG materials.

We have also studied 2D disordered systems which are periodic on the average. The corresponding periodic systems consist of cylinders forming a square lattice and embedded in a different dielectric medium. By introducing disorder in these periodic systems, the higher gaps, which are narrow, disappear quickly and the logarithmic average of the transmission, $< ln(T) >$, or the localization length, ℓ, becomes almost constant at relatively high frequencies (ω higher than about c/a). These high frequency values of the localization length depend on the filling ratio and can be as small as 5.2a (a is the lattice constant of the unperturbed periodic system) for the cases that we have studied. On the other hand, for low frequencies, $< ln(T) >$ is not affected by the disorder and it is close to zero which corresponds to very high localization lengths. At intermediate frequencies, there are large drops in the $< ln(T) >$ which correspond to the lowest gaps of the periodic cases. The wider the gaps of the periodic cases, the higher the amount of disorder needed to close these gaps. The gaps of the s-polarized waves are generally wider and survive a high amount of disorder, in contrast with the gaps of the p-polarized waves which are destroyed easily by the disorder. A systematic study of the optimum conditions for the appearance of the gaps has shown these conditions are fulfilled for cylinders of high dielectric material with filling ratio around 0.25 for the s-polarized waves.

In these cases, the gaps are wider and they survive even a high amount of disorder, resulting to localization lengths smaller than 5a.

Finally, we have also studied 3D disordered systems which are periodic on the average. Preliminary results show that for structures with network topology, the gaps disappear easily by breaking the network. In contrast, introducing positional disorder but keeping the network topology, the gaps survive a high amount of disorder and the corresponding localization length at frequencies close to the gap can be small.

7. Acknowledgments

We want to thank E. Ozbay, G. Tuttle, and S. McCalmont for discussions about their experimental results and R. Biswas for useful discussions. Ames Laboratory is operated for the U. S. Department of Energy by Iowa State University under contract No. W-7405-ENG-82. This work was supported by the Director of Energy Research, Office of Basic Energy Sciences and NATO Grant No. CRG 940647.

References

[1] See the special issue of the J. Opt. Soc. Amer. **B 10**, 208-408 (1993) on *Development and Applications of Materials Exhibiting Photonic Band Gaps.*

[2] See the proceedings of the NATO ARW, *Photonic Band Gaps and Localization,* ed. C. M. Soukoulis, (Plenum, New York, 1993).

[3] K. M. Ho, C. T. Chan, and C. M. Soukoulis, Phys. Rev. Lett. **65**, 3152 (1990).

[4] C. T. Chan, K. M. Ho, and C. M. Soukoulis, Europhys. Lett. **16**, 563 (1991).

[5] E. Yablonovitch, T. J. Gmitter, and K. M. Leung, Phys. Rev. Lett. **67**, 2295 (1991).

[6] H. S. Sozuer, J. W. Haus, and R. Inguva, Phys. Rev. **B45**, 13962 (1992); J. Opt. Soc. Am. **B10**, 296 (1993).

[7] P. R. Villeneuve and M. Piche, Phys. Rev. **B46**, 4969 (1992); **B46**, 4973 (1992).

[8] R. D. Meade, K. D. Brommer, A. M. Rappe, and J. D. Joannopoulos, Appl. Phys. Lett. **61**, 495 (1992).

[9] M. Plihal, A. Shambrook, A. A. Maradudin, and P. Sheng, Opt. Commun. **80**, 199 (1991); M. Plihal and A. A. Maradudin, Phys. Rev. **B44**, 8565 (1991).

[10] E. Yablonovitch, T. J. Gmitter, R. D. Meade, A. M. Rappe, K. D. Brommer, and J. D. Joannopoulos, Phys. Rev. Lett. **67**, 3380 (1991).

[11] R. D. Meade, K. D. Brommer, A. M. Rappe, and J. D. Joannopoulos, Phys. Rev. **B44**, 13772 (1991).

[12] S. L. McCall, P. M. Platzman, R. Dalichaouch, D. Smith, and S. Schultz, Phys. Rev. Lett. **67**, 2017 (1991).

[13] W. Robertson, G. Arjavalingan, R. D. Meade, K. D. Brommer, A. M. Rappe, and J. D. Joannopoulos, Phys. Rev. Lett. **68**, 2023 (1992).

[14] J. B. Pendry and A. MacKinnon, Phys. Rev. Lett. **69**, 2772 (1992); J. B. Pendry, J. Mod. Opt. **41**, 209 (1994).

[15] M. M. Sigalas, C. M. Soukoulis, E. N. Economou, C. T. Chan and K. M. Ho, Phys. Rev. **B48**, 14121 (1993).

[16] K. M. Ho, C. T. Chan, C. M. Soukoulis, R. Biswas, and M. Sigalas, Solid State Commun. **89**, 413 (1994).

[17] E. Ozbay, E. Michel, G. Tuttle, M. Sigalas, R. Biswas, and K. M. Ho, Appl. Phys. Lett. **64**, 2059 (1994).

[18] E. Ozbay, E. Michel, G. Tuttle, R. Biswas, and K. M. Ho, J. Bostak, and D. M. Bloom, Opt. Lett. **19**, 1155 (1994).

[19] M. M. Sigalas, C. M. Soukoulis, C. T. Chan and K. M. Ho, Phys. Rev. **B49**, 11080 (1994).

[20] A.R. McGurn and A.A. Maradudin, Phys. Rev. **B 48**, 17576 (1993).

[21] D. R. Smith, S. Shultz, N. Kroll, M. Sigalas, K. M. Ho and C. M. Soukoulis, Appl. Phys. Lett. **65**, 645 (1994).

[22] J. D. Jackson, *Classical Electrodynamics*, (Wiley, New York, 1975).

[23] C. M. Soukoulis, I. Webman, G. S. Grest, and E. N. Economou, Phys. Rev. **B26**, 1838 (1982).

[24] A. MacKinnon and B. Kramer, Z. Phys. **B53**, 1 (1983).

[25] E. Ozbay, A. Abeyta, G. Tuttle, M. Tringides, R. Biswas, C. M. Soukoulis, C. T. Chan, and K. M. Ho, Phys. Rev. **B50**, 1945 (1994).

[26] R. Pettit, editor, *Electromagnetic Theory of Gratings*, (Springer-Verlag, Berlin, 1980).

[27] W. Lamb, D. M. Wood, and N. W. Ashcroft, Phys. Rev. **B21**, 2248 (1980).

[28] E. N. Economou and M. M. Sigalas, Phys. Rev. **B48**, 13434 (1993).

[29] N. Stefanou and A. Modinos, J. Phys.:Condens. Matter **3**, 8135 (1991).

[30] H. S. Sozuer and J. W. Haus, J. Opt. Soc. Am. **B10**, 296 (1993); H. S. Sozuer and J. P. Dowling, J. Mod. Opt. **41**, 231 (1994).

[31] E. Yablonovitch (private communication).

[32] E. D. Palik, in *Handbook of Optical Constants of Solids*, edited by E. D. Palik (Academic, London, 1985), p. 434.

[33] P. Sheng, B. White, Z.-Q. Zhang, and G. Papanicolaou, in Ref. 2; P. Sheng, B. White, Z.-Q. Zhang, and G. Papanicolaou, Phys. Rev. **B 34**, 4757 (1986).

[34] J.M. Frigerio, J. Rivory and P. Sheng, Optics Commun. **98**, 231 (1993).

[35] A.R. McGurn, K.T. Christensen, F.M. Mueller, and A.A. Maradudin, Phys. Rev. B **47**, 13120 (1993).

[36] A. Kondilis and P. Tzanetakis, Phys. Rev. B **46**, 15426 (1992).

[37] C.T. Chan, Q.L. Yu and K.M. Ho, Phys. Rev. B (submitted).

TRANSFER MATRIX TECHNIQUES FOR ELECTROMAGNETIC WAVES

J. B. PENDRY AND P. M. BELL
The Blackett Laboratory, Imperial College,
London SW7 2BZ, United Kingdom

1. Introduction

The concept of a transfer matrix is extremely simple: if we know electric and magnetic fields in the x-y plane at $z=0$, then we can use Maxwell's equations at a fixed frequency to integrate the wavefield and find electric and magnetic fields in the x-y plane at $z=c$. In fact, if we assume that both **B** and **D** have zero divergence, we need only know two components of each field: let us say,

$$F(z = 0) = \left[E_x(z = 0), E_y(z = 0), H_x(z = 0), H_y(z = 0) \right] \tag{1.1}$$

Then,

$$F(z = c) = T(c,0)F(z = 0) \tag{1.2}$$

defines the transfer matrix, **T**.

Since the transfer matrix defines how waves cross a slab of material, it is closely related to the transmission and reflection coefficients of the slab. Also, if the slab is infinitely repeated throughout the material in a periodic array, we can impose the Bloch condition,

$$\exp(iKc)F(z = 0) = F(z = c) = T(c,0)F(z = 0) \tag{1.3}$$

and calculate the band structure, $K(\omega)$, from the eigenvalues of **T**.

Implementation of this strategy on a computer requires Maxwell's equations to be reformulated on a lattice, which has been accomplished in reference [1], the devising of suitable algorithms for construction of **T** and extraction of transmission of reflection coefficients which has been done in reference [2]. The relevant computer codes can be obtained by contacting *Computer Physics Communications* who will forward them by email. The purpose of this paper is to supply detailed documentation of the programs for those who wish to develop them further.

C. M. Soukoulis (ed.), Photonic Band Gap Materials, 203–228.

204

The computational strategy is shown in figure 1.1. The subroutines documented here are named in the figure. The rest are either standard routines or trivial to write.

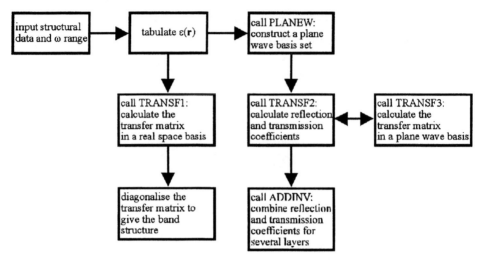

figure 1.1. Flow chart for a **T**-matrix based calculation of the electromagnetic band structure, reflection, and transmission coefficients of an arbitrary object.

2. subroutine PLANEW

objective: We are concerned with calculating the reflection or transmission of an incident plane wave into an outgoing plane wave. To do this we need a plane wave basis set for the reference medium in which our dielectric structure is embedded. The subroutine calculates this basis set as the right and left eigenvectors of the T-matrix for the reference medium.

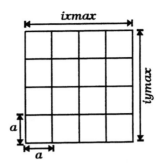

figure 2.1. shows the 2D unit cell of the system as a whole, and its associated lattice. The lattice cell is the unit cell of the reference medium.

given: A, EW, CLIGHT, IXMAX, IYMAX, IXYMAX, EXPAKX, EXPAKY, EMACH, EPSREF

other subroutine calls:
MAXONE   normalises the current vector so that the largest element has modulus unity
CROPRD   take the vector product of two vectors and sets it in a third

output: RVEC, LVEC, EIGF

Formulae

the right eigenvectors
Start from p212 of reference [1]:

$$i\mathbf{k} \times \mathbf{E} = i\omega\,\mathbf{B}, \quad i\mathbf{k} \times \mathbf{H} = -i\omega\,\mathbf{D} \tag{2.1a,b}$$

and approximate,

$$k_x \approx (ia)^{-1}\left[\exp(ik_x a) - 1\right] = \kappa_x^E$$
$$k_y \approx (ib)^{-1}\left[\exp(ik_y b) - 1\right] = \kappa_y^E$$
$$k_z \approx (ic)^{-1}\left[\exp(ik_z c) - 1\right] = \kappa_z^E \tag{2.2}$$

or,

$$k_x \approx -(ia)^{-1}\left[\exp(-ik_x a) - 1\right] = \kappa_x^H$$
$$k_y \approx -(ib)^{-1}\left[\exp(-ik_y b) - 1\right] = \kappa_y^H$$
$$k_z \approx -(ic)^{-1}\left[\exp(-ik_z c) - 1\right] = \kappa_z^H \tag{2.3}$$

then,

$$\kappa^E \times \mathbf{E} = +\omega\mu_0\mathbf{H}, \quad \kappa^H \times \mathbf{H} = -\omega\varepsilon_{ref}\varepsilon_0\mathbf{E} \tag{2.4a,b}$$

or in matrix notation,

$$\begin{bmatrix} \mathbf{0} & -\varepsilon_{ref}^{-1}\varepsilon_0^{-1}\left\{\begin{matrix} 0 & -\kappa_z^H & +\kappa_y^H \\ +\kappa_z^H & 0 & -\kappa_x^H \\ -\kappa_y^H & +\kappa_x^H & 0 \end{matrix}\right\} \\ \mu_0^{-1}\left\{\begin{matrix} 0 & -\kappa_z^E & +\kappa_y^E \\ +\kappa_z^E & 0 & -\kappa_x^E \\ -\kappa_y^E & +\kappa_x^E & 0 \end{matrix}\right\} & \mathbf{0} \end{bmatrix}\begin{bmatrix} \mathbf{E} \\ \mathbf{H} \end{bmatrix} = \omega\begin{bmatrix} \mathbf{E} \\ \mathbf{H} \end{bmatrix} \tag{2.5}$$

or substituting for $\mathbf{H}$:

$$\kappa^H \times \left(\kappa^E \times \mathbf{E}\right) = \omega\mu_0\kappa^H \times \mathbf{H} = -\omega^2\mu_0\varepsilon_0\varepsilon_{ref}\mathbf{E}, \text{ or}$$

$$\left(\kappa^H \cdot\kappa^E\right)\mathbf{E} - \kappa^E\left(\kappa^H \cdot \mathbf{E}\right) = \omega^2\mu_0\varepsilon_0\varepsilon_{ref}\mathbf{E} \tag{2.6a,b}$$

Therefore there are 3 modes which can be found by inspection:

$\omega = 0$ - the longitudinal mode

$$\mathbf{E}^{(L)} = \kappa^E, \quad \mathbf{H}^{(L)} = \kappa^H \tag{2.7a,b}$$

$\omega \neq 0$ - the transverse modes

Defined as having the E-field perpendicular to κ^H, and the H-field to κ^E. Remember that the programs work in terms of the renormalised H-fields and define S and P polarisations relative to the surface normal, $\mathbf{n}$:

$$\mathbf{E}^{(S)} = \kappa^H \times \mathbf{n}, \tag{2.8a}$$

$$\mathbf{H}^{(S)} = \frac{i}{c\omega\varepsilon_0}\mathbf{H}^{(S)}$$

$$= \frac{i}{c\omega^2\varepsilon_0\mu_0}\kappa^E \times \left(\kappa^H \times \mathbf{n}\right) = \frac{ic_0^2}{c\omega^2}\left[\kappa^H\left(\kappa^E \cdot \mathbf{n}\right) - \mathbf{n}\left(\kappa^E \cdot\kappa^H\right)\right] \tag{2.8b}$$

$$\mathbf{H}^{(P)} = \frac{i}{c\omega\varepsilon_0}\mathbf{H}^{(P)} = \kappa^E \times \mathbf{n}, \tag{2.9a}$$

$$\mathbf{E}^{(P)} = ia\varepsilon_{ref}^{-1}\kappa^H \times \left(\kappa^E \times \mathbf{n}\right) \tag{2.9b}$$

The S & P-polarisations are degenerate:

$$\omega^2\mu_0\varepsilon_0\varepsilon_{ref} = \kappa^E \cdot \kappa^H$$

$$= 2a^{-2}\left[1 - \cos(k_xa)\right] + 2b^{-2}\left[1 - \cos(k_yb)\right] + 2c^{-2}\left[1 - \cos(k_zc)\right] \tag{2.10}$$

ie, assuming that we have fixed $\mathbf{k}_x$ and $\mathbf{k}_y$,

$$\cos(k_zc) = 1 - \frac{c^2\omega^2\mu_0\varepsilon_0\varepsilon_{ref}}{2} + \frac{c^2}{a^2}\left[1 - \cos(k_xa)\right] + \frac{c^2}{b^2}\left[1 - \cos(k_yb)\right] \tag{2.11}$$

These formulae have to be evaluated for each of the distinct components of $\mathbf{k}$:

$$VKX = k_x = AKX + 2\pi\left(JX - 1\right)/\left(A*IXMAX\right),$$
$$VKY = k_y = AKY + 2\pi\left(JY - 1\right)/\left(A*IYMAX\right) \tag{2.12}$$

Remember that $\mathbf{g}$ is a reciprocal lattice vector of the 2D *unit cell* not of the lattice on which we quantise the fields, and that we have assumed $a=b=c$. We test whether the Bloch wave decays or carries current in the $+z$ direction by testing either the modulus of the eigenvector, or the group velocity,

$$\frac{d\omega}{dk_z} = +\left(c^2\omega\mu_0\varepsilon_0\varepsilon_{ref}\right)^{-1}\sin(k_zc) \tag{2.13}$$

and adjust the sign of $k_z = VKZ$ to ensure that for our choice the wave *does* decay or carry current in the $+z$ direction. The four eigenvectors are then stored in *EIGF* as follows

EIGF(1)
Waves decaying exponentially or carrying current in the $-z$ direction with sources at $z = +\infty$, and S-polarised.

EIGF(2)
Waves decaying exponentially or carrying current in the $-z$ direction with sources at $z = +\infty$, and P-polarised.

EIGF(3)
Waves decaying exponentially or carrying current in the $+z$ direction with sources at $z = -\infty$, and S-polarised.

EIGF(4)
Waves decaying exponentially or carrying current in the $+z$ direction with sources at $z = -\infty$, and P-polarised.

The eigenvectors are then constructed from the formulae above, with special treatment for normal incidence where we are free to choose the E-field for S polarisation in the -x-direction, and for P in the y-direction.

One word of warning: this basis set is unstable if $IXMAX=4\times$integer, and $IYMAX=4\times$integer'. This choice of dimensions will result in overflows due to our normalisation conditions.

the right eigenvectors
Start from,

$$\begin{bmatrix} E_x^r(\mathbf{r+c}) \\ E_y^r(\mathbf{r+c}) \\ H_x^r(\mathbf{r+c}) \\ H_y^r(\mathbf{r+c}) \end{bmatrix} = \mathbf{TB} \times \mathbf{TA} \begin{bmatrix} E_x^r(\mathbf{r}) \\ E_y^r(\mathbf{r}) \\ H_x^r(\mathbf{r}) \\ H_y^r(\mathbf{r}) \end{bmatrix} = \mathbf{T} \begin{bmatrix} E_x^r(\mathbf{r}) \\ E_y^r(\mathbf{r}) \\ H_x^r(\mathbf{r}) \\ H_y^r(\mathbf{r}) \end{bmatrix} = \mathbf{T}|r\rangle \tag{2.14}$$

where,

$$\mathbf{TA} = \begin{bmatrix} 1, & 0, & \varepsilon_{ref}^{-1}\begin{bmatrix} -e^{-i\theta_y}+1 \\ +e^{+i\theta_x-i\theta_y} - e^{+i\theta_x} \end{bmatrix}, & \dfrac{a^2\omega^2}{c_0^2}+\varepsilon_{ref}^{-1}\begin{bmatrix} +e^{-i\theta_x}-1 \\ -1+e^{+i\theta_x} \end{bmatrix} \\ 0, & 1, & -\dfrac{a^2\omega^2}{c_0^2}-\varepsilon_{ref}^{-1}\begin{bmatrix} +e^{-i\theta_y}-1 \\ -1+e^{+i\theta_y} \end{bmatrix}, & -\varepsilon_{ref}^{-1}\begin{bmatrix} -e^{-i\theta_x}+1 \\ +e^{+i\theta_y-i\theta_x} - e^{+i\theta_y} \end{bmatrix} \\ 0, & 0, & 1, & 0 \\ 0, & 0, & 0, & 1 \end{bmatrix} \tag{2.15a}$$

$$\mathbf{TB} = \begin{bmatrix} 1, & 0, & 0 & 0 \\ 0, & 1, & 0 & 0 \\ \dfrac{c_0^2}{a^2\omega^2}\begin{bmatrix} +e^{+i\theta_y-i\theta_x} - e^{-i\theta_x} \\ -e^{-i\theta_y}+1 \end{bmatrix}, & \varepsilon_{ref}+\dfrac{c_0^2}{a^2\omega^2}\begin{bmatrix} -1+e^{-i\theta_x} \\ +e^{+i\theta_x}-1 \end{bmatrix}, & 1 & 0 \\ -\varepsilon_{ref}-\dfrac{c_0^2}{a^2\omega^2}\begin{bmatrix} -1+e^{-i\theta_y} \\ +e^{+i\theta_y}-1 \end{bmatrix}, & \dfrac{-c_0^2}{a^2\omega^2}\begin{bmatrix} +e^{+i\theta_x-i\theta_y} - e^{-i\theta_y} \\ -e^{+i\theta_x}+1 \end{bmatrix}, & 0 & 1 \end{bmatrix} \tag{2.15b}$$

where,

$$\exp(+i\theta_x) = \exp\left[i(k_x+g_x)a\right], \qquad \exp(+i\theta_y) = \exp\left[i(k_y+g_y)a\right] \tag{2.16a,b}$$

and examine the *left* eigenvectors,

$$\langle l|\mathbf{TB} \times \mathbf{TA} = \langle l|\exp(ik_z c)$$ (2.17)

Taking the transpose,

$$\mathbf{TA}^T \times \mathbf{TB}^T|l\rangle = \exp(ik_z c)|l\rangle$$ (2.18)

Note that,

$$
\mathbf{TA}^T = \begin{bmatrix}
1, & 0, & 0, & 0 \\
0, & 1, & 0, & 0 \\
\varepsilon_{ref}^{-1}\begin{bmatrix} -e^{-i\theta_y}+1 \\ +e^{+i\theta_x-i\theta_y}-e^{+i\theta_x} \end{bmatrix}, & -\dfrac{a^2\omega^2}{c_0^2}-\varepsilon_{ref}^{-1}\begin{bmatrix} +e^{-i\theta_y}-1 \\ -1+e^{+i\theta_y} \end{bmatrix}, & 1, & 0 \\
+\dfrac{a^2\omega^2}{c_0^2}+\varepsilon_{ref}^{-1}\begin{bmatrix} +e^{-i\theta_x}-1 \\ -1+e^{+i\theta_x} \end{bmatrix}, & -\varepsilon_{ref}^{-1}\begin{bmatrix} -e^{-i\theta_x}+1 \\ +e^{+i\theta_y-i\theta_x}-e^{+i\theta_y} \end{bmatrix}, & 0, & 1
\end{bmatrix}
$$

$$= \mathbf{Q}^{-1}\mathbf{TBQ}$$ (2.19)

where,

$$
\mathbf{Q} = \begin{bmatrix}
0, & 1, & 0, & 0 \\
-1, & 0, & 0, & 0 \\
0, & 0, & 0, & -\dfrac{c_0^2\varepsilon_{ref}}{a^2\omega^2} \\
0, & 0, & \dfrac{c_0^2\varepsilon_{ref}}{a^2\omega^2}, & 0
\end{bmatrix}
$$ (2.20)

Similarly,

$$
\mathbf{TB}^T = \begin{bmatrix}
1, & 0, & \dfrac{c_0^2}{a^2\omega^2}\begin{bmatrix} +e^{+i\theta_y-i\theta_x}-e^{-i\theta_x} \\ -e^{+i\theta_y}+1 \end{bmatrix}, & -\varepsilon_{ref}-\dfrac{c_0^2}{a^2\omega^2}\begin{bmatrix} -1+e^{-i\theta_y} \\ +e^{+i\theta_y}-1 \end{bmatrix} \\
0, & 1, & \varepsilon_{ref}+\dfrac{c_0^2}{a^2\omega^2}\begin{bmatrix} -1+e^{-i\theta_x} \\ +e^{+i\theta_x}-1 \end{bmatrix}, & \dfrac{-c_0^2}{a^2\omega^2}\begin{bmatrix} +e^{+i\theta_x-i\theta_y}-e^{-i\theta_y} \\ -e^{+i\theta_x}+1 \end{bmatrix} \\
0, & 0, & 1, & 0 \\
0, & 0, & 0, & 1
\end{bmatrix}
$$

$$= \mathbf{Q}^{-1}\mathbf{TAQ}$$ (2.21)

Therefore,

$$\mathbf{TA}^T \times \mathbf{TB}^T|l\rangle = \mathbf{Q}^{-1}\mathbf{TBQQ}^{-1}\mathbf{TAQ}|l\rangle = \exp(ik_z c)|l\rangle$$ (2.22)

or,

$$\mathbf{TB\,TAQ}|l\rangle = \exp(ik_z c)\mathbf{Q}|l\rangle \tag{2.23}$$

Hence if,

$$\mathbf{TB} \times \mathbf{TA}
\begin{bmatrix}
\mathbf{E}_x^r(\mathbf{r}) \\
\mathbf{E}_y^r(\mathbf{r}) \\
\mathbf{H}_x^r(\mathbf{r}) \\
\mathbf{H}_y^r(\mathbf{r})
\end{bmatrix}
= \exp(ik_z c)
\begin{bmatrix}
\mathbf{E}_x^r(\mathbf{r}) \\
\mathbf{E}_y^r(\mathbf{r}) \\
\mathbf{H}_x^r(\mathbf{r}) \\
\mathbf{H}_y^r(\mathbf{r})
\end{bmatrix} \tag{2.24}$$

then,

$$\begin{bmatrix} \mathbf{E}_x^l(\mathbf{r}) & \mathbf{E}_y^l(\mathbf{r}) & \mathbf{H'}_x^l(\mathbf{r}) & \mathbf{H'}_y^l(\mathbf{r}) \end{bmatrix} \mathbf{TB} \times \mathbf{TA}$$

$$= \exp(ik_z c)\begin{bmatrix} \mathbf{E}_x^l(\mathbf{r}) & \mathbf{E}_y^l(\mathbf{r}) & \mathbf{H'}_x^l(\mathbf{r}) & \mathbf{H'}_y^l(\mathbf{r}) \end{bmatrix} \tag{2.25}$$

where,

$$\begin{bmatrix} \mathbf{E}_x^l(\mathbf{r}) & \mathbf{E}_y^l(\mathbf{r}) & \mathbf{H'}_x^l(\mathbf{r}) & \mathbf{H'}_y^l(\mathbf{r}) \end{bmatrix}^T =$$

$$\mathbf{Q}^{-1}
\begin{bmatrix}
\mathbf{E}_x^r(\mathbf{r}) \\
\mathbf{E}_y^r(\mathbf{r}) \\
\mathbf{H'}_x^r(\mathbf{r}) \\
\mathbf{H'}_y^r(\mathbf{r})
\end{bmatrix}
=
\begin{bmatrix}
0, & -1, & 0, & 0 \\
1, & 0 & 0, & 0 \\
0, & 0, & 0, & \dfrac{a^2\omega^2}{c_0^2\varepsilon_{ref}} \\
0, & 0, & -\dfrac{a^2\omega^2}{c_0^2\varepsilon_{ref}}, & 0
\end{bmatrix}
\begin{bmatrix}
\mathbf{E}_x^r(\mathbf{r}) \\
\mathbf{E}_y^r(\mathbf{r}) \\
\mathbf{H'}_x^r(\mathbf{r}) \\
\mathbf{H'}_y^r(\mathbf{r})
\end{bmatrix} \tag{2.26}$$

NB there is an apparent paradox in that the left eigenvectors so generated are orthogonal to the right eigenvectors from which they are derived. We resolve the paradox by noting that the eigenvalues are two-fold degenerate with respect to S and P polarisation. Thus the right P eigenvector generates the left S eigenvector and vice-versa.

If k_z is real, the current is calculated from the Poynting vector for which a formula is given in reference [1],

$$J(\mathbf{r}, \mathbf{r} - \mathbf{c}) = (a)^{-1}\Big[+\mathbf{E}_x(\mathbf{r})\mathbf{H}_y(\mathbf{r} - \mathbf{c}) - \mathbf{E}_y(\mathbf{r})\mathbf{H}_x(\mathbf{r} - \mathbf{c}) \Big] \tag{2.27}$$

There is the slight complication here that we are working with complex fields, but all we need to do is to recognise that the real and imaginary parts of the fields are independent solutions of Maxwell's equations and add the two currents together,

$$J(\mathbf{r}+\mathbf{c},\mathbf{r}) = (a)^{-1}\Big[+\Re E_x(\mathbf{r}+\mathbf{c})\Re H_y(\mathbf{r}) - \Re E_y(\mathbf{r}+\mathbf{c})\Re H_x(\mathbf{r})\Big]$$
$$+(a)^{-1}\Big[+\Im E_x(\mathbf{r}+\mathbf{c})\Im H_y(\mathbf{r}) - \Im E_y(\mathbf{r}+\mathbf{c})\Im H_x(\mathbf{r})\Big]$$
$$= (a)^{-1}\Re\Big[+E_x^*(\mathbf{r}+\mathbf{c})H_y(\mathbf{r}) - E_y^*(\mathbf{r}+\mathbf{c})H_x(\mathbf{r})\Big] \tag{2.28}$$

In the programs we actually calculate $\mathbf{H}'$, not $\mathbf{H}$. We are only interested in the sign of J so the constants do not matter, but we must divide out the factor of i. Hence the final expression for the current is,

$$J'(\mathbf{r}+\mathbf{c},\mathbf{r}) = (a)^{-1}\Re\left[+\Big[\exp(iK_z a)E_x(\mathbf{r})\Big]^* H_y(\mathbf{r})/i - \Big[\exp(iK_z a)E_y(\mathbf{r})\Big]^* H_x(\mathbf{r})/i\right] \tag{2.29}$$

and we have used the eigenvalues of the transfer matrix to translate the $\mathbf{E}$-field in the z-direction. The current is then used to renormalise the eigenvectors which carry current.

subroutine PLANEW - variables		
variable	**type**	**description**
IXMAX	integer	number of lattice points in the x-direction. NB do NOT choose 4×int
IYMAX	integer	number of lattice points in the y-direction. NB do NOT choose 4×int'
IXYMAX	integer	number of lattice points in the xy-plane
CURRNT	real*8	used to store the Poynting vector
PI	real*8	the value of π - set in the subroutine
A	real*8	the lattice spacing in Bohr radii - ie the small lattice
EW	real*8	the frequency of the light expressed in atomic energy units (1au = 27.2eV = 2 Rydbergs)
CLIGHT	real*8	the velocity of light in atomic units - 137.0360
EMACH	real*8	machine accuracy
AKX	real*8	x component of the k-vector
AKY	real*8	y component of the k-vector
VKX	real*8	x component of $\mathbf{k}+\mathbf{g}$

VKY	real*8	y component of $\mathbf{k+g}$
RVEC	complex*16 array	the right-eigenvectors of the transfer matrix for the reference medium: the first argument labels the vector which is stored in groups of four, first the two eigenvectors with +ve current (or +ve decay), second the two eigenvectors with -ve current (or -ve decay); one group of four per parallel momentum. The second argument labels the component of the vector in the real space representation of T.
LVEC	complex*16 array	the left-eigenvectors of the transfer matrix for the reference medium complementing RVEC
EIGF	complex*16 array	the eigenvalues of T - stored in the same order as RVEC
VN	complex*16 array	surface normal
VKE	complex*16 array	κ^E
VKH	complex*16 array	κ^H
RVECTP	complex*16 array	6 components of the E/H fields - first subscript labels the polarisation - second the component
VSTORE	complex*16 array	temporary storage
VE	complex*16 array	3 components of the right E eigenvector
VH	complex*16 array	3 components of the right H eigenvector
CONS1	complex*16	$1/(ia)$
CONS2	complex*16	$ic_0^2 / \left[(ew)^2 a \right]$
CONS3	complex*16	$a^2 \times ew^2 / \left(epsref \times c_0^2 \right)$

EXPKX2	complex*16	$\exp(ik_x a + i2\pi n/ixmax)$ - phase variation of the current plane wave across the x-direction of the elementary lattice.
EXPKY2	complex*16	$\exp(ik_y a + i2\pi n/iymax)$ - phase variation of the current plane wave across the y-direction of the elementary lattice.
EPSREF	complex*16	the dielectric constant of the reference medium - NB do *not* confuse this variable with the permittivity of free space, ε_0.

3. subroutine TRNSF1

objective: Calculates the transfer matrix in real space prior to diagonalisation to find the band structure.

given: EPSFIB, MUFIB, A, EW, CLIGHT, IXMAX, IYMAX, IZMAX, IXYMAX, EXPAKX, EXPAKY, EMACH.

other subroutine calls: none

output: T

Formulae

The program calculates matrix elements of the transfer matrix connecting points on the -z side of the sample to those on the +z side. Each xy point is selected in turn, see figure 2.1, and each of the four fields in turn set to a non-zero value in the following order:

$$E_x, E_y, H_x, H_y \tag{3.1}$$

To perform the integration through the *izmax* planes the following formulae are applied to advance the integration through steps if the lattice spacing, c. The T-matrix for the reference medium is defined in reference [1] Given the components of the electric field, **E**, and reduced magnetic field,

$$\mathbf{H}' = \frac{i}{a\omega\varepsilon_0}\mathbf{H} \tag{3.2}$$

on the plane **z**, then the fields on the next plane along, **z**+**c**, are defined by,

$$\mathbf{E}_x(\mathbf{r}+\mathbf{c}) = \mathbf{E}_x(\mathbf{r}) + \frac{c^2\omega^2}{c_0^2}\mu(\mathbf{r})\mathbf{H'}_y(\mathbf{r})$$

$$\frac{+c^2}{a\varepsilon(\mathbf{r})}\left[\begin{array}{l}+a^{-1}\{\mathbf{H'}_y(\mathbf{r}-\mathbf{a})-\mathbf{H'}_y(\mathbf{r})\}\\-b^{-1}\{\mathbf{H'}_x(\mathbf{r}-\mathbf{b})-\mathbf{H'}_x(\mathbf{r})\}\end{array}\right]$$

$$\frac{-c^2}{a\varepsilon(\mathbf{r}+\mathbf{a})}\left[\begin{array}{l}+a^{-1}\{\mathbf{H'}_y(\mathbf{r})-\mathbf{H'}_y(\mathbf{r}+\mathbf{a})\}\\-b^{-1}\{\mathbf{H'}_x(\mathbf{r}+\mathbf{a}-\mathbf{b})-\mathbf{H'}_x(\mathbf{r}+\mathbf{a})\}\end{array}\right]$$

$$(3.3a)$$

$$\mathbf{E}_y(\mathbf{r}+\mathbf{c}) = \mathbf{E}_y(\mathbf{r}) - \frac{c^2\omega^2}{c_0^2}\mu(\mathbf{r})\mathbf{H'}_x$$

$$\frac{c^2}{b\varepsilon(\mathbf{r})}\left[\begin{array}{l}+a^{-1}\{\mathbf{H'}_y(\mathbf{r}-\mathbf{a})-\mathbf{H}_y(\mathbf{r})\}\\-b^{-1}\{\mathbf{H'}_x(\mathbf{r}-\mathbf{b})-\mathbf{H'}_x(\mathbf{r})\}\end{array}\right]$$

$$\frac{-c^2}{b\varepsilon(\mathbf{r}+\mathbf{b})}\left[\begin{array}{l}+a^{-1}\{\mathbf{H'}_y(\mathbf{r}-\mathbf{a}+\mathbf{b})-\mathbf{H'}_y(\mathbf{r}+\mathbf{b})\}\\-b^{-1}\{\mathbf{H'}_x(\mathbf{r})-\mathbf{H'}_x(\mathbf{r}+\mathbf{b})\}\end{array}\right]$$

$$(3.3b)$$

$$\mathbf{H'}_x(\mathbf{r}+\mathbf{c}) = \mathbf{H'}_x(\mathbf{r}) + \varepsilon(\mathbf{r}+\mathbf{c})\mathbf{E}_y(\mathbf{r}+\mathbf{c})$$

$$\frac{-c_0^2}{a\omega^2\mu(\mathbf{r}-\mathbf{a}+\mathbf{c})}\left[\begin{array}{l}+a^{-1}\{\mathbf{E}_y(\mathbf{r}+\mathbf{c})-\mathbf{E}_y(\mathbf{r}-\mathbf{a}+\mathbf{c})\}\\-b^{-1}\{\mathbf{E}_x(\mathbf{r}-\mathbf{a}+\mathbf{b}+\mathbf{c})-\mathbf{E}_x(\mathbf{r}-\mathbf{a}+\mathbf{c})\}\end{array}\right]$$

$$\frac{+c_0^2}{a\omega^2\mu(\mathbf{r}+\mathbf{c})}\left[\begin{array}{l}+a^{-1}\{\mathbf{E}_y(\mathbf{r}+\mathbf{a}+\mathbf{c})-\mathbf{E}_y(\mathbf{r}+\mathbf{c})\}\\-b^{-1}\{\mathbf{E}_x(\mathbf{r}+\mathbf{b}+\mathbf{c})-\mathbf{E}_x(\mathbf{r}+\mathbf{c})\}\end{array}\right]$$

$$(3.3c)$$

$$\mathbf{H'}_y(\mathbf{r}+\mathbf{c})] = \mathbf{H'}_y(\mathbf{r}) - \varepsilon(\mathbf{r}+\mathbf{c})\mathbf{E}_x(\mathbf{r}+\mathbf{c})$$

$$\frac{-c_0^2}{b\omega^2\mu(\mathbf{r}-\mathbf{b}+\mathbf{c})}\left[\begin{array}{l}+a^{-1}\{\mathbf{E}_y(\mathbf{r}+\mathbf{a}-\mathbf{b}+\mathbf{c})-\mathbf{E}_y(\mathbf{r}-\mathbf{b}+\mathbf{c})\}\\-b^{-1}\{\mathbf{E}_x(\mathbf{r}+\mathbf{c})-\mathbf{E}_x(\mathbf{r}-\mathbf{b}+\mathbf{c})\}\end{array}\right]$$

$$\frac{+c_0^2}{b\omega^2\mu(\mathbf{r}+\mathbf{c})}\left[\begin{array}{l}+a^{-1}\{\mathbf{E}_y(\mathbf{r}+\mathbf{a}+\mathbf{c})-\mathbf{E}_y(\mathbf{r}+\mathbf{c})\}\\-b^{-1}\{\mathbf{E}_x(\mathbf{r}+\mathbf{b}+\mathbf{c})-\mathbf{E}_x(\mathbf{r}+\mathbf{c})\}\end{array}\right]$$

$$(3.3d)$$

which translates into,

$$
\begin{bmatrix} \mathbf{E}_x(\mathbf{r}+\mathbf{c}) \\ \mathbf{E}_y(\mathbf{r}+\mathbf{c}) \\ \mathbf{H'}_x(\mathbf{r}) \\ \mathbf{H'}_y(\mathbf{r}) \end{bmatrix} = \mathbf{TA} \begin{bmatrix} \mathbf{E}_x(\mathbf{r}) \\ \mathbf{E}_y(\mathbf{r}) \\ \mathbf{H'}_x(\mathbf{r}) \\ \mathbf{H'}_y(\mathbf{r}) \end{bmatrix}, \qquad \begin{bmatrix} \mathbf{E}_x(\mathbf{r}+\mathbf{c}) \\ \mathbf{E}_y(\mathbf{r}+\mathbf{c}) \\ \mathbf{H'}_x(\mathbf{r}+\mathbf{c}) \\ \mathbf{H'}_y(\mathbf{r}+\mathbf{c}) \end{bmatrix} = \mathbf{TB} \begin{bmatrix} \mathbf{E}_x(\mathbf{r}+\mathbf{c}) \\ \mathbf{E}_y(\mathbf{r}+\mathbf{c}) \\ \mathbf{H'}_x(\mathbf{r}) \\ \mathbf{H'}_y(\mathbf{r}) \end{bmatrix}
$$

(3.4a,b)

therefore,

$$
\begin{bmatrix} \mathbf{E}_x(\mathbf{r}+\mathbf{c}) \\ \mathbf{E}_y(\mathbf{r}+\mathbf{c}) \\ \mathbf{H'}_x(\mathbf{r}+\mathbf{c}) \\ \mathbf{H'}_y(\mathbf{r}+\mathbf{c}) \end{bmatrix} = \mathbf{TB} \times \mathbf{TA} \begin{bmatrix} \mathbf{E}_x(\mathbf{r}) \\ \mathbf{E}_y(\mathbf{r}) \\ \mathbf{H'}_x(\mathbf{r}) \\ \mathbf{H'}_y(\mathbf{r}) \end{bmatrix} = \mathbf{T} \begin{bmatrix} \mathbf{E}_x(\mathbf{r}) \\ \mathbf{E}_y(\mathbf{r}) \\ \mathbf{H'}_x(\mathbf{r}) \\ \mathbf{H'}_y(\mathbf{r}) \end{bmatrix}
$$

(3.5)

In addition we have to specify the boundary conditions at the edges of the 2D unit cell, see figure 2.1, which is done by requiring that the fields obey the 2D Bloch theorem,

$$
\begin{bmatrix} \mathbf{E}_x(\mathbf{r}+a \times ixmax \times \hat{\mathbf{x}}) \\ \mathbf{E}_y(\mathbf{r}+a \times ixmax \times \hat{\mathbf{x}}) \\ \mathbf{H'}_x(\mathbf{r}+a \times ixmax \times \hat{\mathbf{x}}) \\ \mathbf{H'}_y(\mathbf{r}+a \times ixmax \times \hat{\mathbf{x}}) \end{bmatrix} = \exp(ik_x \times ixmax \times a) \begin{bmatrix} \mathbf{E}_x(\mathbf{r}) \\ \mathbf{E}_y(\mathbf{r}) \\ \mathbf{H'}_x(\mathbf{r}) \\ \mathbf{H'}_y(\mathbf{r}) \end{bmatrix}
$$

(3.6a)

$$
\begin{bmatrix} \mathbf{E}_x(\mathbf{r}+a \times iymax \times \hat{\mathbf{y}}) \\ \mathbf{E}_y(\mathbf{r}+a \times iymax \times \hat{\mathbf{y}}) \\ \mathbf{H'}_x(\mathbf{r}+a \times iymax \times \hat{\mathbf{y}}) \\ \mathbf{H'}_y(\mathbf{r}+a \times iymax \times \hat{\mathbf{y}}) \end{bmatrix} = \exp(ik_y \times iymax \times a) \begin{bmatrix} \mathbf{E}_x(\mathbf{r}) \\ \mathbf{E}_y(\mathbf{r}) \\ \mathbf{H'}_x(\mathbf{r}) \\ \mathbf{H'}_y(\mathbf{r}) \end{bmatrix}
$$

(3.6b)

so that when the equations require a field at a point beyond the boundaries of the cell, it can be obtained from the value of a known point within the cell by applying the Bloch theorem. These equations are used to step the field through the medium one layer of cells at a time and the T-matrix element set equal to the fields on the final plane.

subroutine TRNSF1 - variables		
variable	type	description
IXMAX	integer	number of lattice points in the x-direction
IYMAX	integer	number of lattice points in the y-direction

IZMAX	integer	number of lattice points in the z-direction
IXYMAX	integer	number of lattice points in the xy-plane
A	real*8	the lattice spacing in Bohr radii - ie the small lattice
EW	real*8	the frequency of the light expressed in atomic energy units (1au = 27.2eV = 2 Rydbergs)
CLIGHT	real*8	the velocity of light in atomic units - 137.0360
AWC2	real*8	(A*EW/CLIGHT)**2
EX	complex*16 array	x-component of the electric field
EY	complex*16 array	y-component of the electric field
HX	complex*16 array	x-component of the magnetic field
HY	complex*16 array	y-component of the magnetic field
T	complex*16 array	the transfer matrix in a real space representation
EPSFIB	complex*16 array	the dielectric function defined at each lattice point, with different values for x, y, and z components of the field
MUFIB	complex*16 array	the magnetic permeability defined at each lattice point, with different values for x, y, and z components of the field
EXPAKX	complex*16	$\exp(ik_x a*ixmax)$ - phase variation across the x-direction of the unit cell
EXPAKY	complex*16	$\exp(ik_y a*iymax)$ - phase variation across the y-direction of the unit cell

4. subroutine TRNSF2

objective: Calculates the reflection and transmission coefficients for plane waves incident on a layer,

figure 4.1. Shows the definition of the transmission and reflection coefficients.

given: EPSPOT, A, EW, CLIGHT, IXMAX, IYMAX, IZMAX, IXYMAX, AKX, AKY, RVEC, LVEC, EIGF, EMACH.

other subroutine calls:

TRNSF3	Given the fields on one side of a medium, calculates the fields on the other side
ZGE,ZSU	matrix inversion
TRN2T1	tests of the calculation

output: TPP, TPM, TMP, TMM

Formulae

As shown in figure 4.1, there are really two calculations to be made. To calculate *TPP* and *TMP* we start on the right of the sample ($+z$) and integrate backwards through the sample to the $-z$ side. Each emerging beam is specified by one of the plane waves in the reference medium extending to $z=+\infty$. The plane waves are labelled by (i) the reciprocal lattice vector, **g**, (ii) the polarisation. The polarisation state is specified by the eigenvector subroutine in subroutine PLANEW as *S* and *P*. If we specify a plane wave of unit amplitude on the right, then integrating backwards through the sample gives *TPP-1* and *TPP^{-1}TMP* as figure 4.2 shows, providing that we can decompose the wavefield on the $-z$ side into its component Bloch waves.

figure 4.2. Shows the rearrangement of the transmission and reflection coefficients prior to calculation. Note that we can now specify the plane wave amplitude on the right (left) before we know the transmission and reflection coefficients.

So first the program selects backwards integration, $JFORWD = -1$, then selects the reciprocal lattice vector,

$$\mathbf{g} = \left[(KX-1)\frac{2\pi}{ixmax \times a}, \quad (KY-1)\frac{2\pi}{ixmay \times a} \right] \tag{4.1}$$

and the polarisation specified by $KPOL$. To start the integration the program needs the E and H fields on the $+z$ side of the sample. These are obtained for the lattice point at the origin from $RVEC$ and the values at other lattice points in the unit cell (see figure 2.1) differ only by a phase factor,

$$\exp\left[i(\mathbf{k} + \mathbf{g}) \cdot (\mathbf{r}_l - \mathbf{r}_0) \right] = \exp\left[i(k_x + g_x) \cdot (x_l - x_0) + i(k_y + g_y) \cdot (y_l - y_0) \right] \tag{4.2}$$

where $\mathbf{r}_l$ is the lattice point in question. The phase factor is constructed as a series of factors of,

$$EXPKX2 = \exp\left[i(k_x + g_x)a \right], \quad \text{and } EXPKY2 = \exp\left[i(k_y + g_y)a \right] \tag{4.3}$$

stored in $EXPVCX$, $EXPVCY$. Once we have these phase factors we can construct the fields at all $IXMAX*IYMAX$ points in the 2D unit cell.

Next the transfer matrix subroutine $TRNSF3$ is called to integrate the fields backwards through the sample to the $-z$ side where they must then be projected onto the plane wave states. To do this the separate Fourier components are filtered out of the wavefield and then each Fourier component is decomposed into plane waves by projecting onto $LVEC$, the matrix of left-eigenvectors of the plane wave transfer matrix. These vectors are orthogonal to the right-eigenvectors by definition. TPP^{-1} and $TPP^{-1}TMP$ are stored in the matrix, $WORK$.

$$WORK = \begin{bmatrix} TPP^{-1} & * \\ TPP^{-1}TMP & * \end{bmatrix} \tag{4.4}$$

where the stars indicate unused space. The program goes on to extract TPP and TMP.

In the next cycle of the loop $JFORWD=+1$ and we calculate TMM and TPM.

subroutine TRNSF2 - variables		
variable	type	description
IXMAX	integer	number of lattice points in the x-direction
IYMAX	integer	number of lattice points in the y-direction

IZMAX	integer	number of lattice points in the z-direction
IXYMAX	integer	number of lattice points in the xy-plane
JFORWD	integer	determines whether we integrate backwards (-1) or forwards (+1) through the unit cell - backwards to calculate TPP AND TMP, forwards to calculate TMM and TPM.
PI	real*8	the value of π - set in the subroutine
A	real*8	the lattice spacing in Bohr radii - ie the small lattice
EW	real*8	the frequency of the light expressed in atomic energy units (1au = 27.2eV = 2 Rydbergs)
CLIGHT	real*8	the velocity of light in atomic units - 137.0360
EMACH	real*8	machine accuracy
AKX	real*8	x-component of the wavevector
AKY	real*8	y-component of the wavevector
EX	complex*16 array	x-component of the electric field
EY	complex*16 array	y-component of the electric field
HX	complex*16 array	x-component of the magnetic field
HY	complex*16 array	y-component of the magnetic field
TPP	complex*16 array	the matrix of transmission coefficients: transmitted left to right
TPM	complex*16 array	the matrix of reflection coefficients: reflected from right back to right
TMP	complex*16 array	the matrix of reflection coefficients: reflected from left back to left
TMM	complex*16 array	the matrix of transmission coefficients: transmitted right to left

RVEC	complex*16 array	the right-eigenvectors of the transfer matrix for the reference medium: the first argument labels the vector which is stored in groups of four, first the two eigenvectors with +ve current (or +ve decay), second the two eigenvectors with -ve current (or -ve decay); one group of four per parallel momentum. The second argument labels the component of the vector in the real space representation of T.
LVEC	complex*16 array	the left-eigenvectors of the transfer matrix for the reference medium complementing RVEC
EIGF	complex*16 array	the eigenvalues of T - stored in the same order as RVEC
EXPVCX	complex*16 array	phase factors for the current plane wave: relates amplitudes of the plane wave as the lattice site is displace in the x-direction
EXPVCY	complex*16 array	phase factors for the current plane wave: relates amplitudes of the plane wave as the lattice site is displace in the y-direction
EXPAKX	complex*16	$\exp(ik_x a*ixmax)$ - phase variation across the x-direction of the unit cell
EXPAKY	complex*16	$\exp(ik_y a*iymax)$ - phase variation across the y-direction of the unit cell
EXPKX2	complex*16	$\exp(ik_x a+i2\pi n/ixmax)$ - phase variation of the current plane wave across the x-direction of the elementary lattice.
EXPKY2	complex*16	$\exp(ik_y a+i2\pi n/iymax)$ - phase variation of the current plane wave across the y-direction of the elementary lattice.
EPSPOT	complex*16	the dielectric constant stored for each lattice point

5. subroutine TRNSF3

objective: This subroutine, given the fields on one side of a non-uniform medium, EX, EY, HX, HY, calculates the fields on the far side: called by TRNSF2.

given: EX, EY, HX, HY, AWC2, EXPAKX, EXPAKY, EPSPOT, IXMAX, IYMAX, IZMAX, JFORWD *other subroutine calls:* none

output: EX, EY, HX, HY, the fields on the far side of the non-uniform medium.

Formulae

The T-matrix for the reference medium has been defined in equations (3.2)-(3.6). The same code can be used for forward or backward integration by noting that,

$$
\begin{bmatrix} E_x(r+c) \\ E_y(r+c) \\ H'_x(r+c) \\ H'_y(r+c) \end{bmatrix} = TB \times TA \begin{bmatrix} E_x(r) \\ E_y(r) \\ H'_x(r) \\ H'_y(r) \end{bmatrix} = \begin{bmatrix} 1 & 0 \\ B & 1 \end{bmatrix} \times \begin{bmatrix} 1 & A \\ 0 & 1 \end{bmatrix} \begin{bmatrix} E_x(r) \\ E_y(r) \\ H'_x(r) \\ H'_y(r) \end{bmatrix}
$$
(5.1)

and hence, exploiting the readily invertable form of the factor matrices,

$$
\begin{bmatrix} E_x(r) \\ E_y(r) \\ H'_x(r) \\ H'_y(r) \end{bmatrix} = \begin{bmatrix} 1 & -A \\ 0 & 1 \end{bmatrix} \times \begin{bmatrix} 1 & 0 \\ -B & 1 \end{bmatrix} \begin{bmatrix} E_x(r+c) \\ E_y(r+c) \\ H'_x(r+c) \\ H'_y(r+c) \end{bmatrix}
$$
(5.2)

enabling is to step in the reverse direction by changing a few lines in the code.

subroutine TRNSF3 - variables		
variable	type	description
IXMAX	integer	number of lattice points in the x-direction
IYMAX	integer	number of lattice points in the y-direction
IZMAX	integer	number of lattice points in the z-direction
JFORWD	integer	determines whether we integrate backwards (-1) or forwards (+1) through the unit cell - backwards to calculate TPP AND TMP, forwards to calculate TMM and TPM.
AWC2	real*8	(A*EW/CLIGHT)**2 - see *TRNSF2* for these variables
EX	complex*16 array	x-component of the electric field
EY	complex*16 array	y-component of the electric field

HX	complex*16 array	x-component of the magnetic field
HY	complex*16 array	y-component of the magnetic field
EXPAKX	complex*16	$\exp(ik_x a *ixmax)$ - phase variation across the x-direction of the unit cell
EXPAKY	complex*16	$\exp(ik_y a *iymax)$ - phase variation across the y-direction of the unit cell
EPSPOT	complex*16	the dielectric constant stored for each lattice point

6. subroutine EIGCAL

objective: Given the transfer matrix, T, this subroutine calculates the eigenvalues of T and hence the band structure. The band structure is then written to file. Optionally, by setting WANTX = .true., the eigenvectors can be found and are return in WORK by subroutine LZIT.

given: T, IXMAX, IYMAX, IGMAX, EW, EMACH

other subroutine calls: LZHES, LZIT

output: writes the band structure, the eigenvalues, to file

Formulae

T-matrix is assumed to translate the fields through one unit cell in the z-direction. Therefore from Bloch's theorem we have,

$$\mathbf{T}\begin{bmatrix} E_x(\mathbf{r}) \\ E_y(\mathbf{r}) \\ H_x(\mathbf{r}) \\ H_y(\mathbf{r}) \end{bmatrix} = \begin{bmatrix} E_x(\mathbf{r}+\mathbf{c}) \\ E_y(\mathbf{r}+\mathbf{c}) \\ H_x(\mathbf{r}+\mathbf{c}) \\ H_y(\mathbf{r}+\mathbf{c}) \end{bmatrix} = \exp(ik_z c)\begin{bmatrix} E_x(\mathbf{r}) \\ E_y(\mathbf{r}) \\ H_x(\mathbf{r}) \\ H_y(\mathbf{r}) \end{bmatrix} \qquad (6.1)$$

The most 'important' eigenvalues are usually those with modulus close to unity, and a filter can be adjusted to select these from the total number of eigenvalues.

The eigenvalues routines supplied are standard numerical packages not written by us, but which have worked well for many years

subroutine EIGCAL - variables		
variable	type	description
IXMAX	integer	number of lattice points in the x-direction
IYMAX	integer	number of lattice points in the y-direction
IGMAX	integer	4*IXMAX*IYMAX dimensions of the T matrix
ITER	integer array	working space
EW	real*8	the frequency of the light expressed in atomic energy units (1au = 27.2eV = 2 Rydbergs)
EMACH	real*8	machine accuracy
GO	complex*16 array	working space for the eigenvalue subroutines
G1	complex*16 array	working space for the eigenvalue subroutines
WORK	complex*16 array	working space for the eigenvalue subroutines. If the eigenvectors are calculated, they are returned in WORK: WORK(k,i) = kth component of the ith eigenvector
EIGA	complex*16 array	eigenvectors = EIGB/EIGA
EIGB	complex*16 array	eigenvectors = EIGB/EIGA
T	complex*16 array	the transfer matrix
CZ	complex*16	complex zero
CI	complex*16	square root -1
WANTX	complex*16	specifies whether the eigenvectors are to be calculated: WANTX = .true. calculate eigenvectors WANTX = .false. do not calculate eigenvectors

224

7. subroutine ADDINV

objective: this subroutine takes reflection and transmission coefficients for a pair of layers,

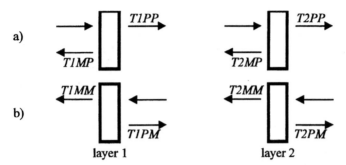

figure 7.1. Shows the definition of the transmission and reflection coefficients for the pair of layers. Case a) for waves incident from the left; case b) for waves incident from the right. The names of variables correspond to their function.

and combines them to give the corresponding quantities for the composite later,

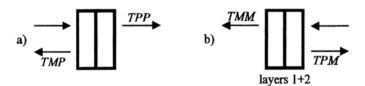

figure 7.2. Shows the definition of the transmission and reflection coefficients for the composite layer. Case a) for waves incident from the left; case b) for waves incident from the right. The names of variables correspond to their function.

If the variable JMAXST>1, then further operations may be performed in which two identical pairs of layers are combined,

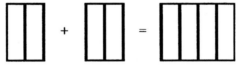

figure 7.3. Shows further operations in which two pairs of layers are combined into a single composite layer.

The process is repeated until a composite of 2**JMAXST is obtained. If it is wished to combine further *non-identical* pairs of layers then the subroutine must be entered with JMAXST=1 and subsequent sequencing organised from outside the subroutine.

Finally the subroutine prints the total transmitted and reflected currents when a wave is incident from the right onto the composite set of layers. It does this for each polarisation of the incident wave. It is assumed that the plane wave eigenvectors for the reference medium have been normalised to unit current if the wave vector is real. The incident wave is defined by the first two plane wave eigenvectors in the reference medium. These waves have the same wave vectors parallel to the surface but have different polarisation.

given: T1PP, T1PM, T1MP, T1MM, T2PP, T2PM, T2MP, T2MM, A, EW, CLIGHT, CPLNCK, IXMAX, IYMAX, IXYMAX, RVEC, EIGF, EMACH, JMAXST

other subroutine calls: ZGE, ZSU (matrix inversion subroutines)

output: TPP, TPM, TMP, TMM, and as printed output: CURTRN, CURREF

Formulae

Scattering from a pair of layers is combined using the multiple scattering formulae,

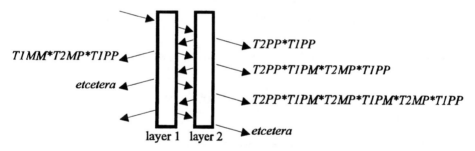

figure 7.4. Multiple scattering between a pair of layers.

which are as follows,

$$TPP = T2PP*T1PP + T2PP*T1PM*T2MP*T1PP$$
$$+ T2PP*T1PM*T2MP*T1PM*T2MP*T1PP + ...$$
$$= T2PP*(1 - T1PM*T2MP)^{-1}*T1PP \qquad (7.1a)$$
$$TMP = T1MM*T2MP*T1PP + T1MM*T2MP*T1PM*T2MP*T1PP + ...$$
$$= T1MM*T2MP*(1 - T1PM*T2MP)^{-1}*T1PP \qquad (7.1b)$$

A further pair of formulae result when a wave is incident from the right,

$$TMM = T1MM*T2MM + T1MM*T2MP*T1PM*T2MM$$

$$+ \: T1MM*T2MP*T1PM*T2MP*T1PM*T2MM + \: ...$$

$$= T1MM*(1 - T2MP*T1PM)^{-1}*T2MM \tag{7.2a}$$

$$TPM \: = T2PP*T1PM*T2MM + T2PP*T1PM*T2MP*T1PM*T2MM + ...$$

$$= T2PP*T1PM*(1 - T2MP*T1PM)^{-1}*T2MM \tag{7.2b}$$

If JMAXST>1 the next step is to replace,

$$T1PP = TPP, \: T1PM = TPM, \: T1MP = TMP, \: T1MM = TMM \tag{7.3a}$$

$$T2PP = TPP, \: T2PM = TPM, \: T2MP = TMP, \: T2MM = TMM \tag{7.3b}$$

and repeat the operations in (5.1, 5.2).

Finally the currents are calculated for waves incident from the right,

$$CURTRN = \sum_{J=1}^{N} FAC \times |TMM(J,2)|^2, \quad \text{where} \quad FAC = 1 \quad \text{if} \quad |EIGF(J)| = 1$$

$$CURREF = \sum_{J=1}^{N} FAC \times |TPM(J,2)|^2, \quad \text{and} \quad FAC = 0 \quad \text{if} \quad |EIGF(J)| \neq 1 \tag{7.4a,b}$$

The calculation is repeated for incident waves of the other polarisation,

$$CURTRN = \sum_{J=1}^{N} FAC \times |TMM(J,2)|^2, \quad \text{where} \quad FAC = 1 \quad \text{if} \quad |EIGF(J)| = 1$$

$$CURREF = \sum_{J=1}^{N} FAC \times |TPM(J,2)|^2, \quad \text{and} \quad FAC = 0 \quad \text{if} \quad |EIGF(J)| \neq 1 \tag{7.5a,b}$$

It is assumed that for the two incident waves considered,

$$|EIGF(1)| = 1, \quad |EIGF(2)| = 1 \tag{7.6a,b}$$

so that they both carry current towards the layers.

subroutine ADDINV- variables		
variable	type	description
IXMAX	integer	number of lattice points in the x-direction
IYMAX	integer	number of lattice points in the y-direction
IXYMAX	integer	number of lattice points in the xy-plane
ITER	integer	working space for matrix inversion subroutines

JMAXST	integer	number of times the layer doubling operation is performed
JSTACK	integer	number of the current layer doubling operation
N	integer	= 2*IXYMAX - dimension of the transmission and reflection matrices
JRVEC, KRVEC	integer	used as labels for the basis set of plane waves incident on/reflected from the layers.
JPOL, KPOL	integer	used as labels for the polarisation of the basis set of plane waves incident on/reflected from the layers.
AWC2	real*8	(A*EW/CLIGHT)**2
CURTRN	real*8	the total current transmitted through the stack of layers when a wave is incident from the right
CURREF	real*8	the total current reflected from the stack of layers when a wave is incident from the right
EMACH	real*8	machine accuracy
A	real*8	the lattice spacing in Bohr radii - ie the small lattice
EW	real*8	the frequency of the light expressed in atomic energy units (1au = 27.2eV = 2 Rydbergs)
CLIGHT	real*8	the velocity of light in atomic units - 137.0360
CPLNCK	real*8	Planck's constant in Hartree-seconds - (if $f=1Hz$, $E=cplnck*f*(27.2eV)$)
TPP	complex*16 array	the matrix of transmission coefficients: left to right
TPM	complex*16 array	the matrix of reflection coefficients: right to right
TMM	complex*16 array	the matrix of transmission coefficients: right to left
TMP	complex*16 array	the matrix of reflection coefficients: left to left

RVEC	complex*16 array	the right-eigenvectors of the transfer matrix for the reference medium: the first argument labels the vector which is stored in groups of four, first the two eigenvectors with +ve current (or +ve decay), second the two eigenvectors with -ve current (or -ve decay); one group of four per parallel momentum. The second argument labels the component of the vector in the real space representation.
EIGF	complex*16 array	the eigenvalues of T - stored in the same order as RVEC
T1PP	complex*16 array	the matrix of transmission coefficients: left to right; left layer
T1PM	complex*16 array	the matrix of reflection coefficients: right to right; left layer
T1MM	complex*16 array	the matrix of transmission coefficients: right to left; left layer
T1MP	complex*16 array	the matrix of reflection coefficients: left to left; left layer
T2PP	complex*16 array	the matrix of transmission coefficients: left to right; right layer
T2PM	complex*16 array	the matrix of reflection coefficients: right to right; right layer
T2MM	complex*16 array	the matrix of transmission coefficients: right to left; right layer
T2MP	complex*16 array	the matrix of reflection coefficients: left to left; right layer
CI	complex*16	square root of -1

8. References

1. Pendry, J.B. (1993) Photonic Band Structures, *J. Mod. Optics* **41**, 209.

2. Bell, P.M., Pendry, J.B., Martín-Moreno L., and Ward, A.J. (1995) A Program for Calculating Photonic Band Structures and Transmission Coefficients of Complex Structures, *Comp. Phys. Comm.*, **85**, 306.

LAYER-BY-LAYER METHODS IN THE STUDY OF PHOTONIC CRYSTALS AND RELATED PROBLEMS

A. MODINOS[1], N. STEFANOU[2], and V. KARATHANOS[1]

[1]*Department of Physics, National Technical University of Athens, Zografou Campus, GR-15780, Athens, Greece*
[2]*Solid State Section, University of Athens, Panepistimioupolis, GR-15784, Athens, Greece*

1. Introduction

The optical properties of a two-dimensional array of metallic particles on a dielectric slab have been investigated by a number of authors, mainly because of the possible applications of such systems in coating technology. In most cases the metallic particles are randomly distributed on the substrate surface and vary both in volume and shape [1], although in some instances a periodic arrangement in space of nearly identical particles has been achieved [2]. Usually, the average diameter of the metallic particles is of the order of 100Å and together the non-overlapping particles cover 30-70% of the substrate surface. The traditional analysis of such experiments is based on the Maxwell-Garnett theory and various extensions of it [1]. Essentially, the particles are replaced by interacting dipoles; an effective (local) field is evaluated by the use of the Clausius-Mossotti equation, and an effective dielectric function for the composite medium is obtained. This analysis (we refer to as the dipolar or electrostatic approximation) breaks down when the wavelength of the incident radiation is relatively small, of the same order of magnitude as the size of the particles and/or the interparticle distance, or when the volume occupied by the particles is about half or more of the total volume. We shall deal with this problem in section 2. We shall represent the particles by spheres characterised by a certain dielectric function and we shall assume they are arranged periodically on a plane. The case of a disordered array of spheres is discussed briefly in section 4.

In section 3 we present a method for the calculation of the transmission, reflection, and absorption of light incident on a slab of non-overlapping spheres, when the slab consists of a succession of planes of spheres with the same periodicity parallel to the surface of the slab. The optical properties of the slab are obtained in terms of the optical properties of the planes which

C. M. Soukoulis (ed.), Photonic Band Gap Materials, 229–251.
© *1996 Kluwer Academic Publishers. Printed in the Netherlands.*

make up the slab. We use this method to describe certain photonic crystals and their properties. The importance of photonic crystals has been emphasised by Yablonovitch and other authors of this volume, so we need not say anything more about it in our introduction.

2. Scattering of Electromagnetic Waves by a Two-Dimensional Ordered Array of Spheres

We assume the spheres are centred on the sites $\{\mathbf{R}_n\}$ of a two-dimensional lattice in the xy-plane, i.e.,

$$\mathbf{R}_n = n_1 \mathbf{a}_1 + n_2 \mathbf{a}_2 \qquad , \qquad (1)$$

where $\mathbf{a}_1$ and $\mathbf{a}_2$ are primitive vectors in the xy-plane and $n_1, n_2 = 0, \pm 1, \pm 2, \ldots$ We assume the spheres, which do not overlap, have the same radius S and are made of the same material which is characterised by a frequency-dependent, in general complex, relative dielectric function $\varepsilon_M(\omega)$. The surrounding medium has a relative dielectric function $\varepsilon(\omega)$. We further assume the spheres and the surrounding medium have the magnetic permeability μ_0 of vacuum.

We now wish to know how a plane electromagnetic wave whose electric field component is given by

$$\mathbf{E}(\mathbf{r}, t) = \text{Re}\{\tilde{\mathbf{E}}(\mathbf{r}) \exp(-i\omega t)\}$$
$$\tilde{\mathbf{E}}(\mathbf{r}) = \mathbf{E}_0(\mathbf{q}) \exp(i\mathbf{q} \cdot \mathbf{r}) \qquad (2)$$

is scattered by the above mentioned plane of spheres. Here and in what follows we do not denote explicitly the magnetic field $\mathbf{H} = (-i / \omega\mu_0)\nabla \times \mathbf{E}$ associated with the electromagnetic field under consideration and we omit the time factor $\exp(-i\omega t)$ also. A notation of plane waves which is more convenient for our purposes can be introduced as follows.

We introduce the reciprocal (to (1)) lattice

$$\mathbf{g} = m_1 \mathbf{b}_1 + m_2 \mathbf{b}_2 \quad , \quad m_1, m_2 = 0, \pm 1, \pm 2, \ldots \qquad (3)$$

where $\mathbf{b}_1$, $\mathbf{b}_2$ are primitive vectors (in $\mathbf{q}_{//}$-space) defined by

$$\mathbf{b}_i \cdot \mathbf{a}_j = 2\pi\delta_{ij} \tag{4}$$

and construct the surface Brillouin zone (SBZ) of (3) in the usual manner. We can write $\mathbf{q}_{//} = (q_x, q_y)$, the component of the incident wavevector on the plane of the spheres, as follows,

$$\mathbf{q}_{//} = \mathbf{k}_{//} + \mathbf{g}' \tag{5}$$

where the reduced wavevector $\mathbf{k}_{//}$ lies in the SBZ of (3) and $\mathbf{g}'$ is one of the reciprocal lattice vectors (3). We note that, according to (3), (4), and (5) $\exp(i\mathbf{q}_{//} \cdot \mathbf{R}_n) = \exp(i\mathbf{k}_{//} \cdot \mathbf{R}_n)$. In what follows we shall write the wavevector of a plane wave as

$$\mathbf{K}_{\mathbf{g}}^{\pm} \equiv \left(\mathbf{k}_{//} + \mathbf{g}, \pm\left[q^2 - \left(\mathbf{k}_{//} + \mathbf{g} \right)^2 \right]^{1/2} \right) \tag{6}$$

where the $+$ $(-)$ sign defines the sign of the z-component of the wavevector, and $q = (\mu_0\varepsilon\varepsilon_0)^{1/2}\omega$ is the magnitude of the wavevector.

We can write a plane wave as a sum of spherical waves about any centre (say about the origin) using a well-known formula [3] which tells us that in a medium characterised by a given $\varepsilon(\omega)$, any field $\tilde{\mathbf{E}}(\mathbf{r})\exp(-i\omega t)$ can be expanded in spherical waves as follows

$$\tilde{\mathbf{E}}(\mathbf{r}) = \sum_{\ell=1}^{\infty} \sum_{m=-\ell}^{\ell} \left(\frac{i}{q} a_{\ell m}^E \nabla \times z_\ell(qr)\mathbf{X}_{\ell m}(\hat{\mathbf{r}}) + a_{\ell m}^H z_\ell(qr)\mathbf{X}_{\ell m}(\hat{\mathbf{r}}) \right) . \tag{7}$$

The function $z_\ell(qr)$ is, in general, a linear combination of the spherical Bessel function $j_\ell(qr)$ and the spherical Hankel function $h_\ell^+(qr)$; the latter has the asymptotic form appropriate to an outgoing spherical wave, i.e., $h_\ell^+(qr) \rightarrow (-i)^\ell \exp(iqr)/(iqr)$ as $r \rightarrow \infty$. The $\mathbf{X}_{\ell m}(\hat{\mathbf{r}})$, where $\hat{\mathbf{r}}$ denotes the angular spherical coordinates (θ, φ) of $\mathbf{r}$, are defined by

$$\sqrt{\ell(\ell+1)}\,\mathbf{X}_{\ell m}(\hat{\mathbf{r}}) = \hat{\mathbf{L}}Y_{\ell m}(\hat{\mathbf{r}}) \quad , \quad \hat{\mathbf{L}} \equiv -i\mathbf{r} \times \nabla \quad , \tag{8}$$

where $Y_{\ell m}$ denotes as usual a spherical harmonic; for given $\ell\,(=1,2,3,...)$, m takes the values $-\ell, -\ell+1,...\ell$. By definition $\mathbf{X}_{00}(\hat{\mathbf{r}}) = \mathbf{0}$. Finally, $a_{\ell m}^{E}$, $a_{\ell m}^{H}$ are coefficients to be determined.

A plane wave ($\tilde{\mathbf{E}}(\mathbf{r})$ is given by (2)) is described by (7) with $z_{\ell}(qr) = j_{\ell}(qr)$; the corresponding coefficients are given by

$$a_{\ell m}^{0E(H)} = \mathbf{A}_{\ell m}^{0E(H)} \cdot \mathbf{E}_0 \qquad , \qquad (9)$$

where $\mathbf{A}_{\ell m}^{0E(H)}$ are certain functions of the wavevector [4].

Because of the two-dimensional periodicity of the structure under consideration, the scattered wave corresponding to the incident plane wave (2) can be written in the following form

$$\tilde{\mathbf{E}}_{sc}(\mathbf{r}) = \sum_{\ell=1}^{\infty} \sum_{m=-\ell}^{\ell} \left(\frac{i}{q} b_{\ell m}^{+E} \mathbf{\nabla} \times \sum_{\mathbf{R}_n} \exp(i\mathbf{k}_{//} \cdot \mathbf{R}_n) h_{\ell}^{+}(qr_n) \mathbf{X}_{\ell m}(\hat{\mathbf{r}}_n) \right.$$

$$\left. + b_{\ell m}^{+H} \sum_{\mathbf{R}_n} \exp(i\mathbf{k}_{//} \cdot \mathbf{R}_n) h_{\ell}^{+}(qr_n) \mathbf{X}_{\ell m}(\hat{\mathbf{r}}_n) \right) \qquad , \qquad (10)$$

with $\mathbf{r}_n \equiv \mathbf{r} - \mathbf{R}_n$. Equation (10) tells us that the scattered wave is a sum of outgoing spherical waves centred on the spheres of the plane, and the wave scattered from the sphere centred at the $\mathbf{R}_n$-site differs from that scattered from the sphere at $\mathbf{R}_n = 0$ (at the origin of coordinates) only by the phase factor $\exp(i\mathbf{k}_{//} \cdot \mathbf{R}_n)$.

Now the coefficients $b_{\ell m}^{+E}$, $b_{\ell m}^{+H}$ which determine the scattered wave from the sphere at the origin are determined from the *total* incident wave on that sphere, which consists of the incident plane wave and the scattered waves from all other spheres described by the terms corresponding to $\mathbf{R}_n \neq 0$ in (10). The incident plane wave is described as we have seen by (7) with $z_{\ell}(qr) = j_{\ell}(qr)$ and the coefficients (9). The incident on the sphere at the origin wave which consists of the scattered waves from all other spheres can also be expanded into a sum of spherical waves about the origin; we have

$$\tilde{\mathbf{E}}'_{sc}(\mathbf{r}) = \sum_{\ell=1}^{\infty} \sum_{m=-\ell}^{\ell} \left(\frac{i}{q} b'^{E}_{\ell m} \nabla \times j_{\ell}(qr) \mathbf{X}_{\ell m}(\hat{\mathbf{r}}) + b'^{H}_{\ell m} j_{\ell}(qr) \mathbf{X}_{\ell m}(\hat{\mathbf{r}}) \right) \quad , \quad (11)$$

where $\tilde{\mathbf{E}}'_{sc}(\mathbf{r})$ is obtained from $\tilde{\mathbf{E}}_{sc}(\mathbf{r})$ by omitting the term corresponding to $\mathbf{R}_n = 0$ in the sum of equation (10). It can be shown that

$$b'^{E}_{\ell'm'} = \sum_{\ell=1}^{\infty} \sum_{m=-\ell}^{\ell} \left(\Omega^{(1)}_{\ell'm';\ell m} b^{+E}_{\ell m} - \Omega^{(2)}_{\ell'm';\ell m} b^{+H}_{\ell m} \right) \quad (12a)$$

$$b'^{H}_{\ell'm'} = \sum_{\ell=1}^{\infty} \sum_{m=-\ell}^{\ell} \left(\Omega^{(2)}_{\ell'm';\ell m} b^{+E}_{\ell m} + \Omega^{(1)}_{\ell'm';\ell m} b^{+H}_{\ell m} \right) \quad , \quad (12b)$$

where $\Omega^{(1)}_{\ell'm';\ell m}$, $\Omega^{(2)}_{\ell'm';\ell m}$ depend on the $\mathbf{k}_{//}$, ω of the incident wave and the geometry (the parameters of the lattice (1)) of the plane of the spheres.

Explicit expressions for the $\Omega^{(1)}_{\ell'm';\ell m}$, $\Omega^{(2)}_{\ell'm';\ell m}$ matrix-elements have been derived in [5]. It turns out these matrices involve quantities which are either identical or very similar to those met in the treatment of electron scattering by a plane of atoms which allows one to use existing programs [6] for the computation of these quantities. We note also that the Ω-matrices have certain symmetry properties which facilitate their computation [4].

The scattered wave from the sphere at the origin, i.e., the coefficients $b^{+E(H)}_{\ell m}$, is determined from the total wave incident on this sphere, i.e., from the coefficients $a^{0E(H)}_{\ell m}$ and $b'^{E(H)}_{\ell m}$, by the requirement that the tangential components of the electric and magnetic fields be continuous on the surface of the sphere (the same from within and outside the sphere); and because of the spherical shape of the scatterer, we obtain

$$b^{+E(H)}_{\ell m} = T^{E(H)}_{\ell}(\omega) \left(a^{0E(H)}_{\ell m} + b'^{E(H)}_{\ell m} \right) \quad . \quad (13)$$

234

Explicit expressions for $T_\ell^{E(H)}(\omega)$ have been derived in [5]. Substituting (12) in (13), we can obtain the coefficients $b_{\ell m}^{+E(H)}$ of the scattered wave (10) in terms of the coefficients $a_{\ell m}^{0E(H)}$ of the incident plane wave. Using standard formulae, we can rewrite the scattered wave (10) as a sum of plane waves as follows [4]

$$\tilde{E}_{sc}(\mathbf{r}) = \sum_g \left[E_{sc}\right]_g^\pm \exp\left(i\mathbf{K}_g^\pm \cdot \mathbf{r}\right) \qquad , \qquad (14)$$

where the superscript $+ (-)$ holds for $z>0$ ($z<0$), respectively. $\mathbf{K}_g^\pm$ are given by (6); they have the same magnitude q and the same reduced wavevector $\mathbf{k}_{//}$ as the incident wave. We see the scattered wave consists, in general, of a number of diffracted beams corresponding to different $\mathbf{g}$-vectors, as indicated schematically in Fig. 1. We note, however, when $q < |\mathbf{g}|$ for $\mathbf{g} \neq \mathbf{0}$, only the $\mathbf{g}=\mathbf{0}$ term will contribute to the scattered wave. The coefficients in (14) are, of course, functions of the $b_{\ell m}^{+E(H)}$ coefficients and through them depend on the incident plane wave. In what follows we shall need to denote this dependence explicitly. We write the three components (i=x, y, z) of the electric field vector of the incident plane wave as

$$\left[E_{in}\right]_{g'i}^s \exp\left(i\mathbf{K}_{g'}^s \cdot \mathbf{r}\right) \qquad , \qquad (15)$$

where $\mathbf{K}_{g'}^s$ is given by (6); $s = + (-)$ corresponds to a wave incident on the plane of spheres from the left (right). Let us for the sake of clarity assume that a plane wave (15) is incident on the plane of spheres from the left as in Fig. 1a. Then the transmitted wave (incident+scattered) on the right of the plane of spheres can be written as

$$\sum_g \left[E_{tr}\right]_{gi}^+ \exp\left(i\mathbf{K}_g^+ \cdot \mathbf{r}\right) \quad , \quad z > 0 \qquad (16a)$$

with

$$\left[E_{tr}\right]_{gi}^+ = \left[E_{in}\right]_{g'i}^+ \delta_{gg'} + \left[E_{sc}\right]_{gi}^+ = \sum_{i'} M_{gi;g'i'}^{++} \left[E_{in}\right]_{g'i'}^+ \qquad (16b)$$

and the reflected wave as

$$\sum_{g}\left[E_{rf}\right]_{gi}^{-}\exp\left(iK_{g}^{-}\cdot r\right) \quad , \quad z < 0 \tag{17a}$$

with

$$\left[E_{rf}\right]_{gi}^{-}=\left[E_{sc}\right]_{gi}^{-}=\sum_{i'}M_{gi;g'i'}^{-+}\left[E_{in}\right]_{g'i'}^{+} \quad . \tag{17b}$$

Similarly, we can define the transmission matrix elements $M_{gi;g'i'}^{--}$ and

reflection matrix elements $M_{gi;g'i'}^{+-}$ starting from a plane wave incident on the plane of spheres from the right (see Fig. 1b).

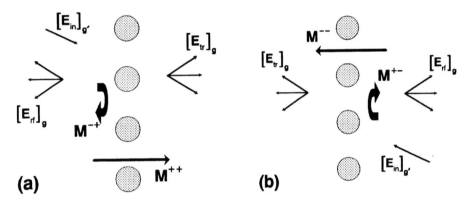

Figure 1. Scattering of a plane wave from a periodic monolayer of spherical particles: (a) plane wave incident from the left; (b) plane wave incident from the right.

The transmittance T and reflectance R are defined as the ratios of the fluxes of the transmitted and reflected waves, respectively, to the flux of the incident wave. Integrating the Poynting vector over the xy-plane on both sides of the plane of spheres and taking the average over a period, we obtain

$$T = \sum_{g,i}\left[E_{tr}\right]_{gi}^{+}\left(\left[E_{tr}\right]_{gi}^{+}\right)^{*}K_{gz}^{+}\bigg/\sum_{i}\left[E_{in}\right]_{g'i}^{+}\left(\left[E_{in}\right]_{g'i}^{+}\right)^{*}K_{g'z}^{+} \tag{18}$$

$$R = \sum_{g,i} \left[E_{rf}\right]_{gi}^{-} \left(\left[E_{rf}\right]_{gi}^{-}\right)^{*} K_{gz}^{+} \bigg/ \sum_{i} \left[E_{in}\right]_{g'i}^{+} \left(\left[E_{in}\right]_{g'i}^{+}\right)^{*} K_{g'z}^{+} \quad , \quad (19)$$

where we have assumed an incident wave from the left. The requirement of energy conservation implies that the absorbance of the plane of spheres is given by

$$A = 1 - T - R \qquad . \tag{20}$$

We used the above formalism to calculate the relative transmittance of a square array of gold spheres on a sapphire substrate with respect to the transmittance of the bare substrate for normally incident light. Our results for the transmittance as a function of the wavelength of the incident light are shown in Fig. 2 for different values of S, the radius of the spheres. The formulae required for the calculation of the transmittance through a system consisting of a periodic array of spheres on a slab of uniform dielectric constant have been derived in [4]. The experimental data in Fig. 2 are due to Craighead and Niklasson [2]. In the actual experiments the gold particles

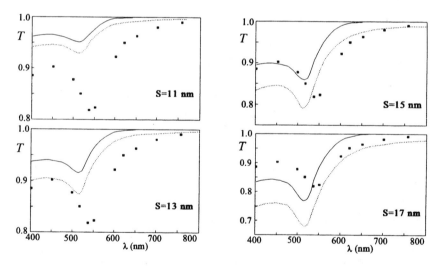

Figure 2. Relative transmittance T of a square array of gold spheres on a sapphire substrate with respect to the transmittance of the bare substrate for normally incident light of wavelength λ (full line), for various sphere radii S. The lattice constant is equal to 50nm. The transmittance (dotted line) of the coating array of gold particles alone and the experimental data (squares) are also shown.

were nearly spherical with diameters between 20 and 35nm. We see the calculated transmittance agrees reasonably well with the experimental data when 15nm<S<17nm. The main discrepancy with the experimental data relates to the position of the minimum of the transmittance (corresponding to a maximum of the absorbance) whose calculated value of 512nm lies below its experimental value of 540nm. We should mention that in the case of gold on sapphire and in the range of the wavelengths considered, our method gave essentially the same results as the electrostatic dipole approximation treatment of the same problem. However, this is not generally true. We have shown [7] that in certain cases ℓ-pole terms beyond the dipole ($\ell = 1$) are very important and change considerably the calculated transmittance and absorbance of a two-dimensional array of particles. This is demonstrated in Fig. 3 which shows the absorbance of a periodic array of plasma particles described by

$$\varepsilon_M(\omega) = 1 - \omega_p^2 \Big/ \omega\big(\omega + i\tau^{-1}\big) \qquad , \qquad (21)$$

where $\hbar\omega_p = 6.93\text{eV}$ and $\hbar\tau^{-1} = 0.158\text{eV}$. The ℓ_{max} in this figure tells us that in the ℓ-expansion of the EM field we have kept all terms with $\ell \le \ell_{max}$. In the dipole approximation, $\ell_{max} = 1$.

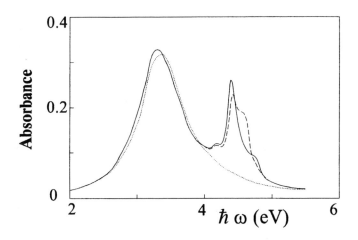

Figure 3. Absorbance of light of angular frequency ω, incident normally on a square array of silver spheres of radius S=80Å with lattice constant a=200Å: full line ℓ_{max}=6; dashed line, ℓ_{max}=4; dotted line, ℓ_{max}=1.

3. Photonic Crystals

3.1. PHOTONIC COMPLEX BAND-STRUCTURE

A crystalline slab of spheres is constituted as a stack of identical planes of spheres parallel to the xy-plane, so the (n+1)th plane along the positive z direction is obtained from the (n)th plane by a simple translation described by a vector $\mathbf{a}_3$ (Fig. 4). When the stack of planes extends over all space (from $z \to -\infty$ to $z \to \infty$), we have an infinite crystal and $\{\mathbf{a}_1, \mathbf{a}_2, \mathbf{a}_3\}$ is a set of primitive vectors for this crystal. We define a reduced $\mathbf{k}$-zone $(\mathbf{k}_{//}, k_z)$ corresponding to it as follows: $\mathbf{k}_{//} = (k_x, k_y)$ extends over the SBZ of the plane under consideration and $-|\mathbf{b}_3|/2 < k_z < |\mathbf{b}_3|/2$, where $\mathbf{b}_3 = 2\pi \mathbf{a}_1 \times \mathbf{a}_2 / (\mathbf{a}_1 \cdot (\mathbf{a}_2 \times \mathbf{a}_3))$. The complex band structure of the EM field associated with the given surface is obtained in the same way, as for the Schrödinger wavefield of electrons in an ordinary crystal [6, 8].

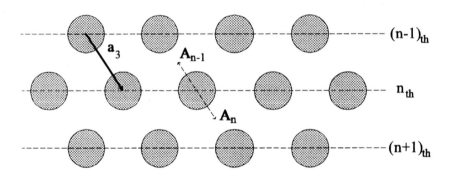

Figure 4. A composite crystal as a stack of successive layers parallel to a crystallographic plane.

In the region between the (n)th and the (n+1)th planes of spheres the wavefield of given ω and $\mathbf{k}_{//}$ has the form

$$\sum_g \left\{ \mathbf{E}_g^+(n) \exp\left(i\mathbf{K}_g^+ \cdot (\mathbf{r} - \mathbf{A}_n)\right) + \mathbf{E}_g^-(n) \exp\left(i\mathbf{K}_g^- \cdot (\mathbf{r} - \mathbf{A}_n)\right) \right\} \quad , \quad (22)$$

where $\mathbf{A}_n$ is the midpoint between the (n)th and the (n+1)th plane of spheres (see Fig. 4). The vector coefficients $\mathbf{E}_g^{\pm}(n)$ are related to the $\mathbf{E}_g^{\pm}(n+1)$ coefficients through the scattering properties of the (n+1)th plane. We have

$$E_{gi}^-(n) = \sum_{g',i'} Q_{gi;g'i'}^{--} E_{g'i'}^-(n+1) + \sum_{g',i'} Q_{gi;g'i'}^{-+} E_{g'i'}^+(n) \qquad (23a)$$

$$E_{gi}^+(n+1) = \sum_{g',i'} Q_{gi;g'i'}^{++} E_{g'i'}^+(n) + \sum_{g',i'} Q_{gi;g'i'}^{+-} E_{g'i'}^-(n+1) \quad , \qquad (23b)$$

where

$$Q_{gi;g'i'}^{--} = \exp\left(-\frac{i}{2}\left(\mathbf{K}_g^- + \mathbf{K}_{g'}^-\right) \cdot \mathbf{a}_3\right) M_{gi;g'i'}^{--}$$

$$Q_{gi;g'i'}^{-+} = \exp\left(-\frac{i}{2}\left(\mathbf{K}_g^- - \mathbf{K}_{g'}^+\right) \cdot \mathbf{a}_3\right) M_{gi;g'i'}^{-+}$$

$$Q_{gi;g'i'}^{++} = \exp\left(\frac{i}{2}\left(\mathbf{K}_g^+ + \mathbf{K}_{g'}^+\right) \cdot \mathbf{a}_3\right) M_{gi;g'i'}^{++}$$

$$Q_{gi;g'i'}^{+-} = \exp\left(\frac{i}{2}\left(\mathbf{K}_g^+ - \mathbf{K}_{g'}^-\right) \cdot \mathbf{a}_3\right) M_{gi;g'i'}^{+-} \qquad . \qquad (24)$$

The phase factors in the above equations arise from the fact that the waves in (22) refer to an origin half-way between two successive planes, whereas the $M_{gi;g'i'}^{ss'}$ matrix elements, as defined in section 2, refer to an origin of coordinates at the centre of the plane of spheres.

Now it is in the nature of a generalised Bloch wave that

$$E_g^{\pm}(n+1) = \exp(i\mathbf{k} \cdot \mathbf{a}_3) E_g^{\pm}(n) \qquad (25)$$

$$\mathbf{k} = \left(\mathbf{k}_{//}, k_z(\mathbf{k}_{//}, \omega)\right) \qquad , \qquad (26)$$

where k_z is a function, to be determined, of $\mathbf{k}_{//}$ and ω.

Substituting (25) into (23) leads, after some algebra, to the following system of equations

$$
\begin{pmatrix}
\mathbf{Q}^{++} & \mathbf{Q}^{+-} \\
-\left[\mathbf{Q}^{--}\right]^{-1}\mathbf{Q}^{-+}\mathbf{Q}^{++} & \left[\mathbf{Q}^{--}\right]^{-1}\left[1-\mathbf{Q}^{-+}\mathbf{Q}^{+-}\right]
\end{pmatrix}
\begin{pmatrix}
\mathbf{E}^{+}(n) \\
\mathbf{E}^{-}(n+1)
\end{pmatrix} =
$$

$$
\exp(i\mathbf{k}\cdot\mathbf{a}_3)
\begin{pmatrix}
\mathbf{E}^{+}(n) \\
\mathbf{E}^{-}(n+1)
\end{pmatrix}
\qquad , \qquad (27)
$$

where $\mathbf{E}^{\pm}$ are column vectors with components: $E^{\pm}_{g_1x}, E^{\pm}_{g_1y}, E^{\pm}_{g_1z}, E^{\pm}_{g_2x}, E^{\pm}_{g_2y}, E^{\pm}_{g_2z}, \ldots$ In practice we keep N **g**-vectors in the calculation (those of smallest magnitude) in which case $\mathbf{E}^{\pm}$ has 3N components. The enumeration of the **g**-vectors implied in the definition of $\mathbf{E}^{\pm}$ allows us to write the Q-matrix elements in a compact form as matrices $\mathbf{Q}^{++}$, $\mathbf{Q}^{+-}$ etc., each has $3N \times 3N$ elements; 1 is the $3N \times 3N$ unit matrix. This means that (27) is a system of 6N homogeneous equations with 6N unknowns. Equation (27) is the EM field equivalent to McRae's equation for the Schroedinger wavefield of an electron in a crystal [6,8]. For given $\mathbf{k}_{//}$ and ω, we obtain 6N eigenvalues of k_z from the solvability condition of (27). Looked upon as functions of ω, these define, for each $\mathbf{k}_{//}$, 6N different lines $k_z(\mathbf{k}_{//}, \omega)$ in the complex k_z-space. We shall not be concerned here with the general properties of these so-called real-frequency lines. It is sufficient for our purposes to note the following. For given $\mathbf{k}_{//}$ and ω, one-third of the corresponding modes (eigensolutions of (27)) are dropped because the divergence of the electric field associated with them is not zero, as it should be in a free of charge space. Of the remaining, none or at best a few are propagating waves corresponding to real eigenvalues of k_z, and the rest are evanescent waves corresponding to complex eigenvalues of k_z. These have an amplitude which increases exponentially in the positive or negative z-direction and do not exist as physical entities in an infinite crystal. However, they are useful in constructing physical solutions of the EM field in a semi-infinite crystal (extending from $z \to -\infty$ to z=0) or in a slab of finite thickness (see section 3.2).

A region of frequencies where propagating waves do not exist for a given $\mathbf{k}_{//}$ constitutes a frequency gap of the EM field for the given $\mathbf{k}_{//}$. If the crystal is not infinite but of finite extension in the z-direction (a slab), the transmission of EM waves with ω within the gap is determined essentially by the eigenmode, of the given $\mathbf{k}_{//}$ and ω, with the smallest imaginary component of k_z. For the reasons pointed out by Yablonovitch [9], more

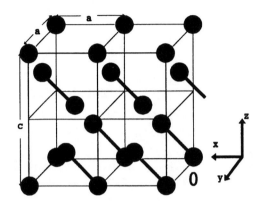

Figure 5. A photonic crystal of tetragonal symmetry. Each layer parallel to the xy-plane consists of two planes of spheres.

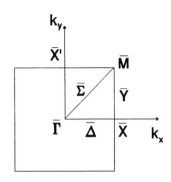

Figure 6. The surface Brillouin zone of the crystal (xy plane)

important are the regions of frequency, so-called absolute frequency gaps, over which no propagating Bloch waves exist no matter the value of $\mathbf{k}_{//}$. Using the formalism we have described, we were able to prove the existence of an absolute frequency gap in the photonic crystal shown in Fig. 5. The crystal, with a centred tetragonal lattice, is described by the primitive vectors $\mathbf{a}_1 = (a,0,0)$ and $\mathbf{a}_2 = (0,a,0)$ in the xy-plane, and $\mathbf{a}_3 = (a/2,a/2,c/2)$ and has two spheres per unit cell at $(0,0,0)$ and $(a/2,0,c/4)$. We note that when $c = a\sqrt{2}$, one obtains the well-known diamond structure. The spheres, of radius S=0.265a, have a dielectric constant $\varepsilon_M = 12.96$, and the surrounding medium has a dielectric constant equal to unity.

We calculated the photonic complex band structure for the (100) surface of the above crystal for $\mathbf{k}_{//} = (k_x, k_y)$ along the symmetry lines of the SBZ shown in Fig. 6 and at other points as well. For given $\mathbf{k}_{//}$, we obtained the real frequency lines $k_z(\mathbf{k}_{//}, \omega)$ as functions of ω, with real values of k_z over certain regions of ω and complex values elsewhere. The sections of real

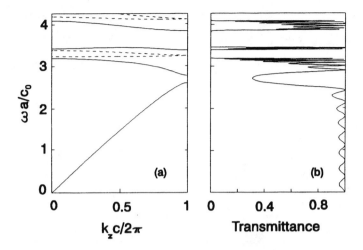

Figure 7. (a) Frequency bands for $k_{//}=0$ in a crystal of tetragonal structure (Fig.5 with $c/a=1.7$) consisting of non-overlapping spheres ($\varepsilon_M=12.96$) in a mediun of dielectric constant equal to unity. c_0 denotes the speed of light. (b) Transmittance of light incident normally on a slab of the crystal consisting of eight layers.

$k_z(k_{//}, \omega)$ of these lines define the frequency bands and the sections of complex $k_z(k_{//}, \omega)$ the frequency gaps for the given $k_{//}$. In Fig. 7a we show the frequency bands for $k_{//}=0$. After calculating the frequency bands for a sufficient number of $k_{//}$-points in the SBZ, we obtained the projection of the frequency band structure on the SBZ shown in Fig. 8. The shaded areas on this diagram correspond to frequency gaps. We see that for the given

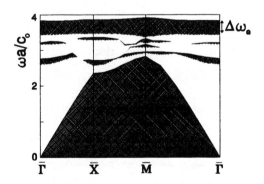

Figure 8. Projection of the photonic band structure on the SBZ. The gaps in the frequency spectrum are represented by the shaded areas. We observe the existence of an absolute gap of width $\Delta\omega_G$

structure there is a region of frequencies where no propagating wave exists in the crystal regardless the value of $\mathbf{k}_{//}$. This absolute gap extends over a region of $\Delta\omega_G$ around the midgap frequency ω_{MG}. A discussion of the dependence of the gap on the geometry of the structure can be found in [10].

3.2. TRANSMISSION OF LIGHT THROUGH A SLAB

We assume the slab is a stack of a finite number of sphere planes with the same periodic structure parallel to the xy-surface. In this case the transmission and absorption of incident light of given $\mathbf{k}_{//}$ and ω can be obtained by matching the wave outside the slab (incident+reflected beams on one side, transmitted beams on the other side of the slab) to the wavefield inside the slab, which can be written as a superposition of Bloch waves (propagating and evanescent) of the same $\mathbf{k}_{//}$, ω of the corresponding infinite crystal. An alternative method, the so-called doubling-layer method [6,8], consists of constructing the scattering matrix elements of the slab from those of the individual planes which make up the slab. We proceed as follows. Combine the matrix elements of two successive planes, as shown schematically in Fig. 9, to obtain those of the double plane.

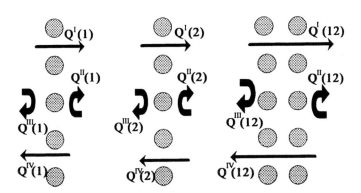

Figure 9. The Q-matrix elements for two successive layers are obtained from those of the individual layers (schematic description).

The matrices $\mathbf{Q}^{++}, \mathbf{Q}^{+-}, \mathbf{Q}^{-+}, \mathbf{Q}^{--}$ for the first plane are denoted by $\mathbf{Q}^{I}(1), \mathbf{Q}^{II}(1), \mathbf{Q}^{III}(1), \mathbf{Q}^{IV}(1)$, respectively. The corresponding matrices for the second plane are denoted by $\mathbf{Q}^{I}(2), \mathbf{Q}^{II}(2)$, etc. Of course, all matrices refer to the same $\mathbf{k}_{//}$ and ω, and the waves are expressed with respect to origins defined in each region as in Fig. 4. For example, the waves incident from the left on the double plane refer to an origin which lies where the midpoint would be between the first plane and that plane preceding it in the corresponding infinite crystal. It is easy to show that

$$\mathbf{Q}^{\mathrm{I}}(12) = \mathbf{Q}^{\mathrm{I}}(2)\left[1 - \mathbf{Q}^{\mathrm{II}}(1)\mathbf{Q}^{\mathrm{III}}(2)\right]^{-1}\mathbf{Q}^{\mathrm{I}}(1)$$

$$\mathbf{Q}^{\mathrm{II}}(12) = \mathbf{Q}^{\mathrm{I}}(2)\mathbf{Q}^{\mathrm{II}}(1)\left[1 - \mathbf{Q}^{\mathrm{III}}(2)\mathbf{Q}^{\mathrm{II}}(1)\right]^{-1}\mathbf{Q}^{\mathrm{IV}}(2) + \mathbf{Q}^{\mathrm{II}}(2)$$

$$\mathbf{Q}^{\mathrm{III}}(12) = \mathbf{Q}^{\mathrm{IV}}(1)\mathbf{Q}^{\mathrm{III}}(2)\left[1 - \mathbf{Q}^{\mathrm{II}}(1)\mathbf{Q}^{\mathrm{III}}(2)\right]^{-1}\mathbf{Q}^{\mathrm{I}}(1) + \mathbf{Q}^{\mathrm{III}}(1)$$

$$\mathbf{Q}^{\mathrm{IV}}(12) = \mathbf{Q}^{\mathrm{IV}}(1)\left[1 - \mathbf{Q}^{\mathrm{III}}(2)\mathbf{Q}^{\mathrm{II}}(1)\right]^{-1}\mathbf{Q}^{\mathrm{IV}}(2) \qquad , \qquad (28)$$

where argument 12 refers to the $\mathbf{Q}$-matrices of the two planes. Similarly, we can combine the $\mathbf{Q}$-matrices of two double planes to obtain those for four planes and so on. We should also emphasise that the method applies equally well (at the expense of further computation) to slabs consisting of different planes (e.g., planes of different spheres) as long as these have the same periodic structure parallel to the xy-surface.

We used the above method to calculate the transmittance of a slab parallel to the xy- surface of the crystal shown in Fig. 5. The results for light incident normally ($\mathbf{k}_{//}=0$) on a slab which consists of eight layers of spheres (each layer consists of two planes of spheres) are shown in Fig. 7b, next to the frequency band structure of the same $\mathbf{k}_{//}$. The relationship between the two is obvious: the transmittance practically vanishes or is very small for frequencies within gaps and is non zero (close to unity) for frequencies within bands that couple to the incident field. The small transmittance for frequencies within the lowest gap is due to the fact that the imaginary part of $k_z(\mathbf{k}_{//} = 0, \omega)$, which determines the attenuation of the EM wave, is quite small in this case leading to an attenuation length comparable with the thickness of the slab. The broken lines in Fig. 7a correspond to normal modes of the EM field in the crystal which are not activated by the incident radiation and therefore do not allow the transmission of light. For an explanation of this phenomenon, see [10]. The oscillations of the transmission coefficient in the lowest frequency range, where an effective medium approximation is valid, results from interference between waves reflected at the two surfaces of the slab [10,11].

3.3. INTERFACE STATES OF THE ELECTROMAGNETIC FIELD

In Fig. 10a we show the transmission coefficient of light incident normally on a slab which has the tetragonal structure (c/a=1.7) of Fig. 5. The slab consists of nine layers of spheres; all spheres have the same radius S=0.265a, except those of the second plane of the fifth layer (we count as first the layer on which the light is incident) which have a smaller radius $S_i = rS$. We refer to this plane as the impurity plane. All the spheres in the slab, including those of the impurity plane, have the same dielectric constant $\varepsilon_M = 12.96$. The dielectric constant of the host material equals unity. We see a sharp resonance in the transmittance curve appears within the absolute gap (see Fig. 8) of the crystal and this moves to higher frequencies the smaller the radius of the impurity spheres. We infer from this, the existence of interface states of the EM field (states localised on the impurity plane) with reduced wavevector $\mathbf{k}_{//} = 0$ whose eigenfrequencies within the photonic gap correspond to the above resonances. It appears a reduction of the region of

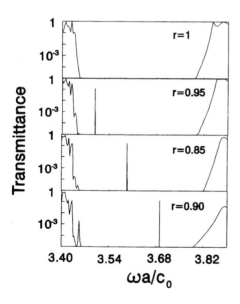

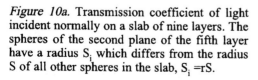

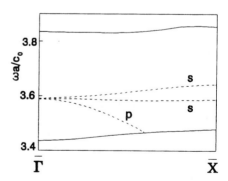

Figure 10a. Transmission coefficient of light incident normally on a slab of nine layers. The spheres of the second plane of the fifth layer have a radius S_i which differs from the radius S of all other spheres in the slab, S_i =rS.

Figure 10b. Bands of interface states (localised on the impurity plane of a nine-layer slab) along the $\overline{\Gamma}\overline{X}$ symmetry line. These are shown by dashed lines. The solid lines represent the edges of the frequency gap. The slab is constructed as in Fig. 10a with r=0.9.

higher dielectric constant within the unit cell of the photonic crystal pushes out of the lower band of frequencies into the gap one or more states (normal modes) of the EM field. If we had an isolated impurity sphere, a "pushed-up" state would have been localised on that sphere. In this respect, the phenomenon is analogous to that of acceptor-type impurity states in a semiconductor. There the introduction of a less attractive centre in place of an existing stronger one pushes an electron state off the valence band into the energy gap of the semiconductor. In our example, we do not have isolated impurity spheres but an impurity plane, and we expect a band, or bands, of interface states to be associated with it. Moreover, we expect a dispersion of these bands because of multiple scattering within the plane. This is indeed the case as seen in Fig. 10b. The dispersion curves of this figure were obtained from corresponding transmittance curves of light incident on a slab of the crystal at an angle corresponding to the given $k_{//}$. The bands denoted by "s" are activated by s-polarised light (the electric field is parallel to the plane of the spheres) and the band denoted by "p" is activated by p-polarised light (the electric field lies in the plane of incidence and has a component normal to the plane of the spheres). For a more extensive discussion of interface states of the EM field see [12].

3.4. OPTICALLY ACTIVE PHOTONIC CRYSTALS

In this section we deal with a property of photonic crystals which is intimately connected with the vector nature of the EM field and therefore has no analogue in the electron physics of semiconductors. The question is: how does a photonic crystal (a slab of it) affect the plane of polarisation of an EM wave transmitted through it? We can put the question differently: can one design optically active photonic crystals, i.e., crystals which rotate the plane of polarisation of an EM wave transmitted through them? We know the optical activity of a material is often related to a helicoidal symmetry in the structure of the material and we have shown this is also true for photonic crystals. We considered, as an example, the crystal shown in Fig. 11. The periodicity of the layers parallel to the xy-plane is described by a square lattice defined by

$$\mathbf{a}_1 = (a,0,0) \, , \, \mathbf{a}_2 = (0,a,0) \qquad . \qquad (29)$$

The (n+1)th layer along the z-axis is obtained from the (n)th layer by the primitive translation

$$\mathbf{a}_3 = (0,0,c) \qquad (30)$$

and each layer consists of four planes defined by the four-point basis at $(0,0,0)$, $(b,0,c/4)$, $(b,b,c/2)$ and $(0,b,3c/4)$. The helicoidal configuration of the spheres along the z-axis is indicated by the broken lines in Fig. 11. The spheres are all the same: of the same dielectric constant $\varepsilon_M = 9.3636$, of the same radius S, and are embedded in a medium of dielectric constant $\varepsilon=1$.

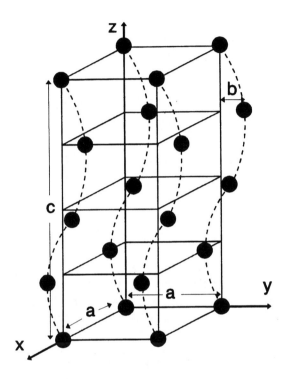

Figure 11. A photonic crystal exhibiting optical activity.

In Fig. 12 we show the angle of rotation φ of an EM wave transmitted through a slab of the above crystal 30-layers thick. The incident and transmitted waves are plane waves with wavevectors normal to the surface of the slab. We see at the lower frequencies the angle of rotation is positive (meaning an anticlockwise rotation) and increases slowly with ω. At higher frequencies (above the frequency gap indicated by the shaded region), the variation of φ with ω is more complicated; it can be positive or negative and become very large, which suggests there can be photonic crystals with optical activity much greater than those of the naturally occurring, optically active materials. A more extensive discussion of optical activity and of the related phenomenon of optical anisotropy, in relation to photonic crystals, can be found in [13].

248

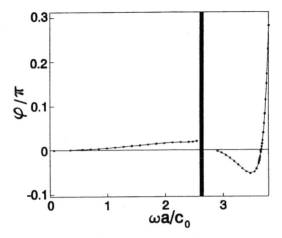

Figure 12. Variation of the angle of rotation of the plane of polarization with incident frequency. The wave is transmitted through a slab 30-layers thick. The parameter values (see Fig. 11) are c/a=2, b/a=0.25 and S/a=0.25.

4. The Treatment of Disorder

We shall consider only the scattering of light by a disordered two-dimensional array of spheres. We have already mentioned in our introduction that when an array of metallic particles is deposited on a dielectric slab, the particles (spheres) are not situated on a regular lattice but usually are randomly distributed in space, and they may also have different sizes. The easiest way to introduce disorder into the calculation is to assume the metallic spheres occupy randomly a fraction of the sites of a regular lattice, e.g., a square lattice. We can then proceed to calculate the scattering of an EM wave from such an array of spheres using methods analogous to those used in the treatment of electron scattering in substitutionally disordered alloys [14]. The simplest approximation, known as the average T-matrix approximation (ATA) [15], takes the following form in the EM problem. We note that the T-matrix elements appearing in (13) depend of course on the size and on the dielectric function of the given scatterer. If there is more than one type of scatterer, these will be described by different

T-matrices $\mathbf{T}^A, \mathbf{T}^B$, etc. In the ATA method we replace every one of the scatterers by an average T-matrix defined by

$$\langle \mathbf{T}(\omega) \rangle = c_A \mathbf{T}^A(\omega) + c_B \mathbf{T}^B(\omega) + \dots \quad , \tag{31}$$

where c_A, c_B,... are the concentrations of scatterers A, B,... We have

$$c_A + c_B + \ldots = 1 \qquad (32)$$

In the special case when a fraction c_A of the sites is occupied by a scatterer A and the rest of the sites are empty we obtain

$$\langle T(\omega) \rangle = c_A T^A(\omega) \qquad (33)$$

With each lattice site occupied by the same average scatterer, the calculation proceeds as with the ordered array of scatterers.

In a more sophisticated method, known as the coherent potential approximation (CPA) [16], one introduces the average CPA scatterer in such a way that the correction to the scattering due to the difference of the actual scatterers from the CPA scatterer vanishes on the average. The formulae necessary for the evaluation of the CPA scatterer have been derived in [17]. In Fig. 13 we show the absorbance as a function of frequency of p-polarised light incident at an angle $\theta=\pi/4$ ($\mathbf{k}_{//}=(k_{//},0)$) on a disordered array of plasma spheres occupying randomly 25% of the sites of a square lattice, with a lattice constant a=100Å, as calculated by the ATA method and by the CPA method. In the same figure we show the absorbance by an ordered array of spheres on a square lattice, whose lattice constant $a' = a/\sqrt{c_A} = 182.57$Å is such that the coverage (the number of spheres per unit area) is the same as in the disordered array. The spheres have the same radius S=50Å and are characterised by the same dielectric function, given by (21). The low-energy peak of the absorbance curve corresponds to a parallel-mode resonance (the induced charge oscillates parallel to the plane of the spheres) and is excited by the component of the incident electric field parallel to the plane. The high-energy peak corresponds to a normal-mode resonance (the induced charge oscillates normal to the plane of the spheres) and is excited by the normal component of the incident electric field. We see the disorder induces a shift of the parallel-mode absorbance peak to lower frequencies and a shift of the normal-mode peak to higher frequencies. We can understand why the peak is broader for the parallel-mode resonance as follows. In this case the induced dipoles on, say, two neighbouring spheres may be parallel, normal, or any angle to the axis (straight line) joining the centres of the spheres, whereas in the normal mode the induced dipoles are always normal to the line joining the centres of the spheres. The interaction between two dipoles depends on the orientation of the dipoles relative to the axis joining them and therefore we obtain a spread of resonance frequencies corresponding to the different orientations of the dipoles. In a disordered array of spheres where different cluster formations are possible, we expect a

250

correspondingly greater broadening of the above absorbance peak. The CPA method takes into account the above effect by the way the CPA scatterer is constructed. The ATA method fails in this respect, as one can see from the results of Fig. 13.

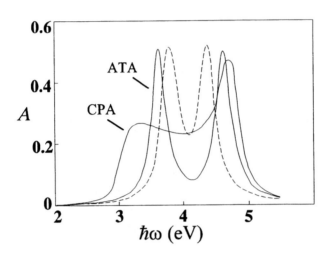

Figure 13. Absorbance *A* of p-polarized light of angular frequency ω incident at an angle θ=π/4 on a disordered array of plasma spheres of radius S=50Å occupying randomly 25% of the sites of a square lattice of lattice constant a=100Å (solid line) and on an ordered square array of plasma spheres of radius S=50Å with lattice constant a=182.57Å (dashed line).

References

1. Abelès, F., Borensztein, Y., and López-Rios, T. (1984) Optical properties of discontinuous thin films and rough surfaces of silver, *Festkörperprobleme (Advances in Solid State Physics)* vol. XXIV (Braunschweig: Vieweg), 93-117.
2. Craighead, H.G., and Niklasson, G.A. (1984) Characterization and optical properties of arrays of small gold particles, *Appl. Phys. Lett.* **44** 1134-1136.
3. Jackson, J.D. (1975) *Classical Electrodynamics*, John Wiley & Sons, Inc., New York.
4. Stefanou, N. and Modinos, A. (1991) Scattering of light from a two-dimensional array of spherical particles on a substrate, *J. Phys.: Condens. Matt.* **3**, 8135-8148.
5. Modinos, A. (1987) Scattering of electromagnetic waves by a plane of spheres-formalism, *Physica* A **141**, 575-588.
6. Pendry, J.B. (1974) *Low Energy Electron Diffraction*, Academic Press, London.
7. Stefanou, N. and Modinos, A. (1991) Optical properties of thin discontinuous metal films, *J. Phys.: Condens. Matt.* **3**, 8149-8157.
8. Modinos, A. (1984) *Field, Thermionic and Secondary Electron Emission Spectroscopy*, Plenum Press, New York.
9. Yablonovitch, E. (1993) Photonic band-gap crystals *J. Phys.: Condens. Matt.* **5**, 2443-2460.

10. Stefanou, N., Karathanos, V., and Modinos, A. (1992) Scattering of electromagnetic waves by periodic structures, *J. Phys.: Condens. Matt.* **4**, 7389-7400.

11. Born, M. and Wolf, E. (1975) *Principles of Optics*, Pergamon Press, Oxford.

12. Karathanos, V., Modinos, A., and Stefanou, N. (1994) Planar defects in photonic crystals, *J. Phys.: Condens. Matt.* **6**, 6257-6264.

13. Karathanos, V., Stefanou, N., and Modinos, A. (1995) Optical activity of photonic crystals, *J. Mod.Optics* **42**, 619-626.

14. Gyorffy, B.L. and Stocks, G.M. (1979) First principles band theory for random metallic alloys, in P. Phariseau, B.L. Gyorffy and L. Scheine (eds.), *Electrons in Disordered Metals and at Metallic Surfaces*, Plenum Press, New York, 89-192.

15. Beeby, J.L. (1964) The electronic structure of disordered systems, *Proc. R. Soc. (London)* **A279**,82-97.

16. Soven, P. (1967) Coherent-potential model of substitutional disordered alloys, *Phys. Rev.* **156**, 809-813.

17. Stefanou, N. and Modinos, A. (1993) Scattering of electromagnetic waves by a disordered two-dimensional array of spheres, *J. Phys.: Condens. Matt.* **5**, 8859-8868.

ELECTROMAGNETIC FIELD DISTRIBUTIONS IN COMPLEX DIELECTRIC STRUCTURES

P. M. BELL
Condensed Matter Theory Group, The Blackett Laboratory
Imperial College, Prince Consort Road, London SW7 2BZ, UK

L. MARTIN MORENO
Instituto de Ciencia de Materiales (Sede B)
Consejo Superior de Investigaciones Científicas
Universidad Autónoma de Madrid, 28049 Madrid, Spain

AND

F. J. GARCIA VIDAL AND J. B. PENDRY
Condensed Matter Theory Group, The Blackett Laboratory
Imperial College, Prince Consort Road, London SW7 2BZ, UK

Abstract. Recently there has been renewed interest in the electromagnetic (EM) properties of complex dielectric materials. One outcome of this interest has been the application of novel methods in solving Maxwell's equations. One of these methods, the transfer matrix technique, is particularly suitable for the study of the propagation of EM waves. In this work we show how it is possible to apply it to the study of several properties, such as energy loss of charged particles passing close to corrugated surfaces, absorption of light in metallic systems and enhanced Raman scattering in rough surfaces. As an example we study these properties in an array of metallic cylinders and explore how they vary as a function of the cylinders packing fraction.

1. Introduction

Following the successful experiments of Yablonovitch [1], there has been much interest in the area of photonics because of the potential application to areas as diverse as the inhibition of spontaneous emission and,

253

C. M. Soukoulis (ed.), Photonic Band Gap Materials, 253–260.
© 1996 *Kluwer Academic Publishers. Printed in the Netherlands.*

ultimately, the design of optical computers. Following hard on the heels of Yablonovitch, several theoretical approaches were developed [2-5] which can be used to study the behaviour of proposed materials before the immensely difficult task of fabrication is attempted. However, in this paper we show that the theoretical techniques which have been developed are not just restricted to materials design, but can be used to provide new insights into problems where the interaction of radiation with complicated metallic structures gives rise to phenomena which are not well understood. The problems we consider here are electron energy loss effects and surface enhanced Raman scattering (SERS). In both cases, it is the effect of coupling between the metallic objects which is dominant, but provides enormous difficulties to the standard mathematical approaches; using the transfer matrix technique for EM waves [6, 7] we are able to treat this coupling with no difficulty and, perhaps more importantly, we are able to gain extra information because we can map out the electric field in real space.

Perhaps the simplest system which displays all these interesting effects is the close packed chain of metallic cylinders and this is the system we study here, Figure 1. Interesting behaviour has already been observed in

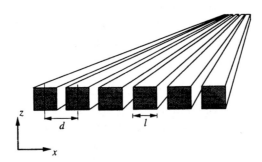

Figure 1. The system of metallic cylinders. The rods are infinitely long in the y-direction, and we study a single layer which forms an infinite chain in the x-direction. We keep the width constant, $l = 50\text{Å}$ and vary the spacing between cylinders, d, over a wide range.

similar systems [8], but in this paper we explore how the behaviour of the system evolves with the packing fraction.

2. Electron Energy Loss Effects

The problem of determining the energy loss of fast moving electrons to metallic surfaces, which are rough on the nanometre scale, has proven an interesting and complicated problem for some time. Some beautiful experiments conducted by Howie and Walsh [9] on aluminium colloids showed a broad loss spectrum which could not be interpreted by the then available analytic models. In principle, mechanisms for the electron energy loss are well understood: evanescent waves emitted from the moving charge can scatter from the surface of the material and undergo processes such as refraction or diffraction. These processes can allow the wavevector to become real and thus the possibility arises for the wave to carry energy away from the electron to the material, exciting plasmon modes which are characteristic of the type and shape of the material. However, analytical models considering losses to a single sphere [10] or to two coupled spheres [11] or employing an effective medium approach proved unable to describe the complexity of the spectra to any great precision.

Pendry and Martín Moreno [12] described how to use the EM transfer matrix method as a tool for the study of energy loss in these systems. Given details of the structure, the transfer matrix method will calculate the reflection matrix, $\mathbf{R}(\mathbf{k}, \sigma; \mathbf{k}', \sigma')$ where $\mathbf{k}$ labels the wavevector and σ labels the polarisation of the wave [6, 7]. With this information, the total E–field at the surface of the material (defined to lie in the xy-plane) can be calculated as a sum of the incident waves from the electron (constructed using a Lorentz transformation of the Coulomb field of the particle at rest) and the reflected waves from the surface

$$\mathbf{E}_{\text{total}}(\omega) = \sum_{\mathbf{k}, \sigma} \left(\mathbf{E}_{\text{inc}}(\mathbf{k}, \sigma, \omega) + \sum_{\mathbf{k}', \sigma'} \mathbf{R}(\mathbf{k}, \sigma; \mathbf{k}', \sigma'; \omega) \mathbf{E}(\mathbf{k}', \sigma', \omega) \right), \quad (1)$$

where ω is the frequency and similarly for the H–field. The fields are propagated in free space from the surface of the material to a plane below the electron, where the flow of energy per unit time from the particle to the material is evaluated $\mathcal{E} = \int S_z(\omega) dx dy$, where S_z is the z-component of the Poynting vector, $\mathbf{S}(\omega) = \mathbf{E}_{\text{total}} \times \mathbf{H}_{\text{total}}$. Thus the energy loss of the electron can be evaluated with no assumptions made about the nature of the interactions between the constituent objects in the material.

In a recent paper [13], we applied this technique to an ordered array of aluminium nanospheres and obtained excellent agreement between our numerical loss curves and those obtained by Howie and Walsh in their colloid experiments [9]. Here we consider the simpler situation of a chain of cylinders, shown in Figure 1. Although it is a very simple configuration, it is ideal for probing the nature of the interactions between cylinders. We model the metal using a frequency dependent dielectric of the form $\varepsilon(\omega) = 1 - \omega_p^2 / [\omega(\omega + i\gamma)]$, where ω is the frequency of the incident radiation, ω_p is the plasma frequency of the metal and γ is the loss. For aluminium, we use the typical values of $\omega_p = 15\text{eV}$ and $\gamma = 1\text{eV}$. Initially we consider the situation where the cylinders touch, forming a slab (*i.e.*, $d = l$). The loss curves are shown in Figure 2. In this case we see there are two peaks in the spectrum at approximately 4 and 13eV, corresponding to the upper and

256

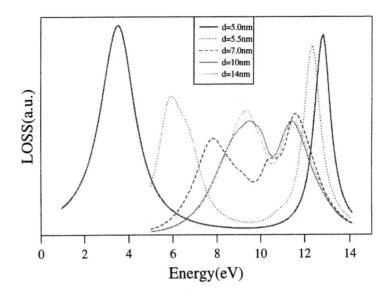

Figure 2. The energy loss experienced by the fast moving electron as the spacing between the cylinders (with diameter $l = 50\text{Å}$) is changed.

lower surface plasmons, as expected. As we move the cylinders apart, we observe the two peaks move towards one another. Eventually we observe the two peaks in the spectrum at approximately 9 and 11.5eV, when the separation between the cylinders is $d = 100\text{Å} = 2l$ or more. Beyond this separation, the spectrum does not change at all. Thus we see that for the chain, there is no coupling between the cylinders when they are more than twice their diameter apart and the loss in this case must be due to the plasmon modes of an isolated cylinder. However, at separations below 100Å, the coupling between the plasmons on each cylinder profoundly affects the loss spectrum.

3. Absorption of Light and SERS

When molecules are adsorbed onto certain metal surfaces which are rough on the nanometre scale (colloidal dispersions, island films, electrode surfaces, *etc.*) [14], the molecules show enormous enhancements (as much as 10^6) in the intensity of the Raman signal in comparison with the same molecules in solution. This discovery was made in the mid 1970s and since then SERS has become a widely used technique for studying the excitation spectra of molecules. However, although the technique is now well established, there remains a controversy about the overall enhancement for the two mechanisms that can enhance the Raman signal from a rough surface. The two ways which can change the signal from the molecules are: 1) a chemical effect, where the surface can modify the initial and final states of the Raman process through a transfer of charge between the surface and the molecule, changing the molecule's polarizability or 2) an EM effect, where the incident light can couple to the plasmon resonances of the surface, enhancing the electric field at the position of the molecule.

One of the problems in the quantitive comparison between both mechanisms has been that the usual EM theories have relied upon the modes of isolated nanometre objects. But, as a careful reading of the bibliography shows, SERS is an effect that appears in surfaces composed of closely interacting structures with individual dimensions much smaller than the wavelength of light. Moreover, in most of the experimental setups, the SERS maxima appear more broadened and at frequencies lower than those predicted for an isolated particle of the same shape.

In order to do a proper EM theory calculation of SERS, we apply the transfer matrix technique and calculate the transmission and reflection matrices for the scattering of an incident plane wave with momentum k and energy ω with our surface. Having obtained these quantities, we reconstruct the total E-field outside the surface using Eq. (1) and then we propagate the fields in real space through the structure, obtaining a detailed picture of the E-field around the constituent objects of our system. The E-field is important for SERS because, to a first approximation, the EM-enhancement of the signal is

$$\rho\left(\omega\right) = \left| \frac{E_{\text{total}}(\mathbf{r},\omega)}{E_{\text{inc}}(\omega)} \right|^4, \tag{2}$$

where $\mathbf{r}$ is the position at which the molecule sits and E_{inc} is the electric field associated with the incident plane wave.

We model a rough surface using a chain of silver cylinders, Figure 1. Silver is well known as a good enhancer in SERS experiments [14]. With this simple model we address two fundamental questions: 1) a quantitive measure of the importance of the interaction between the cylinders and 2) the relation between the enhancement of the Raman signal and the optical properties of the system.

The dielectric function for silver is modelled by $\varepsilon(\omega) = 5.7 - \omega_p^2/\omega^2 + i\gamma$, where $\omega_p = 9$eV and $\gamma = 0.4$ [15] and again we use cylinders with a 50Å side. We assume that the Raman active molecules are adsorbed on the surface of the cylinders and, therefore, we average Eq. (2) over the entire surface of a cylinder. In Figure 3 we present our results for this quantity and compare them with the absorption of our system (calculated using the reflection and transmission coefficients that we obtain). Figure 3 demonstrates that, as

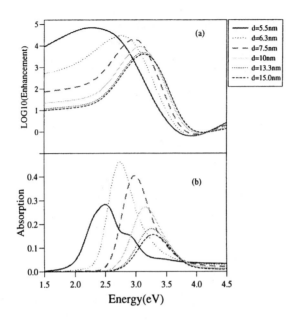

Figure 3. (a) The Raman Enhancement as a function of the frequency of the incident light for different values of *d*. (b) The corresponding absorption of the system.

with the electron energy loss, the interaction between cylinders is important when they are brought closer than twice the diameter of the individual cylinders. Below this separation the effect of the interaction grows rapidly. Another point to notice is the obvious relationship between the absorption and the enhancement in figure 3: the maxima of both quantities are at the same frequency and as the cylinders approach, the maxima tend towards lower energies. The system is clearly absorbing the energy to create the large E-fields. We see the effect of interaction between the objects can account for the shifting of the SERS maxima to lower frequencies as observed in most of the experiments. Not only are the positions of the maxima affected by the coupling, but the average enhancement grows from 10^3 for isolated cylinders to 10^5 for the chain where the cylinders are almost touching. This value for the enhancement is very close to the maximum enhancement observed in experiments.

Hence, using with our simple model for surface roughness, we have shown how the full interaction between the different metallic structures can modify, in a quantitive and qualitative way, the SERS spectra of an isolated particle.

4. Real Space Mapping of the E-field

As we stated in the introduction, the transfer matrix technique allows us to map the electric field in real space. This information can help address some fundamental questions. For example, it would be very interesting to know at which locations on the surface of the silver cylinders the largest enhancement of the Raman signal is found. Further, from a theoretical point of view, it is important to gain information about the particular plasmon modes which are excited by the incident EM waves and are responsible for the huge enhancements in the case of SERS and the loss mechanisms in the electron energy loss experiments. As a first attempt at providing some answers, we present some initial calculations of the E-field and its divergence (related to the induced charge) in the system of silver cylinders, where the separation is very small ($d = 55$Å and $l = 50$Å) and where the separation is large so the cylinders are isolated ($d = 150$Å), Figure 4. In Figure 4 (a) which corresponds to the maximum enhancement in the

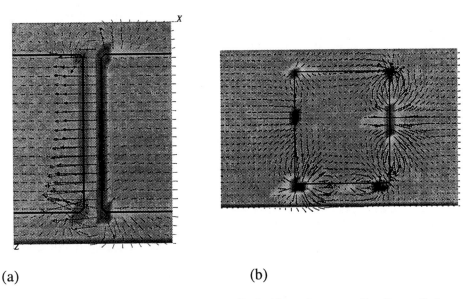

(a)   (b)

Figure 4. The E-fields generated by a normally incident plane wave for silver cylinders. (a) shows the mode responsible for the SERS maximum for a separation of $d = 55$Å ($\omega = 2.5$eV) and (b) shows the multipole mode responsible for the SERS maximum for isolated cylinders, $d = 150$Å and $\omega = 3.3$eV. The shading corresponds to regions of positive and negative charge and the outline of the solid metal cylinders is shown.

SERS calculation, we see the incident radiation is exciting a localised dipole plasmon between the cylinders, creating a huge E-field in this location. This seems to indicate that these types of positions are responsible for the enhanced signal observed in some rough surfaces. By comparison, in Figure

4 (b) the incident radiation is exciting a multipole plasmon on the surface of the isolated cylinder, which does not couple to the neighbouring objects. This mode also corresponds to a SERS maximum, but the enhancement is considerably lower in this case where the cylinders are far apart, $d = 150\text{Å}$.

5. Conclusions

We have demonstrated how the transfer matrix technique for EM waves can be applied successfully to problems which, at first sight, appear unrelated to photonics. In both cases, we have been able to improve upon previous theories by including the effects of interaction between nanometre scale metal objects. We have also presented some initial results on the real space mapping of the E-field in such systems and we hope this will become a useful tool for probing the fields in complex dielectric structures.

6. Acknowledgements

FJGV acknowledges a postdoctoral fellowship of the Spanish Ministerio de Educación y Ciencia. LMM is supported by the Comisión Interministerial de Ciencia y Tecnología of Spain under contract No. MAT94-0058-C02-01. This work was supported in part by the Acciones Integradas research programme.

References

1. E. Yablonovitch, J. Phys.: Condens. Matter 5, 2443 (1993).
2. K.M. Ho, C.T. Chan and C.M. Soukoulis, Phys. Rev. Lett. 65 , 3152 (1990).
3. K.M. Leung and Y.F. Liu, Phys. Rev. Lett. 65, 2646 (1990).
4. Ze Zhang and Sashi Satpathy, Phys. Rev. Lett. 65, 2650 (1990).
5. J.B. Pendry and A. MacKinnon, Phys. Rev. Lett. 69, 2772 (1992).
6. J.B. Pendry, J. Mod. Opt. 41, 209 (1994)
7. P.M. Bell, J.B. Pendry, L. Martín Moreno and A.J. Ward, Comp. Phys. Commun. 85, 306 (1995)
8. M. Sigalas, C.M. Soukoulis, E.N. Economou, C.T. Chan and K.M. Ho, Phys. Rev. B 48, 14121 (1993).
9. A. Howie and C.A. Walsh, Microsc. Microanal. Microstruc. 2, 171 (1991).
10. P.M. Echenique, A. Howie and D.J. Wheatley, Philos. Mag. B 56, 335 (1987).
11. M. Schmeits and L. Dambly, Phys. Rev. B. 44, 12706 (1991).
12. J.B. Pendry and L. Martín Moreno, Phys. Rev. B. 50, 5062 (1994).
13. L. Martín Moreno, P.M. Bell and J.B. Pendry, Journal of Microspcopy, In Press (1995).
14. Martin Moskovits, Rev. Mod. Phys. 57, 783 (1985).
15. P.B. Johnson and R.W. Christy, Phys. Rev. B. 6, 4370 (1972).

PHOTONIC BAND STRUCTURES AND RESONANT MODES

P. J. ROBERTS, P. R. TAPSTER AND T. J. SHEPHERD

Defence Research Agency, St. Andrews Road,
Great Malvern, Worcs. WR14 3PS, UK.

Abstract. This paper reports on schemes for the accurate determination of the electromagnetic properties of photonic crystals with arbitrary underlying lattice symmetry, using a real space transfer matrix method. The schemes are applied to hexagonal crystals and diamond-symmetry crystals, and results are compared with those obtained using a plane wave expansion method. The transfer matrix method is then applied to systems which comprise stacked finite-thickness photonic crystals with different but overlapping photonic stop bands, between which there can exist a planar cavity. Such ultra wide band gap structures can display scattering characteristics attributable to the presence of resonant modes at frequencies within the intersection of the stop band frequency ranges of the individual crystals. An initial study is presented of two stacked hexagonal crystals whose invariant axes are parallel. Results for stacked diamond-symmetry photonic crystals are imminent.

1. Introduction

The electromagnetic properties of periodic arrangements of dielectric materials have been a subject of interest for some time. Initially, one dimensionally (1D) periodic stacks of dielectric layers were considered; this system — the Bragg stack — has by now found many optical applications. The current interest in dielectric "crystals" of higher dimensionality was initiated by Yablonovitch [1, 2], who succeeded in fabricating a 3D crystal which possesses a gap in its photon density-of-states over all directions at microwave frequencies. The presence of such a photonic band gap (PBG) has many important potential applications [3, 4]. At microwave length-scales, structures exhibiting a PBG can be exploited in the design of efficient antennae

C. M. Soukoulis (ed.), Photonic Band Gap Materials, 261–270.
© 1996 *Kluwer Academic Publishers. Printed in the Netherlands.*

and electromagnetic filtering devices. The most technologically significant potential application of PBG materials occurs at optical or infra-red wavelengths; they can be exploited in the design of highly efficient light-emitting diode and laser devices.

Although a complete PBG is obtainable only for systems in which the dielectric function varies in all three spatial dimensions, much attention has been paid to systems of 2D periodicity, due primarily to the relative ease of fabrication of such structures at the important optical and near infra-red length scales. For sufficiently high dielectric contrast, these 2D systems can still be completely reflective to propagative incident radiation for all angles of incidence in vacuum and arbitrary polarisation [5]: over certain ranges of frequency, the modes of the crystal can only be coupled to by fields which are evanescent in the vacuum. Thus in many filtering applications, where it is propagative waves which are of interest, 2D periodicity can be entirely sufficient.

The purpose of the present paper is to accurately apply a transfer matrix method to photonic crystals possessing symmetries of most interest — those which can give rise to broad photonic band gaps. The particular examples of hexagonal symmetry in 2D and diamond symmetry in 3D are focused upon, and schemes for the necessary discretisation of a primitive unit cell described. Results for the obtained band gap frequency ranges are compared with those calculated using a plane wave expansion method.

The interest in the transfer matrix method is primarily due to its ability to calculate the scattering matrix and Green function of crystals with finite thickness. In the current paper, the method is applied in an initial numerical study of the electromagnetic properties of two stacked photonic crystals between which there can exist a planar cavity. Cavities formed between tetragonal arrangements of dielectric spheres have been investigated in reference [6] using an LKKR technique, and the existence of sharp cavity modes clearly demonstrated. Stacked dielectric crystals with FCC underlying lattice symmetry have been experimentally investigated [7] in the context of antennae substrates, and resonance modes between non-adjacent crystals were observed. In the present work, stacked 2D crystals displaying hexagonal symmetry are considered. Results for stacked diamond crystals are imminent. The quantities calculated are the transmission intensity of the stacked system, and the diagonal Green function evaluated in the cavity formed between the crystals.

2. Numerical Method

The primary numerical method which has been employed is the transfer matrix approach introduced by Pendry and MacKinnon [8-10]. The method is based on a direct real space discretisation of Maxwell's equations: the wavefields and dielectric function are sampled only at a discrete set of points which themselves form a lattice. The equations can be arranged to integrate the wavefield through the system by a linear recursive technique. The matrix which specifies this recursion is known as a transfer matrix. The transfer matrix relates fields on a plane of mesh-points to those on an adjacent plane.

In practice, numerical instabilities arise after a certain number of steps. The problem can be remedied by inclusion of scattering matrix steps into the algorithm [9]. The system is divided into sub-regions within which the transfer matrix method can be applied safely. A scattering matrix description of the multiple scattering is then used to combine the sub-regions and relate the fields across the entire system.

The method relates fields at a fixed frequency (on frequency-shell), and can be used to obtain dispersion surfaces (constant frequency surfaces within the Brillouin zone) of periodic crystals [11].

For a determination of the dispersion surfaces of a periodic crystal, it is necessary only to integrate the wave-field through a unit cell of the material: Bloch's theorem can be directly invoked in the form of a boundary condition at the unit cell surfaces. In order to apply the transfer matrix method in a straightforward manner, it is necessary the cross-section remain constant at each transfer matrix step. The simplest 3D geometry which can be described in this way is therefore a cuboid, within which the fields and dielectric function are sampled on a cuboidal grid. Systems with a cuboidal primitive unit cell, i.e., simple orthorhombic lattices (which includes simple cubic and simple tetragonal as special cases) are therefore most efficiently described by this sort of discretisation. The Bravais lattices of most interest — those which can give rise to wide photonic band gaps — are not of these symmetries, however. To achieve accurate results for 3D crystals, it is essential that only a region with the volume of a *primitive* unit cell be considered in the discretisation. Thus, for a good description of non simple-orthorhombic lattices, a conventional primitive unit cell must be discretised in a non-orthogonal co-ordinate scheme. A discretisation of Maxwell's equations along non-orthogonal axes has been performed, and the resulting linear set of equations cast in the form of an iterative transfer matrix equation. The transfer matrix in a non-orthogonal discretisation scheme is not as sparse as that stemming from a conventional orthogonal discretisation, but the added complexity in practise does not significantly increase computer time. The numerical stability is generally comparable to the orthogonal case, although it can happen that in the very near vicinity of one or more special lines within the Brillouin zone, stability is worsened due to shape of the dispersion surface of the discretised vacuum [9].

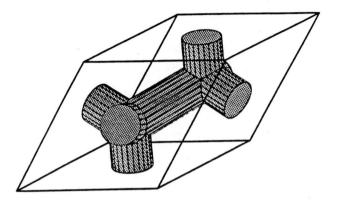

Figure 1. Conventional unit cell of a diamond photonic crystal.

Thus far, the non-orthogonal discretisation scheme has been applied to the FCC Bravais lattice in 3D, and the hexagonal lattice in 2D. The FCC symmetry is of particular interest because a large class of dielectric distributions which possess a complete photonic band gap in their electromagnetic response fall into this symmetry group [12-15]. The achievable in-plane-of-periodicity band gap of crystals with hexagonal symmetry is also known to be particularly wide [16-19].

Figure 1 shows the conventional unit cell of a crystal showing full diamond symmetry (diamond has an underlying FCC Bravais lattice). The crystal comprises cylinders joining nearest-neighbour points in a diamond arrangement, and its mode structure has been numerically investigated using the plane wave method [15]. It has been found to possess a particularly wide band gap. This structure is analysed in section 4 below.

An alternative strategy for obtaining the dispersion of non simple-orthorhombic crystals involves integrating the wavefield through a non-conventional primitive unit cell, which is cuboidal and can thus be conventionally discretised along orthogonal directions. Figure 2 illustrates how such a cell matches on to its neighbours for the diamond crystal whose conventional cell is in Figure 1. The sides of the unconventional cell can be taken as $ac_1 = (-1, 1, 0)a/4$, $ac_2 = (-1, -1, 2)a/4$ and $ac_3 = (1, 1, 1)a/3$, where a is the cubic unit cell side length. If the centre of a reference cell is taken as the origin, the centres of all its adjoining cells are at the positions of the primitive vectors of the FCC lattice. As shown in Figure 2, the cells adjoining the surfaces of the reference cell with normals ac_2 and ac_3 must therefore be displaced. If the transfer matrix iteration is performed along the ac_3 direction, the boundary matching can still be applied using Bloch's

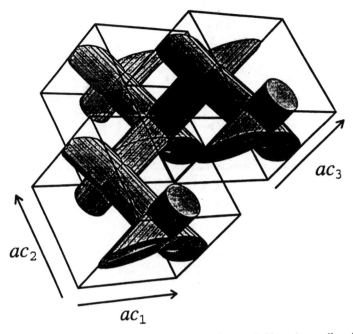

Figure 2. Unconventional unit cell showing how neighbouring cells adjoin.

theorem at each plane.

Since the fields are integrated through a system at a fixed frequency, the transfer matrix allows for the determination of the scattering matrix or Green function of photonic crystals with finite thickness. Using the primitive unit cells shown in Figures 1 and 2, the truncation surfaces of such a finite crystal are necessarily (111) planes. Other choices of primitive cell can be made, so that crystals with different truncation surfaces can be described.

To provide a comparison with the results from the real space technique, a plane wave method code based on the Fourier expansion of the **H**-field together with the inverse of the dielectric function [12, 14] has also been implemented.

3. Results for 2D Structures

The 2D periodic system which shows optimum band gap properties comprises a hexagonal arrangement of parallel cylinders of low dielectric strength embedded in a high dielectric strength host medium [16-19]. In the present paper, this structure is considered with the cylinders taken to be air and the high dielectric medium to have a dielectric constant of 13.24. The broadest in-plane-of-periodicity band gap occurs at this dielectric contrast when the air filling fraction is close to 0.8 [18]. The lowest band gap frequency-ranges

266

Figure 3. An asymmetric cavity between hexagonal crystals.

for the two polarisations calculated using the transfer matrix method with a mesh of 24×24 points agrees to within about a percent with those obtained from the plane wave method.

An asymmetrical cavity was formed between a crystal with an air filling fraction (f) of 0.8 and one with an air filling fraction of 0.75. Both crystals have identical lattice constants, and were taken to have a thickness of 5 unit cells. The truncation surfaces are as depicted in Figure 3. The cavity was assumed to be filled with air, and to be of thickness $0.57a$. Figure 4 shows the transmission intensity at normal incidence for the two crystals taken separately, and for the combined cavity system. The H-field of the incident field was taken as linearly polarised along the direction of the cylinder axes (H-polarisation).

The transmission intensities of the individual crystals, as expected, decrease sharply within the band gaps accessible to the normally incident polarised wave (H-polarised gaps along the Γ-X direction in the Brillouin zone), and show the familiar oscillatory behaviour associated with the repeated reflections at the surfaces of the crystals. The cavity system shows a stop band which is the union of the stop bands of the individual crystal, but with a sharp cavity mode resonance at a frequency $2.4a/c$ which is within the intersection of the stop band regions.

The interference effects within the cavity which give rise to the resonance are particularly evident in the intra-cavity Green function. Figure 5 shows the imaginary part of the diagonal Green function $G_{\mathrm{HH}}(\omega; \mathbf{k}_{\|} = \mathbf{0}, z, \mathbf{k}_{\|} = \mathbf{0}, z)$ evaluated at the centre of the cavity plotted against angular frequency. Here, $\mathbf{k}_{\|}$ is the k-vector component in the plane of the surfaces, z is the real space co-ordinate normal to the cavity faces, and the subscript HH indicates that a projection onto the H-polarisation direction has been performed.

The behaviour at normal incidence is completely analogous to that of

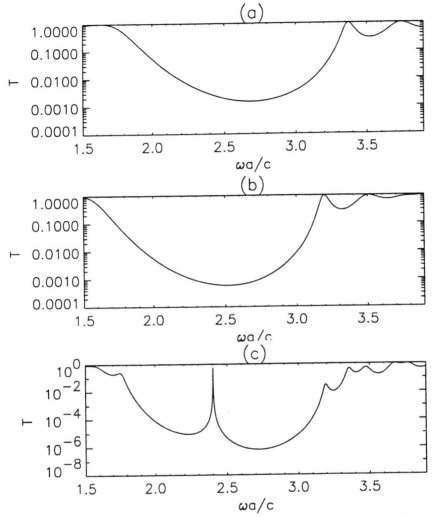

Figure 4. Transmission intensities at normal incidence for (a) 5 layers of f=0.8 crystal, (b) 5 layers of f=0.75 crystal and (c) the two crystals stacked as in Figure 3 separated by a planar cavity of length 0.57a. H-polarisation data is shown.

an asymmetrical cavity formed between two 1-dimensional Bragg stacks, except for polarisation dependence. The interest in systems of higher dimensionality stems from the angular cut-off to the stop bands necessarily suffered by the Bragg stack system. Within a narrow frequency range, it is possible for a 2D crystal to reflect all propagative incident radiation for all angles and polarisation states. Thus, a cavity formed to have a resonance within such a range of frequencies is not subject to loss over a range of solid angles. The angular variation of the cavity modes is currently under investigation.

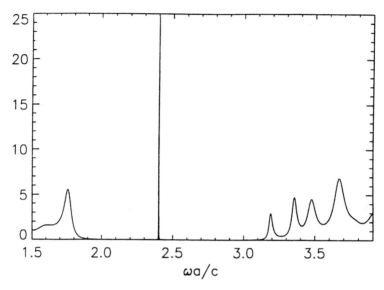

Figure 5. Imaginary part of diagonal Green function evaluated at the centre of the cavity for H- polarisation.

4. Results for 3D Structures

The dispersion of the diamond structure whose conventional primitive unit cell is shown in Figure 1 has been determined using the two real space discretisation approaches discussed in section 2, as well as using the plane wave method. Thus far, grids up to a size $8 \times 8 \times 8$ within each primitive unit cell have been used with the real space approach, and 1067 plane waves have been used in the k-space method. The complete photonic band gaps obtained using the two real space procedures agree very well with each other, and typically differ by a few percent from those found from the plane wave method; the real space approaches appear to give a slightly lower and slightly narrower gap than the plane wave approach. As an example, for a system of air cylinders in a host medium of dielectric constant 12.96, with an air filling fraction of 0.8, the real space approaches give the range $3.2a/c$-$4.05a/c$ for the gap, whereas the plane wave method gives $3.35a/c$-$4.35a/c$. A thorough investigation of the convergence as a function of the number of included grid points is currently underway.

Both discretisation procedures discussed in section 2 do not fully maintain the diamond symmetry. As a result, the dispersion surfaces (constant frequency surfaces in the Brillouin zone) can be expected to show some deviation from this symmetry. This is indeed observed, particularly near band gap edges, but the deviation obtained using the cuboidal cell of Figure 2 is very slight. The dispersion surfaces determined using the rhomboidal unit cell of Figure 1 show larger deviations, however.

At present, the band gap ranges are determined by evenly scanning over frequency and two k-vector components. A more efficient algorithm for their determination is currently being developed.

5. Conclusions and Outlook

The electromagnetic properties of hexagonal and diamond photonic crystals can be efficiently calculated using a real space discretisation approach. The results for the dispersion curves of the infinite crystal agree to within about a percent with results determined from the plane wave approach for the 2D structure, and the band gap frequency regions to within a few percent for the 3D structure. Accurate convergence tests in the latter case are currently under study. Extensions to accurately treat systems with tetragonal symmetry, such as the recently studied layer-by-layer structures [20, 21] which show particularly wide gaps, are currently in progress.

The real space method allows for the calculation of the scattering matrix and Green function of a finite thickness sample. This has been applied to the case of two stacked finite hexagonal photonic crystals between which there can exist a cavity. These structures show ultra-wide stop bands, within which sharp resonant cavity modes can be present. Results for stacked diamond symmetry crystals are imminent.

Thus far, only cavities formed between crystals of the same lattice constant have been considered in the numerical study. Extensions to treat the more interesting case of stacked crystals with differing lattice constants, such as the systems experimentally studied in [7], would be interesting but technically more demanding if accurate results are to be obtained.

6. Acknowledgements

P. J. Roberts would like to acknowledge Professor John Pendry for useful discussions. This work was funded by DRA Malvern.

References

1. E. Yablonovitch, *Phys. Rev. Lett.* **58**, 2059 (1987).
2. E. Yablonovitch, T. J. Gmitter, and K. M. Leung, *Phys. Rev. Lett.* **67**, 2295 (1991).
3. E. Yablonovitch, *J. Opt. Soc. Am.* **10**, 283 (1993).
4. P. St.J. Russell, T. A. Birks, and F. D. Lloyd-Lucas, to be published in "Confined Electrons and Photons: New Physics and Applications", E. Burstein & C. Weisbuch (editors), (Plenum Press).
5. P. J. Roberts, T. A. Birks, P. St.J. Russell, T. J. Shepherd, and D. M. Atkin, submitted to *Opt. Lett.*.
6. V. Karathanos, A. Modinos, and N. Stefanou, *J. Phys: Condens. Matter* **6**, 6257 (1994).

7. K. Agi, E. R. Brown, O. B. McMahon, C. Dill III, and K. J. Malloy, *Electron. Lett.* **30**, 2166 (1994).

8. J. B. Pendry and A. MacKinnon, *Phys. Rev. Lett.* **69**, 2772 (1992).

9. J. B. Pendry, J. Mod. Opt. **41**, 209 (1994).

10. P. M. Bell, J. B. Pendry, L. Martin-Moreno, and A. J. Ward, *Comp. Phys. Commun.* **85**, 306 (1995).

11. A. J. Ward, J. B. Pendry, and W. J. Stewart, *J. Phys: Condens. Matter* **7**, 2217 (1995).

12. K. M. Leung and Y. F. Liu, *Phys. Rev. Lett.* **65**, 2646 (1991).

13. Z. Zhang and S. Zatpathy, *Phys. Rev. Lett.* **65**, 2650 (1991).

14. K. M. Ho, C. T. Chan, and C. M. Soukoulis, *Phys. Rev. Lett.* **65**, 3152 (1991).

15. C. T. Chan, K. M. Ho, and C. M. Soukoulis, *Europhys. Lett.* **16**, 563 (1991).

16. P. R. Villeneuve and M. Piché, *Phys. Rev.* **B46**, 4969 (1992).

17. R. D. Meade, K. D. Brommer, A. M. Rappe, and J. D. Joannopoulos, Appl. Phys. Lett. **61**, 495 (1992).

18. R. Padjen, J. M. Gerard, and J. Y. Marzin, *J. Mod. Opt.* **41**, 295 (1994).

19. J. N. Win, R. D. Meade, and J. D. Joannopoulos, *J. Mod. Opt.* **41**, 257 (1994).

20. K. M. Ho, C. T. Chan, C. M. Soukoulis, R. Biswas, and M. Sigalas, *Solid State Commun.* **89**, 413 (1994).

21. E. Ozbay, E. Michel, G. Tuttle, R. Biswas, M. Sigalas, and K. M. Ho, *App. Phys. Lett.* **64**, 2059 (1994).

PHOTONIC BAND STRUCTURES OF SYSTEMS WITH COMPONENTS CHARACTERIZED BY FREQUENCY-DEPENDENT DIELECTRIC FUNCTIONS

A. A. MARADUDIN
Department of Physics and Astronomy
University of California, Irvine CA 92717 USA

V. KUZMIAK
Institute of Radio Engineering and Electronics
Czech Academy of Sciences
Chaberska 57, 182 51 Praha 8, Czech Republic

AND

A. R. MCGURN
Department of Physics
Western Michigan University, Kalamazoo, MI 49008 USA

1. Introduction

The great majority of the existing calculations of photonic band structures have been limited to periodic structures, either discrete or continuous, fabricated from a dielectric characterized by a positive, real, frequency-independent dielectric constant ϵ_A, embedded in a dielectric matrix characterized by a dielectric constant ϵ_B that is also, positive, real, and frequency-independent [1]. The restriction to components characterized by such dielectric constants seems overly restrictive, since one can envision periodic two- and three-dimensional structures fabricated from metal wires or metal particles, or from rods and particles of polar semiconductors or ionic crystals, all of which are characterized by frequency-dependent dielectric functions, that can be negative in certain frequency ranges. It might be expected that the dispersion curves for electromagnetic waves propagating through such systems (their photonic band structures) may display interesting features,

C. M. Soukoulis (ed.), Photonic Band Gap Materials, 271–318.
© *1996 Kluwer Academic Publishers. Printed in the Netherlands.*

especially in the frequency ranges in which the dielectric functions of the embedded components are negative.

The calculation of photonic band structures of systems containing components with frequency-dependent dielectric functions is more difficult than it is for systems consisting only of components characterized by positive, real, frequency-independent dielectric constants. This is because the use of standard methods, such as expansions of the electromagnetic field components in plane waves, which reduce the calculation of the photonic band structure of a system of the latter type to the diagonalization of a matrix, does not allow such a reduction for a system of the former type without some effort.

In these lectures we present the results of some of our recent work in which computational methods, based on the plane wave method, have been developed for the determination of the photonic band structures of periodic dielectric structures containing components with frequency-dependent dielectric functions [2,3]. We restrict ourselves here to dielectric functions that are real, therefore effects of dissipation are not taken into account. They will be discussed elsewhere in these Proceedings [4].

We will consider structures which contain a metallic component that is characterized by a dielectric function that has the simple, free-electron metal form

$$\epsilon(\omega) = 1 - (\omega_p^2/\omega^2), \tag{1.1}$$

where ω_p is the plasma frequency of the conduction electrons, and usually lies in the ultraviolet region of the optical spectrum. This form for $\epsilon(\omega)$ is negative in the frequency range $0 \leq \omega < \omega_p$. We will also consider structures which contain a component that is fabricated from a cubic, diatomic, polar semiconductor or ionic crystal, characterized by a dielectric function of the form

$$\epsilon(\omega) = \epsilon_\infty \frac{\omega_L^2 - \omega^2}{\omega_T^2 - \omega^2}, \tag{1.2}$$

where ϵ_∞ is the optical frequency dielectric constant, and ω_L and ω_T are the frequencies of the longitudinal and transverse optical vibration modes of infinite wavelength, respectively. These frequencies typically lie in the infrared region of the optical spectrum. The dielectric function (1.2) is negative in the frequency range $\omega_T < \omega < \omega_L$. In all the numerical calculations in these lectures based on Eq. (1.2) we will use the values $\epsilon_\infty = 10.9, \omega_T = 8.12$ THz, and $\omega_L = 8.75$ THz appropriate to GaAs.

In keeping with the tutorial nature of these lectures we begin, in Section 2, by considering a simple, one-dimensional, periodic system consisting of alternating layers of vacuum and a medium characterized by a dielectric function of the form given by Eq. (1.1) or Eq. (1.2). Four different methods for calculating the photonic band structure of this model system will be described, and the results obtained by their use will be compared. In Section 3, several of these methods will be used to obtain the photonic band structures of two-dimensional periodic structures consisting of infinite, parallel, metallic or semiconducting cylinders in vacuum, whose intersections with a perpendicular plane form a two-dimensional Bravais lattice. One of these four approaches will be used in Section 4 to obtain the photonic band structure of a face centered cubic array of metal spheres embedded in vacuum. A discussion of the results obtained, and some directions for future work, will be presented in Section 5.

2. A One-Dimensional System

The one-dimensional system we study in this section is depicted in Fig. 1. It consists of alternating layers of a metal or semiconductor of thickness a separated by vacuum layers of thickness b. The volume filling fraction of the metallic subsystem is therefore $f = a/(a+b)$. We consider the propagation of an s-polarized electromagnetic field through it in the x_1-direction, i.e., normal to the vacuum-dielectric interface. The electric vector in this case has the form

$$\vec{E}(\vec{x};t) = (0, E(x_1|\omega), 0) \exp(-i\omega t), \tag{2.1}$$

while the magnetic vector is given by

$$\vec{H}(\vec{x};t) = (0, 0, H(x_1|\omega)) \exp(-i\omega t). \tag{2.2}$$

Finally, the electric displacement vector is

$$\vec{D}(\vec{x};t) = (0, D(x_1|\omega), 0) \exp(-i\omega t). \tag{2.3}$$

From the Maxwell curl equations

$$\frac{d}{dx_1} E(x_1|\omega) = \frac{i\omega}{c} H(x_1|\omega) \tag{2.4a}$$

$$\frac{d}{dx_1} H(x_1|\omega) = \frac{i\omega}{c} D(x_1|\omega), \tag{2.4b}$$

and the fact that $D(x_1|\omega) = \epsilon(\omega)E(x_1|\omega)$ in each dielectric medium, while $D(x_1|\omega) = E(x_1|\omega)$ in each vacuum region, we find that the amplitude function $E(x_1|\omega)$ satisfies

$$\left(\frac{d^2}{dx_1^2} + \epsilon(\omega)\frac{\omega^2}{c^2} \right) E(x_1|\omega) = 0 \tag{2.5}$$

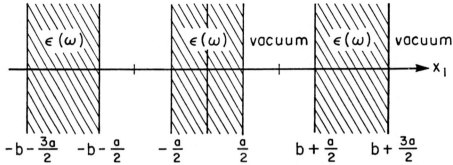

Fig. **1.** The one-dimensional periodic structure studied in this paper. It consists of alternating layers of a metal or semiconductor of thickness a separated by vacuum layers of thickness b.

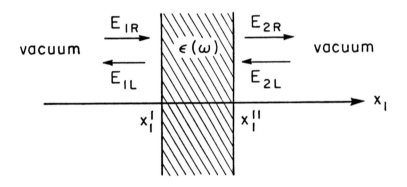

Fig. **2.** The structure, consisting of a single dielectric layer, used in obtaining the transfer matrix for the periodic structure depicted in Fig. 1.

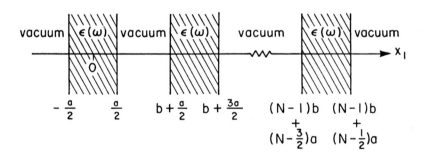

Fig. **3.** A periodic array of N dielectric layers, each of thickness a, and separated from its neighbors by vacuum layers of thickness b.

in each dielectric medium, and the equation

$$\left(\frac{d^2}{dx_1^2} + \frac{\omega^2}{c^2}\right) E(x_1|\omega) = 0 \tag{2.6}$$

in each vacuum region. The boundary conditions satisfied at each interface require the continuity of $E(x_1|\omega)$ and $dE(x_1|\omega)/dx_1$ across it. We now proceed to obtain the dispersion relation for the propagation of such electromagnetic waves through this structure by several different approaches.

2.1. THE TRANSFER MATRIX METHOD

Let us consider first the transmission and reflection of electromagnetic waves incident on a single dielectric layer from the right and from the left (Fig. 2). The layer fills the region $x_1' < x_1 < x_1''$. In the region $x_1 < x_1'$ the solution of Eq. (2.6) is

$$E(x_1|\omega) = E_{1R} e^{i\frac{\omega}{c}x_1} + E_{1L} e^{-i\frac{\omega}{c}x_1}, \tag{2.7}$$

while in the region $x_1 > x_1''$ the solution of Eq. (2.6) is

$$E(x_1|\omega) = E_{2R} e^{i\frac{\omega}{c}x_1} + E_{2L} e^{-i\frac{\omega}{c}x_1}. \tag{2.8}$$

Inside the dielectric layer $x_1' < x_1 < x_1''$ the solution of Eq. (2.5) has the form

$$E(x_1|\omega) = A e^{i\frac{\omega}{c}\sqrt{\epsilon(\omega)}x_1} + B e^{-i\frac{\omega}{c}\sqrt{\epsilon(\omega)}x_1}, \tag{2.9}$$

where $Re\sqrt{\epsilon(\omega)} > 0, Im\sqrt{\epsilon(\omega)} > 0$. The boundary conditions at $x_1 = x_1'$ become

$$E_{1R} e^{i\frac{\omega}{c}x_1'} + E_{1L} e^{-i\frac{\omega}{c}x_1'} = A e^{i\frac{\omega}{c}\sqrt{\epsilon(\omega)}x_1'} + B e^{-i\frac{\omega}{c}\sqrt{\epsilon(\omega)}x_1'} \tag{2.10a}$$

$$i\frac{\omega}{c}\left(E_{1R} e^{i\frac{\omega}{c}x_1'} - E_{1L} e^{-i\frac{\omega}{c}x_1'}\right)$$

$$= i\frac{\omega}{c}\sqrt{\epsilon(\omega)}\left(A e^{i\frac{\omega}{c}\sqrt{\epsilon(\omega)}x_1'} - B e^{-i\frac{\omega}{c}\sqrt{\epsilon(\omega)}x_1'}\right), \tag{2.10b}$$

while the boundary conditions at $x_1 = x_1''$ are

$$A e^{i\frac{\omega}{c}\sqrt{\epsilon(\omega)}x_1''} + B e^{-i\frac{\omega}{c}\sqrt{\epsilon(\omega)}x_1''} = E_{2R} e^{i\frac{\omega}{c}x_1''} + E_{2L} e^{-i\frac{\omega}{c}x_1''} \tag{2.11a}$$

$$i\frac{\omega}{c}\sqrt{\epsilon(\omega)}\left(A e^{i\frac{\omega}{c}\sqrt{\epsilon(\omega)}x_1''} - B e^{-i\frac{\omega}{c}\sqrt{\epsilon(\omega)}x_1''}\right)$$

$$= i\frac{\omega}{c}\left(E_{2R} e^{i\frac{\omega}{c}x_1''} - E_{2L} e^{-i\frac{\omega}{c}x_1''}\right). \tag{2.11b}$$

By eliminating A and B from this set of equations we obtain the following relation between the amplitudes $E_{2R,L}$ and the amplitudes $E_{1R,L}$:

$$\begin{pmatrix} E_{2R} \\ E_{2L} \end{pmatrix} = \overleftrightarrow{T}(x_1''; x_1') \begin{pmatrix} E_{1R} \\ E_{1L} \end{pmatrix},$$
(2.12)

where the 2×2 *transfer matrix* $\overleftrightarrow{T}(x_1''; x_1')$ is given by

$$\overleftrightarrow{T}(x_1''; x_1') = \begin{pmatrix} T_{11}(x_1''; x_1') & T_{12}(x_1''; x_1') \\ T_{21}(x_1''; x_1') & T_{22}(x_1''; x_1') \end{pmatrix}$$
(2.13)

with

$$T_{11}(x_1''; x_1') = e^{-i\frac{\omega}{c}(x_1''-x_1')}[\cos\frac{\omega}{c}\sqrt{\epsilon(\omega)}(x_1'' - x_1')$$

$$+ \frac{i}{2}(\sqrt{\epsilon(\omega)} + \frac{1}{\sqrt{\epsilon(\omega)}})\sin\frac{\omega}{c}\sqrt{\epsilon(\omega)}(x_1'' - x_1')]$$
(2.14a)

$$T_{12}(x_1''; x_1') = \frac{i}{2}e^{-i\frac{\omega}{c}(x_1''+x_1')}\left(\sqrt{\epsilon(\omega)} - \frac{1}{\sqrt{\epsilon(\omega)}}\right)\sin\frac{\omega}{c}\sqrt{\epsilon(\omega)}(x_1'' - x_1')$$

(2.14b)

$$T_{21}(x_1''; x_1') = -\frac{i}{2}e^{i\frac{\omega}{c}(x_1''+x_1')}\left(\sqrt{\epsilon(\omega)} - \frac{1}{\sqrt{\epsilon(\omega)}}\right)\sin\frac{\omega}{c}\sqrt{\epsilon(\omega)}(x_1'' - x_1')$$

(2.14c)

$$T_{22}(x_1''; x_1') = e^{i\frac{\omega}{c}(x_1''-x_1')}[\cos\frac{\omega}{c}\sqrt{\epsilon(\omega)}(x_1'' - x_1')$$

$$- \frac{i}{2}(\sqrt{\epsilon(\omega)} + \frac{1}{\sqrt{\epsilon(\omega)}})\sin\frac{\omega}{c}\sqrt{\epsilon(\omega)}(x_1'' - x_1')].$$
(2.14d)

We note that

$$\det \overleftrightarrow{T}(x_1''; x_1') = 1,$$
(2.15)

a result which holds whether $\sqrt{\epsilon(\omega)}$ is real or pure imaginary.

Now let us consider an array of N dielectric layers each of thickness a, each of which is separated from its neighbors by vacuum layers of thickness b (Fig. 3). The dielectric layers thus occupy the intervals of the x_1-axis defined by $((n-1)b+(n-\frac{3}{2})a, (n-1)b+(n-\frac{1}{2})a)$ with $n = 1, 2, 3, \ldots, N$. We assume that for $x_1 > (N-1)b+(N-\frac{1}{2})a$ the electric field has the form $E(x_1|\omega) = e^{-i\frac{\omega}{c}x_1} + re^{i\frac{\omega}{c}x_1}$, while for $x_1 < -\frac{a}{2}$ it has the form $te^{-i\frac{\omega}{c}x_1}$. Thus, r is the reflection amplitude and t is the transmission amplitude for the wave of unit amplitude incident on our structure from the right. By using the preceding results we find

$$
\begin{pmatrix} r \\ 1 \end{pmatrix} = \begin{pmatrix} T_{11}^{(N)} & T_{12}^{(N)} \\ T_{21}^{(N)} & T_{22}^{(N)} \end{pmatrix} \begin{pmatrix} 0 \\ t \end{pmatrix}, \tag{2.16}
$$

where the 2×2 matrix $\overleftrightarrow{T}^{(N)}$ is given by

$$
\overleftrightarrow{T}^{(N)} = \prod_{n=1}^{N} \overleftrightarrow{T}\left((n-1)b+(n-\frac{1}{2})a; (n-1)b+(n-\frac{3}{2})a\right). \tag{2.17}
$$

This product of 2×2 transfer matrices can easily be evaluated by a computer. It follows from Eq. (2.16) that r and t are given by

$$
r = \frac{T_{12}^{(N)}}{T_{22}^{(N)}}, \quad t = \frac{1}{T_{22}^{(N)}}. \tag{2.18}
$$

Below we will present results for the transmissivity $T = |t|^2$ of this N-layer system as a function of frequency, for comparison with the results of calculations of the photonic band structure of this system in the limit as $N \to \infty$, when it becomes a periodic system.

To obtain this band structure we note that, according to the Bloch-Floquet theorem, the values of the electric field at $x_1 = -\frac{a}{2}$ and at $x_1 = d-\frac{a}{2}$ are related by

$$
E(d - \frac{a}{2}|\omega) = e^{ikd} E(-\frac{a}{2}|\omega). \tag{2.19}
$$

Similarly, the values of the derivative of the field at these two values of x_1 are related by

$$
E'(d - \frac{a}{2}|\omega) = e^{ikd} E'(-\frac{a}{2}|\omega). \tag{2.20}
$$

With the use of Eqs. (2.7) and (2.8) these relations become

$$
E_{2R}e^{i\frac{\omega}{c}(d-\frac{a}{2})} + E_{2L}e^{-i\frac{\omega}{c}(d-\frac{a}{2})} = e^{ikd}(E_{1R}e^{-i\frac{\omega}{c}\frac{a}{2}} + E_{1L}e^{i\frac{\omega}{c}\frac{a}{2}}) \tag{2.21}
$$

and

$$\frac{i\omega}{c}(E_{2R}e^{i\frac{\omega}{c}(d-\frac{a}{2})} - E_{2L}e^{-i\frac{\omega}{c}(d-\frac{a}{2})}) = e^{ikd}\frac{i\omega}{c}(E_{1R}e^{-i\frac{\omega}{c}\frac{a}{2}} - E_{1L}e^{i\frac{\omega}{c}\frac{a}{2}}).\tag{2.22}$$

These equations can be rearranged into

$$\begin{pmatrix} E_{2R} \\ E_{2L} \end{pmatrix} = e^{ikd} \begin{pmatrix} e^{-i\frac{\omega}{c}d} & 0 \\ 0 & e^{i\frac{\omega}{c}d} \end{pmatrix} \begin{pmatrix} E_{1R} \\ E_{1L} \end{pmatrix}$$

$$= \overleftrightarrow{T}\left(\frac{a}{2}; -\frac{a}{2}\right) \begin{pmatrix} E_{1R} \\ E_{1L} \end{pmatrix},\tag{2.23}$$

where we used Eq. (2.12) with $x_1'' = \frac{a}{2}$ and $x_1' = -\frac{a}{2}$ in writing the second equation. The solvability condition for the homogeneous system of equations (2.23) is

$$det \begin{pmatrix} T_{11}(\frac{a}{2}; -\frac{a}{2}) - e^{ikd-i\frac{\omega}{c}d} & T_{12}(\frac{a}{2}; -\frac{a}{2}) \\ T_{21}(\frac{a}{2}; -\frac{a}{2}) & T_{22}(\frac{a}{2}; -\frac{a}{2}) - e^{ikd+i\frac{\omega}{c}d} \end{pmatrix} = 0.\tag{2.24}$$

On evaluating the determinant with the aid of Eq. (2.15) we obtain the dispersion relation for electromagnetic waves in our structures in the form

$$\cos kd = \frac{1}{2}\left[e^{i\frac{\omega}{c}d}T_{11}(\frac{a}{2}; -\frac{a}{2}) + e^{-i\frac{\omega}{c}d}T_{22}(\frac{a}{2}; -\frac{a}{2})\right]$$

$$= \cos\frac{\omega}{c}\sqrt{\epsilon(\omega)}a \cos\frac{\omega}{c}b - \frac{1}{2}(\sqrt{\epsilon(\omega)} + \frac{1}{\sqrt{\epsilon(\omega)}})\sin\frac{\omega}{c}\sqrt{\epsilon(\omega)}a$$

$$\times \sin\frac{\omega}{c}b.\tag{2.25}$$

All of the distinct solutions of Eq. (2.25) are obtained if kd is restricted to the range $0 \le kd \le \pi$, i.e. to the right hand half of the first Brillouin zone of the periodic structure depicted in Fig. 1.

Insight into the photonic band structure of one dimensional periodic systems of the type depicted in Fig. 1, when the component characterized by the frequency-dependent dielectric function is a metal (Eq. (1.1)) or a polar semiconductor (Eq. (1.2)) can be gained from an examination of the dispersion relation (2.25). At first glance it might appears is if interesting features should occur in the band structure in the vicinity of the frequencies at which $\epsilon(\omega)$ goes to zero, due to the presence of $\sqrt{\epsilon(\omega)}$ in the denominator of the second term on the right hand side of this equation. However, this factor is cancelled by the factor $\sin \frac{\omega}{c}\sqrt{\epsilon(\omega)}a$ multiplying it, so that nothing unusual should be observed in the band structures for frequencies in the vicinity of ω_p in the case of metallic components, or ω_L in the case of polar semiconductor components. The situation is quite different, however, for frequencies in the vicinity of the values at which $\epsilon(\omega)$ becomes infinite, viz. in the vicinity of $\omega = 0$ in the case of a metal, and $\omega = \omega_T$ in the case of a polar semiconductor. Here it is necessary to differentiate between the case in which $\epsilon(\omega)$ becomes infinite through negative values, as occurs when $\omega \to 0+$ in the case of a metal, and when $\omega \to \omega_T+$ in the case of a polar semiconductor, and the case in which $\epsilon(\omega)$ becomes infinite through positive values, as occurs when $\omega \to \omega_T-$ in the case of a polar semiconductor. In the former case the trigonometric functions containing $\sqrt{\epsilon(\omega)}$ in their arguments on the right hand side of Eq. (2.25) become hyperbolic functions and hence monotonically increasing functions of frequency as the singularity of $\epsilon(\omega)$ is approached instead of oscillatory functions. The limiting value of the right hand side of Eq. (2.25) is positively infinite as the singularity is reached for all filling fractions in the metal case, and for all filling fractions greater than 0.5 in the polar semiconductor case; it is negatively infinite for all filling fractions smaller than 0.5 in the latter case. An important consequence of this result is that when $kd = 0$ the lowest frequency solution has a nonzero value of ω in the case of the metal system, while it is zero in the case of the polar semiconductor system. In other words, there is a gap below the lowest frequency band in the photonic band structure of the metallic system, but not in the case of the polar semiconductor system. However, in the latter system there is a gap above ω_T before the first band is encountered, at $kd = 0$, for all filling fractions greater than 0.5, and at $kd = \pi$ for all filling fractions smaller than 0.5. A discrete set of bands above these lowest frequency bands is then found in both systems in the frequency range in which $\epsilon(\omega)$ is negative. In contrast, when $\epsilon(\omega)$ becomes infinite through positive values, as when $\omega \to \omega_T-$ in the polar semiconductor system, the second term on the right hand side behaves like $x^{-1}\sin(x^{-1})$ as $x \to 0+$, where $x = (\omega_T - \omega)^{\frac{1}{2}}$. Thus, the right hand side of Eq. (2.25) oscillates more and more rapidly with increasing amplitude as $\omega \to \omega_T-$. The result is an infinite number of bands in the frequency range just below ω_T in the polar semiconductor system. These bands are nearly dispersionless, since the slopes of $x^{-1}\sin(x^{-1})$ are nearly infinite for values of this function in the interval $(-1,1)$ determined by the left hand side of Eq. (2.25). Consequently, there is very little dependence on k of the frequencies of these bands. Such nearly dispersionless bnands will not be present in the band structure of the metallic system, because there is no way in this case for $\epsilon(\omega)$ to become infinite through positive values.

These qualitative conclusions are confirmed by the results displayed in Figs. 4 and 5, in which the left and right hand sides of Eq. (2.25) are plotted as functions of frequency, for $kd = 0, \pi/2, \pi$ for the metallic and polar semiconductor systems, respectively.

2.2. THE DETERMINANTAL APPROACH

The dielectric function of the system depicted in Fig. 1 is clearly a periodic function of x_1 with a period $d = a + b$. We can write it in the form $\epsilon(x_1|\omega)$. The two Maxwell equations (2.1) and (2.2) can then be combined into the single equation

$$\left(\frac{d^2}{dx_1^2} + \epsilon(x_1|\omega)\frac{\omega^2}{c^2} \right) E(x_1|\omega) = 0. \qquad (2.26)$$

In view of its periodicity we can expand $\epsilon(x_1|\omega)$ in a Fourier series according to

$$\epsilon(x_1|\omega) = \sum_{n=-\infty}^{\infty} \hat{\epsilon}_n(\omega)e^{i\frac{2\pi n x_1}{d}}, \qquad (2.27)$$

where

$$\hat{\epsilon}_n(\omega) \quad = \quad \frac{1}{d}\int_{-\frac{d}{2}}^{\frac{d}{2}} dx_1 \epsilon(x_1|\omega)e^{-i\frac{2\pi n x_1}{d}}$$

$$= \quad 1 - f(1 - \epsilon(\omega)) \qquad\qquad n = 0 \qquad (2.28a)$$

$$= \quad -f(1 - \epsilon(\omega))\frac{\sin n\pi f}{n\pi f} \qquad n \neq 0. \qquad (2.28b)$$

We also expand $E(x_1|\omega)$ in a form

$$E(x_1|\omega) = \sum_{n=-\infty}^{\infty} \hat{E}_n(k, \omega)e^{i(k+\frac{2\pi n}{d})x_1}, \qquad (2.29)$$

that satisfies the Bloch-Floquet condition required by the periodicity of the system:

$$E(x_1 + d|\omega) = e^{ikd}E(x_1|\omega). \qquad (2.30)$$

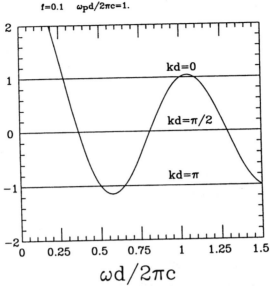

f=0.1   ωpd/2πc=1.

$$\omega d / 2\pi c$$

Fig. **4.** The left and right hand sides of Eq. (2.25) are plotted as functions of frequency for $kd = 0, \pi/2, \pi$, for the one-dimensional periodic system depicted in Fig. 1, when the material characterized by the frequency-dependent dielectric function is a metal. E-polarization. $\omega_p d/2\pi c = 1$, $f = 0.1$.

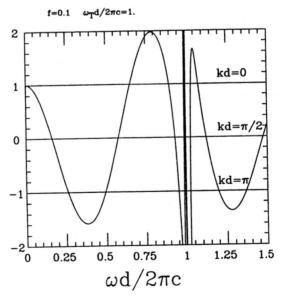

f=0.1   ωTd/2πc=1.

$$\omega d / 2\pi c$$

Fig. **5.** The same as Fig. 4, except that the material characterized by the frequency-dependent dielectric function is GaAs. E-polarization, $\omega_T d/2\pi c = 1$, $f = 0.1$.

When we substitute Eqs. (2.27) and (2.29) into Eq. (2.26), we obtain as the equation satisfied by the Fourier coefficients $\{\hat{E}_n(k,\omega)\}$

$$\sum_{n=-\infty}^{\infty} M_{mn}(k,\omega)\hat{E}_n(k,\omega) = 0 \qquad m = 0, \pm 1, \pm 2, \ldots, \qquad (2.31a)$$

where

$$M_{mn}(k,\omega) = \delta_{mn}(k + \frac{2\pi n}{d})^2 - \frac{\omega^2}{c^2}\hat{\epsilon}_{m-n}(\omega). \qquad (2.31b)$$

The solvability condition for this system of equations,

$$det\, M_{mn}(k,\omega) = 0, \qquad (2.32)$$

is the dispersion relation for electromagnetic waves in our system.

It can be shown from Eqs. (2.31) that any solution $\omega(k)$ is an even function of k, $\omega(-k) = \omega(k)$, that is also a periodic function of k with period $2\pi/d$, $\omega(k + (2\pi/d)) = \omega(k)$. It follows that we can restrict k to the interval $0 \le k \le \pi/d$ in obtaining all of the distinct solutions of Eq. (2.32).

In practice the infinite determinant in Eq. (2.32) is approximated by the $(2N+1) \times (2N+1)$ determinant which is obtained by restricting m and n to the values $0, \pm 1, \pm 2, \ldots, \pm N$. N is then increased until convergence is achieved for all the solutions of interest. The solution of Eq. (2.32) is carried out by choosing a value of k in the interval $(0, \pi/d)$ and increasing ω from zero to some arbitrarily chosen upper limit in equal steps of $\Delta\omega$, and looking for changes in the sign of the determinant. If a sign change occurs in going from $n\Delta\omega$ to $(n+1)\Delta\omega$, the determinant is then calculated at $(n + \frac{1}{2})\Delta\omega$. If the sign of this determinant is the same as at $n\Delta\omega$, the zero lies in the interval $((n + \frac{1}{2})\Delta\omega, (n+1)\Delta\omega)$; if it is not, the zero lies in the interval $(n\Delta\omega, (n + \frac{1}{2})\Delta\omega)$. By a repetition of this interval-halving procedure the zeros of the determinant can be obtained very accurately. This determinantal approach can be used to calculate the photonic band structure when $\epsilon(\omega)$ has either of the forms given by Eqs. (1.1) and (1.2).

The determinantal approach described above can also be applied to the equation satisfied by the amplitude of the single, nonzero component of the electric displacement vector, $D(x_1|\omega)$, and it is instructive to do so. From Eqs. (2.4) and the relation $E(x_1|\omega) = D(x_1|\omega)/\epsilon(x_1|\omega)$ we find that $D(x_1|\omega)$ satisfies the equation

$$\frac{d^2}{dx_1^2}\left(\frac{1}{\epsilon(x_1|\omega)}D(x_1|\omega)\right) + \frac{\omega^2}{c^2}D(x_1|\omega) = 0. \qquad (2.33)$$

In view of its periodicity, we can expand $1/\epsilon(x_1|\omega)$ in a Fourier series according to

$$\frac{1}{\epsilon(x_1|\omega)} = \sum_{n=-\infty}^{\infty} \hat{\kappa}_n(\omega) e^{i\frac{2\pi n x_1}{d}}, \tag{2.34}$$

where

$$\hat{\kappa}_n(\omega) = \frac{1}{d} \int_{-\frac{d}{2}}^{\frac{d}{2}} dx_1 \frac{e^{-i\frac{2\pi n x_1}{d}}}{\epsilon(x_1|\omega)}$$

$$= 1 + f\left(\frac{1}{\epsilon(\omega)} - 1\right) \qquad n = 0 \tag{2.35a}$$

$$= f\left(\frac{1}{\epsilon(\omega)} - 1\right) \frac{\sin n\pi f}{n\pi f} \qquad n \neq 0. \tag{2.35b}$$

We also expand $D(x_1|\omega)$ in a form that satisfies the Bloch-Floquet condition,

$$D(x_1|\omega) = \sum_{n=-\infty}^{\infty} \hat{D}_n(k,\omega) e^{i(k+\frac{2\pi n}{d})x_1}. \tag{2.36}$$

When Eqs. (2.34) and (2.36) are substituted into Eq. (2.33) we find that the equation satisfied by the Fourier coefficients $\{\hat{D}_n(k,\omega)\}$ is

$$\sum_{n=-\infty}^{\infty} (k + \frac{2\pi m}{d})^2 \hat{\kappa}_{m-n}(\omega) \hat{D}_n(k,\omega) = \frac{\omega^2}{c^2} \hat{D}_m(k,\omega). \tag{2.37}$$

A more symmetric equation is obtained if we make the replacement

$$\hat{D}_n(k,\omega) = (k + \frac{2\pi n}{d}) \hat{A}_n(k,\omega), \tag{2.38}$$

viz.

$$\sum_{n=-\infty}^{\infty} N_{mn}(k,\omega) \hat{A}_n(k,\omega) = 0 \qquad m = 0, \pm 1, \pm 2, \cdots, \tag{2.39a}$$

where

$$N_{mn}(k,\omega) = \left(k + \frac{2\pi m}{d}\right) \hat{\kappa}_{m-n}(\omega) \left(k + \frac{2\pi n}{d}\right) - \frac{\omega^2}{c^2} \delta_{mn}. \tag{2.39b}$$

Thus, the dispersion for electromagnetic waves in our system takes the equivalent form

$$\det N_{mn}(k,\omega) = 0. \tag{2.40}$$

We will compare the dispersion curves obtained from Eq. (2.40) with those obtained from Eq. (2.32).

2.3. REDUCTION TO A STANDARD EIGENVALUE PROBLEM

In the particular case that the dielectric function $\epsilon(\omega)$ has the free electron form (1.1) , Eq. (2.31) can be reduced to a standard eigenvalue problem for a real symmetric matrix. To see this we begin by rewriting Eq. (2.31) in the form

$$\hat{E}_m k_m^2 - \frac{\omega^2}{c^2} \hat{\epsilon}_0(\omega) \hat{E}_m - \frac{\omega^2}{c^2} \sum_n' \hat{\epsilon}_{m-n}(\omega) \hat{E}_n = 0 \tag{2.41}$$

where, to simplify the notation, we have set $k_m = (k + \frac{2\pi m}{d})$, and where the prime on the sum indicates that the term with $n = m$ is omitted. When we use the explicit expression for $\hat{\epsilon}_n(\omega)$ given by Eq. (2.28), Eq. (2.41) becomes

$$\hat{E}_m k_m^2 - \frac{\omega^2}{c^2} \hat{E}_m + f \frac{\omega^2}{c^2}(1 - \epsilon(\omega)) \left[\hat{E}_m + \sum_n' \frac{\sin(m-n)\pi f}{(m-n)\pi f} \hat{E}_n \right] = 0. \tag{2.42}$$

If we note that $\sin p\pi f/(p\pi f) = 1$ for $p = 0$, and use Eq. (1.1), Eq. (2.30) can be rewritten as

$$\sum_n [\delta_{mn} k_n^2 + f \frac{\omega_p^2}{c^2} \frac{\sin(m-n)\pi f}{(m-n)\pi f}] \hat{E}_n = \frac{\omega^2}{c^2} \hat{E}_m. \tag{2.43}$$

This is a standard eigenvalue problem for a real symmetric matrix.

The result that the photonic band structure can be calculated through the diagonalization of a real symmetric matrix in this case is due solely to the particular form (1.1) for the dielectric function assumed.

In practice, the infinite matrix equation (2.43) is approximated by a $(2N + 1) \times (2N + 1)$ matrix equation by restricting the indices m and n to the values $0, \pm 1, \pm 2, \ldots, \pm N$. Again, N is increased until convergence is obtained for all the eigenvalues of interest.

2.4. REDUCTION TO A SET OF EIGENVALUE PROBLEMS

When the dielectric function for the dielectric layers has the form (1.2), the determination of the photonic band structure of the resulting periodic system cannot be reduced to the solution of a standard eigenvalue problem. However, we can proceed in the following manner. We return to Eq. (2.26) and rewrite it as

$$\frac{1}{\epsilon(x_1|\omega)} \frac{d^2}{dx_1^2} E(x_1|\omega) + \frac{\omega^2}{c^2} E(x_1|\omega) = 0. \tag{2.44}$$

When Eqs. (2.29) and (2.34) are substituted into Eq. (2.44), the equation for the amplitudes $\{\hat{E}_n(k,\omega)\}$ takes the form

$$\sum_n \hat{\kappa}_{m-n}(\omega) k_n^2 \hat{E}_n - \frac{\omega^2}{c^2} \hat{E}_m = 0 \qquad m = 0, \pm 1, \pm 2, \ldots. \tag{2.45}$$

The replacement

$$\hat{E}_n(k,\omega) = \frac{\hat{F}_n(k,\omega)}{k_n} \tag{2.46}$$

yields the more symmetric equation,

$$\sum_n k_m \hat{\kappa}_{m-n}(\omega) k_n \hat{F}_n - \frac{\omega^2}{c^2} \hat{F}_m = 0. \tag{2.47}$$

If we now use the explicit form for $\hat{\kappa}_n(\omega)$ given by Eqs. (2.35), and again interpret $\sin \pi p f / (\pi p f)$ as unity when $p = 0$, we can rewrite Eq. (2.47) as

$$\sum_n \left[\delta_{mn} \left(k_m^2 - \frac{\omega^2}{c^2} \right) - f \left(1 - \frac{1}{\epsilon(\omega)} \right) k_m \frac{\sin \pi(m-n)f}{\pi(m-n)f} k_n \right] \hat{F}_n = 0. \tag{2.48}$$

With the substitution of Eq. (1.2) into Eq. (2.48), and the introduction of the notation

$$\frac{\omega^2}{c^2} = \mu, \tag{2.49}$$

the resulting equation can be written in matrix form as

$$\left(\mu^2 \overleftrightarrow{I} - \mu \overleftrightarrow{M} + \overleftrightarrow{N} \right) \vec{F} = 0, \tag{2.50}$$

where the matrices $\overleftrightarrow{M}(k)$ and $\overleftrightarrow{N}(k)$ have the elements

$$M_{mn}(k) = \delta_{mn} \left(k_m^2 + \frac{\omega_L^2}{c^2} \right) - f \left(1 - \frac{1}{\epsilon_\infty} \right) k_m \frac{\sin \pi(m-n)f}{\pi(m-n)f} k_n \tag{2.51a}$$

$$N_{mn}(k) = \delta_{mn} \frac{\omega_L^2}{c^2} k_m^2 - f \left(\frac{\omega_L^2}{c^2} - \frac{\omega_T^2}{\epsilon_\infty c^2} \right) k_m \frac{\sin \pi(m-n)f}{\pi(m-n)f} k_n. \tag{2.51b}$$

Equation (2.50) can be written in the factored form

$$\left(\mu \overset{\leftrightarrow}{I} - \overset{\leftrightarrow}{B}\right)\left(\mu \overset{\leftrightarrow}{I} - \overset{\leftrightarrow}{C}\right) \vec{F} = 0, \tag{2.52}$$

where the matrices $\overset{\leftrightarrow}{B}$ and $\overset{\leftrightarrow}{C}$ satisfy the pair of equations

$$\overset{\leftrightarrow}{B} + \overset{\leftrightarrow}{C} = \overset{\leftrightarrow}{M}, \qquad \overset{\leftrightarrow}{B}\overset{\leftrightarrow}{C} = \overset{\leftrightarrow}{N}. \tag{2.53}$$

If we write

$$\overset{\leftrightarrow}{C} = \overset{\leftrightarrow}{M} - \overset{\leftrightarrow}{B}, \tag{2.54}$$

the equation satisfied by the matrix $\overset{\leftrightarrow}{B}$ becomes

$$\overset{\leftrightarrow}{B} = \overset{\leftrightarrow}{N}\overset{\leftrightarrow}{M}{}^{-1} + \overset{\leftrightarrow}{B}{}^2 \overset{\leftrightarrow}{M}{}^{-1}. \tag{2.55}$$

We solve this equation by iteration, by first rewriting it as

$$\overset{\leftrightarrow}{B} = \epsilon \overset{\leftrightarrow}{N}\overset{\leftrightarrow}{M}{}^{-1} + \overset{\leftrightarrow}{B}{}^2 \overset{\leftrightarrow}{M}{}^{-1}, \tag{2.56}$$

expressing the solution as an expansion in powers of ϵ,

$$\overset{\leftrightarrow}{B} = \sum_{n=1}^{\infty} \epsilon^n \overset{\leftrightarrow}{B}_n, \tag{2.57}$$

and setting $\epsilon = 1$ at the end of the calculation. The matrices $\{\overset{\leftrightarrow}{B}_n\}$ can be calculated recursively:

$$\overset{\leftrightarrow}{B}_1 = \overset{\leftrightarrow}{N}\overset{\leftrightarrow}{M}{}^{-1} \tag{2.58a}$$

$$\overset{\leftrightarrow}{B}_n = \sum_{m=1}^{n-1} \overset{\leftrightarrow}{B}_{n-m}\overset{\leftrightarrow}{B}_m\overset{\leftrightarrow}{M}{}^{-1} \qquad n \geq 2. \tag{2.58b}$$

Once $\overset{\leftrightarrow}{B}$ has been calculated, $\overset{\leftrightarrow}{C}$ is obtained from Eq. (2.54). In this way, the problem of calculating the photonic band structure is reduced to the diagonalization of the two matrices $\overset{\leftrightarrow}{B}$ and $\overset{\leftrightarrow}{C}$.

In practice the infinite matrices $\overset{\leftrightarrow}{M}$ and $\overset{\leftrightarrow}{N}$ are approximated by $(2N+1)\times(2N+1)$ matrices by restricting the indices m and n in Eqs. (2.51) to take the values $0, \pm 1, \pm 2, \ldots, \pm N$. Then enough terms have to be kept in the expansion (2.57) of the matrix $\overset{\leftrightarrow}{B}$ (with $\epsilon = 1$) to ensure a converged result for it. Finally, N is increased until the bands of interest have converged.

2.5. RESULTS

In this Section we present results for the photonic band structure of the simple one-dimensional system depicted in Fig. 1, that we obtain by each of the methods described in Sections 2.1 - 2.4. From a comparison of the results obtained by each of these methods, we can access the accuracy and reliability of these several approaches to the calculation of the photonic band structures containing components characterized by frequency-dependent dielectric functions. The ability to do so is particularly important, since there are no comparably simple models of analogous two- and three-dimensional systems whose photonic band structures can be used to judge the quality of different computational approaches to the determination of the band structures of realistic, and hence more complicated, systems.

We begin by considering the results obtained by the transfer matrix method. These results will be used as the benchmarks against which the results of the other approaches will be compared. This is because the transfer matrix method yields an explicit, analytic dispersion relation, Eq. (2.25), whose solutions are free from the concerns about convergence that are always present when the plane wave method, and variants of it, are employed.

In the left-hand half of Fig. 6(a) we present the photonic band structure for the system in Fig. 1, when the material characterized by the frequency-dependent dielectric function is a metal whose dielectric function is given by Eq. (1.1). We have chosen $\omega_p d/2\pi c = 1$, and $f = 0.1$. From this result we see that allowed bands exist in the frequency range $\omega d/2\pi c < 1$ in which the dielectric function of the metallic component is negative. Perhaps the most interesting feature of the band structure in this case is the gap below the lowest frequency band which exists for any nonzero filling fraction. The width of this gap is found to increase monotonically with increasing filling fraction.

In Fig. 6(b) we plot the frequency dependence of the transmissivity of a system of 20 slabs of a metal characterized by the dielectric function (1.1) embedded in a vacuum, in the case where $\omega_p d/2\pi c = 1$ and $f = 0.1$. A comparison of this result with the photonic band structure shows that calculating the transmissivity as a function of frequency is an effective way of obtaining the frequency ranges of the allowed and forbidden bands, but it is incapable of providing the detailed information about the photonic band structure contained in the dispersion curves.

When the dielectric function characterizing the medium with a frequency-dependent dielectric function has the form (1.2), the resulting photonic band structure possesses no gap below the lowest frequency band, and posseses a series of bands in the frequency range $1 < \omega d/2\pi c < 1.07759$ in which $\epsilon(\omega)$ is negative, and a series of nearly dispersionless bands in the

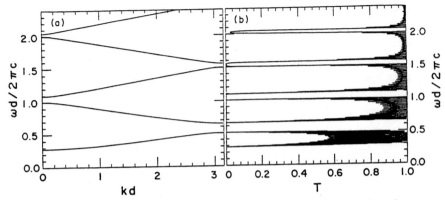

Fig. **6.** (a) The photonic band structure of the one-dimensional system depicted in Fig. 1, when the material characterized by the frequency-dependent dielectric function is a metal. The transfer matrix method (Eq. (2.25)) has been used to obtain it. E-polarization. $\omega_p d/2\pi c = 1$, $f = 0.1$. (b) The frequency dependence of the transmissivity T of a periodic system of 20 metal layers depicted in Fig. 3. E-polarization. $\omega_p d/2\pi c = 1$, $f = 0.1$. The transfer matrix method (Eq. (2.18)) has been used in obtaining this result.

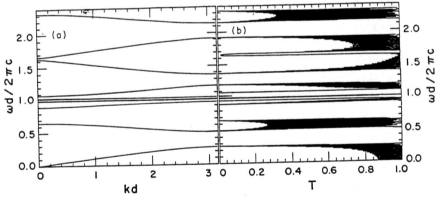

Fig. **7.** (a) The photonic band structure of the one-dimensional system depicted in Fig. 1, when the material characterized by the frequency-dependent dielectric function is GaAs. The transfer matrix method (Eq. (2.25)) has been used to obtain it. E-polarization. $\omega_T d/2\pi c = 1$, $f = 0.1$. (b) The same as Fig. 4(b), but for a periodic system of 20 GaAs layers. E-polarization. $\omega_T d/2\pi c = 1$, $f = 0.1$.

frequency range just below the frequency $\omega d/2\pi c = 1$. The former bands are also nearly dispersionless because there are several of them and they are confined to a rather narrow frequency range. This is shown in Fig. 7 (a), which has been obtained for the case that $\omega_T d/2\pi c = 1$ and $f = 0.1$.

In Fig. 7(b) we plot the frequency dependence of the transmissivity of a system of 20 slabs of a medium characterized by the frequency-dependent dielectric function (1.2) embedded in a vacuum. The filling fraction is $f = 0.1$ and we assumed $\omega_T d/2\pi c = 1$ in obtaining this result. Again, a comparison of the frequency dependence of the transmissivity with the corresponding band structure shows that it predicts the positions of the allowed bands, including the nearly dispersionless bands, and the band gaps, but lacks the details present in the dispersion curves.

The detailed form of the band structures displayed in Figs. 6 and 7 are in complete agreement with the predictions of the qualitative analysis of Eq. (2.25) presented in Section 2.1.

We now look at the determinantal method applied to the metal system with $\omega_p d/c = 1$ and to the GaAs system with $\omega_T d/c = 1$. For both systems we assume $f = 0.001$. We take smaller values for $\omega_p d/c$ and $\omega_T d/c$ than the values used in obtaining Figs. 4 and 5 because for these values $\epsilon(\omega)$ is negative at the frequencies on the lowest branch of the dispersion relation near the center of the Brillouin zone. The dispersion relation in this region is simple and hence easy to map using the determinantal method. The small filling fraction $f = 0.001$ is chosen in the hope that fewer plane waves will be required to represent the electric field in the metal or GaAs slabs accurately.

We find that the determinantal method based on Eq. (2.26) converges quickly to the transfer matrix results for the GaAs and the metal systems, but the determinantal method based on Eq. (2.33) is much less satisfactory when applied to these systems. We have studied the mode frequencies (zeros of the determinants) at a fixed wavevector versus N for the modes computed from a truncated $N \times N$ matrix version of Eq. (2.26) or Eq. (2.33). In Fig. 8(a) we plot mode frequencies, $\omega d/2\pi c$, versus N for the results from Eq. (2.26) (crosses) and Eq. (2.33) (squares) for the GaAs system. The results shown are computed for the wave vector $kd = 0.628$. The transfer matrix results for GaAs with $f = 0.001$ and $kd = 0.628$ indicate a lowest frequency mode at $\omega d/2\pi c = 0.1$ and a flat band at $\omega d/2\pi c = 0.1592$. These modes are both observed in the determinantal method based on Eq. (2.26). The determinantal method based on Eq. (2.33) gives the mode at $\omega d/2\pi c = 1$ accurately, but for $N = 2000$ has still not yielded a completely converged flat band at $\omega d/2\pi c = 0.1592$.

In the transfer matrix solution for the metal, the frequency of the lowest frequency mode for $kd = 0.628$ is given by $\omega d/2\pi c = 0.1$, and no flat

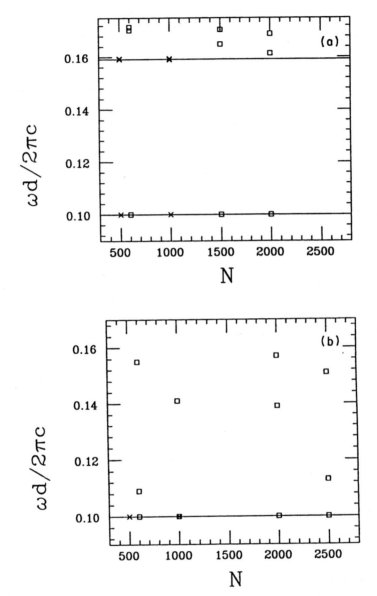

Fig. 8. (a) The frequencies corresponding to the smallest zeros of Eq. (2.32) (×) and Eq. (2.40) (□) for the one-dimensional system depicted in Fig. 1, when the material characterized by the frequency-dependent dielectric function is GaAs, obtained from $N \times N$ matrix approximations to the matrices $\overleftrightarrow{M} (k,\omega)$ and $\overleftrightarrow{N} (k,\omega)$ in these equations, when $kd = 0.628$. $\omega_T d/c = 1$, $f = 0.001$. (b) The same as (a), except that the material characterized by the frequency-dependent dielectric function is a metal. $\omega_p d/c = 1$, $f = 0.001$. In both cases, the horizontal lines give the lowest frequencies obtained by the transfer matrix method (Eq. (2.25)).

bands are present. In Fig. 8(b) we present determinantal results from Eq. (2.26) (crosses) and Eq. (2.33) (squares) for this system. We plot mode frequencies $\omega d/2\pi c$ versus N for modes computed from truncated $N \times N$ matrix versions of Eq. (2.26) or Eq. (2.33). While the method based on Eq. (2.26) converges quickly to the mode at $\omega d/2\pi c = 0.1$, the method based on Eq. (2.33) converges to the mode at $\omega d/2\pi c = 0.1$ but, in addition, exhibits a series of spurious modes which change a great deal as N is increased. The method based on Eq. (2.33) has not completely converged for $N = 2000$, which is the largest value of N for which we have computed the band structure.

We turn now to the method developed in Section 2.3, in which the calculation of the photonic band structure is transformed into the solution of a standard eigenvalue problem. The result obtained by this approach, with only NG = 25 plane waves used in the expansion of the electric field, is presented in Fig. 9, for a periodic structure of metallic slabs. It is indistinguishable from the one obtained by the transfer matrix approach which has been plotted in Fig. 4(a). This agreement validates the method developed in Section 2.3, at least in those cases in which it is $\epsilon(x_1|\omega)$ that is expanded in a Fourier series rather than $1/\epsilon(x_1|\omega)$. In Fig. 10 we plot the width of the gap below the lowest frequency band as a function of the filling fraction for this type of structure. We see that as the filling fraction approaches unity, the gap increases to cover the entire frequency range $0 < \omega < \omega_p$ in which the dielectric function (1.1) is negative, a not unexpected result.

Finally, in Fig. 11 we plot the photonic band structure for a periodic structure of GaAs slabs obtained by the method developed in Section 2.4, in which the calculation of the photonic band structure in this case is transformed into the solution of a pair of eigenvalue problems. A total of NG = 151 plane waves was used in the expansion of the electric field in these calculations. A comparison of the results presented in Figs. 5(a) and 11 shows very good agreement, including the presence of nearly dispersionless bands in the frequency range $\omega_T d/2\pi c \cong 1$ where $\epsilon(\omega)$ is negative. This result validates the method developed in Section 2.4 for the calculation of photonic band structures for periodic systems containing components characterized by the dielectric function in Eq. (1.2).

We now apply several of these methods to the determination of the photonic band structures of periodic two-and three-dimensional systems containing components with frequency-dependent dielectric functions of the forms given by Eqs. (1.1) and (1.2).

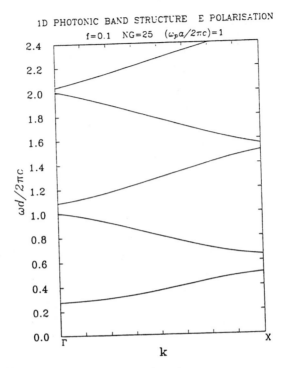

Fig. 9. The photonic band structure for the same system assumed in obtaining Fig. 6(a), but now calculated on the basis of the approach described in Section 2.3 (Eq. (2.43)). $NG = 25$.

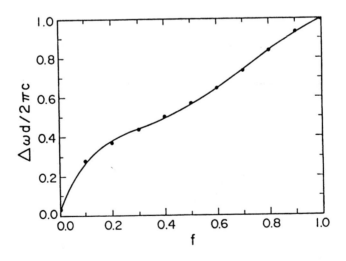

Fig. 10. The width of the gap below the lowest frequency band in the photonic band structure plotted in Figs. 6(a) and 9, as a function of the filling fraction f.

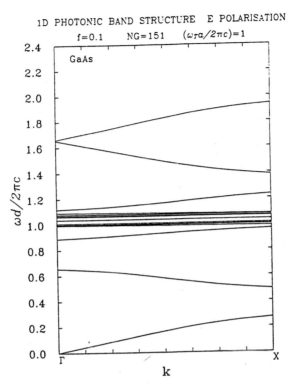

1D PHOTONIC BAND STRUCTURE E POLARISATION

f=0.1   NG=151 $(\omega_T a/2\pi c)=1$

GaAs

Fig. **11.** The photonic band structure for the same system assumed in obtaining Fig. 7(a), but now calculated on the basis of the approach described in Section 2.4. NG = 99.

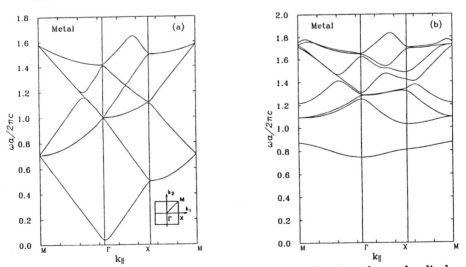

Fig. **12.** The photonic band structure of a square lattice of metal cylinders in vacuum. E-polarization. $\omega_p a/2\pi c = 1$. (a) $f = 0.001$. (b) $f = 0.7$. A band gap is present in the latter structure. NG = 197.

3. Two-Dimensional Systems

In this Section we study the photonic band structures for electromagnetic waves of E- and H-polarization propagating in a system consisting of an infinite array of indentical, infinitely long, parallel, metal or semiconductor cylinders of circular cross-section, embedded in a vacuum, whose intersections with a perpendicular plane form one of the five two-dimensional Bravais lattices.

We assume that the axes of the cylinders are parallel to the x_3-axis. The translation vectors of the Bravais lattice are

$$\vec{x}_\parallel(\ell) = l_1 \vec{a}_1 + l_2 \vec{a}_2, \tag{3.1}$$

where $\vec{a}_1$ and $\vec{a}_2$ are two, noncollinear, primitive translation vectors of the lattice, while l_1 and l_2 are arbitrary integers we denote collectively by l. The area a_c of a primitive unit cell of this lattice is given by

$$a_c = |\vec{a}_1 \times \vec{a}_2|. \tag{3.2}$$

The lattice reciprocal to the direct lattice whose points are defined by Eq. (3.1) is defined by the translation vectors

$$\vec{G}_\parallel(h) = h_1 \vec{b}_1 + h_2 \vec{b}_2, \tag{3.3}$$

where the primitive translation vectors $\vec{b}_1$ and $\vec{b}_2$ of the reciprocal lattice are the solutions of the equations

$$\vec{a}_i \cdot \vec{b}_j = 2\pi \delta_{ij} \qquad i, j = 1, 2, \tag{3.4}$$

and h_1 and h_2 are arbitrary integers we denote collectively by h. The dielectric function of this system is position-dependent, $\epsilon(\vec{x}_\parallel|\omega)$, and is a periodic function of $\vec{x}_\parallel$ with the periodicity of the two-dimensional Bravais lattice defined by Eq. (3.1), $\epsilon(\vec{x}_\parallel + \vec{x}_\parallel(\ell)|\omega) = \epsilon(\vec{x}_\parallel|\omega)$. The same property is possessed by the reciprocal of the dielectric function, $1/\epsilon(\vec{x}_\parallel + \vec{x}_\parallel(\ell)|\omega) = 1/\epsilon(\vec{x}_\parallel|\omega)$.

3.1. E-POLARIZATION

We now turn to a calculation of the photonic band structure for electromagnetic waves propagating in a plane perpendicular to the rods. In the case of E-polarization, we seek solutions of Maxwell's equations which have the forms

$$\vec{E}(\vec{x}; t) = (0, 0, E_3(\vec{x}_\parallel|\omega)) \exp(-i\omega t) \tag{3.5a}$$

$$\vec{H}(\vec{x}; t) = (H_1(\vec{x}_\parallel|\omega), H_2(\vec{x}_\parallel|\omega), 0) \exp(-i\omega t). \tag{3.5b}$$

The Maxwell curl equations for the three nonzero field components are

$$\frac{\partial H_2}{\partial x_1} - \frac{\partial H_1}{\partial x_2} = -i\frac{\omega}{c}D_3 = -i\frac{\omega}{c}\epsilon(\vec{x}_\parallel|\omega)E_3 \qquad (3.6a)$$

$$\frac{\partial E_3}{\partial x_1} = -i\frac{\omega}{c}H_2 \qquad (3.6b)$$

$$\frac{\partial E_3}{\partial x_2} = i\frac{\omega}{c}H_1. \qquad (3.6c)$$

When we eliminate H_1 and H_2 from these equations, we obtain as the equations satisfied by E_3

$$\left(\frac{\partial^2}{\partial x_1^2} + \frac{\partial^2}{\partial x_2^2}\right)E_3 + \epsilon(\vec{x}_\parallel|\omega)\frac{\omega^2}{c^2}E_3 = 0. \qquad (3.7)$$

To solve Eq. (3.7) we expand $\epsilon(\vec{x}_\parallel|\omega)$ and $E_3(\vec{x}_\parallel|\omega)$ according to

$$\epsilon(\vec{x}_\parallel|\omega) = \sum_{\vec{G}_\parallel}\hat{\epsilon}(\vec{G}_\parallel)e^{i\vec{G}_\parallel\cdot\vec{x}_\parallel} \qquad (3.8)$$

$$E_3(\vec{x}_\parallel|\omega) = \sum_{\vec{G}_\parallel}B(\vec{k}_\parallel|\vec{G}_\parallel)e^{i(\vec{k}_\parallel+\vec{G}_\parallel)\cdot\vec{x}_\parallel}, \qquad (3.9)$$

where $\vec{k}_\parallel = (k_1, k_2, 0)$ is the two-dimensional wave vector of the wave. When these expansions are substituted in Eq. (3.7) we obtain as the equation satisfied by the coefficients $\{B(\vec{k}_\parallel|\vec{G}_\parallel)\}$

$$(\vec{k}_\parallel + \vec{G}_\parallel)^2 B(\vec{k}_\parallel|\vec{G}_\parallel) = \frac{\omega^2}{c^2}\sum_{\vec{G}_\parallel}\hat{\epsilon}(\vec{G}_\parallel - \vec{G}_\parallel')B(\vec{k}_\parallel|\vec{G}_\parallel'). \qquad (3.10)$$

Alternatively, we can rewrite Eq. (3.7) as

$$\frac{1}{\epsilon(\vec{x}_\parallel|\omega)}\left(\frac{\partial^2}{\partial x_1^2} + \frac{\partial^2}{\partial x_2^2}\right)E_3 + \frac{\omega^2}{c^2}E_3 = 0. \qquad (3.11)$$

To solve this equation we expand $\epsilon^{-1}(\vec{x}_\parallel|\omega)$ according to

$$\frac{1}{\epsilon(\vec{x}_\parallel|\omega)} = \sum_{\vec{G}_\parallel}\hat{\kappa}(\vec{G}_\parallel)e^{i\vec{G}_\parallel\cdot\vec{x}_\parallel}, \qquad (3.12)$$

and $E_3(\vec{x}_\parallel|\omega)$ as in Eq. (3.9). When we substitute these expansions into Eq. (3.11), we obtain as the equation satisfied by the coefficients $\{B(\vec{k}_\parallel|\vec{G}_\parallel)\}$

$$\sum_{\vec{G}_\parallel} \hat{\kappa}(\vec{G}_\parallel - \vec{G}_\parallel')(\vec{k} + \vec{G}_\parallel')^2 B(\vec{k}_\parallel|\vec{G}_\parallel') = \frac{\omega^2}{c^2} B(\vec{k}_\parallel|\vec{G}_\parallel). \tag{3.13}$$

However, the replacement

$$C(\vec{k}_\parallel|\vec{G}_\parallel) = |\vec{k}_\parallel + \vec{G}_\parallel| B(\vec{k}_\parallel|\vec{G}_\parallel) \tag{3.14}$$

yields the more symmetric equation

$$\sum_{\vec{G}_\parallel'} |\vec{k}_\parallel + \vec{G}_\parallel| \hat{\kappa}(\vec{G}_\parallel - \vec{G}_\parallel') |\vec{k}_\parallel + \vec{G}_\parallel'| C(\vec{k}_\parallel|\vec{G}_\parallel') = \frac{\omega^2}{c^2} C(\vec{k}_\parallel|\vec{G}_\parallel). \tag{3.15}$$

For a structure consisting of metal rods whose cross-section is a circle of radius R, the Fourier coefficients $\{\hat{\epsilon}(\vec{G}_\parallel)\}$ are

$$\hat{\epsilon}(\vec{G}_\parallel) = 1 - f\frac{\omega_p^2}{\omega^2} \qquad\qquad \vec{G}_\parallel = 0 \tag{3.16a}$$

$$= -f\frac{\omega_p^2}{\omega^2}\frac{2J_1(G_\parallel R)}{(G_\parallel R)} \qquad \vec{G}_\parallel \neq 0, \tag{3.16b}$$

where $J_1(x)$ is a Bessel function. The corresponding Fourier coefficients $\{\hat{\kappa}(\vec{G}_\parallel)\}$ are

$$\hat{\kappa}(G_\parallel) = 1 + f\frac{\omega_p^2}{\omega^2 - \omega_p^2} \qquad\qquad \vec{G}_\parallel = 0 \tag{3.17a}$$

$$= f\frac{\omega_p^2}{\omega^2 - \omega_p^2}\frac{2J_1(G_\parallel R)}{(G_\parallel R)} \qquad \vec{G}_\parallel \neq 0. \tag{3.17b}$$

For a structure consisting of GaAs rods of circular cross-section, these two sets of Fourier coefficients are

$$\hat{\epsilon}(\vec{G}_\parallel) = 1 + f\frac{(\epsilon_\infty\omega_L^2 - \omega_T^2) - (\epsilon_\infty - 1)\omega^2}{\omega_T^2 - \omega^2} \qquad \vec{G}_\parallel = 0 \tag{3.18a}$$

$$= f\frac{(\epsilon_\infty\omega_L^2 - \omega_T^2) - (\epsilon_\infty - 1)\omega^2}{\omega_T^2 - \omega^2}\frac{2J_1(G_\parallel R)}{(G_\parallel R)} \quad \vec{G}_\parallel \neq 0, \tag{3.18b}$$

and

$$\hat{\kappa}(\vec{G}_\parallel) \;=\; 1 + f\frac{\omega_T^2 - \epsilon_\infty\omega_L^2 + \omega^2(\epsilon_\infty - 1)}{\epsilon_\infty(\omega_L^2 - \omega^2)} \qquad \vec{G}_\parallel = 0 \quad (3.19a)$$

$$\;=\; f\frac{\omega_T^2 - \epsilon_\infty\omega_L^2 + \omega^2(\epsilon_\infty - 1)}{\epsilon_\infty(\omega_L^2 - \omega^2)}\frac{2J_1(G_\parallel R)}{(G_\parallel R)} \qquad \vec{G}_\parallel \neq 0. \quad (3.19b)$$

The use of the result for the $\{\hat{\epsilon}(\vec{G}_\parallel\}$ given by Eq. (3.16) in Eq. (3.10) transforms the latter into

$$\sum_{\vec{G}'_\parallel}\left\{(\vec{k}_\parallel + \vec{G}_\parallel)^2\delta_{\vec{G}_\parallel,\vec{G}'_\parallel} + f\frac{\omega_p^2}{c^2}\frac{2J_1(|\vec{G}_\parallel - \vec{G}'_\parallel|R)}{(|\vec{G}_\parallel - \vec{G}'_\parallel|R)}\right\} B(\vec{k}_\parallel|\vec{G}'_\parallel)$$

$$= \frac{\omega^2}{c^2}B(\vec{k}_\parallel|\vec{G}_\parallel), \qquad (3.20)$$

which has the form of a standard eigenvalue problem for a real, symmetric matrix. In solving Eq. (3.20) it should be kept in mind that $2J_1(x)/x$ equals unity at $x = 0$.

The use of the results given by Eqs. (3.19) transforms Eq. (3.15) into

$$(\omega_L^2 - \omega^2)\left[\frac{\omega^2}{c^2} - |\vec{k}_\parallel + \vec{G}_\parallel|^2\right] B(\vec{k}_\parallel|\vec{G}_\parallel) = \frac{f}{\epsilon_\infty}[\omega_T^2 - \epsilon_\infty\omega_L^2$$

$$+(\epsilon_\infty - 1)\omega^2]\sum_{\vec{G}'_\parallel}|\vec{k}_\parallel + \vec{G}_\parallel|\,|\vec{k}_\parallel + \vec{G}'_\parallel|\frac{2J_1(|\vec{G}_\parallel - \vec{G}'_\parallel|R)}{(|\vec{G}_\parallel - \vec{G}'_\parallel|R)}B(\vec{k}_\parallel|\vec{G}'_\parallel). \,(3.21)$$

At this point we define

$$\frac{\omega^2}{c^2} = \mu \qquad (3.22)$$

and rewrite Eq. (3.21) in the form

$$(\mu^2\,\overleftrightarrow{I} - \mu\,\overleftrightarrow{M} + \overleftrightarrow{N})\vec{B} = 0, \qquad (3.23)$$

where the elements of the matrices $\overleftrightarrow{M}$ and $\overleftrightarrow{N}$ are given by

$$\overleftrightarrow{M}(\vec{G}_\parallel|\vec{G}'_\parallel) = \delta_{\vec{G}_\parallel,\vec{G}'_\parallel}\left[\frac{\omega_L^2}{c^2} + |\vec{k}_\parallel + \vec{G}_\parallel|^2\right]$$

$$-f\left(1 - \frac{1}{\epsilon_\infty}\right) |\vec{k}_\| + \vec{G}_\| ||\vec{k}_\| + \vec{G}_\|'| \frac{2J_1(\vec{G}_\| - \vec{G}_\|')|R)}{(|\vec{G}_\| - \vec{G}_\|'|R)} \qquad (3.24a)$$

$$\overleftrightarrow{N}(\vec{G}_\| | \vec{G}_\|') = \delta_{\vec{G}_\|, \vec{G}_\|'} |\vec{k}_\| + \vec{G}_\||^2 \frac{\omega_L^2}{c^2}$$

$$+f\left[\frac{1}{\epsilon_\infty}\frac{\omega_T^2}{c^2} - \frac{\omega_L^2}{c^2}\right] |\vec{k}_\| + \vec{G}_\| ||\vec{k}_\| + \vec{G}_\|'| \frac{2J_1(|\vec{G}_\| - \vec{G}_\|'|R)}{(|\vec{G}_\| - \vec{G}_\|'|R)}. \qquad (3.24b)$$

Equation (3.23) can be written in the factored form

$$\left(\mu \overleftrightarrow{I} - \overleftrightarrow{B}\right)\left(\mu \overleftrightarrow{I} - \overleftrightarrow{C}\right)\vec{B} = 0, \qquad (3.25)$$

where the matrices $\overleftrightarrow{B}$ and $\overleftrightarrow{C}$ satisfy the pair of equations

$$\overleftrightarrow{B} + \overleftrightarrow{C} = \overleftrightarrow{M} \qquad \overleftrightarrow{B}\overleftrightarrow{C} = \overleftrightarrow{N}. \qquad (3.26)$$

When we use the iterative approach described in Section 3.4 the problem of calculating the photonic band structure can be reduced to the diagonalization of the two matrices $\overleftrightarrow{B}$ and $\overleftrightarrow{C}$.

3.2. H-POLARIZATION

In the case of H-polarization, we seek solutions of Maxwell's equations which have the forms

$$\vec{H}(\vec{x}; t) = (0, 0, H_3(\vec{x}_\||\omega))\exp(-i\omega t) \qquad (3.27a)$$

$$\vec{E}(\vec{x}; t) = (E_1(\vec{x}_\||\omega), E_2(\vec{x}_\||\omega), 0)\exp(-i\omega t). \qquad (3.27b)$$

The Maxwell curl equations in this case are

$$\frac{\partial E_2}{\partial x_1} - \frac{\partial E_1}{\partial x_2} = i\frac{\omega}{c}H_3 \qquad (3.28a)$$

$$\frac{\partial H_3}{\partial x_1} = i\frac{\omega}{c}D_2 = i\frac{\omega}{c}\epsilon(\vec{x}_\||\omega)E_2 \qquad (3.28b)$$

$$\frac{\partial H_3}{\partial x_2} = -i\frac{\omega}{c}D_1 = -i\frac{\omega}{c}\epsilon(\vec{x}_\||\omega)E_1. \qquad (3.28c)$$

When we eliminate E_1 and E_2 from these equations, we obtain the equation satisfied by H_3, which we write in the form

$$\frac{\partial}{\partial x_1}\left(\frac{1}{\epsilon(\vec{x}_{\|}|\omega)}\frac{\partial H_3}{\partial x_1}\right) + \frac{\partial}{\partial x_2}\left(\frac{1}{\epsilon(\vec{x}_{\|}|\omega)}\frac{\partial H_3}{\partial x_2}\right) + \frac{\omega^2}{c^2}H_3 = 0. \qquad (3.29)$$

To solve this equation we expand $\epsilon^{-1}(\vec{x}_{\|}|\omega)$ according to Eq. (3.12) and $H_3(\vec{x}_{\|}|\omega)$ according to

$$H_3(\vec{x}_{\|}|\omega) = \sum_{\vec{G}_{\|}} A(\vec{k}_{\|}|\vec{G}_{\|})e^{i(\vec{k}_{\|}+\vec{G}_{\|})\cdot\vec{x}_{\|}}. \qquad (3.30)$$

When we substitute the expansions (3.12) and (3.30) into Eq. (3.29), we obtain the equation satisfied by the coefficients $\{A(\vec{k}_{\|}|\vec{G}_{\|})\}$

$$\sum_{\vec{G}'_{\|}}(\vec{k}_{\|} + \vec{G}_{\|})\cdot(\vec{k}_{\|} + \vec{G}'_{\|})\hat{\kappa}(\vec{G}_{\|} - \vec{G}'_{\|})A(\vec{k}_{\|}|\vec{G}'_{\|}) = \frac{\omega^2}{c^2}A(\vec{k}_{\|}|\vec{G}_{\|}). \qquad (3.31)$$

The use of the results given by Eqs. (3.17) in Eq. (3.31) transforms the latter into

$$\frac{(\omega^2 - \omega_p^2)}{\omega_p^2}\left[\frac{\omega^2}{c^2} - |\vec{k}_{\|} + \vec{G}_{\|}|^2\right]A(\vec{k}_{\|}|\vec{G}_{\|})$$

$$= f\sum_{\vec{G}'_{\|}}(\vec{k}_{\|} + \vec{G}_{\|})\cdot(\vec{k}_{\|} + \vec{G}'_{\|})\frac{2J_1(|\vec{G}_{\|} - \vec{G}'_{\|}|R)}{(|\vec{G}_{\|} - \vec{G}'_{\|}|R)}A(\vec{k}_{\|}|\vec{G}'_{\|}). \qquad (3.32)$$

At this point we define

$$\frac{\omega^2}{\omega_p^2} = \mu \qquad (3.33)$$

and rewrite Eq. (3.32) in the form

$$(\mu^2\, \overleftrightarrow{I} - \mu\, \overleftrightarrow{M} + \overleftrightarrow{N})\vec{A} = 0, \qquad (3.34)$$

where the elements of the matrices $\overleftrightarrow{M}$ and $\overleftrightarrow{N}$ are given by

$$\overleftrightarrow{M}(\vec{G}_{\|}|\vec{G}'_{\|}) = \delta_{\vec{G}_{\|},\vec{G}'_{\|}}\left[1 + \frac{c^2|\vec{k}_{\|} + \vec{G}_{\|}|^2}{\omega_p^2}\right] \qquad (3.35a)$$

$$\overset{\leftrightarrow}{N}(\vec{G}_{\|}|\vec{G}_{\|}') = \delta_{\vec{G}_{\|},\vec{G}_{\|}'} \frac{c^2|\vec{k}_{\|} + \vec{G}_{\|}|^2}{\omega_p^2}$$

$$-f\frac{c^2}{\omega_p^2}(\vec{k}_{\|} + \vec{G}_{\|}) \cdot (\vec{k}_{\|} + \vec{G}_{\|}') \frac{2J_1(|\vec{G}_{\|} - \vec{G}_{\|}'|R)}{(|\vec{G}_{\|} - \vec{G}_{\|}'|R)}. \qquad (3.35b)$$

Equation (3.34) can be written in the factored form

$$(\mu \overset{\leftrightarrow}{I} - \overset{\leftrightarrow}{B})(\mu \overset{\leftrightarrow}{I} - \overset{\leftrightarrow}{C})\vec{A} = 0, \qquad (3.36)$$

where the matrices $\overset{\leftrightarrow}{B}$ and $\overset{\leftrightarrow}{C}$ satisfy the pair of the equations (3.26). The solution of the latter is then carried out by means of the iterative approach described in Section 2.4.

The use of the results for the Fourier coefficients $\{\hat{\kappa}(\vec{G}_{\|})\}$ given by Eq. (3.19) transforms Eq. (3.31) into

$$(\omega_L^2 - \omega^2)\left[\frac{\omega^2}{c^2} - (\vec{k}_{\|} + \vec{G}_{\|})^2\right] A(\vec{k}_{\|}|\vec{G}_{\|})$$

$$= \frac{f}{\epsilon_\infty}[\omega_T^2 - \epsilon_\infty\omega_L^2 + (\epsilon_\infty - 1)\omega^2]\sum_{\vec{G}_{\|}'}(\vec{k}_{\|} + \vec{G}_{\|}) \cdot (\vec{k}_{\|} + \vec{G}_{\|}')$$

$$\times \frac{2J_1(|\vec{G}_{\|} - \vec{G}_{\|}'|R)}{(|\vec{G}_{\|} - \vec{G}_{\|}'|R)} A(\vec{k}_{\|}|\vec{G}_{\|}'). \qquad (3.37)$$

At this point we define

$$\frac{\omega^2}{c^2} = \mu \qquad (3.38)$$

and rewrite Eq. (3.37) in the form

$$(\mu^2 \overset{\leftrightarrow}{I} - \mu \overset{\leftrightarrow}{P} + \overset{\leftrightarrow}{Q})\vec{A} = 0, \qquad (3.39)$$

where the elements of the matrices $\overset{\leftrightarrow}{P}$ and $\overset{\leftrightarrow}{Q}$ are given by

$$\overset{\leftrightarrow}{P}(\vec{G}_{\|}|\vec{G}_{\|}') = \delta_{\vec{G}_{\|},\vec{G}_{\|}'}\left[\frac{\omega_L^2}{c^2} + |\vec{k}_{\|} + \vec{G}_{\|}|^2\right]$$

$$-f\left(1 - \frac{1}{\epsilon_\infty}\right)(\vec{k}_{\|} + \vec{G}_{\|}) \cdot (\vec{k}_{\|} + \vec{G}_{\|}') \frac{2J_1(|\vec{G}_{\|} - \vec{G}_{\|}'|R)}{(|\vec{G}_{\|} - \vec{G}_{\|}'|R)} \qquad (3.40a)$$

$$\overleftrightarrow{Q}\left(\vec{G}_{\parallel}|\vec{G}_{\parallel}'\right) = \delta_{\vec{G}_{\parallel},\vec{G}_{\parallel}'}|\vec{k}_{\parallel} + \vec{G}_{\parallel}|^2\frac{\omega_L^2}{c^2}$$

$$+f\left[\frac{1}{\epsilon_\infty}\frac{\omega_T^2}{c^2} - \frac{\omega_L^2}{c^2}\right]\frac{2J_1(|\vec{G}_{\parallel} - \vec{G}_{\parallel}'|R)}{(|\vec{G}_{\parallel} - \vec{G}_{\parallel}'|R)}(\vec{k}_{\parallel} + \vec{G}_{\parallel})\cdot(\vec{k}_{\parallel} + \vec{G}_{\parallel}').\text{(3.40b)}$$

Equation (3.39) can be written in the factored form

$$\left(\mu\overleftrightarrow{I} - \overleftrightarrow{D}\right)\left(\mu\overleftrightarrow{I} - \overleftrightarrow{E}\right)\vec{A} = 0, \tag{3.41}$$

where the matrices $\overleftrightarrow{D}$ and $\overleftrightarrow{E}$ satisfy the pair of equations

$$\overleftrightarrow{D} + \overleftrightarrow{E} = \overleftrightarrow{P} \qquad \overleftrightarrow{D}\overleftrightarrow{E} = \overleftrightarrow{Q}. \tag{3.42}$$

By the use of the iterative approach described in Section 2.4, the problem of calculating the photonic band structure is reduced to the diagonalization of the two matrices $\overleftrightarrow{D}$ and $\overleftrightarrow{E}$.

3.3. RESULTS

We apply the preceding results first to the calculation of the photonic band structure of a periodic array of parallel metallic rods characterized by the dielectric function $\epsilon(\omega)$ given by Eq. (1.1). We consider the case of a simple square lattice of rods of lattice parameter a for which the primitive translation vectors are

$$\vec{a}_1 = a(1,0), \qquad \vec{a}_2 = a(0,1), \tag{3.43a}$$

while the primitive translation vectors of the reciprocal lattice are

$$\vec{b}_1 = \frac{2\pi}{a}(1,0), \qquad \vec{b}_2 = \frac{2\pi}{a}(0,1). \tag{3.43b}$$

In this case the filling fraction $f = \pi R^2/a^2$ is restricted to the range $0 \leq f \leq \pi/4 = 0.7854$, if the cylinders are not to overlap.

In Fig. 12(a) we present the photonic band structure for E-polarized electromagnetic waves in a square lattice of metallic rods when the filling fraction of the rods is $f = 0.001$. A total of $NG = 197$ plane waves was used in the solution of Eq. (3.2) in obtaining this result. We have taken $\omega_p a/2\pi c = 1$, which is the value of $\omega a/2\pi c$ at which the change in sign of $\epsilon(\omega)$ occurs. For small filling fractions, namely up to $f = 0.1$, we have obtained a band structure that is not significantly different from the dispersion curves for electromagnetic waves in a vacuum. However, for higher values of the filling fraction, the band structure differs substantially from the dispersion relation for electromagnetic waves in a vacuum and reveals the existence of an absolute band gap between the first and second bands, which appears for values of the filling fraction $f \gtrsim 0.25$. The presence of the band gap is illustrated in Fig. 12(b), where the photonic band structure for E-polarization when $f = 0.7$ is shown. The variation of the width of the band gap with the filling fraction is nonmonotonic. We found that the optimal filling fraction, which is defined as the value of f that gives the largest width of the band gap, is $f = 0.7$. In Fig. 13(a) we plot the width ofthe band gap as a function of the filling fraction. In Fig. 13(b) we plot the ratio of the width of the band gap to the frequency at the center of the band gap as a function of the filling fraction. From the latter figure we see that this ratio can be as large as 17% for a filling fraction $f \cong 0.65$. A total of 197 plane waves was used in obtaining the results for all values of the filling fraction. It should be noted that the convergence of these calculations is rapid, and relatively small matrices are required for an accurate determination of the photonic band structure for the case of E-polarization.

A notable feature of the results for the photonic band structures for electromagnetic waves of E-polarization in a square lattice of metallic cylinders in vacuum presented in Fig. 12 is the presence of a band gap below the lowest frequency band. The existence of this gap is a consequence of the metallic nature of the cylinders, and has its origin in the second term inside the braces on the left-hand side of Eq. (3.20). Its width is seen to increase with increasing filling fraction f.

In Fig. 14 we present the photonic band structure for the case of H-polarization for the square lattice when the filling fraction of the rods is $f = 0.001$. A total of 529 plane waves was used in solving Eq. (3.36) to obtain these results, and the first 12 terms in the expansion (2.57) for $\overleftrightarrow{B}$ (with $\epsilon = 1$) were kept. In contrast to the dispersion curves for E-polarized waves depicted in Fig. 13, at this filling fraction no absolute band gap is present in the frequency range investigated. At and below ω_p a number of very flat bands of electromagnetic modes are observed to be superimposed on a dispersion relation which is otherwise a slightly perturbed version of the dispersion curves for electromagnetic waves in a vacuum. The convergence of the dispersive part of the photonic band structure is rapid, and relatively small matrices are required for an accurate determination of the photonic band structures for the case of H-polarization. We found that the nearly dispersionless part of the photonic band structure converges slowly, and we used up to 1417 plane waves, together with an extrapolation procedure, to obtain converged results.

We next apply the results of Sections 3.1 and 3.2 to the calculation of the photonic band structures of a square lattice of parallel rods fabricated from

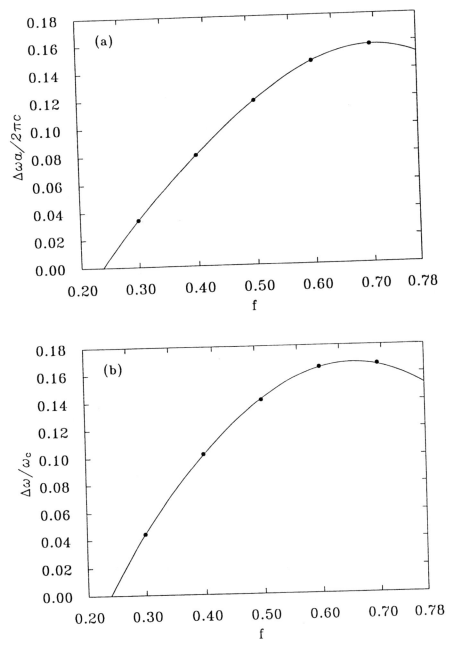

Fig. **13.** (a) The width of the band gap between the first and second bands present in the results displayed in Fig. 12(b) is plotted as a function of the filling fraction f. (b) The ratio of the width of this band gap $\Delta\omega$ to the frequency at the center of the gap ω_c is plotted as a function of the filling fraction f.

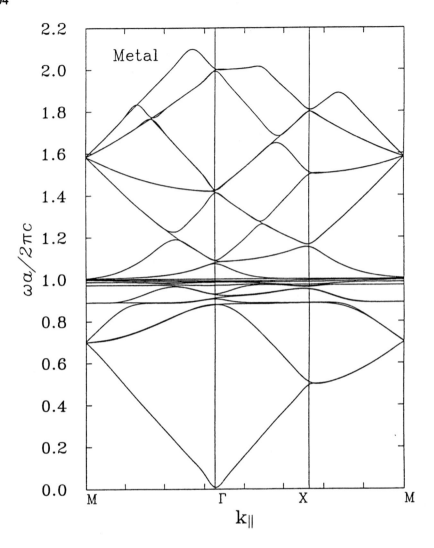

Fig. **14.** The photonic band structure of a square lattice of metal cylinders in a vacuum calculated by diagonalizing the $\overset{\leftrightarrow}{B}$ and $\overset{\leftrightarrow}{C}$ matrices. H-polarization. $\omega_p a/2\pi c = 1$, $f = 0.001$. NG = 529.

GaAs, which represents a polar crystal material characterized by the dielectric function $\epsilon(\omega)$ given by Eq. (1.2). In Fig. 15(a) we present the photonic band structure for electromagnetic waves of E-polarization in this square lattice, when the filling fraction of the rods is $f = 0.001$, and $\omega_T a/2\pi c = 1$. The method of Section 3.1 (Eq. (3.25)) was used to obtain this result. The calculation of the photonic band structure is transformed into the solution of a pair of eigenvalue problems. A total of $NG = 529$ plane waves was used in the expansion of the electric field in these calculations, and the first 12 terms in the expansion (2.57) for $\overleftrightarrow{B}$ (with $\epsilon = 1$) were kept. We again see that for small filling fractions, namely up to $f = 0.01$, the photonic band structure is a slightly perturbed version of the dispersion relation for electromagnetic waves in a vacuum. For larger values of the filling fraction, the band structure differs substantially from the dispersion relation for electromagnetic waves in a vacuum, and reveals the existence of a band gap between the first and second bands. The presence of the band gap is illustrated in Fig. 15(b), where the photonic band structure for E-polarization when $f = 0.1$ is shown. The variation of the width of the band gap with the filling fraction is nonmonotonic, as is seen from Fig. 16, where the ratio of the width of the band gap $\Delta\omega$ to the midgap frequency ω_c is plotted as a function of f. The largest value of this ratio, $\Delta\omega/\omega_c = 0.36$, occurs for a filling fraction $f \cong 0.15$. In addition to the "vacuum-like" part of the band structures, we also found a number of nearly dispersionless bands in the frequency range $\omega_T < \omega < \omega_L$, and in the frequency range $\omega \lesssim \omega_T$. These very flat bands of electromagnetic modes are observed to be superimposed on a photonic band structure which is otherwise a slightly perturbed version of the dispersion curves of electromagnetic waves in a vacuum. The convergence of the calculation of the band structure was monitored by increasing the number of the plane waves used in the expansion (3.9), and by increasing the number of terms retained in the expansion (2.57) for $\overleftrightarrow{B}$ (with $\epsilon = 1$). For small values of the filling fraction $f \simeq 0.01$, we used a modest number of plane waves ~ 100 and the first 12 terms in the expansion for $\overleftrightarrow{B}$ to obtain converged results. For large values of the filling fraction, the convergence of the bands is slower and we have used up to $\simeq 530$ plane waves and an extrapolation procedure to obtain accurate values of the bands.

In Fig. 17(a) we present the photonic band structure of a square lattice of GaAs rods, for H-polarization, when the filling fraction of the rods is $f = 0.001$. A total of 97 plane waves was used in obtaining this result and we have taken $\omega_T a/2\pi c = 1$. For small filling fractions, namely up to $f = 0.01$, the band structures do not differ substantially from the dispersion curves for electromagnetic waves in a vacuum. For larger values of the filling fraction the band structures depart significantly from the dispersion curves

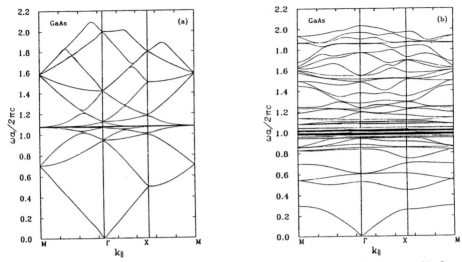

Fig. **15.** The photonic band structure of a square lattice of GaAs cylinders in a vacuum. E-polarization. $\omega_T a/2\pi c = 1$. (a) $f = 0.001$ (b) $f = 0.1$. Band gaps are present in the latter structure. NG = 529.

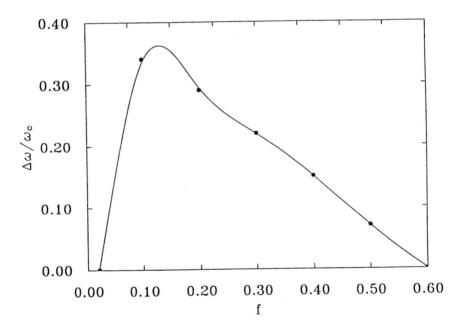

Fig. **16.** The ratio of the width of the band gap between the first and second bands present in the results displayed in Fig. 15(b) to the midgap frequency is plotted as a function of the filling fraction f.

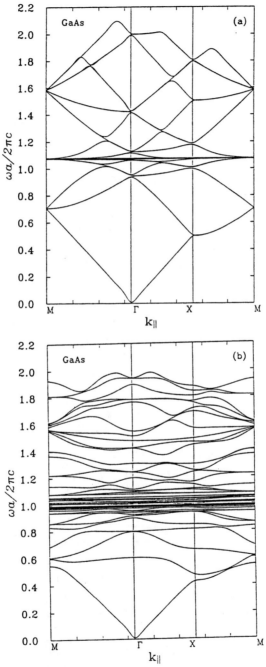

Fig. **17.** The photonic band structure of a square lattice of GaAs cylinders in a vacuum. H-polarization. $\omega_T a/2\pi c = 1$. (a) $f = 0.001$. (b) $f = 0.1$. NG = 529.

for electromagnetic waves in a vacuum, as is shown in Fig. 17(b), where the photonic band structure for a square lattice when $f = 0.1$ is plotted. In contrast to the case of E-polarization, the photonic band structure reveals no energy band gaps when the value of the filling fraction f is increased. This result clearly demonstrates the different natures of the band structures, depending on the polarization of the electromagnetic waves, of the same periodic, two-dimensional, dielectric system. The convergence of the calculation of the band structure was monitored by increasing the number of the plane waves used in the expansion (3.30), and by increasing the number of terms retained in the expansion for $\overleftrightarrow{D}$ (with $\epsilon = 1$). Since the convergence of these calculations for small values of the filling fraction $f \simeq 0.01$ is rapid, relatively small matrices $N \sim 100$ are required for an accurate determination of the photonic band structure for the case of H-polarization. However, for larger values of the filling fraction the convergence of the frequencies of the bands with an increasing number of plane waves is slower and we have used up to $\simeq 530$ plane waves and an extrapolation procedure to obtain converged bands.

Another interesting feature of the photonic band structures of electromagnetic waves propagating through a periodic array of GaAs rods is the absence of a band gap below the lowest frequency band, in contrast with the band structures of systems containing metallic rods.

In addition to dispersive bands, we also have found a nearly dispersionless part of the band structure in the frequency range $\omega_T < \omega < \omega_L$. We observed that the eigenvalues of the matrix $\overleftrightarrow{D}$ yield the flat bands and the dispersive bands below the frequency ω_T of transverse optical vibration modes, while the eigenvalues of $\overleftrightarrow{E}$ produce the dispersive bands in the frequency range $\omega > \omega_L$ which for small filling fractions are not significantly different from the dispersion curves for electromagnetic waves in a vacuum. The complete band structure is then the superposition of the band structures associated with each of the eigenvalue problems. The convergence of the calculation of the band structure was monitored by increasing the number of the plane waves used in the expansion for $\overleftrightarrow{D}$ (with $\epsilon = 1$). Since the convergence of these calculations for small values of the filling fraction $f \simeq 0.01$ is rapid, relatively small matrices $N \sim 100$ are required for an accurate determination of the photonic band structures for the case of H-polarization. However, for larger values of the filling fraction the convergence of the frequencies of the bands with an increasing number of plane waves is slower and we have used up to $\simeq 530$ plane waves and an extrapolation procedure to obtain converged bands.

4. A Three-Dimensional System

Of the four methods for calculating photonic band structures described in Section 2, the determinantal method is the only one to have been applied to three-dimensional periodic structures containing components with frequency-dependent dielectric functions[2]. In this section we apply it to the determination of the photonic band structure of a face centered cubic array of non-overlapping spheres or radius R formed from a dielectric material characterized by the dielectric function (1.2).

Our starting point is the Maxwell equation for the magnetic vector in an arbitrary periodic three-dimensional structure, which we write in the form

$$\nabla \times \left(\frac{1}{\epsilon(\vec{x}|\omega)} \nabla \times \vec{H}(\vec{x}|\omega) \right) - \frac{\omega^2}{c^2} \vec{H}(\vec{x}|\omega) = 0, \tag{4.1}$$

where $\epsilon(\vec{x}|\omega)$ is the spatially periodic dielectric function of the structure. We expand $1/\epsilon(\vec{x}|\omega)$ in a Fourier series according to

$$\frac{1}{\epsilon(\vec{x}|\omega)} = \sum_{\vec{G}} \hat{\kappa}(\vec{G}|\omega) e^{i\vec{G}\cdot\vec{x}}, \tag{4.2}$$

where the $\{\vec{G}\}$ are the translation vectors of the lattice reciprocal to the one defined by the structure. The magnetic vector is expanded in plane waves in a form that satisfies the Bloch-Floquet theorem

$$\vec{H}(\vec{x}|\omega) = \sum_{\vec{G}} \vec{C}(\vec{k} + \vec{G}) e^{i(\vec{k}+\vec{G})\cdot\vec{x}}. \tag{4.3}$$

When Eqs. (4.2) and (4.3) are substituted into Eq. (4.1) the following equation for the coefficient vector $\vec{C}(\vec{k} + \vec{G})$ is obtained:

$$\sum_{\vec{G}'} \hat{\kappa}(\vec{G} - \vec{G}')\{[(\vec{k} + \vec{G}) \cdot \vec{C}(\vec{k} + \vec{G})](\vec{k} + \vec{G}')$$

$$-(\vec{k} + \vec{G}) \cdot (\vec{k} + \vec{G}')\vec{C}(\vec{k} + \vec{G}')\} + \frac{\omega^2}{c^2} \vec{C}(\vec{k} + \vec{G}) = 0. \tag{4.4}$$

We can simplify this equation somewhat by using the Maxwell equation $\nabla \cdot \vec{H}(\vec{x}|\omega) = 0$ to express $C_3(\vec{k} + \vec{G})$ as a linear combination of $C_1(\vec{k} + \vec{G})$ and $C_2(\vec{k} + \vec{G})$. The result is the pair of equations

$$\sum_{\vec{G}} \sum_{\beta=1}^{2} M_{\alpha\beta}(\vec{k} + \vec{G}; \vec{k} + \vec{G}')C_\beta(\vec{k} + \vec{G}') = 0 \qquad \alpha = 1, 2, \tag{4.5}$$

where

$$M_{\alpha\beta}(\vec{k} + \vec{G}; \vec{k} + \vec{G}') = \hat{\kappa}(\vec{G} - \vec{G}')[\delta_{\alpha\beta}(\vec{k} + \vec{G}) \cdot (\vec{k} + \vec{G}')$$

$$-(k_\alpha + G'_\alpha)(k_\beta + G_\beta) + \frac{k_3 + G_3}{k_3 + G'_3}(k_\alpha + G'_\alpha)(k_\beta + G'_\beta)]$$

$$-\frac{\omega^2}{c^2}\delta_{\alpha\beta}\delta_{\vec{G},\vec{G}'}.$$

(4.6)

The dispersion relation for electromagnetic waves in this structure is therefore

$$det\ M_{\alpha\beta}(\vec{k} + \vec{G}; \vec{k} + \vec{G}') = 0.$$ (4.7)

For the case of a three-dimensional array of spheres of radius R whose centers form a face-centered cubic lattice, the translation vectors of the system are

$$\vec{x}(\ell) = \ell_1\vec{a}_1 + \ell_2\vec{a}_2 + \ell_3\vec{a}_3 \qquad \ell_1, \ell_2,\ \ell_3 = 0, \pm1, \pm2, \ldots,$$ (4.8)

where the primitive translation vectors are

$$\vec{a}_1 = \frac{a_0}{2}(0, 1, 1),\ \vec{a}_2 = \frac{a_0}{2}(1, 0, 1),\ \vec{a}_3 = \frac{a_0}{2}(1, 1, 0).$$ (4.9)

The translation vectors of the reciprocal lattice are

$$\vec{G}(h) = h_1\vec{b}_1 + h_2\vec{b}_2 + h_3\vec{b}_3 \qquad h_1, h_2, h_3 = 0, \pm1, \pm2, \ldots,$$ (4.10)

where the primitive translation vectors are

$$\vec{b}_1 = \frac{2\pi}{a_0}(-1, 1, 1),\ \vec{b}_2 = \frac{2\pi}{a_0}(1, -1, 1),\ \vec{b}_3 = \frac{2\pi}{a_0}(1, 1, -1).$$ (4.11)

The volume of a primitive unit cell is $a_0^3/4$, so that the filling fraction is $f = \frac{16}{3}\pi R^3/a_0^3$. The Fourier coefficients $\{\hat{\kappa}(\vec{G})\}$ are then given by

$$\hat{\kappa}(\vec{G}) = 1 - f\left(1 - \frac{1}{\epsilon(\omega)}\right) \qquad\qquad \vec{G} = 0 \quad (4.12a)$$

$$= -f\left(1 - \frac{1}{\epsilon(\omega)}\right)\frac{3(\sin GR - GR\cos GR)}{(GR)^3} \quad \vec{G} \neq 0. \ (4.12b)$$

In Figs. 18(a) and 18(b) we present results for the photonic band structure of a face centered cubic lattice of GaAs spheres in vacuum with volume filling fractions $f = 0.001$ and $f = 0.01$, respectively. It has been assumed that $\omega_T a_0/c = 1$ in both cases. The calculations converged reasonably rapidly as the number of plane waves in the expansion (4.3) was increased. A total of NG = 749 plane waves was used in obtaining the results for both filling fractions. However, we note that when the equation for the electric vector $\vec{E}(\vec{x}_\parallel|\omega)$ was used instead of Eq. (4.1), so that $\epsilon(\vec{x}_\parallel|\omega)$ rather than $1/\epsilon(\vec{x}_\parallel|\omega)$ appeared in this equation, we failed to obtain converged results for the photonic band structure.

For the most part, the band structures resemble the dispersion curves for electromagnetic waves in a vacuum, except for the presence of nearly dispersionless bands in the frequency range between ω_T and ω_L in which $\epsilon(\omega)$ is negative. The density of these flat bands increases with increasing filling fraction f, spreading from the high frequency (ω_L) to the low frequency (ω_T) end of the range in which $\epsilon(\omega)$ is negative.

5. Discussion and Conclusions

In these lectures we have considered the calculation of the photonic band structures of one-, two-, and three-dimensional periodic systems containing components characterized by dielectric functions that are frequency dependent, and can be negative in some frequency range. We have presented four different approaches to such calculations in the context of a simple one-dimensional model problem, where the analysis is particularly transparent, and have then applied several of these approaches to the calculations of photonic band structures of two- and three-dimensional period systems. With the exception of the transfer matrix method, whose use was restricted to the one-dimensional model, the remaining methods are based on the use of a position- and frequency-dependent dielectric function, and the expansion of the electromagnetic field components in plane waves. With this approach it is always possible to obtain the photonic band structure of a particular system by searching for the zeros of the determinant obtained from the matrix equation for the coefficients in the plane wave expansions of the field components. However, this method is computationally intensive, and has been found to work well only for small filling fractions. Thus, we have developed alternative approaches that exploit the specific forms of the dielectric functions (1.1) and (1.2) to reduce the determination of photonic band structures of two-dimensional periodic systems with metallic or semiconducting components to the solution of standard eigenvalue problems. These alternative approaches have proved to be accurate and computationally tractable, and yield accurate results, subject to qualifications discussed

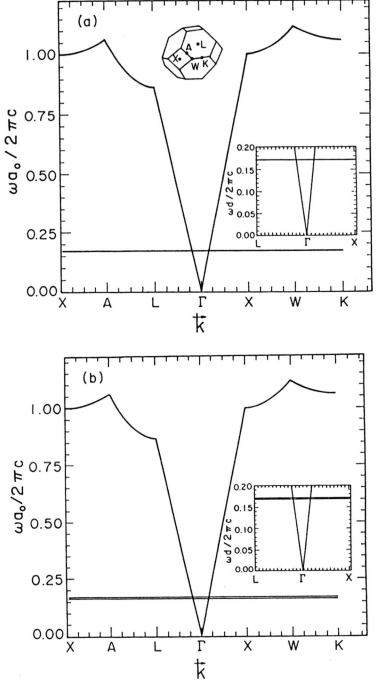

Fig. **18.** The photonic band structure of a face centered cubic array of GaAs spheres in a vacuum. $\omega_T a_0/2\pi c = 1$. (a) $f = 0.001$; (b) $f = 0.01$. NG = 749.

below.

Among the interesting features displayed by the photonic band structures of the two-dimensional systems studied here, is the presence of an absolute band gap below the lowest frequency band in the band structure for E-polarized electromagnetic waves propagating in a plane perpendicular to a periodic array of metal cylinders. A similar gap is found in our one-dimensional example as well. An analogous gap is not found in the band structure for H-polarized electromagnetic waves. It is also not found in the band structure of electromagnetic waves of either polarization when the metal cylinders are replaced by GaAs cylinders.

The photonic band structures of periodic systems with metallic or polar semiconductor components also display nearly dispersionless bands in the frequency range in which the dielectric function of these components is negative, and just below this range in the case that the periodic system contain components fabricated from a polar semiconductor. The origin of these flat bands in the case of the one-dimensional periodic structure fabricated from GaAs layers, and their absence in the case that the layers are metallic, considered here has been traced to the presence of a pole at a nonzero frequency (ω_T) in the dielectric function (1.2) of GaAs.

These flat bands have been found in H-polarization in the case that the dielectric function has the metallic form (1.1), and in both E- and H-polarizations when the dielectric function has the polar semiconductor form (1.2), in the photonic band structures of two-dimensional periodic structures. They have earlier been shown to be present in the photonic band structures of periodic three-dimensional systems fabricated from metal spheres [2,5], and in the present work they are also found in the band structures of such systems fabricated from GaAs spheres.

In discussing the photonic band structures of two-dimensional periodic systems formed from metallic cylinders, we first note that the dielectric function (1.1) diverges as the frequency ω tends to zero and vanishes at $\omega = \omega_p$. This behavior of $\epsilon(\omega)$ results in an asymmetry in the nature of the results obtained, depending on whether it is $\epsilon(\vec{x}_{\parallel}|\omega)$ that appears in Maxwell's equations or $1/\epsilon(\vec{x}_{\parallel}|\omega)$. It should be kept in mind that, in contrast with the one-dimensional case, the independence of the E- and H-polarized waves in the two-dimensional systems studied here is reflected in the profoundly different natures of their photonic band structures, and also plays an important role in the interpretation of the results obtained.

Specifically, the use of $\epsilon(\vec{x}_{\parallel}|\omega)$ in Maxwell's equations in the case of E-polarization leads to a standard eigenvalue problem for the determination of the corresponding photonic band structure. For small filling fractions, namely up to $f \simeq 0.15$, the resulting band structure is not significantly different from the dispersion relation for electromagnetic waves in vacuum.

For larger values of the filling fraction, the band structure differs substantially from the dispersion relation for electromagnetic waves in a vacuum, and reveals the existence of a band gap between the first and second bands. It also displays a band gap below the lowest frequency band, whose width increases with increasing filling fraction. The existence of this gap is a consequence of the metallic nature of the cylinders.

Our first approach to the determination of the photonic band structure for waves of H-polarization also employed $\epsilon(\vec{x}_{\|}|\omega)$ rather than $1/\epsilon(\vec{x}_{\|}|\omega)$ [3], and resulted in a standard eigenvalue problem for obtaining the band structure. The resulting band structure does not possess a band gap in the frequency range studied for large values of the filling fraction, and does not possess a gap below the lowest frequency band, in contrast to the band structure for electromagnetic waves of E-polarization. It also possesses a number of nearly dispersionless bands in the frequency range $\omega < \omega_p$, effectively superimposed on the dispersive band structure. However, the slow convergence of this method, especially for obtaining the flat bands, prompted us to develop the approach outlined in Section 3.2, in which it is $1/\epsilon(\vec{x}_{\|}|\omega)$ that appears in Maxwell's equations.

The use of $1/\epsilon(\vec{x}_{\|}|\omega)$ in Maxwell's equations, which constitutes an alternative approach to the calculation of the photonic band structure for waves of both E- and H-polarization, leads to the necessity of solving a pair of eigenvalue problems. In this case, in addition to the vacuum-like part of the photonic band structure nearly dispersionless bands are present in the frequency range $\omega < \omega_p$. In the limit as the filling fraction f tends to zero, these flat bands degenerate into a single flat band at $\omega = \omega_p$. This additional solution, superimposed on the dispersion curves for electromagnetic waves in a vacuum, must be regarded as spurious for waves of both polarizations.

The origin of the flat bands in the frequency range in which $\epsilon(\omega)$ is negative ($\omega < \omega_p$) is believed to be the weak overlap of the fields of electromagnetic excitations associated with each cylinder in isolation, that are characterized by discrete frequencies, when an infinite number of cylinders is brought together to form the two-dimensional periodic structures considered here. The overlap of these fields broadens the discrete frequencies into narrow bands. The frequencies of the electromagnetic modes of H- and E-polarization of a dielectric cylinder of radius R, characterized by a dielectric function $\epsilon(\omega)$, embedded in vacuum are the roots of [6]

$$D_n^{(H)}(\omega) = I_n(|\epsilon(\omega)|^{\frac{1}{2}}\frac{\omega}{c}R)' H_n^{(1)}\left(\frac{\omega}{c}R\right)$$

$$+|\epsilon(\omega)|^{\frac{1}{2}} I_n(|\epsilon(\omega)|^{\frac{1}{2}}\frac{\omega}{c}R) H_n^{(1)}\left(\frac{\omega}{c}R\right)' = 0$$

$$n = 0, 1, 2, \ldots \qquad (5.1)$$

and

$$D_n^{(E)}(\omega) = I_n(|\epsilon(\omega)|^{\frac{1}{2}} \frac{\omega}{c} R) H_n^{(1)} \left(\frac{\omega}{c} R\right)'$$

$$-|\epsilon(\omega)|^{\frac{1}{2}} I_n(|\epsilon(\omega)|^{\frac{1}{2}} \frac{\omega}{c} R)' H_n^{(1)} \left(\frac{\omega}{c} R\right) = 0,$$

$$n = 0, 1, 2, \ldots \qquad (5.2)$$

respectively, where $H_n^{(1)}(z)$ is a Hankel function, $I_n(z)$ is a modified Bessel function, and the prime denotes differentiation with respect to argument. Our notation emphasizes the fact that we are seeking frequencies in the range in which $\epsilon(\omega)$ is negative. We have also assumed that the electromagnetic field in the region outside the cylinder has the form of an outgoing wave at infinity, as would be the case in the scattering of a plane wave propagating perpendicularly to the axis of the cylinder. The solutions of both of these dispersion relations for each n are complex, $\omega_{n,s} = \omega_{n,s}^{(R)} - i\omega_{n,s}^{(I)}$ ($n = 0, 1, 2, \ldots, s = 1, 2, 3, \ldots$). These solutions were found by plotting $|D_n^{(H,E)}(\omega)|^2$ as functions of ω in the range in which $\epsilon(\omega)$ is negative. Such plots will have Lorentzian peaks centered at $\omega_{n,s}^{(R)}$, whose half-width at half-maximum is $\omega_{n,s}^{(I)}$. We have found that such peaks occur in the frequency range in which $\epsilon(\omega)$ is negative in H-polarization for a metallic cylinder, but not in E-polarization. This result is consistent with the fact that flat bands are present in the photonic band structure of H-polarized waves, but not in that of E-polarized waves.

To summarize the preceding discussion, we see that the photonic band structure for waves of H-polarization possess flat bands in the frequency range $\omega < \omega_p$ for nonzero filling fractions, irrespective of whether its calculation is based on the use of $\epsilon(\vec{x}_{\parallel}|\omega)$ or $1/\epsilon(\vec{x}_{\parallel}|\omega)$ in Maxwell's equations. Their origin has been identified as the overlap of the fields of electromagnetic excitations associated with each of the cylinders comprising the system in isolation. The differences between the band structures obtained by the use of $\epsilon(\vec{x}_{\parallel}|\omega)$ and $1/\epsilon(\vec{x}_{\parallel}|\omega)$ in Maxwell's equations in the case of E-polarization are more striking than in the case of H-polarization, since the special form of $\epsilon(\omega)$ allows transforming the solution of Maxwell's equations in the former case to the solution of a standard eigenvalue problem. The resulting band structure possesses no flat bands in the frequency range $\omega < \omega_p$. The alternative approach based on the use of $1/\epsilon(\vec{x}_{\parallel}|\omega)$, however, introduces flat bands in the frequency range $\omega < \omega_p$. In view of the fact that these flat bands were not found in the results based on the use of $\epsilon(\vec{x}_{\parallel}|\omega)$, and because no electromagnetic modes of E-polarization associated with an isolated metal cylinder have been found in this frequency range, we can regard the flat bands as spurious in the case of E-polarization. Their presence in the photonic band structures of periodic two-dimensional systems containing metallic components in H-polarization is genuine, and is attributed to the electromagnetic excitations of this polarization associated with isolated metal cylinders.

Turning now to periodic two-dimensional systems formed from GaAs cylinders, we note that, in addition to the dispersive part, a number of nearly dispersionless bands is present in the band structures of electromagnetic waves of both E- and H-polarizations in such systems. The frequencies of these flat bands lie just below ω_T and in the frequency range $\omega_T < \omega < \omega_L$ in which $\epsilon(\omega)$ is negative. The existence of the former cluster of flat bands is associated with the pole of the dielectric function (1.2) at $\omega = \omega_T$. The origin of the flat bands whose frequencies are in the range in which $\epsilon(\omega)$ is negative is again believed to be the weak overlap of the electromagnetic fields of E- and H-polarized excitations associated with each GaAs cylinder in isolation, when an infinite number of these cylinders is brought together to form a periodic two-dimensional system. We have found that both Eqs. (5.1) and (5.2) have solutions in the frequency range $\omega_T < \omega < \omega_L$ when the dielectric function $\epsilon(\omega)$ has the form (1.2), which is consistent with this explanation.

Flat bands are also present in the photonic band structures of three-dimensional periodic systems for both forms of the dielectric function characterizing the component with the frequency-dependent dielectric function. The origin of these flat bands appears to be the weak overlap of the electromagnetic fields associated with the electromagnetic modes of an isolated sphere when they are brought together to form a lattice. These bound states exist only in the frequency range in which $\epsilon(\omega) < 0$, viz. $\omega_T < \omega < \omega_L$, in the case where the spheres are fabricated from a polar semiconductor.

As we have noted above, we have used a transfer matrix approach only for the calculation of the photonic band structure of the one-dimensional model system. However, it should be pointed out that extensions of this simple transfer matrix approach, which can be used for the calculation of photonic band structures of two- and three-dimensional periodic systems containing components characterized by frequency-dependent dielectric functions, have recently been published[5]. This development may be most significant for calculations of the photonic band structures of three-dimensional periodic systems of this type, for which the only other existing approach, viz. the determinantal method, is computationally intensive. As is the case with the one-dimensional transfer matrix method described here, these higher-dimensional extensions of the transfer matrix method can also be used for calculating the transmissivity of finite two- and three-dimensional periodic systems.

The methods described in this paper should be applicable to the determination of the photonic band structures of other two- and three-dimensional systems containing components characterized by frequency-dependent dielectric functions. Examples of such systems which have not been studied until now include a periodic array of metal cylinders or metal spheres embedded in a background medium characterized by a dielectric constant ϵ_b instead of vacuum; a periodic array of cylindrical holes drilled in a metallic or polar crystal matrix; dielectric cylinders or dielectric spheres embedded in a metallic matrix; polar crystal rods embedded in a background medium fabricated from a different polar crystal; and three-dimensional systems in which the lattice sites of a space lattice, e.g., an fcc or diamond structure, are joined by nearest neighbor bonds of metal wires. The goal of such calculations would be to find systems which possess well-defined band gaps, to understand better the factors leading to gap formation, and to find new systems which may prove to be suitable for existing and emerging applications.

The same approaches provide the photonic band structures in the case that dissipation is introduced into the metallic or polar crystal components of periodic two-dimensional systems through the use of dielectric functions of the form

$$\epsilon(\omega) = 1 - \frac{\omega_p^2}{\omega(\omega + i\gamma)}$$

and

$$\epsilon(\omega) = \epsilon_\infty + \frac{(\epsilon_0 - \epsilon_\infty)\omega_T^2}{\omega_T^2 - \omega^2 - i\omega\gamma},$$

respectively, where ϵ_0 is the static dielectric constant. In this case the reduction of the band structure calculation to the solution of a single, standard eigenvalue problem is not possible, even in E-polarization, but it can be transformed into a problem requiring the solution of a set of standard eigenvalue problems that yield complex values of the frequency, $\omega(\vec{k}_\parallel) = \omega_R(\vec{k}_\parallel) - i\omega_I(\vec{k}_\parallel)$, for each value of the two-dimensional wave vector $\vec{k}_\parallel$. The imaginary part of each frequency characterizes the attenuation of the corresponding electromagnetic wave due to dissipation, as it propagates through the periodic system. The extension of this approach to the calculation of the photonic band structures of three-dimensional periodic systems containing lossy components should also be feasible.

The study of such more complex systems should reward their investigators with additional unexpected and interesting properties of their photonic band structures.

Acknowledgments

The work of A.A.M. was supported in part by National Science Foundation Grant No. DMR93-19404. The work of A.R.M. was supported in part by National Science Foundation Grant No. DMR92-13793. The work of V.K. was supported in part by the Grant of the Czech Academy of Sciences No. 167108. This research was also supported by the University of California, Irvine, through an allocation of computer time.

References

[1] Recent accounts of this work can be found in the book *Photonic Band Gaps and Localization*, ed. C. M. Soukoulis (Plenum, New York, 1993); and in the following special journal issues: J. Opt. Soc. Am. B10 (2) (1993) and J. Mod. Optics 41 (2) (1994).

[2] A. R. McGurn and A. A. Maradudin, Phys. Rev. B48, 17576 (1993).

[3] V. Kuzmiak, A. A. Maradudin, and F. Pincemin, Phys. Rev. B50, 16835 (1994).

[4] See the paper by V. Kuzmiak and A. A. Maradudin in these Proceedings.

[5] J. B. Pendry, J. Mod. Optics 41, 209 (1994).

[6] R. Englman and R. Ruppin, J. Phys. C. 1, 1515 (1968).

PHOTONIC BAND STRUCTURES OF 1D AND 2D PERIODIC SYSTEMS WITH METALLIC COMPONENTS IN THE PRESENCE OF DISSIPATION

V.KUZMIAK

Institute of Radio Engineering and Electronics
Czech Academy of Sciences
Chaberska 57, 182 51 Prague 8, Czech Republic

AND

ʳ A. A. MARADUDIN

Department of Physics and Astronomy
University of California, Irvine, CA 92715 USA

Abstract. We present a new approach for calculating the dispersion curves of electromagnetic waves in periodic media which contain metallic components characterized by a complex, frequency-dependent dielectric function.The formalism is based on the use of a position-dependent dielectric function and the plane wave technique. Because of the complex form of the dielectric function the reduction of the band structure calculation to the solution of a single standard eigenvalue problem is not possible. Instead, a generalized eigenvalue problem has to be solved. At low filling fractions of the metallic components ($f \leq 1\%$) the generalized eigenvalue problem is reduced to the problem of solving sets of nonlinear simultaneous equations which correspond to the diagonal terms of the matrix equation in the plane wave representation, with the non-diagonal elements taken into account perturbatively. The resulting complex band structure yields besides the dispersion curves also the attenuation of each mode as it propagates through the system. The method has been applied to the calculation of the photonic band structures of electromagnetic waves propagating through both 1D and 2D periodic systems. The real part of the photonic band structure for small filling fractions is not significantly affected by the presence of the dissipation.Both parts of the complex photonic band structure exhibit different behaviour depending on the polarization of the electromagnetic waves.

C. M. Soukoulis (ed.), Photonic Band Gap Materials, 319–339.
© 1996 *Kluwer Academic Publishers. Printed in the Netherlands.*

1. Introduction

In recent years the propagation of electromagnetic (EM) waves in periodic dielectric structures has received much experimental and theoretical attention.The photonic band structures of these systems exhibit intervals of frequencies in which EM waves are forbidden – photonic band gaps – which can open up under favorable circumstances.The existence of photonic band gaps can lead to a variety of interesting phenomena of both fundamental and practical interest, and have potential applications in many scientific and technical areas[1].

To date, the theoretical calculations of the dispersion relation for propagation of EM waves in 2D and 3D periodic media have been carried out for purely dielectric media, whose components are characterized by dielectric functions that are real, positive, and frequency-independent[2]-[23]. Interest in the nature of the photonic band structures of periodic systems containing components fabricated from metallic and semiconducting materials has led recently to several theoretical investigations[24]-[28]. Both the plane-wave technique and the transfer matrix approach[10] were applied to the calculation of the photonic band structures of EM waves propagating in 2D and 3D periodic systems, whose components are characterized by a real, frequency-dependent dielectric function, and in systems constructed from dispersive and highly absorptive materials characterized by a dielectric function which is frequency-dependent and has a non-negligible imaginary part.

The use of the plane-wave technique in the study of dispersive photonic-band-gap materials presents a more challenging problem than does the case of purely dielectric materials, since it does not lead to the solution of a standard eigenvalue problem, but rather requires some additional effort to incorporate the specific nature of the frequency-dependent dielectric function[27]. The aim of this paper is to present a new approach, developed within the framework of the plane-wave technique, which allows calculating photonic band structures of periodic systems containing metallic components, which are characterized by a complex, frequency-dependent dielectric function, and to explore how the photonic band structures in 1D and 2D periodic systems with metallic components are affected by the presence of dissipation. If the dissipation is introduced into the metallic components through the use of a complex dielectric function of the form

$$\epsilon(\omega) = 1 - \frac{\omega_p^2}{\omega(\omega + i\gamma)}, \tag{1.1}$$

where ω_p is the plasma frequency of the conduction electrons and $\gamma = 1/\tau_e$ is an inverse electron relaxation time, the problem of obtaining the photonic band structure cannot be reduced to the solution of a single standard eigenvalue problem, but it can be transformed into a generalized type of eigen-value problem.At low filling fractions of the metallic component ($f \leq 1\%$) the generalized eigenvalue problem is further reduced to a problem requiring the solution of sets of nonlinear simultaneous equations which correspond to the diagonal terms of the matrix equation in the plane wave representa-tion, with the non-diagonal elements taken into account perturbatively. The resulting complex band structure yields in addition to the dispersion curves also the attenuation of each mode as it propagates through the system.

We start, in Section 2, by considering a model system represented by a one-dimensional, periodic array of alternating layers of vacuum and metal-lic slabs characterized by the complex dielectric function described above.In Section 3 we apply this method to obtain the photonic band structures of an infinite array of parallel, infinitely long, metallic rods whose intersec-tions with a perpendicular plane form a simple square lattice. The results obtained by the use of this method in both 1D and 2D systems considered in this paper are presented in Section 4. In Section 5 we discuss the results obtained, summarize the conclusions of this work, and present possible di-rections for future research.

2. One-dimensional system - metallic slabs in the presence of dissipation

We begin by formulating the problem of obtaining the photonic band struc-tures of one-dimensional periodic structures.The physical system we con-sider consists of alternating layers of vacuum and a metal characterized by the frequency-dependent dielectric function $\epsilon(\omega)$ given by Eq.(1.1). We consider s-polarized(TE) waves with the electric vector perpendicular to the plane of incidence, which are assumed to propagate along the x_1-axis The intersections of the axes of the slabs with the x_1-axis form a one-dimensional lattice whose sites are given by the points $x(l) = la$, where a is the lattice constant, while l is an arbitrary positive or negative integer, or zero.The ratio of the thickness of the metallic layers to the period of the lattice is the filling fraction $f = 2R/a$, where $d = 2R$ is the thickness of the metal layer, and takes values in the range $(0, a)$. Because the dielectric function $\epsilon(x_1|\omega)$ of this system is a position-dependent, periodic function of x_1 with the period given by the lattice constant a,

$$\epsilon(x_1 + a|\omega) = \epsilon(x_1|\omega), \tag{2.1}$$

it may be expanded in terms of the reciprocal lattice vectors $G = 2\pi n/a$, where $n = 0, \pm 1, \pm 2, \ldots$, in a one-dimensional Fourier series according to

$$\epsilon(x_1|\omega) = \sum_{G} \hat{\epsilon}(G) \, e^{iGx_1}, \tag{2.2}$$

where

$$\hat{\epsilon}(G) = 1 - f\frac{\omega_p^2}{\omega(\omega + i\gamma)} \qquad\qquad G = 0 \qquad (2.3a)$$

$$= -f\frac{\omega_p^2}{\omega(\omega + i\gamma)}\frac{\sin(G\,R)}{(G\,R)} \qquad\qquad G \neq 0. \qquad (2.3b)$$

In the case of E-polarization we seek solutions of the Maxwell equations which have the forms

$$\vec{E}(\vec{x}; t) = (0, E(x_1|\omega), 0)\, exp(-i\omega t) \qquad (2.4a)$$

$$\vec{H}(\vec{x}; t) = (0, 0, H(x_1|\omega))\, exp(-i\omega t). \qquad (2.4b)$$

The Maxwell curl equations for the two nonzero field components are

$$\frac{d}{dx_1}E(x_1|\omega) = i\frac{\omega}{c}H(x_1|\omega) \qquad (2.5a)$$

$$\frac{d}{dx_1}H(x_1|\omega) = i\frac{\omega}{c}D(x_1|\omega) = i\frac{\omega}{c}\epsilon(x_1|\omega)E(x_1|\omega). \qquad (2.5b)$$

When we eliminate $H(x_1|\omega)$ from these equations we obtain the equation satisfied by $E(x_1|\omega)$, which we write in the form

$$\frac{d^2}{dx_1^2}E(x_1|\omega) + \epsilon(x_1|\omega)\frac{\omega^2}{c^2}E(x_1|\omega) = 0. \qquad (2.6)$$

Since $\epsilon(x_1|\omega)$ is unchanged by translation through a lattice constant a, we can also expand $E(x_1|\omega)$ in the form

$$E(x_1|\omega) = \sum_G B(k|G)e^{i(k+G)x_1}. \qquad (2.7)$$

To solve Eq. (2.6) we substitute the expansions (2.2) and (2.7) in Eq. (2.6), and we obtain as the equation satisfied by the coefficients $\{B(k|G)\}$

$$(k + G)^2 B(k|G) = \frac{\omega^2}{c^2}\hat{\epsilon}(0)B(k|G) + \frac{\omega^2}{c^2}\sum_{G'}{}'\hat{\epsilon}(G - G')B(k|G'), \qquad (2.8)$$

where the prime on the sum over G' indicates that the term with $G' = G$ is omitted. At this point we define a complex variable

$$\mu = \frac{\omega}{c}, \qquad (2.9)$$

and use the results for $\hat{\epsilon}(G)$ given by Eqs. (2.3) to transform Eq. (2.8) into

$$\left(\mu^3 + i\frac{\gamma}{c}\mu^2 - \mu\left[(k+G)^2 + f\frac{\omega_p^2}{c^2}\right] - i\frac{\gamma}{c}(k+G)^2\right)B(k|G)$$

$$-\mu\frac{\omega_p^2}{c^2}\sum_{G'}'f\frac{sin(|G-G'|\ R)}{(|G-G'|\ R)}B(k|G') = 0, \qquad (2.10)$$

which has the form of a generalized eigenvalue problem for a complex matrix. For sufficiently small values of the filling fraction, $f \leq 1\%$, we can treat the nondiagonal terms in Eq. (2.10) as a perturbation, and we proceed in two steps as follows: we first seek the zero-order eigenvalues given by the solutions of the equations which correspond to the diagonal terms of the matrix equation (2.10) for each value of the reciprocal lattice vector G used in the expansions given by Eqs.(2.2) and (2.7). In the second step we substitute the zero-order eigenvalues and the nondiagonal terms in Eq.(2.10), $-\mu f(\omega_p^2/c^2)sin(|G-G'|\ R)/(|G-G'|\ R)$, into a standard first-order perturbation formula to calculate the corrected eigenvalues.

We can write the complex variable μ in the form $\mu = \omega_R/c + i\omega_I/c$, where ω_R represents the real part of the frequency and ω_I determines the lifetime of the wave according to the definition

$$\frac{1}{\tau} = -2\omega_I. \qquad (2.11)$$

It is obvious that in terms of ω_R/c and ω_I/c Eq.(2.10) can be replaced by a pair of coupled equations for the real and imaginary parts of the complex variable μ for each value of G of the form

$$\omega_R^3 - \omega_R\left[3\omega_I^2 + 2\omega_I\gamma + c^2(k+G)^2 + f\omega_p^2\right] = 0 \qquad (2.12a)$$

$$\omega_I^2 + \gamma\omega_I^2 - \omega_I\left[3\omega_R^2 - c^2(k+G)^2 - f\omega_p^2\right] - \gamma\left[\omega_R^2 - c^2(k+G)^2\right] = 0. \qquad (2.12b)$$

If we assume that $\omega_R \neq 0$, we can eliminate ω_R by substituting ω_R^2 given by

$$\omega_R^2 = 3\omega_I^2 + 2\omega_I\gamma + c^2(k+G)^2 + f\omega_p^2 \qquad (2.13)$$

into Eq.(2.12b), which gives

$$\omega_I^3 + \omega_I^2\gamma + \omega_I\frac{1}{4}\left[c^2(k+G)^2 + f\omega_p^2 + \gamma^2\right] + \frac{\gamma}{8}\omega_p^2 f = 0. \qquad (2.14)$$

The latter equation can be transformed into the form

$$\omega_Z^2 + p\omega_Z + q = 0, \qquad (2.15)$$

by the substitution $\omega_Z = \omega_I + \gamma/3$, where

$$p = \frac{1}{4}\left[c^2(k+G)^2 + f\omega_p^2\right] - \frac{\gamma^2}{12} \tag{2.16a}$$

$$q = \frac{2\gamma^3}{27} - \frac{\gamma}{12}\left[c^2(k+G)^2 + f\omega_p^2 + \gamma^2\right] + \frac{\gamma}{8}\omega_p^2 f. \tag{2.16b}$$

Then we retain the roots of Eqs.(2.12) which correspond to the solutions which yield both a positive frequency and a positive lifetime for each EM mode. Finally, the substitution of the zero-order eigenvalues given by the roots $\mu_G^{(0)}$ of Eq.(2.12), and the nondiagonal terms of Eq.(2.10), into the standard first-order perturbation formula gives the corrected eigenvalues μ_G,

$$\mu_G = \mu_G^{(0)} + \sum_{G'} \frac{Q_{GG'}Q_{G'G}}{\mu_G^{(0)} - \mu_{G'}^{(0)}}, \tag{2.17}$$

where $\mu_G^{(0)} = \omega_R/c + i\omega_I/c$ is the zero-order eigenvalue for each value of the reciprocal lattice vector G, and $Q_{GG'}$ are the nondiagonal elements

$$Q_{GG'} = -\mu_G^{(0)} f\frac{\omega_p^2}{c^2}\frac{sin(|G-G'|\,R)}{(|G-G'|\,R)}. \tag{2.18}$$

3. Two-dimensional system - metallic rods in the presence of dissipation

The method outlined in the preceding section can be readily generalized to two dimensions. Specifically, we consider a system consisting of an array of parallel, infinitely long metallic cylinders of circular cross section surrounded by a vacuum, whose intersections with a perpendicular plane form a simple square lattice. We assume that the axes of the cylinders are parallel to the x_3-axis and the positions of the sites of this lattice are given by the vectors

$$\vec{x}_{\parallel}(l) = l_1\vec{a}_1 + l_2\vec{a}_2, \tag{3.1}$$

where $\vec{a}_1$ and $\vec{a}_2$ are the two, noncolinear, primitive translation vectors of the lattice, while l_1 and l_2 are arbitrary integers which we denote collectively by l. The area a_c of a primitive unit cell of this lattice is given by

$$a_c = |\vec{a}_1 \times \vec{a}_2|. \tag{3.2}$$

The lattice reciprocal to the direct lattice whose points are defined by Eq. (3.1) is defined by the translation vectors

$$\vec{G}_{\parallel}(h) = h_1\vec{b}_1 + h_2\vec{b}_2, \tag{3.3}$$

where $\vec{b}_1$ and $\vec{b}_2$ are the primitive translation vectors of the reciprocal lattice, and h_1 and h_2 are arbitrary integers which we denote collectively by h. The dielectric function $\epsilon(\vec{x}_{\|}|\omega)$, of this system is a position-dependent, periodic function of $\vec{x}_{\|}$ with the periodicity of the Bravais lattice defined by Eq. (3.1),

$$\epsilon(\vec{x}_{\|} + \vec{x}_{\|}(l)|\omega) = \epsilon(\vec{x}_{\|}|\omega). \tag{3.4}$$

It can therefore be expanded in a two-dimensional Fourier series according to

$$\epsilon(\vec{x}_{\|}|\omega) = \sum_{\vec{G}_{\|}} \hat{\epsilon}(\vec{G}_{\|}) \ e^{i \ \vec{G}_{\|} \cdot \vec{x}_{\|}}. \tag{3.5}$$

In the particular case of cylinders whose cross section is a circle of radius R we obtain for the Fourier coefficients $\hat{\epsilon}(\vec{G}_{\|})$

$$\hat{\epsilon}(\vec{G}_{\|}) = 1 - f\frac{\omega_p^2}{\omega(\omega + i\gamma)} \qquad\qquad \vec{G}_{\|} = \vec{0} \tag{3.6a}$$

$$\hat{\epsilon}(\vec{G}_{\|}) = -f\frac{\omega_p^2}{\omega(\omega + i\gamma)}\frac{2J_1(G_{\|} R)}{(G_{\|} R)} \qquad\qquad \vec{G}_{\|} \neq \vec{0}, \tag{3.6b}$$

where $f = \pi R^2/a_c$ is the filling fraction, i.e. the fraction of the total volume occupied by the rods, and $J_1(x)$ is a Bessel function.

The inverse dielectric function $\epsilon(\vec{x}_{\|}|\omega)^{-1}$ can also be expanded in a two-dimensional Fourier series according to

$$\frac{1}{\epsilon(\vec{x}_{\|}|\omega)} = \sum_{\vec{G}_{\|}} \hat{\kappa}(\vec{G}_{\|}) \ e^{i \ \vec{G}_{\|} \cdot \vec{x}_{\|}}. \tag{3.7}$$

The Fourier coefficients $\hat{\kappa}(\vec{G}_{\|})$ in the case of metallic rods assumed to have a circular cross section of radius R are then given by

$$\hat{\kappa}(\vec{G}_{\|}) = 1 + f\frac{\omega_p^2}{\omega^2 - \omega_p^2 + i\gamma\omega} \qquad\qquad \vec{G}_{\|} = \vec{0} \tag{3.8a}$$

$$= \frac{\omega_p^2}{\omega^2 - \omega_p^2 + i\gamma\omega} f\frac{2J_1(G_{\|} R)}{(G_{\|} R)} \qquad\qquad \vec{G}_{\|} \neq \vec{0}. \tag{3.8b}$$

We now apply these results to the determination of the photonic band structures of E and H-polarized electromagnetic waves in the system described by the dielectric function (1.1).

3.1. E - POLARIZATION

In the case of E-polarization, we seek solutions of the Maxwell equations which have the forms

$$\vec{E}(\vec{x};t) = (0, 0, E_3(\vec{x}_{\|}|\omega))\, exp(-i\omega t) \qquad (3.9a)$$

$$\vec{H}(\vec{x};t) = (H_1(\vec{x}_{\|}|\omega), H_2(\vec{x}_{\|}|\omega), 0)\, exp(-i\omega t). \qquad (3.9b)$$

The Maxwell curl equations for the three nonzero field components are

$$\frac{\partial H_2}{\partial x_1} - \frac{\partial H_1}{\partial x_2} = -i\frac{\omega}{c}D_3 = -i\frac{\omega}{c}\epsilon(\vec{x}_{\|}|\omega)E_3 \qquad (3.10a)$$

$$\frac{\partial E_3}{\partial x_1} = -i\frac{\omega}{c}H_2 \qquad (3.10b)$$

$$\frac{\partial E_3}{\partial x_2} = i\frac{\omega}{c}H_1. \qquad (3.10c)$$

When we eliminate H_1 and H_2 from these equations, we obtain as the equation satisfied by E_3

$$\left(\frac{\partial^2}{\partial x_1^2} + \frac{\partial^2}{\partial x_2^2}\right) E_3 + \epsilon(\vec{x}_{\|}|\omega)\frac{\omega^2}{c^2}E_3 = 0. \qquad (3.11)$$

To solve Eq. (3.11) we use the expansion (3.5) and write $E_3(\vec{x}_{\|}|\omega)$ in the form

$$E_3(\vec{x}_{\|}|\omega) = \sum_{\vec{G}_{\|}} B(\vec{k}_{\|}|\vec{G}_{\|})e^{i\,(\vec{k}_{\|}+\vec{G}_{\|})\cdot\vec{x}_{\|}}, \qquad (3.12)$$

where $\vec{k}_{\|} = (k_1, k_2, 0)$ is the two-dimensional wave vector of the wave. When these expansions are substituted in Eq. (3.11) we obtain as the equation satisfied by the coefficients $\{B(\vec{k}_{\|}|\vec{G}_{\|})\}$

$$(\vec{k}_{\|} + \vec{G}_{\|})^2 B(\vec{k}_{\|}|\vec{G}_{\|}) = \frac{\omega^2}{c^2}\hat{\epsilon}(\vec{0})B(\vec{k}_{\|}|\vec{G}_{\|}) + \frac{\omega^2}{c^2}{\sum_{\vec{G}_{\|}'}}'\hat{\epsilon}(\vec{G}_{\|} - \vec{G}_{\|}')B(\vec{k}_{\|}|\vec{G}_{\|}'),$$

$$(3.13)$$

where the prime on the sum over $\vec{G}_{\|}'$ indicates that the term with $\vec{G}_{\|}' = \vec{G}_{\|}$ is omitted. We now use the definition of μ given by Eq.(2.9) and the results for $\hat{\epsilon}(\vec{G}_{\|})$ given by Eqs. (3.6) to transform Eq. (3.13) into

$$\left(\mu^3 + i\frac{\gamma}{c}\mu^2 - \mu\left[(\vec{k}_{\|} + \vec{G}_{\|})^2 + f\frac{\omega_p^2}{c^2}\right] - i\frac{\gamma}{c}(\vec{k}_{\|} + \vec{G}_{\|})^2\right) B(\vec{k}_{\|}|\vec{G}_{\|})$$

$$-\mu \frac{\omega_p^2}{c^2} \sum_{\vec{G}_\|}' f \frac{2J_1(|\vec{G}_\| - \vec{G}_\|'| \, R)}{(|\vec{G}_\| - \vec{G}_\|'| \, R)} B(\vec{k}_\| | \vec{G}_\|') = 0, \qquad (3.14)$$

which has the form of a generalized eigenvalue problem for a complex matrix. To solve this matrix equation for sufficiently small values of the filling fraction we proceed in the same way as in the preceding section: we first solve separately the equations which correspond to the diagonal terms of the matrix equation (3.14) to obtain the zero-order eigenvalues $\mu_{G_\|}^{(0)}$ for each value of the reciprocal lattice vector $\vec{G}_\|$ used in the expansions given by Eqs.(3.5) and (3.12), and then substitute the non-diagonal terms of Eq.(3.14), $-\mu f(\omega_p^2/c^2)2J_1(|\vec{G}_\| - \vec{G}_\|'| \, R)/(|\vec{G}_\| - \vec{G}_\|'| \, R)$, into the standard first-order perturbation formula to calculate the corrected eigenvalues.

3.2. H - POLARIZATION

In the case of H-polarization, we seek solutions of the Maxwell equations which have the forms

$$\vec{H}(\vec{x}; t) = (0, 0, H_3(\vec{x}_\| | \omega)) \, exp(-i\omega t) \qquad (3.15a)$$

$$\vec{E}(\vec{x}; t) = (E_1(\vec{x}_\| | \omega), E_2(\vec{x}_\| | \omega), 0) \, exp(-i\omega t). \qquad (3.15b)$$

The Maxwell curl equations in this case are

$$\frac{\partial E_2}{\partial x_1} - \frac{\partial E_1}{\partial x_2} = i\frac{\omega}{c} H_3 \qquad (3.16a)$$

$$\frac{\partial H_3}{\partial x_1} = i\frac{\omega}{c} D_2 = i\frac{\omega}{c} \epsilon(\vec{x}_\| | \omega) E_2 \qquad (3.16b)$$

$$\frac{\partial H_3}{\partial x_2} = -i\frac{\omega}{c} D_1 = -i\frac{\omega}{c} \epsilon(\vec{x}_\| | \omega) E_1. \qquad (3.16c)$$

When we eliminate E_1 and E_2 from these equations we obtain the equation satisfied by H_3, which we write in the form

$$\frac{\partial}{\partial x_1} \left(\frac{1}{\epsilon(\vec{x}_\| | \omega)} \frac{\partial H_3}{\partial x_1} \right) + \frac{\partial}{\partial x_2} \left(\frac{1}{\epsilon(\vec{x}_\| | \omega)} \frac{\partial H_3}{\partial x_2} \right) + \frac{\omega^2}{c^2} H_3 = 0. \qquad (3.17)$$

To solve this equation we expand $H_3(\vec{x}_\|)$ according to

$$H_3(\vec{x}_\| | \omega) = \sum_{\vec{G}_\|} A(\vec{k}_\| | \vec{G}_\|) e^{i \, (\vec{k} + \vec{G}_\|) \cdot \vec{x}_\|}. \qquad (3.18)$$

When we substitute the latter expansion together with the expansion of $\epsilon^{-1}(\vec{x}_\parallel|\omega)$ given by Eq.(3.7) into Eq. (3.17), we obtain as the equation satisfied by the coefficients $\{A(\vec{k}_\parallel|\vec{G}_\parallel)\}$

$$\sum_{\vec{G}_\parallel{}'}(\vec{k}_\parallel+\vec{G}_\parallel)\cdot(\vec{k}_\parallel+\vec{G}_\parallel') \, \hat{\kappa}(\vec{G}_\parallel-\vec{G}_\parallel')A(\vec{k}_\parallel|\vec{G}_\parallel') = \frac{\omega^2}{c^2}A(\vec{k}_\parallel|\vec{G}_\parallel) . \quad (3.19)$$

Now we use the definition of μ given by Eq. (2.9) and the Fourier coefficients $\{\hat{\kappa}(\vec{G}_\parallel)\}$ given by Eqs. (3.8) in Eq. (3.19) to obtain

$$\left(\mu^4 + i\frac{\gamma}{c}\mu^3 - \mu^2\left[(\vec{k}_\parallel+\vec{G}_\parallel)^2 + \frac{\omega_p^2}{c^2}\right]\right.$$

$$\left. -i\mu\frac{\gamma}{c}(\vec{k}_\parallel+\vec{G}_\parallel)^2 + (1-f)\frac{\omega_p^2}{c^2}(\vec{k}_\parallel+\vec{G}_\parallel)^2\right)A(\vec{k}_\parallel|\vec{G}_\parallel)$$

$$-f\frac{\omega_p^2}{c^2}\sum_{\vec{G}_\parallel'}{}'\frac{2J_1(|\vec{G}_\parallel-\vec{G}_\parallel'|\,R)}{(|\vec{G}_\parallel-\vec{G}_\parallel'|\,R)}(\vec{k}_\parallel+\vec{G}_\parallel)\cdot(\vec{k}_\parallel+\vec{G}_\parallel')A(\vec{k}_\parallel|\vec{G}_\parallel') = 0, \quad (3.20)$$

which has the form of a generalized eigenvalue problem for a complex matrix. To solve this matrix equation we use the same approach as we applied in the case of E-polarization: we first solve the equations which correspond to the diagonal terms of the matrix equation (3.20) to obtain the zero-order eigenvalues $\mu_{\vec{G}_\parallel}^{(0)}$ for each value of the reciprocal lattice vector $\vec{G}_\parallel$ used in the expansions given by Eqs.(3.7) and (3.18), and then we use perturbation theory to obtain the corrected eigenvalues by taking the non-diagonal terms given by the last term in the latter equation, $-f(\omega_p^2/c^2)(\vec{k}_\parallel+\vec{G}_\parallel)\cdot(\vec{k}_\parallel+\vec{G}_\parallel')\,2J_1(|\vec{G}_\parallel-\vec{G}_\parallel'|\,R)/(|\vec{G}_\parallel-\vec{G}_\parallel'|\,R)$, as a perturbation.

The diagonal terms of the matrix equation (3.20) can be rewritten as a pair of coupled nonlinear equations in terms of the components ω_R/c and ω_I/c of the complex variable μ given by Eq.(2.9) for each $\vec{G}_\parallel$:

$$\omega_R^4 - \omega_R^2\left[6\omega_I^2 + 3\omega_I\gamma + c^2(\vec{k}_\parallel+\vec{G}_\parallel)^2 + \omega_p^2\right] + \omega_I^4 + \gamma\omega_I^3 + \omega_I^2\left[c^2(\vec{k}_\parallel+\vec{G}_\parallel^2\omega_p^2\right]$$

$$+\omega_I\gamma c^2(\vec{k}_\parallel+\vec{G}_\parallel)^2 + (1-f)\omega_p^2c^2(\vec{k}_\parallel+\vec{G}_\parallel)^2 = 0 \quad (3.21a)$$

$$\omega_I^3 + \frac{3\gamma}{4}\omega_I^2 + \omega_I\{\frac{1}{2}\left[c^2(\vec{k}_\parallel+\vec{G}_\parallel)^2 + \omega_p^2\right] - \omega_R^2\} - \omega_R^2\frac{\gamma}{4} + \frac{\gamma}{4}c^2(\vec{k}_\parallel+\vec{G}_\parallel)^2 = 0. \quad (3.21b)$$

We solved these coupled nonlinear equations numerically for each vector $\vec{G}_\parallel$ used in the expansions given by Eqs.(3.7) and (3.18) by using the computational procedure from $MATHEMATICA^{TM}$ [30], and we retained only the solutions of physical interest i.e. those for which $Re[\mu_{G_\parallel}^{(0)}(\vec{k}_\parallel)] \geq 0$, $Im[\mu_{G_\parallel}^{(0)}(\vec{k}_\parallel)] \leq 0$.

4. Results

All results presented in this section were obtained for a structure consisting of metallic components characterized by the dielectric function $\epsilon(\omega)$ given by Eq. (1.1). A typical value of the plasma frequency for metals is $\omega_p \sim 10^{15}s^{-1}$, while a typical value of the electron lifetime is $\tau_e \sim 10^{-13}s$. Therefore, we used the value of $\gamma = 0.01\omega_p$ in obtaining these results. We took as a normalization condition $\omega_p a/2\pi c = 1$, which determines the value of $\omega_R a/2\pi c$ at which the change in sign of $Re[\epsilon(\omega)]$ occurs when $\omega_R = \omega_p(1 - \gamma^2/\omega_p^2)^{\frac{1}{2}}$.

In Figs. 1(a) and 1(b) we present the five lowest frequency dispersion curves for a 1D lattice formed by metallic slabs in a vacuum when the filling fraction of the slabs is $f = 0.001$ and $f = 0.01$, respectively. A total of 91 plane waves was used to obtain these results. The presented results, which correspond to the real part of the complex photonic band structure at the filling fraction of the metallic slabs considered, are a slightly perturbed version of the dispersion curves of EM waves in a vacuum, and are in good agreement with the results presented elsewhere in these Proceedings [29], where the simple free-electron form of the dielectric function without damping was employed. The dispersion curves exhibit the existence of a band gap below the lowest band at the Γ-point, whose width increases as the filling fraction is increased. The imaginary part of the complex photonic band structures is represented by the lifetimes of the modes as determined from the imaginary part ω_I of the normalized complex frequency $\mu = \omega_R/c + i\omega_I/c$ according to the definition (2.11).

In Figs. 2(a) and 2(b), we plot the lifetimes of the modes on a logarithmic scale as functions of the wave vector k for a 1D lattice formed by metallic slabs in a vacuum when the filling fraction is $f = 0.001$ and $f = 0.01$, respectively. As shown in Figs. 2(a) and 2(b), the lifetime of the modes in a given band increases with increasing values of the wave vector k, and displays a minimum at the Γ-point, where the strongest attenuation of the modes associated with the lowest band occurs. We display the lifetime associated with the 12 lowest bands to demonstrate the fact that the lifetimes saturate to a finite value in the limit of large wave vector k. The value of the lifetime at the Γ-point is equal to the electron relaxation

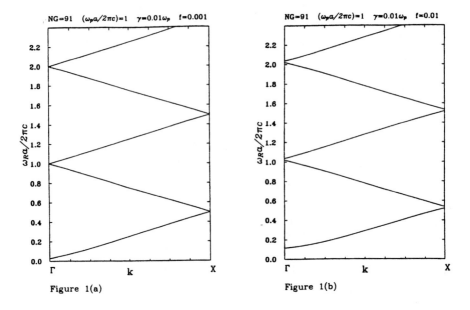

Figure 1(a)   Figure 1(b)

Fig. 1. The photonic band structure of a 1D lattice consisting of metallic slabs in a vacuum. E-polarization:(a) f= 0.001 (b) f=0.01. The number of plane waves used in these calculations was NG = 91.

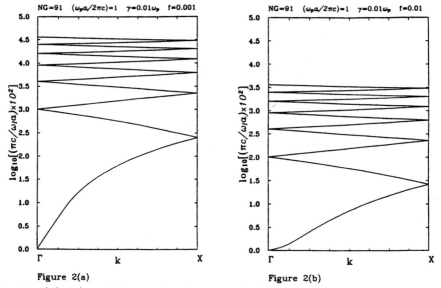

Figure 2(a)   Figure 2(b)

Fig. 2. The lifetimes of the modes associated with the photonic band structure of a 1D lattice consisting of metallic slabs in a vacuum. E-polarization: (a) f= 0.001 (b) f=0.01. NG = 91.

time τ_e given by the damping constant γ used in the dielectric function $\epsilon(\omega)$. The lifetimes associated with the modes at and above the plasma frequency are $10^2 - 10^3$-times larger than the electron relaxation time, and are of the order of 10^{-11} and $10^{-10}s$ for the the filling fractions $f = 0.001$ and $f = 0.01$, respectively. A study of the dependence of the band structure on the damping constant γ indicates that the imaginary part is directly proportional to the value of the damping constant γ, unlike the real part, which does not vary significantly when the damping constant is changed. The convergence of the calculation of the band structure was monitored by increasing the number of plane waves used in the expansion (2.2). For small values of the filling fraction, $f \leq 1\%$, the use of a modest number of plane waves (~ 100) was sufficient to produce converged results.

In Figs. 3(a) and 3(b), we present the real part of the photonic band structures in the case of E-polarization for a two-dimensional system consisting of metallic rods arrayed in a simple square lattice, when the filling fraction of the rods is $f = 0.001$ and $f = 0.01$, respectively. A total of 113 plane waves was used to obtain these results. For both filling fractions considered, $f = 0.01$ and $f = 0.001$, we have obtained a band structure that is a slightly perturbed version of the dispersion curves in a vacuum, and which is in quantitative agreement with the results obtained by the present authors when a simple free electron dielectric function $\epsilon(\omega)$ was used[27]. The calculated band structure also reveals the existence of a band gap below the lowest frequency band as a consequence of the metallic nature of the cylinders. The width of this gap is seen to increase with increasing filling fraction f.

In Figs. 4(a) and 4(b) we plot the lifetimes of these modes on a logarithmic scale as functions of the wave vector $\vec{k}_\parallel$, when the filling fraction $f = 0.001$ and $f = 0.01$, respectively. As shown in these figures the lifetimes of the modes as functions of the wave vector $\vec{k}_\parallel$ resemble the behaviour of the associated real part of the photonic band structure and displays the strongest attenuation for the lowest band at the $\overline{\Gamma}$-point. The value of the lifetime at this point is equal to the electron relaxation time τ_e given by the damping constant γ used in the dielectric function $\epsilon(\omega)$. For frequencies ω above ω_p, the lifetimes are $10^3 - 10^2$-times larger than the electron relaxation time, and are of the order of 10^{-10} and $10^{-11}s$ for the filling fractions $f = 0.001$ and $f = 0.01$, respectively. The 2D photonic band structures display the same behaviour as observed in the 1D case with respect to their dependence on the damping constant γ – the imaginary part is found to be directly proportional to the damping constant γ, unlike the real part, which does not vary when the damping constant is changed. The convergence of the calculation of the band structure was monitored by increasing the number of plane waves used in the expansion (3.12). For small values

332

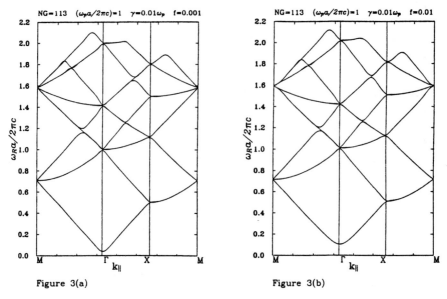

Figure 3(a)

Figure 3(b)

Fig. 3. The photonic band structure of a square lattice of metal cylinders in a vacuum. E-polarization:(a) f= 0.001 (b) f=0.01. The number of plane waves used in these calculations was NG = 113.

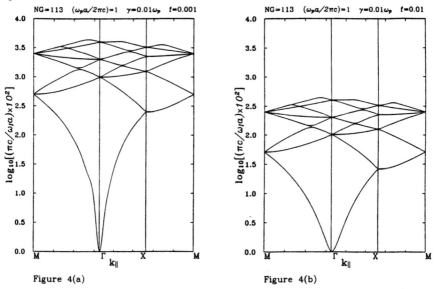

Figure 4(a)

Figure 4(b)

Fig. 4. The lifetimes of the modes associated with the photonic band structure of a square lattice of metal cylinders in a vacuum. E-polarization:(a) f= 0.001 (b) f=0.01. NG = 113.

of the filling fraction, $f \leq 1\%$, the use of a modest number of plane waves (~ 100) was sufficient to produce converged results.

In Fig. 5(a) we present the real part of the photonic band structure for the case of H-polarization for a square lattice when the filling fraction of the metal rods is $f = 0.001$. By solving the sets of nonlinear equations (3.21a) and (3.21b) we have found two independent solutions for each band for a given wave vector $\vec{k}_{\parallel}$. The two solutions represent two parts of the photonic band structure: a nearly dispersionless one which is consistent with the existence of the flat bands in the region $0 < \omega < \omega_p$ reported recently [27], and a dispersive part which resembles the dispersion curves for electromagnetic waves in a vacuum. The convergence of the calculation of the band structure was monitored by increasing the number of plane waves used in the expansion (3.18). Since the convergence of these calculations for small values of the filling fraction, $f < 1\%$, is rapid, relatively small matrices were required for an accurate determination of the dispersive part of the photonic band structure. In fact, a total of 113 plane waves was used to obtain these results. As is known from our earlier calculations [24], [27], the convergence of the flat bands is substantially slower than the convergence of the dispersive part of the photonic band structure, and the use of a large number of plane waves, together with an extrapolation procedure, was required to obtain accurate results. Taking into account that the eigenvalues converge asymptotically as the number of plane waves used is increased, and the inherent inaccuracy associated with the perturbative approach used in our method, the calculated flat bands should be regarded as only indicative of the true flat band structure that exists in the frequency range $0 < \omega < \omega_p$.

In Figs. 5(b) and 5(c) we plot the lifetimes of these modes on a logarithmic scale as functions of the wave vector $\vec{k}_{\parallel}$, for modes with frequencies greater and smaller than ω_p, respectively. We see that the lifetimes of the modes whose frequencies are greater than ω_p increase with increasing values of the wave vector $\vec{k}_{\parallel}$. In contrast, the lifetimes of the modes whose frequencies are smaller than ω_p increase as the wave vector $\vec{k}_{\parallel}$ decreases, and a traveling free space wave with zero attenuation occurs in the lowest frequency band at the $\bar{\Gamma}$ point. The minimum of the lifetime is associated with the flat bands, which begin to occur at the plasma frequency ω_p.

In Fig. 6(a) we present the real part of the photonic band structure for the case of H-polarization for a square lattice, when the filling fraction of the metal rods is $f = 0.01$. In Figs. 6(b) and 6(c) we plot the lifetimes of these modes on a logarithmic scale as functions of the wave vector $\vec{k}_{\parallel}$, for modes with frequencies smaller and greater than ω_p, respectively. The calculations whose results are presented in these figures were carried out in exactly the same way as were the calculations whose results are presented

334

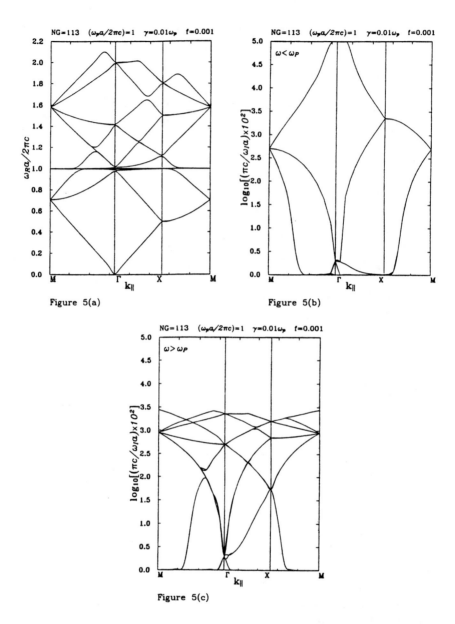

Figure 5(a)

Figure 5(b)

Figure 5(c)

Fig. 5. (a) The photonic band structure of a square lattice of metal cylinders in a vacuum. H-polarization, $f = 0.001$. (b) The lifetimes of the modes associated with this band structure whose frequencies are smaller than ω_p. (c) The lifetimes of the modes associated with this band structure whose frequencies are larger than ω_p. NG = 113.

335

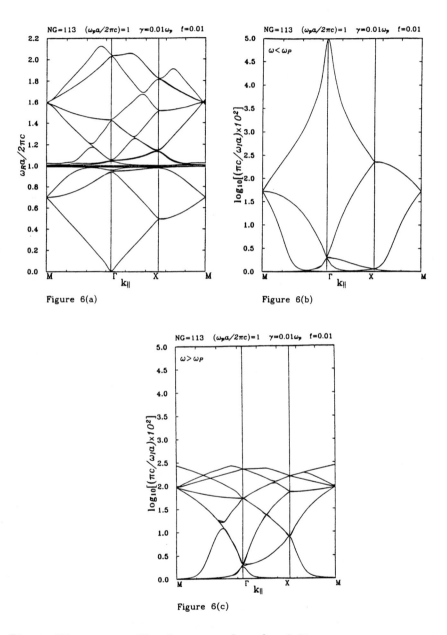

Figure 6(a)

Figure 6(b)

Figure 6(c)

Fig. **6.** The same as Fig. 5, except that $f = 0.01$.

in Fig. 5, and the results are qualitatively very similar to those for a lower filling fraction. Quantitatively, the lifetimes of the modes corresponding to the filling fraction $f = 0.01$ are smaller than the lifetimes of the modes calculated for $f = 0.001$, which is consistent with the expected fact that the modes are more attenuated as the filling fraction of the slabs (rods) increases.

5. Discussion and Conclusions

In this paper we have presented a new approach, developed within the framework of the plane-wave method, which provides a computationally viable approach for calculating the photonic band structures of both one- and two-dimensional, periodic arrays containing metallic components in the presence of dissipation. The calculated complex photonic band structure yields both the dispersion curves and the lifetimes of the modes associated with the bands. The dispersion curves determined from the real part of both the one- and two-dimensional photonic band structures of the periodic arrays of metallic slabs(rods) embedded in a vacuum are not significantly different from the dispersion curves for EM waves in a vacuum, and qualitatively agree with the results obtained by using a simple free-electron model dielectric function without damping. The photonic band structures of the EM waves of E-polarization, in addition to structural gaps between the higher frequency bands, possess an absolute band gap below the lowest frequency band, whose dispersion curve does not tend to zero frequency at the $\Gamma(\overline{\Gamma})$-point. The width of this gap increases with increasing filling fraction. Such a gap is not observed in the 1D structures consisting of dielectric slabs, and in 2D systems when the metal cylinders are replaced by dielectric cylinders, and is a consequence of the metallic nature of the components. The photonic band structures of EM waves of H-polarization propagating in a system of metallic rods in the presence of dissipation do not possess a band gap below the lowest frequency, and exhibit additional, nearly dispersionless bands in the frequency range $\omega < \omega_p$, effectively superimposed on the dispersive part of the band structures. These features demonstrate the differences between the band structures depending on the polarization of the waves, which reflects the fact that the system of conducting cylinders can act as a polarizer which transmits EM waves of H-polarization. From the imaginary part of the resulting photonic band structure we have determined the attenuation of the modes as they propagate through the one- and two-dimensional lattices. The results for E-polarized EM waves in both the 1D and 2D periodic systems indicate that the the lowest band is most strongly attenuated, and its lifetime tends to the value of the conduction electron relaxation time τ_e at the $\Gamma(\overline{\Gamma})$-point. The lifetimes of the modes as functions of the wave vector $k(\vec{k}_{\|})$ resemble the behaviour of the associated real part of the photonic band structure, and display the strongest attenuation for the lowest band at the $\overline{\Gamma}$-point. The difference between the lifetime of the lowest band at zero frequency and the lifetimes of the modes at frequencies ω above ω_p decreases as the filling fraction of the slabs(cylinders) increases. A quite different dependence is observed for the attenuation of the modes as a function of the wave vector $\vec{k}_{\|}$ for H-polarized EM waves propagating through an array of metal rods. The attenuation of the flat bands also exhibits nearly dispersionless behaviour, and the corresponding lifetimes of the modes are found to be close to the electron relaxation time τ_e, which corresponds to the bands at the plasma frequency ω_p. The lifetimes of the modes in the dispersive part of the photonic band structure whose frequencies ω are larger than ω_p increase with increasing values of the wave vector $\vec{k}_{\|}$. The lifetimes of the modes associated with the bands below ω_p, however, also increase with decreasing values of the wave vector $\vec{k}_{\|}$, and yield a travelling free space wave with zero attenuation at the $\overline{\Gamma}$-point.

Studies in progress focus on the photonic band structures of a periodic array of metallic slabs(rods) embedded in a background medium characterized by a dielectric constant ϵ_a , cylindrical holes drilled in a metallic or cubic polar crystal matrix, dielectric rods embedded in a metallic matrix, and polar crystal rods embedded in a background medium fabricated from a different cubic polar crystal. Such structures have not been investigated up to now. We also plan to explore the propagation of the EM waves through two-dimensional square and triangular lattices of cylinders fabricated from a lossy polar semiconductor whose dielectric function has the form

$$\epsilon(\omega) = \epsilon_\infty + (\epsilon_0 - \epsilon_\infty)\frac{\omega_T^2}{\omega_T^2 - \omega^2 - i\Gamma\omega}, \tag{5.1}$$

where ϵ_0 is the static dielectric constant and Γ the damping constant. In this case the Maxwell equations can be transformed into a generalized eigenvalue problem, and for structures with small filling fractions the approach requiring the solution of sets of nonlinear equations developed in this paper can be used.

It is expected that the photonic band structures of EM waves propagating through such systems will display new and interesting features associated with the specific nature of the (complex) frequency-dependent dielectric functions.

338

6. Acknowledgments

The work of V.K. was supported in part by the Grant of Czech Academy of Sciences No.167108. The work of A. A. M. was supported in part by NSF Grant No. DMR93-19404. This research was also supported by the University of California, Irvine, through an allocation of computer time.

References

1. See the special issue of J. Opt. Soc. Am. B **10**, 208-408 (1993).
2. E. Yablonovitch, Phys. Rev. Lett. **58**, 2059 (1987).
3. M. Plihal, A.Shambrook, A. A. Maradudin, and P.Sheng, Opt.Commun. **50**, 199 (1991).
4. M. Plihal and A. A. Maradudin, Phys. Rev. B **44**, 8565 (1991).
5. P.R.Villeneuve and M.Piché, J. Opt. Soc. Am. A **8**, 1296 (1991).
6. S. L. McCall, P. M. Platzman, R. Dalichaouch, D. R. Smith, and S. Schultz, Phys. Rev. Lett. **67**, 14 2017 (1991).
7. R. D. Meade, K. D. Brommer, A. M. Rappe, and J. D. Joannopoulos, Appl. Phys. Lett. **61**, 495 (1992).
8. P.R.Villneuve and M. Piché, Phys. Rev. B **46**, 4969 (1992).
9. P.R.Villneuve and M. Piché, Phys. Rev. B **46**, 4973 (1992).
10. J.B.Pendry, and A.MacKinnon, Phys. Rev. Lett. **69**, 2772 (1992).
11. A. A. Maradudin and A. R. McGurn, J. Opt. Soc. Am., B **10**, 307 (1993).
12. D. R. Smith, R. Dalichaouch, N. Kroll, S. Schultz, S. L. McCall, and P. M. Platzman, J. Opt. Soc. Am., B **10**, 314 (1993).
13. R. D. Meade, A. M. Rappe, K. D. Brommer, and J. D. Joannopoulos, J. Opt. Soc. Am., B **10**, 328 (1993).
14. D. L. Bullock, C. - C. Shih, and R. S. Margulies, J. Opt. Soc. Am., B **10**, 399 (1993).
15. K. M. Leung and Y. Qiu, Phys. Rev. B **48**, 7767 (1993).
16. R. D. Meade, A. M. Rappe, K. D. Brommer, J. D. Joannopoulos, and O. L. Alerhand, Phys. Rev. B **48**, 8434 (1993).
17. A. A. Maradudin and A. R. McGurn, in *Photonic Band Gaps and Localization*, edited by C. M. Soukoulis (Plenum, New York, 1993), p.247.
18. J. N. Winn, R. D. Meade, and J. D. Joannopoulos, J. Mod. Opt. **41**, 257 (1994).
19. A. A. Maradudin and A. R. McGurn, J. Mod. Opt. **41**, 275 (1994).
20. R. D. Meade, A. Devenyi, J. D. Joannopoulos, O. L. Alerhand, D. A. Smith, and K. Kash, J. Appl. Phys. **75**, 4753 (1994).
21. W. Robertson, G. Arjavalingam, R. D. Meade, K. D. Brommer, A. M. Rappe, and J. D. Joannopoulos, Phys. Rev. Lett. **68**, 2033 (1992).
22. W. Robertson, G. Arjavalingam, R. D. Meade, K. D. Brommer, A. M. Rappe, and J. D. Joannopoulos, J. Opt. Soc. Am., B **10**, 333 (1993).
23. S. Schultz and D. R. Smith, in *Photonic Band Gaps and Localization*, edited by C. M. Soukoulis (Plenum, New York, 1993), p.305.
24. A. R. McGurn and A. A. Maradudin, Phys. Rev. B **48**, 17567 (1993).
25. M. M. Sigalas, C. M. Soukoulis, E. N. Economou, C. T. Chan, and K. M. Ho, Phys. Rev. B **48**, 14 121 (1993).
26. M. M. Sigalas, C. M. Soukoulis, C. T. Chan, and K. M. Ho, Phys. Rev. B **49**, 11080 (1994).
27. V. Kuzmiak, A. A. Maradudin, and F. Pincemin, Phys. Rev. B **50**, 16835 (1994).
28. J. B. Pendry, J. Mod. Opt. **41**, 209 (1994).

29. See the paper by A. A. Maradudin, V. Kuzmiak, and A. R. McGurn in these Proceedings

30. S. Wolfram, *MATHEMATICA*^{TM} (Addison-Wesley Publishing Company, Inc., 1988)

BAND STRUCTURE AND TRANSMISSION OF PHOTONIC MEDIA: A REAL-SPACE FINITE-DIFFERENCE CALCULATION WITH THE R-MATRIX PROPAGATOR

J. MERLE ELSON and PHUC TRAN
Code 474400D
Research and Technology Division
Naval Air Warfare Center Weapons Division
China Lake, CA 93555-6001 USA

1. Introduction

In a recent paper, Elson and Tran [1] applied a multilayer modal method with an **R**-matrix propagator to calculate diffraction intensity from gratings and dispersion in photonic media. This calculation was done in Fourier space where the fields and photonic media are manifest as Fourier transforms of spatially variable quantities. With regard to photonic media, the results of this method [1] agree with experimental data for transmission through a seven-row deep square arrangement of round cylinders. In the present work, a similar calculation is done in real space, where modal expansion solutions are not required. Like the previous work, [1] the method presented here divides the photonic medium into sublayers and the **R**-matrix algorithm is applied to propagate the solutions through the inhomogeneous media. Unlike the previous work, we do not need to compute eigenvalues and eigenvectors to find modal solutions. This can be an important advantage since such computations can be time consuming.

First described in the field of chemical physics [2], the **R**-matrix propagation algorithm is comparatively free of numerical instability that exists for certain **T**-matrix propagation schemes. Unlike the **T**-matrix approach which seeks a matrix that relates the electric and magnetic field on one side of a sublayer to the electric and magnetic field on the other side of the sublayer, the **R**-matrix method relates the electric field of both sides of a sublayer to the magnetic field on both sides. A recursive formula is derived to allow adding successive sublayers to form a global **R**-matrix for any structure of interest. This approach is very simple and allows characterization of arbitrary photonic media structure.

Recently, there has been much interest in photonic crystal with band gaps in the band structure [3-5] and there are many techniques for bulk dispersion calculation, [6-13]. However, a transmission calculation can also play an important role. It may be desirable to limit the size of the crystal for practical applications, and a transmission calculation offers

341

C. M. Soukoulis (ed.), Photonic Band Gap Materials, 341–354.
© 1996 *Kluwer Academic Publishers. Printed in the Netherlands.*

the ability to examine when the bulk limit is reached. Furthermore, the search for such structures is done experimentally by measuring the transmissivity through several layers of crystal. Direct comparison with experimental data is very good for designing photonic crystals.

In addition to transmission, we also apply our method to direct calculation of the dispersive properties of bulk photonic media. The method is similar to that discussed by Sigalas, *et. al.* [14] and Pendry [15] but naturally follows from our **R**-matrix results. We compare our dispersion results with those of Robertson, *et. al.* [16]

Section 2 outlines the theory used in this work. In Section 3 the numerical results are given. We calculate the transmission of square arrays of cylinders with round and square cross section. This is done for 1, 7, and 50 rows of cylinders. These transmission results are compared to the dispersion curve results.

2. Method of Calculation

From our basic formalism, we calculate transmission through finite thickness photonic structure and dispersion of bulk photonic media. Initially we describe the transmission calculation where the photonic structure is periodic in the $\hat{x}$ direction, uniform in the $\hat{y}$ direction, with the $\hat{z}$ direction encountering a finite-thickness region consisting of N identical sections. When considering transmission, there are three regions of interest: the homogeneous region containing the incident and reflected wave, the homogeneous region containing the transmitted wave, and the region containing the structure, which is described by a spatially variable permittivity $\varepsilon(\mathbf{r})$, where $\mathbf{r} = (x, z)$. In Fig. 1, an example of one section is shown with nomenclature associated with a periodic array of infinitely long cylinders. In the case of dispersion in bulk photonic media, we let $N \rightarrow \infty$ and apply the Floquet theorem, but we defer discussion regarding this calculation until the end of this section.

As shown in Fig. 1, we subdivide a section into n_z sublayers each having thickness Δz and further subdivide the x-coordinate over one period into n_x discrete points each separated by Δx. Considering a sublayer bounded by $z = z_1$ and $z = z_1 + \Delta z$, we now write the two Maxwell's equations $\nabla \times \mathbf{E}(\mathbf{r}) = i(\omega/c)\mathbf{H}(\mathbf{r})$ and $\nabla \times \mathbf{H}(\mathbf{r}) = -i(\omega/c)\varepsilon(\mathbf{r})\mathbf{E}(\mathbf{r})$ and the constitutive relation $\mathbf{D}(\mathbf{r}) = \varepsilon(\mathbf{r})\mathbf{E}(\mathbf{r})$ as finite difference equations. We eliminate the z-component fields which yields a set of equations relating the $\hat{x}$ and $\hat{y}$ field components.

Equation (1) is valid at a given discrete point x in the set $\mathbf{X}$ of n_x discrete x-coordinates. Also, Eq. (1) is quite general since the only part that depends on the photonic structure is the permittivity $\varepsilon(x, z)$. This allows complex structures to be treated and different structures typically require minor changes to computer programs.

$$\frac{E_x(x, z_1 + \Delta z) - E_x(x, z_1)}{\Delta z} = \frac{i\omega}{c} H_y(x, z_1)$$

$$+ \frac{ic}{\omega(\Delta x)^2} \left\{ \frac{H_y(x + \Delta x, z_1) - H_y(x, z_1)}{\varepsilon(x + \Delta x/2, z_1)} - \frac{H_y(x, z_1) - H_y(x - \Delta x, z_1)}{\varepsilon(x - \Delta x/2, z_1)} \right\} \qquad (1a)$$

$$\frac{E_y(x, z_1 + \Delta z) - E_y(x, z_1)}{\Delta z} = -\frac{i\omega}{c} H_x(x, z_1 + \Delta z) \qquad (1b)$$

$$\frac{H_x(x, z_1 + \Delta z) - H_x(x, z_1)}{\Delta z} = -\frac{i\omega}{c} \varepsilon(x, z_1) E_y(x, z_1)$$

$$+ \frac{ic}{\omega(\Delta x)^2} \{ 2E_y(x, z_1) - E_y(x + \Delta x, z_1) - E_y(x - \Delta x, z_1) \} \qquad (1c)$$

$$\frac{H_y(x, z_1 + \Delta z) - H_y(x, z_1)}{\Delta z} = \frac{i\omega}{c} \varepsilon(x, z_1 + \Delta z) E_x(x, z_1 + \Delta z) \qquad (1d)$$

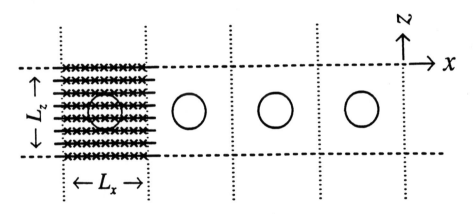

Figure 1. Cross section schematic of the round cylinder model with nomenclature. The two horizontal dash lines represent the boundaries of one section, which is the same as the period, L_z, in the z-direction. Four cylinders are shown, although the cylinders repeat infinitely in the x-direction. The vertical dotted lines denote the period in the x-direction, L_x. For the area around one cylinder, the × marks denote locations of discrete x-coordinate points which are each separated by distance Δx. The horizontal solid lines indicate the sublayer divisions of one section and the thickness of a sublayer is Δz. In this example, the number of discrete x-points is $n_x = 9$ and the number of sublayers $n_z = 7$.

In Eq. (1), if we now include all discrete x-points covering one period, denoted by $\mathbf{X}$, the resulting $4n_x$ equations can be concisely written in matrix form as

$$\begin{pmatrix} \tilde{\mathbf{E}}(\mathbf{X}, z_1) \\ \tilde{\mathbf{E}}(\mathbf{X}, z_1 + \Delta z) \end{pmatrix} = \mathbf{r}(\Delta z) \begin{pmatrix} \tilde{\mathbf{H}}(\mathbf{X}, z_1) \\ \tilde{\mathbf{H}}(\mathbf{X}, z_1 + \Delta z) \end{pmatrix} \tag{2}$$

For brevity we have defined a column vector with a *tilde* notation as

$$\tilde{\mathbf{E}} = \begin{pmatrix} E_x \\ E_y \end{pmatrix} \quad \text{and} \quad \tilde{\mathbf{H}} = \begin{pmatrix} H_x \\ H_y \end{pmatrix}$$

In forming Eq. (2) from Eq. (1), certain terms of Eqs. (1a) and (1c) require special attention. We note that for extreme x-values within a period, some E_y and H_y terms containing $x \pm \Delta x$ will fall outside the dimension of one period. These terms are "wrapped around" by using the Floquet relationship (similar to Eq. (12) below, except in the x-direction).

The elements of $\mathbf{r}(\Delta z)$, which is the sublayer r-matrix can be obtained from Eq. (1). Note that, by definition, the r-matrix is such that the electric and magnetic fields are on opposite sides and we have emphasized the Δz argument for reasons which will be apparent below. Equation (2) pertains to a given sublayer bounded from $z_1 \rightarrow z_1 + \Delta z$, but we also assume that an analogous equation exists for an arbitrary thickness, $z_1 \rightarrow z_2$, which satisfies

$$\begin{pmatrix} \tilde{\mathbf{E}}(\mathbf{X}, z_1) \\ \tilde{\mathbf{E}}(\mathbf{X}, z_2) \end{pmatrix} = \mathbf{R}(z_2 - z_1) \begin{pmatrix} \tilde{\mathbf{H}}(\mathbf{X}, z_1) \\ \tilde{\mathbf{H}}(\mathbf{X}, z_2) \end{pmatrix} \tag{3}$$

where

$$\mathbf{R}(z_2 - z_1) = \begin{pmatrix} \mathbf{R}_{11}(z_2 - z_1) & \mathbf{R}_{12}(z_2 - z_1) \\ \mathbf{R}_{21}(z_2 - z_1) & \mathbf{R}_{22}(z_2 - z_1) \end{pmatrix} \tag{4}$$

is the global R-matrix. Using Eqs. (2) and (3) and the continuity of $\tilde{\mathbf{E}}$ and $\tilde{\mathbf{H}}$ across any sublayer boundary, we obtain the following recursive relationships

$$\mathbf{R}_{11}(z_2 + \Delta z - z_1) = \mathbf{R}_{11}(z_2 - z_1) + \mathbf{R}_{12}(z_2 - z_1)\left[\mathbf{r}_{11}(\Delta z) - \mathbf{R}_{22}(z_2 - z_1)\right]^{-1}\mathbf{R}_{21}(z_2 - z_1) \tag{5a}$$

$$\mathbf{R}_{12}(z_2 + \Delta z - z_1) = -\mathbf{R}_{12}(z_2 - z_1)\left[\mathbf{r}_{11}(\Delta z) - \mathbf{R}_{22}(z_2 - z_1)\right]^{-1}\mathbf{r}_{12}(\Delta z) \tag{5b}$$

$$\mathbf{R}_{21}(z_2 + \Delta z - z_1) = \mathbf{r}_{21}(\Delta z)\left[\mathbf{r}_{11}(\Delta z) - \mathbf{R}_{22}(z_2 - z_1)\right]^{-1}\mathbf{R}_{21}(z_2 - z_1) \tag{5c}$$

$$\mathbf{R}_{22}(z_2 + \Delta z - z_1) = \mathbf{r}_{22}(\Delta z) - \mathbf{r}_{21}(\Delta z)\left[\mathbf{r}_{11}(\Delta z) - \mathbf{R}_{22}(z_2 - z_1)\right]^{-1}\mathbf{r}_{12}(\Delta z) \tag{5d}$$

With Eq. (5) we have a prescription for obtaining a global R-matrix for any structure of interest by adding successive sublayers where Eq. (1) can be used to get the r-matrix of each sublayer. To initialize the recursion relations in Eq. (5), we see from Eq. (2) that we may set $\mathbf{R}(\Delta z) = \mathbf{r}(\Delta z)$. It is evident from Eq. (1) that the r-matrix depends on the difference in Δz and not on the particular value of z_1. It follows that the R-matrix must depend only on the difference $z_2 - z_1$. When the structure has N identical sections, each of thickness $z_2 - z_1$, the R-matrix that relates the tangential field components at both boundaries of a given section will be identical for all sections since they only depend on the differences in z-coordinate. We therefore only need to calculate the R-matrix for one section using the recursive algorithm in Eq. (5). With this, it is a simple matter to calculate the R-matrix for N identical sections, $\mathbf{R}_N$, by using the same recursive algorithm $N-1$ times with the R-matrix for a single section in place of the r-matrix. This is done in this work for $N = 7$ and 50 sections.

2.1 TRANSMISSION OF FINITE THICKNESS PHOTONIC MEDIA

Having found the R-matrix for a given structure, as in Eq. (3), we can proceed to determine the reflection and transmission coefficients. At this point we have established a relationship between the electric and magnetic fields across a finite thickness structure. To calculate transmission or reflection, we transform to Fourier space in order to find a relationship between the electric and magnetic fields. First, we define a discrete Fourier transform matrix $\mathbf{F}(\mathbf{K}, \mathbf{X})$ where

$$\mathbf{F}(\mathbf{K}, \mathbf{X})\begin{pmatrix} \bar{\mathbf{E}}(\mathbf{X}, z_1) \\ \bar{\mathbf{E}}(\mathbf{X}, z_2) \end{pmatrix} = \begin{pmatrix} \bar{\mathbf{E}}(\mathbf{K}, z_1) \\ \bar{\mathbf{E}}(\mathbf{K}, z_2) \end{pmatrix} \tag{6}$$

and similarly for the magnetic field. The vector $\mathbf{K}$ denotes the set of wave vectors in Fourier space. We apply this $\mathbf{F}$ matrix to Eq. (3) which yields

$$\begin{pmatrix} \bar{\mathbf{E}}(\mathbf{K}, z_1) \\ \bar{\mathbf{E}}(\mathbf{K}, z_2) \end{pmatrix} = \overline{\mathbf{R}}(z_2 - z_1)\begin{pmatrix} \tilde{\mathbf{H}}(\mathbf{K}, z_1) \\ \tilde{\mathbf{H}}(\mathbf{K}, z_2) \end{pmatrix} \tag{7}$$

where

$$\overline{\mathbf{R}}(z_2 - z_1) = \mathbf{F}\mathbf{R}(z_2 - z_1)\mathbf{F}^{-1} \tag{8}$$

Since the z_1 and z_2 are the substrate and superstrate boundaries, respectively, we can use the boundary conditions to relate the substrate and superstrate fields. This yields from Eq. (7)

$$\begin{pmatrix} \bar{\mathbf{E}}'(\mathbf{K}, z_1) \\ \bar{\mathbf{E}}'(\mathbf{K}, z_2) + \bar{\mathbf{E}}^{inc}(\mathbf{K}^{inc}, z_2) \end{pmatrix} = \overline{\mathbf{R}}(z_2 - z_1)\begin{pmatrix} \tilde{\mathbf{H}}'(\mathbf{K}, z_1) \\ \tilde{\mathbf{H}}'(\mathbf{K}, z_2) + \tilde{\mathbf{H}}^{inc}(\mathbf{K}^{inc}, z_2) \end{pmatrix} \tag{9}$$

where K^{inc} is the $(\hat{x}, \hat{y})$ component of the incident beam wave vector and the incident field is described by $\tilde{E}^{inc}$ and $\tilde{H}^{inc}$. The superstrate and substrate media are homogeneous and contain the reflected and transmitted fields, denoted by r and t, respectively. As discussed elsewhere, [1] the $\tilde{E}$ and $\tilde{H}$ fields in Eq. (9) can be related by using

$$\tilde{H}^j(K, z) = Z^j \tilde{E}^j(K, z) \tag{10}$$

where $j = r, t$ or inc. Using Eq. (10) in Eq. (9) yields

$$\begin{pmatrix} I - \bar{R}_{11} Z^r & -\bar{R}_{12} Z^t \\ -\bar{R}_{21} Z^r & I - \bar{R}_{22} Z^t \end{pmatrix} \begin{pmatrix} E^r(K, z_1) \\ E^t(K, z_2) \end{pmatrix} = \begin{pmatrix} 0 & \bar{R}_{12} \\ -I & \bar{R}_{22} \end{pmatrix} \begin{pmatrix} E^{inc}(K^{inc}, z_2) \\ H^{inc}(K^{inc}, z_2) \end{pmatrix}$$

$$= \begin{pmatrix} \bar{R}_{12} Z^{inc} E^{inc}(K^{inc}, z_2) \\ [\bar{R}_{22} Z^{inc} - I] E^{inc}(K^{inc}, z_2) \end{pmatrix} \tag{11}$$

where I and 0 are the identity and null matrices. Equation (11) can be solved for the x- and y-components of the reflected and transmitted electric fields. From these results it is straightforward to obtain the diffraction efficiencies.

2.2 DISPERSION OF BULK PHOTONIC MEDIA

If we assume that the photonic array has x and z periods of L_x and L_z, respectively, and the number of sections $N \to \infty$, we may invoke the Floquet theorem and obtain a dispersion relation. In Eq. (3), we set $z_2 = z_1 + L_z$ and use

$$\tilde{E}(X, z_1 + L_z) = \tilde{E}(X, z_1) e^{iK_z L_z} \quad \text{and} \quad \tilde{H}(X, z_1 + L_z) = \tilde{H}(X, z_1) e^{iK_z L_z} \tag{12}$$

With this, and after some manipulation, Eq. (3) yields

$$E(X, z_1) = \left(R_{11} + R_{12} e^{iK_z L_z} \right) \left(R_{22} + R_{21} e^{-iK_z L_z} \right)^{-1} E(X, z_1) \tag{13}$$

If the photonic medium is lossless, then we can obtain the dispersion relation from Eq. (13) as follows. Consider first the case where $K_x = 0$. We choose a frequency ω, calculate the $R_{i,j}$ matrices, vary K_z and look for eigenvalues $\lambda = 1$ of the matrix product in Eq. (13). Values of K_z which yield $\lambda = 1$ yield points (K_z, ω) on the dispersion curve.

In the event $K_x \neq 0$, we seek the dispersion curve which has direction (K_x, K_z). Choosing a nonzero value for K_x affects certain field terms in Eq. (1) which overlap into adjoining x-periods. When overlap occurs and $K_x \neq 0$, we must apply the Floquet relationship in the

Floquet relationship in the x-direction. One possibility is to fix (K_x, K_z), which corresponds to the desired direction of propagation, and vary ω until an eigenvalue $\lambda = 1$ is found. Alternatively, we can fix (K_x, ω), vary K_z, and look for eigenvalues $\lambda = 1$. However, this does not necessarily yield the desired direction of propagation. It may then be necessary to repeat the process having made adjustments in ω and or K_x.

Over a range of ω, diagonalization of Eq. (13) may not yield any eigenvalues which are unity and this implies a band gap. For an eigenvalue $\lambda = 1$ obtained from Eq. (13), we use the associated eigenvector $\bar{E}(X, z_1)$ to determine the polarization.

3. Numerical Results

3.1 TRANSMISSION AND DISPERSION OF CYLINDRICAL ARRAYS

In this section, we give numerical results for transmission of two examples of square periodic arrays of cylinders having round and square cross section. We consider transmission of arrays having 1, 7, and 50 sections (rows) of cylinders. For the 50 section transmission data, points on the corresponding dispersion curves are superimposed. All transmission calculations are done for normal incidence with polarization parallel or perpendicular to the cylinder axes. The period in the x-direction is $L_x = 1.87$mm. The diameter of the round cylinders and side length of the square cylinders is 0.74mm. The cylinders have permittivity (8.8804,0.0) and are surrounded by vacuum. The z-distance between rows of cylinder arrays is $L_z = 1.87$mm and this is also the thickness of a section. The parameters for the round cylinders correspond to those given by Robertson, *et. al.* [16]

For these calculations, we used 75 discrete x values ($n_x = 75, \Delta x = 0.025$mm) over the x-period L_x and 41 z-subdivisions ($n_z = 41, \Delta z = 0.046$mm) over the thickness L_z of one section. Most of the computational effort is done in calculating through the first section. After that, calculating the 7 and 50 section transmission data is accomplished much more quickly by using the recursion method to add one whole section at a time.

3.1.1 *Square Cross Section Cylinders*

Figure 2 shows parallel polarization transmission versus frequency for square cylinder arrays with 1 and 7 sections. Figure 3 shows the same graphs except for perpendicular polarization.

Figures 4 (parallel polarization) and 5 (perpendicular polarization) show calculated transmission versus frequency for 50 sections. Also, superimposed on these data are calculated points from 5 bands on the dispersion curve. Much correlation between the transmission and dispersion band gaps is evident. However, points of band 4, perpendicular polarization, lie where the calculated transmission is zero. Therefore, this band would be measurable by normal incidence transmission. Examination of the eigenvectors (electric field, see Eq. (13)) associated with the points of this band show asymmetry over one period in the *x*-direction. It follows that the field associated with this band does not couple with symmetric incident field.

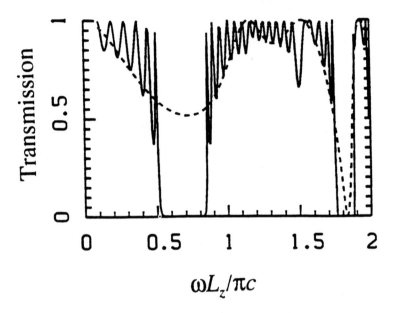

Figure 2. Transmission versus $\omega L_z/\pi c$ for 1 (dash line) and 7 (solid line) sections of square cylinders. The polarization is parallel to the cylinder axes.

349

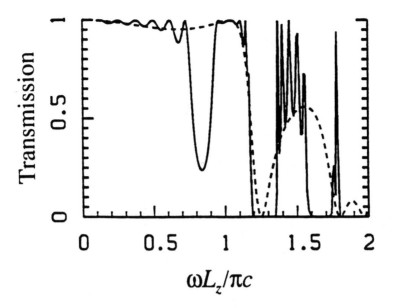

Figure 3. Transmission versus $\omega L_z/\pi c$ for 1 (dash line) and 7 (solid line) sections of square cylinders. The polarization is perpendicular to the cylinder axes.

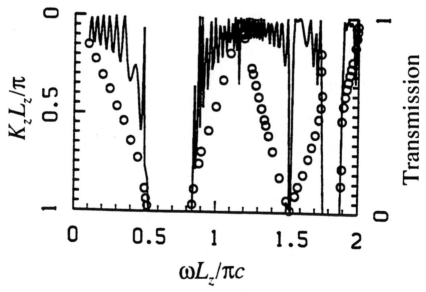

Figure 4. Transmission versus $\omega L_z/\pi c$ (50 sections) and dispersion curve versus $K_z L_z/\pi$ for square cylinders. The polarization is parallel to the cylinder axes. The solid line is transmission and the open circles are calculated points for bands one through five (from left vertical axis) on the dispersion curve. To view the dispersion curves in normal fashion, turn figure 90 degrees counterclockwise.

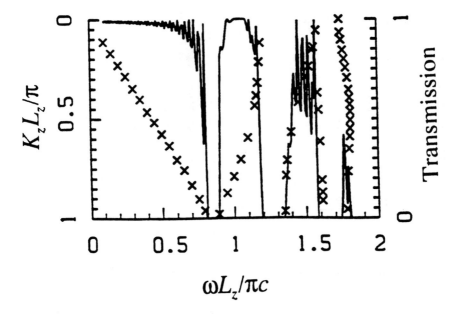

Figure 5. Transmission versus $\omega L_z/\pi c$ (50 sections) and dispersion curve versus $K_z L_z/\pi$ for square cylinders. The polarization is perpendicular to the cylinder axes. Analogous to Fig. 4, the solid line is transmission and the × are calculated points for bands one through five on the dispersion curve.

3.1.2 Round Cylinders

For round cylinders, we show in Fig. 6 parallel polarization transmission versus frequency for 1 and 7 sections. Likewise, for perpendicular polarization, Fig. 7 shows similar graphs. These results are analogous to Figs. 2 and 3. For 50 sections, we show in Fig. 8 the parallel polarization transmission versus frequency results and compare this with the corresponding dispersion curve. Figure 9 shows analogous results for perpendicular polarization. Like Figs. 4 and 5, the agreement between transmission and dispersion band gaps is good. There are points on band 3 (parallel polarization) and band 4 (perpendicular polarization) which fall within or near a region of zero transmission. It turns out that the fields associated with these bands are asymmetric and do not couple to the incident field. That these dispersion bands would not be detected by transmission measurement is in agreement with the experimental measurements of Robertson, et. al., [16] and their arguments regarding asymmetry.

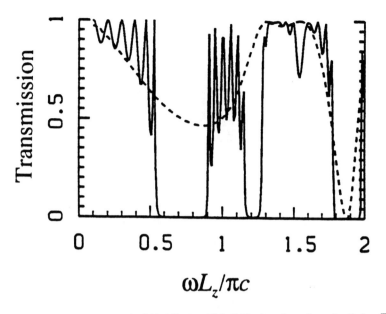

Figure 6. Transmission versus $\omega L_z/\pi c$ for 1 (dash line) and 7 (solid line) sections of round cylinders. The polarization is parallel to the cylinder axes.

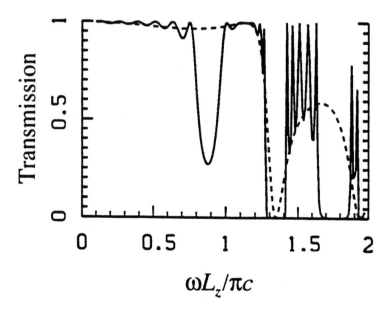

Figure 7. Transmission versus $\omega L_z/\pi c$ for 1 (dash line) and 7 (solid line) sections of round cylinders. The polarization is perpendicular to the cylinder axes.

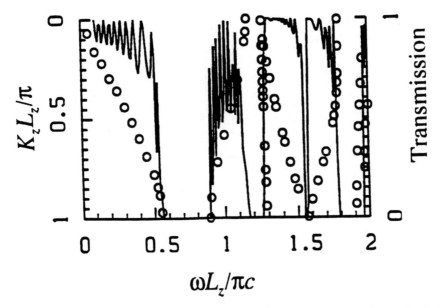

Figure 8. Transmission versus $\omega L_z/\pi c$ (50 sections) and dispersion curve versus $K_z L_z/\pi$ for round cylinders. The polarization is parallel to the cylinder axes. The solid line is transmission and the open circles are calculated points for bands one through five (from left vertical axis) on the dispersion curve. To view the dispersion curves in normal fashion, turn figure 90 degrees counterclockwise.

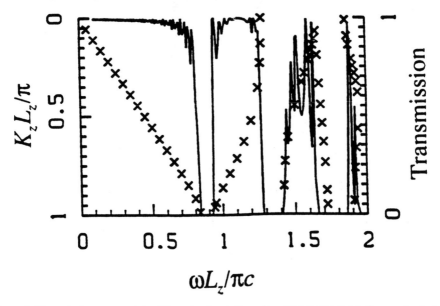

Figure 9. Transmission versus $\omega L_z/\pi c$ (50 sections) and dispersion curve versus $K_z L_z/\pi$ for round cylinders. The polarization is perpendicular to the cylinder axes. The solid line is transmission and the × are calculated points on the dispersion curve. Analogous to Fig. 8, the solid line is transmission and the × are calculated points for bands one through five on the dispersion curve.

4. Conclusions

We have presented a finite difference theory, which uses an R-matrix algorithm propagator, for calculation of transmission and dispersion of photonic media. The method is quite general in that complex three-dimensional periodic structures can be treated. It is straightforward to treat structures which are periodic in the x-y plane and of arbitrary structure in the z-direction by dividing this structure into sublayers. The numerical stability of this R-matrix algorithm is superior to that of the T-matrix and this has been recently documented. [1,17,18]

We have good agreement between calculated transmission spectra and dispersion. Some higher frequency dispersion points fall within regions where the predicted transmission is zero. See band 4 in Figs. 4, 5 and 9, and band 3 in Fig. 8. It follows that some portions of these bands would not be seen in transmission measurements. Also, the round cylinder dispersion calculations agree with measurement [16], where band 3 (Fig. 8) and band 4 Fig. (9) were predicted, but not seen in transmission.

We calculate the transmission spectra of a single row of square or round cylinders. From this result, we use a recursive scheme to calculate transmission spectra for arrays that are 7 and 50 rows deep. This is consistent with the numerical stability of the R-matrix algorithm. This calculation method is quite versatile and is easy to implement. The numerical stability of the R-matrix algorithm is perhaps more evident in deep grating calculations. Although numerical results are easily obtained for much deeper gratings, we have obtained numerical results for 17-wave-deep grating depths with a total of $n_x = 281$ (this is the same as the number of diffracted orders) and $n_z = 400$. Also, the grating material was an absorbing metal. This study on deep gratings is currently in progress.

5. References

1. Elson, J.M. and Tran, P. (1995) Dispersion and diffraction in photonic media: A different modal expansion for the R-matrix propagation technique, *J. Opt. Soc. Am. A*, accepted for publication.

2. Zvijac, D.J. and Light, J.C. (1976) R-matrix theory for collinear chemical reactions, *Chem. Phys.* 12, 237-251; Light, J.C. and Walker, R.B. (1976) An R-matrix approach to the solution of coupled equations for atom-molecule reactive scattering, *J. Chem. Phys.* 65, 4272-4282.

3. Botten, L.C., Craig, M.S., Mcphedran, R.C., Adams, J.L., and Andrewartha, J.R. (1981) The dielectric lamellar diffraction grating, *Opt. Acta* 28, 413-428; Botten, L.C., Craig, M.S., Mcphedran, R.C., Adams, J.L., and Andrewartha, J.R. (1981) The finitely conducting lamellar diffraction grating, *Opt. Acta* 28, 1087-1102.

4. Yablonovitch, E. (1987) Inhibited spontaneous emission in solid-state physics and electronics, *Phys. Rev. Lett.* 58, 2059.

5. John, S. (1987) Strong localization of photons in certain disordered dielectric super lattices, *Phys. Rev. Lett.* 58, 2486.

6. John, S., and Wang, J. (1990) Quantum electrodynamics near a photonic band gap: photon bound states and dressed atoms, *Phys. Rev. Lett.* 64, 2418.

7. Leung, K.M. and Liu, Y.F. (1990) Full vector wave calculation of photonic band structures in face-centered-cubic dielectric media, *Phys. Rev. Lett.* 65, 2646.

8. Zhang, Z. and Satpathy, S. (1990) Electromagnetic wave propagation in periodic structures: Bloch wave solutions of Maxwell's equations, *Phys. Rev. Lett.* 65, 2650.

9. Ho, K.M., Chan, C.T., Soukoulis, C.M. (1990) Existence of a photonic gap in periodic dielectric structure, *Phys. Rev. Lett.* 65, 3152.

10. Leung, K.M., and Qiu, Y. (1993) Multiple scattering calculation of the two-dimensional photonic band structure, *Phys. Rev. B* **48**, 7767-7771.

11. Wang, X., Zhang, X.G., Yu, Q., and Harmon, B.N. (1992) Multiple scattering theory for electromagnetic waves, *Phys. Rev. B* **47**, 4161-4167.

12. Pendry J.B. and MacKinnon, A. (1992) Calculations of photon dispersion relations, *Phys. Rev. Lett.* **69**, 2772-2775.

13. Johnson, N.F. and Hui, P.M. (1993) Theory of propagation of scalar waves in periodic and disordered composite structures, *Phys. Rev. B* **48**, 10118-10123.

14. Sigalas, M., Soukoulis, C.M., Economu, E.N., Chan, C.T., and Ho, K.M. (1993) Photonic band gaps and defects in two dimensions: Studies of the transmission coefficient, *Phys. Rev. B* **48** 14121-14126.

15. Pendry, J.B. (1994) Photonic band structures, *J. Mod. Opt.* **41**, 209-230.

16. Robertson, W.M., Arjavalingam, G., Meade, R.D., Brommer, K.D., Rappe, A.M., and Joannopoulos, J.D. (1992) Measurement of photonic band structure in a two-dimensional periodic dielectric array, *Phys. Rev. Lett.* **68** 2023-2026.

17. Li, L. (1993) Multilayer modal method for diffraction gratings of arbitrary profile, depth, and permittivity, *J. Opt. Soc. Am. A* **10**, 2581-2591.

18. Montiel, F. and Neviere, M. (1994) Differential theory of gratings: extension to deep gratings of arbitrary profile and permittivity through the R-matrix propagation algorithm, *J. Opt. Soc. Am. A* **11**, 3241-3250.

MICROWAVE APPLICATIONS OF PHOTONIC CRYSTALS

E. R. BROWN, O. B. MCMAHON, C. D. PARKER, C. DILL III
Lincoln Laboratory, Massachusetts Institute Of Technology,
Lexington, Ma   02173-9108 USA

K. AGI and K. J. MALLOY
Center for High Technology Materials
University of New Mexico,
Albuquerque, NM   87131 USA

Abstract

This paper reviews three applications we have investigated using conventional (i.e., all dielectric) photonic crystals at frequencies up to about 30 GHz: (1) microwave mirrors, (2) substrates for planar antennas, and (3) photonic-crystal heterostructures. In each case, an important characteristic of the photonic crystal is that the reflection at frequencies in the stop band is distributed over at least one lattice constant in depth. Thus, the heat generated by residual dielectric absorption is distributed over a much larger volume than the heat generated by surface losses in a metal mirror, enabling a lower operating temperature. An additional characteristic of the photonic crystal, essential to the antenna application, is that its stop band is three-dimensional and thus rejects the majority of power radiated by an antenna mounted on its surface. This makes the planar antenna much more efficient than the same antenna placed on a homogeneous substrate made from the same dielectric material as the photonic crystal. A key factor in the ultimate practicality of these applications is the development of new types of photonic crystals that are superior structurally to conventional crystals or that display enhanced stop-band characteristics. To widen the stop band, we have studied a photonic-crystal heterostructure consisting of a stack of monoperiodic sections having different lattice constants. The resulting structure is shown to have a stop band of nearly one octave.

C. M. Soukoulis (ed.), Photonic Band Gap Materials, 355–375.

1. Introduction

Photonic crystals are two- or three-dimensional periodic dielectric or metallic structures that display a stop band in their electromagnetic transmission characteristics. If this stop band is omnidirectional, then the photonic crystals have a forbidden region or band gap in their ω-k dispersion relation. As such, photonic crystals are well suited to a number of microwave and millimeter-wave applications for which conventional materials and components are unsatisfactory. Paramount among these are applications that require high reflectivity without the presence of metal.

The first photonic crystal to exhibit a three-dimensional stop band was a (111)-oriented face-centered-cubic (fcc) structure with nonspherical air atoms. The crystal was fabricated at Bellcore from the synthetic dielectric material Stycast [1], and displayed a band stop in the microwave range between 13 and 16 GHz. More recently, it has been shown that superior band-stop characteristics can be obtained from the diamond crystal structure, and such a crystal can be fabricated by stacking dielectric rods in a "woodpile" structure [2,3]. The technique of silicon micromachining has enabled the fabrication of diamond photonic crystals having stop bands up to 450 GHz [4].

At Lincoln Laboratory we have devised a (111)-oriented fcc structure whose repeat unit is a stack of three slabs, each slab containing a two-dimensional triangular lattice of cylindrical air atoms [5]. Because the atomic shape in this fcc structure is cylindrical, one does not expect the band-stop properties (or the band structure) to be the same at common-symmetry points (i.e., the 6 X points) in the fcc Brillouin zone. Nevertheless, the structure displays a sizable stop band along the axis normal to the top facet (corresponding to one of the L points in the Brillouin zone) and over a large angle about the normal. Furthermore, the triangular lattice constituting each slab is known to have good two-dimensional band-stop properties in the plane [6]. Hence, this fcc structure is very useful as a substrate for planar antennas or transmission lines operating at frequencies near the center of the first stop band.

Our choice of photonic crystal exemplifies the strategy we have generally applied in the investigation of microwave applications. That is, the presence of an omnidirectional stop band or, equivalently, a band gap, is usually not as important an issue as it is in most theoretical studies of photonic crystals. The more important issues are: (1) the stop-band properties along the directions involved in the application at hand, and (2) the mechanical compatibility of the photonic crystal with the metallic antennas or transmission lines that must be fabricated on its surface. Mechanical compatibility bears on ruggedness and cost — both of which often determine whether or not an application becomes popular.

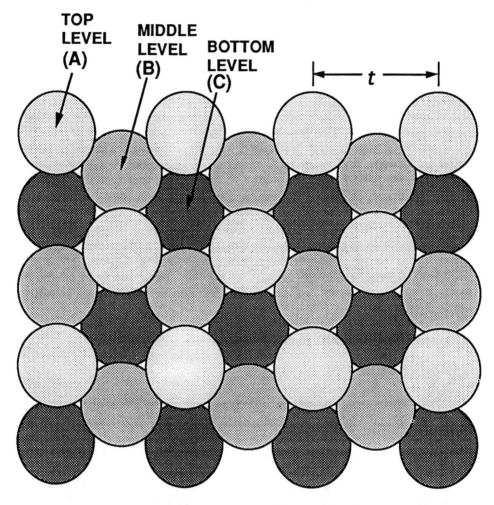

Fig. 1. Top view of Lincoln Laboratory (111)-oriented photonic crystal. The shaded circles represent the cylindrical atoms in three successive layers of the fcc repeat unit.

2. Crystal Fabrication

The vertical repeat unit of our fcc photonic crystal consists of three slabs (A,B,C) in which the middle layer (B) is stacked below the top layer A in such a way that each atom lies directly below the center of a triangular unit cell in A. The bottom layer (C) is aligned so its atoms lie directly below the remaining unit cells in (A). The top view of this stacking arrangement (ABC) is shown in Fig. 1. If the atoms were spherical and there was no intervening dielectric material, this stacking arrangement would result in the (111)-oriented fcc close-packed lattice

[7]. The (111) fcc lattice requires the triangular lattice constant t be related to the slab thickness s by $t = (3/2)^{1/2}s$. The fcc conventional cubic lattice constant a is then given by $a = 2^{1/2} t$.

All of our conventional photonic crystals were fabricated from Stycast, a synthetic dielectric material consisting of titanium dioxide grit immersed in a polymeric binder. The dielectric constant of the Stycast has nominally been either 10, 12, or 13, depending on the concentration of titanium dioxide. The Stycast was obtained commercially in 1x1 ft by 1/4-inch-thick plates, and fabricated into 6x6x1/4-inch triangular-lattice slabs by sawing and numerical routing. The router is a custom-built computer-controlled machine with a three-axis control system. The x, y, and z position stages are driven by stepper motors having 2000 steps/revolution, corresponding to a linear displacement of 0.0005 inch/step and a bidirectional repeatability of 0.0008 inch/inch. The x and y stepper motors are controlled in velocity and acceleration over a total travel of 12 inches. The z stage offers 2 inches of travel. The router speed is adjustable from 0 to 22,000 RPM. The control programs are written and executed on an IBM 386 in a Microsoft Windows environment.

3. Transmission and Reflection Characteristics

3.1. MEASUREMENT TECHNIQUES

After fabrication, the microwave transmission or reflection through a photonic crystal is routinely measured from 5 to 26 GHz with an HP 8510 network analyzer in the setup shown in Fig. 2. A set of three pyramidal feedhorns is used in sequence at the generator and transmission ports to transform the radiation from coaxial cable to free space. Each horn is used only within its designated bandwidth (less than 1 octave) to ensure good beam characteristics incident on the sample. As in all microwave antennas, the beam from a feedhorn is characterized by its emitted or collected intensity vs polar angle q in at the electric E plane (coincident with the polarization) and in the orthogonal H plane. The full width of the beams is defined by the difference in q between the points where the collected radiation is down 10 dB (10%) relative to that at broad side. For the feedhorns used in our experiments the 10-dB full widths were 30° in the E plane and 22° in the H plane. As displayed by other feedhorns of this type, the E-plane pattern is broader than the H-plane pattern because of small sidelobes.

An important factor in the transmission or reflection measurements is the spot size of the incident microwave beam at the surface of the photonic crystal. Our experience is this spot size must be substantially greater than the separation of nearest-neighbor lattice sites to see well-defined stop bands with precipitous band edges. With

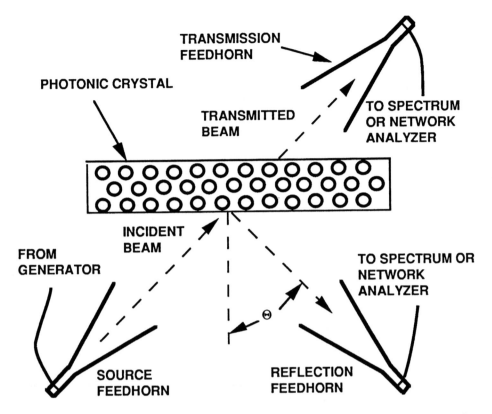

Fig. 2. Schematic diagram of experimental setup used to measure the transmission and reflection properties of photonic crystals.

the 22 to 30° Gaussian-like beam patterns of our feedhorns, this is accomplished by separating the sample from the feedhorn by roughly 20-cm or more. For the 20-cm separation, the spot diameter on the crystal is about 8 cm, which is roughly 10 times the nearest-neighbor separation in a photonic crystal having its first band stop in the 10-to-20-GHz range.

In the transmission measurements, the sample is placed between the feedhorns with its front and back facet at a specific angle Θ relative to the beam propagation direction. The most interesting angles in our measurements have been the following four high-symmetry propagation directions in the conventional cubic unit cell: [111], [100], [1, 1/2, 0], and [110]. These directions correspond to vectors in the first fcc Brillouin zone that pass through the following points: (1) Γ to L, (2) Γ to X, (3) Γ to W, and (4) Γ to K. The Γ to W direction is particularly interesting from a scientific standpoint since this the direction in which the fcc crystal with spherical atoms fails to demonstrate a strong stop band [8].

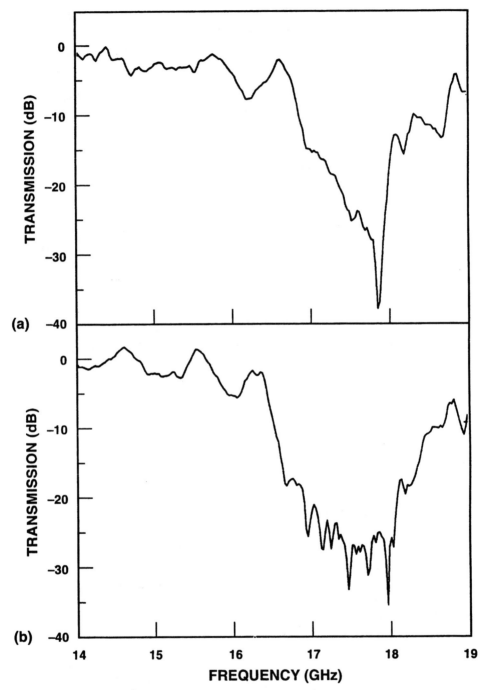

Fig. 3. Experimental transmission through (111)-oriented fcc photonic crystal for radiation incident along the [210] direction. (a) TE polarization. (b) TM polarization.

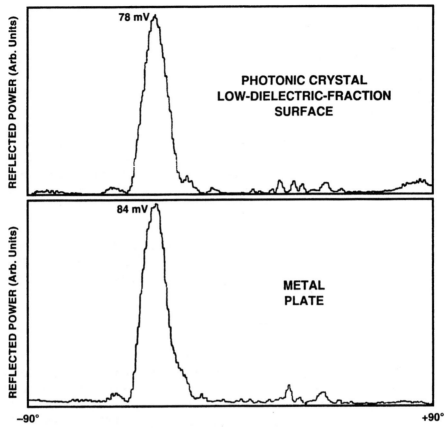

Fig. 4. Reflection from fcc photonic crystal for incident radiation at 39° and the reflected radiation measured over the front hemisphere. The incident frequency was fixed at the center of the stop band.

3.2. TRANSMISSION AND REFLECTION RESULTS

Shown in Figs. 3(a) and (b) are the transmission spectrum through 3 repeat units (9 slabs) of our fcc crystal for radiation incident along the [210] direction and polarized either transverse to the plane of incidence (TE polarization) or parallel to the plane of incidence (TM) polarization. For TE polarization, the stop band occurs between roughly 16.8 and 18.0 GHz. For TM polarization, the stop band occurs between about 16.5 and 18.0 GHz. The dominant feature of the TE spectrum is the deep notch centered around 17.8 GHz. The rejection associated with this notch is approximately 35 dB. Aside from this feature, the maximum rejection is about 25 dB, corresponding to a rejection of 7.5 dB per cubic lattice constant. For the TM polarization in Fig. 3(b), the maximum rejection outside of the small notches is approximately 25 dB at 17.6 GHz.

The set-up in Fig. 2 allows measurement of the reflection coefficient by moving the feedhorn from the back side of the crystal to the front side.

Our experience has been that the reflection is a better indicator of subtleties in the electromagnetic propagation through the photonic crystal than the transmission. A good example is shown in Fig. 4 which was obtained with the generator feedhorn oriented for propagation along the [210] direction 39° away from the normal and the reflection feedhorn varied in angle Θ over the entire front hemisphere. Both feedhorns were set for TE polarization. The frequency of the generator is set at the center of the photonic band stop along this direction. Clearly, the reflection from the photonic crystal is not quite specular, because the reflected signal amplitude is lower than the generator amplitude by about 7%. Although part of this difference is certainly caused by imperfect specular alignment of the feedhorns and losses in the coaxial cables, there is a significantly increased strength of the reflection from the photonic crystal at $\Theta = +90°$ (in the plane of the crystal, 129° away from the generator feedhorn). This is probably caused by the excitation of surface modes, which are thought to exist at the surfaces of all photonic crystals, particularly those having high dielectric surface density (i.e., the fraction of the surface area that is dielectric).

4. Planar Antennas on Photonic-Crystal Substrates

The application of the photonic crystal as an antenna substrate is important from a historical perspective because it was the first known microwave application in which the omnidirectional stop band of the three-dimensional photonic crystal was essential to the performance. Therefore, in this section we start with a thorough conceptual review.

4.1. PHYSICAL PRINCIPLES

Planar antennas play the important role in microwave and millimeter-wave integrated circuits of radiating signals off chip into free space. When fabricated monolithically on a semiconductor substrate such as silicon, such circuits greatly enhance the performance and functional density compared to the alternative hybrid circuits. High-resistivity ($\approx 10^4$ Ω cm) silicon has weak enough electromagnetic absorption ($\alpha < 0.2$ cm^{-1}) to be useful up to millimeter-wave frequencies (> 30 GHz) [9]. However, silicon has a high dielectric constant, which generally makes the performance of planar antennas inferior to that of metallic feedhorns. At the present time, feedhorns are used in the majority of applications at frequencies above 20 GHz.

The problems associated with planar antennas on semiconductor substrates originate in the fundamental electromagnetics of a conductor on a dielectric surface. Shown in Fig. 5 is the conductor-substrate interface for a generic metallic planar antenna on a uniform substrate

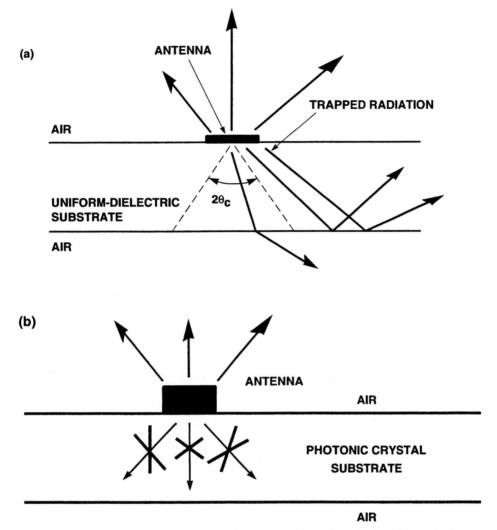

Fig. 5. (a) Cross-sectional view of propagation of radiation from a planar antenna on a uniform-dielectric substrate. (b) Same view for planar antenna on photonic-crystal substrate.

having a purely real dielectric function (i.e., no electromagnetic attenuation) represented by the dielectric constant ε. Independent of its shape, the antenna has a tendency to radiate more power into the substrate than into the free space above the substrate. The ratio of the power into the substrate to the power into free space increases with e. For example, an infinitesimal planar dipole radiates approximately $\varepsilon^{3/2}$ more power into the substrate than into free space [10]. Thus, a dipole on a silicon substrate ($\varepsilon = 11.8$) radiates 40 times more power into the substrate. A second problem is that the power radiated into the substrate at angles greater than $\theta_c = \sin^{-1}\varepsilon^{-1/2}$ is totally internally reflected at the top and bottom substrate-air interfaces. For silicon, this occurs at $\theta_c = 17°$, so that in many antenna structures the vast majority of the radiated power is trapped in the substrate, as shown in Fig. 5(a).

An elegant way to alleviate these problems is to use a three-dimensional photonic crystal as the antenna substrate, as depicted in Fig. 5(b). If the driving frequency of the antenna lies within the photonic band gap, one expects that no power will be radiated into the substrate at any angle, since at every point along the conductor-substrate interface there is no propagation over the full hemisphere on the substrate side. However, it is not clear what fraction of the driving power will be radiated into the air side, since evanescent modes still exist at the air-substrate interface [11], and impedance mismatch can reflect power back to the generator.

4.2. FABRICATION AND MEASUREMENT OF ANTENNAS

Two types of planar antennas have been studied to date: bow-ties (i.e., long, tapered dipoles) and resonant dipoles. The bow-tie was studied first because of its nonresonant nature and well-known impedance properties on uniform dielectric substrates. It was the first antenna type to demonstrate the nearly complete rejection of radiation by the Bellcore fcc photonic crystal [12]. However, its radiation pattern was always highly scalloped, partly because of the complicated nature of the antenna pattern of the bow-tie even in free space, and partly because it was fabricated with thin copper tape that was easily twisted and lifted off the substrate by the coaxial feed lines. To start with a simpler antenna pattern, we then investigated planar dipole antennas [13], which will be the only results displayed here. In this case, the antennas were fabricated with either thin copper tape stuck directly on the surface or free-standing metal shimstock abutted to the crystal. As will be shown later, the stiffness and maneuverability of the shimstock dipole yielded the most desirable antenna patterns that we have observed to date.

After fabrication, the radiation patterns from planar antennas are measured with the compact, scalar antenna test range shown schematically in Fig. 6. The photonic crystal is mounted on one end of the range in a plastic yoke designed to rotate in azimuth and elevation by mechanical coupling to metallic gear assemblies. The gear assemblies are shielded from the photonic crystal by microwave absorbing foam. The electric E-plane pattern of a dipole antenna, for example, is measured by rotating the mounting yoke through 180° in elevation with the crystal and feedhorn configured as in Fig. 6. The mounting yoke is then rotated 180° in azimuth to obtain the magnetic H-plane pattern. The separation between the crystal and the mouth of the feedhorn is approximately 1.4 m.

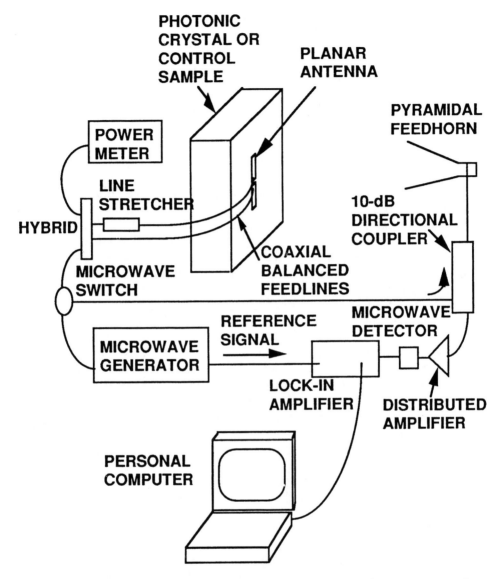

Fig. 6. Diagram of compact antenna test range used to measure the radiation pattern from planar antennas on photonic crystals.

The source of microwave drive radiation is an amplitude-modulated generator with a coaxial output port. To drive the planar antenna in a balanced manner, the oscillator power is divided equally by a 3-dB, 180° hybrid coupler into two thin (2.2-mm diameter) coaxial feed lines. The feed lines are routed to the planar antenna with a line-stretcher-type phase shifter added in one line to achieve a balanced drive at the end of the lines. To determine how balanced the drive is, the power is monitored in the unused port of the hybrid and the line stretcher is adjusted to minimize this power. The radiated power is measured with the scalar receiver of Fig. 6. The receiver consists of a pyramidal feedhorn connected to a standard-height waveguide, waveguide-to-coaxial transition, distributed amplifier, and microwave power detector. To enhance the sensitivity, the output of the microwave detector is synchronously detected with a lock-in amplifier tuned to the modulation frequency of the generator.

4.3. PLANAR-DIPOLE RESULTS

For the sake of brevity, the experimental results presented here will be limited to the E-plane patterns of a copper-tape dipole on the Bellcore fcc crystal and the shimstock dipole on the Lincoln Laboratory fcc crystal, since these are the most educational results obtained to date. The Bellcore crystal is particularly interesting because it afforded two different surface dielectric densities by virtue of its nonspherical and noncylindrical atoms. Figure 7(a) shows the E-plane pattern for a dipole mounted on the high-dielectric surface and the driving point positioned at the center of the dielectric rib between two nearest-neighbor air holes. The zero of the polar angle defines the zenith. The pattern is characterized by multiple narrow peaks and a lack of symmetry about the zenith. In addition, it shows a significant amount of power at one of the horizons and some radiation into the back hemisphere. Since the photonic crystal cannot propagate bulk electromagnetic modes into the back side at the given drive frequency, this is a clue that surface waves are playing a role. Surface waves propagate from the dipole to the edge of the crystal, and can then radiate into the back-side hemisphere because of the abrupt discontinuity in the dielectric properties at the edge.

Figure 7(b) shows the E plane pattern of the same dipole mounted on the surface with low dielectric density, shown in Fig. 7(b). In this case, the driving point of the dipole was placed over an air atom between dielectric mesas. The antenna pattern is the most desirable that we observed from the Bellcore crystal. It peaked near the zenith and contained practically no power in the back-side hemisphere. However, it was somewhat asymmetric about the zenith, displaying a predominant side lobe near $\smile := -70°$. The cause of this asymmetry was not understood, although distortion of the dipoles by the coaxial drive lines was suspected.

The experimental results for the shimstock dipole on the Lincoln Laboratory fcc crystal are shown in Fig. 8. The overall length of the dipole was equal to one-half of a free-space wavelength at the drive frequency of 17.4 GHz. Several orientations of the dipole antenna were attempted relative to the unit cell of the fcc lattice. The best pattern, by

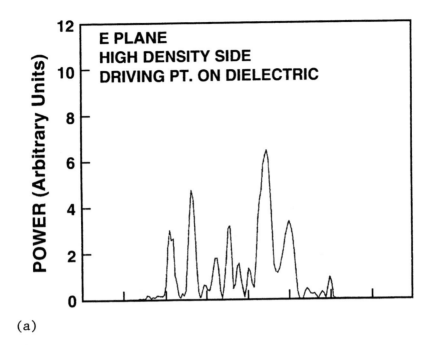

(a)

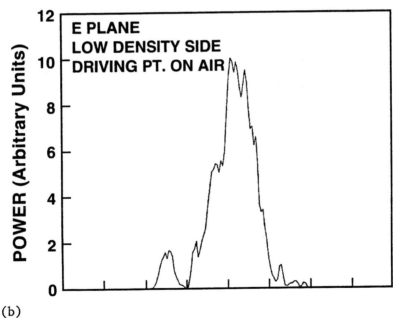

(b)

Fig. 7. E-plane radiation patterns from copper-tape dipole antenna mounted on two difference surfaces of the Bellcore fcc photonic crystal. (a) High dielectric density. (b) Low dielectric density.

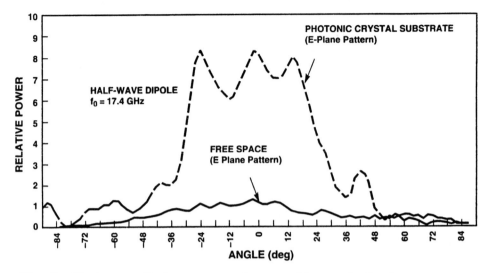

Fig. 8. E-plane radiation pattern from a shimstock-dipole antenna on Lincoln Laboratory fcc photonic crystal and the same dipole in free space.

far, was obtained for the dipole antenna mounted with its driving point located directly over a cylindrical air atom along a line through nearest neighbors of the surface triangular lattice. The resulting E-plane pattern displays a strong central lobe consisting of three local peaks and sharp skirts that occur at roughly ±30° in polar angle. Secondary lobes occur at roughly ±42°, outside of which the power drops to 10 dB below the main lobe at roughly ±48°.

A big advantage of the free-standing shimstock dipole is that it affords calibration of the antenna pattern by removal of the photonic crystal from the scene. Figure 8 also plots the E-plane radiation pattern of the same shimstock dipole in free space. As expected, the pattern of the free-space dipole is rather featureless with a broad peak at the zenith and decrease with polar angle that goes roughly as $\cos^2\theta$. Clearly, over the entire central lobe, the dipole on the fcc sample radiates with much more intensity than it does in free space. At the zenith, the ratio of the intensities is a factor of 6.8.

5. High-Reflectivity Mirrors

A straightforward application of photonic crystals in both the optical and microwave regions is high reflectivity mirrors. In the optical regime, such a mirror has the attribute of suppressing the three-dimensional density of photon states, which was the reason for the expression "photonic crystal" and is the basis for the optical applications low-threshold lasers and very efficient LEDs [14]. In the microwave region we have confirmed that aside from small surface-mode generation, the reflection is specular provided the impinging free-space beam is extended laterally over many (at least ten) interatomic lengths.

5.1. PHYSICAL CONCEPT

In the microwave region, omnidirectional reflection is readily provided by a standard metal mirror with the following limitation. Because of the skin effect, the small fraction of power absorbed during the reflection of a high-power beam is dissipated in a small depth of metal. Because of the finite thermal conductivity of metals at room temperature, this small fraction of absorbed power can still result in high surface temperatures and ultimately mirror damage or destruction. The solution offered by the photonic-crystal reflector is to distribute this absorbed heat over a much greater volume and, hence, render much lower operating temperatures at the surface. In addition, many of the high-permittivity dielectric materials that are well suited to conventional photonic crystals (e.g., sapphire) are highly refractory and have comparable thermal conductivity to mirror metals at room temperature, so they tend to withstand absorbed power better in any case.

To demonstrate this concept quantitatively, we start with the conventional flat metal mirror having electrical resistivity ρ and subject to a powerful incident microwave beam of intensity I_i and frequency 10 GHz. In this case, the penetration or skin depth of the radiation at the surface of the metal is $\delta = (2\rho/\mu\omega)^{1/2}$ [MKSA Units], the surface impedance is $Z_s \approx \rho/\delta$, and the power reflectivity $R = [(Z_s - Z_0)/(Z_s + Z_0)]^2$, where Z_0 is the characteristic impedance of free space (377 Ω). The power absorbed per unit area by the mirror is $(1 - R)I_{inc}$ and the power absorbed per unit volume and averaged over the skin depth, $\overline{P_{abs}}$, is roughly $(1.0 - R)I_{inc}/\delta$. For example, when the metal is high-purity copper, we have $\rho \approx 2 \times 10^{-6}$ Ω–cm, $\delta \approx 0.7$ μm at 10 GHz, $Z_s \approx 0.028$ Ω, and $R \approx 0.9997$. This leads to the rough approximation $\overline{P_{abs}} \approx 4.2\ I_i$ [W cm^{-3}]. Given this absorbed power, the surface temperature can rise to a very high level, potentially leading to mirror damage. The surface temperature is determined largely by the thermal conductivity of copper, which, in pure form, is approximately 4.0 W-cm^{-1} K^{-1} at 300 K.

Unlike the metal, the absorbed power in the photonic crystal is a parasitic mechanism, unrelated to the distributed reflection. Intuitively, we expect the intensity of a broad beam (compared to a lattice constant) will decay in the photonic crystal as $I(z) = I_0\exp[-(\alpha_r + \alpha_a)z]$, where α_r is the decay constant associated with the (lossless) reflection, α_a is the decay constant associated with the real absorption in the dielectric material, and $z = 0$ is at the air photonic-crystal interface. The decay of power per unit volume (caused by reflection and absorption) is given by $-dI/dz = (\alpha_r + \alpha_a)I_0\exp[-(\alpha_r + \alpha_a)z] = (\alpha_r + \alpha_a)I_0$. The portion of this decay associated with absorption is given by $\alpha_a I_0 \equiv P_{abs}$, so the net power absorbed per unit area down to a depth z is

$$I_{abs} \equiv \int_0^z P_{abs}dz' = \frac{\alpha_a I_0}{\alpha_a + \alpha_r}\left[1 - \exp(-[\alpha_a + \alpha_r]z)\right], \qquad (1)$$

and the average absorbed power per unit volume is

$$\overline{P_{abs}} = z^{-1}\int_0^z P_{abs}dz' = \frac{\alpha_a z^{-1}I_0}{\alpha_a + \alpha_r}\left[1 - \exp(-[\alpha_a + \alpha_r]z)\right]. \quad (2)$$

Note, in the usual case where $\alpha_r \gg \alpha_a$, $\overline{P_{abs}}$ saturates much more rapidly with z than it does in bulk dielectric material having loss α_a. This illustrates the fact that little power can be absorbed after most of it is already reflected from the crystal.

For example, we consider a photonic crystal made from high-resistivity ($\rho \approx 1\times10^4$ Ω-cm) silicon. In bulk form, this material has an absorption coefficient α_0 of approximately 0.1 cm^{-1} [9]. The value of α_a in the photonic crystal should be less than this, since the volumetric fraction of the silicon will be much less than unity. In the present analysis, we make a worst case estimate of $\alpha_a = \alpha_0$. For consistency, we also assume that the photonic crystal has an overall thickness that yields approximately the same reflection (0.9997) as the copper mirror. In a typical photonic crystal at 10 GHz, $\alpha_r \approx 2.0$ cm^{-1} and a ≈ 1 cm, so the same reflectivity as the copper is achieved with roughly four lattice constants, 4a. Substituting these parameters into Eq. (2), we find $\overline{P_{abs}}$(z=4 cm) $\approx 0.012 \cdot I_0$ W-cm^{-3}. This is 0.3% of the same quantity estimated above for the copper mirror. This result coupled with the fact that the thermal conductivity of silicon (1.5 W-cm^{-1}-K^{-1} [15]) is within a factor of three of copper at room temperature means the surface temperature of the silicon photonic-crystal mirror is probably much lower for a given incident intensity. Similar reasoning would suggest other high-permittivity dielectric materials with good thermal conductivity may be even better because of their higher electrical resistivity and, hence, lower absorption coefficient. One notable material would be sapphire ($\varepsilon \approx 9.4$), whose resistivity is so high that the residual alternative is probably caused by a mechanism other than free-carrier absorption at microwave frequencies.

6. Photonic-Crystal Heterostructures

Most conventional photonic crystals studied to date provide a limited stop-bandwidth of about 15% of the center frequency. To overcome this limit, we have investigated a composite structure fabricated by stacking monolithic sections of our (111)-oriented fcc crystal as shown in Fig. 9 [16]. By making a discrete change in the lattice constant between adjacent monolithic sections, the net stop-bandwidth can be extended well beyond that of a given section.

One of the features that makes the composite crystal attractive is the high transmission through the individual monolithic sections *outside* the stop band. This was exemplified in Fig. 3, where we saw the experimental transmission just below the stop band approached 90%. This means the net reflection can be very high at frequencies within the stop band of sections lying well below the top of the heterostructure. Without this property, large standing waves would likely be established between the photonic band gap buried section and the partially reflecting upper layer or layers above it. This would lead to a resonantly high net reflection at some frequencies, but very low reflection at others. Of course, such behavior would be undesirable for most applications.

6.1. CRYSTAL DESIGN

In the absence of standing waves between the individual sections, the net reflection from the photonic-crystal heterostructure can be determined by an incoherent addition of the power reflected from each section. This leads to the following expression for the total reflection,

$$R_T \approx R_1 + T_1 R_2 T_1 + T_1 R_2^2 R_1 T_1 + T_1 R_2^3 R_1^2 T_1 + ... = R_1 + \frac{T_1^2 R_2}{1 - R_1 R_2}, \qquad (3)$$

where R_1 (T_1) and R_2 are the reflections (transmissions) from the upper- and lower-frequency, respectively, associated with each heterojunction.

In a rough sense, the maximum overall bandwidth will result from a contiguous arrangement of the band gaps of the adjacent sections in the composite structure. Assuming the band gap of each section has a constant fractional width d about the center frequency, one can derive (by mathematical induction) the following expressions for the overall fractional width Δ,

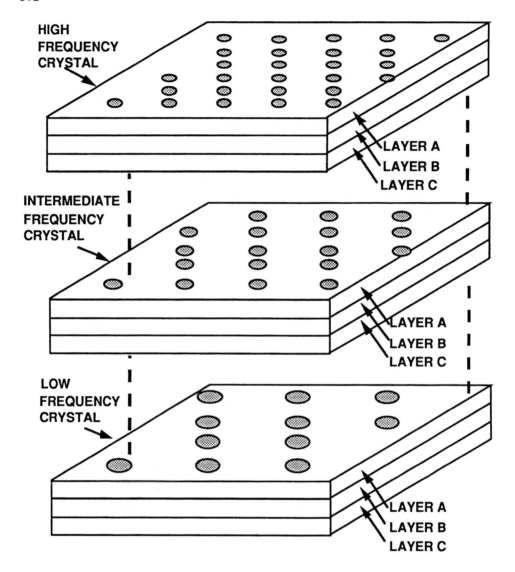

Fig. 9. Photonic crystal heterostructure consisting of a vertical stack of fcc crystals having different lattice constants.

$$\Delta = \frac{(1+\delta/2)^{N} - (1-\delta/2)^{N}}{(1+\delta/2)^{N} + (1-\delta/2)^{N}}, \tag{4}$$

and the ratio r of the maximum and minimum frequencies of the overall band gap, $r = (1 + \delta/2)^{N}/(1 - \delta/2)^{N}$. The values of r for various values

Table I. Ratio of maximum to minimum frequency of the stop band of a heterostructure consisting of N sections of monolithic photonic crystal, each having $\delta = 15\%$

N	1	2	3	4	5	6	7	8	9	10
r	1.16	1.35	1.57	1.82	2.12	2.46	2.86	3.33	3.87	4.49

of N are listed in Table I, assuming a δ of 15%. Notice a gap width of over one octave ($r = 2$) can be obtained with just 5 sections and over two octaves with 10 sections.

6.2. EXPERIMENTAL RESULTS

A photonic crystal heterostructure consisting of three different sections of the Lincoln Laboratory fcc crystal has been constructed and tested in the range of 15 to 25 GHz. Figures 10(a) through 10(d) show the experimental transmission at normal incidence through the heterostructure, the high-frequency component crystal, the middle-frequency component crystal, and the low-frequency component crystal. The heterostructure has a maximum rejection of approximately 48 dB at 22.5 GHz, a minimum rejection of approximately 7 dB at 17.9 GHz, and a stop band that extends from 16 to at least 25 GHz. The low rejection at 17.9 GHz occurs from too little overlap between the component band stops as seen in Figs. 10(c) and 10(d). Clearly, the rejection of the heterostructure around 22.5 GHz is being assisted by the 2nd stop band of the low-frequency crystal, as seen in Fig. 10(d). The fact that it is much deeper than the 1st stop band is typical for the Lincoln Laboratory fcc crystal.

7. Summary

The existence of a three-dimensional stop band in photonic crystals enables several interesting applications, such as high-power mirrors and planar-antenna substrates. The mirror application utilizes the fact that the reflection process is distributed in space over at least one lattice constant, so the heat generated by residual absorption density (power absorbed per unit volume) in the dielectric component of the crystal is much smaller than it is in conventional, all-metal mirrors. The heat problem is also alleviated by the fact that the high-permittivity materials required for conventional photonic crystals tend to have high thermal conductivity. The application of photonic crystals as substrates for planar antennas is unique in the sense that it requires the presence of the three-dimensional

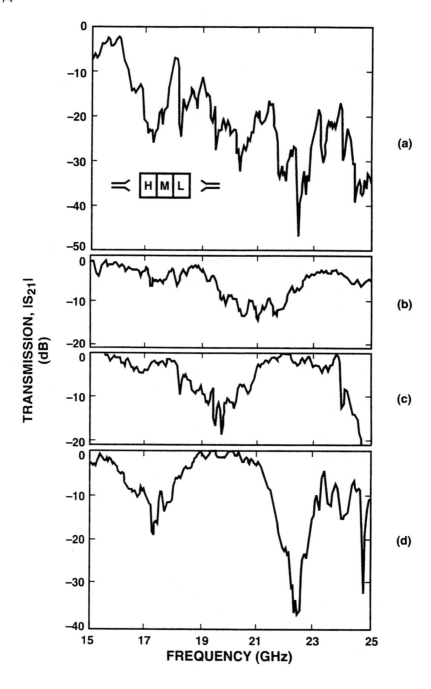

Fig. 10. (a) Magnitude of transmission for three-section photonic-crystal heterostructure over the range of 15 to 25 GHz. (b) Also shown is the isolated transmission through the (b) high-frequency section, (c) mid-frequency section, and (d) low-frequency section.

stop band, antennas naturally being three-dimensional radiators. In addition, the fact that the reflection is distributed means that the antenna is not shorted out at the driving point, as are planar antennas mounted directly on metal. Finally, we have explored a way of increasing the spectral bandwidth of photonic crystals in these applications. The technique is to stack sections of photonic crystal having different lattice constants. This technique, which is analogous to constructing heterostructures in electronic semiconductors, is well suited to the high-power mirror application.

References

[1] E. Yablonovitch, T. J. Gmitter, and K. M. Leung, Phys. Rev. Lett. **67**, 2295 (1991).
[2] K. M. Ho, C. T. Chan, C. M. Soukoulis, R. Biswas, and M. Sigalas, Solid. State Commun. **89**, 413 (1994).
[3] H. S. Sözüer and J. P. Dowling, J. Mod. Opt. **41**, 231 (1994).
[4] E. Özbay, E. Michel, G. Tuttle, R. Biswas, M. Sigalas, and K.-M. Ho, Appl. Phys. Lett. **64**, 2059 (1994).
[5] E. R. Brown, K. Agi, C. D. Dill III, C. D. Parker, and K. J. Malloy, Microw. Opt. Tech. Lett. **7**, 777 (1994).
[6] R. D. Meade, K. D. Brommer, A. M. Rappe, and J. D. Joannopoulos, Appl. Phys. Lett. **61**, 495 (1992).
[7] N.W. Ashcroft and N.D. Mermin, *Solid State Physics* (Saunders College, Philadelphia, 1976).
[8]. K. H. Ho, C. T. Chan, and C. M. Soukoulis, Phys. Rev. Lett. **65**, 3152 (1990).
[9] M. N. Afsar and K. J. Button, Proc. IEEE **73**, 131 (1985).
[10] D. B. Rutledge, D. P. Neikirk, and D. P. Kasilingam, "Integrated-Circuit Antennas," in *Infrared and Millimeter Waves,* Vol. 10 (Academic, Orlando, 1983), p. 1.
[11] R. D. Meade, K. D. Brommer, A. M. Rappe, and J. D. Joannopoulos, Phys. Rev. B **44**, 10961 (1991).
[12] E. R. Brown, C. D. Parker, and E. Yablonovitch, J. Opt. Soc. Am. B **10**, 404 (1993).
[13] E. R. Brown, C. D. Parker, and O. B. McMahon, Appl. Phys. Lett. **64**, 3345 (1994).
[14] E. Yablonovitch, Phys. Rev. Lett. **58**, 2059 (1987).
[15] S. Sze, *Physics of Semiconductor Devices* , 2nd Ed. (Wiley, New York, 1981), p. 850.
[16] K. Agi, E. R. Brown, O. B. McMahon, C. Dill III, and K. J. Malloy, Electron. Lett. **30**, 2166 (1994).

OPTIMIZED ANTENNAS ON PHOTONIC BAND GAP CRYSTALS

R. BISWAS[†‡], S. D. CHENG[†], E. OZBAY[*†], S. McCALMONT[†], W. LEUNG[†], G. TUTTLE[†] and K.-M. HO[†‡]

[†]*Microelectronics Research Center and Ames Laboratory - USDOE, Iowa State University, Ames, IA 50011.*

[‡] *Department of Physics and Astronomy, Iowa State University, Ames, IA 50011.*

[*]*Department of Physics, Bilkent University, Bilkent, Ankara 06533, Turkey*

Abstract

The radiation patterns of a dipole antenna on a photonic crystal substrate are measured and compared with dipole antennas on dielectric substrates. The layer-by-layer photonic crystal with a gap between 12 and 15 GHz is utilized. The angular distribution of radiated power is optimized by varying the position, orientation and driving frequency of the antenna. Virtually no radiated power is lost to the photonic crystal resulting in gains and radiation efficiencies larger than on other conventional dielectric substrates.

C. M. Soukoulis (ed.), Photonic Band Gap Materials, 377–390.

1.   Introduction

Photonic band gap crystals are a novel class of periodic dielectric structures where propagation of electromagnetic (EM) waves is forbidden for all frequencies in the photonic band gap.[1,2] With the fabrication of periodic dielectric structures that have full three-dimensional photonic band gaps in the microwave[3,4] and the millimeter wave regime[5-7], one important direction has been to design novel applications for these photonic band gap (PBG) materials. The three-cylinder structure with the symmetry of diamond was fabricated with drilling techniques by Yablonovitch et al[3] and was the first structure to have[3] a full photonic bandgap, at microwave frequencies. The Iowa State group then designed[4] an alternative layer-by-layer crystal structure that is easy to fabricate and has a full three-dimensional PBG, and was fabricated at microwave[4] (13 GHz) and millimeter wave frequencies (90-500 GHz)[5-7].

This paper demonstrates a simple application of the layer-by-layer PBG crystal: an efficient directional antenna, that was first proposed by Brown, Parker and Yablonovitch[8]. In microelectronics applications it is often necessary to integrate the antenna and the electronics on the same wafer. Such integrated circuit antennas on a standard semi-infinite semiconductor substrate (with dielectric constant ε) have the drawback that the power radiated into the substrate is a factor $\varepsilon^{3/2}$ larger than the power into free space.[9] Hence, antennas on GaAs or Si ($\varepsilon \approx 12$), radiate only about 2-3% of their power into free space. Of the power radiated into the substrate, a large fraction is in the form of trapped waves propagating at angles larger than the critical angle.[9] Brown, Parker and Yablonovitch[8] proposed that by fabricating the antenna on a PBG crystal with a driving frequency in the stop band, no power should be transmitted into the photonic crystal and all power should be radiated in free space- if there are no evanescent surface modes. Brown et al[8] were the first to demonstrate this by fabricating a bow-tie antenna on their 3-cylinder PBG crystal and found a complex radiation pattern confined to free space. Brown, Parker and McMahon[10] then improved the antenna directionality by placing a dipole antenna on high- and low-dielectric surfaces of their PBG crystal,[10] and found that the radiation pattern was much more directional for dipole antennas on the low-dielectric surface.

In this paper we examine ways by which the dipole antenna performance can be enhanced by placing the dipole antenna on the surface of our layer-by-layer PBG crystal, constructed with stacked alumina rods,[5] that has a full 3-dimensional photonic band gap between 12 and 14 GHz. This frequency range was chosen for ease of measurement, and results found here can be scaled to other frequencies in the microwave and millimeter wave regime.

Practical applications of antennas in the millimeter wave regime rely on the transmission bands around 35, 94 and 140 GHz where the atmospheric attenuation is a minimum. The 94 GHz band has attracted much interest for high-bandwidth military communications. The band around 60 GHz which has high attenuation has much promise for short-range secure communications such as in vehicular communications, as well as in satelite-to-satelite communications. The lower frequency band of 0.8-3 GHz is undergoing rapid growth with cellular telephones and personal communications systems.

2. Results

A thin copper dipole antenna was photolithographically patterned on a 0.031" thick Duroid/5880 sheet (ε_r=2.2). The experimental setup (Fig. 1), was very similar to Ref. [8], consisting of a *Ku*-band synthesizer generating the signal that was divided by a 3 dB hybrid coupler into two components 180° out of phase. Each component was routed through adjustable phase shifters and 50 Ω coaxial cables that were soldered to center feed points of the dipole. *Ku*-band measurements were performed with an HP8510B network analyzer, with the dipole as a rotating source and a stationary pyramidal feedhorn as a receiving antenna. In later measurements, the hybrid coupler and phase shifters were also mounted

380

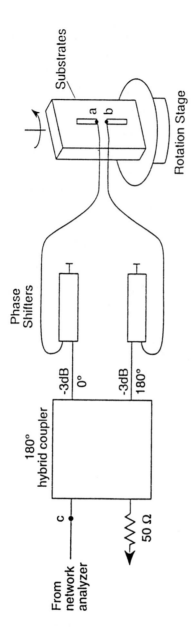

Fig. 1.   Schematic experimental setup for antenna measurements.

on the rotary table so that the feed wires would rotate with the dipole source, and minimize the changes in phase caused by movement or twisting of feed wires.

The radiation of the dipole on a 0.25" thick lexan dielectric (ε=2.56) sheet was first characterized, since the low ε makes it similar to a dipole in free space. The dipole length (1.1 cm) was designed for a good impedance match to the coaxial feed and the reflection coefficient (S_{11}) was small in magnitude (<15 dB). Measured radiation patterns at 13 GHz (Fig. 2) displayed an approximately equal division of power between the front and back sides of the dielectric similar to a free space dipole. A large surface wave near the slab edges (Fig. 2), is caused by the trapped radiation in the dielectric slab emerging near the edges of the dielectric. The H-plane is more noisy since the trapped waves in the dielectric are much larger in the H-plane and can produce interference effects.

The far-field radiation of the dipole placed above a finite dielectric slab, was calculated[11] by expanding the spherical dipole fields in plane waves, and utilizing Fresnel reflection and transmission coefficients. The calculation assumes the slab is infinite in the two lateral directions and hence does not permit the trapped wave in the dielectric to emerge from the slab. This provides the difference between calculation and measurement (Fig. 2) for the surface radiation. The remainder of the pattern, exhibits good agreement between measurement and theory, especially for the E-plane, indicating that the fabricated dipole is close to the half-wavelength design. The plane wave method also predicts that the trapped waves in the dielectric have much larger power in the H-plane - accounting for the differences between calculation and measurement in the H-plane measurement (Fig. 2).

We have also performed similar measurements on the high dielectric stycast substrates ($\varepsilon \approx 12$) and found a similar division of radiated power between the air and substrate hemispheres. However the radiation patterns were less reliable since i) small changes in the thickness of such slabs led to large changes in the phase of waves reflected from the slab, and ii) variations in dielectric constant of the stycast slab led to noise in the radiation patterns.

382

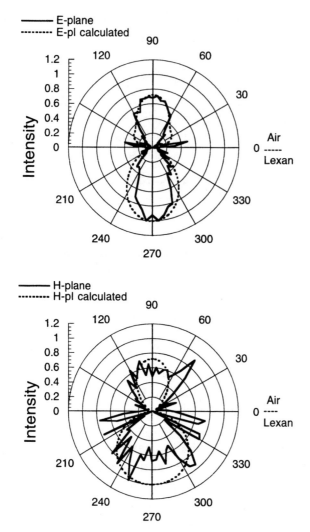

Figure 2. Measured E- and H-plane radiation patterns for the dipole antenna on a 0.25" lexan dielectric sheet (solid) compared to calculations (dashed).

This dipole was then mounted on the stacking surface of our PBG crystal. We drove the dipole at frequencies within the band gap and found virtually no radiation propagating through the PBG crystal. As expected the radiation is confined to the air side. However, the angular distribution of the radiated power strongly depended on the driving frequency and the dipole position. Accordingly, we systematically varied the position, orientation and the driving frequency (12-15 GHz) to optimize the radiation pattern.

We obtained the most directional radiation pattern (Fig. 3a) when the dipole was parallel to the second layer rod and placed above the intersection of the first two layer rods. Both E- and H-planes have central lobes over a broad 0.6 GHz frequency band ($\approx$12.5-13.1 GHz). The measured E-plane pattern is narrower than the H-plane radiation.

The pattern changes when the dipole is rotated by 90°, so that it is parallel to the first layer rods, with its center at the same position as in Fig. 3a. The measured radiation (Fig. 3b) is concentrated in the E-plane in very strong, narrow peaks at 60°-70° from the normal. The weaker H-plane radiation consists of a broad central lobe in addition to weaker side lobes.

Other dipole positions in the surface unit cell, in particular the placement of the dipole above the intersection of the third and fourth layer rods (Fig. 4) led to all the power radiated to the front air side and in two strong side peaks in the H-plane, and a central lobe in the E-plane. This is essentially a reversal of the E- and H-planes in Fig. 3b with most of the power concentrated in the side lobes in the H-plane.

We have verified the results in Fig. 3a and 3b by performing measurements in a state-of-art anechoic chamber at Rockwell where the dipole was used as a receiving antenna, that was rotated in a precisely controlled automated setup.

The radiation patterns on the PBG are similar to a dipole placed a distance above a dielectric substrate, which have similar angular dependence of the patterns. Frequency dependence arises since the phase of reflected plane waves from the PBG crystal varies with frequency in the band gap. Although the bulk of the photonic crystal behaves as a perfect reflector, the surface layers may act as normal periodic dielectric

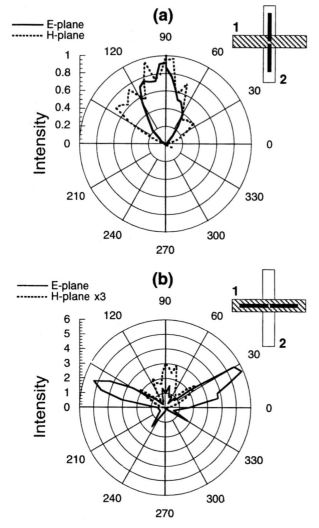

Figure 3. Radiation patterns of the dipole antenna on the photonic bandgap crystal for the two different surface positions drawn in the insets.The dipole is above the intersection of the first (1) and second layer rod (2) as shown in the inset. The E- and H-plane intensities are very different in (b).

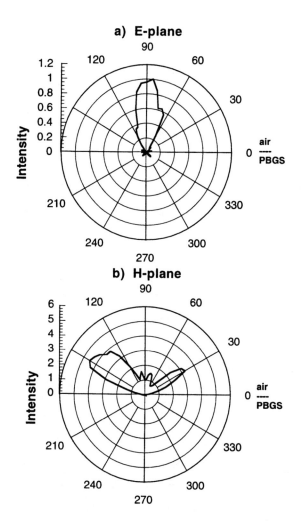

Figure 4. Radiation patterns of the dipole antenna in the E and H- planes when placed above intersection of third and fourth layer rods and with the dipole parallel to the first layer rods. This is the 'open' position in the surface unit cell. The scales are different for the E- and H-plane intensities.

scatterers, and hence result in a complex radiation pattern. The quantitative modeling of these patterns will require the transfer matrix method where plane wave EM fields are propagated in time through a complex structure. Such modeling will account for effect of surface fields on the radiation patterns.

We extracted gains and radiation efficiencies with the standard three-antenna technique and the Friis transmission equations. [12] The sum of the gain of the dipole antenna (G_i) and gain of the horn receiver (G_j) is proportional to the ratio of the peak received power to transmitted power (P_r/P_t), and a space factor depending on the separation R between transmitter and receiver and the wavelength λ,

$$G_i + G_j = 20 \log (4\pi R/\lambda) + 10 \log (P_r/P_t). \qquad (1)$$

Here

$$P_r/P_t = S_{12}/ (1 - S_{11}) = (P_r/P_{inc}) / (\{P_{inc} - P_{refl}\}/P_{inc}). \qquad (2)$$

In practice, the transmission measurements are first done for a pair of standard gain horn antennas (i=j), to extract the gain Gj of the horn antenna as a function of frequency which we find to be about 15 dB. The measurements of transmitted power S_{12} and reflected power S_{11} (back to the source) are then performed for the dipole-antenna and horn antenna combination. This provides the ratio of the powers P_r/P_t and the gain G_i of the dipole. In our definition (2) we subtract out the reflected power P_{refl} back to the network analyzer since this does not determine radiation efficiency, rather the matching back to the source.

The transmitted power was referred to input point c (Fig. 1) so that dipole, feed cables, phase shifters and the hybrid coupler were treated as a composite unit with gain G_c. Alternatively, we subtracted losses from these components (estimated with calibrations to be about - 5.4 dB), to extract the input power delivered to the antenna terminals points a and b (Fig. 1) and the gain of the dipole G_{ab}. The directivities are inferred from radiation pattern measurements, while the corresponding efficiencies (η_c, η_{ab}) are the ratios of the gain and directivities (Table 1).

TABLE 1. The gains and radiation efficiencies for the dipole antenna on different substrates at 13 GHz. These are either referenced (Fig. 1) to point c which includes components, or to the dipole feed points a,b taking out the losses in the system components.

Substrate	position	Directivity	with coupler		without coupler	
		D	Gain G_c	Efficiency η_c	Gain G_{ab}	Efficiency η_{ab}
lexan 0.5"		1.5	-8.4 dB	10%	-3.0 dB	33 %
Stycast 2"		≈1.5	-11.8 db	4%	-5.6 dB	18 %
Microstrip	metal-back	≈3.0	-5.3 dB	10%	0.1 dB	34%
PBG	Fig. 3(a)	3.2	-4.8 dB	10%	0.6 dB	36%
PBG	Fig. 3(b)	3.4	-1.4 dB	21%	4.0 dB	74 %

The antenna on the PBG in Fig. 3a has a higher gain but a similar efficiency to that on the lexan. The gain and efficiency of the dipole (η_c or η_{ab}) on the PBG crystal in Fig. 3b, is much larger by a factor of 2, than the lexan, and the efficiency η_{ab} exceeded 70%. We also performed measurements (Table 1) for i) a stycast substrate ($\varepsilon \approx 12$) and found much poorer gains and efficiencies, and ii) a microstrip geometry with a metal back-plane behind the duroid and found antenna performance similar to lexan but inferior to the PBG of Fig. 3(b).

3. Conclusions

These antenna results require a full-three-dimensional photonic band gap to prevent the transmission of waves in any direction in the PBG crystal. Simpler structures with two-dimensional PBG's would not be adequate since radiation would propagate along the third dimension of the 2-D structures and lead to losses in the substrate. Slot antennas, whose fields are the complement of the dipole antenna, also show enhanced gains when combined with the perfectly reflecting photonic band gap substrate.

The highly reflecting and dissipation-less PBG crystal, together with the optimized antenna radiation as a function of position on the surface, are promising for applications to high power and directional antennas and antenna arrays. These results are readily scaleable to antennas on the higher frequency millimeter-wave PBG crystals that have the same layer-by-layer structure. These PBG crystal antennas would need to be compared with the performance of microstrip antennas which are commonly used at microwave frequencies but become more lossy at higher millimeter wave frequencies.

We thank R. Weber, M. Sigalas, C.M. Soukoulis, and E. Brown for helpful discussions. We thank J. West and S. Oglesby for performing measurements at Rockwell-Collins. This work was supported by the Department of Commerce through the Center for Advanced Technology (CATD) and the Director for Energy Research, Office of Basic Energy Sciences and Advanced Energy Projects. The Ames Laboratory is operated for the U.S. Department of Energy by Iowa State University under Contract No. W-7405-Eng-82.

REFERENCES

1. For a recent review see C.M. Soukoulis, ed., *Photonic bandgaps and Localization* (Plenum, New York, 1993).

2. K.M. Ho, C.T. Chan and C.M. Soukoulis, Existence of a photonic band gap in periodic dielectric structures, *Phys. Rev. Lett.* **65**, 3152 (1990).

3. E. Yablonovitch, T.J. Gmitter and K.M. Leung, Photonic band structure: the face-centered-cubic case employing nonspherical atoms, *Phys. Rev. Lett.* **67**, 2295 (1991).

4. K.M. Ho, C.T. Chan, C.M. Soukoulis, R. Biswas and M. Sigalas, Photonic band gaps in three dimensions: New layer-by-layer periodic structures, *Solid State Comm.* **89**, 413 (1994).

5. E. Ozbay, A. Abeyta, G. Tuttle, M. Tringides, R. Biswas, M. Sigalas, C.M. Soukoulis, C.T. Chan and K. M. Ho, Measurement of three-dimensional photonic band gap in new crystal structure made of dielectric rods, *Phys. Rev. B.* **50**, 1945 (1994).

6. E. Ozbay, G. Tuttle, R. Biswas, M. Sigalas, and K.M. Ho, Micromachined millimeter wave photonic band gap crystals, *Appl. Phys. Lett.* **64**, 2059 (1994).

7. E. Ozbay, E. Michel, G. Tuttle, R. Biswas, K.M. Ho, J. Bostak, and D.M. Bloom, Terahertz spectroscopy of three-dimensional photonic band gap crystals, *Optics Letters* **19**, 1155 (1994).

8. E.R. Brown, C.D. Parker, and E.J. Yablonovitch, Radiation properties of a planar antenna on a photonic-crystal substrate, *J. Opt. Soc. Am. B* **10**, 404 (1993).

9. D.B. Rutledge, D.P. Neikirk, and D.P. Kasilingam, in *Infrared and Millimeter Waves* (Academic, Orlando, 1983), Vol 10, p. 1, and references therein.

10. E.R. Brown, C.D. Parker, and O.B. McMahon, Effect of surface composition on the radiation pattern from a photonic crystal planar-dipole antenna, *Appl. Phys. Lett.* **64**, 3345 (1994).

11. S.D. Cheng, M.S. thesis, Iowa State University (1995) and to be published.

12. C.A. Balanis, Antenna Theory, Analysis and Design, (Harper and Row, New York, 1982) p. 716.

DESIGN CONSIDERATIONS FOR A 2-D PHOTONIC BAND GAP ACCELERATOR CAVITY

D. R. SMITH, N. KROLL, S. SCHULTZ

Department of Physics, University of California, San Diego
9500 Gilman Drive, La Jolla, California, 92093-0319, USA

We discuss recent progress in our effort to develop a high gradient accelerator cavity based on Photonic Band Gap (PBG) concepts. Our proposed cavity consists of a two-dimensional (2-D) photonic lattice, composed of either dielectric or metal scatterers, bounded in the third dimension by flat conducting (or superconducting) plates. A defect introduced to the lattice, usually a removed scatterer, produces a defect mode with fields concentrated at the defect site and decaying exponentially in all directions away from the defect site. The defect mode is designed to resonate at frequencies in the 2-20 GHz range, where metals can still be used to confine the energy with minimal loss. We present in this paper some of the technical considerations which have arisen relevent to this application, and to PBG structures in general. In particular, we focus on measurements and calculations carried out for a 2-D metal PBG cavity.

1. Introduction

Since 1992, we have been pursuing the possibility of incorporating photonic lattices in high-Q, high gradient accelerator cavities. This idea was prompted by observing experimentally obtained field maps of a defect mode in a 2-D photonic lattice composed of dielectric cylinders and bounded by metal plates [1]. It was noted that the electric field had an antinode in the center, was uniformly polarized along the cylinder axis, and had a localization length on the order of one or two lattice constants; thus, the photonic lattice and bounding plates produced a mode analogous to, for example, the TM_{010} mode of a cylindrical "pill-box" cavity, but without using metal for the cavity side-walls. The accelerating mode of the photonic lattice, however, had several potentially advantageous features over the typical accelerating cavity.

(i) The set of resonant frequencies of the PBG cavity was entirely different from those associated with usual cavity structures. In contrast to typical metal cavities (such as the aforementioned pill-box cavity), the defect modes of PBG structures were sparse; that is, in a given frequency range the number of bound modes was very small. In fact, as we will discuss, some structures can be found which have only one bound mode in the entire spectrum; all other modes are propagating, and, in principle, can be damped by placing absorber around the periphery of the structure. An absence of higher order modes is a desirable property for accelerator cavities, since beam coupling to such modes leads to energy spread and (potentially catastrophic) emittance growth.

C. M. Soukoulis (ed.), Photonic Band Gap Materials, 391–410.
© *1996 Kluwer Academic Publishers. Printed in the Netherlands.*

(ii) Since the spatial distribution of the field for the PBG defect mode fell off exponentially from the center of the defect site, material constraints would be lessened toward the boundary of the structure. This could be an important issue when the bounding plates were superconducting, since the quality of a superconductor is difficult to maintain across large distances. Also, only the low loss dielectric scatterers nearest the defect would need to be of the highest purity.

(iii) It was determined that structures fabricated utilizing suitable low loss sapphire cylinders and superconductor should be able to achieve Q values $> 10^9$. In such a structure, the only superconducting material needed would be in the form of flat sheets, with no bends, joints, or welds. Such a structure might be particularly suited to high gradient applications.

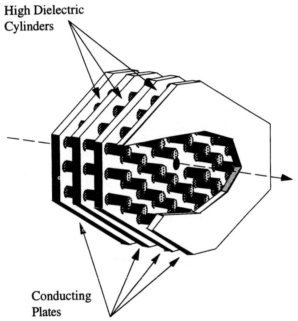

High Dielectric
Cylinders

Conducting
Plates

Figure 1. A schematic view of a proposed 2π accelerator unit. The unit here consists of three triangular photonic lattices composed of low loss dielectric cylinders, separated by conducting or superconducting sheets. Each of the lattices has a cylinder removed to allow the formation of a monopole defect mode. Holes drilled through the conducting plates allow a particle beam to be accelerated through the unit.

A conceptual diagram of a proposed accelerating unit is shown in Fig. 1. In this example, we have (arbitrarily) chosen a triangular dielectric photonic lattice as the basic component of each of the subunits, and have grouped together three subunits to make a single accelerating unit. In this example, the three cavities would be coupled, either via the aperatures or by some other means, and the appropriate mode chosen such that a suitably phased bunched beam entering the unit would experience acceleration as it passed through each of the subunits.

We began a program to optimize the properties of the dielectric PBG cavity using numerical analysis. It was found during the process of this optimization, by us and by other groups, that the best (1/e) decay length of the fields which could be obtained with a dielectric structure was on the order of the lattice constant. Thus, to build a useful cavity with reasonable Q operating at typical microwave frequencies would have required on the order 500+ high dielectric, low-loss tangent cylinders. Building numerous dielecric PBG models for testing would therefore be an expensive proposition, due to the cost and difficulty of machining dielectrics.

While we maintain the eventual goal of producing a dielectric-based PBG cavity for accelerator purposes, we have turned our attention to the metal PBG cavity. This structure uses metal rather than dielectric cylinders, but retains most of the essential properties inherent to PBG structures (e. g., the existence of pass bands, stop bands, and defect modes). The metal structure is easier and less expensive to fabricate, so that numerous test models can be constructed. Furthermore, the metal PBG cavity was found to have unique properties which also warranted study, and may be advantageous for accelerator designs. We feel that the investigation of the properties of the metal PBG cavity will make our program in characterizing the dielectric PBG cavities more efficient.

In order to characterize various properties of the PBG cavities, we have found it necessary to develop accurate numerical codes for simulating various aspects of the structures. We have applied three different techniques; the well-known plane wave method, a method based on finite differences, and the transfer matrix method. We describe each of these in some detail in Section 2, followed by some of the significant results in Section 3. In Section 4 we discuss the Q factors of 2-D PBG cavities, and in Section 5 we report on experiments performed on 2-D metal PBG test cavities. Some concluding remarks are made in Section 6.

2. Numerical Methods

2.1. PLANE WAVE METHOD

Throughout our work, we have been primarily concerned with 2-D modes, those in which the fields do not vary as a function of one of the coordinates. Since our experimental configurations always utilize metal plates to confine the fields in one direction, only TM modes, in which the electric field is parallel to the cylinder axis, satisfy this requirement. For this case, Maxwell's equations reduce to a single scalar wave equation, which can be written in Cartesian coordinates as

$$\frac{\partial^2 E}{\partial x^2} + \frac{\partial^2 E}{\partial y^2} = -\varepsilon(x,y)\frac{\omega^2}{c^2}E, \tag{2.1}$$

where the dielectric function $\varepsilon(x,y)$ is periodic in x and y, and uniform in z. Expanding both the dielectric function and the field in plane waves, we can rewrite Eq. (2.1) as

$$\frac{c^2}{\omega_k^2}u_G^k = \sum_{|G'|<|G_c|}\frac{\left(\delta_{G,G'}+\chi_{G-G'}\right)}{\sqrt{(k+G)\cdot(k+G)}\sqrt{(k+G')\cdot(k+G')}}u_{G'}^k, \tag{2.2}$$

where **k** is a Bloch vector lying within the first Brillouin zone, and G and G' are reciprocal lattice vectors. We have written Eq. (2.2) as a finite rank eigenvalue equation, with size determined by a plane wave cutoff vector $\mathbf{G}_C$. The details of the spatial dielectric function are contained in $\chi(\mathbf{x}) = [\varepsilon(\mathbf{x}) - 1]/4\pi$, from which can be found the Fourier space coefficients

$$\chi_G = \frac{1}{a_c} \int\limits_{unit\,cell} \chi(\mathbf{x}) e^{i\mathbf{G}\cdot\mathbf{x}} dA.$$

(2.3)

a_c is the area of a unit cell, and **x** is a two-dimensional position vector. The integral in Eq. (2.3) can either be done analytically, if the geometry is simple enough, or numerically (e. g., using an FFT) for more complicated structures. Equation (2.2) is generally solved to find the frequencies of the lowest bands of interest; this is known as the Plane Wave (PW) method [1-4]. Convergence is improved as more plane waves are included in the calculation; the standard criteria for convergence is that the band structure in the range of interest remains stable as more and more plane waves are added, although care must be taken to avoid regions of quasi-stability [5]. Finally, the eigenfields can be determined from

$$E_{\mathbf{k}} = \sum_{\mathbf{G}} u_{\mathbf{G}}^{\mathbf{k}} e^{i(\mathbf{k}+\mathbf{G})}.$$

(2.4)

Solving Eq. (2.2), we can find not only the band structure, which provides information as to the position and size of the band gaps, but also the *complex* photonic band structure [6], which can be determined by allowing **k** to have complex components. When **k** has complex components, however, Eq. (2.2) is no longer Hermitian, and the eigenfrequencies will no longer be necessarily real-valued; in this situation, we select all solutions which do have purely real frequencies as composing the complex band structure. We can use the complex band structure for information about the rate of decay of evanescent modes occurring at frequencies corresponding to the band gap frequencies. These modes, forbidden in infinitely periodic structures, are allowed in terminated photonic lattices and lattices containing defects. For many applications, a lattice configuration can be immediately excluded from consideration based on the rate of decay of the fields from a defect. The complex band structure is a quick means of obtaining this information.

2.2. FINITE DIFFERENCE METHOD

The plane wave method, which initially proved reliable for simple periodic systems, is usually found to be insufficient for more intricate structures, such as structures with defects, or terminated lattices, or even structures with metal included. As an alternative to the plane wave method, authors began applying techniques based on real-space methods [7,8]. Instead of evaluating Maxwell's equations in k-space, where the differential operators are treated as exact but the dielectric function is approximate, the real space calculations approximate the differential operators. The precision of the dielectric function is limited only by the precision of the spatial discretization. We developed the finite difference (FD) method as a means of accurately calculating the defect modes in photonic lattices; the quantitative understanding of the frequencies and field patterns of such modes was crucial in the design of the cavities.

Since the FD method has proved so easy to implement, we choose to present the formulation for a *truncated* photonic lattice, a term introduced by the authors of Ref. [9]. The geometry of the truncated two-dimensional PBG structure is the following: an array of cylinders (or like objects which are uniform in one direction) with dielectric tensor ε_{ij}, periodically spaced in the x and y (transverse) directions, bounded on top and bottom by metal plates with infinite conductivity. The separation between the plates is h, the cylinder repeat distance is d, and the radius of a cylinder is r. We consider here a dielectric tensor with diagonal elements only, with one value for the transverse directions and another for the longitudinal direction, corresponding to the situation when the dielectric is a uniaxial crystal (such as pure sapphire); the following procedure, however, can be easily extended to include a general dielectric tensor satisfying $\varepsilon_{ij}=\varepsilon_{ji}$, with real ε_{ij}.

We follow the authors of Ref. [9] and write down the coupled equations for the electric field in the region of the cavity:

$$-\frac{\partial^2 E_x}{\partial y^2} - \frac{\partial^2 E_x}{\partial z^2} + \frac{\partial^2 E_y}{\partial x \partial y} + \frac{\partial^2 E_z}{\partial x \partial z} = \frac{\omega^2}{c^2} \varepsilon^T(x,y) E_x$$

$$-\frac{\partial^2 E_y}{\partial x^2} - \frac{\partial^2 E_y}{\partial z^2} + \frac{\partial^2 E_x}{\partial x \partial y} + \frac{\partial^2 E_z}{\partial y \partial z} = \frac{\omega^2}{c^2} \varepsilon^T(x,y) E_y \qquad (2.5)$$

$$-\frac{\partial^2 E_z}{\partial x^2} - \frac{\partial^2 E_z}{\partial y^2} + \frac{\partial^2 E_y}{\partial y \partial z} + \frac{\partial^2 E_x}{\partial x \partial z} = \frac{\omega^2}{c^2} \varepsilon^L(x,y) E_z$$

to which we add the boundary conditions

$$E_x(x+d,y+d,z) = e^{ik_x d} e^{ik_y d} E_x(x,y,z)$$

$$E_y(x+d,y+d,z) = e^{ik_x d} e^{ik_y d} E_y(x,y,z) \qquad (2.6)$$

$$E_z(x+d,y+d,z) = e^{ik_x d} e^{ik_y d} E_z(x,y,z)$$

and

$$E_x(x,y,0) = E_x(x,y,h) = 0$$

$$E_y(x,y,0) = E_y(x,y,h) = 0 \qquad (2.7)$$

The boundary condition Eq. (2.6) is a statement of Bloch's theorem, while the boundary condition in Eq. (2.7) is the requirement that the tangential fields vanish on the conducting surfaces. The above boundary conditions enable us to write explicit forms for the longitudinal dependence of the fields as

$$E_x(x,y,z) = \tilde{E}_x(x,y)\sin qz$$
$$E_y(x,y,z) = \tilde{E}_y(x,y)\sin qz, \qquad (2.8)$$
$$E_z(x,y,z) = \tilde{E}_z(x,y)\cos qz$$

where $q_n = n\pi/h$, and $n=0,1,2,....$ As can be seen from Eq. (2.8), modes with $n=0$ are TM, and have no variation along the z-axis. Modes with $n>0$ have variation along the z-axis, and we will refer to such modes as higher order modes. Substituting Eq. (2.8) into Eq. (2.5) yields

$$-\frac{\partial^2 \tilde{E}_x}{\partial y^2} + q^2 \tilde{E}_x + \frac{\partial^2 \tilde{E}_y}{\partial x \partial y} - q\frac{\partial \tilde{E}_z}{\partial x} = \frac{\omega^2}{c^2}\varepsilon^T(x,y)\tilde{E}_x$$

$$-\frac{\partial^2 \tilde{E}_y}{\partial x^2} + q^2 \tilde{E}_y + \frac{\partial^2 \tilde{E}_x}{\partial x \partial y} - q\frac{\partial \tilde{E}_z}{\partial y} = \frac{\omega^2}{c^2}\varepsilon^T(x,y)\tilde{E}_y. \qquad (2.9)$$

$$-\frac{\partial^2 \tilde{E}_z}{\partial x^2} - \frac{\partial^2 \tilde{E}_z}{\partial y^2} + q\frac{\partial \tilde{E}_y}{\partial y} + q\frac{\partial \tilde{E}_x}{\partial x} = \frac{\omega^2}{c^2}\varepsilon^L(x,y)\tilde{E}_z$$

With the substitution

$$\tilde{\mathbf{E}} = \left[\frac{1}{\sqrt{\varepsilon}}\right]\mathbf{f}, \qquad (2.10)$$

Eq. (2.9) becomes a symmetric matrix eigenvalue equation relating the three field components.

We now must approximate the derivatives in Eq. (2.9) as finite differences; but this process is complicated by the appearance of the first order derivatives which can be approximated by different schemes, some of which will not produce meaningful results. We follow here the procedure outlined by Pendry [10], in which Eq. (2.5) is Fourier transformed, the dispersion relations approximated (for example, ik is replaced by $(\exp(ik\ell)-1)/\ell$, where ℓ is the discretization length, and we assume a square discretization grid with equal spacing in both directions), and the resulting equations transformed back to real space yielding a set of correct real space finite difference formulas. Applying these finite difference formulas to Eq. (2.9) with the substitution of Eq. (2.10) yields

$$\left[(2+q^2\ell^2)\frac{f^x_{i,j}}{\varepsilon^T_{i,j}} - \frac{f^x_{i,j-1}}{\sqrt{\varepsilon^T_{i,j}\varepsilon^T_{i,j-1}}} - \frac{f^x_{i,j+1}}{\sqrt{\varepsilon^T_{i,j}\varepsilon^T_{i,j+1}}}\right]$$

$$-\left[\frac{f^y_{i+1,j-1}}{\sqrt{\varepsilon^T_{i,j}\varepsilon^T_{i+1,j-1}}} - \frac{f^y_{i,j-1}}{\sqrt{\varepsilon^T_{i,j}\varepsilon^T_{i,j-1}}} - \frac{f^y_{i+1,j}}{\sqrt{\varepsilon^T_{i,j}\varepsilon^T_{i+1,j}}} + \frac{f^y_{i,j}}{\varepsilon^T_{i,j}}\right]$$

$$-q\ell\left[\frac{f^z_{i+1,j}}{\sqrt{\varepsilon^T_{i,j}\varepsilon^L_{i+1,j}}} - \frac{f^z_{i,j}}{\sqrt{\varepsilon^T_{i,j}\varepsilon^L_{i,j}}}\right] = \frac{\omega^2\ell^2}{c^2}f^x_{i,j}$$

$$\left[(2+q^2\ell^2)\frac{f^y_{i,j}}{\varepsilon^T_{i,j}} - \frac{f^y_{i-1,j}}{\sqrt{\varepsilon^T_{i,j}\varepsilon^T_{i-1,j}}} - \frac{f^y_{i+1,j}}{\sqrt{\varepsilon^T_{i,j}\varepsilon^T_{i+1,j}}}\right]$$

$$-\left[\frac{f^x_{i-1,j+1}}{\sqrt{\varepsilon^T_{i,j}\varepsilon^T_{i-1,j+1}}} - \frac{f^x_{i-1,j}}{\sqrt{\varepsilon^T_{i,j}\varepsilon^T_{i-1,j}}} - \frac{f^x_{i,j+1}}{\sqrt{\varepsilon^T_{i,j}\varepsilon^T_{i,j+1}}} + \frac{f^x_{i,j}}{\varepsilon^T_{i,j}}\right] \quad (2.11)$$

$$-q\ell\left[\frac{f^z_{i,j+1}}{\sqrt{\varepsilon^T_{i,j}\varepsilon^L_{i,j+1}}} - \frac{f^z_{i,j}}{\sqrt{\varepsilon^T_{i,j}\varepsilon^L_{i,j}}}\right] = \frac{\omega^2\ell^2}{c^2}f^y_{i,j}$$

$$\left[4\frac{f^z_{i,j}}{\varepsilon^L_{i,j}} - \frac{f^z_{i-1,j}}{\sqrt{\varepsilon^L_{i,j}\varepsilon^L_{i-1,j}}} - \frac{f^z_{i+1,j}}{\sqrt{\varepsilon^L_{i,j}\varepsilon^L_{i+1,j}}} - \frac{f^z_{i,j-1}}{\sqrt{\varepsilon^L_{i,j}\varepsilon^L_{i,j-1}}} - \frac{f^z_{i,j+1}}{\sqrt{\varepsilon^L_{i,j}\varepsilon^L_{i,j+1}}}\right]$$

$$-q\ell\left[\frac{f^x_{i-1,j}}{\sqrt{\varepsilon^L_{i,j}\varepsilon^T_{i-1,j}}} - \frac{f^x_{i,j}}{\sqrt{\varepsilon^L_{i,j}\varepsilon^T_{i,j}}}\right]$$

$$-q\ell\left[\frac{f^y_{i,j-1}}{\sqrt{\varepsilon^L_{i,j}\varepsilon^T_{i,j-1}}} - \frac{f^y_{i,j}}{\sqrt{\varepsilon^L_{i,j}\varepsilon^T_{i,j}}}\right] = \frac{\omega^2\ell^2}{c^2}f^z_{i,j}$$

Equations (2.11) together with the boundary conditions in Eqs. (2.6) and (2.7) form an infinite Hermitian eigenvalue equation for the frequencies. We solve numerically the finite matrix formed by limiting the number of lattice points to N per direction, or N^2 per cell. We can then write Eq. (2.11) as

$$\mathbf{M}_{\alpha,\beta,m,n}f^\beta_n = \frac{\omega^2\ell^2}{c^2}f^\alpha_m, \quad (2.12)$$

where the Greek symbols refer to the polarization, and m,n index the spatial coordinate (m=(j-1)*N+i, for example). Thus, the final matrix **M** to solve has rank $3N^2$, if a higher order mode is being computed, or has rank N^2, if an n=0 mode is being computed. For typical applications, the rank of M is usually on the order of ten or twenty thousand elements, and would be nearly impossible to solve with typical eigenvalue/eigenvector packages. However, from the form of Eq. (2.11), we note the matrix is actually sparse, and thus specialized techniques can be applied to find the eigenvalues and eigenvectors of the system. The specific algorithm we used for the subsequent calculations is publically available, and is called "Lanczos Algorithm for Large Symmetric Hermitian Eigenvalue Computations" [11]. The procedure is based on an iterative process, and stores only a number of elements equal to the rank of the matrix. The user supplies the range where the program searches for eigenvalues, the maximum number of iterations, the desired accuracy, and a subroutine which efficiently computes the matrix-vector product. The numerical procedure proved very efficient; even the most complicated structures we studied could be simulated on a standard 486 personal computer in about an hour.

To improve the convergence for dielectric structures, the dielectric tensor is averaged according to the formula

$$\varepsilon_{m,ij} = \bar{\varepsilon}_m n_i n_j + \tilde{\varepsilon}_m e_{nli} e_{kjl} n_k n_n, \tag{2.13}$$

where **n** is the unit normal to the dielectric interface, and ε_{ijk} is the Levi-Civita pseudotensor. The values $\bar{\varepsilon}$ and $\tilde{\varepsilon}$ at the grid point indexed by m are determined by averaging the inverse of the dielectric constant and the dielectric constant, respectively, over the cell which the point represents. This averaging was introduced by the authors of Ref. [12], who found that averaging the dielectric constant improved convergence for TE modes, but worsened convergence for TM modes, while averaging the inverse of the dielectric had the opposite effect. We approximated these averages by

$$\bar{\varepsilon} = \sum_{n.n.} \varepsilon$$

$$\frac{1}{\tilde{\varepsilon}} = \sum_{n.n.} \frac{1}{\varepsilon} \tag{2.14}$$

where the sums were taken over the nearest neighbor grid points. We found this filtering of the dielectric function adequate, and we did not pursue more sophisticated averaging schemes.

In computing the modes for metal PBG structures, the tangential electric field was assumed to vanish at the surface and in the interior of a metal scatterer. Our approach assumes perfect conductors, and cannot take into account skin depths, losses, or other effect associated with real metals. For truncated metal PBG structures, we have that the divergence of the electric field vanishes everywhere within the structure. In this case, Eq. (2.11) decouples into an equation for E_z, and two coupled equations for E_x and E_y; thus, there can only be purely TM or TE modes for the metal PBG cavity. The n>0 mode frequencies can be found from

$$\frac{\omega_n^2}{c^2} = \frac{\omega_0^2}{c^2} + \frac{n^2 \pi^2}{h^2}, \qquad (2.15)$$

where ω_0 is the frequency of a given n=0 mode. The transverse components of fields corresponding to n>0 modes are thus trivial repetitions of the transverse components of n=0 modes, modulated by a sin(qz) term which describes the variation in fields as a function of the longitudinal coordinate.

2.3. TRANSMISSION CALCULATIONS

While the knowledge of the eigenfrequencies and eigenfields of a PBG system are essential for design purposes, it is also useful to have a quantity which can be calculated and measured directly. Recently, the problem of computing the transmission coefficient has been successfully formulated and demonstrated to be reliable [7,8]. The method, known as the transfer matrix method, has the additional advantage that it can be used for systems with finite conductivity, losses, or frequency dependent dielectric functions.

We have developed a version of the transfer matrix method, applicable to our 2-D, n=0 systems. Starting with the discretized scalar wave equation for the electric field, we write

$$E_{i+1,j} = \left[4 - \frac{\omega^2 \ell^2}{c^2} \varepsilon_{i,j} \right] E_{i,j} - E_{i,j+1} - E_{i,j-1} - E_{i-1,j}, \qquad (2.16)$$

where i indexes points in the direction of propagation, while j indexes points perpendicular to the direction of propagation. Thus the section of lattice can be viewed as being broken up into strips, with Eq. (2.16) relating the i+1 strip to the i and i-1 strips. Eq. (2.16) can be written formally as

$$F_{i+1} = \begin{pmatrix} M & -I \\ I & 0 \end{pmatrix} F_i = TF_i, \qquad (2.17)$$

where

$$M_{j,j'} = \left[4 - \frac{\omega^2 \ell^2}{c^2} \varepsilon_{i,j} \right] \delta_{j,j'} - \delta_{j+1,j'} - \delta_{j-1,j'} \qquad (2.18)$$

and

$$F_i = \begin{pmatrix} E_i \\ E_{i-1} \end{pmatrix}. \qquad (2.189)$$

We then follow exactly the same procedure as in Pendry [10], to determine the transmittance for a section of lattice with a finite width along the propagation direction. Since microwave absorber plays a critical role in all of our structures, the transfer matrix method may provide us with the means to study some of the effects due to the interaction of the PBG structure with the absorber.

3. Results of Calculations

3.1. BAND STRUCTURE CALCULATIONS

In our search for a suitable photonic lattice to form the basis of the accelerator cavity, we systematically computed the photonic band structures for a wide range of material and lattice parameters. Eventually, however, our dielectric constant (for dielectric structures)

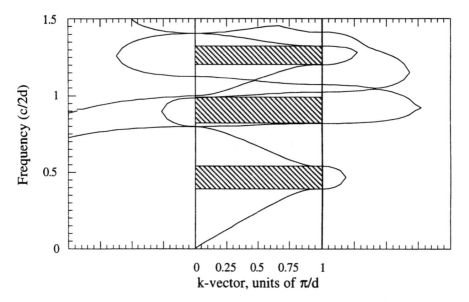

Figure 2. The complex photonic band structure for the (10) surface of the square lattice. The cylinders have dielectric constant $\varepsilon=9$, radii $r/d=0.375$. The central portion of the plot corresponds to k taking values from the Γ point ($k_r=0$) to the X point ($k_r=\pi/d$); the transverse component of **k** is zero. The left hand portion of the graph shows the real frequency lines as a function of k_i, with $k_r=0$, while the right hand portion of the graph shows the real frequency lines with $k_r=\pi/d$. The horizontal scale for the regions where k is complex is the same as the labled center region.

is constrained by the requirement that the material also have very low losses; one of the few materials we can take advantage of is sapphire, which has a dielectric constant ~9. A representative band structure of a square photonic lattice is shown in Fig. 2. The parameters for this lattice were r/d=0.375 and ε=9. We have shown the bands corresponding to the [10] direction of propagation in the center of the figure, while on either side we show the bands corresponding to modes which are evanescent along the [10] direction. Note that for frequencies in the band gaps, only modes which are evanescent can occur.

In addition to knowing the gap structure of the n=0 modes, it is also of importance to have some idea of the gap structure of the higher order modes, those modes associated with the truncated structure. In Fig. 3, we present the n=1 band structure for the dielectric square lattice with the same parameters as used in Fig. 1, with the added parameter of h/d=0.8. Typical of higher n band structures, there is a cutoff frequency at which point n=1 pass bands begin. In general, the larger the n value, the higher the cutoff frequency for the propagating modes; so for the frequency range of our experimental system, only the n=0 and n=1 modes were of interest. As can be seen from the Fig. 3, an absolute band gap does exist for this set of parameters.

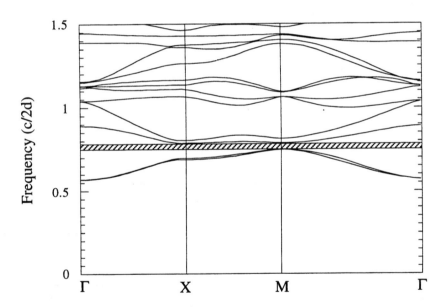

Figure 3. The n=1 hybrid band structure for a truncated square lattice with dielectric cylinders. The cylinders have dielectric constant ε=9 for both polarizations, and radii r/d=0.375. The plate separation h/d=0.8. These parameters correspond to those used extensively in our experiments. An absolute frequency band gap is highlighted in the figure by shading. Analogous somewhat to the metal PBG lattice, the frequency region from zero frequency to the cutoff frequency (at ~0.6) can also be considered a region of band gap.

In Fig. 4, we present the n=0 TM band structure corresponding to a lattice of metal cylinders, with r/d=0.187. This calculation was computed using the finite difference method described above, but has since been verified by both plane wave [13] and transfer matrix [14] calculations. The band structure exhibits a single absolute PBG, which starts at zero frequency and extends up to a cutoff frequency, where the pass bands begin. This is the only absolute PBG gap in the structure. A separate calculation of the complex band structure (not shown) demonstrates that the rate of energy attenuation at near the center of the gap is over three orders of magnitude per lattice constant.

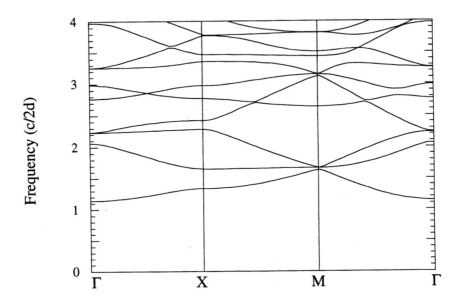

Figure 4. The photonic band structure for a lattice of metal cylinders, r/d=0.187. The calculation was done using the finite difference method.

3.2. DEFECT CALCULATIONS

To initially test the accuracy of the finite difference method, we simulated the n=0 defect modes of a square dielectric PBG lattice, with the same parameters as those used in the band structure calculation of Fig. 2. These parameters also coincide with those of lattices which were studied extensively in Ref. [15], experimentally. The simulated supercell consisted of a 4 x 4 array of cylinders in which the radius r of the central cylinder was varied. The supercell was discretized with a 113 x 113 point grid, for an effective matrix rank of 12,769. As the radius of the defect cylinder is changed from r/a=1 to r/a=0, defect modes move across the gap, as shown in Fig. 5. Experimental measurements were only available for defect modes in the second gap, so the figure highlights only the second gap data. Over the range of defects applied, the gap was found via the simulation to have two types of defect modes, one monopole, the other with dipole symmetry. The predicted defect frequencies were found to agree well with the measured points, also shown on the figure, except toward the band edges where such supercell calculations are expected to fail. The predicted symmetries were also confirmed by phase measurements.

In addition to finding bound defect modes in the n=0 band structure, we also found a defect mode in the small PBG in the n=1 band structure. Fig. 6 shows the frequency of the mode as a function of r/d of the defect cylinder. While it was beyond the scope of our program to fully investigate the types of higher order defect modes which occur in the truncated dielectric structures, it is anticipated that knowledge of these modes will be important at the time we design a dielectric PBG cavity, since they represent a potential source of degradation for accelerating beams.

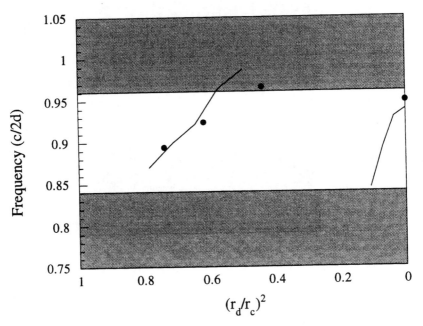

Figure 5. Calculation of the evolution of defect frequencies in the second gap of the square lattice with cylindrical scatterers. The dielectric constant of the cylinders is ε=9, and the radii of the cylinders is r/d=0.375. A supercell consisting of 4 x 4 unit cells was discretized with a 113 x 113 point grid. The radius of one of the cylinders was varied from r/d=0 to r/d=1 in steps of constant area. Two modes are found to move across the gap as the radius of the defect cylinder is decreased; the first which occurs is a doubly degenerate dipole mode, while the second is a monopole mode. The solid circles were experimentally measured points. The agreement between the calculation and the measurements is quite good, except for the point near the upper band edge where the supercell calculation is expected to yield poor results.

The range of defects is somewhat more restricted in the case of the metal photonic lattice. Since the calculation assumes perfectly conducting cylinders, we cannot alter the material properties to create defects; furthermore, any defect configuration must allow for

404

a beam to pass through the central region. Therefore, we performed a calculation of a 5 x 5 array of metal cylinders with the central cylinder completely removed. The parameters were the same as in Fig. 4, but the supercell was discretized by a 161 x 161 point grid for a matrix rank of 25,921. We found that a defect mode, with monopole symmetry, did indeed move into the gap; field plots (Fig. 7) confirmed that the attenuation rate was greater than 20 dB per lattice constant.

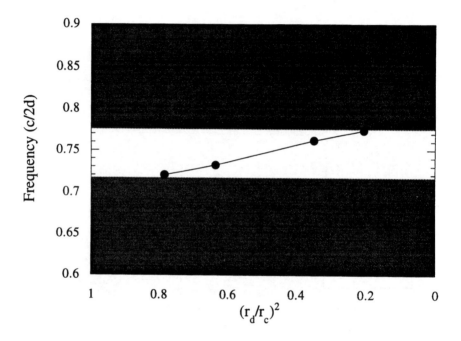

Figure 6. The defect frequency as a function of defect radius. A 5 x 5 supercell of lattice with the same parameters as in Fig. 3 was used for the calculation. The calculations were performed for a number of different radii of the defect cylinder; solid circles indicate the defect frequencies calculated. The mode shown is doubly degenerate, indicating it has dipole symmetry with respect to rotations in the plane of the lattice. Note: This defect occurs in the n=1 band structure.

4. Q of PBG Cavities

The cavity, as we have proposed it, is an open structure. There are thus three possible components to the unloaded Q_0 of the cavity. These are: $Q_{leakage}$, set by leakage out of the structure; Q_{walls}, set by the resistive losses in the metal or superconductor; and $Q_{dielectric}$, if there is dielectric present. These factors are related by the equation

$$\frac{1}{Q_0} = \frac{1}{Q_{leakage}} + \frac{1}{Q_{walls}} + \frac{1}{Q_{dielectric}}. \tag{4.1}$$

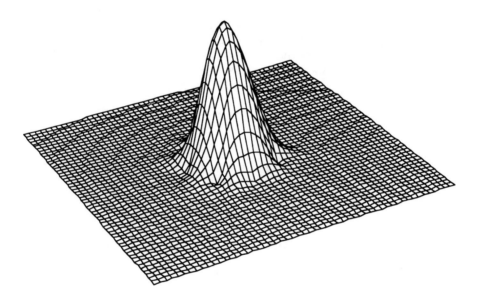

Figure 7. Electric energy density of a defect mode in a square PBG lattice with metal cylinders. The cylinders had radii of $r/d=0.187$. The supercell used for the calculation consisted of a 5 x 5 array of unit cells, discretized by a 161 x 161 point grid. The defect consisted of a cylinder being completely removed from the center unit cell.

The lateral size of the cavity will be dictated by the smallest tolerable $Q_{leakage}$. This quantity can be estimated by using the generated field vs. position data from the defect simulations described above. For the n=0 TM defect modes, the stored energy is

$$U = \frac{1}{8\pi} \int_{\substack{volume \\ of\ cavity}} E_z^2(\mathbf{x})dV, \tag{4.2}$$

while the power lost out the sides is

$$P = \int_{\substack{surface \\ of\ cavity}} \mathbf{S} \cdot d\mathbf{A}, \tag{4.3}$$

where $\mathbf{S}$ is the Poynting vector. The periodic boundary condition which was applied in the simulations is equivalent to a magnetic mirror surrounding the sides of the structure; that is, the derivative of the electric field goes to zero everywhere around the boundary. In this situation, the defect mode has standing wave character, and the time-averaged Poynting vector vanishes everywhere. Nevertheless, a reasonable estimate of the power lost in a lattice terminated by absorber can be obtained by calculating the surface integral of

$$|\mathbf{S}| = \frac{c}{4\pi}|E_z|\sqrt{H_x^2 + H_y^2} \tag{4.4}$$

around the perimeter of the cell, using half the magnitude of the calculated fields. For the defect mode of the metal PBG lattice shown in Fig. 7, $Q_{leakage}$ was calculated to be $\sim 10^6$ for the 5 x 5 cell.

The power dissipated in the metal walls of the structure is related to the surface magnetic fields by

$$P = R_s \left(\frac{c}{4\pi}\right)^2 \int_{\substack{metal \\ surfaces}} H^2 dS, \tag{4.5}$$

where R_s is the surface impedance, and is roughly proportional to $1/\sigma\delta$ where σ is the conductivity of the metal and δ is the classical skin depth. In the dielectric PBG cavity, in which the only metal surfaces are the bounding flat plates, we have for n=0 modes,

$$Q_{walls} = \frac{2\omega_0}{\pi R_s c^2} h \tag{4.6}$$

and

$$Q_{dielectric} = \frac{1}{f \tan\delta}, \tag{4.7}$$

where f is the fraction of the electric energy density within the dielectric region, and $\tan(\delta)$ is the loss tangent of the dielectric. For all metal structures, $Q_{dielectric}$ will not contribute to the total Q_0, and Q_{walls} will be more complicated due to the losses on the scatterer surfaces.

To measure the Q of a cavity, power must be coupled into the structure by some means, and this coupling defines a Q_{ext}. If the coupling circuit is assumed to be lossless, and if the measurement is made "at match" (when $Q_0 = Q_{ext}$), then the measured Q will be one-half of the unloaded Q (Q_0) of the cavity. In practice, there are always losses in the coupling circuit, and it is generally more convenient to make the measurement somewhat undercoupled (when $Q_0 < Q_{ext}$); for our measurements, we followed the procedure for measuring Q outlined in Ref. [16].

5. Experimental Results on Metal Test Cavities

Several experimental metal PBG cavities were designed and built to investigate some of the technical issues. The first cavity built was designed to have the frequency of the defect mode occurring at 9.0 GHz. The cavity consisted of a 7 x 7 array of copper tubes with 3/8" diameter brazed into two copper plates space 0.4" apart. The tubes were spaced a distance of 0.5" apart, with the center tube removed to form the defect. Measurements of reflected power vs. frequency were made using an HP8756A microwave network analyzer, from which the resonance frequency of the defect mode could be found. The mode was excited by an antenna into the unit through a small hole drilled in one of the plates, centered above the defect. The frequency of the defect mode was predicted by the supercell simulation to occur at 8.96 GHz; the measured frequency was found to be 8.94 GHz.

Two identical metal PBG cavities were subsequently constructed, scaled to have the defect mode occur at 3.3 GHz. This fundamental frequency was chosen to be low enough that a frequency region of at least four times the fundamental frequency could be probed easily with our equipment. The 3.3 GHz test unit consisted of a 5 x 5 lattice of copper tubes brazed into copper plates, with parameters d=1.33", r=0.25", and h=0.8". Microwave absorber was placed around the periphery of the unit to minimize reflection into the structure. Measurements of reflected power over the frequency range 3-14 GHz were made using the microwave network analyzer; microwaves were coupled to the cavity again by means of an antenna which entered via a coaxial line through one of several small coupling holes in one plate of the cavity. The coupling could be varied by moving the coaxial line slightly further into or out of the the cavity. Coupling holes just off the center of the defect were also made so that nodal modes (e. g., dipole, quadrupole) could be probed.

In Fig. 8 the resonance curve of the 3.3 GHz copper test cavity is shown. The unloaded Q was measured to be ~4,000 from the data. In Fig. 9 an experimental sweep of reflected power vs. frequency is displayed over a frequency span which is three times the fundamental frequency. It is immediately clear from this trace there are numerous higher order resonances. The Q values for these resonances were found to be much lower (~100-200) than the fundamental. Since the calculated Q due to leakage out the sides of the defect mode was determined to be on the order of a million, the experimentally observed Q is most likely dominated by the losses of room temperature copper. The Q of the resonances, however, is most likely dominated by leakage to the surrounding absorber. It should be noted that the fundamental is not observed in this trace because the coupling was very strong to identify the low Q resonances.

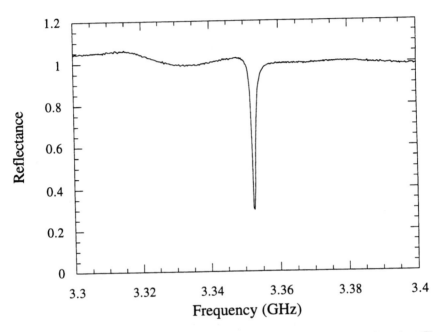

Figure 8. Reflectance vs. frequency for the defect mode of the 3.3 GHz metal PBG cavity. The unloaded Q of the cavity was measured to be ~4,000.

We have labeled the observed resonances in the spectrum of Fig. 9 by an r^n_j notation, where j indicates (arbitrarily) the number of the resonance and n is the mode index, as used above. For the test unit measured, the height of h=0.8" leads to a cutoff frequency of fc=7.4 GHz, above which modes with n=1 longitudinal dependence can exist. The cutoff frequency for n=2 modes is 14.8 GHz, so we do not expect to measure such modes in the frequency range selected. For accelerator applications, the low Q higher order resonances are undesirable. In principle, however, all of these resonances lie in the pass bands, and it should be possible to change their frequencies and Qs by making changes to the defect region, without adversely affecting the desired properties of the defect mode.

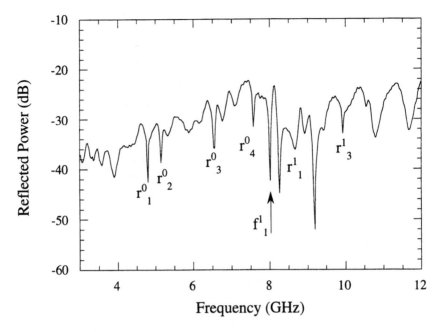

Figure 9. Reflected power vs. frequency for the 3.3 GHz metal PBG cavity. The fundamental mode, labeled as f^0_1, does not appear on the trace due to the degree of coupling which was needed for the measurement. The mode f^1_1, set off by an arrow below the resonance curve, corresponds to the n=1 defect mode. The other n=0 and n=1 low-Q resonances have frequencies placing them in the pass bands of the band structure.

6. Conclusion and Future Work

If the metal PBG cavity is to be used in accelerator applications, the low Q higher order resonances are undesirable. It is troubling, in fact, that the resonances lie in the upper pass bands, where the simulations suggest only propagating modes should occur; this implies the same trouble may also affect dielectric PBG cavities. We have begun to study the origin of these resonances, both experimentally and numerically. We have found, for example, that it is possible to make changes in the region of the defect which will significantly alter the Q's and frequencies of the resonances, while only slightly perturbing the fundamental defect mode. Furthermore, we have found there can be a very large reflection from the absorber/lattice interface, which may account for some of the resonances we see; by making changes to the shape of the absorber (e.g., by cutting "teeth" into the absorber layer) we can substantially reduce this reflection. We have confirmed these observations also by transmission calculations. These initial experiments provide us with the hope that we may be able to eliminate, or at least substantially reduce, the higher order resonances.

410

We are currently designing an all-niobium PBG cavity suitable for testing at liquid helium temperatures. The design frequency is 11.424 GHz, a standard operating frequency at several high energy accelerator test sites. Such testing will help us assess the bottom line practicality of using PBG based structures as accelerator cavities. From low temperature, high power testing, we can expect to learn the ultimate Q limitations and maximum accelerating field. Further studies in which a beam will be introduced into the accelerating cavity are also expected to provide important data and insight.

7. Acknowledgments

This research has been supported by the DOE, contracts DE-FG-03-93ER40793 and DE-AC-03-76SF00515. AccSys Technologies fabricated the test cavities used in this study. We thank M. Sigalas and C. M. Soukoulis for helpful discussions.

8. References

1. S. L. McCall, P. M. Platzman, R. Dalichaouch, David Smith , S. Schultz, *Phys. Rev. Lett.* **67**, 2017 (1991).
2. M. Plihal, A. Shambrook, A. A. Maradudin, Sheng Ping, *Opt. Comm.* **80**, 199 (1991).
3. M. Plihal, A. A. Maradudin, *Phys. Rev. B* **45**, 13962 (1992).
4. R. D. Meade, K. D. Brommer, A. M. Rappe, J. D. Joannopoulos, *App. Phys. Lett.* **61**, 495 (1992).
5. H. S. Sozuer, J. W. Haus, R. Inguva, *Phys. Rev. B* **45**, 13962 (1992).
6. S. Schultz, D. R. Smith, N. Kroll, *Proc. 1993 Particle Accelerator Conference*, Washington D. C., USA **4**, 2559 (1994).
7. J. B. Pendry, A. MacKinnon, *Phys. Rev. Lett.* **69**, 2772 (1992).
8. M. Sigalas, C. M. Soukoulis, E. N. Economou, C. T. Chan, K. M. Ho, *Phys Rev. B* **48**, 14121 (1993).
9. A. A. Maradudin, A. R. McGurn, *J. Opt. Soc. Am B* **10**, 314 (1993).
10. J. B. Pendry, *J. Mod. Opt.* **41**, 209 (1994).
11. J. K. Cullum, R. A. Willoughby, *Lanczos Algorithms for Large Symmetric Eigenvalue Computations, Vol. II Programs*, Birkhauser, Boston (1985).
12. R. D. Meade, A. M. Rappe, K. D. Brommer, J. D. Joannopoulos, *Phys. Rev. B* **48**, 8434 (1993).
13. T. Suzuki, private communication (1994).
14. D. R. Smith, S. Schultz, N. Kroll, M. Sigalas, K. M. Ho, C. M. Soukoulis, *Appl. Phys. Lett.* **65**, 645 (1994).
15. D. R. Smith, S. Schultz, S. L. McCall, P. M. Platzman, *J. Mod. Opt.* **41**, 395 (1994).
16. E. D. Reed, "Measurement of Q by reflected power", *Bell Telephone Technical Publication* (1953).

MICROCAVITIES IN CHANNEL WAVEGUIDES

PIERRE R. VILLENEUVE, SHANHUI FAN AND
J. D. JOANNOPOULOS
Department of Physics,
Massachusetts Institute of Technology,
Cambridge, MA 02139, U.S.A.

AND

KUO-YI LIM, JERRY C. CHEN, G. S. PETRICH,
L. A. KOLODZIEJSKI AND RAFAEL REIF
Department of Electrical Engineering and Computer Science,
Massachusetts Institute of Technology,
Cambridge, MA 02139, U.S.A.

Abstract. We introduce and analyse a new type of resonant microcavity consisting of a channel waveguide and a one-dimensional photonic crystal. A band gap for the guided modes is opened and a state is created within the gap by adding a single defect in the periodic system. An analysis of the eigenstates in the system shows that a strong field confinement of the defect state can be achieved with a modal volume less than half of a cubic half-wavelength. The coupling efficiency to this mode will be shown to exceed 80%. As a proof of concept, we present a feasibility study for the fabrication of these microcavities in an air-bridge configuration with micron-sized features using semiconductor materials.

1. Introduction

Atomic radiative dynamics can be strongly affected by boundary conditions. For instance, electromagnetic cavities with perfectly reflecting walls have the ability of altering the density of allowed states for radiative transitions. By scaling down the dimensions of the cavity to the atomic transition wavelength, the density of states is reduced to a spectrally discrete set of modes. If there are no modes available at the atomic transition frequency, atomic radiative decay can be essentially suppressed.

C. M. Soukoulis (ed.), Photonic Band Gap Materials, 411–426.
© *1996 Kluwer Academic Publishers. Printed in the Netherlands.*

At microwave frequencies, electromagnetic cavities can be fabricated by using highly reflecting metallic walls. However, at optical frequencies, metals become very lossy and one needs to turn to other materials. In recent years, low-loss dielectric materials have been used for the fabrication of optical microcavities. High-Q modes have been observed in one-dimensional Fabry-Perot resonators using distributed Bragg reflectors and in microdisks using total internal reflection [1]. It has also been suggested that "photonic crystals" could be used for the fabrication of optical microcavities, since they have the ability to eliminate radiation losses along every direction in space [2-4].

In this manuscript, we introduce a novel class of coplanar microcavities which use index guiding to confine light along two dimensions and a one-dimensional photonic crystal to confine light along the third. Conceptually, this approach is analogous to that of one-dimensional Fabry-Perot resonators, but it differs in that it has both a coplanar geometry and the ability to give rise to strong field confinement. An analysis of the eigenstates of the microcavity will show that the modal volume of the cavity state is less than half of a cubic half-wavelength. As a proof of concept, we will also demonstrate that these microcavities can be fabricated using conventional lithographic techniques applied to technologically important semiconductor-based materials.

2. Computational Methods

To design a microcavity, we need to answer two questions: what are the modes in the cavity, and how can we couple energy in and out of these modes? To answer these questions, we use two different computational approaches. The first one solves Maxwell's equations in the frequency domain while the second solves Maxwell's equations in the time domain. Each of these methods will teach us something different about the cavity. The first one will allow us to find every eigenstate of the device — the frequency, polarization, symmetry and field distribution of the modes — and the second will allow us to determine its temporal behavior. By looking at the evolution of the fields in time, we will be able to determine the coupling efficiency, the scattering and the quality factor of the microcavity.

2.1. FREQUENCY DOMAIN

In the first method, we expand the fields into a set of harmonic modes and write the wave equation for the magnetic field in the form

$$\nabla \times \left\{ \frac{1}{\epsilon(r)} \nabla \times H(r) \right\} = \frac{\omega^2}{c^2} H(r) \ . \tag{1}$$

This equation is an eigenvalue problem and can be rewritten as

$$\Theta H_n = \lambda_n H_n \tag{2}$$

where Θ is a Hermitian differential operator, and λ_n is the nth eigenvalue, proportional to the squared frequency of the mode. We solve Eq. (2) by using a variational approach where each eigenvalue is computed separately by minimizing the functional $< H_n|\Theta|H_n >$. This method is described in more details in references [4] and [5]. To find the minimum, we use the conjugate gradient method, keeping H_n orthogonal to the lower states. The conjugate gradient method has the advantage of being more effective than the traditional method of steepest descents, in that it requires less iterations to reach convergence. In order to minimize the functional, we need to calculate

$$\Theta H_n(r) = \left\{ \nabla \times \frac{1}{\epsilon(r)} \nabla \times \right\} H_n(r) . \tag{3}$$

We compute Eq. (3) by using several Fast Fourier Transform (FFT) steps. The motivation is the following: since the curl is a diagonal operator in reciprocal space, and $1/\epsilon(r)$ is a diagonal operator in real space, we carry out each operator in the space where it is diagonal by going back and forth between real and reciprocal space. This way, we can diagonalize the operator without storing every element of the $N \times N$ matrix; instead, we need only store the N elements of H_n. In turn, we will be able to consider structures of very large dimensions.

Equation (3) is computed in the following way: first, we take the FFT of $H_n(r)$ to get $H_n(G)$. Then, we compute the curl of $H_n(G)$ and define $C_n(G) = (k+G) \times H_n(G)$. We transform $C_n(G)$ into real space and devide it by the dielectric function. Finally, we define $F_n(r) = C_n(r)/\epsilon(r)$, transform it back into reciprocal space, and compute the last curl operator.

2.2. TIME DOMAIN

The second method solves Maxwell's equations in real space where, this time, the explicit time dependency of the equations is maintained. The equations for the electric and magnetic fields can be written as

$$\frac{\partial}{\partial t} H(r, t) = -\nabla \times E(r, t) \tag{4}$$

$$\epsilon(r) \frac{\partial}{\partial t} E(r, t) = \nabla \times H(r, t) . \tag{5}$$

We discretize these equations on a simple cubic lattice [6] where space-time points are separated by fixed units of time and distance. We approximate the derivatives at each lattice point by a corresponding centered difference, which gives rise to finite-differences equations. By solving these equations, we can determine the temporal response of the microcavity.

3. Microcavity Design

The microcavities are made of high-index channel waveguides in which a strong periodic variation of the refractive index is added along the axial direction. The periodic index is introduced by etching a series of holes vertically through the guide and the microcavity is formed by breaking the perfect translational symmetry of the array of holes. In order to understand the mechanism for field confinement in the microcavity, it is useful to investigate first the effect of the holes on the properties of the guided modes.

3.1. GUIDED MODES

The modes in the waveguide are computed by using the frequency domain approach. In our model, we use the supercell approximation where a waveguide with holes is placed into a large supercell and is repeated periodically in space. We show in Fig. 1 the dispersion relation of a standard channel waveguide with holes, for the modes with even symmetry with respect to the xy-plane. The dielectric waveguide was chosen to be made of GaAs with a refractive index of 3.37 at $1.55 \mu m$ [7]. The dispersion relation was constructed by computing the lowest-frequency eigenstates (solid circles) of the waveguide for different values of the wavevector along the direction of propagation. The waveguide is shown in the inset of Fig. 1. The width and height of the guide are $1.2\,a$ and $0.4\,a$, respectively, where a is the distance from center to center between neighboring holes. The holes have a diameter of $0.6\,a$. These parameters were chosen so as to achieve good field confinement in the microcavity, as we will show in the next section.

We plot the dispersion relation in the first Brillouin zone. States above the light line (shaded area) are not guided by the dielectric, and form a continuum of radiation modes. States below the light line are evanescent along the x and z directions; they are guided in the dielectric. These guided modes undergo multiple scattering by the periodic array of holes and destructive interference opens a gap between the first and second guided-mode bands. This gap[1] plays a vital role in our ability to localize a state in the waveguide, since it allows a mode (or group of modes) to be introduced within the gap by adding a defect in the periodic structure. The size of the

---

[1]The gap does not extend over the entire Brillouin zone. The gap exists only for the guided modes.

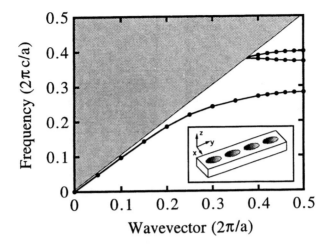

Figure 1. Dispersion relation computed for modes with even symmetry with respect to the xy-plane in the channel waveguide shown in the inset. The distance from center to center between neighboring holes is the lattice constant a. The diameter of the holes is $0.6\,a$ and the width and height of the guide are $1.2\,a$ and $0.4\,a$, respectively. The dielectric constant of the guide is 11.36. Only the lowest three bands are shown.

gap is determined by the dielectric constant of the waveguide and by the size of the holes. In the case shown in Fig. 1, the size of the gap is 27% of the midgap frequency.

To gain some insight into the nature of the gap, we examine the field pattern of the two lowest bands. Figure 2 shows a vector plot of the electric field in the xy-plane passing through the middle of the guide, for a wavevector at the edge of the Brillouin zone. Since the modes are shown at the zone edge, the fields alternate in sign from one period to the next. We see that the field is confined along the waveguide, and decays exponentially away from it, characteristic of guided modes. In Fig. 2(a), the electric field is strongly concentrated in the dielectric regions surrounding each air hole. This contrasts strongly with Fig. 2(b) which shows the electric field to be concentrated inside the holes. Since the frequency of light for a given wavelength is lower in a dielectric material than it is in air, it seems reasonable that the mode of the first band has a lower frequency than that of the second band. This simple observation explains the large splitting between the two bands [4].

It is instructive, at this point, to compare the periodic waveguides with standard channel waveguides. In a standard waveguide (with no holes), modes are guided if $\omega < k_{||}$ and extended if $\omega > k_{||}$. On the other hand, in a periodic waveguide, $k_{||}$ is constrained between $-\pi/a$ and π/a, where a is the lattice constant along the waveguide. Modes with a frequency $\omega > \pi/a$ automatically satisfy the extended-mode condition. Thus, spacial periodicity

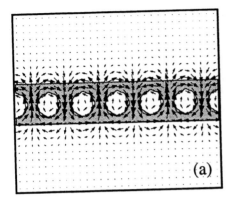 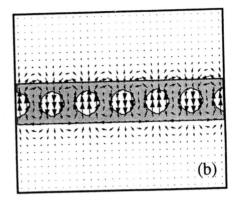

Figure 2. Vector plot of the electric field associated with (a) the first and (b) the second bands of Fig. 1. The field is shown in the xy-plane passing through the middle of the guide, for a wavevector at the edge of the Brillouin zone. The waveguide is shown in gray.

imposes an upper-frequency cutoff for guided modes.

3.2. RESONANT MODES

We now introduce a defect in the waveguide shown in Fig. 1 by adding a local phase shift in the perfect array of holes. If the defect consists of extra dielectric material between two of the holes, then a defect state is "pulled down" from one of the above guided-mode bands. The state appears initially at the top of the gap and is pulled deeply into the gap by increasing the size of the defect. Since this defect state can be expanded primarily in terms of the guided modes, its projection onto the continuum is very small. This leads to a resonant mode which is highly confined within the vicinity of the defect. It can only couple its energy to the waveguide mode through the evanescent fields across the array of holes.

We show in Fig. 3 the resonant mode of a cavity made up of six holes, with three holes on either side of the defect. The frequency of this mode is $f = 0.31\,c/a$ where c is the speed of light in vacuum. As expected, the frequency lies within the gap shown in Fig. 1. The holes act like Bragg mirrors. By increasing the number of holes, the reflectivity of the mirrors is increased, along with the field confinement in the cavity. However, as we increase the number of holes, we also reduce our ability to couple light in and out of the cavity. We will discuss this issue further in Section 4. A good compromise can be reached by reducing the number of holes on only one side of the cavity.

By changing the different parameters of the microcavity, we can adjust the number of resonant modes in the gap, the frequency of these modes

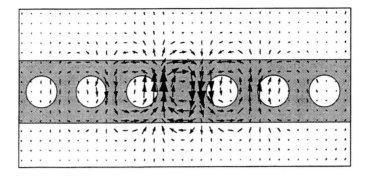

Figure 3. Vector plot of the electric field in the xy-plane passing through the center of the guide. The dielectric structure is shown in gray.

and their confinement in the cavity. The different parameters are the width and height of the waveguide, the radius of the holes, the distance between neighboring holes, and the size of the defect. Since the microcavity can be scaled to any wavelength simply by scaling every parameter, it is convenient to choose one parameter and scale the other ones with respect to it. In this manuscript, we have chosen to scale every parameter with respect to a. In Fig. 3, the defect had dimensions $d = 1.5\,a$, where d is the distance from center to center between the holes on either side of the defect. This had the effect of introducing a single resonant state in the gap. A defect size of $1.5\,a$ corresponds to a quarterwave phase shift and yields the strongest field confinement of the resonator mode. Unlike Fabry-Perot resonators, the quarterwave phase shift in the microcavity does not introduce a mode in the middle of the gap. Instead, the mode appears about a third of the way up from the bottom of the gap. The defect state has the same symmetry as the modes of the second band. The electric field is polarized mostly in the xy-plane while the magnetic field (not shown) is mostly polarized along the z-direction. The electric field has a nodal point at the center of the cavity. The fields decay rapidly; the modal volume is smaller than half of a cubic half-wavelength.

The mode shown in Fig. 3 is the lowest-order resonator mode. Since the introduction of the defect may cause several states to be "pulled down" from the upper bands, more than one state could potentially appear in the gap. In the structure presented above, the first higher-order defect state appears at a frequency of $0.37\,c/a$ which is aligned with the second band outside the gap. Although they are aligned, the defect state and the guided modes of the second band do not couple with each other, since they do not have the same symmetry. Indeed, the first higher-order mode has the symmetry of the third band. The mode remains well confined around the

418

defect. The large separation between the fundamental mode and the higher-order modes $(\Delta f / f_\circ > 19\%)$ will allow these microcavities to be operated with a single-mode output.

Instead of creating a defect by moving the holes apart, it would also have been possible to move them closer together. This would have had the effect of "pushing" a mode up from the lower band into the gap. A computational analysis of this structure has shown that strong field confinement and single-mode operation could also have been achieved using this approach [8].

4. Dynamic Analysis

We now turn our attention to the dynamic behavior of the modes in the microcavity. We want to determine the coupling efficiency between the guided mode in a standard channel waveguide and the resonant mode in the cavity. We also want to determine whether the holes will cause severe scattering of the guided modes, and also compute the quality factor of the cavity. The dynamic analysis is carried out with the time-domain method presented in Section 2.2.

4.1. COUPLING EFFICIENCY

To couple light from the waveguide into the cavity, the waveguide mode must have a component of the same symmetry as that of the defect state. Since TE modes[2] have the same symmetry with respect to the xy-plane, they are more likely to couple energy efficiently into the cavity mode.

We consider a long single-mode channel waveguide, and introduce a series of six holes halfway down the guide, with a single defect of size $d = 1.5\,a$ located at the center of the array. We send a pulse down the waveguide and study its behavior as it propagates through the holes. The pulse has a Gaussian frequency profile centered at $f = 0.30\,c/a$ with a waist of $0.10\,c/a$. The electric field is polarized along the x-axis. We compute the transmission through the holes and normalize it with respect to the incident amplitude. Results are shown in Fig. 4.

Figure 4 shows a wide gap in transmission. The gap extends from $0.28\,c/a$ to $0.37\,c/a$, in agreement with results shown in Fig. 1. The modes inside the gap are strongly attenuated and cannot propagate through the holes. They are reflected back. On the other hand, the modes outside the gap can be transmitted efficiently; some frequencies have a transmission coefficient close to unity. This suggests that the modes remain guided as

[2]TE modes are defined in a slab waveguide as the modes for which the electric field is polarized parallel to the slab and perpendicular to the direction of propagation. In the waveguide presented above (width to height aspect ratio of 3:1), the modes are not purely TE or purely TM, but rather TE-like and TM-like.

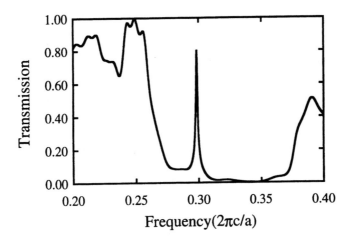

Figure 4. Normalized transmission through the cavity as a function of frequency.

they propagate through the holes, and undergo very little scattering or reflection. Figure 4 also shows the presence of a sharp resonance inside the gap. The coupling efficiency from the waveguide mode to the resonance exceeds 80%.

4.2. QUALITY FACTOR

The quality factor Q is a measure of the losses in the resonator. To compute Q, we choose to use an approach which first involves pumping energy into the cavity and then watching it decay. We recall that the quality factor is defined as [9]

$$Q = \frac{\omega_o E}{P} = -\frac{\omega_o E}{dE/dt} \,, \tag{6}$$

where E is the stored energy, ω_o is the resonant frequency, and $P = -dE/dt$ is the dissipated power. A resonator can therefore sustain Q oscillations before its energy decays by a factor of $e^{-2\pi}$ (or approximately 0.2%) of its original value. If we excite the resonant mode, we can monitor the total energy as a function of time, and compute the number of optical cyles required for the energy to decay.

Before presenting the results for Q, it is worth mentioning here that we could have chosen to compute Q using a different method. Indeed, we recall that Q could also have been defined as $\omega_o/\Delta\omega$, where $\Delta\omega$ is the width of the resonance. By computing the width of the resonance directly from Fig. 4, we could have estimated the value of Q. This method, however, would have led to larger uncertainties, especially for large values of Q.

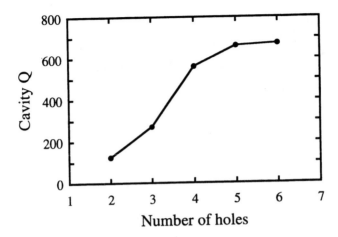

Figure 5. Quality factor as a function of the number of holes on either side of the the defect.

In order to excite the resonance efficiently, the initial condition should be chosen such that the pump mode overlaps with the resonant mode as much as possible. We have chosen to initialize the system by exciting two dipoles in the cavity. The dipoles were excited for a few optical cycles, and were oriented along the x-axis and located at the maxima of the modal profile shown in Fig. 3. After turning off the dipoles, the energy inside the cavity was measured over time. During the initial stages of the decay, every mode — except the high-Q one — quickly radiated away. Then, the only energy which remained in the cavity was that of the resonant mode. The mode continued its slow exponential decay. The Q value was computed to be equal to 271.

Figure 5 shows the value of Q as a function of the number of holes on either side of the defect. In the example above, Q was found to be equal to 271 when three holes were made on either side of the defect. If the number of holes is increased, the losses through the mirrors decrease. This has the effect of increasing Q. However, since the resonant mode is coupled to the continuum — the gap in Fig. 1 does not extend along every direction in space — Q does not increase indefinitely. Indeed, if too many holes are added, the coupling with the continuum becomes dominant and there will be an important outflow of energy in the transverse directions. This will cause Q to saturate. In the case shown in Fig. 5, Q saturates around 700. In order to couple light in and out of the cavity, it is essential that Q not be saturated. If Q is saturated, all the energy coupled to the cavity will simply radiate away in the transverse directions.

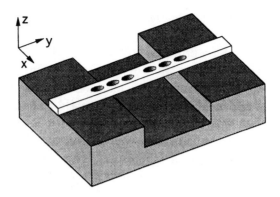

Figure 6. Suspended coplanar microcavity.

5. Fabrication

In order to achieve good confinement of the radiation in the microcavity, it is necessary for the index contrast between the cavity and its surroundings to be as large as possible. We present in this section two possible geometries to achieve good field confinement; first, the air-bridge geometry where the cavity is completely surrounded by air, and second, the insulator geometry where the cavity is grown directly on top of a low-index substrate.

5.1. AIR-BRIDGE GEOMETRY

The first case that we consider is that of the air-bridge. In this geometry, the cavity is completely surrounded by air, allowing for maximum field confinement. The air-bridge can be fabricated by growing the waveguide on a substrate, then removing part of the substrate under the cavity [10]. The resulting structure is shown in Fig. 6. Again, we have chosen to show a microcavity with three holes on either side. In the specific case shown in Fig. 6, two of these holes are in the suspended section of the guide while the other hole extends into the substrate. By etching the third hole deeply below the surface of the substrate, we can optimize the backscattering of the guided modes which extend under the waveguide.

The eigenstates of the suspended microcavity are computed using the same method presented in Section 3.2. We choose a supercell made of the structure shown in Fig. 6 and its mirror image in the xy-plane at the bottom of the substrate — each supercell contains two microcavities — in order for the supercell to have inversion symmetry. The use of a supercell with inversion symmetry significantly increases the efficiency of our computations [5]. By choosing the substrate sufficiently large, the interaction between the two microcavities can be made small; in all our calculations, we have chosen the thickness of the substrate such that the splitting between the eigenvalues converged to within 0.5%. The size of the supercell was chosen to include three holes on either side of the microcavity and the total thickness of the substrate was chosen to be equal to $4a$.

The resonant mode was found to be identical — same frequency, polarization, symmetry and field distribution — to that shown in Fig. 3. The microcavity material was chosen to be GaAs with a refractive index of 3.37 at $1.55\mu m$ [7] and the substrate was chosen to be $Al_{0.3}Ga_{0.7}As$ with a refractive index of 3.07 also at $1.55\mu m$ [7]. The small index contrast between these materials would not have allowed for strong confinement of the guided modes had the cavity been placed directly on top of the substrate. However, we have chosen these materials to demonstrate that a suspended structure can allow for strong field confinement within a distance as little as $2a$ on either side of the defect and radiation losses into the substrate can be negligible in spite of the small index contrast.

As a proof of concept, we have begun exploring the possibilities of fabricating these microcavities with silicon-based materials and with III-V semiconductor-based materials. Our main objective is to demonstrate the feasibility of building suspended structures with micron-sized features.

In the silicon-based system, a $1.0\,\mu m$ thick layer of sacrificial SiO_2 was deposited by low-temperature chemical vapor deposition onto a Si substrate. A $0.5\,\mu m$ thick layer of amorphous Si was deposited on top of the oxide by low-pressure chemical vapor deposition. The bridge pattern was defined using photolithography and the two deposited layers were then anisotropically removed by reactive ion etching. Finally, the sacrificial oxide was selectively etched away with dilute hydrofluoric acid. A scanning electron microscope (SEM) photograph of the bridge structure is shown in Fig. 7(a). The Si bridge is $10\,\mu m$ long and $4\,\mu m$ wide. Four holes with a diameter of $2\,\mu m$ are placed periodically along the bridge and are separated by $3\,\mu m$, center to center.

In the case of the III-V semiconductor materials, a $1.0\,\mu m$ thick layer of sacrificial AlAs and a $0.5\,\mu m$-thick layer of GaAs were grown sequentially by a gas source molecular beam epitaxy on a GaAs substrate. The bridge pattern was again defined using photolithography and the two layers were anisotropically removed by reactive ion etching. The sacrificial material was then selectively etched away by a chemical etch. Figure 7(b) shows an SEM photograph of a series of GaAs bridges. These bridges are $20\,\mu m$ in length and range from 3.5 to $5.5\,\mu m$ in width. The holes are all $2.5\,\mu m$ in diameter and are separated by $7.5\,\mu m$ from center-to-center in the structures shown in the foreground, and $5\,\mu m$ for those in the background. Optical measurements of these and other structures will be performed in future experiments.

 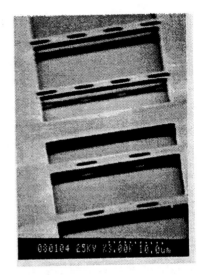

Figure 7. (a) SEM photograph of the Si bridge. The bridge has a length of 10 μm and a width of 4 μm. The holes are 2 μm in diameter and are separated by 3 μm from center to center. (b) SEM photograph of GaAs bridges. The bridges are 20 μm in length and range from 3.5 to 5.5 μm in width. The holes are all 2.5 μm in diameter and are separated from center-to-center by 7.5 μm (foreground) and 5 μm (background).

5.2. INSULATOR GEOMETRY

Instead of having the cavity suspended in mid-air, it may be possible to grow it directly on top of a low-index substrate and still achieve good field confinement. Since the penetration of the field into the substrate will depend on the index contrast between the guide and the substrate, it will be necessary for the index contrast to be as large as possible. However, for any choice of material, the field will always extend more into the substrate than it will into the air. Therefore, we need to use additional schemes to help confine the mode into the cavity.

The first scheme is shown in Fig. 8. It uses a pedestal (or monorail) geometry to elevate the cavity above the surface of the substrate. This has the effect of isolating the cavity from the substrate, hence of reducing the losses. The second scheme calls for the etching of deep holes into the waveguide. Since the field extends into the substrate, only a fraction of the field "sees" the holes. However, by etching the holes through the waveguide and into the substrate, a larger fraction of the field can overlap with the holes and be affected by the periodic variation of the index. This has the effect of increasing the reflectivity of the mirrors, hence of reducing the losses.

Several materials can be used for the fabrication of microcavities on

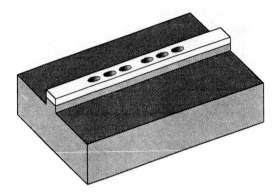

Figure 8. High-index microcavity on a low-index substrate.

insulators. Materials such as silicon and silicon dioxide offer a large index contrast — Si has an index of 3.48 while that of its oxide is 1.44, both at 1.55μm [7] — and Si-on-SiO$_2$ technology is well known today. The fabrication of the microcavity would rely only on standard silicon processing technologies with no need for multilayer epitaxy.

We have computed the Q value for the structure shown in Fig. 8, using the same parameters as those presented in Fig. 6. The guide was made of Si, and rested on a SiO$_2$ substrate. The holes were etched into the substrate; the total depth of the holes was equal to twice the thickness of the waveguide. The cavity Q was computed to be equal to 172. This and other similar structures have been fabricated at M.I.T. and are currently being tested [11].

6. TM Resonators

In the previous sections, we studied microcavities made of channel waveguides with vertical holes. These cavities were designed such that the resonant mode in the cavity would couple efficiently with TE modes in the waveguide. Modes with TM polarization would simply not couple to the cavity mode, nor would they be reflected by the array of holes. In order to design a resonator for TM modes, we need to find a structure that will open a large gap for modes polarized in the *vertical* direction.

We recall that a large gap is opened in the dispersion relation of a waveguide when two bands show a large difference in field concentration between the air and dielectric regions. A structure that gives rise to a large gap for TM modes is shown in Fig 9. It is made of a periodic array of long dielectric rods aligned in the vertical direction. The electric field of the lowest band is concentrated in the rods; it goes up one rod, out the top

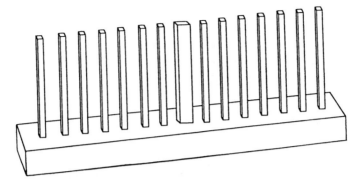

Figure 9. Microcavity for guided modes polarized in the vertical direction.

and down the next. The field in the second band has a nodal plane through each rod, and the antinodes occur in the air region between the rods. A two-dimensional analysis of this structure can be found in reference [12].

In order to design a structure more amenable to microfabrication, it may be necessary to reduce the height of the rods. However, as the rods are shorten, the component of the electric field in the longitudinal direction increases. Another improvement could be made by connecting the top of all the rods together. In doing so, the electric field would no longer have to pass through air as it "jumps" from rod to rod, and would concentrate the field of the first band into the dielectric. This would have the effect of increasing the size of the band gap. If the rods are both shorten and connected, then the structure would resemble the air-bridge microcavity presented in Section 5, simply tilted on its side! In addition to having a large component of the electric field in the vertical direction, the cavity mode in this structure would also have a large component in the longitudinal direction. Variations of this structure could include the use of square holes instead of circular ones — this would make the structure more amenable to microfabrication while having little effect on the overall performance of the cavity — and also the fabrication of wider structures, which could have the effect of introducing additional modes in the cavity.

7. Conclusion

We have presented and analysed a new class of resonant microcavities integrated directly along the plane of the substrate. We found that these microcavities give rise to strong field confinement and allow for efficient coupling into channel waveguides. By concentrating the field fluctuations into a small volume, we expect the recombination rate of carriers will increase. This could lead to the enhancement of spontaneous emission and could allow the microcavities to be modulated at very high speeds. These new microcavities offer exciting possibilities for the fabrication of high density and high speed optical interconnects, ultra-low threshold microlasers, and sharp transmission optical filters.

8. Acknowledgements

We acknowledge helpful contributions by J. N. Damask, J. S. Foresi, L. C. Kimerling, and L. C. West. This work was supported in part by the Army Research Office grant number DAAH04-93-G-0262 and by the MR-SEC Program of the NSF under Award Number DMR-9400334.

References

1. Y. Yamamoto, and R. E. Slusher, Physics Today, June, 66 (1993).
2. E. Yablonovitch, J. Opt. Soc. Am. B **10**, 283 (1993).
3. P. R. Villeneuve and M. Piché, Prog. Quantum Electron. **18**, 152 (1994).
4. J. D. Joannopoulos, R. D. Meade, and J. N. Winn, *Photonic Crystals* (Princeton, New York, 1995).
5. R. D. Meade, A. M. Rappe, K. D. Brommer, and J. D. Joannopoulos, Phys. Rev. B **48**, 8434 (1993).
6. K. S. Yee, IEEE Trans. Antennas Propag. **AP-14**, 302 (1996).
7. *Handbook of Optical Constants of Solids*, edited by E. D. Palik (Academic, New York, 1985).
8. P. R. Villeneuve, S. Fan, J. C. Chen, and J. D. Joannopoulos (unpublished).
9. A. Yariv, *Optical Electronics* (Saunders, Philadelphia, 1991).
10. P. R. Villeneuve, S. Fan, J. D. Joannopoulos, Kuo-Yi Lim, G. S. Petrich, L. A. Kolodziejski, and R. Reif, Appl. Phys. Lett. **67**, 167 (1995).
11. J. S. Foresi, L. C. Kimerling, P. R. Villeneuve, S. Fan, and J. D. Joannopoulos (unpublished).
12. S. Fan, J. N. Winn, A. Devenyi, J. C. Chen, R. D. Meade, and J. D. Joannopoulos, J. Opt. Soc. Am. B **12**, 1267 (1995).

EXPLORING THE TWO-DIMENSIONAL PHOTONIC BANDGAP IN SEMICONDUCTORS

THOMAS F.KRAUSS AND RICHARD M.DE LA RUE
Optoelectronics Group,
Department of Electronics and Electrical Engineering
Glasgow University, Glasgow, G12 8LT, Scotland, U.K.

Abstract.
    We discuss the fabrication of two-dimensional photonic bandgap (PBG) lattices in a GaAs/AlGaAs semiconductor and present a waveguide structure that allows their optical characterisation. The one-dimensional case has been used as a vehicle for exploring, theoretically and experimentally, the buildup of a PBG in a semiconductor-air photonic lattice. By designing a lattice with a high semiconductor fill-fraction, the scattering loss has been reduced to a tolerable level and true PBG effects have been observed.

1. Introduction

One of the ultimate aims that drives research into photonic bandgap structures is the quest for threshold-less laser devices. In order to achieve this goal, the active emitter must be incorporated in a photonic lattice which suppresses radiation into all but a selected few modes. Despite several reports of the fabrication of two-dimensional [1, 2, 3, 4] and three-dimensional [5] lattices of suitable periodicity, there is an evident lack of quantitative optical frequency characterisation data for structures of this kind.

    Ideally, one would like to incorporate active optical emitters into a photonic lattice and study the change of emission as a function of lattice parameters; this approach, however, has the added complication of photo or electropumping the emitters and is limited to the available spectrum of emission wavelengths. We believe that, in the one- and two-dimensional case, incorporating the photonic lattice into a waveguide structure and studying its passive properties is a simpler and more versatile method of

427

C. M. Soukoulis (ed.), Photonic Band Gap Materials, 427–436.
© *1996 Kluwer Academic Publishers. Printed in the Netherlands.*

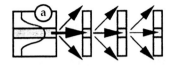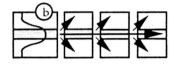

Figure 1. Out-of-plane scattering in semiconductor waveguide photonic lattices, (a) low semiconductor fill-factor leads to high scattering loss, (b) high fill-factor leads to low scattering loss, but only a polarisation-specific bandgap can be achieved

characterisation. The lattice can then be examined via direct measurements and, using a tunable Ti:Sapphire laser, a relatively large wavelength range is accessible. Furthermore, PBG waveguide structures deserve attention in their own right as highly selective filters [6] or, due to the strong light-matter interaction, novel and compact integrated optical elements.

In this paper, we discuss the fabrication of 2-D PBG lattices, show the smallest features achieved so far (to our knowledge), and present a waveguide structure that allows the optical assessment of such lattices. We also present theoretical and experimental data showing the buildup of a one-dimensional PBG as the number of elements increases.

2. Design considerations for 2-D PBG structures

Careful study and interpretation of "PBG maps" as in [7] yields important information for the design and fabrication of photonic lattices. For a square or triangular lattice, a full, polarisation-independent bandgap can only be achieved with a honeycomb structure, i.e., one with "air-pillars" where the high-index region is continuous. The optimum fill-factor of around 80% air implies that, in the case of cylindrical holes, the walls between adjacent air-pillars are less than 10nm wide (for a PBG centre wavelength of around 900nm). Accurately controlling the size of such structures seems very difficult, if not impossible, which is why we have used hexagonally shaped features throughout our work. Hexagons allow higher filling ratios with relatively thick (20-30nm) sidewalls, as we have already demonstrated [3]. For a hexagonal lattice, as has been shown very recently [8], it is also possible to obtain a full, polarisation - independent 2-D PBG if the lattice is composed of high-index pillars. As in the case of the honeycomb, however, this is only possible for a low fill-fraction of semiconductor (20-30%) in the lattice.

In a simple etched semiconductor waveguide structure (Fig.1), the light is only guided in the semiconductor and not while travelling through the airspace, hence diffraction loss and scattering into the third dimension (i.e., out of the plane of the waveguide) is inevitable. This loss can be reduced by increasing the fill-fraction of semiconductor, for example by designing a lattice of pillars with narrow airgaps or a semiconductor honeycomb structure with small air-pillars (Fig.1b). The principle of this approach is that light can "hop" across the narrow gaps without suffering excessive scattering loss, while still experiencing the full refractive index contrast between semiconductor and air.

The aforementioned map [7] shows that for such a semiconductor-rich structure, a PBG opens up for a semiconductor fill-factor below approximately 80%, for both a pillar and honeycomb-type lattice and for one polarisation only. The first order bandgap occurs for a periodicity of approximately 20% of the PBG centre wavelength, i.e., between 160nm and 180nm for a centre wavelength of 800 to 900nm. In our present work, we have concentrated on semiconductor-pillar lattices, because, unlike honeycombs, they offer the practical possibility of sizeable second and third-order PBG's, allowing for more process latitude. The fact that the bandgap is polarisation-specific has no serious consequences for the semiconductor laser applications that this work is aimed at, because the polarisation state of the quantum well emission can be engineered into the structure, e.g. by including strained layers.

3. Fabrication

The fabrication process is wholly based on electron-beam lithography (EBL), because this method affords a high degree of flexibility and both the periodicity of the pattern and the feature size can be accurately controlled. In particular, a Leica Beamwriter is employed. Its 100kV capability makes it very suitable for writing two-dimensional gratings, because the deleterious proximity effect, caused by electrons that are backscattered from the substrate, is less pronounced at higher acceleration voltages. The machine uses a 16×16 bit pattern generator that can cover fields as big as $160 \mu m \times 160 \mu m$ with 5nm nominal resolution, which is sufficient for many purposes.

Instead of a metal mask deposited by lift-off, as is commonly used for the fabrication of nanostructures [9], we employ a silica mask. The pattern, initially generated in electron-beam resist (PMMA, 100nm thick), is then transferred into the silica by dry etching with C_2F_6. The GaAs/AlGaAs multilayer is etched by RIE with $SiCl_4$ using special low-damage conditions [10], which allow etch depths of up to $1.5 \mu m$ with a mask thickness of 120nm. Figure 2a shows the detail of a $150 \mu m \times 150 \mu m$ two-dimensional array which has been produced with this technology, note the periodicity of 150nm and the 20–30nm gaps between the pillars. Although somewhat distorted, the hexagonal shape of each pillar can still be recognised, which emphasises the resolution capability of our EBL system.

It is an unfortunate but natural characteristic of the dry etching process that very narrow gaps cannot be etched to the same depth as bigger features. We found that it is difficult to etch narrow gaps with near-vertical

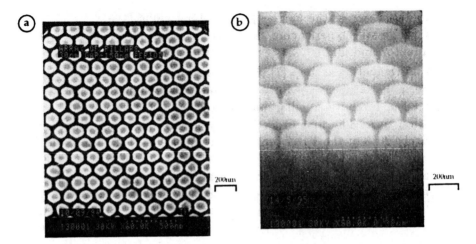

Figure 2. (a) Detail of a 150μm square photonic lattice with 150nm periodicity and 30nm airgaps.(b) Etching narrow gaps. The micrograph shows a lattice similar to (a), but with 300nm periodicity. The airgaps are 30nm wide and the etch depth seems limited to approximately 0.35μm.

sidewalls beyond a width:depth ratio of approximately 1:10; for example, a depth of 0.35μm was the maximum we achieved for 30nm gaps (Fig.2b). We therefore decided to move to higher-order lattices, in particular third order, where a high semiconductor:air ratio is still possible for airgaps between 60 and 100nm [7]. Airgaps of this size allow a sufficient etch depth of nearly 1.0μm.

4. One-dimensional PBG Waveguide

4.1. DESIGN

Referring to a "normal" grating as a one-dimensional photonic bandgap structure may sound somewhat exaggerated, but one should keep in mind that "normal" gratings, e.g., as in DFB lasers, usually employ refractive index contrasts of at most 1%, whereas photonic bandgap structures demand a contrast of greater than 2:1 [11]. Despite the fact that reports of optical devices employing gratings are numerous, there is, to our knowledge, no published investigation of semiconductor waveguide structures with a refractive index contrast as high as 2:1 or even 3:1. The question of diffraction losses (i.e., coupling to radiation modes) requires serious consideration, particularly since the third dimension (conveniently assumed infinite in most 2-D calculations) only extends to at most 1μm in semiconductor waveguide structures.

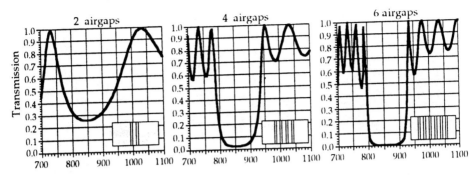

Figure 3. Buildup of a photonic bandgap in a third order one-dimensional structure. The period is 400nm and and the airgaps are 80nm wide. The horizontal scale denotes wavelength in nanometres, the vertical scale relative transmission, and the inset indicates the physical structure.

Two-dimensional structures produced by pulling fibre-bundles [12, 13] provide an alternative approach, but these structures do not readily allow incorporation of efficient light emitters and are therefore only suitable for passive applications. III-V semiconductors, in contrast, are the most well-developed, the most efficient and, therefore, the most natural electroluminescent source material, which makes them so interesting for this study and for application in PBG structures in general.

4.2. GRATING LENGTH

One of the questions that arise when studying these high contrast lattices is the physical length, i.e., how many grating periods are required. A reasonable approximation can be found by studying 1-D gratings with infinite boundaries using a simple transmission matrix routine [14]. Such a simplified simulation does not accurately represent the real structure, because it neglects boundary effects which are undoubtedly present, but can still be expected to show important trends.

The simulation shows that a bandgap starts to become evident with as little as two periods, is almost fully developed with four, and changes very little once the number of periods exceeds six (Fig.3). In comparison with near-ideal PBG structures, which have been estimated to attenuate at 16dB/unit cell [15], our lattice only attenuates at approximately 3dB/unit cell. The reasons for this lower attenuation are that, in order to minimise the diffraction loss as quoted above, our structures are third order and deviate strongly from the ideal of having an optical pathlength of $\lambda/4$ in both the high and the low index regions. The interaction, however, is still strong enough for the bandgap to build up over a distance as short as a few microns.

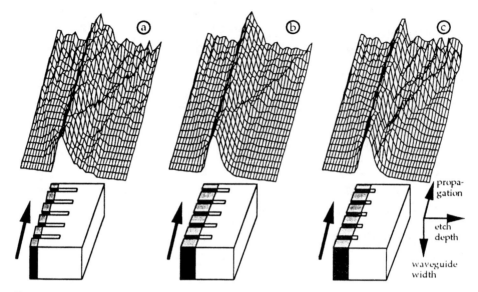

Figure 4. Indication of the scattering loss out of the waveguide plane. The propagation through the waveguide (shaded) was simulated with the beam popagation method (BPM). The period in all cases is 425nm (third order at 870nm) and the propagation is through 5 airgaps of 100nm width as indicated by the physical structure sketched underneath. (a) 0.2μm surface waveguide, etched 0.8μm deep. Transmission: 65%. (b) 0.4μm thick surface waveguide, etched 0.8μm deep. Transmission: 82%. (c) 0.4μm thick surface waveguide, etched 0.5μm deep. Transmission: 43%.

4.3. PROPAGATION LOSS

We used another simple and approximate method to study the diffraction (scattering) loss that the modes experience when propagating through our waveguides. We examined the problem by means of the beam propagation method (BPM), which is commonly used in the analysis of integrated optics waveguide structures. Its operating principle is to propagate a given field through the structure while assuming energy conservation in the forward direction, hence the method cannot cope with reflections and considers a single pass only. This approach allows a qualitative investigation of different types of waveguide, and is a useful tool for the derivation of important design rules.

The results are summarised in Fig.4. Figure 4a shows a single mode waveguide, 0.2μm thick, with a refractive index difference between core and cladding of approximately 0.2 (3.5-3.3). Waveguides of this thickness and index contrast are commonly used in semiconductor lasers. The guiding layer was positioned at the top of the structure to minimise the required etch depth, since the etch depth:gapwidth ratio was limited to approximately 10:1, as discussed previously. An etch depth of 0.8μm and 100nm wide gaps were chosen in order to ensure that the structure could be fabricated.

After passing through 5 airgaps, the power still contained in the waveguide dropped to 65% (Fig. 4a), a significant loss considering that the result has been calculated for a single pass; in a real structure, multiple reflections would occur, increasing the effective pathlength of the mode within in the structure and hence increasing the loss further. This result can be improved to 82% by maintaining the refractive index difference, gap-width and etch depth, and increasing the waveguide depth from $0.2\mu m$ to $0.4\mu m$ (Fig.4b). The throughput, somewhat surprisingly, is not improved further by increasing the refractive index difference within the range accessible to GaAs/AlGaAs semiconductors; the best result was obtained for the strongly guiding single-mode waveguide shown in Fig.4b.

Altering another parameter, the width of the airgap, to 50nm, reduced the scattering loss to 10% (i.e., 90% throughput) which supports our model of light "hopping" across the narrow gap with reduced ability of coupling to radiation modes.

Reducing the etch depth, on the other hand, from $0.8\mu m$ to $0.5\mu m$ as shown in Fig.4c, leads to a substantial increase of the scattering loss and a reduction in throughput to 43%, which underlines the need to etch the airgaps as deeply as possible.

4.4. EXPERIMENTAL RESULTS

The waveguide of Fig.4b was grown by MBE and a deep grating etched. Figure 5a depicts the pattern, here 6×100nm airgaps of $0.9\mu m$ depth etched into the waveguide. The width of the waveguide is $2\mu m$ and the patterns seen above and below are identical structures. The lattices were designed to show a third order bandgap centred at 870nm.

Light is launched into the waveguide, propagates for approximately $500\mu m$, interacts with the grating and propagates for another $500\mu m$ before reaching the detector. The available tuning range of the laser is between 820nm and 930nm, so that the lower limit coincides approximately with the absorption edge of the 12% AlGaAs material in the waveguide core. Figure 5b is a normalised graph of the grating transmission vs. wavelength for the two different periodicities of 420nm and 470nm examined here.

Both curves clearly show a sharp drop in transmission at 840nm and 910nm, respectively, which can be identified as a photonic bandedge. This interpretation is supported by the theoretical simulation of identical structures (Fig.5c) which shows a similar drop in transmission at the same wavelengths. The simulation also reveals that the corresponding rise in transmission, i.e., the "high-frequency bandedge", cannot be detected, be-

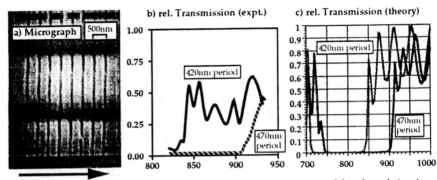

Figure 5. Experimental results of a third order 1-D PBG structure and comparison with simulation. (a) Physical structure, 420nm period, 100nm airgaps, etched 0.9μm deep. (b) Normalised transmission through two different waveguide structures (420 and 470nm period) measured with a tunable Ti:Sapphire laser. (c) Simulation of the properties of the same structure. The wavelength range accessible in our measurement is highlighted.

cause the bandgap is wider than the wavelength range accessible to our measurement. The fringes in the 420nm curve (Fig.5b) are a combination of the fringes shown in Fig.5c, which correspond to localised states near the bandedge[16], and alignment problems that occured because the laser beam/waveguide coupling varied as the laser wavelength was changed.

In order to match the wavelength of the observed and the simulated bandedges, a refractive index of 3.15 was used for the high-index region (i.e., the semiconductor part of the lattice), which is considerably lower than the effective waveguide index of 3.42 in an unetched waveguide. We believe that boundary problems (light leaking out and travelling in the airspace above and beside the grating) are not sufficient to explain this discrepancy, but see it as evidence for changes in the material properties such as dry etch damage and partial oxidation.

5. Conclusion and Outlook

We have described design concepts, presented important aspects of the fabrication, and shown experimental evidence for the feasibility of photonic bandgap structures in III-V semiconductor material at optical frequencies. By departing from the standard approach of a honeycomb lattice with high air-filling fraction and using a semiconductor-rich lattice instead, we were able to show the first optical characterisation of a photonic lattice embedded in a waveguide structure, which shows that the waveguide approach is a versatile tool for characterising one and two-dimensional photonic microstructures. Using the same approach, we hope to demonstrate a two-dimensional PBG shortly and plan to explore the inclusion of photonic defects. We also believe that it is easier to make photonic lattices with very small periodicity and both high (Fig.2) and low [3] semiconductor fill fraction by using hexagonal pillars instead of the commonly assumed circular cylinders.

Intuitively, it is not obvious that significant transmission can occur in this type of structure because of diffraction (scattering loss) and mismatch between the modes in the waveguide and the Bloch modes in the photonic lattice [17]. The throughput of approximately 50% that we have measured in the wavelength region outwith the stopband is hence very encouraging for a first attempt and we hope to increase it significantly in due course.

In a simple semiconductor-air structure, however, some residual scattering loss will always be present, even if the mode confinement is significantly stronger than in our all-semiconductor design [18]. In order to eliminate this loss completely, a different approach is therefore required. The reason for the loss is that light is only guided in the semiconductor region; when travelling through the airgaps, there is no confinement and some diffraction occurs. This diffraction can be suppressed if a guiding structure is also created in the airgap, e.g., by backfilling the gap with dielectrics of suitable refractive index, and creating a perfect overlap between the guided mode in both the high and the low-index region.

Furthermore, our results are immediately relevant to laser applications, because the all-semiconductor waveguide can be easily redesigned to include quantum wells and allow current injection. By using strained layers, the polarisation of the quantum well emission can be adapted to the specific properties of the photonic lattice.

We believe that these are significant milestones towards the realisation of novel optoelectronic devices based on the new concept of photonic microstructures.

6. Acknowledgements

We are very grateful for the technical support of the Nanoelectronics Group and the Dry Etching Facility, also for the epitaxially grown waveguide structure supplied by the MBE Group at Glasgow University. TFK would like to acknowledge a personal fellowship by the Engineering and Physical Sciences Research Council of the U.K.

References

1. Gerard, J.M., Izrael,A., Marzin,J.Y., and Padjen,R. (1994), Photonic bandgap of two-dimensional dielectric crystals, Solid State Electronics, **37**, 1341-1344.
2. Wendt,J.R., Vawter,G.A., Gourley,P.L., Brennan,T.M., and Hammons,B.E. (1993), Nanofabrication of photonic lattice structures in GaAs/AlGaAs, Journ. Vac. Sci. Tech. B., **11**, 2637-2640.

436

3. Krauss, T., Song, Y.P., Thoms, S., Wilkinson, C.D.W., DeLaRue, R.M. (1994), Fabrication of 2-D photonic bandgap structures in GaAs/AlGaAs, Electron.Lett., **30**, 1444-1446.
4. Baba, T and Matsuzaki, T. (1995), 2-Dimensional photonic crystals with multiple refractive index steps, this volume.
5. Cheng, C.C., Sherer, A., Arbet-Engels,V., and Yablonovitch, E. (1995), Fabrication of Photonic Band-gap Structures, 39th Electron, Ion,and Photon Beam Meeting, to be published in Journ. Vac. Sci. Tech. B, Nov/Dec 1995.
6. Bullock, D.L., Shih, C-C, and Margulies, R.S. (1993), Photonic band structure investigation of two-dimensional Bragg reflector mirrors for semiconductor laser mode control, Journ. Opt. Soc. Am. (B), **10**, 399-403.
7. Padjen, R., Gerard, J.M., and Marzin, J.Y. (1994), Analysis of the filling pattern dependence of the photonic bandgap for two-dimensional systems, Journ.Mod.Optics, **41**, 295-310.
8. Cassagne, D., Jouanin, C., Bertho, D. (1995), Two-dimensional photonic band gaps: New hexagonal structures, this volume.
9. Beaumont,S.P., Tamamura,T., and Wilkinson, C.D.W., (1981) A two-layer resist system for efficient lift-off in very high resolution electron beam lithography, Proc.Microcircuit Engineering, (Delft University Press, Amsterdam), 381-388.
10. Murad, S.K., Wilkinson, C.D.W., Wang, P.D., Parkes, W., Sotomayor-Torres, C.M. and Cameron, N. (1993), Very low damage etching of GaAs, J.Vac.Sci.Tech.B **11**, 2237-2243.
11. Meade, R.D., Brommer, K.D., Rappe, A.M., and Joannopoulos, J.D.(1992), Existence of a photonic band gap in two dimensions, Appl.Phys.Lett., **61**, 495-497.
12. Tonucci, R.J., Justus, B.L., Campillo, A.J., Ford, C.E. (1992), Nanochannel array glass, Science, **258**, 783-785.
13. Inoue, K., Wada, M., Sakoda, K., Yamanaka, A., Hayashi, M.,and Haus, J.W. (1994), Fabrication of two-dimensional photonic band structure with near-infrared band gap, Jpn.J.Appl.Phys., **33**, L1463-1465.
14. T.Tamir, Guided-Wave Optoelectronics, (Springer, New York, 1990), 43-50.
15. Ozbay, E., Michel,E., Tuttle, E., Biswas,R., and Ho, K.M. (1994), Terahertz spectroscopy of three-dimensional photonic band-gap crystals, Opt.Lett., **19**, 1155-1157.
16. John, S. (1995), Collective Phenomena in a photonic band gap, this volume.
17. Russell, P.St.J. (1984), Novel thick-grating beam-squeezing device corrugated waveguide, Electron. Lett., **20** 72-73.
18. Fan, S.H., Villeneuve, P.R., Meade, R.D., Joannopoulos, J.D. (1995), Design of 3-dimensional photonic crystals at submicron length scales, Appl. Phys. Lett. **65**, 1466-1468; also this volume.

2D PHOTONIC BAND GAP STRUCTURES IN FIBRE FORM

Band Gaps by the Kilometre

T.A. BIRKS, D.M. ATKIN, G. WYLANGOWSKI, P.ST.J. RUSSELL
Optoelectronics Research Centre
University of Southampton, Southampton SO17 1BJ, United Kingdom
P.J. ROBERTS
Defence Research Agency
St Andrew's Road, Malvern, Worcs., WR14 3PS, United Kingdom

1. Two-Dimensional PBG Structures

A structure with a refractive index that varies periodically in three dimensions can fail to transmit light of certain "forbidden" frequencies, despite being non-absorbing. This is because Bragg diffraction, constructive multiple-beam interference from a periodic array of scatterers, takes place for all directions; only an exponentially-decaying evanescent field is supported. These frequencies ω are the full photonic band-gaps (PBGs) of the structure. Light that attempts to enter the structure is expelled and returns to the medium from which it came. If that medium is free space, the structure behaves as a totally-reflecting mirror; if the medium is a buried excited atom wanting to emit a photon at a forbidden frequency, the spontaneous emission is suppressed[1].

In any 3D periodic structure, the distance between successive scatterers changes somewhat with direction. In other words, the Brillouin zone is not spherical. The contrast in the refractive index within each unit cell must therefore be large if Bragg diffraction is to occur for all directions. Furthermore, the repeat period must be of the order of the optical wavelength. Such structures are exceedingly hard to build for visible and near infrared frequencies, though they have been constructed for microwave frequencies[1].

This has led researchers to consider instead structures that are periodic in just two dimensions, and uniform in the third dimension[2-5], Figure 1. Successful techniques include reactive ion etching of semiconductors[2] and the etching of one component of two-component glass arrays[3]. This gives an array of air holes in a high index host material. Theoretical analysis is also simpler, and when propagation is restricted to the transverse plane full PBGs are possible. Such structures have been constructed and optically probed, and their transmission spectra accord with theory. However, they do

437

C. M. Soukoulis (ed.), Photonic Band Gap Materials, 437–444.
© *1996 Kluwer Academic Publishers. Printed in the Netherlands.*

not have full PBGs because in reality light is free to propagate out of the transverse plane; the restriction to the plane is artificial. Spontaneous emission within the structure can always escape. To understand the behaviour of 2D periodic structures fully, it is necessary to consider waves with a longitudinal (z axis) wave-vector component β, something which few researchers have done[5]. Perhaps this is because existing methods of construction are limited to samples that are very thin in the z direction.

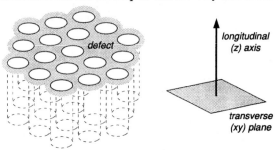

Figure 1. A two-dimensional PBG structure, with rods of one material arranged in a honeycomb array within a different material. A defect (in this case a missing rod) is shown. The structure is uniform in the z direction.

2. Propagation Along a 2D PBG Structure - Optical Fibres

Although there are no full 3D PBGs in a 2D periodic structure, there can still be photonic stop bands of a restricted sort. For a given ω (which can always be imposed by using a light source of that frequency), there can be forbidden values of β for which no light propagates with real transverse wave-vector components k_x and k_y.

In a uniform medium of refractive index n, only values of β greater than kn (where $k = \omega / c$) are forbidden, so it can be said to have a stop band for such β values. Indeed, this gives rise to the phenomenon of total internal reflection upon which conventional optical fibres are based, Figure 2.

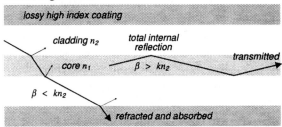

Figure 2. Light in a conventional optical fibre is guided in the core by total internal reflection if $\beta > kn_2$. Otherwise the light refracts into the cladding and is absorbed by the coating.

A core of index n_1 is surrounded by a cladding of lower index n_2. Light can propagate in the core if the longitudinal wave-vector component β is less than kn_1. If $\beta > kn_2$, the light cannot pass through the cladding and so can only propagate along the core. In contrast, light with smaller β values can escape from the core, but real optical fibres are coated with a polymer that has a high index and absorbs light. Hence, all waves not confined to the core are eventually absorbed in this coating, leaving just the confined wave in the core to emerge from the far end of a length of fibre. Optical fibres are usually made from silica glass with very low losses and in lengths of several kilometres[6].

In a 2D periodic structure also, all β values greater than a certain maximum are forbidden. However, there can also be stop bands for smaller β values, separated from each other and the ultimate large-β stop band by bands of allowed β (the pass bands). The β positions of the stop bands can be varied by adjusting the pitch or absolute scale of the structure.

Light with β within a stop band cannot propagate within the periodic structure. However, if the periodicity is broken in a small part of the cross-section by deliberately introducing a thread-like "defect", the light can propagate along the defect. It provides an allowed mode for the light, which is confined to the defect by Bragg reflection in the surrounding periodic structure. The defect can take many forms, like the simple absence of a refractive index component in one unit cell, as depicted in Figure 1. If a portion of the periodic structure is surrounded by an absorbing high-index coating, then all the waves filling the structure are absorbed, and only defect modes will propagate from one end of the structure to the other. The structure therefore acts like an optical fibre ("PBG fibre") that guides light by Bragg reflection instead of total internal reflection. The defect is the fibre core, and the periodic structure is the cladding. Unlike a conventional fibre, the PBG fibre can be made to guide light with any β value, including values approaching zero, simply by appropriately engineering the structure to give a stop band for the appropriate β. Unlike the 2D structures considered elsewhere, this fibre can be made in lengths of several kilometres, thus giving a plentiful supply of PBG material. Such a fibre is the goal of our project.

The fibre is of interest as a test-bed of PBG ideas. However, it also has practical applications. One example is the optical delay line, because in principle a PBG fibre can be designed with a group velocity as small as necessary. A second application is in environmental or bio-sensors. A silica PBG fibre can be designed to give β/k any value lower than the index of silica, for example 1.33 to permit a strong evanescent interaction with an aqueous solution.

3. Photonic Band Calculations

The photonic band properties of a structure periodic in 1D are very easy to calculate. Indeed, for a multilayer stack of two materials, the solutions are known in closed form[7,8]; even the eigenvectors giving the field distributions are expressible analytically if a sensible co-ordinate system is defined[8]. However, numerical methods are required for calculating the photonic bands of a 2D (or 3D) structure. A real-space finite-element method[9] was used. The unit cell of the structure is divided into a lattice of points, on which Maxwell's equations are discretised. The field in one layer of lattice points is related to that in the previous layer by a translation matrix. The matrix for the entire unit cell can be found by concatenating such single-layer matrices. The translation matrix method is fast but is susceptible to numerical instabilities; in practice, it can only be applied over a limited number of layers at a time.

A scattering matrix method, which relates transmission and reflection coefficients between layers, is more stable but slower. To optimise accuracy and speed, the two methods are combined, using the translation matrix method over groups of lattice layers small enough to retain accuracy, and relating the fields in adjacent groups using the scattering matrix method[9].

The photonic band properties of a hexagonal array of circular cylindrical "rods" of air ($n = 1.00$) in a silica matrix ($n = 1.45$), like that of Figure 1, were calculated. An air filling fraction of 0.45 within the structure was assumed, which is consistent with the experimental techniques to be described below. This calculation gave the graph shown in Figure 3, where the normalised frequencies $k\Lambda$ of the stop band edges are plotted against the normalised wave constant $\beta\Lambda$, where Λ is the spacing of nearest-neighbour air holes. In addition to the ultimate large-β stop band that would be expected in a uniform structure, finite stop bands for relatively low β can also be seen. For light with a free-space wavelength of 1550 nm, the corresponding hole spacing is $\Lambda \approx 2\,\mu$m. The aim of the experimental investigation is to make a fibre with this structure, including a defect to guide the light and a high index jacket to strip off all unconfined waves.

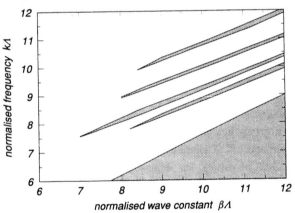

Figure 3. The stop bands (shaded) of the 2D structure described in the text.

4. Construction of a PBG Fibre

To make a PBG fibre, we need to provide a refractive index variation across a glass structure, and we need that variation to repeat periodically. From earlier theoretical work, we know it is advantageous to maximise the index variation within the structure, a relatively large proportion of the structure should be of the lower index material, it is better to have isolated low index regions in a connected high index matrix than *vice versa*, and a hexagonal honeycomb lattice has maximum symmetry and is likely to give the widest stop bands for given materials[4].

We have started our investigation with a silica structure. Silica is easier to work with because its viscosity varies relatively gradually with temperature, making the fibre drawing process easier to control. The required index variation is provided by the inclusion of air gaps, giving an index ratio of 1.45:1. We decided to make the 2D periodic fibre by a multi-step process, because the size reduction ratio required (from 30 mm to 2 μm) is so large. This process is depicted in Figure 4.

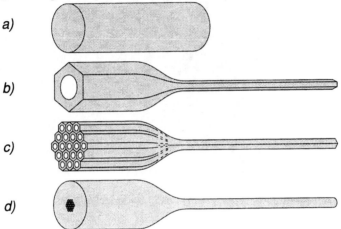

Figure 4. The fabrication of a PBG fibre, starting from a silica rod (a). A central hole is drilled and flats milled on the outside, and the rod is then drawn into hex cane (b). This is stacked and the stack is drawn into stack cane (c). Finally, the stack cane is jacketed with silica and drawn to form the final fibre (d).

The first question to be answered is whether the fabrication process is successful mechanically: is stacking possible, do the canes fuse together, do they remain in the desired array as they do so, and are the air holes retained without filling in or enlarging? We therefore did our first experiments with a structure that lacked a deliberate defect. All three drawing steps were carried out at relatively low furnace temperatures and large trawing tensions, to preserve the shapes of the canes and prevent the air holes filling in under surface tension.

The unit cell of the structure was made on a macroscopic scale from a solid cylindrical silica rod of diameter 30 mm and length 250 mm. An ultrasonic mill was used to drill a central hole of diameter 16 mm to provide the air region, and mill flats on the outside to provide a hexagonal cross-section. The rod was drawn down to a width of about 1 mm on a fibre drawing tower, to give tens of metres of *hex cane* (cane is simply fibre with a large enough diameter to be stiff). This was cut into 250 mm lengths, and 331 pieces were stacked in a honeycomb array with 11 canes along each side, as shown in Figure 5 (at this stage, a defect could be provided by, for example, leaving one cell empty, or inserting a cane without an air hole into one cell).

Figure 5. A photograph of the complete stack of canes, showing the simple holder used to support it.

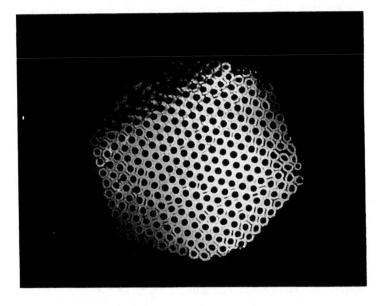

Figure 6. The cross-section of a piece of stack cane as seen under the microscope. It is about 900 μm wide.

The stack was tied together using soft tantalum wire, and drawn down again to form *stack cane*, again of diameter ~1 mm. Figure 6 is a photograph of the stack cane as seen under a microscope.

A piece of stack cane was surrounded by a silica jacket, and this assembly was pulled down one last time to form the *final fibre*. The silica jacket is necessary to make the final fibre big enough to be handleable. The number of canes in a stack is limited by the bore of the furnace (33 mm) and the minimum size of a cane (the canes must be big enough to be stiff, and to be individually manipulated during stacking). The silica jacket still permits light not confined to a defect to be stripped off by index-matching fluid. The jacket was provided by solid hexagonal canes stacked around the central stack cane. These were made of silica with a high water content (1200 ppm OH instead of 120 ppm). The intention was for the corresponding lower viscosity to allow them to fuse intimately with each other and with the central cane, without the holes in the central cane filling in.

An early result is shown in Figure 7. Although the jacket did not fuse together as required, it can be seen that the holes in the central cane survived. The centre-to-centre separation of these holes was about 2.5 μm, which is only slightly greater than the target value of 2 μm. Lengths of fibre with hole separations of 1.7 μm have since been drawn. These result show that our goal, a PBG fibre in silica and air, is possible.

Figure 7. The cross-section of a trial PBG fibre. Each cane is about 50 μm wide across corners. Although the outer solid canes have not fused together, the air holes in the central cane have clearly survived.

5. Conclusions

We have described a 2D photonic band-gap structure made of "rods" of air arranged periodically in a honeycomb lattice within a silica matrix. Although there is no full PBG because light of any frequency can always pass through the structure for certain values of the longitudinal wave vector component β, there can be certain values of β for which propagation is forbidden. With an appropriate defect in one unit cell to act as an optical fibre core, it should be possible to guide light along the fibre by Bragg reflection instead of total internal reflection.

The structure has been analysed numerically by a real-space discretisation method. We have shown that it is possible to draw lengths of fibre with an array of air holes about 1 μm in diameter using silica fibre fabrication techniques, without the holes collapsing. The next steps in this project will be to produce lengths of fibre like that in Figure 7, except with close fusion of the canes. A deliberate defect will be introduced into the central structure by omitting one of the air holes. The resulting fibre can then be characterised optically, with a demonstration of wave guidance by Bragg reflection instead of total internal reflection. This will be the first time that PBG material will have been produced in quantity.

6. Acknowledgement

This project is supported by the U.K. Defence Research Agency at Malvern.

7. References

1. Yablonovitch, E. (1993) *Photonic band-gap structures*, J. Opt. Soc. Am. B **10**, 283-295.

2. Wendt, J.R., Vawter, G.A., Gourley, P.L., Brennan, T.M., and Hammons, B.E. (1993) *Nanofabrication of photonic lattice structures in GaAs / AlGaAs*, J. Vac. Sci. Technol. B **11**, 2637-2640.

3. Inque, K., Wada, M., Sakoda, K., Yamanaka, A., Hayashi, M., and Haus, J.W. (1994) *Fabrication of two-dimensional photonic band structure with near-infrared band gap*, Jpn. J. Appl. Phys. **33**, L1463-L1465. Similar structures, though not for PBGs, are described in Tonucci, R.J., Justus, B.L., Campillo, A.J., and Ford, C.Z. (1992) *Nanochannel array glass*, Science **258**, 783-785.

4. Villeneuve, P.R., and Piché, M. (1992) *Photonic band gaps in two-dimensional square and hexagonal lattices*, Phys. Rev. B **46**, 4969-4972.

5. Maradudin, A.A., and McGurn, A.R. (1994) *Out of plane propagation of electromagnetic waves in a two-dimensional periodic dielectric medium*, J. Mod. Opt. **41**, 275-284.

6. Senior, J.M. (1985) *Optical Fiber Communications*, Prentice Hall International, London.

7. Yeh, P., Yariv, A., and Hong, C.-S. (1977) *Electromagnetic propagation in periodic statified media*, J. Opt. Soc. Am. **67**, 423-448.

8. Russell, P.St.J., Birks, T.A., and Lloyd-Lucas, F.D. (1994) Photonic Bloch waves and photonic band gaps, in E. Burstein and C. Weisbuch (eds.), *Confined Electrons and Photons: New Physics and Applications*, Plenum Press.

9. Pendry, J.B. (1994) *Photonic band structures*, J. Mod. Opt. **41**, 209-229.

DISPERSION, TUNABILITY AND APPLICATIONS OF DEFECT MODES IN PHOTONIC BAND-GAP STRUCTURES

R.D. PECHSTEDT, P.ST.J. RUSSELL and T.A. BIRKS
Optoelectronics Research Centre, University of Southampton,
Southampton SO17 1BJ, U.K.

1. Introduction

Electronic surface states, whose energies lie within the semiconductor band-gap, are normally regarded as a nuisance to be avoided in semiconductor lasers. This is particularly true of micro-pillar laser arrays where the surface-area-to-volume ratio is large. Similarly, intra-*photonic*-band-gap interface or surface states will reduce the effectiveness of a photonic bandgap material by introducing intra-band-gap states into which unwanted spontaneous emission and lasing can occur [1,2]. These *photonic defect states* are not, however, always undesirable. Examples include: i) the DFB laser mode supported by a structural defect at the centre of a uniform Bragg mirror - the resonant frequency of this mode lies within the photonic bandgap of the Bragg mirror [3]; ii) the surface-guided Bloch modes (SGBM) confined at the surface of multilayer stacks [4]; and iii) Bragg waveguide modes (BWGM) in which total internal reflection is replaced by Bragg reflection between two multilayer stacks [4]. A general feature of defect modes is a phase velocity that is highly sensitive both to optical frequency and to the "strength" of the local aperiodicity that defines the defect. Small compositional and structural changes can radically alter the position of the mode within the stop-band, providing an effective tuning mechanism. Owing to these and other unique properties, defect modes may provide the basis for the development of a versatile new family of optoelectronic devices.

   In this paper we discuss some properties of SGBM's within the context of photonic bandgaps and show how they can be used to realize a tunable narrow-band fibre tap and a mode-selective fibre coupler.

2. The Band Structure

For completeness and clarity we now outline the results of the transfer matrix analysis, given in full in [4,5]. Let us consider a dielectric multilayer stack consisting of alternating planar layers with constant refractive indices n_1, n_2 and widths h_1, h_2, respectively. The stack period Λ equals $h_1 + h_2$ and a Cartesian system of co-ordinates is used with y (z) normal (parallel) to the layers; there is no field variation with x. Due

C. M. Soukoulis (ed.), Photonic Band Gap Materials, 445–452.
© 1996 *Kluwer Academic Publishers. Printed in the Netherlands.*

to the planar geometry the solutions for the electromagnetic field within the stack separate into TE and TM polarization states with $H_x = E_y = E_z = 0$ and $E_x = H_y = H_z = 0$, respectively. In each case, all field components can be expressed in terms of the remaining x component, denoted by F. The normal modes of a periodic medium are Bloch waves in the same sense as planar waves are the normal modes of a homogeneous medium. Due to the translational symmetry of the medium, Bloch waves obey the Floquet-Bloch theorem and the following relationship holds:

$$F_{\pm}(y+\Lambda) = F_{\pm}(y)\exp(\pm ik_y\Lambda) , \qquad (1)$$

where k_y is the Bloch wavevector given by

$$\cos(k_y\Lambda) = \cos(p_1h_1)\cos(p_2h_2) - \frac{1}{2}(r+1/r)\sin(p_1h_1)\sin(p_2h_2) , \qquad (2)$$

where $r = p_1\xi_1/p_2\xi_2$ and $p_j = (k^2n_j^2 - \beta^2)^{1/2}$ is the wavevector component normal to the interface in the j-th layer. Here k denotes the vacuum wavevector at optical frequency $\omega/2\pi$, β is the propagation constant along the z direction parallel to the layers, and ξ_j is a polarization parameter equal to 1 and $1/n_j^2$ for TE and TM polarization, respectively. The field $F(y)$ is obtained by solving Helmholtz's equation in each layer and matching the solutions in two adjacent layers via the boundary conditions. The general field in the structure (for given ω, β, and polarization) is expressible as a superposition of the two Bloch waves in Eq.(1).

If k_y is real, the Bloch waves in Eq.(1) are progressive and may transport energy normal to the layers as well as along them. If, however, values of ω and β exist for which the magnitude of the RHS of Eq.(2) exceeds 1, k_y has an imaginary part. In this case, the Bloch waves are evanescent, growing or decaying exponentially from period to period normal to the layers, while progressing along them. If the structure is infinite in the y-direction, these waves cannot be supported and no real states exist; the ranges of ω and β where this occurs are known as *stop-bands*. The band edges between real and virtual states (i.e., travelling and evanescent Bloch waves) occur when $\cos(k_y\Lambda) = \pm 1$.

By introducing a defect into the structure, real states within the stop-band can occur. In this case the Bloch waves play the role of tunnelling fields, resulting in field localization near the defect. The most radical defect is a truncation. In a truncated structure, photonic surface waves can form when the external field is evanescent and matches to a Bloch wave decaying into the stack. This is the case of a SGBM, the properties of which will be considered in more detail in the next section. On the other hand, by changing the thickness of only one layer in a particular period or replacing it with a layer of different refractive index, a structural or compositional defect can easily be introduced into an infinite structure. This type of defect mode corresponds to a BWGM and the degree of spatial localization is entirely determined by its position within the stop-band via the imaginary part of k_y. If a defect layer is introduced into a semi-infinite stack supporting a SGBM, hybrid modes between a SGBM and a BWGM can form.

3. Surface-Guided Bloch Modes

3.1. DISPERSION

The energy (wavevector) of Bloch waves can be quantized by truncating the medium in which they exist. This results in a reduction of the spectral density of states (i.e., the number of states per unit frequency in a given volume), caused by localizing the waves between the boundaries. Here the simplest case of a semi-infinite periodic medium with a common interface to an isotropic external medium with refractive index n_{ext} will be considered, leading to a single SGBM. It propagates within the plane parallel to the layers of the stack and the associated field is localized near the interface with the external medium. The variety of possible localized states in periodic structures is much richer than in isotropic media due to the fact that complete reflection can be produced within ranges of β (the *stop-bands*) where the structure would be expected to be transparent when considering the average index only. Further examples of localized states supported by multilayer stacks are given in [5].

The dispersion relation of the SGBM may be shown to take the form [5]:

$$p_j \frac{h_j}{2} - \arctan\left(\frac{\xi_e p_e \Lambda a_j + b_j}{\xi_j p_j \Lambda a_j - (\xi_e p_e / \xi_j p_j) b_j}\right) = m\pi \,, \tag{3}$$

where the parameters with subscripts j are those belonging to the last partial layer of the stack, $p_e = (\beta^2 - k^2 n_{ext}^2)^{1/2}$ is the amplitude decay rate into the external medium, and $\xi_e = 1 \ (1/n_{ext}^2)$ for TE (TM) polarization. The constants a_j and b_j determine the field amplitude and can be expressed via the components of the transfer matrix.

SGBM's can exist for both TE and TM polarizations. This is in contrast to other electromagnetic surface excitations like surface plasmons (oscillations of surface charges located at the interface with a metal) which can only form for TM polarization; in the TE case, the electric field component normal to the interface E_y equals zero and hence no surface charges are generated. A SGBM, however, is confined to the interface by different mechanisms that apply to both polarization states: total internal reflection from the external medium and Bragg reflection from the multilayer stack.

Figure 1 shows the effective propagation index $n_{eff} = \beta/k$ (solving Eq.(3) for β) and the band edges as a function of wavelength $\lambda = 2\pi/k$. For all numerical simulations, a stack consisting of alternating layers of $Al_{0.1}Ga_{0.9}As$ ($n_1 = 3.58$ at 830nm, $h_1 = 65.6nm$) and AlAs ($n_2 = 2.99$ at 830nm, $h_2 = 75.4nm$) was assumed. The final untruncated layer is made of AlAs, followed by a topmost layer of $Al_{0.1}Ga_{0.9}As$ of thickness $h_{fl} = 7.3nm$. As can be seen from Fig.1, SGBM's exhibit a considerable dispersion with wavelength. This is due to the fact that the formation of a SGBM relies on Bragg reflection, the effect of which is determined by the mutual phase shifts between the waves partially backreflected from each layer of the stack. In other words, the phase velocity is highly sensitive to the optical frequency, which is a general feature of defect modes.

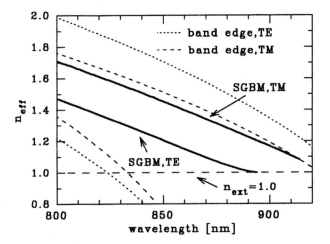

Figure 1. Stop-band edges and dispersion of SGBM's.

3.2. LOCATION WITHIN STOP-BAND

The phase velocity of defect modes is also strongly dependent on structural changes of the aperiodicity which defines the defect; an example is provided by changing the thickness of the topmost layer h_{f1} (or h_{f2}, if the topmost truncated layer consists of AlAs). The high sensitivity of SGBM's to h_{fj}, $j = 1,2$, is a result of the condition that the round-trip phase between two evanescent reflections at the interface must be an integral multiple of 2π in order to form a guided mode (this condition is equivalent to

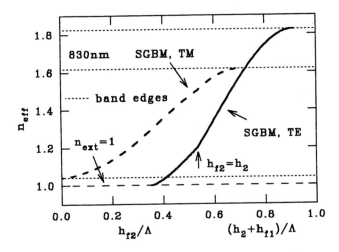

Figure 2. Effective propagation index of SGBM's versus thickness of the topmost layer.

the dispersion relation). By choosing the appropriate thickness of this layer, it is possible to place the SGBM anywhere within the stop-band, provided that n_{eff} remains larger than n_{ext} to ensure an evanescent behaviour within the external medium. This can be seen in Fig.2, where n_{eff} is plotted as a function of h_{fj}, starting with medium 2 up to h_2, followed by medium 1 up to h_1. The kink apparent in the graph for TE-polarization occurs at the position where the completed layer of medium 2 finishes (i.e., $h_{f2} = h_2$) and the next layer of medium 1 begins. Because $n_1 > n_2$, the change in round-trip phase per unit layer thickness is larger in medium 1, resulting in a steeper graph. Due to the considerable width of the stop-bands obtainable with these structures, a substantial range of quasi-single mode operation exists. For a certain range of thickness, this effect also can be used to completely suppress one or both states of polarization.

3.3. TUNING

In addition to a high sensitivity to structural changes, defect modes are also highly sensitive to compositional changes. For example, by changing the refractive index of the external medium, it is possible to tune n_{eff} as the associated field tail extends into the external medium and hence senses the refractive index n_{ext}. In Fig.3 the effective index is plotted as a function of the external index for both TE and TM-polarized SGBM's.

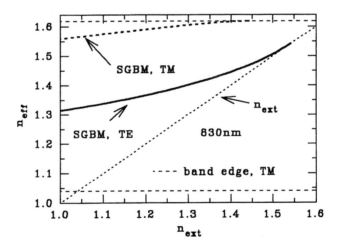

Figure 3. Effective propagation index of SGBM's versus refractive index of external medium.

The rate of change of n_{eff} with n_{ext} increases with increasing n_{ext} in the TE case. The reason is that for larger n'_{eff} the difference ($n_{eff} - n_{ext}$) becomes small, so the evanescent field extends more into the external medium according to $L_{decay} \propto 1/(n_{eff}^2 - n_{ext}^2)^{1/2}$. In the TM case the increase in n_{eff} is restricted by the presence of the band edge, resulting in a reduced tunability.

4. Applications

4.1. TUNABLE NARROW-BAND FIBRE TAP

The considerable dispersion suggests that SGBM's would be useful in designing a narrow-bandwidth filter, in which a multilayer stack is brought in close contact with a side-polished section of single-mode fibre, separated from it by a layer of material (e.g., oil) with an alterable refractive index n_{oil} and a thickness h_{oil} (see Fig.4). A remaining cladding thickness of 1 μm is assumed to separate the fibre core from the oil layer. In practice, n_{oil} can be varied thermally.

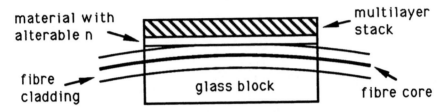

Figure 4. Generic fibre tap geometry.

Power exchange from the fibre mode to the SGBM supported by the stack occurs only if the modes are phase-matched. Due to the large difference in dispersion, the range of considerable coupling is restricted to a narrow band of wavelengths. In Fig.5 the maximum coupled power P_{max} (normalized to the input power P_o carried by the fibre mode) is shown as a function of the wavelength. In order to calculate P_{max}, the normal modes of the coupler were obtained first, modelling the fibre as a planar waveguide supporting a single mode with an appropriate effective index. Using the relationship between the normal-mode approach and coupled-mode theory [6], the coupled power follows in a straightforward manner.

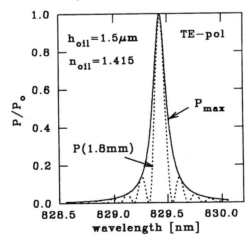

Figure 5. Maximum coupled power and power coupled at propagation length $L = 1.8$mm.

The maximum coupled power corresponds to the envelope of the graph obtained when calculating the power coupled to the SGBM at a fixed position $z = L$ along the fibre. An example is given by the dashed line for a coupling length equal to 1.8mm (which is the propagation distance needed to completely couple the power P_0 to the SGBM at the phase-matching point). A larger spacing h_{oil} between the fibre and stack leads to weaker coupling, resulting in a narrower bandwidth and a longer coupling length. However, the product of both remains fixed and can be used as a figure of merit to characterize the device performance; in the example considered it equals 0.3 nm mm. Due to the high sensitivity of the SGBM to the index of the intervening layer n_{oil}, good tunability is predicted. This is demonstrated in Fig.6, where the change in centre wavelength is shown as a function of n_{oil}.

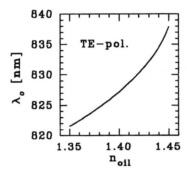

Figure 6. Centre wavelength of fibre tap versus refractive index of the oil layer.

4.2. MODE-SELECTIVE FIBRE COUPLER

Another application of SGBM's is in mode-selective coupling, where the aim is to selectively couple different spatial modes out of a multimode fibre. The device geometry is the same as in the previous section, with the single-mode fibre replaced

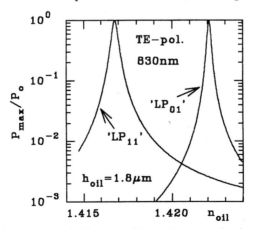

Figure 7. Maximum power coupled from each fibre mode versus refractive index of the oil layer.

by a multimode fibre. Coupling to different fibre modes is accomplished by phase matching them to the SGBM by tuning n_{oil}. Earlier experimental attempts to realize a mode-selective coupler employing surface plasmons supported by a thin metal film gave unacceptably high cross-talk from the non-phase-matched higher-order mode when phase-matched to the fundamental mode [7]. The reason is a deeper penetration of the modal field into the fibre cladding as the mode index increases, leading to a much stronger coupling. In addition, unavoidable ohmic losses in the metal layer cause a broadening of the resonances. The consequence was a trade-off between loss and selectivity, severely limiting the overall device performance. The replacement of the surface-plasmon by a virtually lossless SGBM with its higher sensitivity to the external medium results in lower cross-talk figures and better tunability. This can be seen from Fig.7, where the maximum power coupled from the modes of a two-moded fibre is shown as a function of n_{oil}. When tuned to the fundamental mode, the cross-talk from the non-phase-matched higher-order mode is $\sim -26\text{dB}$. When tuned to the higher-order mode, the cross-talk is $\sim -36\text{dB}$. These values compare favourably with the theoretical predictions for an optimized device based on surface plasmons [7], where the cross-talk figures are $\sim -10\text{dB}$ (tuned to LP_{01}) and $\sim -27\text{dB}$ (tuned to LP_{11}), respectively. As with the fibre tap, cross-talk figures can be decreased at the expense of a larger coupling length.

In conclusion, surface-guided Bloch modes on dielectric multilayer stacks can be regarded as a special type of defect modes in photonic band-gap structures. Owing to their unique properties they offer distinctive advantages in applications such as tunable narrow-band fibre taps and mode-selective fibre couplers.

5. Acknowledgements

The work was partly supported by the U.S. Department of the Air Force through the European Office of Aerospace Research and Development, London. The Optoelectronics Research Centre is an Interdisciplinary Research Centre of the U.K. Engineering and Physical Sciences Research Council.

6. References

1. Meade, R.D., Brommer, K.D., Rappe, A.M., and Joannopoulos, J.D. (1991) Electromagnetic Bloch waves at the surface of a photonic crystal, *Phys. Rev. A* **44**, 10961–10964.
2. Smith, D.R., Dalichaouch, R., Kroll, N., Schultz, S., McCall, S.L., and Platzman, P.M. (1993) Photonic band structure and defects in one and two dimensions, *J. Opt. Soc. Am. B* **10**, 314–321.
3. Stanley, R.P., Houdré, R., Oesterle, U., Ilegems, M., and Weisbuch, C. (1993) Impurity modes in one-dimensional systems: The transition from photonic band gaps to microcavities, *Phys. Rev. A* **48**, 2246–2249.
4. Yeh, P., Yariv, A., and Hong, Ch-S. (1977) Electromagnetic propagation in periodic stratified media. I. General theory, *J. Opt. Soc. Am.* **67**, 423–438.
5. Russell, P.St.J., Birks, T.A., and Lloyd-Lucas, F.D. (1995) Photonic Bloch Waves And Photonic Band Gaps, in E. Burstein and C. Weisbuch (eds.), *Confined Electrons and Photons: New Physics and Applications*, Plenum Press, New York, pp.585–633.
6. Vasallo, C. (1991) *Optical Waveguide Concepts*, Elsevier, Amsterdam, pp.27–35.
7. Barcelos, S., Zervas, M.N., and Russell, P.St.J. (1995) Selective excitation of fiber modes using surface plasmons, *IEEE Photon. Technol. Lett.* **7**, No.9.

FABRICATION OF 2-D INFRARED PHOTONIC CRYSTALS IN MACROPOROUS SILICON

U. GRÜNING AND V. LEHMANN
Siemens Corp. Research and Development,
Dep. ZFE T ME, 81730 Munich, Germany

Abstract. The fabrication of photonic band gap materials at micron and submicron lengthscales still poses some problems. Macroporous silicon is a way to meet the requirements of regularity and high index contrast for a 2-D infrared photonic material. The controlled formation of pores in n-type silicon by light-assisted electrochemical etching in hydrofluoric acid can lead to a regular pattern of uniform holes with minimal changes of the pore diameter both between neighboring pores and with depth. To show the feasibility of the approach, we etched a 340 μm deep 2-D square lattice of circular air rods with a lattice constant of 8 μm in an n-type silicon substrate. This structure possesses individual photonic gaps for both polarizations in the infrared region between 250 and 500 cm^{-1} (20 - 40 μm). The transmission spectra between 50 and 650 cm^{-1} were in good agreement with the theoretical calculated structure. The pore formation technique should allow the fabrication of photonic lattices with a complete 2-D band gap in the middle and near infrared.

1. Introduction

Recently theoretical and experimental interest has been drawn on material with modified electromagnetic properties[1]-[4]. The propagation of electromagnetic radiation in periodic dielectric structures can be described similar to the case of electrons propagating in a crystal. If the wavelength is in the order of the dimensions of the periodic structure, a photonic band gap - a frequency range where no electromagnetic modes exist - can open up in two or three dimensions and lead to interesting phenomena such as frequency

C. M. Soukoulis (ed.), Photonic Band Gap Materials, 453–464.

selective mirrors or the inhibition of spontaneous emission from an atom inside the gap[1].

Photonic band structures have first been mechanically fabricated in two[5]-[7]- and three[8]-[10]- dimensions, in the centi- and millimeter range, and allowed the verification of theoretically calculated band structures[11]-[15]. As the dimensions of the lattice must be in the same order of magnitude as the desired band gap wavelength, the scaling down to the interesting optical and infrared frequencies has posed problems due to the demanded uniformity and regularity of the photonic lattice. Additionally, material with a high refractive index is demanded for realizing large gaps. Two-dimensional photonic band gaps are technologically easier to realize and sufficient for many applications. Regular cylindrical hole structures on a submicron scale have been fabricated in AlGaAs and GaAs by reactive ion etching[16, 17]. However, extreme process control is necessary even for lattice depths of less than a micron, and problems with the uniformity still exist. Very recently, Inoue et al.[18] reported the first fabrication and experimental verification of a 2-D photonic band gap material in the near infrared in PbO- glass. However, in this material photonic gaps can only be realized for one polarization due to the low refractive index of PbO-glass.

Microporous silicon has recently received much attention due to its optical properties. A significant blueshift of the absorption edge[19] and visible room-temperature photoluminescence[20] are attributed to the existence of a porous, spongelike silicon skeleton with dimensions of a few nanometers in which quantum confinement effects are present. In contrast to this spongelike microporous structure, ordered arrays of macropores can be etched by both modifying slightly the formation conditions and changing the doping density of the substrate. *Macroporous silicon*[21] exhibits pores with a diameter in the micrometer region and a depth of several hundreds of micrometers. Consequently, no quantum size effects exist in macroporous silicon. The controlled formation of these pores can lead to a regular pattern of uniform holes with minimal changes of the pore diameter both between neighboring pores and with depth.

To show the feasibility of this approach for the experimental realization of photonic crystals, we fabricated a two-dimensional photonic band gap structure in macroporous silicon with polarization-dependent gaps between 250 and 500 cm^{-1} (20-40 μm)[22].

2. Electrochemistry of Silicon and Pore Formation Mechanism

First, the basics of electrochemistry of silicon electrodes will be briefly presented and the mechanism for pore formation and morphology will be discussed. Then the technique for controlled formation of regular two- dimensional arrays of macropores will be shortly explained, as details are published elsewhere[21].

Pore formation is a well-known feature of many metal and semiconductor electrodes under anodic conditions in different electrolytes. Though the pore formation process is often poorly understood, it has been technologically used for decades; for example, to enhance the performance of aluminum capacitors. Electrochemical pore formation in a solid-state electrode occurs if the pore walls are passivated in some way, while dissolution continues at the pore tips.

Microporous silicon was first observed by Uhlir and Turner in 1956 during anodic oxidation in an aqueous hydrofluoric acid solution (HF)[23]. Crystalline silicon, which is usually inert to HF, is attacked if the wafer is anodically biased. It is generally accepted that electronic holes are required for the dissolution reaction which must be supplied by the silicon. Holes are majority carriers in p-type silicon and must be created by illumination in n-type silicon. The exact dissolution mechanism and hence the surface morphology produced by this dissolution depend critically on whether the diffusion of reactants in the electrolyte or charge supply from the electrode is the rate-limiting step. If there is a lack of holes at the surface, but a high concentration of chemical reactants, i.e., HF molecules, every hole reaching the surface will be immediately consumed in the dissolution reaction. As the electric field lines are focused at surface depressions, the dissolution is enhanced there, leading to pore formation. If the reaction is limited by the diffusion of HF in the electrolyte to the electrode, and not by the hole supply, a surface charge of holes builds up. Convex parts of the silicon electrode which are more exposed to the electrolyte will then dissolve faster. This results in a smoothening of the surface which is known as electropolishing. Thus, two distinct regimes can be distinguished in the anodic part of the current-voltage plot, observable by a peak and a change of slope, as shown for a flat intensely illuminated silicon electrode in Fig. 1. For current densities higher than a critical current density J_{PS}, electropolishing is observed. For current densities lower than J_{PS}, pores form under hydrogen evolution. J_{PS} marks the point of steady state between ionic transfer and charge supply and is dependent on temperature and electrolyte concentration.

As holes are necessary for the reaction, the availability of holes determines the location of dissolution and is the origin of the substrate-doping dependent morphology of the porous structure. In n-type silicon, the anodic hole current can be adjusted independently of the applied voltage by photogeneration of holes. Thus, every point right of the I-V curve of a n-type electrode, as shown in Fig. 1, can be realized by increasing the applied bias. The critical current density J_{PS} depends on substrate orientation and is highest in $< 100 >$ direction. Macropore formation is found for the sha-

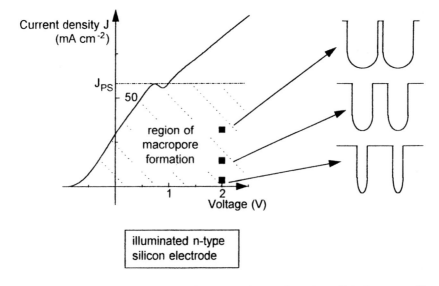

Figure 1. The anodic current density J versus voltage plot of a polished n-type silicon electrode in 5% aqueous HF under intense illumination is plotted as a solid line. Formation of macropores is observed for current densities J within the dashed region of the plot which can be realized by adjusting the illumination intensity. The dependence of pore diameter and current density J for a regular array of pores is sketched for three different values of J.

ded region in Fig. 1. If the sample is illuminated, either with red light at the front surface or with blue light at the backside, holes are generated in the bulk of the silicon electrode. The flow of holes will then be focused mainly to pore tips for two reasons: i) pore tips are located closest to the source of holes and ii) the electric field lines are focused at them, as illustrated in Fig. 2. The pore growth velocity will be limited by the HF transfer rate to the pore tips. Thus, a steady state condition between HF transfer and current density exists at the pore tips which is specified by the critical current density J_{PS}. The ratio of the total etching current density J to the critical current density J_{PS} equals the porosity of the macroporous layer. If this process is applied to a polished silicon substrate, the pores will grow in a random pattern. The average pore diameter is in the same order of magnitude as the space charge layer which forms at the semiconductor surface in contact with the electrolyte, thus increasing with substrate resistivity up to several μm. However, if etch pits, for example, generated by standard lithography and alkaline etching, are present, pore formation will initiate at these pits and regular patterns can be realized. The porosity of a square lattice with lattice constant a of circular pores with diameter d is given by $[\pi \cdot d^2/(4 \cdot a^2)]$. Thus the pore diameter d can be adjusted by the photogenerated current density according to

$$d = a \cdot \sqrt{\frac{4}{\pi} \cdot \frac{J}{J_{PS}}}, \tag{1}$$

with a being the lattice constant, J the etching current density and J_{PS} the critical current density which depends on HF concentration and temperature. This relation is illustrated in the insets of Fig. 1 where pore diameters are shown for three different current densities $J < J_{PS}$.

As the range of diameters for stable pore growth is strongly influenced by the extension of the space charge layer, the pore size can be varied over a wide range by changing the substrate resistivity. Macro pores have been grown between 100 nm and several tens of μm. For pore diameters below 100 nm, other charge transfer mechanisms like electrical breakdown and tunneling will become important. Degenerately doped n-material, for example, shows a porous, however, less ordered structure with features sizes in the 10 nm region. Therefore, todays limits of the macropore formation technique are set by the resolution of lithography for the initial pattern

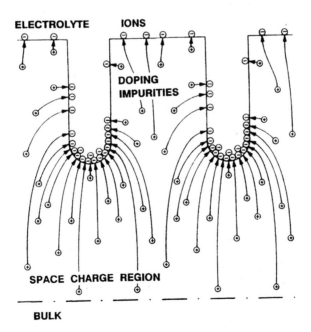

Figure 2. A schematic view of the etching process is shown. The electric field lines in the space charge layer at the n-type substrate/electrolyte interface are focused at the pore tips. Thus, holes arriving from the substrate are mainly collected at the pore tips, and pore walls are protected from dissolution.

definition. In addition, the minimal wavelength for PBG applications of macroporous Si is given by its band gap energy (1.1 eV).

3. Sample Preparation

The sample preparation will be only briefly explained as details of the experimental setup are published elsewhere[24]. First, a square lattice of etch pits with lattice constant of 8 μm is predefined on a 350 μm thick, 40 Ωcm n-type < 100 > silicon sample by oxidation, a standard photolithography, and subsequent alkaline etching. Then 340 μm deep macropores are grown on an area of about 2.6 cm^2 using the experimental setup sketched in Fig. 3. During etching, the front side of the sample is in contact with the HF electrolyte and kept in the dark. An ohmic contact at the edge of the sample and a platinum wire in the electrolyte were used to apply a bias. The back side of the wafer is illuminated and the hole current density is adjusted according to Eq. (1) to grow pores of a diameter of approximately 6.2 μm. In Fig. 4, a SEM micrograph of the etched structure is shown. In order to show the regularity of the photonic lattice over the depth, the sample was cleaved and one edge was polished to show a plane 45° inclined to the sample surface.

4. Theoretical Band Structure

The photonic band structure for this 2-D square lattice has been theoretically investigated by several groups[15, 25]. Two polarizations of the electromagnetic field are possible for electromagnetic waves propagating in the plane perpendicular to the axes of the pores. If the electric (magnetic) field is polarized in the z-direction, parallel to the axes of the pores, it is called E-polarized (H-polarized). The Brillouin zone of the square lattice is shown in the inset of Fig. 5. According to the approach of Plihal et al.[15], we calculated the band structure by the plane wave method using a total number of 289 plane waves. The results for a lattice constant a of 8 μm, a pore diameter d of 6.2 μm and a dielectric constant ϵ of Si of 11.7 are shown in Fig. 5 for both polarizations. A small photonic band gap exists between 274 and 279 cm^{-1} for the E-polarization and two larger gaps between 329 and 358 cm^{-1} and between 462 and 522 cm^{-1} for the H-polarization. However, as the gaps do not overlap, there is no common gap for both polarizations.

5. Transmission Measurements

As the transmission must be measured from the side surface of the wafer in order to reveal the photonic structure, small pieces of 1 mm x 5 mm are broken out of the etched sample along the Γ-X or along the Γ- M directions

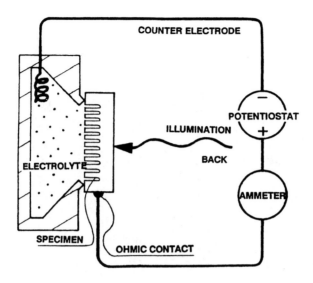

COUNTER ELECTRODE

POTENTIOSTAT

ILLUMINATION

BACK

ELECTROLYTE

AMMETER

SPECIMEN

OHMIC CONTACT

Figure 3. Sketch of the experimental setup for the electrochemical etching of macropores in silicon. The front side of the sample is in contact with the HF electrolyte and kept in the dark. An ohmic contact at the edge of the sample and a platinum wire in the electrolyte were used to apply a bias. The back side of the wafer is illuminated and the hole current density adjusted to provide the desired pore diameter.

of the photonic lattice (see arrows in Fig. 4). Several pieces with the same lattice orientation are then stacked in z-direction, resulting in two batches of about 1x5x3 mm^3 (x-y-z-direction). Transmission is measured along the x-direction between 50 and 650 cm^{-1} (15-200 μm) by a FT-IR spectrometer (Bruker IFS 113v), composed of a mercury lamp, a black polyethylene filter, a 3.5 μm thick mylar beamsplitter, and a silicon bolometer which is cooled to 4 K. The light is weakly focused on the sample. A Perkin-Elmer gold wire grid polarizer was used for polarization-resolved measurements between 200 and 650 cm^{-1}. A metal cone is used for increased efficiency in collecting the transmitted radiation behind the sample. Measurements are taken with a resolution of 4 cm^{-1} in a vacuum chamber at a pressure of 3 Pa. The spectrum of the empty sample holder is used as reference for normalization of the transmission spectra. The transmission spectra of the sample are shown in Fig. 6(a) for E- and in Fig. 6(b) for H-polarization for both the wave vector $\vec{k}$ parallel to the Γ-X (10)-direction (solid lines) and $\vec{k}$ parallel to the Γ-M (11)- direction (dotted lines).

First, we consider the structure of the transmission spectra. For E-polarized light in (10)-direction (solid line in Fig. 6(a)), an opaque region

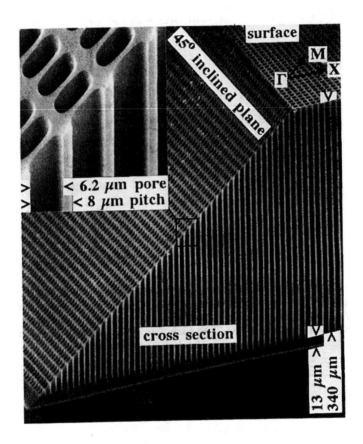

Figure 4. Scanning electron micrograph of the cleaved edge of a macroporous silicon sample with pores forming a two-dimensional square lattice. The lattice was etched 340 μm deep in a 353 μm substrate. One edge of the sample was polished to show a plane 45° inclined to the sample surface. The inset shows a magnification of the structure to reveal the pore dimensions. The lattice constant a is 8 μm and the pore diameter approximately 6.2 μm.

is observed between 225 and 275 cm^{-1}, corresponding to the calculated gap in (10)-direction between 211 and 279 cm^{-1}. A second drop in the transmittance is observed between 430 and 490 cm^{-1} with an additional dip around 455 cm^{-1}. According to the band diagram in Fig. 5, radiation

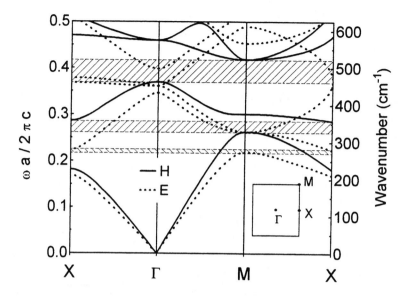

Figure 5. Photonic band structure for E- (dotted line) and H- (solid line) polarizations, for a two-dimensional (2-D) square lattice of circular holes (ϵ=1) in Si (ϵ=11.7) with a lattice constant a of 8 μm and a pore diameter of 6.2 μm. The left axis shows the angular frequency in units of $2\pi c/a$, whereas the right axis displays the corresponding wavenumbers. The inset shows the first Brillouin zone of the 2-D square lattice.

should be transmitted in this region via the third and fourth band for E-polarized light. However, a similar observation has already been made by Robertson et al.[6] when they found that E-polarized radiation was not transmitted by the third band of a square lattice of dielectric cylinders. They suggested certain bands could not be efficiently excited by plane waves due to symmetry constraints. In (11)-direction (dotted line in Fig. 6(a)), two small dips in the transmittance are observed, the first around 280 cm^{-1}, which may be ascribed to the band edges calculated at 274 and 327 cm^{-1}, even if it is somewhat narrower than expected. The second dip around 460 cm^{-1} may again be due to the missing (or inefficient) excitation of the third band[26]. Thus the spectra seem to indicate the existence of the small gap for E-polarization centered around 275 cm^{-1}, even if it is not completely resolved.

For H-polarization in (10)-direction (solid line in Fig. 6(b)), a drop of the transmitted intensity is observed between 230 and 350 cm^{-1} and for wave numbers higher than 450 cm^{-1}. This corresponds to the photonic band edges calculated at 227 and 358 cm^{-1} and at 462 and 575 cm^{-1}. In (11)-direction (dotted line in Fig. 6(b)), a sharp drop in the transmittance can be observed around 340 cm^{-1}, corresponding to the beginning of the first gap, calculated between 329 and 377 cm^{-1}. However, we observed no increase in the transmission between the first and the second gap (theoretically expected between 462 and 522 cm^{-1}). Apparently, the second band for H-polarization does not efficiently transmit radiation in this direction, though no obvious symmetry constraints exist for this band. Again, this observation is similar to those made by Robertson et al.[26]. Thus, there is evidence for two gaps for H-polarized light, the first centered at 345 cm^{-1} and the second starting around 450 cm^{-1} with its upper end experimentally not clearly resolved.

Now, having shown the principle agreement of the theoretically expected with the experimentally observed structure of the transmission spectra, we discuss the values of the absolute transmittance. Within the band gaps, the transmittance should drop to zero. The small intensities which we still observed are most likely due to some leakage paths around the photonic lattice, as there is about 13 μm of unetched substrate left which represents nearly 4% of the sample height. Neglecting the gap regions, the overall transmittance decreases with increasing wave numbers from about 20% at 50 cm^{-1} to 2% at 500 cm^{-1}. As already suggested by Inoue et al.[18], this decrease is not due to the available density of transmitting states which,

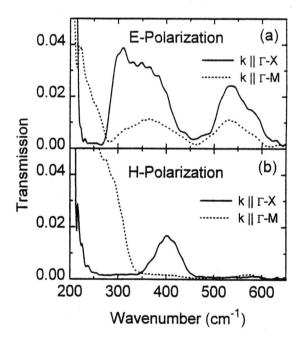

Figure 6. Transmission spectra between 200 and 650 cm^{-1} (a) for E-polarized (electric field parallel to the pore axes) and (b) for H-polarized radiation propagating along the Γ-X (solid lines) and the Γ-M direction (dotted lines).

in principle, increases with frequency, but can be understood in terms of scattering. The porous structure behaves like an effective medium for wavelengths much larger than the pore dimensions, whereas the pore walls act as strong scatterers for shorter wavelengths. This is especially true for the interfaces between the photonic lattice and the vacuum. As the cleaving edges of our samples were not smoothed (in order to avoid filling of the pores by polishing remnants), the surfaces were very rough, thus causing considerable scattering of the incident and transmitted radiation. To confirm this, the metal cone behind the sample which served to collect some of the divergent emanating radiation was removed. For the Γ-X direction, the transmitted intensity in the region between 290 and 390 cm^{-1} decreased by a factor of 3, while it stayed approximately constant inside the gap between 230 and 275 cm^{-1}.

6. Summary and Conclusion

We have succeeded in fabricating a two-dimensional photonic band gap structure in macroporous silicon with individual gaps for both E- and H-polarized radiation in the infrared region between 20 and 40 μm. A small opaque region common to both polarizations is observed around 460 cm^{-1} which may be due to inefficient excitation of some bands by plane waves. The transmittance spectra are in good agreement with theoretical predictions. Thus, it has been demonstrated that macroporous silicon can meet the requirements for a photonic band gap material. Further work will be focused on scaling down the gap to higher frequencies in the infrared. As the size of the pores can be adjusted by the substrate resistivity, the down-scaling seems to be limited only by the available lithography for the pattern definition. The use of a two-dimensional triangular lattice instead of the square lattice will lead to a large bandgap common to both polarizations[7, 27]. As the location of pore growth is defined by lithography, small disorder or defects can be easily and precisely introduced in the photonic lattice, allowing the investigation of localized photon modes. Thus, macroporous silicon provides an excellent medium in the infrared region for investigating new phenomena related to the existence of photonic band gaps.

7. Acknowledgments

We are grateful to C.M. Engelhardt and Prof. M. Stutzmann from the Walter-Schottky-Institut at Garching for the use of and assistance with the FT-IR spectrometer.

464

References

1. E. Yablonovitch, Phys. Rev. Lett. **58**, 2059 (1987).
2. S. John, Phys. Rev. Lett. **58**, 2486 (1987).
3. *Photonic Bandgaps and Localization*, edited by C.M. Soukoulis (Plenum, New York, 1993).
4. see also the special issues of the J. Opt. Soc. Am. B **10**, No. 2 (1993), and of the J. Mod. Optics **41**, No. 2 (1994).
5. S.L. McCall, P.M. Platzman, R. Dalichaouch, D. Smith, and S. Schultz, Phys. Rev. Lett. **67**, 2017 (1991).
6. W.M. Robertson, G. Arjavalingam, R.D. Meade, K.D. Brommer, A.M. Rappe, and J.D. Joannopoulos, Phys. Rev. Lett. **68**, 2024 (1992).
7. R.D. Meade, K.D. Brommer, A.M. Rappe, and J.D. Joannopoulos, Appl. Phys. Lett. **61**, 495 (1992).
8. E. Yablonowitch, T.J. Gmitter, and K.M. Leung, Phys. Rev. Lett. **67**, 2295 (1991).
9. E. Yablonowitch, T.J. Gmitter, R.D. Meade, A.M. Rappe, K.D. Brommer, and J.D. Joannopoulos, Phys. Rev. Lett. **67**, 3380 (1991).
10. E. Özbay, E. Michel, G. Tuttle, R. Biswas, M. Sigalas, and K.-M. Ho, Appl. Phys. Lett. **64**, 2059 (1994).
11. K.M. Leung and Y.F. Liu, Phys. Rev. Lett. **65**, 2646 (1990).
12. Z. Zhang and S. Satpathy, Phys. Rev. Lett. **65**, 2650 (1990).
13. K.M. Ho, C.T. Chan, and C.M. Soukoulis, Phys. Rev. Lett. **65**, 3152 (1990).
14. J.B. Pendry and A. MacKinnon, Phys. Rev. Lett. **69**, 2772 (1992).
15. M. Plihal, A. Shambrook, A.A. Maradudin, and P. Sheng, Opt. Commun. **80**, 199 (1991).
16. J.R. Wendt, G.A. Vawter, P.L. Gourley, T.M. Brennan, and B.E. Hammons, J. Vac. Sci. Technol. B **11**, 2637 (1993).
17. J.M. Gerard, A. Izraël, J.Y. Marzin, R. Padjen, and F.R. Ladan, Sol. Stat. Electr. **37**, 1341 (1994).
18. K. Inoue, M. Wada, K. Sakoda, A. Yamanaka, M. Hayashi, and J.W. Haus, Jpn. J. Appl. Phys. **33**, **L** 1463 (1994).
19. V. Lehmann and U. Gösele, Appl. Phys. Lett. **58**, 856 (1991).
20. L.T. Canham, Appl. Phys. Lett. **57**, 1046 (1990).
21. V. Lehmann, J. Electrochem. Soc. **140**, 2836 (1993).
22. U. Grüning, C.M. Engelhardt, and V. Lehmann, Appl. Phys. Lett. **66**, 3254 (1995)
23. A. Uhlir, Bell Syst. Techn. J. **36**, 333 (1956); D.R. Turner, J. Electrochem. Soc. **105**, 402 (1958).
24. V. Lehmann and H. Föll, J. Electrochem. Soc. **137**, 653 (1990).
25. P.R. Villeneuve and M. Piché, Phys. Rev. B **46**, 4973 (1992).
26. W.M. Robertson, G. Arjavalingam, R.D. Meade, K.D. Brommer, A.M. Rappe, and J.D. Joannopoulos, J. Opt. Soc. Am. B **10**, 322 (1993).
27. M. Plihal and A.A. Maradudin, Phys. Rev. B **44**, 8565 (1991).

TECHNIQUES FOR BANDSTRUCTURES AND DEFECT STATES IN PHOTONIC CRYSTALS

K. BUSCH[1,2], C. T. CHAN,[1,3] AND C. M. SOUKOULIS[1]

[1] *Ames Laboratory and Department of Physics and Astronomy, Iowa State University, Ames, IA 50011*
[2] *Institut für Theorie der Kondensierten Materie Universität Karlsruhe, 76128, Karlsruhe, Germany*
[3] *Physics Department, Hong Kong University of Science and Technology, Clear Water Bay, Hong Kong*

1. Introduction

In recent years experimental and theoretical studies of artificially manufactured, lossless periodic dielectric media called Photonic Band Gap (PBG) Materials or Photonic Crystals, have experienced a dramatically increased amount of interest[1, 2]. This can be attributed to their highly unusual, yet rather easily controllable properties which, firstly, offer exciting and challenging new problems for basic research and, secondly, give rise to numerous device applications.

Under favorable circumstances (high dielectric contrast and appropriate lattice structures, such as diamond[3] or rod-stacking structures [4]) a range of frequencies for which electromagnetic (EM) wave propagation is forbidden, irrespective of propagation direction, may be found. Furthermore, introducing deviations from periodicity can lead to localized modes inside these forbidden frequency ranges. It is precisely this absence of zero-point fluctuations (" The Photonic Crystal is quieter than even the vacuum", E. Yablonovitch in [2]) and the possibility of introducing well localized defect modes that may be utilized for studies of bound photon-atom states, the single-mode light emitting diode, high Q-cavities in frequency ranges not accessible to metal cavities, improving the performance of planar antennas and many more.

EM waves propagating in Photonic Crystals are in many ways analogous to electron waves propagating in "real" crystals, and, therefore, the famil-

465

C. M. Soukoulis (ed.), Photonic Band Gap Materials, 465–485.
© *1996 Kluwer Academic Publishers. Printed in the Netherlands.*

iar nomenclature of electronic crystals, such as the notions of Reciprocal Space, Brillouin Zones, Bandstructure, Bloch functions, etc., will be carried over to the EM case. From a theoretical viewpoint, it is highly desirable to have several independent methods for calculating the bandstructure and defect states of Photonic Crystals, thereby testing very thoroughly our understanding of wave propagation. Furthermore, from a practical point of view, some methods may be more efficient than the others in calculating a desired quantity. Thus the engineering of Photonic Crystals can benefit greatly by having at hand several techniques for "calculating" its properties. Since Maxwells Equations are essentially exact, we expect not only a good (naturally) agreement between these methods but also between theory and experiment.

In this paper we present various techniques which can be used to determine the bandstructure and defect modes of Photonic Crystals. In order to focus on the basic concepts and not to be bothered with notational and technical complications, we restrict ourselves mainly to one-dimensional (1D) systems, occasionally quoting two (2D) - and three (3D) dimensional results and/or giving a brief outline how to generalize the 1D case to 2D and 3D. First, we consider the most intuitive method, the so-called Transfer Matrix Method which allows for an analytical solution in 1D and has recently been extended to higher dimension. The analytical solution obtained by the Transfer Matrix Method is then used as a test for two more sophisticated techniques, the Plane Wave Method and the Multiple Scattering Technique. Since the proposal for PBG materials in 1987 by S. John, and E. Yablonovitch, the Plane Wave Method has become the standard technique to attack the bandstructure and single defect problems and has proved to be very accurate and efficient. Until now only a few publications used the Multiple Scattering Technique for bandstructure calculations. This is somewhat surprising, since it is known from electronic problems that it is a very powerful tool, although, admittedly, conceptual also the most difficult one.

2. The Layered Model System

To become specific, we start with a description of the layered system which will serve as our model system to apply the various techniques to: Consider a time harmonic EM wave with frequency ω propagating normal to the interfaces of an infinitely extended periodic arrangement of alternating layers A and B, as seen in Fig. 1.

We define the direct lattice $\tilde{R} = [z \cdot a \,|\, z \in Z]$ and the reciprocal lattice $\tilde{G} = [z \cdot (2\pi/a) \,|\, z \in Z]$. The experimentally controllable parameters are the dielectric constants ϵ_b and ϵ_a of the "high" and "low" dielectric material, the lattice constant a and the filling ratio $f = d/a$, where d is the thickness

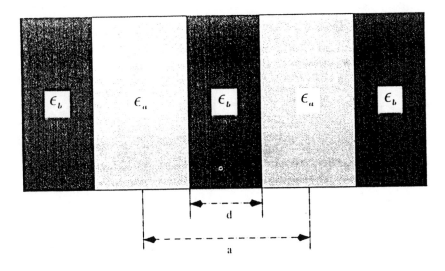

Figure 1. A schematic view of the infinite layered model system consting of alternating layers of dielectric materials A and B, having dielectric constants ϵ_a and ϵ_b with thicknesses a and a−d, respectively.

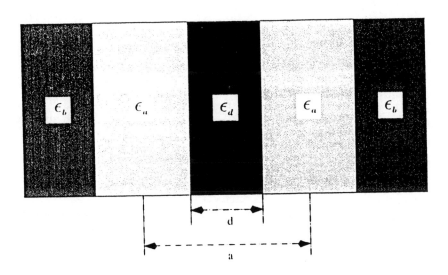

Figure 2. A schematic view of the defect model system obtained by replacing the B-layer of the periodic model at the origin by a defect layer D having the same thickness as the B-layer but dielectric constant $\epsilon_d = \epsilon_b + \Delta\epsilon$

of the B-layers.

The primary goal of all techniques lies in finding the frequencies ω for which propagation is allowed, i.e., mathematically speaking, we are interested in the (bounded) solutions of

$$\left(\partial_x^2 + \frac{\omega^2}{c^2}\epsilon_p(x)\right)E(x) = 0, \tag{1}$$

where $\epsilon_p(x) = \epsilon_a + (\epsilon_b - \epsilon_a)\sum_{R\in\bar{R}} S(x-R)$. The function $S(x)$ equals one if $|x| \leq d/2$ and zero else. Since Eq.(1) comprises a differential equation with periodic coefficients, its solutions obey the Bloch-Floquet Theorem

$$E(x+a) = e^{iqa}E(x), \tag{2}$$

where q is the "wave vector" labeling the solutions.

In order to study defect modes in this system, we replace the B-layer centered at $x = 0$ by a defect layer D having dielectric constant $\epsilon_d = \epsilon_b + \Delta\epsilon$, as seen in Fig. 2. In complete analogy to the electronic case, we expect that some band states will be pushed into the forbidden frequency ranges (gaps) of the periodic structure for both $\Delta\epsilon > 0$ ("donor modes") and $\Delta\epsilon < 0$ ("acceptor modes").

Consequently, the frequencies of the defect states are to be determined from

$$\left(\partial_x^2 + \frac{\omega^2}{c^2}\epsilon_p(x)\right)E(x) = -\frac{\omega^2}{c^2}\Delta\epsilon\, E(x). \tag{3}$$

3. Transfer Matrix Method

The starting point of the Transfer Matrix Method (TMM) is the observation that the solutions of Eq.(1), (3) inside each layer l can be written in terms of plane waves

$$E(x) = A_1^l e^{ik_l x} + A_2^l e^{-ik_l x},$$

where $k_l = \frac{\omega}{c}\sqrt{\epsilon_l}$ and ϵ_l is the dielectric constant of the l-th layer. The coefficients A_1^l and A_2^l have to be determined from the boundary conditions that both the E-field and its first derivative are continuous across an interface. It is now very convenient to write the field in form of a "formal" vector

$$\vec{E}(x) = \begin{pmatrix} A_1^l e^{ik_l x} \\ A_2^l e^{-ik_l x} \end{pmatrix}. \tag{4}$$

With this notation it is easily shown that the field in layer $l-1$ at x_{l-1} is related to the field at x_l in layer l, the layer to the right of layer $l-1$, via

$$\vec{E}(x_l) = \mathcal{P}^l(\Delta x_l)\,\mathcal{M}^{l-1,l}\,\mathcal{P}^{l-1}(\Delta x_{l-1})\,\vec{E}(x_{l-1})$$

where $\Delta x_l = |x_l - d_{l-1,l}|$ and $\Delta x_{l-1} = |x_{l-1} - d_{l-1,l}|$ are the distances from x_l and x_{l-1} to the interface of layers $l-1$ and l located at $x = d_{l-1,l}$. The matrices $\mathcal{P}$ and $\mathcal{M}$ are given by

$$\mathcal{P}^l(x) = \begin{pmatrix} e^{ik_l x} & 0 \\ 0 & e^{-ik_l x} \end{pmatrix} \tag{5}$$

$$\mathcal{M}^{l-1,l} = \begin{pmatrix} \frac{1}{2}(1 + k_{l-1}/k_l) & \frac{1}{2}(1 - k_{l-1}/k_l) \\ \frac{1}{2}(1 - k_{l-1}/k_l) & \frac{1}{2}(1 + k_{l-1}/k_l) \end{pmatrix}. \tag{6}$$

The physical meaning of these matrices is that $\mathcal{P}$ "propagates" the E-field a distance x in a uniform medium, whereas $\mathcal{M}$ takes the E-field from one side of an interface to the other.

In order to solve for the bandstructure of the periodic crystal, we have - according to Eq.(2) - to relate the fields at points separated by a lattice constant. Let these points be $x = -a/2$ and $x = a/2$. We then find

$$\vec{E}(-a/2) = \mathcal{M}\,\vec{E}(a/2) \tag{7}$$

$$\mathcal{M} = \mathcal{P}^a((a-d)/2)\,\mathcal{M}^{b/a}\,\mathcal{P}^b(d)\,\mathcal{M}^{a/b}\,\mathcal{P}^a((a-d)/2)$$

where the matrix elements of $\mathcal{M}$ are given by

$$\mathcal{M}_{\infty\infty} = \mathcal{M}_{\in\in}^* = e^{ik_a(a-d)}\left(\cos(k_b d) + \frac{1}{2}i\left(\frac{k_a}{k_b} + \frac{k_b}{k_a}\right)\sin(k_b d)\right)$$

$$\mathcal{M}_{\infty\in} = \mathcal{M}_{\in\infty}^* = \frac{1}{2}i\left(\frac{k_a}{k_b} - \frac{k_b}{k_a}\right)\sin(k_b d)$$

A comparison between Eq.(2) and Eq.(7) shows that we are looking for eigenvalues of $\mathcal{M}$ of the form e^{iqa}. Indeed, it is known from linear algebra, that the eigenvalues $\lambda_{1/2}$ of a 2×2 unimodal matrix such as $\mathcal{M}$, i.e., $\det[\mathcal{M}] = 1$, are given by $\lambda_{1/2} = e^{\pm iqa}$ where $\cos(qa) = \frac{1}{2}(\mathcal{M}_{11} + \mathcal{M}_{22})$. Explicitly, this reads as

$$\cos(qa) = \cos(k_a(a-d))\cos(k_b d) - \frac{1}{2}\left(\frac{k_a}{k_b} + \frac{k_b}{k_a}\right)\frac{\sin(k_a(a-d))}{\sin(k_b d)}, \tag{8}$$

which is the desired bandstructure equation. Obviously, those frequencies for which the right-hand side of Eq.(8) becomes greater than 1 or less than -1 are not allowed and, therefore, form the Photonic Band Gaps. A typical bandstructure using $\epsilon_a = 1$, $\epsilon_b = 13$ and $f = 0.5$ is shown in Fig. 3.

In order to proceed with the calculation of defect frequencies, we note that the periodic crystal together with the defect possesses inversion symmetry, i.e., the solutions are symmetric or antisymmetric with respect to inversion.

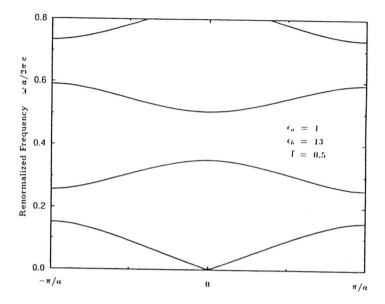

Figure 3. The bandstructure of the periodic model system with $\epsilon_a = 1$, $\epsilon_b = 13$ and f = d/a = 0.5 in terms of the renormalized frequency $\omega\,a/2\pi\,c$. The bandstructure was obtained with all three methods.

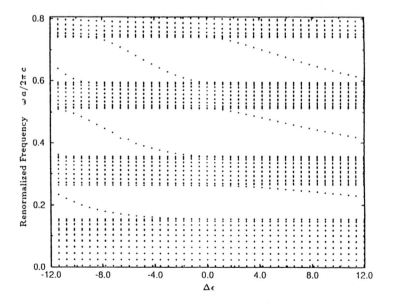

Figure 4. The defectfrequencies $\omega\,a/2\pi\,c$. as a function of $\Delta\epsilon$. The parameters of the underlying periodic lattice are: $\epsilon_a = 1$, $\epsilon_b = 13$ and f = 0.5

This fact can most conveniently be taken into account by specifying the fields at $x = 0$ to be

$$\vec{E}(0) = \begin{pmatrix} 1 \\ \pm 1 \end{pmatrix}$$

Next, since we are dealing with localized solutions, we have to replace the Bloch condition, Eq.(2) by

$$\vec{E}(x+a) = \pm e^{-\gamma a} E(x), \qquad |x| \geq a/2, \qquad \gamma > 0. \tag{9}$$

The E-field at $x = a/2$ then is

$$\begin{aligned}
\vec{E}(a/2) &= \mathcal{D} \vec{E}(0) \\
\mathcal{D} &= \mathcal{P}^a((a-d)/2) \, \mathcal{M}^{d/a} \, \mathcal{P}^d(d/2)
\end{aligned}$$

Finally, using Eq.(9) with $x = a/2$, we arrive at

$$(\mathcal{M} \mp e^{-\gamma a} \, \mathcal{E}) \, \mathcal{D} \, \vec{E}(0) = 0, \tag{10}$$

and $\mathcal{E}$ is the 2×2 unit matrix. Thus, we have obtained two equations, the second being the complex conjugate of the first. Rephrasing Eq.(10) as

$$\left(\mathcal{M}_{11} + \mathcal{M}_{12} \frac{(\mathcal{D}_{12} \pm \mathcal{D}_{11})^*}{(\mathcal{D}_{11} \pm \mathcal{D}_{12})} \right) \mp e^{-\gamma a} = 0, \tag{11}$$

leads, upon taking the imaginary part, to the desired defect mode equation

$$\tan(k_a(a-d)) = -\frac{1}{2} \tan(k_b d) \left(\left(\frac{k_a}{k_b} + \frac{k_b}{k_a} \right) - \left(\frac{k_a}{k_b} - \frac{k_b}{k_a} \right) \phi^{\pm}(k_d) \right) \tag{12}$$

$$\begin{aligned}
\phi^{\pm}(k_d) &= \frac{\cos(k_a(a-d)) \, \Lambda_1^{\pm}(k_a, k_d) \mp 2 \frac{k_d}{k_a} \sin(k_a(a-d)) \sin(k_d d)}{\Lambda_2^{\pm}(k_a, k_d) \cos(k_a(a-d))} \\
\Lambda_1^{\pm}(k_a, k_d) &= \left(1 - \left(\frac{k_d}{k_a} \right)^2 \right) \pm \left(1 + \left(\frac{k_d}{k_a} \right)^2 \right) \cos(k_d d) \\
\Lambda_2^{\pm}(k_a, k_d) &= \left(1 + \left(\frac{k_d}{k_a} \right)^2 \right) \pm \left(1 - \left(\frac{k_d}{k_a} \right)^2 \right) \cos(k_d d)
\end{aligned}$$

and the sign of the right-hand side of Eq.(8) determines whether to use ϕ^+ or ϕ^-. For the simplest case, $k_d = k_a$, i.e., replacing the B-layer at the origin by an A-layer, $\phi^{\pm}$ simplifies considerably

$$\phi^{\pm}(k_a) = \pm \frac{\cos(k_a d)}{\cos(k_a(a-d))},$$

which is the result given by D. Smith et al. [5].

Obviously, the TMM is a very physical approach for 1D systems and, in addition, can be used to numerically study disordered systems [6], [7]. Generalization of the TMM to higher dimensions is non trival, but numerical formulation has been given recently [8], [9] and has already been applied succesfully to many problems of interest [8] - [11].

4. Plane Wave Method

The Plane Wave Method (PWM) is based on the use of the Bloch-Floquet Theorem in the form which displays the lattice periodicity of the Bloch waves

$$E_q(x) = e^{iqx} u_q(x), \tag{13}$$

$$u_q(x) = u_q(x+R) \quad \forall R \in \tilde{R}. \tag{14}$$

Here, q is the (unrestricted) "wave vector" of the Bloch functions $E_q(x)$. Now, q may be restricted to lie within the first Brillouin Zone (1.BZ) $[-\pi/a, \pi/a]$, any points outside providing no new information. Then, however, for any given $q \in$ 1.BZ, we have several discrete solutions which we will label by the band index n. And finally, for purposes of computing the defect states, it is most convenient to impose the Born-v.Karman (cyclic) boundary condition

$$E_{nq}(x) = E_{nq}(x+Na) \quad N \gg 1, \tag{15}$$

from which it follows that q is of the form

$$q = \frac{m}{N} \cdot \frac{2\pi}{a}, \quad -N/2 \leq m \leq N/2, \quad m \in Z, \tag{16}$$

i.e., the exponential factors occurring in Eq.(13) are nothing but the irreducible representations of the finite (abelian) translation group defined via Eq. (2) and Eq.(15).

The PWM proceeds in expanding the lattice periodic $u_{nq}(x)$ into Fourier series, leading to the following standard symmetric eigenvalue problem to be solved for the allowed (eigen-) frequencies and respective plane wave expansion coefficients E_G^{nq}

$$u_{nq}(x) = \sum_{G \in \tilde{G}} E_G^{nq} e^{iGx}, \tag{17}$$

$$\sum_{G' \in \tilde{G}} \frac{\epsilon_{G-G'}}{(2\pi/a)|q + G| (2\pi/a)|q + G'|} C_{G'}^{nq} = \left(\frac{2\pi c}{\omega_{nq} a}\right) C_G^{nq}, \tag{18}$$

$$\epsilon_G = \frac{1}{a} \int_{-a/2}^{a/2} dx\, e^{-iGx}\, \epsilon_p(x) \quad , \quad E_G^{nq} = C_G^{nq}/|q+G|.$$

When doing the numerics, this infinite system of linear equationshas to be truncated, i.e, using a finite number of reciprocal lattice "vectors" G. We find that excellent agreement with the analytical solution Eq. (8) is already obtained for a fairly small number of reciprocal lattice "vectors" of 37 in the case of the bandstructure siplayed in Fig. 3, i.e. the parameters being $\epsilon_a = 1$, $\epsilon_b = 13$ and $f = 0.5$.

There are numerous other ways to obtain an equivalent eigenvalue problem. For example, one can rewrite Eq.(1) in such a way that the inverse of the dielectric function occurs [12]. Also, one can use the corresponding equation satisfied by the H-field rather than the one satisfied by the E-field [3], [13].

In 2D the two possible polarizations can be decoupled, effectively leading to two distinct scalar equations of the Bloch-type. The band gaps created in this way do not always overlap [13], thus opening a vast field for interesting applications. In 3D, the polarizations cannot be separated any more and one has to solve the full vector problem [3], since the scalar approximation [14], [15] turns out to agree rather poorly with experiment. Very recently, there has been growing interest in periodic structures in which one material posseses a frequency dependent dielectric function such as in a metal or polar semiconductor. To address this kind of Photonic Crystals a modification of the PWM is needed [16].

The most straightforward way to investigate the single defect problem in the framework of the PWM consists in employing what is called the supercell method [17]. The approximation made, lies in replacing the single defect problem by an artificially periodic problem that uses a "super unit cell" (cp. Eq.(15)) "constructed" of the defect and a huge portion of the embedding periodic crystal. This leads to a bandstructure containing very narrow "impurity bands" inside the gaps of the original, purely periodic, system, thus representing the defect frequencies. The reasoning behing that is the fact that once the "supercell" is large enough, the interaction of the defects in neighbouring cells can be neglected. However, since this method deals with very complicated unit cells, the convergence in 3D is rather slow and a supercomputer has to be used to do the necessary computations.

A more elegant and numerically more convenient way to solve the defect problem, Eq.(3), uses the fact that the Greens function of Eq.(1) is given by

$$G(x, x'|\omega^2) = \frac{1}{Na} \sum_n \sum_q \frac{E_{nq}^*(x')\, (\omega_{nq}^2/c^2)\, E_{nq}(x)}{\omega_{nq}^2/c^2 - \omega^2/c^2}, \tag{19}$$

where the q are determined by Eq.(16) and the number of bands has later on to be specified such that the emerging results are reasonably converged. With the help of Eq.(19) we can recast Eq.(3) in the form of an integral equation

$$E(x) = \frac{\omega^2}{c^2} \int_{defect} dx' \, G(x, x'|\omega^2) \, \Delta\epsilon \, E(x'). \tag{20}$$

As can be seen from Eq.(20), this method -like the supercell method- allows us to deal with more general defect problems, such as a spatially varying $\Delta\epsilon(x)$.

Discretizing the real space integral in Eq.(20) yields the following condition on ω:

$$\det\left(\mathcal{E} - \frac{\omega^2}{c^2}\Delta\epsilon\,\mathcal{G}\,\mathcal{W}\right) = 0, \tag{21}$$

where $\mathcal{E}$ is the unit matrix. The matrix elements of $\mathcal{G}$ and $\mathcal{W}$ are $\mathcal{G}_{ij} = G(x_i, x_j|\omega^2)$ and $\mathcal{W}_{ij} = w_j$; $x_i,.., x_p$, and $w_1,..,w_p$ being the discretization points and respective weights of a p-point discretization scheme applied to the integral in Eq.(20). Solving Eq.(21) using Eq.(18), (19) for our model defect problem gives excellent agreement with the exact solution using the TMM.

Furthermore, the PWM allows us to compute the actual defect profile. To this end, we expand the E-field in Eq.(20) into Bloch waves

$$E(x) = \sum_n \sum_q f_{nq} \, E_{nq}(x), \tag{22}$$

where, again, we use the finite number of q points given by Eq.(16) and use a high enough number of bands to get converged results. The orthonormality relation (which is a consequence of the normalization of the C_G^{nq} obtained in Eq.(18))

$$\frac{1}{Na} \int_{-N/2a}^{N/2a} dx \, E_{n'q'}^*(x) \, \epsilon_p(x) \, E_{nq}(x) = \frac{c^2}{\omega_{nq}^2} \delta_{nn'} \, \delta_{qq'}, \tag{23}$$

between any two Bloch waves can be utilized to project Eq.(20) onto a single Bloch wave. The final result is a standard Hermitian eigenvalue problem for the expansion coefficients f_{nq} of the E-field

$$\left(\frac{2\pi c}{\omega a}\right)^2 f_{nq} = \sum_{n'} \sum_{q'} \left(<nq|\Delta\epsilon|n'q'> + \delta_{nn'} \, \delta_{qq'} \left(\frac{2\pi c}{\omega_{nq}a}\right)^2\right) f_{n'q'} \tag{24}$$

$$<nq|\Delta\epsilon|n'q'> = \left(\frac{2\pi}{a}\right)^2 \frac{1}{Na} \int_{-N/2a}^{N/2a} dx \, E_{nq}^*(x) \, \Delta\epsilon \, E_{n'q'}(x). \tag{25}$$

Here, again, we note in passing that within the PWM, Eq.(24), it is possible to deal with spatially varying defects. Furthermore, if the defect posses inversion symmetry -as in our model problem- Eq.(24) reduces to a standard (real) symmetric eigenvalue problem. In Fig. 4. we present a plot of the defect frequencies obtained with Eq.(24) versus the defect strength $\Delta\epsilon$ where the parameters of the perfect lattice are $\epsilon_a = 1$, $\epsilon_b = 13$ and $f = 0.5$. Again, the defect frequencies obtained using the PWM agree with the analytical solution Eq. (12) of the TMM within the numerical accuracy for 67 reciprocal lattice vectors, 21 q-points and 8 bands. In this way, we get a feeling for the number of reciprocal lattice point G_{max}, q-points q_{max} and bands n_{max} that are necessary to get a fully converged result. Typical defect profiles for $\Delta\epsilon = -10$ and $\Delta\epsilon = 10$ in the 1st and 3rd gap, obtained with the PWM ,i.e, Eq.(24), are shown in Fig. 5, 6, respectively. They clearly display the localized character of the defect states inside the gaps.

The conversion of the defect procedure to higher dimensions is rather straightforward [12], however, in 2D we run into serious convergence problems using the 2D-analogue of Eq.(24). It appears that the expansion of a localized defect mode into extended Bloch waves is like forcing square peg into a round hole, i.e., bad convergence properties may be inherent in this apporach, becoming more pronounced in 2D and 3D. Therefore, it would be interesting to implement a theory based on Wannier functions [18], each of which is localized around a single lattice point.

5. Multiple Scattering Technique

The Multiple Scattering Technique (MST) is in some sense the most powerful and, consequently, the most complicated technique. MST allows for analytical solutions in 1D, is applicable to higher dimensions, gives both the defect frequencies and field patterns and, finally, can be used to study disordered systems. The price to be paid for such a powerful method are quite cumbersome calculations involving many indices. We want to mention that our exposition of the theory follows very closely the work of Butler [24].

Since the MST was originally developed for 3D and, for radial symmetric "potentials", is based on expansions of all quantities of interest into spherical harmonics, we start by setting up the notion of the 1D "spherical" harmonics

$$Y_l(x) = \frac{1}{\sqrt{2}} \left[\text{sign}(x) \right]^l, \quad l = 0, 1. \tag{26}$$

Just like in the case of 3D spherical harmonics there exists an orthonormality and a closure relation as well as an addition theorem between the 1D "spherical" harmonics. Furthermore, we define the 1D "spherical" Bessel

476

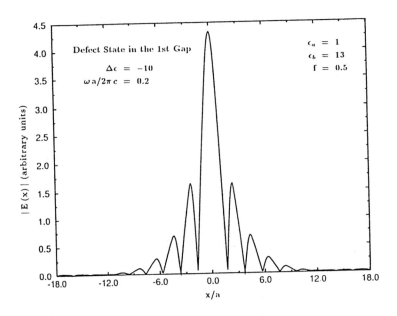

Figure 5. Field pattern of a defect state inside the first gap. The parameters are: $\Delta\epsilon = -10$, $\omega\, a/2\pi\, c = 0.2$, $\epsilon_a = 1$, $\epsilon_b = 13$ and $f = 0.5$

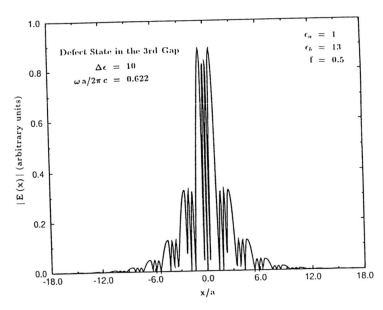

Figure 6. Field pattern of the defect state inside the third gap. The parameters are: $\Delta\epsilon = 10$, $\omega\, a/2\pi\, c = 0.622$, $\epsilon_a = 1$, $\epsilon_b = 13$ and $f = 0.5$

functions according to

$$j_l(z) = \cos(z - l\pi/2), \quad n_l(z) = \sin(z - l\pi/2) \tag{27}$$
$$h_l^+(z) = j_l(z) + i n_l(z), \quad l = 0, 1. \tag{28}$$

where, again in complete analogy to the 3D spherical Bessel functions, we have a Wronskian and a sum rule

$$j_l(z) n_l'(z) - j_l'(z) n_l(z) = 1 \tag{29}$$
$$\sum_{l=0,1} j_l^2(z) = 1 \tag{30}$$

It is an easy exercise to expand a plane wave and the free Greens function $G_0(x, x')$ into 1D "spherical" harmonics

$$e^{ik_a x} = 2 \sum_l i^l j_l(k_a|x|) Y_l(k_a) Y_l(x) \tag{31}$$

$$G_0(x, x') = \frac{e^{ik_a|x-x'|}}{2ik_a} = \frac{1}{ik_a} \sum_l j_l(k_a r_<) h_l(k_a r_>) Y_l(x) Y_l(x'). \tag{32}$$

$r_<$ and $r_>$ denote the lesser and greater of $|x|$ and $|x'|$, respectively.

Let us first consider the single barrier problem, i.e., a layer of dielectric constant ϵ_b and thickness d at the origin separating two halfspaces of dielectric constant ϵ_a. To solve the scattering problem, we use the following expansions for the fields $u_{in}(x)$ and $u_{out}(x)$ inside and outside the B-layer:

$$u_{in}(x) = 2 \sum_l i^l A_l j_l(k_b r) Y_l(k_b) Y_l(x), \tag{33}$$

$$u_{out}(x) = 2 \sum_l i^l [j_l(k_a|x|) + D_l h_l(k_a|x|)] Y_l(k_a) Y_l(x), \tag{34}$$

in this way introducing the scattering coefficients A_l and D_l which can be calculated by matching the boundary conditions. This ansatz clearly reflects our expectation that a plane wave impinging on the B-layer leads to a outgoing "spherical wave". However, for 1D systems it is more usual to discuss the single barrier problem in terms of transmission and reflection coefficients. Indeed, after a little algebra, it is easily shown that $t-1 = D_0 + D_1$ and $r = D_0 - D_1$, t and r being the transmission and reflection amplitudes, respectively. With those scattering coefficients, the main ingredients of MST, namely the scattering phaseshifts δ_l and the single scatterer t-matrix t_l, are given by

$$D_l = i e^{i\delta_l} \sin(\delta_l) \tag{35}$$

$$t_l = i k_a D_l \tag{36}$$

$$D_l = \frac{j_l(k_b d/2) j_l'(k_a d/2) - (k_b/k_a) j_l'(k_b d/2) j_l(k_a d/2)}{(k_b/k_a) j_l'(k_b d/2) h_l(k_a d/2) - j_l(k_b d/2) h_l'(k_a d/2)}. \tag{37}$$

The single barrier Greensfunction with outgoing-wave-boundary conditions may now be determined by standard procedures [19]:

$$G^{(1)}(x, x') = \frac{1}{k_a} \sum_l R_l(r_<) F_l(r_>) Y_l(x) Y_l(x'), \tag{38}$$

where $F_l(z) = N_l(z) - i R_l(z)$. The functions R_l and N_l for $|x| > a/2$ are given by $R_l(x) = \cos(\delta_l) j_l(k_a|x|) - \sin(\delta_l) n_l(k_a|x|)$ and $N_l(x) = \cos(\delta_l) j_l(k_a|x|) + \sin(\delta_l) n_l(k_a|x|)$. For $|x| < a/2$ they are given by linear combinations of $j_l(k_b|x|)$ and $n_l(k_b|x|)$ such that $G^{(1)}(x, x')$ is continuous across the "potential step", $R_l(x) \simeq x^l$ as $l \to 0$ and their Wronskian equals k_a:

$$R_l(x) F_l'(x) - R_l'(x) F_l(x) = k_a.$$

The central idea of MST is now to express the Greensfunction of an arbitrary assembly of scatterers using only the positions $\{R_m\}$ and t-matrices $\{t_l^m\}$ of the single scatterers. In operator notation can be written as [20]:

$$\hat{G} = \hat{G}_0 + \hat{G}_0 \hat{T} \hat{G}_0 \tag{39}$$

$$\hat{T} = \sum_m \hat{t}^m + \sum_m \sum_{n \neq m} \hat{t}^m \hat{G}_0 \hat{t}^n +$$

$$\sum_m \sum_{n \neq m} \sum_{p \neq n} \hat{t}^m \hat{G}_0 \hat{t}^n \hat{G}_0 \hat{t}^p + \cdots \tag{40}$$

Formally, this equations can be solved in real space by introducing the Scattering Path Operator $\tau_{l_1 l_2}^{mn}(\{R_m\})$ introduced by Gyorffy and Stott [21] which, in some sense, is the analogue to the Locator Greens function known from Tight-Binding electronic structure calculations [22]:

$$G(x, x') = G_0(x, x') +$$

$$\sum_{m,n} \sum_{l_1, l_2} \left(\frac{1}{ik_a}\right) h_{l_1}(k_a|x - R_m|) Y_{l_1}(x - R_m) *$$

$$\tau_{l_1 l_2}^{mn}(\{R_m\}) \left(\frac{1}{ik_a}\right) h_{l_2}(k_a|x' - R_n|) Y_{l_2}(x' - R_n) \tag{41}$$

$$\tau_{l_1 l_2}^{mn}(\{R_m\}) = t_{l_1}^m \delta_{mn} \delta_{l_1 l_2} +$$

$$\sum_{l_3} \sum_{p \neq m} t_{l_1}^m \tilde{G}_{l_1 l_3}(R_m - R_p) \tau_{l_3 l_2}^{pn}(\{R_m\}). \tag{42}$$

$\tilde{G}_{l_1 l_2}(x)$ is related to the free Greens function. Details are given in the appendix, Eq. (66). The density of states (DOS) $N(\omega)$ can be extracted from the imaginary part of $G(x, x)$ via [23]

$$N(\omega) = -\frac{2\omega}{\pi c^2} \int dx \operatorname{Im}[G(x, x)].$$

Consequently, the poles of τ determine the eigenfrequencies of the sytem. Our attempts, therefore, must be directed towards evaluating the scattering path operator τ in Eq. (42). For the periodic case we accomplish this task by means of a lattice Fourier transform.

$$t_{l_1}^m \equiv t_{l_1} \quad \forall m \tag{43}$$

$$\breve{G}(q) = \sum_{p \neq 0} e^{iqpa}\, \tilde{G}(pa)\,, \quad \tilde{G}(0) = 0\,, \tag{44}$$

$$\breve{\tau}_{l_1 l_2}(q) = \lim_{N \to \infty} \frac{1}{N} \sum_m \sum_n e^{iq(m-n)a}\, \tau_{l_1 l_2}^{mn}(\{R_m\}) \tag{45}$$

which leads to the solution

$$(\breve{\tau}^{-1})_{l_1 l_2}(q) = t_{l_1}^{-1}\delta_{l_1 l_2} + \breve{G}_{l_1 l_2}(q) \tag{46}$$

Therefore the eigenfrequencies of the periodic system, i.e. the bandstructure is given by the zeroes of the determinant of $\breve{\tau}^{-1}$. After some algebraic manipulations [24] we then find

$$\cos(qa) = \frac{\cos(k_a a + \delta_0 + \delta_1)}{\cos(\delta_0 - \delta_1)} \tag{47}$$

We note that k_a as well as the scattering phaseshifts δ_0 and δ_1 depend on the frequency ω. The solutions coming from this bandstructure equation are identical to those determined by Eq.(8) and Eq.(18).

However, in order to treat the single defect problem, a multicenter expansion of $G(x, x')$ as in Eq.(41) is not very useful. We are maily interested in the Greens function $G(x, x')$ when x and x' are both within the same Wigner-Seitz cell. Let us suppose that x and x' are both within the Wigner-Seitz cell at the origin (but for the time being still in the A-material). We use the two center expansions Eq.(62)-(64) and the relations Eq.(69)-(71) to express all terms in Eq.(41) as 1D "spherical harmonics" about the origin, the final result being [24]

$$G(x, x') = G^{(1)}(x, x') + \tag{48}$$
$$\sum_{l_1 l_2} R_{l_1}(k_a|x|)\, e^{i\delta_{l_1}^0}\, Y_{l_1}(x)\, G_{l_1 l_2}^{00}\, R_{l_2}(k_a|x'|)\, e^{i\delta_{l_2}^0}\, Y_{l_2}(x') \tag{49}$$

where we have introduced the $G_{l_1 l_2}^{00}$ as

$$G_{l_1 l_2}^{00} = (t_{l_1}^0)^{-1}(\tau_{l_1 l_2}^{00} - t_{l_1}^0 \delta_{l_1 l_2})(t_{l_2}^0)^{-1} \tag{50}$$

Our derivation applies only in the A-material region of the cell at the origin. However, using the definitions of $R_l(x)$ for the B-material, Eq.(48) satisfies the requirements on the Greens function also everywhere inside the cell and has the correct boundary conditions so that it is in fact valid at all points within the cell at the origin. In addition, we note that the imaginary part of the Greens function is

$$\text{Im}\left[G(x, x')\right] = \sum_{l_1 l_2} \frac{R_{l_1}(k_a|x|)\, R_{l_2}(k_a|x'|)}{\sin(\delta_{l_1})\,\sin(\delta_{l_2})}\, \text{Im}\left[\tau_{l_1 l_2}^{00}\right] Y_{l_1}(x)\, Y_{l_2}(x') \tag{51}$$

Thus, the density of states (DOS) is determined directly by τ^{00}.
Consequently, we have to proceed with computing the scattering path operator $\bar{\tau}^{00}$ for a single defect in an otherwise periodic lattice. Defining, cp. Eq.(44), $\tilde{G}_{l_1 l_2}(x=0)=0$ and using the equation of motion of τ, Eq.(42), it is easy to show that the site diagonal elements τ^{mm} can be written in the following "angular momentum"-matrix form

$$\tau^{mm} = ((t^m)^{-1} - \Delta^m)^{-1}, \tag{52}$$

where Δ^m depends on all t^n EXCEPT t^m. Consequently, $\bar{\tau}^{00}$ is given in terms of τ^{00}, the scattering path operator for the periodic system, by

$$\bar{\tau}^{00} = \tau^{00}\,(\mathcal{E} + \tau^{00}((t_d)^{-1} - (t_b)^{-1}))^{-1} \tag{53}$$

leaving us with the need to calculate τ^{00}.
Fortunately, this can be done using the Brillouin zone integral of $\breve{\tau}_{l_1 l_2}(k)$ as given in Eq.(46)

$$\tau_{l_1 l_2}^{00} = \frac{1}{2\pi} \int_{-\pi}^{\pi} d\theta\, \tilde{\tau}_{l_1 l_2}(\theta) \tag{54}$$

The result of which turns out to be [24]

$$\tau_{10}^{00} = \tau_{01}^{00} = 0 \tag{55}$$

$$\tau_{00}^{00} = -k_a \frac{\tan(\delta_0)}{1 + \tan(\delta_0)\tan(\delta_1)}\, A(x, f_1) \tag{56}$$

$$\tau_{11}^{00} = -k_a \frac{\tan(\delta_1)}{1 + \tan(\delta_0)\tan(\delta_1)}\, A(x, f_0) \tag{57}$$

where we used the following nomenclature

$$x = \frac{\cos(k_a a + \delta_0 + \delta_1)}{\cos(\delta_0 - \delta_1)} \tag{58}$$

$$f_0 = \cos(k_a a) - \tan(\delta_0)\sin(k_a a) \tag{59}$$

$$f_1 = \cos(k_a a) - \tan(\delta_1)\sin(k_a a) \tag{60}$$

$$A(x, f) = 1 + \frac{x - f}{(x^2 - 1)^{1/2}} \tag{61}$$

It is now a straightforward procedure to obtain the DOS for the single defect problem. Fig. 7 shows the DOS of a periodic 1D photonic crystal with parameters $\epsilon_a = 1$, $\epsilon_b = 13$ and $f = 0.5$ as well as the DOS when replacing the layer at the origin by defect layers having $\epsilon_d = 9.0$. As can be seen from Fig. 4 the agreement to the previously discussed methods is excellent .

In order to obtain the field pattern of the defect states, we would have to chose a different approach, namely to start from the wavefields rather than from the Greens function. This has been done for the electronic case [25] and follows pretty much the same lines as the present approach. Once the single defect problem has been solved, substitutionally disordered systems can be studied in the framework of the Coherent Potential Approximation (CPA). Again, until now this has been done only for the electronic case (cp. the references given in [22]).

Bandstructure calculations for photonic systems employing MST for 2D [26] and 3D [27] have shown excellent agreement with the plane wave results. Until now, these works are the only ones that use MST in connection with photonic problems.

6. Discussion

In conclusion, we have shown how the TMM, PWM and MST can be applied to PBG materials and have obtained excellent agreement between these methods. The advantages and limitations have also been discussed.

We did not discuss the various finite difference approaches that are used to directly solve Maxwells equation in either the time- or frequency domain, since this would go beyond the scope of this article. It is fair to say that the bandstructures obtained with PWM and MST can be regarded as very accurate and reliable. The single defect problem, especially in 2D and 3D, has received much less attention and therefore a lot of work needs to be done in this direction. The Wannier function approach for defects should be investigated more carefully. Furthermore, it will be very interesting to develop a theory of weakly disordered PBG structures, since imperfections are inevitable the PGB production processes and the stability of the bandstructure with respect to such imperfections are crucial to any application. Another problem of interest is non-linear (e.g. Kerr type) defects in an otherwise periodic linear medium. This problem is of great technological interest since it could provide us with tunable and/or bistable defect modes, the key ingriedients for switching devices. One way to approach this, is by using an iterative PWM, such that once the defect field pattern is computed, it allows us to compute a new, spatially varying, defect dielectric "constant", which can serve as starting point for the next iteration. The hope is that

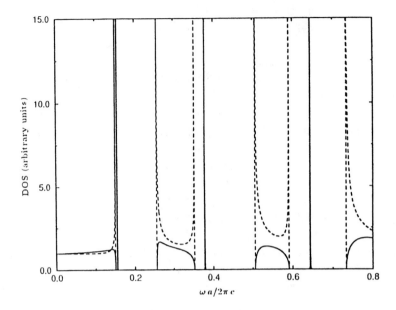

Figure 7. The Density of States (DOS) as a function of the renormalized frequency $\omega a/2\pi c$, for the periodic lattice (dashed line) and the single defect lattice (solid line) with parameters $\Delta\epsilon = -4$, $\epsilon_a = 1$, $\epsilon_b = 13$ and $f = 0.5$.

this procedue will converge and, thus, giving us the non-linear defectmodes.

7. Acknowledgments

K. Busch acknowledges the financial support of a "DAAD Doktoranden-stipendium aus Mitteln des zweiten Hochschulsonderprogramms HSPII/AUFE". We want to thank E. N. Economou and P. Wölfle for useful discussions. Ames Laboratory is operated for the U. S. Department of Energy by Iowa State University under contract No. W-7405-ENG-82. This work was supported by the Director of Energy Research, Office of Basic Energy Sciences and NATO Grant No. CRG 940647.

A. Two center expansions and Structural Greens function

In this appendix we give the formulas needed to convert the multicenter expension Eq.(41) into an expansion around the origin Eq.(48).

$$
\begin{aligned}
j_l(|x-y|)\, Y_l(x-y) &= 2\sum_{l_1 l_2} i^{l_1-l_2-l}\, C(l\,l_1\,l_2)\, * \\
&\quad j_{l_1}(|x|) j_{l_2}(|y|) Y_{l_1}(x) Y_{l_2}(y) \quad (62)
\end{aligned}
$$

$$
\frac{i}{ik_a}\, h_{l_1}(|x-y|)\, Y_{l_1}(x-y) = \sum_{l_2} \tilde{G}_{l_1 l_2}(y)\, j_{l_2}(|x|)\, Y_{l_2}(x) \quad (63)
$$

$$
\begin{aligned}
G_0(x-x'+y) &= \sum_{l_1 l_2} \tilde{G}_{l_1 l_2}(y)\, j_{l_1}(|x|)\, Y_{l_1}(x)\, * \\
&\quad j_{l_2}(|x'|)\, Y_{l_1}(x') \quad (64)
\end{aligned}
$$

where Eq.(63) only holds for $|y| > |x|$ and Eq.(64) is valid only if $|y| > |x|$ and either $|x'| < |x-y|$ or $|x| < |x'-y|$. Furthermore, in complete analogy to the 3D case, the 1D Gaunt numbers (Clebsch Gordon coefficients) $C(l\,l_1\,l_2)$ are given by

$$
C(l\,l_1\,l_2) = \sum_{x=\pm 1} Y_{l_1}(x)\, Y_{l_2}(x)\, Y_l(x) = \frac{1}{\sqrt{2}}(1 - \delta_{l_1 l_2})\delta_{l\,1} + \frac{1}{\sqrt{2}}\delta_{l_1 l_2}\delta_{l\,0} \quad (65)
$$

and $\tilde{G}_{l_1 l_2}(x)$ is defined as

$$
\tilde{G}_{l_1 l_2}(x) = \frac{2}{ik_a}\sum_{l_3} i^{l_1-l_2+l_3}\, C(l_1\,l_2\,l_3)\, h_{l_3}(|x|)\, Y_{l_3}(x), \quad (66)
$$

which can be evaluted to yield

$$
\tilde{G}(x) = \frac{e^{ik_a|x|}}{ik_a}\begin{pmatrix} 1 & -i\,\mathrm{sign}(x) \\ i\,\mathrm{sign}(x) & 1 \end{pmatrix}. \quad (67)
$$

The so-called structural Greens function $\breve{G}(q)$ which contains all the information about the lattice is nothing but the lattice Fourier transform of $\tilde{G}(x)$ and reads as

$$
\begin{aligned}
\breve{G}(q) &= \sum_{p \neq 0} e^{iqpa}\, \tilde{G}(pa) \\
&= \frac{1}{ik_a} \left(
\begin{array}{cc}
-1 + i\frac{\sin(\phi)}{\cos(\theta)-\cos(\phi)} & \frac{\sin(\theta)}{\cos(\theta)-\cos(\phi)} \\
-\frac{\sin(\theta)}{\cos(\theta)-\cos(\phi)} & -1 + i\frac{\sin(\phi)}{\cos(\theta)-\cos(\phi)}
\end{array}
\right)
\end{aligned}
\tag{68}
$$

where $\phi = k_a a$ and $\theta = qa$. Finally, from the definition of $G_{l_1 l_2}^{00}$, Eq.(50), and the equation of motion for τ, Eq. (42) the following relations are easily derived

$$
\sum_{p \neq 0} \sum_{l_3} \tilde{G}_{l_1 l_3}(R_p)\, \tau_{l_3 l_2}^{p0}(\{R_m\}) = G_{l_1 l_2}^{00}\, t_{l_1}^{0}
\tag{69}
$$

$$
\sum_{p \neq 0} \sum_{l_3} \tau_{l_1 l_3}^{0p}(\{R_m\})\, \tilde{G}_{l_3 l_2}(R_p) = t_{l_1}^{0}\, G_{l_1 l_2}^{00}
\tag{70}
$$

$$
\sum_{m \neq 0} \sum_{n \neq 0} \sum_{l_3 l_4} \tilde{G}_{l_1 l_3}(R_m)\, \tau_{l_3 l_4}^{mn}(\{R_m\})\, \tilde{G}_{l_4 l_2}(R_n) = G_{l_1 l_2}^{00}
\tag{71}
$$

References

[1] J. Opt. Soc. Am B, **10** , (February 1993).

[2] Photonic Band Gaps and Localization, Ed. by C. M. Soukoulis, NATO ASI series B, Vol. 308, Plenum (1993).

[3] K.M. Ho, C.T. Chan and C.M. Soukoulis, Phys. Rev. Lett. **65**, 3152 (1990).

[4] E. Ozbay, E. MIchel, G. Tuttle, M. Sigalas, R. Biswas, and K.M. Ho, Appl. Phys. Lett. **64**, 2059 (1994)

[5] D. R. Smith, R. Dalichaouch, N. Kroll, S. Schultz, S.L. McCall, and P.M. Platzman, J. Opt. Soc. Am. B **10**, 314 (1993).

[6] A. Kondilis and P. Tzanetakis, Phys. Rev. B. **46**, 15426 (1992).

[7] J.M. Frigerio, J. Rivory and P. Sheng, Opt. Com., **98**, 231 (1993).

[8] J.B. Pendry and A. MacKinnon, Phys. Rev. Lett., **89**, 2772 (1992).

[9] J.B. Pendry, J. Mod. Opt., **41**, 209 (1994).

[10] M. Sigalas, C.M. Soukoulis, E.N. Economou, C.T. Chan and K.M. Ho, Phys. Rev B **48**, 14221 (1993).

[11] M. Sigalas, C.M. Soukoulis, C.T. Chan and K.M. Ho, Phys. Rev B **49**, 11080 (1993).

[12] A.A. Maradudin and A.R. McGurn in [1].

[13] R.D. Maede, K.D. Brommer, A.M. Rappe, and J.D. Joannopoulos, Appl. Phys. Lett., **61**, 495 (1992)

[14] S. Satpathy, Z. Zhang and M.R. Salehpour, Phys. Rev. Lett. **64**, 1239 (1990).

[15] K.M. Leung and Y.F. Liu, Phys. Rev. B, **41**, 10188 (1990)

[16] V. Kuzmiak, A.A. Maradudin and F. Pincemin, Phys. Rev. B, **50**, 16835 (1994).

[17] E. Yablonovitch, T.J. Gmitter, R.D. Maede, K.D. Brommer, A.M. Rappe, and J.D. Joannopoulos, Phys. Rev. Lett., **67**, 3380 (1991).

[18] K.M. Leung in [1]

[19] P.W. Anderson and W.L. McMillan, Proceedings of the International School of Physics ENRICO FERMI, Course XXXVII, p. 50 (1967)

[20] E.N. Economou, "Greens Functions in Quantum Physics", 2nd Edition, Springer (1983)

[21] B.L Gyorffy and M.J. Stott, Solid State Com., **9**, 613 (1971)

[22] A. Gonis, "Green Functions for Ordered and Disordered Systems". North Holland (1992)

[23] P. Sheng, "Introduction to Wave Scattering, Localization and Mesoscopic Phenomena", Academic Press (1995)

[24] W.H. Butler, Phys. Rev. B, **14**, 468 (1976)

[25] N.A.W. Holzwarth, Phys. Rev. B, **11**, 3718 (1975)

[26] K.M. Leung and Y. Qiu, Phys. Rev. B, **48**, 7767 (1993)

[27] X. Wang, X.-G. Zhang, Q. Yu and B. Harmon, Phys. Rev. B, **47**, 4161 (1993)

IMPURITY MODES FROM FREQUENCY DEPENDENT DIELECTRIC IMPURITIES IN PHOTONIC BAND STRUCTURES

ARTHUR R. MCGURN and MICHAEL KHAZHINSKY
Department of Physics
Western Michigan University
Kalamazoo, Michigan 49008 U.S.A.

1. Introduction

A recent problem of interest in the study of photonic band structures has been that of a single dielectric impurity in an otherwise periodic photonic band structure. This problem has been studied using supercell methods [1], transfer matrix methods [2,3], and Green's function methods [4-6]. In this paper we will look at the exact solution of the single impurity problem using Green's function methods. Results will be presented for one-dimensional (layers) and two-dimensional (periodic array of rods) systems for impurities with frequency dependent dielectric constants substituted into periodic dielectric media composed of frequency independent dielectric materials. The impurities will be substitutional impurities (replacement of a single slab or rod by that of another dielectric composition). Particular attention will be given to obtaining theoretical results using dielectric constants appropriate to materials which have been used in recent experimental studies of photonic systems.

In Section 2, the Green's function theory and some results for a layered system with an impurity will be given. In Section 3, the two-dimensional system will be discussed and results for this system will be presented. Results for frequency-dependent dielectric media will be presented for both one- and two-dimensional systems using parameters appropriate to GaAs and aluminum composites. In Section 4, our conclusions will be presented.

2. Theory and Results in One-Dimension

We consider a layered array of dielectric slabs described by a periodic dielectric function of the form

C. M. Soukoulis (ed.), Photonic Band Gap Materials, 487–495.
© 1996 *Kluwer Academic Publishers. Printed in the Netherlands.*

$$\epsilon(x_1) = \begin{cases} 1 & , \ n(a+b) \le x_1 \le n(a+b) + a \\ \epsilon & , \ n(a+b) + a \le x_1 \le (n+1)(a+b) \end{cases} \tag{1}$$

where n ranges over the integers. For linearly polarized electromagnetic waves propagating parallel to the x_1-axis

$$-\frac{\partial}{\partial x_1}\left(\frac{1}{\epsilon(x_1)}\frac{\partial B_2}{\partial x_1}\right) = \left(\frac{\omega}{c}\right)^2 B_2 \tag{2}$$

$$-\frac{\partial^2}{\partial x_1^2} E_3 = \left(\frac{\omega}{c}\right)^2 \epsilon(x_1) E_3 \tag{3}$$

with B_2 along the x_2-axis and E_3 along the x_3-axis. Solutions of Eqs. (2) and (3) are then of the form

$$B_2(x_1) = e^{ikx_1} \sum_{n,\ell} a_n^{(\ell)}(k) \ e^{i\frac{2\pi n}{a+b}x_1} \tag{4}$$

$$E_3(x_1) = \frac{ic}{\omega\epsilon(x_1)}\frac{\partial}{\partial x_1} B_2$$

$$= -\frac{c}{\omega\epsilon(x_1)} e^{ikx_1} \sum_{n,\ell}\left(\frac{2\pi n}{a+b} + k\right) a_n^{(\ell)}(k) \ e^{i\frac{2\pi n}{a+b}x_1} \quad , \tag{5}$$

where $-\dfrac{\pi}{a+b} \le k \le \dfrac{\pi}{a+b}$, ℓ is a band index,

$$\sum_n \left[k + \frac{2\pi m}{a+b}\right]\hat{K}_{m-n}\left[k + \frac{2\pi}{a+b}n\right] a_n^{(\ell)}(k) = \left(\frac{\omega}{c}\right)^2 a_m^{(\ell)}(k) \tag{6}$$

and

$$\hat{K}_m = \frac{\sin\left(\dfrac{m\pi a}{a+b}\right)}{m\pi} \left[1 - \frac{1}{\epsilon}\right] e^{-i\frac{m\pi a}{a+b}} \qquad (m \neq 0) \tag{7}$$

$$= \frac{1}{a+b}\left[a + \frac{b}{\epsilon}\right] \qquad (m = 0) \qquad .$$

A solution of the impurity problem in which one of the dielectric slabs of our periodic array is replaced by a defect slab or a slab containing a defect can be obtained by studying a modified version of Eq. (3). Specifically, in the presence of an impurity we consider

$$-\frac{\partial^2}{\partial x_1^2} E_3 = \left(\frac{\omega}{c}\right)^2 [\epsilon(x_1) + \delta\epsilon(x_1)] E_3 \qquad , \tag{8}$$

where $\epsilon(x_1)$ is given in Eq. (1) and $\delta\epsilon(x_1)$ represents an additional contribution to the total dielectric constant of the system due to the impurity. As we shall now see below, using the electromagnetic modes of the periodic system in Eq. (3), we can rewrite Eq. (8) into the form of an integral equation. This integral equation can then be solved using standard numerical techniques.

The modes of the periodic system are the normalized eigenvectors of

$$-\frac{\partial^2}{\partial x_1^2} \Psi_k^{(\ell)}(x_1) = \lambda_k^{(\ell)} \epsilon(x_1) \Psi_k^{(\ell)}(x_1) \tag{9}$$

and are of the form

$$\Psi_k^{(\ell)}(x_1) = \frac{i}{\epsilon(x_1)\left[N(a+b)\lambda_k^{(\ell)}\right]^{1/2}} e^{ikx_1}$$

$$* \sum_n \left(\frac{2\pi n}{a+b} + k\right) a_n^{(\ell)}(k) e^{i\frac{2\pi n}{a+b}x_1} \qquad , \tag{10}$$

where N → ∞ is the number of unit cells in the direct lattice, the $\{a_n^{(\ell)}(k)\}$ are solutions of

Eq. (6) for $\left(\dfrac{\omega}{c}\right)^2 = \lambda_k^{(\ell)}$ normalized such that $\sum_n |a_n^{(\ell)}(k)|^2 = 1$, and

$$\int dx \left(\Psi_k^{(\ell)}(x)\right)^* \Psi_{k'}^{(\ell')}(x) \in (x) = \delta_{k,k'} \, \delta_{\ell,\ell'} \tag{11}$$

Using these modes we can expand $E_3(x)$ in Eq. (8) so that

$$E_3(x_1) = \sum_\ell \sum_k b_k^{(\ell)} \Psi_k^{(\ell)}(x_1) \tag{12}$$

Substituting Eq. (12) into Eq. (8) and using the orthogonality relation in Eq. (11) we then find the integral equation

$$E_3(x_1) = \left(\dfrac{\omega}{c}\right)^2 \int dx_1' \; G(x_1, x_1') \, \delta \in (x_1') \, E_3(x_1') \tag{13}$$

for $E_3(x)$ where the Green's function $G(x,x')$ is given by:

$$G(x,x') = \sum_\ell \sum_k \dfrac{\Psi_k^{(\ell)}(x) \, \Psi_k^{(\ell)*}(x')}{\lambda_k^{(\ell)} - \left(\dfrac{\omega}{c}\right)^2} \tag{14}$$

The integral equation in Eq. (13) can be discretized into the form of a matrix equation which is then treated numerically by computer methods to study the impurity modes located in the frequency band gaps. Before we present the results from such a study we discuss some further simplifications which can be applied to Eq. (13).

Some simplification of the numerical solution of Eq. (13) occurs if we write $\delta \in (x_1) = \Delta \, f(x_1)$ where Δ is independent of x_1, so that

$$E_3(x_1) = \Delta \left(\dfrac{\omega}{c}\right)^2 \int dx_1' \; G(x_1, x_1') \, f(x_1') \, E_3(x_1') \tag{15}$$

For a specified function $f(x_1)$, Eq. (15) can be treated as an eigenvalue problem, giving the eigenvalue Δ as a function of the impurity mode frequency $\left(\dfrac{\omega}{c}\right)$. This procedure greatly simplifies the determination of the impurity levels in our system without significantly restricting the generality of the impurity problems which can be addressed.

In the results presented below, we consider two types of impurities. In the first type we take a delta-function form $\delta \epsilon (x_1) = \Delta \delta (x_1 - x_0)$, where $x_0 = a + b/2$. (This choice for x_0 centers the impurity in the slab of dielectric constant ϵ located between $a \leq x \leq a + b$.) When substituted into Eq. (15), the delta-function form yields a single solution

$$\Delta = \left[\left(\frac{\omega}{c} \right)^2 G (x_0 | x_0) \right]^{-1} \tag{16}$$

for Δ as a function of ω in the band gaps. In the second type, we take $\delta \epsilon (x_1) = \Delta H (x_1)$ where $H(x_1) = 1/(2t)$ for $|x_1 - x_0| \leq t$ and $H(x_1) = 0$ otherwise. Again the impurity is centered in the $a \leq x \leq a + b$ slab. The integral equation in Eq. (15) then becomes

$$E_3 (x) = \frac{\Delta}{2t} \left(\frac{\omega}{c} \right)^2 \int_{-t+x_0}^{t+x_0} dx' \, G (x,x') E_3 (x') \tag{17}$$

for $|x - x_0| \leq t$. This equation can be discretized into the form of a matrix eigenvalue problem for Δ. Equation (17) then yields many values of Δ for each ω in the band gaps.

2.1 RESULTS

We have evaluated Eqs. (16) and (17) for a periodic system with $\epsilon = 9$ in Eq. (1). The value $\epsilon = 9$ corresponds to an aluminum composite which has recently been used in some experimental studies of photonic band systems [7]. For simplicity in the presentation of results, we let $a = b$ in Eq. (1) for the results presented below.

In Fig. (1) results are presented for Δ versus $\omega a_0/2\pi c$ in the lowest frequency band gap of our one-dimensional system. (All lengths in Fig. (1) are measured in units of the lattice constant $a_0 = 2a$.) Results are shown from Eq. (16) (dashed line) for a delta-function impurity located at the center of an ϵ slab and from Eq. (17) (solid lines) for the replacement of an entire single slab of dielectric constant ϵ by one with dielectric constant $\Delta/2t + \epsilon$ where $2t = 1/2$ in units of a_0. Only the two solutions of Eq. (17) for Δ closest to the solution of Eq. (16) for Δ are presented. It is seen that the delta-function results for Δ are very similar to the results for one of the Δ from Eq. (17).

In obtaining the results in Fig. (1), we have truncated the matrix eigenvalue problem in Eqs. (6) and (7) by using the 101 smallest reciprocal lattice vectors (101 smallest n in Eq. (6)). The Green's function in Eq. (14) was obtained using 20 values of k in a Gaussian quadrature integration over the first Brillouin zone. The integral in Eq. (17) was obtained using a 20-point Gaussian quadrature. (Gaussian quadrature, unlike the trapezoidal rule, does not use a uniformly spaced set of points to evaluate integrals. The Gaussian quadrature, however, yields a more accurate approximation than does the trapezoidal rule for equal numbers of points in the integration mesh [8].)

3. Two-Dimensional Systems

The mathematics of the Green's function formulation for the impurity problem in two-dimensions (i.e., a two-dimensional array of infinitely long rods) has been presented elsewhere [6]. We shall use Eqs. (41) and (34) in Ref. 6 to study a square lattice array of cylindrical rods of dielectric constant $\epsilon = 9$, circular cross-section and filling fraction $f = 0.4488$, surrounded by a vacuum. We only consider modes for which the electric field is polarized parallel to the axes of the rods. The axes of the rods are perpendicular to the x - y plane and centered at $\vec{r}_\| = n\, a_0\, \hat{i} + m\, a_0\, \hat{j}$ where n and m are integers. Equation (41) of Ref. 6 is used to study a single delta function impurity defined by $\delta\, \epsilon\, (\vec{r}_\|) = \Delta\, \delta\, (\vec{r}_\|)$ (see Eqs. (32) and (35a) of Ref. 6) which is placed at the center of the $\vec{r}_\| = 0$ cylinder. In addition, Eq. (34) of Ref. 6 is used to study the impurity (see Eqs. (32) and (35c) of Ref. 6) defined by

$$
\begin{aligned}
\delta\, \epsilon\, (\vec{r}_\|) &= \Delta/4t^2 &&, \; |x_\|| \le t \text{ and } |y_\|| \le t \\
&&&\text{for } \vec{r}_\| = x_\| \hat{i} + y_\| \hat{j} \\
&= 0 &&, \text{ otherwise}
\end{aligned}
\tag{18}
$$

located about the center of the rod at $\vec{r}_\| = 0$.

In Fig. 2 we present results in the second band gap $0.414 \le \dfrac{\omega\, a_0}{2\pi c} \le 0.468$ for the two-dimensional square array of cylindrical rods for Δ versus $\dfrac{\omega\, a_0}{2\pi c}$. (In Fig. (2) all lengths are measured in units of the two-dimensional lattice constant a_0.) Results are shown for the delta-function (solid line) and square cross-section impurity defined in Eq. (18) with $t = 0.10\, a_0$ (square boxes) and $t = 0.05\, a_0$ (triangles). Again, the delta-function results are single valued. The results for the impurity defined by Eq. (18) are multiple valued and we only show for this case the solution nearest to the results for the delta-function impurity.

3.1. FREQUENCY DEPENDENT IMPURITIES

Given the results in Figs. (1) and (2) for the impurity modes of our respective one- and two-dimensional systems, it is easy to obtain the frequencies of the impurity modes formed

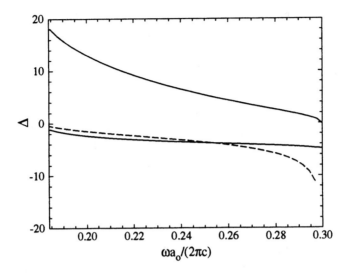

Figure 1. Plot of Δ versus ωa$_0$/(2πc) for the system of slabs. All lengths are measured in units of a$_0$.

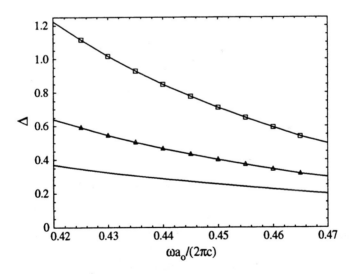

Figure 2. Plot of Δ versus ωa$_0$/(2πc) for the two-dimensional system of cylindrical rods. Results are shown for the delta-function impurity (solid line) and for square cross-section impurity with t = 0.10a$_0$ (square boxes) and *t* = 0.05a$_0$ (triangle boxes). All lengths are measured in units of a$_0$.

from the addition to these periodic systems of frequency dependent dielectric material with dielectric constant $\epsilon_{imp}(\omega)$. For such impurity material the frequency modes, ω, of the impurity levels must satisfy

$$\Delta(\omega) = [\epsilon_{imp}(\omega) - \epsilon] f_i \qquad , \qquad (19)$$

where $\Delta(\omega)$ is from Figs. (1) or (2), f_i is the filling fraction of the impurity in its primitive lattice cell, and ϵ is the frequency independent dielectric constant of the original periodic array. In particular, let us consider GaAs impurities [3]

$$\epsilon_{imp}(\omega) = \epsilon_\infty \frac{\omega_L^2 - \omega^2}{\omega_T^2 - \omega^2} \qquad , \qquad (20)$$

where $\epsilon_\infty = 10.9$, $\omega_L = 8.75$ THz and $\omega_T = 8.12$ THz. We find that to have an impurity mode of frequency ω_0 for $\Delta(\omega_0)$ in a band gap, ω_T must satisfy

$$\frac{\omega_T}{c} = \left[\frac{f_i(10.9 - \epsilon) - \Delta(\omega_0)}{f_i(12.7 - \epsilon) - \Delta(\omega_0)} \right]^{1/2} \frac{\omega_0}{c} \qquad , \qquad (21)$$

where $\epsilon = 9$ for the systems in Figs. (1) and (2).

As an example of the use of Eq. (21), we see in Fig. (1) that $\Delta\left(\frac{\omega a_0}{2\pi c} = 0.240\right) = -3.55$ for one of the impurity slab modes with $f_i = 0.5$. From Eq. (21) we find for this mode that $\frac{\omega_T a_0}{2\pi c} = 0.219$. Consequently, the lattice constant $a_0 = 5.1 \times 10^{-3}$ cm for our system to exhibit this particular impurity mode in its band gap.

4. Conclusions

Using Green's function methods, the solutions of dielectric impurity modes in one- and two-dimensional periodic systems are reduced to the study of integral equation eigenvalue problems. These problems are solved numerically yielding curves for Δ versus $\omega a_0/2\pi c$, which can be used to study the impurity modes due to both frequency independent and frequency dependent dielectric impurities. Numerical results are presented for systems with dielectric constants appropriate to materials recently used in experimental [7] and theoretical [3,4] studies on photonic band structures. The results presented here are currently being modified to treat the geometry of the particular two-dimensional system formed from frequency independent dielectric materials which has been the focus of some recent work [4,7] on impurities formed by the complete removal of a single rod.

Acknowledgment

This work is supported in part by NSF grant DMR 92-13793.

References

1. Yablonovitch, E., Gmilter, T.J., Meade, R.D., Rappe, A.M., Bommer, K.D. and Joannopoulos (1991) Donor and Acceptor Modes in Photonic Band Structure, *Phys. Rev. Lett.* 67, 3380 - 3383.

2. Smith, D.R., Dalichaouch, R., Kroll, N., Schultz, S., McCall, S.L. and Platzman, P.M. (1993) Photonic Band Structure and Defects in One and Two Dimensions, *J. Opt. Soc. Am. B*10, 314-321.

3. Sigalas, M., Soukoulis, C.M. Economou, E.N., Chan, C.T. and Ho, K.M. (1993) Photonic Band Gaps and Defects in Two Dimensions: Studies of the Transmission Coefficient, *Phys. Rev.* 48, 14121-14126.

4. Maradudin, A.A. and McGurn, A.R. (1993) *Photonic Band Gaps and Localization*, ed. C.M. Soukoulis, Plenum, New York, 247.

5. Leung, K.M. (1993) Defect Modes in Photonic Band Structures: A Green's Function Approach Using Vector Wannier Functions, *J. Opt. Soc. Am.* 10, 303-306.

6. Algul, H.G., Khazhinsky, M., McGurn, A.R. and Kapenga, J. (1995) Impurity Modes from Impurity Clusters in Photonic Band Structures, *J. Phys. Condens. Matt.* 7, 447-462.

7. McCall, S.L., Platzman, P.M., Dalichaouch, R., Smith, D. and Schultz, S. (1991) Microwave Propagation in Two-Dimensional Dielectric Lattices, *Phys. Rev. Lett.* 67, 2017-2020.

8. Garcia, A.L. (1994) *Numerical Methods for Physics*, Prentice Hall, New Jersey, chp. 9.

TWO-DIMENSIONAL PHOTONIC BAND GAPS:
NEW HEXAGONAL STRUCTURES

D. CASSAGNE, C. JOUANIN, AND D. BERTHO

Groupe d'Etude des Semiconducteurs, CC074,
Université de Montpellier II,
Place E. Bataillon, 34095 Montpellier Cedex 05, France

Abstract. Periodic dielectric structures have been recently proposed to inhibit spontaneous emission in semiconductors. From this suggestion, the new concepts of photonic band gaps and photonic crystals have been developed. Zero-threshold lasers, waveguides, and polarizers are promising applications. A new class of two-dimensional periodic dielectric structures with hexagonal symmetry is investigated in order to obtain photonic band gap materials. This set has the hexagonal symmetry and contains, in particular, several structures previously discussed. The photonic band gap structure is related to the basic properties of the materials and some features of the opening of the gaps are explained. By varying the crystal pattern, we show how band gaps common to E and H polarizations appear for a new design of two-dimensional periodic dielectric structures. The dependence of the widths of these gaps with the filling patterns is studied and potential application for the creation of photonic crystals in the optical domain is discussed.

Introduction

Since the suggestion in 1987 [1, 2] that the creation of a periodicity in dielectric materials could prevent the propagation of electromagnetic waves with certain frequencies, many works both theoretical and experimental have been performed [3, 4]. The periodicity of the dielectric constant induces the removal of degeneracies of the free-photon states at Bragg planes and provokes a range of forbidden energies for the photons. This has led to the

C. M. Soukoulis (ed.), Photonic Band Gap Materials, 497–505.
© *1996 Kluwer Academic Publishers. Printed in the Netherlands.*

new concept of photonic band gap (PBG) which underscores the analogy between electrons in semiconductors and photons in photonic crystals. Since the periodicity of the medium must be comparable to the wavelength of the electromagnetic waves to inhibit their propagation, PBGs in the optical or infrared domain require submicrometer structures. The fabrication of photonic crystals with three-dimensional (3D) periodicity is still a challenge at this scale. 2D crystals consisting of infinitely long parallel cylinders are easier to realize and the feasibility of such structures is demonstrated at the submicrometer lengths [5-7]. In 2D PBG materials, the electromagnetic waves propagating into the plane perpendicular to the cylinders can be separated in two polarizations according to the electric (E polarization) or magnetic (H polarization) field, parallel to the cylinder axis. The band gaps occurring in each case must overlap to form an absolute band gap which prevents the propagation of the light of any polarization. To date, the only already known 2D structure which exhibits a large absolute PBG is the triangular structure [8-11] of cylindrical holes in dielectric material when the holes are closely to touch each other.

In this paper, we propose a new class of structures with absolute PBGs. First, we expose the main lines of the calculation method and we introduce the different studied structures. We discuss the passage from the triangular structure to the graphite structure [12] by the introduction of a second cylinder into the unit cell. Finally, we present, in detail, the graphite structure of dielectric rods in air which appears to be the most interesting case. We found that two absolute PBGs can be achieved for systems far from the close-packed configuration. This opportunity can allow an easier realization of photonic crystals by avoiding the etching of thin semiconductor layers necessary in the triangular structure.

1. Model and Calculation Method

Using the plane-wave method, we investigate the in-plane propagation of electromagnetic waves in 2D structures of cylinders perpendicular to the xy plane of the lattice. Two different polarizations must be studied. We have two eigenvalue equations:

$$\det\left(\mathcal{H} - \frac{\omega^2}{c^2}\right) = 0, \tag{1}$$

where

$$\mathcal{H}(\mathbf{G}, \mathbf{G}') = |\mathbf{k} + \mathbf{G}||\mathbf{k} + \mathbf{G}'|\eta(\mathbf{G} - \mathbf{G}') \tag{2}$$

for the E polarization and:

$$\mathcal{H}(\mathbf{G}, \mathbf{G}') = (\mathbf{k} + \mathbf{G}) \cdot (\mathbf{k} + \mathbf{G}')\eta(\mathbf{G} - \mathbf{G}') \tag{3}$$

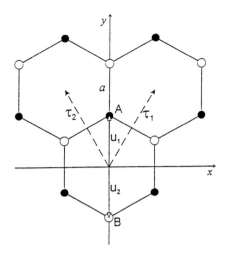

Figure 1. Two-dimensional triangular, graphite, and boron nitride structures.

for the H polarization. Here, $\mathbf{k}$ is a wave vector in the Brillouin zone, $\mathbf{G}$ are 2D reciprocal lattice vectors, and $\eta(\mathbf{G})$ is the Fourier transform of the inverse of $\varepsilon(\mathbf{r})$. For 2D hexagonal lattice, we define in Fig. 1 two sites A and B occupied by cylinders with radius ρ_1 and ρ_2, respectively. α is the radius ratio ρ_1/ρ_2. If $\alpha = 0$, we obtain the well-known triangular structure [8-11]. If $\alpha = 1$, all the sites are occupied by identical cylinders and form a two-dimensional arrangement of hexagons. By analogy with the crystal structure of the graphite, we call it graphite structure [12]. When α varies between 0 and 1, an entire class of configurations with the crystal structure of the boron nitride (BN) is obtained. This class allows the variation of a new parameter to optimize PBGs because the radii, ρ_1 and ρ_2, can be modified independently. The distance between the centers of two nearest neighbor cylinders is a for the graphite and BN structures. Because of the removal of one kind of cylinders in the triangular structure, it becomes $a\sqrt{3}$ in triangular configuration. The first Brillouin zone turns out to be a hexagon. To calculate the photonic band structures, we must first perform the Fourier transforms of $\eta(\mathbf{r})$ when the cylinders are filled with a material of dielectric constant ε_a and embedded in a background of dielectric constant ε_b. The Fourier transform of $\eta(\mathbf{r})$ is:

$$\eta(\mathbf{G}) = \varepsilon_b^{-1} \delta_{\mathbf{G}0} + \sum_{i=1}^{2} \eta^{(i)}(\mathbf{G}) \, e^{-i\mathbf{G}\cdot\mathbf{u}_i} \qquad (4)$$

For cylinders with circular cross-section, the Fourier transform, $\eta^{(i)}(\mathbf{G})$, only depends on $G = |\mathbf{G}|$. If the cylinders are not overlapping, we obtain:

$$\eta^{(i)}(G) = \left(\varepsilon_a^{-1} - \varepsilon_b^{-1}\right) \beta_i \frac{2J_1(G\rho_i)}{G\rho_i} \tag{5}$$

with $\beta_i = \pi\rho_i^2/S_{\text{cell}}$. S_{cell} is the surface of the unit cell and $J_1(x)$ is the Bessel function of the first order. The graphite structure contains two identical cylinders located at $\mathbf{u}_1$ and $\mathbf{u}_2 = -\mathbf{u}_1$. So, $\eta(\mathbf{G})$ can be expressed as:

$$\eta(\mathbf{G}) = \varepsilon_b^{-1}\delta_{\mathbf{G0}} + 2\cos(\mathbf{G} \cdot \mathbf{u}_1)\, \eta^{(1)}(G). \tag{6}$$

Whereas, by choosing the origin on the cylinder axis, we obtain for triangular structure:

$$\eta(G) = \varepsilon_b^{-1}\delta_{G0} + \eta^{(1)}(G). \tag{7}$$

We define the filling factor β of these structures as the fraction of the cell area occupied by cylinders. These formulas underscore the effect of the arrangement and of the form of the cylinders in the unit cell. Changes in the profile of the cross-section modify $\eta^{(i)}(\mathbf{G})$, the presence of two cylinders in the unit cell acts on the structure factors by introducing modulation depending on reciprocal lattice vectors.

2. Photonic Band Gaps

We have considered an entire class of BN structures. These structures can be characterized by two parameters, the filling factor β and the radius ratio $\alpha = \rho_1/\rho_2$. We have studied the evolution of PBGs as functions of α for β given values, which allows the description of structures with the same average dielectric constant ε_{av}. Configurations with different filling factors have been examined and we present here the most significant results. We only consider here the case of non-overlapping cylinders. The maximum filling factor for the graphite structure $\beta = 60\%$ is reached when the cylinder diameter is equal to the distance between the centers of two nearest neighbor cylinders. This is smaller than for the triangular case, $\beta = 91\%$, because the closed-packed graphite is obtained from the triangular one by removing the cylinders at hexagon centers. We examine the two following complementary configurations. The first consists of air cylinders ($\varepsilon_a = 1$) in GaAs ($\varepsilon_b = 13.6$). The second is formed by cylinders of GaAs in air. GaAs has been chosen because it presents interesting optical properties in the infrared domain and is representative of many semiconductors. This limitation is not essential and calculations on other systems of the same nature give similar results. We use 475 plane waves in the calculations, which ensures sufficient convergence for the frequencies of interest for studied structures.

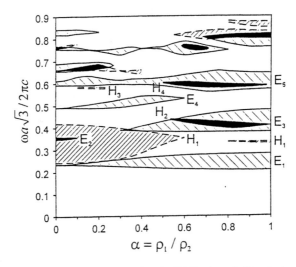

Figure 2. Photonic band gaps for E polarization (solid line) and H polarization (dashed line) of boron nitride structures of air cylinders in GaAs for $\beta = 60\%$. The absolute band gaps are represented in black.

2.1. BORON NITRIDE STRUCTURE OF AIR CYLINDERS IN GaAs.

We first consider the BN structure of air cylinders in GaAs at $\beta = 60\%$, which corresponds to the graphite close-packed configuration. We show in Fig. 2, the evolution of PBGs with respect to the radius ratio. Many gaps appear for E polarization. As predicted by symmetry analysis [13], an E_1 gap which cannot exist in the triangular structure opens up for $\alpha > 0$ and reaches its maximum width for the graphite structure at $\alpha = 1$. The E_2 gap evolves in a complementary way. It is maximum for $\alpha = 0$ and closes over 0.1. New gaps due to the lower symmetry of BN structure appear only at intermediate values of α. For H polarization, only one H_1 gap occurs in the triangular structure. Its width decreases for increasing value of α. This gap closes for $\alpha = 0.6$ and opens again for $0.8 < \alpha < 1$. There are three absolute PBGs in this configuration. The first is due to the overlap of E_2 and H_1 gaps and its greater width is obtained for the triangular configuration, whose existence has been previously reported by some authors [8-11]. Two other gaps E_3-H_2 and E_6-H_4 occur for $0.5 < \alpha < 1$. They are characteristic of the BN structure and their maximum widths are 7 and 4% respectively. At $\beta = 60\%$, the cylinders are in close-packed configuration and these structures will be probably rather difficult to realize because the dielectric layers between the cylinders are very thin. We have investigated the evolution of the PBGs for smaller values of the filling factor. Gap widths strongly depend on the filling factor. All the gaps

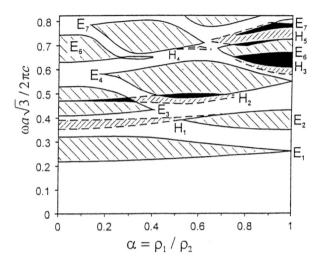

Figure 3. Photonic band gaps for the boron nitride structure of GaAs cylinders in air for $\beta = 30\%$.

shrink and the narrowest ones, like E_2, H_2, and H_4, are suppressed. Thus, absolute PBGs will be only observed for large filling factors. Consequently, the most promising hexagonal structure of air cylinders in GaAs is the triangular structure for which the absolute band gap E_2-H_1 can be widened by increasing the filling factor up to 0.91. In fact, to keep sufficiently rather thick layers between the cylinders, β must be limited to 70%.

2.2. BORON NITRIDE STRUCTURE OF GaAs CYLINDERS IN AIR.

We have also examined the case of BN structures formed of GaAs rods in air. In Fig. 3, we present the dependence of PBGs on the radius ratio for $\beta = 30\%$. This value has been chosen because in this neighborhood, the absolute band gaps are the largest. For E polarization, E_1 and E_3 gaps close when going from triangular to graphite structure, whereas the E_2 gap opens. Some gaps also appear for H polarization. The lowest frequency gap H_1 shrinks for increasing α values and closes at respect $\alpha = 0.7$. The H_2 gap only exists for BN structures with $0.1 < \alpha < 0.7$. More interesting is the appearance of H_3 and H_5 gaps above $\alpha = 0.7$. These gaps widen when α increases. The E_3-H_2 gap exists for $0.1 < \alpha < 0.4$, however its width remains small. The most important result is the appearance of two absolute band gaps, E_6-H_3 and E_7-H_5, which are not present in the triangular structure of GaAs rods in air [11]. As their widths are maximum for $\alpha = 1$, the graphite structure appears as the most interesting configuration to get

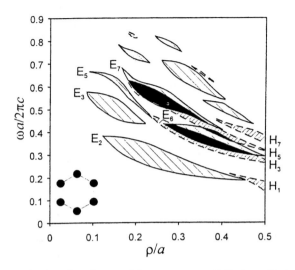

Figure 4. Photonic band gaps for the graphite structure of GaAs cylinders in air as functions of the ratio ρ/a.

optimal absolute band gap for lattices of GaAs rods in air.

2.3. GRAPHITE STRUCTURE OF GaAs CYLINDERS IN AIR.

In Fig. 4, we study the evolution of the gaps as functions of the cylinder radius ρ. Two absolute band gaps appear. They exist in a limited frequency range and their widths present maximums which can be large. The lowest gap resulting from the overlap of E_6 and H_3 gaps is centered near $\omega a/2\pi c = 0.37$ for $\rho = 0.35\, a$ ($\beta = 30\%$) with a 10% relative width. A second absolute band gap of the same width is obtained for $\omega a/2\pi c = 0.55$ for $\rho = 0.25\, a$ ($\beta = 15\%$) by the superposition of E_7 and H_5 gaps. Finally, a smaller gap due to E_6 and H_5 gaps lies in the neighborhood of the latter one. The appearance of these gaps is essential and shows the graphite structure of dielectric rods is a good candidate to the realization of the PBG materials. These values can be compared to the results obtained for the triangular structure of hole cylinders in GaAs [11]. If we take into account the fabrication limits, the largest absolute PBG is centered for this structure near $\omega a/2\pi c \approx 0.40$ for $\beta \approx 70\%$ (where a is here the lattice parameter). As for lattices of air cylinders in GaAs, the proportion of dielectric material is equal to $1 - \beta$. This proportion is close to the value obtained for the center of the E_6-H_3 absolute band gap in the graphite structure of dielectric cylinders and the average dielectric constant is the same in the two structures. In the submicrometer range, it is possible to center the E_7-H_5 absolute band gap in the near infrared at $\lambda = 0.9$ μm by realizing graphite structure of GaAs rods of $\rho = 0.12$ μm distant by $a = 0.50$ μm. The same analysis for the E_6-H_3 absolute band gap gives $a = 0.34$ μm and $\rho = 0.12$ μm. In these two structures, the diameter of the rods, equal to 0.24 μm, is large and does not require etching of thin dielectric layers. The existence of two photonic band gaps gives a further flexibility to optimize the dimension of the lattice parameter for a given value of the wavelength. We hope the fabrication and the experimental studies of such photonic crystals will be realized to compare with the theoretical results of this study.

Conclusions

We have presented a new class of structures for 2D PBG materials. This class allows the variation of a new parameter to optimize PBGs. Numerous large gaps are obtained for E and H polarizations. Two particular cases are especially attractive because of the overlap between gaps for both polarizations, the triangular structure of cylindrical holes in dielectric and the graphite structure of dielectric rods in air.

Acknowledgments

The Groupe d'Etude des Semiconducteurs is "Unité de Recherche Associée au Centre National de la Recherche Scientifique No. 357." We thank the "Centre National Universitaire Sud de Calcul de Montpellier" for a grant of computer time.

Note: During the NATO ASI on Photonic Band Gap materials, we learnt of the existence of similar results for the graphite structure from the group at M.I.T. [14].

References

1. E. Yablonovitch, Phys. Rev. Lett. **58**, 2059 (1987).
2. S. John, Phys. Rev. Lett. **58**, 2486 (1987).
3. For recent reviews, see the articles in *Photonic bandgaps and Localization*, edited by C. Soukoulis (Plenum, New York, 1993) and in the special issue of J. Opt. Soc. Am. **B 10** (1993) edited by C. M. Bowden, J.P. Dowling and H.O. Everitt.
4. P. R. Villeneuve and M. Piché, Prog. Quant. Elect. **18**, 153 (1994)
5. P. L. Gourley, J. R. Wendt, G. A. Vawter, T. M. Brennan, and B. E. Hammons, Appl. Phys. Lett. **64**, 687 (1994).
6. J. M. Gérard, A. Izraël, J. Y. Marzin, R. Padjen, and F. R. Ladan, Solid-State Electronics, **37**, 1341 (1994).
7. K. Inoue, M. Wada, K. Sakoda, A. Yamanaka, M. Hayashi, and J. W. Hus, Jpn. J. Appl. Phys. **33**, L1463 (1994).
8. M. Plihal and A. A. Maradudin, Phys. Rev. **B44**, 8565 (1991).

9. R. D. Meade, K. D. Brommer, A. M. Rappe, and J. D. Joannopoulos, Appl. Phys. Lett. **61**, 495 (1992).
10. P. R. Villeneuve and M. Piché, Phys. Rev. **B46**, 4969 (1992).
11. R. Padjen, J. M. Gérard, and J. Y. Marzin, J. Modern Optics, **41**, 295 (1994).
12. D. Cassagne, C. Jouanin, and D. Bertho, Phys. Rev. **B52**, 2217 (1995).
13. D. Cassagne, C. Jouanin, and D. Bertho, unpublished.
14. J. D. Joannopoulos, R. D. Meade, and J. N. Winn, in *Photonic crystals, Molding the flow of light*, Princeton University Press (1995).

PHOTONIC BAND GAPS IN COMPLEX-UNIT SYSTEMS AND QUASI ONE-DIMENSIONAL WAVEGUIDES

R. AKIS, P. VASILOPOULOS, AND F. SEZIKEYE

Concordia University, Department of Physics
Montreal, Quebec, Canada H3G 1M8

1. Introduction

The existence of a photonic band gap (PBG), in materials where the refractive index varies periodically, gives rise to many interesting and potentially useful properties, including the localization of light[1], the inhibition of radiation [2], etc., see Ref. 3 and references cited therein. These properties become more pronounced when the photonic gap is made large. Accordingly, the search for crystals with large PBGs has been extensive[1-4]. However, to our knowledge this search has had a limited success as it has identified structures with significant PBGs described by a gap-to-midgap frequency ratio of only about 20%.

We searched for large PBGs, tunable in size and frequency range, in i) complex one-dimensional (1D) and 2D structures, and ii) periodically modulated, in both refractive index and geometry, quasi 1D waveguides. The motivation for i) came from a method for tuning the electronic superlattice miniband stucture[5] by adding potential barriers (wells) in the wells (barriers) of the superlattice. Translated to the photonic case, this means adding *dielectric* contrast to the unit cell of a 1D two-layered system or of 2D lattice and expecting it to lead to multiple destructive (constructive) reflections of the waves and thus alter the photonic band structure. The results show that certain PBGs can increase by a factor of two to three relative to those of the literature. For ii) the motivation came from a method for tuning the conductance output in a waveguide using superlattices of electronic stub tuners[6] and exploiting the interference of waves propagating along the main waveguide with those reflected from the stubs[6-8]. Accordingly, we have considered stubless waveguides with a periodic dielectric contrast along the direction of propagation as well as superlattices of photonic stubs *with* and *without* dielectric contrast. In both cases the PBGs can become more than *ten times larger* than those of the 1D quarter-wave (QW) gap for perfect confinement and about *five times larger* for dielectric confinement. We also consider defects and show how one of them, a double stub, can give rise to both resonances and antiresonances when inserted in a waveguide.

In the next section, we present the 1D and 2D results and in Sec. 3 we concentrate on our most important results obtained in quasi 1D waveguides modulated in the manner mentioned above. A summary follows in the last section.

507

C. M. Soukoulis (ed.), Photonic Band Gap Materials, 507–514.

2. Structures With a Complex Unit Cell

2.1. ONE DIMENSIONAL STRUCTURES

The general unit cell of the periodic structures we consider is the conventional two-layered system with one or more dielectric layers inserted in it.The unit with one extra layer, with width d_3 and refractive index $n_3 = n_1$, is shown in the inset of Fig. 1 (a). We *fix* $L=d_1+d_2+d_3+d_4$, the length of the unit cell, so that in the *original* structure (no inserted layer) L would be equal to d_1+d_2, with d_1 and d_2 adjusted to the QW condition $d_1n_1=d_2n_2$. The dispersion relation, obtained by the transfer-matrix method, is given by $\cos(KL) = tr(M)/2$, where tr stands for the trace and M is a 2 by 2 transfer matrix, obtained by multiplying the matrices describing the four layers. Defining $\omega_0 = \pi c / n_1 L_1 = \pi c / n_2 L_2$ and $\omega_L = \pi c / L$, the QW gap is:

$$\Delta\omega_{QW} = \frac{4\omega_0}{\pi}\frac{|n_2-n_1|}{n_2+n_1} = \frac{2\omega_L}{\pi}\frac{|n_2-n_1|}{n_1n_2} . \tag{1}$$

In Fig. 1 (a), we plot the gaps $\Delta\omega_i, i=1,2,...,$ as a function of d_3, for transverse electric (TE) waves and normal incidence. At $d_3=0$, only the *odd* gaps are present. Note that while $\Delta\omega_1$ decreases steadily with d_3, $\Delta\omega_3$ can increase by nearly a factor of two relative to its QW value, $\Delta\omega_{QW} \cong 0.119\,\omega_L$, for $d_3=0$. A similar result occurs for higher order odd gaps. The 2nd and 4th gaps also display a maximum significantly above the QW value. To appreciate the changes brought about by the extra layer, we show in Fig. 1 (b) $\Delta\omega_i$ vs width d_1 for the usual two-layered system ($d_3=0$). The QW condition corresponds to $d_1=.455L$ and gives a maximum in $\Delta\omega_1$. As can be seen, the higher order gaps, $\Delta\omega_i$, $i>1$, all display maxima larger than $\Delta\omega_{QW}$, but the effect is miniscule compared to that obtained with the inserted layer.

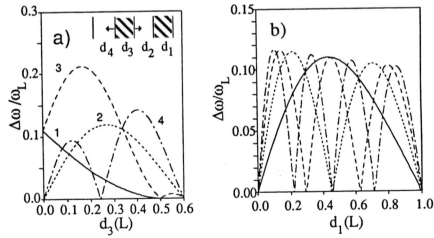

Figure 1. (a) The gaps as a function of the length of the inserted layer, 1st gap (solid) 2nd gap (dotted), 3rd gap (dashed), 4th gap (dash-dotted). (b) The gaps as a function of d_1, with $d_3=0$. For both (a) and (b) we have used $n_1=3.59$ and $n_2=3.0$.

2.2. TWO-DIMENSIONAL STRUCTURES

Several authors have already evaluated the PBG structure of 2D lattices using a plane-wave expansion[4] of the electric and magnetic fields. The Fourier coefficients E_G and H_G of these fields satisfy the following equations

$$(k+G) \times (k+G) \times E_G + (\omega^2/c^2) \sum_{G'} \varepsilon_{GG'} E_G = 0 \quad , \tag{2a}$$

$$\sum_{G'} \eta_{GG'} (k+G) \times (k+G') \times H_G + (\omega^2/c^2) H_G = 0 \quad , \tag{2b}$$

where $\varepsilon_{GG'}$ is the Fourier transform of the dielectric constant, $\varepsilon(r)$, and $\varepsilon = 1/\eta$. Equations 2 (a) and 2 (b) are solved numerically using standard procedures.

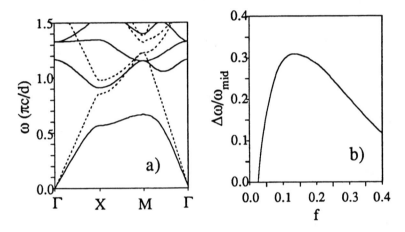

Figure 2. (a) the dispersion for a 2D square lattice of square rods with filling fraction, f=0.09. (b) The first gap in terms of the midgap frequency as a function of f.

We optimized the results as a function of the filling fraction by considering simple 2D structures involving materials A and B with $area_A \times \varepsilon_A = area_B \times \varepsilon_B$. This is the 2D analogue of the QW condition of the previous subsection. This gave the largest gap when, e.g., *high-index* rods ($\varepsilon_{rod} = 9$) are inserted in a *low-index* background ($\varepsilon = 1$). Figure 2 (a) shows the results for TE (solid curves) and transverse magnetic (TM) (dashed curves) waves pertaining to a square lattice of square rods. Our calculated first TE gap is about 3 times larger than that of Ref. 10 for the same index contrast and the gap relative to the midgap frequency, $\Delta\omega/\omega_{mid}$ is larger than 30%. In Fig. 2 (b), we plot $\Delta\omega/\omega_{mid}$ vs filling fraction, f. The maximum at $f \cong 0.15$ corresponds to the 2D QW condition. Away from this value, the gap decreases. The reverse situation, when *low-index* rods are inserted in a *high-index* background is much poorer as $\Delta\omega/\omega_{mid}$ is about 1.6% for areas optically equal and decreases for other values of f.

In an attempt to obtain larger PBGs we have also considered 2D structures with added interstitial rods of the same index at i) the center and ii) the sides of the square unit.

With the parameters of Ref. 10 we obtained $\Delta\omega / \omega_{mid} \cong 28\%$ for i) and $\Delta\omega / \omega_{mid} \cong 20\%$ for ii).

3. Quasi One-Dimensional Waveguides

In this section we evaluate the PBG in waveguides with a refractive index or geometry that vary periodically. The general unit cell of these waveguides, shown in Fig. 3 (a), consists of two segments with refractive indices n_1 and n_2, and corresponding lengths L_1 and L_2. The two stubs have dimensions s_1 and s_2, the total length is $L=L_1+L_2$ and $h=a+s_1+s_2$. This structure is similar to the *corrugated* waveguides, known from optoelectronics, which generally involve only a *single* dielectric constant ($\Delta n = 0$) in the guiding material and the variations in the width of the waveguide are usually quite *small* so that approximate treatments are typically employed. Here, we consider s_1 and s_2 of the same order as the width of the n_1 waveguide, a, and perform *full* calculations.

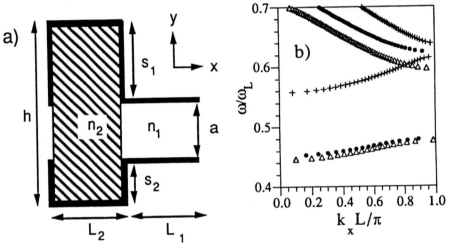

Figure 3. (a) the periodically repeated unit cell of the waveguide superlattice. (b) dispersion for the structure in (a) using the parameters given in the text.

We consider quasi-1D waveguides in which only *one* mode propagates in the stubless parts and only present results for TE waves, as those for TM waves are qualitatively similar. We consider two types of confinement: i) perfectly conducting boundaries, so the normal components of the electromagnetic waves vanish at the walls, and ii) dielectric confinement, with $n_1, n_2 < n_{enc}$, where n_{enc} is the refractive index of the enclosing material. For type ii), the waveguide modes decay exponentially outside the boundaries. For type i), the wave numbers in the y direction are restricted to discrete values and the modes can be expressed in terms of sine functions. For TE waves, we have $E(r) = (0,0,E)$, $H(r) = (H_x, H_y, 0)$ and the scalar wave equation for E reads:

$$\partial^2 E / \partial x^2 + \partial^2 E / \partial y^2 + (\omega^2 n^2 / c^2)E = 0 . \tag{3}$$

For type i), the solution of Eq. (3) for x<0 and x>L2 in Fig. 3 (a) has the form[7]
$E_1 = \sum_m (u_{1m}e^{i\alpha_m x} + \bar{u}_{1m}e^{-i\alpha_m x})\phi_m^w(y)$, where $\alpha_m = \sqrt{(n_1\omega/c)^2 - (\pi m/a)^2}$ is the wave number of the mth mode and $\phi_m^w(y) = \sqrt{2/a}\, sin[m\pi(y-a/2)/a]$. In the stub region the solution is $\sum_k [f_k \xi_{k0}(x,y) + \bar{f}_k \xi_{0k}(x,y)]$. The functions ξ_{k0} and ξ_{0k} are Fourier expansions of $X_l(x)\phi_l^s(y)$ and obey appropriate boundary conditions[7] at x=0 and x=L and $X_l(x)$ is given in Ref. 7. Matching the E field and its derivative at the boundaries[6,7] leads to the matrix equation $M_{cell}u^{\pm} = e^{ik_x L}u^{\pm}$ for the superlattice. Solving it gives the dispersion relation. For a *finite* superlattice, assuming that waves are incident (*in*) from the left and exiting (*out*) from the right, the transmittance is $T = \sum_{mn}|u_{out,n}|^2(\alpha_n/\alpha_m)$. For type ii), it is advantageous to use the discrete version of Eq. (3) which reads[9]

$$\left(4 - \omega^2\Delta^2 n_{i,j}^2 / c^2\right) E_{i,j} - E_{i-1,j} - E_{i,j-1} - E_{i+1,j} - E_{i,j+1} = 0 \qquad (4)$$

where i and j denote, respectively, positions on the x and y axes, and Δ is the distance between neighboring points in the 2D discrete mesh. Transfer matrices can be derived from Eq. (4), with the modes now expressed as vectors.

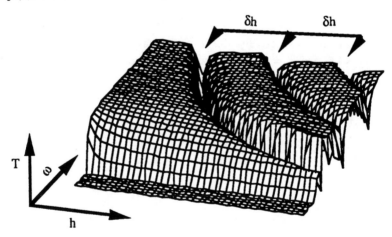

Figure 4. Transmittance T vs frequency ω and width h for a ten-unit superlattice. The period, δh, for the spacing between the transmittance minima, is given by Eq. (5).

In Fig. 3 (b), we plot the first two branches of the dispersion ω vs k_x for waveguides of type i), each with *fixed* L, $L_1 = 0.5114L = a$ and $L_2 = 0.4885L$ chosen to satisfy the 1D QW condition when $n_1 = 3.4$ and $n_2 = 3.6$. The triangles are for a waveguide with stubs ($s_1 = s_2 = 0.3068L$, h= 1.2273L) but *with* $n_1 = n_2 = 3.6$, the solid dots for one with the same stubs but *with* $n_1 = 3.4$ and $n_2 = 3.6$, so that $\Delta n = 0.2$, and the crosses for a stubless waveguide (a=h=0.5114 L) *with* contrast. As can be seen, the first gap for a $\Delta n = 0$ stubbed waveguide is approximately *six* times larger than the gap of the stubless waveguide for this small Δn. As expected, the gap with both stubs and $\Delta n = 0.2$ (solid dots) is larger. Higher gaps behave in a similar way, e.g., the second gap is about *four* times larger than the corresponding gap for the stubless waveguide. In analogy with electronic stub tuners explored recently [6-8], the origin of the gap when $n_1 = n_2$ in type i) or ii) waveguides is the *destructive* interference between waves propagating along the main waveguide and those reflected from the stubs. This interference occurs for $\Delta n \neq 0$ as well. To make it transparent, we show in Fig. 4 a three-dimensional plot of the transmittance T vs frequency ω and width h for a type i) wave-guide with $n_1 = 3.4$, $n_2 = 3.6$, and a, L_1, and L_2 as in Fig. 1. For fixed ω, T varies *periodically with h* and the gaps ($T \approx 0$) occur for a change in width δh given by

$$\delta h = 2\pi / \sqrt{(n_2 \omega / c)^2 - (\pi / L_2)^2} = 2\pi / k_s \quad . \tag{5}$$

This is just the condition for the *destructive* interference in the stub, $k_s \delta s_1 = k_s \delta s_2 = \pi$ ($\delta h = \delta s_1 + \delta s_2$) as the wavelength in the stub is $\lambda_s = 2\pi / k_s$.

The gap with $\Delta n \neq 0$ in Fig. 3 (b) can be increased with a higher Δn and eventually become larger than the $\Delta n = 0$ gap. In all cases, however, there is an almost constant increase in the contribution to the gap from the stubs, $\Delta \omega_s = \Delta \omega - \Delta \omega_{h=a}$, relative to the 1D QW value, defined by Eq. (1). This is shown in Fig. 5 (a) where we plot $\Delta \omega_s / \Delta \omega_{QW}$ as a function of h/a for $a/L = 0.5114$. The solid, long-dashed, and short-dashed curves are, respectively, for type i) waveguides with $n_1 = 3.4, n_2 = 3.6$, $n_1 = 3.0, n_2 = 3.6$, and $n_1 = n_2 = 3.6$. The dotted curve is for a type ii) waveguide with $n_1 = 3.4, n_2 = 3.6$; the stubs still increase the gap but the effect is less pronounced.

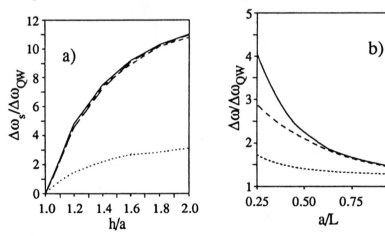

Figure 5. (a) Stub contribution to the first gap vs h/a. (b) The first gap for stubless waveguides vs a/L.

Figure 5 (b) shows the first gap as a function of a/L for $\Delta n = 3.6 - 3.4 = 0.2$ (solid curve) and $\Delta n = 1.6 - 1.0 = 0.6$ (long-dashed curve) for fixed $h = a$ (no stubs) and L_1 and

L_2 adjusted to the QW condition. In both cases the gap can increase significantly, by a factor of *three to four* for the lowest value of a/L. Combining these results with those of Fig. 5a), we see that the gap, when stubs are present, can be more than *one order of magnitude* larger than the 1D QW gap, the unit of the gaps in the figure. Importantly, for comparable lattice constants, refractive indices, and frequency ranges, the 1D QW gap is about 25% larger than the largest PBG reported so far for 2D systems[11]. The gaps we obtain are also large when compared to the midgap frequency. In the $n_1 = 3.4, n_2 = 3.6$ example with stubs, the gap to midgap ratio is 24%, despite the small dielectric difference. These results apply for type i) waveguides. The short-dashed curve is for a type ii) waveguide with $n_1 = 3.4, n_2 = 3.6$ and $n_{enc} = 1$. In that case, the gap at a/L=0.51 with stubs included is $\Delta\omega = 4.55\Delta\omega_{QW}$, which is still large.

We now consider defects. In line with previous studies in 1D and 2D, substitutional defects introduce states in the gaps, i.e., transmittance peaks. A qualitatively different defect is a double stub inserted in a *stubless* waveguide with alternating dielectric segments. Fixing n_1=2.0, n_2=3.6, L_1=0.643L = a, L_2=0.357L, in Fig. 6, we compare the transmittance for an eight unit stubless superlattice with one in which the n_2 portion of the 4th unit has the form of a double stub, h=3.25a. Apart from the usual resonance peaks in the gaps, transmittance *antiresonances* (dips) are now apparent in the first and second band. This simultaneous appearance of peaks and dips, by the same defect is understood as follows. When one mode is allowed in the main waveguide and a resonance or antiresonance occurs, the solution in the stub is a standing wave obeying $\sin(k_s h) = 0$ or $k_s h = \pi q$, q being an odd integer. Odd q's correspond to a wave with even symmetry with respect to the y axis. In the n_1 regions, there is only one propagating mode so $q=1$. The peak corresponds to $q=5$ and the two antiresonances satisfy this condition for $q=3$ and $q=7$, respectively, in the stub. The T minima can be interpreted as occurring due to *resonant reflection* from a bound state, a phenomenon noted in the electronic case[6]. There is no coupling to odd symmetry (q even) waves due to orthogonality, unless the stub is made asymmetric. When more stubs are inserted in the middle of the waveguide, the resonances and antiresonances widen into bands in the gaps and gaps in the bands, respectively. These same results follow for type ii).

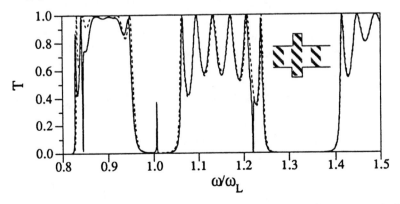

Figure 6. The transmittance, T, for a finite superlattice of 8 units. The dotted line is for a stubless superlattice. For the solid curve, a stubbed defect is present at the 4th unit cell. The inset depicts a unit cell with this type of defect sandwiched between two normal unit cells.

514

4. Conclusions

We have shown that the PBGs can be modulated in 1D and 2D structures with complex basis by approximately a factor of two. In 1D structures this is most easily achieved by varying the width of an extra layer inserted into a QW structure and in 2D structures by considering equal optical areas or by including interstitial elements. In quasi 1D periodically modulated waveguides, by changing their width and adding stubs, the size of the PBGs can be enlarged by *more than one order of magnitude* for perfect confinement and by about five times for dielectric confinement. Moreover, the PBGs exist even when $n_1 = n_2$ as the result of *destructive* interference between waves propagating along the main waveguide and those reflected from the stubs. Finally, if a stub is inserted into a stubless waveguide, *resonances in the gaps and antiresonances within the bands can occur simultaneously* that widen into *new* bands and gaps as more stubs are inserted. A similar effect can occur in a waveguide with stubs, if the inserted stubs are *asymmetric* .

5. References

1. John, S. (1991) Localization of light, *Physics Today*, May, 32-40; Meade, D. R., Bromer, D. K., Rappe, M. A., and Joannopoulos, D. J. Electromagnetic Bloch waves at the surface of a photonic crystal, Phys. Rev. B **44**, 10961-10969-10972 (1991).
2. Yablonovitch, E. (1987) Inhibited spontaneous emission in solid-state physics and electronics, *Phys. Rev. Lett.* **58**, 2059-2062.
3. C. M. Soukoulis (ed.) (1993) Photonic Band Structure and Localization, Plenum, New York; J. Opt. Soc. Am. B. **10**, issue 2.
4. Ho, K.M., Chan, C.T. and Soukoulis, C.M. (1990) Existence of a photonic band gap in periodic dielectric structures, *Phys. Rev. Lett.* **65**, 3152-3160; Villeneuve, P.R. and Piche, M. (1992) Photonic band gaps in two-dimensional square and hexagonal lattices, *Phys. Rev. B* 46,4969-4972; Ozbay, E.,Michel, E., Tuttle, G., Biswas, R., Ho, K. M., Bostak, J., Bloom, D. M., (1994) Double etch geometry for millimeter-wave photonic band gap crystals, Appl. Phys. Lett. 65, 1617-1619.
5. Vasilopoulos, P., Peeters, F.M., and Aitelhabti, D. (1990) Quantum tunability of superlattice minbands, *Phys. Rev. B* **41**, 10021-10027.
6. Akis, R., Vasilopoulos, P., and Debray, P. (1995) Ballistic transport in electron stub tuners: shape and temperature dependence, tuning of the conductance output and resonant tunnelling, *Phys. Rev. B* **52** in press.
7. Wu, H., Sprung, D. W. L. , Martorell, J. and Klarsfeld, S. (1991) Quantum wire with periodic serial structure, *Phys. Rev. B* **44**, 6351-6360 .
8. Sols, F. , Macucci, M., Ravaioli, U. and Hess, K. (1989) Theory for a quantum modulated transistor, *J. Appl. Phys.* 66, 3892-3906.
9. Sigalas, M., Soukoulis, C. M., Economou, E. N., Chan, C. T. and Ho, K. M. (1993) Photonic band gaps and defects in two dimensions: Studies of the the transmission coefficient, *Phys. Rev. B* **48**, 14121-14126.
10. McCall, S. L. and Platzman, P. M. (1991) Microwave propagation in two-dimensional dielectric lattices, *Phys. Rev. Lett.* **67**, 2017-2020 .
11. Plihal, M. and Maradudin, A. A. (1991) Photonic band structure of two-dimensional systems: The triangular lattice, *Phys. Rev. B* **44**, 8565-85770.

THEORY OF LIGHT SCATTERING THROUGH A TWO-DIMENSIONAL PERIODIC ARRAY

D. ANDRÉ[1], A. CASTIAUX[1], A. DEREUX[1], J.-P. VIGNERON[1] AND
C. GIRARD[2]
[1] *Laboratoire de Physique du Solide,*
Facultés Universitaires Notre-Dame de la Paix,
Rue de Bruxelles, 61, B-5000 Namur, Belgium
[2] *Laboratoire de Physique Moléculaire,*
Université de Franche-Comté,
F-25030 Besançon cedex, France

Abstract. Within the framework of the classical electromagnetic scattering theory, this paper develops a practical evaluation scheme to accurately compute the light scattered by a two-dimensional array of particles, with finite extension in the direction perpendicular to the periodicity.

1. Introduction

In the last few years, the interest in the study of light scattering by mesoscopic and microscopic scatterers was renewed in different contexts. The development of scanning near-field microscopy [1] has evidenced the need for accurate methods to compute the scattering of electromagnetic waves by arbitrary structures.

There also has been a great deal of interest in the study of mesoscopic structures exhibiting photonic band gaps [2]. Two-dimensional periodic media were investigated because full band gaps are achieved more easily if the structure displays a translational invariance along a single direction.

It was shown that band gaps appeared in some triangular and square lattices of infinitely long dielectric rods [3, 4]. Some studies were also performed on finite-size periodic systems : the band structure of an array of rods between two perfectly conducting plates was computed [5], with rods perpendicular to the plates. Recently, the transmission coefficient through a slab in which dielectric rods were embedded has been calculated [6]. How-

C. M. Soukoulis (ed.), Photonic Band Gap Materials, 515–520.
© *1996 Kluwer Academic Publishers. Printed in the Netherlands.*

ever, in most of these studies the electromagnetic wave was travelling in the plane of periodicity. Maradudin and McGurn computed band structures of a periodic array of infinite dielectric rods in the case of a wave propagating out of the plane of periodicity [7].

In this article, we develop a perturbed Green's dyadic method to study the scattering of light by a periodic layer of particles. This method is quite general and can be applied to the case of out of plane propagation.

2. Scattering of Electromagnetic Waves

This section summarizes the general theory of electromagnetic wave scattering. One rapidly realizes that the framework detailed in this section extends the well-known quantum theory of scattering to the case of vector waves. Assuming the $\exp(-i\omega t)$ time dependence, the most general vector wave equation obtained from Maxwell equations

$$-\vec{\nabla} \times \vec{\nabla} \times \vec{E}(\vec{r}) + \frac{\omega^2}{c^2} \, \epsilon(\vec{r}) \, \vec{E}(\vec{r}) = 0 \tag{1}$$

may be written as

$$-\vec{\nabla} \times \vec{\nabla} \times \vec{E}(\vec{r}) + \frac{\omega^2}{c^2} \, \epsilon_{ext} \, \vec{E}(\vec{r}) = \mathbf{V}(\vec{r}) \, \vec{E}(\vec{r}). \tag{2}$$

ϵ_{ext} is the isotropic dielectric function of the external medium of which we know the analytical solution $\vec{E}_0(\vec{r})$ satisfying

$$-\vec{\nabla} \times \vec{\nabla} \times \vec{E}_0(\vec{r}) + \frac{\omega^2}{c^2} \, \epsilon_{ext} \, \vec{E}_0(\vec{r}) = 0 \tag{3}$$

and the associated Green's dyadic $\mathbf{G}_0(\vec{r}, \vec{r'})$ defined by

$$-\vec{\nabla} \times \vec{\nabla} \times \mathbf{G}_0(\vec{r}, \vec{r'}) + \frac{\omega^2}{c^2} \, \epsilon_{ext} \, \mathbf{G}_0(\vec{r}, \vec{r'}) = \mathbf{1} \, \delta(\vec{r} - \vec{r'}). \tag{4}$$

The analytical form of $\mathbf{G}_0(\vec{r}, \vec{r'})$ in a homogeneous media is known from past works [8]. Its most general form may be written as

$$\mathbf{G}_0(\vec{r}, \vec{r'}) = \int d\vec{k} \left[\mathbf{1} - \frac{1}{q^2} \, \vec{k} \otimes \vec{k} \right] \frac{e^{i\vec{k} \cdot (\vec{r} - \vec{r'})}}{8\pi^3 (q^2 - k^2)} \tag{5}$$

where

$$q^2 = \frac{\omega^2}{c^2} \, \epsilon_{ext} \tag{6}$$

and $\otimes$ denotes a tensor product. When the source point and the observation point are the same, one must be careful to properly normalize $\mathbf{G}_0(\vec{r}, \vec{r'})$ as described in Appendix A.

Any complicated behavior due to the geometrical shape of the original dielectric tensor profile $\epsilon(\vec{r})$ is described as a difference relatively to the external system:

$$\mathbf{V}(\vec{r}) = \frac{\omega^2}{c^2} \left(1 \, \epsilon_{ext} - \epsilon(\vec{r})\right). \tag{7}$$

The solution of (2) can then be obtained from the implicit Lippmann-Schwinger's equation

$$\vec{E}(\vec{r}) = \vec{E}_0(\vec{r}) + \int_D d\vec{r'} \, \mathbf{G}_0(\vec{r}, \vec{r'}) \, \mathbf{V}(\vec{r'}) \, \vec{E}(\vec{r'}), \tag{8}$$

where D is limited to teh region of space where $\mathbf{V}(\vec{r'})$ is not zero.

3. Scattering by an Array of Nanoparticles

We now proceed to the introduction of a self-supported layer of periodically distributed particles. This will imply that the dielectric tensor of the perturbation, and so $\mathbf{V}(\vec{r})$ has a two-dimensional periodicity in the xy plane, that is

$$\mathbf{V}(\vec{r}_\parallel + z\vec{e}_z) = \mathbf{V}(\vec{\rho} + \vec{l} + z\vec{e}_z) \tag{9}$$

where $\vec{l}$ is a vector of the two-dimensional direct lattice

$$\vec{l} = l_x \vec{e}_x + l_y \vec{e}_y, \qquad l_x \text{ and } l_y \text{ integers.} \tag{10}$$

We note $\vec{r}_\parallel = \vec{\rho} + \vec{l}$, with $\vec{\rho}$ being a vector in the Wigner-Seitz cell. In the reciprocal space, the vectors of the first Brillouin zone will be written $\vec{K}_\parallel$ and the other vectors $\vec{k}_\parallel = \vec{K}_\parallel + \vec{t}$, where $\vec{t}$ is a reciprocal lattice vector.

Using this notation, we can represent the field $\vec{E}(\vec{r})$ as a Bloch wave, with the periodic part expanded into a two-dimensional Fourier series. The use of Bloch's theorem keeps the integration inside the first Brillouin zone:

$$\vec{E}_0(\vec{r}) = \frac{1}{4\pi^2} \int_{BZ} d\vec{K}_\parallel \, \vec{f}_0(\vec{\rho}, z; \vec{K}_\parallel) e^{i\vec{K}_\parallel \cdot (\vec{\rho} + \vec{l})}. \tag{11}$$

The Green's function can be developed similarly to get

$$\mathbf{G}_0(\vec{r}, \vec{r'}) = \frac{1}{4\pi^2} \int_{BZ} d\vec{K}_\parallel \mathbf{\Gamma}_0(\vec{\rho}, \vec{\rho'}; z, z'; \vec{K}_\parallel) e^{i\vec{K}_\parallel \cdot (\vec{\rho} + \vec{l} - (\vec{\rho'} + \vec{l'}))}. \tag{12}$$

Since these equations are, in fact, the equations for the coefficients of a two-dimensional Fourier series, they can be inverted to obtain $\vec{f}(\vec{\rho}, z; \vec{K}_\parallel)$

and $\Gamma(\vec{\rho}, \vec{\rho}'; z, z'; \vec{K}_\parallel)$ as functions of $\vec{E}(\vec{r})$ and $\mathbf{G}(\vec{r}, \vec{r}')$, respectively. We then have

$$\vec{f}_0(\vec{\rho}, z; \vec{K}_\parallel) = \sigma_{cell} \sum_{\vec{l}} \vec{E}_0(\vec{r}) e^{-i\vec{K}_\parallel \cdot (\vec{\rho} + \vec{l})} \qquad (13)$$

$$\Gamma_0(\vec{\rho}, \vec{\rho}'; z, z'; \vec{K}_\parallel) = \sigma_{cell} \sum_{\vec{l} - \vec{l}'} \mathbf{G}_0(\vec{r}, \vec{r}') e^{-i\vec{K}_\parallel \cdot ((\vec{\rho} + \vec{l}) - (\vec{\rho}' + \vec{l}'))}, \qquad (14)$$

where σ_{cell} is the surface of the unit cell. A similar expansion holds for $\vec{f}(\vec{\rho}, z; \vec{K}_\parallel)$ as a function of $\vec{E}(\vec{r})$.

One easily verifies that $\vec{f}(\vec{\rho}, z; \vec{K}_\parallel)$ can be deduced from a Lippman-Schwinger's equation restricted to the Wigner-Seitz cell:

$$\begin{aligned} \vec{f}(\vec{r}, \vec{K}_\parallel) &= \vec{f}_0(\vec{r}, \vec{K}_\parallel) \\ &+ \frac{1}{\sigma_{cell}} \int d\vec{\rho}' \int dz' \, \Gamma_0(\vec{r}, \vec{r}'; \vec{K}_\parallel) \mathbf{V}(\vec{r}') \vec{f}(\vec{r}'; \vec{K}_\parallel), \end{aligned} \qquad (15)$$

where $\vec{r} = \vec{\rho} + z\vec{e}_z$ and $\vec{r}' = \vec{\rho}' + z'\vec{e}_z$.

After an appropriatediscretization, the last formula leads to a linear set of equations solvable by standard numerical procedures. This algorithm allows us to deal with periodic structures having an arbitrary shape.

4. Far-field Scattering

The computation of transmittance and reflectance spectra involves an integration of the scattered field far from the perturbation. In order to calculate it, one carefully separates the effect of the z-coordinate of the observation point. This can be done through an alternative (and equivalent) form of $\Gamma_0(\vec{\rho}, \vec{\rho}'; z, z'; \vec{K}_\parallel)$:

$$\Gamma_0(\vec{\rho}, \vec{\rho}'; z, z'; \vec{K}_\parallel) = \sum_{\vec{t}} \mathbf{g}_0(z, z'; \vec{K}_\parallel + \vec{t}) e^{i\vec{t} \cdot (\vec{\rho} - \vec{\rho}')}, \qquad (16)$$

where

$$\mathbf{g}_0(z, z'; \vec{k}_\parallel) = \left[1 - \frac{1}{q^2} \vec{Q}_s \otimes \vec{Q}_s \right] \frac{e^{ih(z-z')}}{2iQ}, \qquad (17)$$

$Q^2 = q^2 - k_\parallel^2$,

$$\vec{Q}_s = \vec{k}_\parallel + h\vec{e}_z, h = \begin{cases} +Q & \text{if } z > z' \\ -Q & \text{if } z < z' \end{cases} \qquad (18)$$

The scattered field $\vec{E}_s(\vec{r})$ can then be expressed as

$$\vec{E}_s(\vec{r}) = \frac{1}{4\pi^2} \int_{BZ} d\vec{K}_\parallel \vec{F}_s(\vec{\rho}', z'; \vec{K}_\parallel) e^{i\vec{K}_\parallel \cdot (\vec{\rho} + \vec{l})}. \qquad (19)$$

We have defined

$$
\vec{F}_s(\vec{\rho}, z; \vec{K}_\parallel) = \frac{1}{\sigma_{cell}} \int dz' \int d\vec{\rho}' \sum_{\vec{t}} \mathbf{g}_{0s}(z'; \vec{K}_\parallel + \vec{t}) \mathbf{V}(\vec{\rho}' + z'\vec{e}_z)
$$

$$
\vec{f}(\vec{\rho}', z'; \vec{K}_\parallel) e^{ih(\vec{t})z} e^{i\vec{t}\cdot(\vec{\rho} - \vec{\rho}')}, \tag{20}
$$

where $\mathbf{g}_{0s}(z'; \vec{K}_\parallel + \vec{t})$ is the part of $\mathbf{g}_0(z, z'; \vec{k}_\parallel)$ independent of z. Since we are interested only in far-field scattering, that is when z becomes very large, it is obvious that the summation in the reciprocal lattice in (20) can be limited to the few radiative terms.

The computation of the reflectance and the transmittance requires to integrate the z-component of the Poynting vector on a plane of constant z located at the ∞. The reflected power through this plane then becomes

$$
P_r = -\frac{1}{8\pi^2 \omega \mu_0 \sigma_{cell}} \Re \int d\vec{\rho} \int d\vec{K}_\parallel \left| \vec{F}_s(\vec{\rho}, z; \vec{K}_\parallel) \right|^2 \vec{q} \cdot \vec{e}_z, \tag{21}
$$

where the integration is done in a plane $z = -\infty$. Similarly the transmitted power is

$$
P_t = -\frac{1}{8\pi^2 \omega \mu_0 \sigma_{cell}} \Re \int d\vec{\rho} \int d\vec{K}_\parallel \left| \vec{F}_s(\vec{\rho}, z; \vec{K}_\parallel) + \vec{F}_{0s}(\vec{\rho}, z; \vec{K}_\parallel) \right|^2 \vec{q} \cdot \vec{e}_z, \tag{22}
$$

where the integration is done in a plane $z = +\infty$, and $\vec{F}_{0s}(\vec{\rho}, z; \vec{K}_\parallel)$ is for the incident field. The reflection and transmission factors are then obtained by simply renormalizing these powers relatively to the incident power.

5. Conclusion

We developed a method to compute the scattered electromagnetic field of a three-dimensional system with two-dimensional periodicity. The algorithm can deal with structures of arbitrary shape and with various directions of incidence of the light.

Acknowledgments

This work was performed in the framework of the Human Capital and Mobility research network *Near-Field Optics for Nanoscale Science and Technology* initiated by the European Community. The Laboratoire de Physique Moléculaire is Unité Associée au CNRS 772. A. C. acknowledges support from the *Fond National de la Recherche Scientifique*.

A. Electromagnetic Green's Dyadic of an Homogeneous Medium

A solution of Eq. (5) is

$$\mathbf{G}_0(\vec{r}, \vec{r'}) = (1 + \frac{1}{k_0^2}\vec{\nabla} \otimes \vec{\nabla})g(\vec{r}, \vec{r'}), \tag{23}$$

where $g(\vec{r}, \vec{r'})$ is the scalar Green's function of an homogeneous medium [9].

$$g(\vec{r}, \vec{r'}) = -\frac{e^{ik_0 R}}{4\pi R}, \qquad R = \left|\vec{r} - \vec{r'}\right|. \tag{24}$$

Equation (23) leads to

$$\mathbf{G}_0(\vec{r}, \vec{r'}) = -\left\{1 - 1\frac{1 - ik_0 R}{k_0^2 R^2} - \vec{R} \otimes \vec{R}\frac{-3 + 3ik_0 + k_0^2}{k_0^2 R^4}\right\}\frac{e^{ik_0 R}}{4\pi R}. \tag{25}$$

This last equation is singular when $\vec{r} = \vec{r'}$. In this case we integrate the expression on an infinitesimal sphere centered around $\vec{r}$. This leads to a diagonal tensor whose diagonal elements are given by

$$-\frac{2e^{ik_0 a}}{3k_0^2} + \frac{2}{3k_0^2} - 2\frac{ae^{ik_0 a}}{3ik_0}, \tag{26}$$

which at the first order gives $-2a^2/3$. Since practical algorithms need a discretization of space, this never falls to zero. We also have to take into account the depolarization in each discretization cell. In the case of a cube, this leads to [10]

$$\mathbf{G}_0(\vec{r}, \vec{r}) = -\frac{1}{3}(2a^2 + \frac{c^2}{\omega^2}). \tag{27}$$

References

1. For a recent review of scanning near-field optical microscopy, see D. Courjon and C. Bainier, Rep. Prog. Phys. **57**, 989 (1994).
2. See "Photonic Band Gaps and Localization," Ed. by C. M. Soukoulis (Plenum, New York, 1993).
3. P. R. Villeneuve and M. Piché, Phys. Rev. B **46**, 4969 (1992); Phys. Rev. B **46**, 4973 (1992).
4. J. N. Winn, R. D. Meade and J. D. Joannopoulos, J. Mod. Opt. **41**, 257 (1994).
5. A. A. Maradudin and A. R. McGurn, in [2], 247.
6. K. Sakoda, Phys. Rev. B **51**, 4672 (1995).
7. A. A. Maradudin and A. R. McGurn, J. Mod. Opt. **41**, 275 (1994).
8. H. Levine and J. Schwinger, Comm. Pure App. Math. **3**, 355 (1950).
9. E. N. Economou, *Green's functions in quantum physics* (Springer-Verlag, Berlin, 1983).
10. A. D. Yaghjian, Proc. IEEE **68**, 248 (1980).

TWO-DIMENSIONAL GUIDES WITH PHOTONIC BAND GAP BOUNDARIES: MODE STRUCTURES

H. BENISTY

Laboratoire de Physique de la Matière Condensée,
Ecole Polytechnique, F-91128, Palaiseau Cedex, France

Abstract. Guided modes in slab-like structures consisting of dielectric material surrounded by 2D photonic band gap material drilled in the same dielectric are studied by the supercell method. Focusing on E-polarized modes ($\vec{H}$ transverse to the rods), we outline (i) the existence of a guide with mid-gap mode much narrower than the classical quarter-wave layer of the 1D case, (ii) the role of the relative "phases" of the boundary corrugations for contradirectional coupling, and (iii) the symmetries of modes at $k = 0$ and how it appears in the succession of modes.

1. Introduction

Dielectric Photonic Band Gap Crystals (PBGCs) hold great promises for optoelectronic systems due to their ability to forbid any electromagnetic propagation in a sizeable frequency range [1-4]. Of particular interest for integrated optics is the use of PBGC as novel guide boundaries operating at a much narrower width and much shorter bend radius than current technology [5, 6]. Two-dimensional PBGC, although limited in solid angle, are anticipated to work in this case, since most optical losses take place in a reduced cone around the propagation plane.

Here, we develop a computational study of such guides. The chosen two-dimensional PBGC consist of infinite cylindrical air holes in dielectric, centered on a triangular lattice. The guide lies along a dense (11) plane and consists of the same dielectric. This PBGC is known to be feasible [7] up to air filling factors $f = f_{air}$ of order 0.70-0.80 and a full gap of about 10 percent relative width [3, 4]. We keep $\vec{k}$ normal to the cylinder axis and focus on the "E" polarization as a first step.

C. M. Soukoulis (ed.), Photonic Band Gap Materials, 521-528.

In the one-dimensional case, it is well-known that a localized mode appears in the stop band of a periodic dielectric stack when a defect is introduced [8]; namely, the defect reaches mid-gap energy for a quarter-wave (half optical period) additional layer. For defects in PBGC, the strength for which defects do reach band edges and further mid-gap is not obvious[9].

We first focus on the narrowest guide yielding a mid-gap mode. As the guide widens, many modes detach from PBGC bands to cross the gap. In the wide guide limit, the effect of boundary corrugations decreases and the situation is akin to the slightly corrugated planar waveguide case with sharp stop-bands when k reaches zone edges. However, in our case, we will see that the two boundary corrugations "cancel each other." Finally, we will present preliminary results on mode symmetries at $k = 0$ as a first step towards studies of guided modes dispersion relation.

2. Calculations

To model a guide cut along the dense (11) dielectric crystal row, we make use of a rectangular supercell of dimensions a along guide direction y and $a\sqrt{3}\beta$ along x. Denoting $\beta = \beta_0 + \delta\beta$, the triangular PBGC occupies a width $a\sqrt{3}\beta_0$ and the guide, located at mid-cell, a width $a\sqrt{3}\delta\beta$ (see Fig. 1).Modes and their frequencies are obtained through the plane-wave expansion method [3, 4] on reciprocal lattice vectors $\vec{G}$ of electric field $\vec{E}(x,y) = E_z\vec{z}$, for which the following matrix equation applies :

$$\sum_{G'} |\vec{k}+\vec{G'}| \, |\vec{k}+\vec{G}| \, \eta(\vec{G}-\vec{G'}) * \left[E_z(\vec{k},\vec{G})|\vec{k}+\vec{G}|\right] = (\omega/c)^2 * \left[E_z(\vec{k},\vec{G})|\vec{k}+\vec{G}|\right]$$

$$(1)$$

where $\eta(\vec{G})$ is the inverse dielectric constant Fourier transform, obtained by the so-called "Hoś method", i.e., by inverting the matrix of dielectric constant $\epsilon(\vec{G}-\vec{G'})$. $\epsilon(\vec{G})$ is the product of (i) an atomic factor $f(\epsilon_a, \epsilon_b, |G|R)$ where $\epsilon_a = 1$, $\epsilon_b = 12.5$ is a typical value for GaAs, and $R = 0.469a$ is the cylinder radius whose center is located at $\vec{C}_j$; and (ii) a structure factor $S = (2\beta)^{-1} \sum_j exp(-i\vec{G}.\vec{C}_j)$.

In the supercell approach, we actually model parallel guides with a $a\sqrt{3}\beta_0$ wide PBGC barrier. At a given guide width $w = a\sqrt{3}\delta\beta$, coupling vanishes in the large β limit. This is checked through the flatness of dispersion for $\vec{k} = k_y\vec{y}$ normal to the guide. Convergence with respect to the number N of plane waves is very fast. In spite of N values of order of 400, we could also check the convergence vs. N using states well localized in the gap, which are well defined for β_0 as small as 4 or 5. Finally, although band frequencies are slightly overestimated, guided mode frequencies are fairly accurate with respect to band edges.

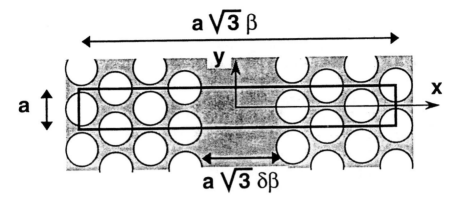

Figure 1. $a\sqrt{3}\beta * a$ rectangular supercell used to model a guide with PBGC boundaries

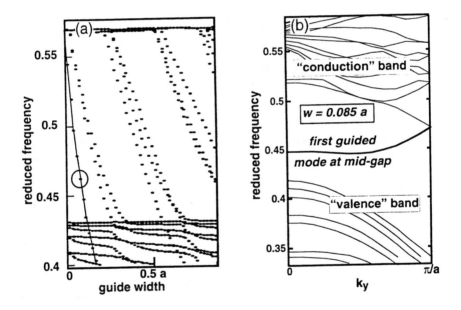

Figure 2. (a) eigenfrequencies as a function of guide width w at $k = 0$. Guided modes detach from the conduction band. The first mode to reach the middle of the full gap is circled at $w = a\sqrt{3}\delta\beta \approx 0.085a$; (b) dispersion relations of all modes (thin lines) that reflect folded bands of uniform PBGC and the guided mode circled in (a) (bold line) .

3. Results

1. Narrowest guide with mid-gap mode. Going to studies of the number
and frequency spacing of guided modes as a function of guide width, we
first focus on the narrowest guide yielding a mid-gap mode at k=0. Mode
reduced frequencies vs. reduced guide width are shown in Fig. 2a. The first
guided mode detached from conduction bands at k=0 crosses the gap very
rapidly as a function of width and, due to its symmetry, does not couple
to valence-band states, unlike the two next modes which smoothly merge
with valence band states.

The first mid-gap guided mode is seen to occur at a particularly narrow
guide width of $w = 0.085a$, hence an optical path $0.085\epsilon_b^{1/2}a \approx a/4$. This is
much less than in the one-dimensional case where a quarter-wave spacer is
$a/2$ thick. The difference partly lies in band structure, since this first defect
mode does not couple to the valence band, it must cross from conduction
band to the lower-lying first E band, of zero frequency at $k = 0$. Thus,
it reaches the middle of the full-gap far before midway across the bands
it actually couples. In view of enhanced light-matter coupling with, e.g.,
semiconductor material active in a relatively narrow energy domain, such
narrowly guided modes might be very desirable.

The dispersion relation of this mode vs. k_y appears in Fig. 2b. It exhibits
a minute downward bending up to $k_y \approx \pi/2a$ and goes upwards next. Its
spatial structure, e.g., at $k = 0$, is depicted by equal amplitude contour
lines in Fig. 3. Notice how narrow is the guide. Main lobes take place at
the thicker ribs of the guide between adjacent air cylinders. This structure
is typical from a conduction band state.

2. Role of relative boundary arrangement. Unlike a naive view, the guides
discussed here exhibit no contradirectional coupling when $2k_y = 2\pi/a$ in
spite of *individual* edge corrugations of period a. Instead of a stop band, a
doubly degenerate state is clearly obtained at $k_y = \pi/a$. This is due to the
$a/2$ relative shift along y between edge patterns. In a classical view, partial
backreflections from each cylinder cancel each other at this k value. From
a Fourier transform viewpoint, we have $\epsilon(\vec{G} = (2\pi/a)\vec{y}) = 0$ due to the
supercell symmetry. Thus, even for more general design of guides bounded
with PBGC, e.g., in the case of tapered or curved guides, care should be
taken of the relative boundary corrugation arrangement.

3. Symmetries of guided mode. Along the same line, all $k = 0$ eigenmodes
exhibit symmetries closely related to the cell mode (detailed dispersion relations are beyond our scope). E modes are by nature odd upon reflection
in the xy plane. The cell has an xz mirror plane whereas the yz reflection
must be combined with an $a/2$ glide along y. In Fig. 4, we show mode reduced frequencies vs. reduced guide width on a broader range than Fig. 1a.

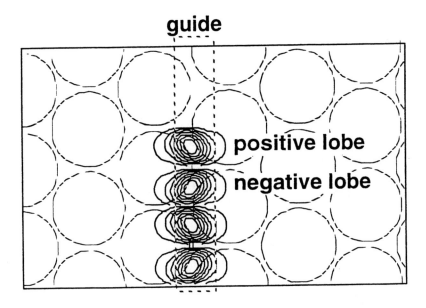

Figure 3. Equal field amplitude contours of the guided mode circled in Fig. 2a. The dashed rectangle indicates the narrow guide location.

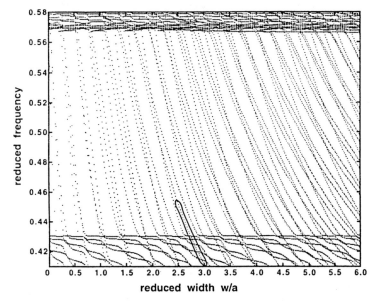

Figure 4. mode reduced frequencies $\omega a/2\pi c$ as a function of reduced guide width w/a at $k = 0$. Notice that one out of every three modes does not couple to the valence band. An example is underlined by a solid line.

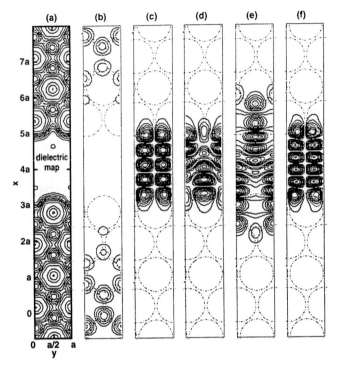

Figure 5. (a) contour map of the dielectric constant as truncated in the plane-wave expansion ; (b) to (f) : amplitude contours of field patterns at increasing frequencies. (b) Anti-guiding valence-like mode; (c) guided mode with xz nodal plane (d) and (e) xz is an antinodal plane ; both modes have opposite parity with respect to the glide-reflection in the yz plane ; (f) same as (c) with one more lobe.

At a given width as pictured in Fig. 5a (here $\delta\beta = 1$), guided states display a systematic succession of symmetries vs. increasing frequency, as shown by equal amplitude patterns : starting from (b) an anti-guided valence-like mode, we have a mode (c) with 4 lobes along x and an xz nodal plane. Next, a mode (d) with an xz antinodal plane and strongest lobes at inner cylinder edges, a mode (e) also with an xz antinodal plane but strongest lobes at ribs between cylinders adjacent to the guide. Finally, (f) is similar to (b) but with an additional lobe along x. (d) and (e) have opposite parity with respect to the glide-reflection. This systematic succession (two xz-antinodal for an xz-nodal mode, and so on) stems from the cost for an additional lobe along x, which results in an opposite parity for the glide-reflection when going from (c) to (f). On the contrary (d) and (e) with xz antinodal planes display both parities with a change in lobe pattern rather than with an additional lobe, albeit with complicated field patterns.

4. Conclusion

We have shown the efficiency of the supercell method in predicting the structures of guided E-polarized modes bounded by two-dimensional PBGCs. In the discussion of our results, some physical insight was presented on (i) enhanced light-matter coupling prospect in very narrow guides, (ii) design rules for guide boundaries, and (iii) symmetries of guided modes.

5. Acknowledgments

We want to thank C. Weisbuch for illuminating discussions and J.-M.Gérard and J.-Y. Marzin for introducing me to plane wave expansion methods. This work was supported by ESPRIT Basic Research Project Number 8447 "SMILES".

References

1. E. Yablonovitch, Phys. Rev. Lett. **58**, 2059 (1987).
2. K. M. Ho, C. T.Chan, and C. M. Soukoulis, Phys. Rev. Lett. **65**, 3152 (1991).
3. M. Plihal and A. A.Maradudin, Phys. Rev. B **44**, 8565 (1991).
4. R. D. Meade, K. D. Brommer, A. M. Rappe and J. D. Joannopoulos, Appl. Phys. Lett. **61**, 495, (1992).
5. R. D. Meade, A. Devenyi, J. D. Joannopoulos, O. L. Alerhand, D. A. Smith, and K.Kash, J. Appl. Phys. **75**, 4753 (1994).
6. R. J. Deri and E. Kapon, IEEE J. Quantum Electron. **QE-27**, 570, 626 (1992) and references therein.
7. J.-M. Gérard, A. Izrael, J.-Y. Marzin, R. Padjen and F. R. Ladan, Solid-State Electronics **37**, 1341 (1994); see also the contribution by T. Krauss and co-workers in this volume.
8. D. R. Smith, R. Dalichaouch, N. Kroll, S. Schultz, S. L. McCall and P. M. Platzmann, J. Opt. Soc. Am. B, **10**, 314 (1993); R. P. Stanley, R. Houdré, U. Oesterle, M. Ilegems,

and C. Weisbuch, Phys. Rev. A **48**, 2246 (1993).

9. We refer the reader to the numerous contributions of the present volume on this matter and references therein.

SECOND HARMONIC SCATTERING FROM SITES OF A CRYSTALLINE LATTICE

JORDI MARTORELL, R. CORBALAN
Departament de Fisica, Universitat Autonoma de Barcelona
08193 Bellaterra (Barcelona), Spain

R. VILASECA, J. TRULL
Departament de Fisica i Enginyeria Nuclear, Universitat Politecnica de Catalunya
C/Colom 11, 08222 Terrassa (Barcelona), Spain

Abstract

Experimental evidence of second harmonic generation in a macroscopically centrosymmetric lattice formed by spherical particles of optical dimensions is presented. Second harmonic light is scattered from the surface of these spherical particles. A simple theoretical model based on scattering in the Rayleigh-Gans approximation indicates constructive interference of light scattered at the second harmonic frequency leads to a plane wave front propagating in the direction of the incident beam. The implications of the addition of defects in a controlled manner will be discussed in the framework of the photonic band gap theory. Along these lines we will discuss in detail the simple case of a truncated periodic lattice in one dimension. We will present additional experimental results showing enhancement and suppression of the radiation of a dipole sheet oscillating at the second harmonic frequency embedded in a 1-dimensional periodic structure with a defect.

1. Introduction

The intrinsic ordering of dielectric material in photonic band gap (PBG) structures may be used in nonlinear processes that require an orientation of the permanent dipole moment. An inversion symmetry breaking is required for second order nonlinear processes, such as second harmonic generation (SHG) or frequency mixing in the dipole approximation. As is well known, Second Harmonic (SH) light may be reflected from the interface of two centrosymmetric media, due to the lack of inversion symmetry of the surface layers [1]. In a PBG structure constructed from dielectric spherical particles suspended in a less dense medium, we encounter a lack of inversion symmetry across the surface of each one of the spheres. Light scattered from different portions of the sphere surface interferes constructively leading to a nonvanishing field, when the radius of the sphere is comparable to the wavelength of the light.

Generation of SH light in structures with a three-dimensional distribution of dielectric material was first considered in sheared colloidal suspensions of latex microspheres by Lawandy et al. [2]. In the first part of this work we will present experimental results that show evidence of SHG in a macroscopic colloidal crystal composed of polystyrene spheres. The mechanism responsible for SHG is based on the scattering of light at the

C. M. Soukoulis (ed.), Photonic Band Gap Materials, 529–534.
© *1996 Kluwer Academic Publishers. Printed in the Netherlands.*

double frequency in the Rayleigh-Gans regime. The SH light scattered at each site of one lattice plane adds coherently and forms a plane wave propagating in the direction of the incident beam. For a continuous growth of the SH field intensity, it is also necessary to phase match the fundamental and SH beams. In a periodic material, such as a crystalline lattice, the necessary mechanism of phase matching is naturally provided by the bending of the photon dispersion curve at the boundary of the forbidden zone.

When the frequency of the SH light is close to the zone boundary, one has the added advantage of a greatly enhanced energy density. Further increasing the energy density is possible in a PBG structure by the introduction of defects or a controlled degree of disorder. In the second part of this paper, we present a preliminary study on the effects of the introduction of a controlled degree of disorder in a PBG structure. Our study is focused on the alteration of the nonlinear interaction in a truncated one-dimensional periodic structure.

2. Second Harmonic Generation in a 3-Dimensional Lattice

In our experiments we used a macroscopically centrosymmetric colloidal crystal composed of negatively charged small spherical particles of polystyrene suspended in water. Large concentrations of these spherical particles are known to arrange themselves in a face centred cubic (fcc) crystal with a lattice constant of optical dimensions [3]. In each spherical particle of the lattice, we encounter a lack of inversion symmetry across any given portion of the sphere surface. To enhance the second order nonlinear interaction present in this boundary, we may adsorb a layer of strongly nonlinear molecules on the surface of each sphere. Stable aqueous suspensions of negatively charged microspheres may be coated by dialysis with a positive chromophore of a dye molecule with a high nonlinear coefficient. The negative surface charge of each sphere helps, via the attractive coulombian interaction, the formation of a layer of nonlinear molecules with a preferred orientation of the permanent dipole moment. In our experiments we coated 0.137 μm in diameter spheres with the chromophore part of Malachite Green. The polystyrene suspension was dialyzed overnight in a 2×10^{-6} Molar aqueous solution of Malachite Green. With the colloidal suspension of dye coated spheres, a single fcc crystal was formed in a 1 mm path length cell. The colloidal suspension was added to a cell (50x10x1 mm) containing in the bottom a mixed bed ion exchange resin. After a couple of days, a single crystal formed when stray ions in solution diffused to the resin.

The prepared sample was excited using an active-passive mode-locked laser emitting 35 ps pulses in the near infrared at a wavelength of 1064 nm. The average energy of the pulse was 5 mJ. The polarization of the incident beam was set to be either parallel or perpendicular to the plane of incidence, and the polarization of the generated SH beam was selected using a Glann-Thompson polarizer. Measurements of the reflected SH intensity from the crystal planes with Miller indices (111) are shown in Figure 1 as a function of the angle of incidence. Both the exciting field and the generated field at the double frequency were polarized parallel to the plane of incidence. A peak of maximum SH intensity was observed at 15 degrees. Additional experiments were performed to determine the passive reflection characteristics of the colloidal crystal

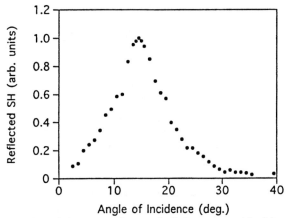

Figure 1. Intensity of the second harmonic light generated inside a single colloidal crystal, measured in a reflection geometry as a function of the angle of incidence, relative to the normal of the (111) planes. The wavelength of the incident laser pulse was 1064 nm. A single crystal was formed in a 1 mm path length cuvette from a concentrated aqueous suspension of 0.137 μm polystyrene spheres in diameter. The spheres were coated in a 2×10^{-6} Molar solution of Malachite Green.

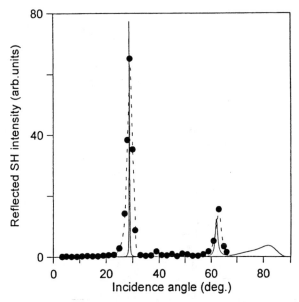

Figure 2. Reflected SH intensity as a function of the angle of incidence, from the multilayer stack described in the text. The angle is given relative to the normal of the multilayer stacks. The black dots indicate the experimental data, while the dashed line is only a guide for the eye. The continuous line corresponds to the numerical prediction of the theoretical analysis.

used. Light at 532 nm, obtained by doubling the frequency of the laser pulse in a KDP crystal, was sent directly to the colloidal crystal surface. Maximum reflection occurred at 19 degrees, where the Bragg condition $2d\cos\theta=\lambda/n$ for the (111) planes was satisfied. Notice that SH light is generated at the smaller angle side of the Bragg stop band centred at 19 degrees. This is consistent with a phase matching of the fundamental and SH beams due to a decrease of 0.02 in the effective index of refraction at the left edge of the stop band. This change in the effective index of refraction for a wave propagating at the SH frequency was sufficient to overcome the phase lag between the fundamental and SH waves introduced by the normal dispersion of water [4].

This behavior, in addition to a nonzero SH reflection at normal incidence, indicates clearly a bulk process responsible for the SHG. This second order nonlinear process at the bulk of a centrosymmetric colloidal crystal is attributed to the nonlinear molecules adsorbed on each sphere. The permanent dipole moment of the molecule was oriented on the average in the radial direction defined from the center of each sphere. Experiments performed four months later showed no significant decrease in the total SH intensity measured, indicating a permanent attachment of the nonlinear molecule to the surface of the sphere. The observed second order nonlinear process was consistent with a theoretical analysis that considered first the SH light scattered from the surface layer of nonlinear material covering each dielectric sphere. The following step considers the coherent addition of the contributions from the spheres located on one of the (111) planes of the fcc crystal. We performed this summation numerically and found a nonzero field polarized parallel to the plane of incidence and propagating in the same direction of the incident beam. This SH field, partially reflected by the periodic distribution of dielectric material, may be detected in either a reflection or transmission geometry. This prediction of the generation of a SH plane wave is in agreement with the experimental observation of an SH field polarized in the same direction as the incident field.

3. Localization of Second Harmonic Light in a 1-dimensional Lattice

In order to investigate the alteration induced in the nonlinear interaction when defects breaking the perfect periodicity of the lattice are included, we considered the simple case of a truncated 1-dimensional lattice or a stack of alternating layers of high and low index of refraction. The truncated periodic structure used in our experiments was formed by two identical multilayer stacks separated by a small air gap. Such a structure strongly reflected light at normal incidence in the wavelength range from 540 to 660 nm. A Bragg stop band or forbidden band was also seen when the wavelength was maintained constant at some value close to the wavelength satisfying the Bragg condition, and the angle of incidence was varied. At 532 nm a gap was found between 15 and 63 degrees when the field was polarized parallel to the plane of incidence. A monolayer of oriented Malaquite Green (MG) chromophores was adsorbed on the surface facing the air gap of only one of the multilayer stacks using the "dipping" technique [5].

The SH intensity reflected from the surface monolayer embedded in the one-dimensional periodic structure described was measured after excitation of the MG molecules with 35 ps pulses from an active-passive mode-locked Nd:YAG laser. The mirror-like periodic distribution of dielectric material on both sides of this monolayer will strongly alter the nonlinear interaction. Experimental measurements of the SH intensity reflected from the surface monolayer of nonlinear molecules through the multilayer stacks as a function of the angle of incidence are shown in Figure 2. These measurements were taken with the incident fundamental field polarized parallel to the plane of incidence. A sharp resonance at 29 degrees corresponding to the excitation of the SH field in a local mode within the forbidden band may be seen in Figure 2, as well as another resonance at the large angle edge of the stop band at 63 degrees. The last experiment was repeated when the multilayer stacks were separated a distance much larger than the coherence length of the laser pulse and all coherent effects disappear. A comparison of these last measurements and the data shown in Figure 2 indicate the generated SH intensity is six times larger when the nonlinear interaction occurs in a mode of the microresonator formed by the two multilayer stacks. The air gap separating the two stacks, equivalent to the addition of dielectric material in the middle period of the 1-dimensional periodic structure, results in the appearance of a defect mode or "impurity level" within the forbidden band. The high energy density of this local mode is responsible for the enhancement of several times the nonlinear interaction in the vicinity of the defect. In the language of PBG theory, this sharp resonance corresponds to the excitation of the SH oscillation in a "donor mode" [3]. In a 1-dimensional lattice this "bound" state exists no matter how small the size of the defect. Its position in the angular spectrum, which is shown in Figure 2, is a function of the size of the defect.

Near the band edge, the bending of the electromagnetic wave dispersion curve slightly above or below the forbidden zone, indicates the group velocity approaches zero, giving rise to an increased effective path length. In solid state electronics, this phenomenon results in a square root singularity for the photon density of states, or a Van Hove singularity of a 1-dimensional lattice. In our experiment this increased effective path length corresponds to an enhancement of the SH radiation at the angle of 63 degrees as shown in Figure 2. This enhancement of the nonlinear interaction at the edge of the forbidden zone, predicted by Bloembergen and Sievers [6], is a result of the periodicity built into the material, and its location in the angular spectrum shown in Figure 2 is essentially independent of the size and position of the defect.

As seen in Figure 2, SHG from a molecular dipole sheet embedded in a 1-dimensional periodic structure is completely suppressed for modes within the forbidden band other than the defect mode. This suppression of the oscillation at the SH frequency corresponds to the inhibition of the radiation from a classical dipole source [7].

4. Concluding Remarks

The observation of SHG from the surface of spherical particles ordered in a three-dimensional lattice opens numerous applications in the development of new nonlinear devices as well as possibilities for research in the basic filed of scattering media. We have also considered the introduction of defects in a periodic structure, in order to alter the nonlinear interaction present in the neighbourhood of the defect. We have experimentally studied the one-dimensional case and found a strong dependence on the environment for SHG from a slab of nonlinear material located at the defect region.

Enhancement of SHG in defect modes could become particularly useful in the implementation of a frequency doubler in vertical surface emitting lasers.

Acknowledgments

We acknowledge support from the Spanish DGICYT, Projects No. PB93-0968 and PB92-0600.

References

1. Fielding Brown and Masahiro Matsuoka, Phys. Rev. **185**, 985 (1969), T. F. Heinz, C. K. Chen, D. Ricard, and Y. R. Shen, Phys. Rev Lett. **48**, 478 (1982).
2. N. M. Lawandy, S. A. Johnston, and Jordi Martorell, Optics Comm. **65**, 425(1988).
3. Roger J. Carlson and Sanford A. Asher, Appl. Spectroscopy **38**, 297 (1984).
4. I. Thormählen, J. Straub, and U. Grigull, J. Phys. Chem. Ref. Data, **14**, 933 (1985).
5. S. R. Meech and K. Yoshihara, Chem. Phys. Lett. **154**, 20 (1989), Stephen R. Meech and Keitaro Yoshihara, J. Phys. Chem. **94**, 4914 (1990).
6. N. Bloembergen and A. J. Sievers, Appl. Phys. Lett. **17**, 483 (1970) 483.
7. Jonathan P. Dowling and Charles M. Bowden, Phys. Rev. A **46**, 46 (1992).

MULTISTABILITY AND SWITCHING IN NONLINEAR LAYERED OPTICAL AND ELECTRONIC MEDIA

N. G. SUN

Center for Nonlinear Science and Department of Physics
University of North Texas, Denton, Texas 76203, USA

AND

G. P. TSIRONIS

Center for Nonlinear Science and Department of Physics
University of North Texas, Denton, Texas 76203, USA
and Physics Department, University of Crete and Research
Center of Crete, P. O. Box 1527, Heraklion 71110, Crete,
Greece

1. Introduction

Photonic crystals that possess a Photonic Band Gap (PBG) are systems with very intersting physical properties and, in addition, are very important in applications. As a result of numerous efforts [1, 2, 3, 4, 5, 6] both in the experimental and theoretical direction it is now well established that PBG systems can be realized in practice. Possible applications include resonance cavities, perfect refractive mirrors, antennas and electromagnetic wave filters with broad frequency range from ultraviolet to microwave. Most PBG studies so far have addressed almost exclusively PGB structures where nonlinearity in the dielectric constant of the constituting matertial is negligible.[7] One reason for this is that in the presence of nonlinearity the superposition principle does not hold any more, rendering the calculations more difficult and also turning familar concepts such as that of bands and gaps essentially meaningless. In the present paper we present recent results obtained through a simple periodic model concerning nonlinear effects in one dimensional PBG crystals and other nonlinear periodic lattice systems.[8, 9] The nonlinear Kronig-Penney (KP) model that we use shows that in the electromagentic case of wave propagation in one dimen-

C. M. Soukoulis (ed.), Photonic Band Gap Materials, 535–545.
© 1996 *Kluwer Academic Publishers. Printed in the Netherlands.*

sion occurence of shifts in the linear stop gaps as a result of nonlinearity, the onset of bistability as well as switching properties.[10] We will show that for reasonable input powers the nonlinear lattice can be used as an optical switch.[10] The nonlinear KP model can be also used for the study of electronic propagation in semiconductor superlattices. We will address the electronic case when in addition to nonlinearity and periodicity there is an external constant electric field applied to the system.[11] The electronic and electromagnetic cases are formally essentially equivalent but there are physical differences between them.[9]

2. The Photonic Wave Equation

The nonlinearity that we consider here arises from the optical Kerr effect producing a field-dependent index of refraction:

$$n(\omega) = n_0(\omega) + n_2|E|^2$$

where n_0 is the linear index of refraction of the medium and n_2 describes the change in the index of refraction effected by the local field intensity $|E|^2$. In what follows, we use the dielectric function ϵ to first order in $|E|^2$, i.e.,

$$
\begin{aligned}
\epsilon &= n^2 = (n_0 + n_2|E|^2)^2 \\
&\approx n_0^2(1 + \frac{2n_2}{n_0}|E|^2)
\end{aligned}
\tag{1}
$$

We study wave propagation in a one-dimensional system that satisfies the following scalar wave equation:

$$\frac{d^2E(z)}{dz^2} + \frac{n^2(z)\omega^2}{c^2}E(z) = 0 \tag{2}$$

Our system consists of a multilayered medium with alternating segments, one of which has only linear dielectric properties whereas the other has also a nonlinear contribution similar to that of Eq. (1). For the linear medium, Eq. (2) reads:

$$E''(z) + k^2E(z) = 0$$

where $k = n_0\omega/c$, and n_0 is the index of refraction in the linear medium. For the second medium that also contains the nonlinear contribution to the dielectric constant we have:

$$\frac{d^2E(z)}{dz^2} + k'^2\left(1 + \frac{2n_2}{n_1}|E|^2\right)E(z) = 0$$

where $k' = n_1 \omega / c$, n_1 is the linear part of the index of refraction of this medium and n_2 determines the nonlinear contribution to the dielectric constant. We will assume that the nonlinear layer is much thinner than the linear one and the effects of the nonlinear medium in the propagation can be represented through a spatially periodic δ-function model with nonlinearly modulated strengths. Using this δ-function type of Kronig-Penney (KP) model, we obtain following equation for the wave propagation in the complete lattice as a fucntion of dimensionless coordinate $x = z/d$:

$$-\frac{d^2 E(x)}{dx^2} + \sum_{m=1}^{N} \delta(x - m) g(x) E(x) = \kappa^2 E(x) \qquad (3)$$

$$g(x) = \kappa^2 - \kappa'^2 - 2\frac{n_2}{n_1}\kappa'^2 |E(x)|^2 \qquad (4)$$

The nonlinear Kronig-Penney model of Eq. (4) can be turned into an effective discrete nonlinear Schrödinger equation[13, 14] that is equivalent to a map representation; it can be studied both analytically and numerically using Hamiltonian methods[9, 12]. In particular this study reveals bistablity and the intensity-dependent transmission phenomena. Most of these properties have been qualitatively and quantitatively addressed for the electronic and photonic cases through the use of a dynamical systems approach.[8, 9, 10, 11, 14, 15, 16]. For the purpose of the present study we will present the physically interesting case of the modification of the linear stop band structure due to the presence of nonlinearity and the subsequent creation of a power switching device. For a Kerr coefficient n_2 of the order of $10^{-11} \text{cm}^2/\text{W}$; this is a typical optical fiber value. We plot in Fig. 1 the transmission-gap transitions on the plane of input laser power and the wavenumber κ. We observe that the amplitude dependent modification to the linear stop gap (that is independent on th input power) occurs well below the GW/cm^2 level leading to the realistic possibility of use of such a fiber system as a nonlinear switch. By examining Fig. 1 we note in particular that for waves with wavenumbers near $\kappa = 4.5$, or wavelengths $\lambda = 400nm$ and $d = 215nm$, a transmission switch is created when the input power reaches the level of $200 MW/cm^2$.

3. Nonlinear Kronig-Penney Model for Electrons

Electric field induced Stark ladder effect, Wannier-Stark (W-S) localization and the electro-optical properties and device applications in the semiconductor superlattices (SL), such as GaAs-AlGaAs, have been studied extensively in recent years [17, 18, 19, 20, 21, 22]. Quantum effects become significant in such systems because the electron wavelength is of the same order as the superlattice constant. Recently, there has been renewed interest in the

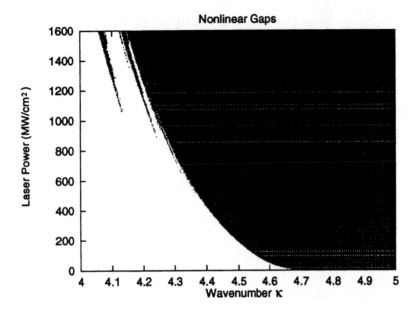

Figure 1. Possible creation of a power switching device. Transmission and gap regions in the power-wavevector plane are shown in this figure. The white areas in the diagram indicates tranmission region, and the black areas represent the gap. At the band edge, when the the laser power is increased, the transmission property of this structure is switched from the transparent region to a gap. The Kerr coefficient $n_2 = 4 \times 10^{-11}\, cm^2/W$. In the absense of nonlinearity, frequency gaps are independent of the amplitude; for relatively weak nonlinearity, the gap reqions become larger as the wave intensity increases.

study of multistability and discontinuity in the current-voltage (I-V) characteristic of doped semiconductor superlattices, both theoretically [23, 24] and experimentally[25]. We propose a simple model which is based on self-consistent potentials and does not include any direct scattering processes, and study the transport of ballistic electrons in terms of transmissions of quantum mechanical waves in a SL heterostructure. We solve the tunneling and transmission problem for the electrons in the quantum well (QW) heterostructure and demonstrate that multistability and discontinuity in the transport of carriers are related to the shrinking and destruction of the miniband structure by the nonlinearity in the doped layers. We assume, for simplicity, that the longitudinal and transverse degrees of freedom are decoupled, resulting thus in an effectively one-dimensional problem. The interaction of an electron with charge accumulation in a doped layer is represented by a nonlinear term, that is seen to arise from a self-consistent potential in that layer [26].

We consider a SL that consists of a square-well/square-barrier semiconductor heterostructure; this is a model of conduction bands representing the mismatch between two component materials of the superlattice. We consider two different models for the location of the doped layers. In the first one we assume that the doped layers coincide with the quantum barriers, whereas in the second model the doped layers are located in the center of the QWs. Following Ref. [26], we write the self-consistent Schrödinger equation for $\psi(x,t)$, the wave function for an electron in the SL, in the absence of an external field, as:

$$i\hbar\frac{\partial}{\partial t}\psi(x,t) = \left[-\frac{\hbar^2}{2m}\frac{\partial^2}{\partial x^2} + V(x) + \int W(t,t';x,x')|\psi(x',t')|^2 dt' dx'\right]\psi(x,t),$$

$$(5)$$

where $V(x)$ is the periodic lattice potential and $W(t,t';x,x')$ is a kernel describing the interaction of the electron with the electrons in the charged layers. We are interested in the time-independent solutions, $\psi(x,t) = \psi(x)e^{iEt}$, and by assuming that the kernel is time-independent we have the integral part of Eq. (5) proportional to the stationary density of charges in the doped layers. If the size of these regions is much smaller than the spatial variations of $\psi(x)$, the integral part of Eq. (5) can be replaced by the summation of the average contributions of the localized charges inside the wells, i.e., $\sum_n \bar{W}b|\psi(x_n)|^2$, where b is the width of the layer and $\bar{W}$ is the average kernel in the well. The latter is proportional to $e^2 n_e/C$, where e is the electron charge, n_e is the charge density in the doped layer and C is the capacitance of that layer. For simplicity, we assume that we have ultrathin doped layers and use δ-function type nonlinear barriers to represent the self-consistent potentials; this is an approximation which makes it possible to obtain a closed form expression for the model. In order for the δ-function model to be qualitatively compatible with the original QW structure, we require that the δ-function strengths are equal to the average barrier height in a cell leading to the replacement of the integral of Eq. (5) with the sum $\sum_n \bar{W}b^2|\psi(x)|^2\delta(x - x_n)$. The problem then reduces to that of studying a Kronig-Penney type model with nonlinear terms [27], [8]-[16]

4. The Electric Field Case

When an external electric field is applied along the growth axis of the superlattice, the most fundamental change that the field makes is the breaking of the translational symmetry. The energy levels of neighboring wells are misaligned, resulting in a W-S localization due to the turning-off of the resonant tunneling between consecutive wells. Wannier-Stark localization has been used to explain shifted absorption edges of photocurrent [17, 18, 19] and widened gap regions in the transmission spectrum[28]. In this case, the time-independent Schrödinger equation for the electron in an external electric field $\mathcal{E}$, with energy E, and approaching a sample of N periodic

540

potential barriers is

$$-\frac{\hbar^2}{2m}\frac{d^2}{dx^2}\psi(x) + \left[\sum_{n=1}^{N} g(|\psi|^2)\delta(x-x_n) - e\mathcal{E}x\right]\psi(x) = E\psi(x) \quad (6)$$

where $g(|\psi|^2) = p(g_0 + g_2|\psi(x)|^2)$, p is the potential strength, g_0 and g_2 are weight factors ($pg_2 = \bar{W}b^2$), representing the linear and self-consistent nonlinear potentials respectively, and $x_n = na$, where a is the lattice constant. We define a characteristic length $l(\mathcal{E}) = (\hbar^2/2me\mathcal{E})^{1/3}$, and a dimensionless parameter $\lambda(\mathcal{E}) = (2m/\hbar^2e^2\mathcal{E}^2)^{1/3}E$. It can be easily shown that in the linear case ($g_2 = 0$), between two adjacent scatters, Eq. (6) is transformed into a *Bessel equation* of order $(1/3)$, whose solution is expressed as a combination of *Hankel functions* of the first and second kind [29]. This solution is also valid in the general nonlinear case ($g_2 \neq 0$) since the nonlinear term is localized; we thus have for the wave function between x_{n-1} and x_n:

$$\psi_n(z) = A_n z^{1/3} H_{1/3}^{(1)}(z) + B_n z^{1/3} H_{1/3}^{(2)}(z) \quad (7)$$

where $H_{1/3}^{(1,2)}(z)$ are the *Hankel functions* of the first and second kind, respectively, and $z(x,\mathcal{E}) = \frac{2}{3}\lambda^{3/2}(\mathcal{E})(1 + x/\lambda(\mathcal{E})l(\mathcal{E}))^{3/2}$, is a dimensionless coordinate. The effect of nonlinearity is included through the amplitude coefficients A_n and B_n which will be determined subsequently through boundary conditions.

We calculate the transmission coefficient for an electron in a SL in the presence of an external electric field, and use Landauer's formula to obtain the corresponding conductance. In order to do that, we must find the wave amplitudes, (A_n, B_n), of Eq. (7). Considering the continuity of $\psi(x_n)$ and the discontinuity of its derivative due to the δ-function at $x = x_n$, a recurrence relation connecting (A_{n+1}, B_{n+1}) with (A_n, B_n) is obtained as follows:[11]

$$\mathcal{M}: \begin{cases} A_{n+1} = \left[1 + w_n(|\psi_n|^2)h_n^{(1)}/h_n^{(0)}\right] A_n + w_n(|\psi_n|^2)h_n^{(2)}/h_n^{(0)} B_n \\ B_{n+1} = \left[1 - w_n(|\psi_n|^2)h_n^{(1)}/h_n^{(0)}\right] B_n - w_n(|\psi_n|^2)h_n^{(3)}/h_n^{(0)} A_n . \end{cases}$$
$$(8)$$

where $w_n = \frac{2ml}{\hbar^2}(\frac{3}{2}z_n)^{-1/3}p(g_0 + g_2|\psi_n(z_n)|^2)$ and all the h_n's are products (or sum of products) of Hankel functions of z_n: $h_n^{(0)} = H_{1/3}^{(2)}(z_n)H_{-2/3}^{(1)}(z_n) - H_{1/3}^{(1)}(z_n)H_{-2/3}^{(2)}(z_n)$, $h_n^{(1)} = H_{1/3}^{(1)}(z_n)H_{1/3}^{(2)}(z_n)$, $h_n^{(2)} = H_{1/3}^{(2)}(z_n)^2$, and $h_n^{(3)} = H_{1/3}^{(1)}(z_n)^2$. When the self-consistent interaction of the electrons in the doped layers is absent, i.e., when $g_2 = 0$, then Eq. (8) becomes independent of amplitudes A_n and B_n, and it essentially represents a transfer matrix-type

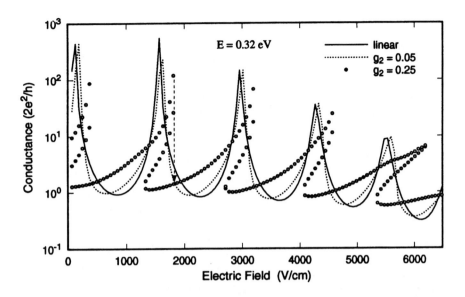

Figure 2. The effects of nonlinearity is shown in the conductance-field $(G\text{-}\mathcal{E})$ diagrams. For small nonlinear parameter g_2, the transmission and conductance curves are tilted and shifted; for large g_2, multiple transmissions and conductions become possible. The energy is chosen in the third transmission band, but multistability is also observed at other energies. An arrow is used to indicate a location of discontinuity. Other parameters are given in the text. The absolute values are used for the field strengths.

of equation. We observe that by using the properties of the δ-functions, the nonlinear Kronig-Penney problem of Eq. (6) is replaced by the simple, invertible nonlinear map $\mathcal{M}$ of Eq. (8). In this map, iteration by one step is equivalent to scattering through a δ-function barrier. In order to analyze the properties of the map $\mathcal{M}$ and calculate the transmission coefficient T for the electrons in the presence of the field $\mathcal{E}$, we use the standard back-propagation approach, i.e., we fix the amplitude of the outgoing wave at site N and iterate backwards to find A_0, the desired input amplitude. We vary the values of A_N and obtain the complete set of A_N's that corresponds to a given A_0. The transmission coefficient through the superlattice with N doped layers is given by $T = |A_N|^2/|A_0|^2$, whereas using the Landauer formula [30, 31], $G = (2e^2/h)T(1-T)^{-1}$, we also obtain the conductance G for the SL.

In Fig. 2, we show the electrical conductance G as a function of the field strength $\mathcal{E}$ for various values of g_2. In the linear case ($g_2 = 0$) and for a moderate electric field, the wave-function of the electrons inside each quantum well is localized (W-S localization) and the transmission is reduced due to the field induced reduction of the resonant tunneling between adjacent wells. As the electric field increases, the electrostatic potential energy of the electrons in each QW is enhanced by the amount of $ea\mathcal{E}$; if this value becomes comparable to ΔE_g, the energy gap between two minibands, enhancement in transmission is expected because of the intersubband resonant tunneling [32]. In the case of many minibands this process of enhanced transmission repeats itself also at higher field values resulting in the oscillatory pattern of the continuous curves in Fig. 2. This oscillatory behavior is a manifestation of the competition between W-S localization and the intersubband resonance-induced delocalization. We note that the delocalization effect is completely absent from a single band model. The effects of W-S localization and the intersubband resonance-induced delocalization can be observed through photon absorption and luminescence [17, 18, 32].

In the case of weak nonlinearity ($g_2 = 0.05$ in Fig. 2), the oscillatory behavior of transmission coefficient and conductance in the field remains similar to the linear case. However, the left and right sides of each peak becomes asymmetric, which means that (a) the W-S localization process is slowed down in the presence of nonlinearity in the doped layers, as shown by the smaller slopes of the increasing curves in Fig. 2; and (b) the widths of the minibands shrink in the presence of moderate nonlinearity, so that the intersubband resonances occur in a narrower range of field values, resulting in the rapid drop after T or G reaches a peak value. Finally, drastic changes are observed in the case of strong nonlinearity ($g_2 = 0.25$ in Fig. 2). We notice that W-S localization process is further slowed down in a increasing field, whereas the minibands structure is totally destroyed by the nonlinearity, resulting in abrupt changes in transmission and conductance, including the occurrences of discontinuity and multistability. In Fig. 2, we use $a = 20\text{Å}$, $N = 40$, and $E = 0.32\text{eV}$ (this energy is roughly at the center of the second miniband of the linear model). For the barrier strength, we use $p = 2.0$ eV Å; $g_2 = 0.05$, and 0.25, respectively, with $g_0 = 1.0 - g_2$.

In Fig. 3, we plot transmission vs. electric field strengths for $E = 0.35eV$ and three different nonlinear strengths or doped densities. We can see that as the nonlinear strength g_2 is increased, multistability is enhanced. We now use the case when the doped layers are placed in the middle of the QWs instead of the barriers. After we obtain the conductance G as in the case of Fig. 2, we use the field strength $\mathcal{E}$ and sample length Na to obtain the voltage $V = Na\mathcal{E}$, we then use the Ohm's law to obtain the current, $I = GV$. The current-field characteristic diagram and possible sweep-up and sweep-down paths for this second model are presented in Fig. 4. We use the following parameters for numerical calculations in Fig. 3: for the barrier potential, $g_0 = 1.0$, $g_2 = 0.0$; and for the doped layers, $g_0 = 0.5$, $g_2 = 0.5$. The rest of the parameters are the same as in Fig. 2. We point out that the behavior depicted in Fig. 4 is in qualitative agreement with

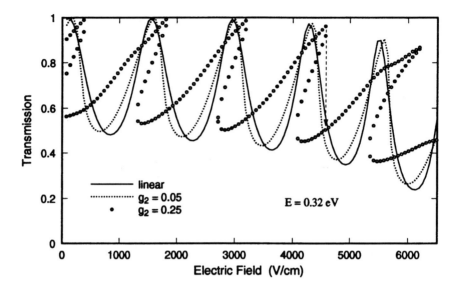

Figure 3. Multistability is shown in the plane of transmission coefficient vs. the electric field. The figure shows three cases with different nonlinear strengths, $g_2 = 0.25, 0.5$ and 1.0. $g_0 = 3.0$ for each case, and the lattice size $N = 50$.

the experimental results of Ref.[25].

5. Conclusions

We have shown that nonlinearity in a one dimensional photonic band gap system changes the properties of the crystal. When nonlinearity is relatively small we can expect that the band picture still holds approximately. In that case, nonlinearity widens the linear stop gaps, enhancing thus the nonpropagating regime.[10] For larger values the band picture changes completely. Furhtermore, nonlinearity induces bistability in the nonlinear dielectric lattice. These properties can be used for optical switching at laser powers that are within the realm of the present technology. On the other hand, we have demonstrated that the occurrence of multistability and discontinuity in the transport processes of electrons can be explained by introducing self-consistent potentials representing the nonlinear space charge effects due to electron accumulation in the doped semiconductor layers. We use a simple model in which the doped layers are assumed to be ultrathin and act nonlinear "kicks" on the wave packets of electrons. The introduction of δ-function type potentials is not essential in obtaining the multistable behavior in transmission.

544

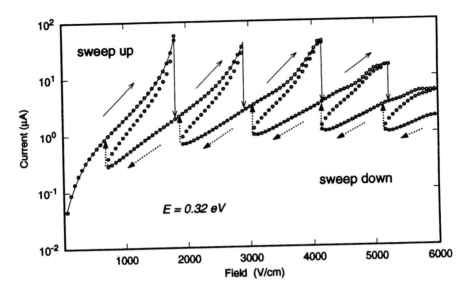

Figure 4. The current-field characteristic for the second model. Possible sweep-up and a sweep-down paths are shown as the field is either increased or decreased. The current values (small circles) are obtained by calculating the conductance under different fields. Parameters are given in the text. The absolute values are used for the field strengths.

Acknowledgement We acknowledge partial support from TARP grant number 003656-073c.

References

1. E. Yablonovitch, Phys. Rev. Lett. **58**, 2059 (1987).
2. S. John, Phys. Rev. Lett. **58**, 2486 (1987).
3. S. John and R. Rangarajan, Phys. Rev. B, **38**, 10101 (1988).
4. K. M. Ho, C. T. Chan and C. M. Soukoulis, Phys. Rev. Lett. **65**, 3152 (1990).
5. M. Sigalas, C. M. Soukoulis, E. N. Economou, C. T. Chan and K. M. Ho, Phys. Rev. B **48**, 121 (1993)
6. J. Joannopoulos, R. D. Meade and J. N. Winn, *Photonic Crystals*, Princeton University Press, Princeton (1995).
7. W. Chen and D. L. Mills, Phys. Rev. Lett. **58**, 160 (1987)
8. D. Hennig, H. Gabriel, G. P. Tsironis and M. I. Molina, Appl. Phys. Lett. **64**, 2934 (1994)
9. D. Hennig, N. G. Sun, G. P. Tsironis and Sp. Pnevmatikos, *Stationary properties of one-dimensional nonlinear lattices*, Erasmus Interuniversity cooperation program publication, Heraklion (1995).
10. N. G. Sun and G. P. Tsironis, submitted to Phys. Rev. Lett.
11. N. G. Sun and G. P. Tsironis, Phys. Rev. B **51**, 11221 (1995).
12. D. Hennig and G. P. Tsironis, in preparation.
13. Y. Wan and C. M. Soukoulis, Phys. Rev. A **41**, 800 (1990).
14. D. Hennig, N. G. Sun, H. Gabriel and G. P. Tsironis, Phys. Rev. E **52**, 255 (1995).
15. N. G. Sun, D. Hennig, M. I. Molina and G. P. Tsironis, J. Phys. Consens. Matter

6, 7741 (1994).

16. D. Hennig, G. P. Tsironis, M. I. Molina and H. Gabriel, Phys. Lett. A **190**, 259 (1994).

17. D. A. B. Miller, D. S. Chemla, T. C. Damen, A. C. Gossard, W. Wiegman, T. H. Wood and C. A. Burrus, Phys. Rev. B **32**, 1043 (1985)

18. E. E. Mendez, F. Agullo-Rueda, and J. M. Hong, Phys. Rev. Lett. **60**, 2426 (1988)

19. Paul Voisin, Surface Sci. **288**, 74 (1990)

20. F. Agullo-Rueda, E. E. Mendez, H. Ohno, and J. M. Hong, Phys. Rev. B **42**, 1470 (1990)

21. M. M. Dignam, and J. E. Sipe, Phys. Rev. Lett. **64**, 1797 (1990)

22. E. E. Mendez, in *Optics of Semiconductor Nanostructures*, ed. by F. Henneberger, S. Schmitt-Rink and E. O. Göbel, Akademie Verlag (1993)

23. B. Laikhtman, and D. Miller, Phys. Rev. B **48**, 5395 (1993)

24. F. Prengel, A. Wacker, and E. Schöll, Phys. Rev. B **50**, 1750 (1994)

25. J. Kastrup, H. T. Grahn, K. Ploog, F. Prengel, A. Wacker, and E. Schöll, Appl. Phys. Lett. **65**, 1808 (1994)

26. C. Presilla, G. Jona-Lasinio and F. Capasso, Phys. Rev. B **43**, 5200 (1991)

27. C. M. Soukoulis, J. V. Jose, E. N. Economou,and P. Sheng, Phys. Rev. Lett. **50**, 764 (1983)

28. N. G. Sun, D. Yuan, and W. D. Deering, Phys. Rev. B, **51**, 4641 (1995).

29. S. Flügge, *Practical Quantum Mechanics*, v.1, Springer-Verlag (1971)

30. R. Landauer, Philos. Mag. **21**, 863 (1970), and J. Phys. Condens. Matt. **1**, 8099 (1989)

31. P. W. Anderson, D. J. Thouless, E. Abrahams, and D.S. Fisher, Phys. Rev. B **22**, 3519 (1980)

32. H. Schneider, H. T. Grahn, K. v. Klitzing, and K. Ploog, Phys. Rev. Lett. **65**, 2720 (1990)

WAVEGUIDES IN PERIODIC STRUCTURES WITH SMOOTHLY VARYING PARAMETERS

V. V. KONOTOP

Department of Physics and Center of Mathematical Sciences, University of Madeira, Praça do Município, 9000 Funchal, Portugal

Abstract.

It is shown that in a periodic medium with smoothly modulated parameters, electromagnetic waves can propagate along effective waveguides. Such waveguides are interpreted in terms of defect modes. Respective theory based on the Wannier function expansion is developed. Some examples and relative effects are considered.

1. Introduction

It is known [1, 2] that in periodic media waveguides are created through the phenomenon of Bragg reflection. These are typically layered dielectric structures in which the periodicity is broken by a boundary, or by the local change of the parameters of layers. Formation of waveguides in these systems can also be interpreted in terms of the so-called *deffect modes* [3, 4]. Indeed, it is well known that deviations of the refractive index profile of a structure from an exactly periodic form can be interpreted as a defect of the periodicity which results in defect levels in the spectrum. Located inside a stop gap, these defect levels correspond to field configurations localized around the defect. If additional periodicity exists only in one direction, then field localization in that direction results in electromagnetic wave propagation along one of the orthogonal directions, giving rise to an effective waveguide.

The defect modes in photonic crystals have been studied experimentally and theoretically [3, 4]. The main emphasis has been defects localized on distances of the order of the underlying structure period and in most theoretical studies the problem was reduced to a numerical analysis.

C. M. Soukoulis (ed.), Photonic Band Gap Materials, 547–554.

Treatment of the defect modes in structures with smooth perturbations of periodicity can be given in terms of the Wannier functions [5]. These functions have been explored in the theory of photonic crystals [4] to reduce the size of matrices determining the defect modes.

In the present paper, the Wannier function representation (Sec. 2) is used in order to develop the theory of the defect modes in a medium where periodicity is smoothly modulated (Sec. 3). In contrast to most situations studied earlier, this case allows a rather complete analytical treatment. In particular, I obtain the mode structure of an effective waveguide in a periodic structure with the modulation having sech-form (Sec. 4) and describe related effects caused by the modulation (Sec. 5).

2. Wannier Functions

Wannier functions have been already used in the propagation theory [4]. Here, they are introduced in the way appropriate for the description of waveguides. The wave equation

$$\frac{\partial^2 E}{\partial x^2} + \frac{\partial^2 E}{\partial z^2} - \frac{\epsilon(x)}{c^2}\frac{\partial^2 E}{\partial t^2} = 0, \tag{1}$$

describes evolution of the electric field $\vec{E} = (0, E(x, z, t), 0)$ in a structure characterized by the dielectric permittivity $\epsilon(x)$. I studied the case where $\epsilon(x)$ allows a representation in the form,

$$\epsilon(x) = [1 + g(x)]\epsilon_0(x), \tag{2}$$

i.e., where there exists modulation $g(x)$ of the dielectric permittivity $\epsilon_0(x)$, periodic in the x-direction with the period L: $\epsilon_0(x + L) = \epsilon_0(x)$. It is assumed that $g(x)$ is bounded for any x.

Since it is easier to work with the discrete spectrum and a finite structure seems to be more realistic for practical applications, let us impose the cyclic boundary conditions

$$E(x, z) = E(x + l_x, z); \quad E(x, z) = E(x, z + l_z), \tag{3}$$

(assume $l_x, l_z \gg L$).

Now, consider the spectral problem

$$-\frac{c^2}{\epsilon_0(x)}\left(\frac{\partial^2}{\partial x^2} + \frac{\partial^2}{\partial z^2}\right)|q; n, k\rangle = \lambda_{qnk}|q; n, k\rangle, \tag{4}$$

where $\epsilon_0(x) > 0$ and the quantum numbers q, n, and k describe quantization in the z-direction (q) and in the x-direction (n, k). The last case n refers to a zone number and k refers to a wave vector inside the reduced Brillouin zone (BZ). A special form of (4) implies the factorization

$$|q; n, k\rangle = \zeta_q(z)\xi_{q,n,k}(x) \qquad (5)$$

and the following orthogonality relations for the normalized eigenfunctions

$$\int_0^{l_z} \bar{\zeta}_{q'}(z)\zeta_q(z)dz = \delta_{q'q}, \qquad (6)$$

$$\int_0^{l_x} \epsilon_0(x)\bar{\xi}_{q,n',k'}(x)\xi_{q,n,k}(x)dx = \delta_{n'n}\delta_{k'k}, \qquad (7)$$

and therefore

$$\langle q'; n', k'|\epsilon_0(x)|q; n, k\rangle =$$
$$\int_0^{l_z} dz \int_0^{l_x} dx\, \epsilon_0(x)\bar{\zeta}_{q'}(z)\zeta_q(z)\bar{\xi}_{q,n',k'}(x)\xi_{q,n,k}(x) = \delta_{q'q}\delta_{n'n}\delta_{k'k}. \qquad (8)$$

The functions introduced by (5) solve the eigenvalue problems as follows

$$\frac{d^2\zeta_q}{dz^2} + \nu_q^2\zeta_q = 0, \qquad (9)$$

$$\frac{d^2\xi_{qnk}}{dx^2} + \left(\lambda_{qnk}\frac{\epsilon_0(x)}{c^2} - \nu_q^2\right)\xi_{qnk} = 0. \qquad (10)$$

It follows from (9) and (3) that $\zeta_q = l_z^{-1/2}e^{i\nu_q z}$ with $\nu_q = \frac{2\pi}{l_z}q$. The gap structure is determined by (10).

For the next consideration, introduce a number N_x of states k. As is well-known N_x coincides with a number of periods and therefore, there exists a relation $l_x = N_x L$. Then, the Wannier functions $W_{qnp}(x)$ are introduced by the formula

$$|q; n, k\rangle = \frac{\zeta_q(z)}{\sqrt{N_x}}\sum_p e^{i\kappa_k L_p}W_{qnp}(x), \qquad (11)$$

where $L_p = pL$ (p is an integer) is a vector of the spatial lattice, while the quantity $\kappa_k = \frac{2\pi}{L}\frac{k}{N_x}$ plays a part of the wave number of the reciprocal space. As is customary [5], one can show that $W_{qnp}(x) \equiv W_{qn}(x - L_p)$, $W_{qn}(x)$ is a function localized near $x = 0$.

It is convenient to define a new ket-vector

$$|q; n, p\} = \frac{1}{\sqrt{N_x}}\sum_k e^{-i\kappa_k L_p}|q; n, k\rangle = \zeta_q(z)W_{qnp}(x). \qquad (12)$$

Holding the definition (8) for the inner product, one easily finds the orthogonality relation

$$\{q'; n', p' | \epsilon_0(x) | q; n, p\} = \delta_{q'q} \delta_{n'n} \delta_{p'p}. \tag{13}$$

3. General Formulas

The Wannier functions introduced above allow the representation of the electric field in the form

$$E = \sum_{q,n,p} f(q, n, p, t) | q; n, p\}. \tag{14}$$

Inserting both this expansion and expression (2) into wave equation (1) and applying $\{q; n, p | \epsilon_0(x)$, deduce the equation for $f(q, n, p, t)$, that is

$$\frac{1}{N_x} \sum_{p',k} f(q, n, p', t) \lambda_{qnk} e^{i\kappa_k (L_p - L_{p'})} + \frac{\partial^2 f(q, n, p, t)}{\partial t^2} +$$

$$\sum_{q',n',p'} \frac{\partial^2 f(q', n', p', t)}{\partial t^2} \{q; n, p | g(x) \epsilon_0(x) | q'; n', p'\} = 0. \tag{15}$$

Now assume that $g(x)$ varies smoothly as compared with the characteristic scale of $W_{qnp}(x)$. Designating a characteristic scale of $g(x)$ by D, the last requirement can be expressed as $L \ll D$. Then, as follows from (13)

$$\{q; n, p | g(x) \epsilon_0(x) | q'; n', p'\} \approx g(L_p) \delta_{q'q} \delta_{n'n} \delta_{p'p},$$

and one obtains the equation

$$\frac{1}{N_x} \sum_{m,k} f(q, n, p - m, t) \lambda_{qnk} e^{i\kappa_k L_m} + \frac{\partial^2 f(q, n, p, t)}{\partial t^2} +$$

$$g(L_p) \frac{\partial^2 f(q, n, p, t)}{\partial t^2} = 0. \tag{16}$$

Introducing a function $f_{qn}(a, t)$ of a continuous dimensionless variable a, which coincides with $f(q, n, p, t)$ in the points $a = L_p / D$ and a function $\lambda_{qn}(\kappa)$ which coincides with λ_{qnk} at $\kappa = \kappa_k$, $\lambda_{qn}(\kappa)$ being periodic in the reciprocal space: $\lambda_{qn}(\kappa + \frac{2\pi}{L}) = \lambda_{qn}(\kappa)$, one arrives at the equation for $f_{qn}(a, t)$ resulting directly from (16),

$$\lambda_{qn} \left(-\frac{i}{D} \frac{\partial}{\partial a} \right) f_{qn}(a, t) + [1 + G(a)] \frac{\partial^2 f_{qn}(a, t)}{\partial t^2} = 0. \tag{17}$$

Here $G(a) = g(aD)$ is a function varying on the unit scale. Finally, using the factorization $f_{qn}(a,t) = T_{qn}(t)A_{qn}(a)$, one obtains the system as follows

$$\frac{d^2 T_{qn}}{dt^2} + \Omega_{qn}^2 T_{qn} = 0, \tag{18}$$

$$\lambda_{qn}\left(-\frac{i}{D}\frac{d}{da}\right)A_{qn} - \Omega_{qn}^2[1 + G(a)]A_{qn} = 0. \tag{19}$$

The spectral parameter Ω_{qn} plays a part of the frequency and the spatial structure of the mode along the x-axis is described by Eq. (19).

Let us concentrate on the case when the propagating wave borders either a boundary or the center of the BZ. It is assumed that the stop gap is large enough. Then, in the center of the BZ one can approximate

$$\lambda_{qn}\left(-\frac{i}{D}\frac{d}{da}\right) = \omega_{qn}^2(0) - \Lambda_{qn}\frac{d^2}{da^2}, \tag{20}$$

where $\omega_{qn}^2(0)$ is the frequency of a stop gap edge in the absence of modulation and the coefficient Λ_{qn} is expressed through the group velocity dispersion $\omega_{qn}''(\kappa) = \frac{d^2\omega_{qn}(\kappa)}{d\kappa^2}$: $\Lambda_{qn} = \omega_{qn}(0)\omega_{qn}''(0)D^{-2}$. Introducing the normalized dimensionless frequency shift Δ_{qn} through the formula

$$\Omega_{qn}^2 = \omega_{qn}^2(0) + \Delta_{qn}\Lambda_{qn} \tag{21}$$

and the designation $\alpha_{qn} = \omega_{qn}^2(0)/\Lambda_{qn}$, one reduces (19) to the eigenvalue problem

$$\frac{d^2 A_{qn}}{da^2} + [\Delta_{qn} + (\Delta_{qn} + \alpha_{qn})G(a)]A_{qn} = 0. \tag{22}$$

By analogy at the boundary of BZ, $\kappa = \pi/L$,

$$\lambda_{qn}\left(-\frac{i}{D}\frac{d}{da}\right) = \omega_{qn}^2(\pi/L) + \tilde{\Lambda}_{qn}\left(\pi\frac{D}{L} + i\frac{d}{da}\right)^2 \tag{23}$$

with $\tilde{\Lambda}_{qn} = \omega_{qn}(\pi/L)\omega_{qn}''(\pi/L)D^{-2}$ and the eigenvalue problem defining the detuning takes the form

$$\frac{d^2 \tilde{A}_{qn}}{da^2} + [\tilde{\Delta}_{qn} + (\tilde{\Delta}_{qn} + \tilde{\alpha}_{qn})g(a)]\tilde{A}_{qn} = 0, \tag{24}$$

where $\tilde{A}_{qn} = A_{qn}\exp(i\pi\frac{D}{L}a)$ and $\tilde{\alpha}_{qn} = \omega_{qn}^2(\pi/L)/\tilde{\Lambda}_{qn}$. The frequency of the mode $\tilde{\Omega}_{qn}^2$ is given by

$$\tilde{\Omega}_{qn}^2 = \omega_{qn}^2\left(\frac{\pi}{L}\right) + \tilde{\Delta}_{qn}\tilde{\Lambda}_{qn}. \tag{25}$$

In a generic situation, a discrete spectrum (associated with a field configuration localized in the x-direction) of the problem exists under definite conditions (see the discussion below). Meantime by direct algebra, one can ensure the problem possesses only a continuum spectrum in the region of parameters corresponding to dispersionless propagation of the wave (i.e., far enough from both the center and the edge of BZ).

4. Modulation of Sech-form and Effective Waveguide

Both limits considered in the preceding section have been formally reduced to the same eigenvalue problem. For the sake of definiteness, I will consider (22). The solution of the last depends on the sign of Λ_{qn}. Indeed, let us illustrate this by a particular example of the modulation of the sech-form

$$G(a) = g_0 \text{sech}^2 a, \tag{26}$$

where g_0 is a characteristic depth of the modulation. Then, the discrete spectrum of (22) exists subject to the condition $\Lambda_{qn} g_0 > 0$. The respective eigenvalue problem is well-known [6] and gives the frequency shift in the form

$$\Omega_{qn}^2 - \omega_{qn}^2(0) = -\frac{\Lambda_{qn}}{4(1+g_0)^2} \times$$
$$\left[(2m+1) - \sqrt{(1+g_0)(1+4g_0\alpha_{qn})} - g_0(2m+1) \right]^2. \tag{27}$$

Here a quantum number m defining the mode is introduced and in the case where $g_0 < 0$, there is one more requirement $|g_0| < 1$ providing positive dielectric permittivity. To describe the BZ edge, one has to replace respectively $\omega_{qn}^2(0)$, α_{qn} and Λ_{qn} by $\omega_{qn}^2(\frac{\pi}{L})$, $\tilde{\alpha}_{qn}$ and $\tilde{\Lambda}_{qn}$.

Thus, the sign of the detuning depends on wether $\omega_{qn}(0)$ is a frequency of the top or the bottom of the stop gap. In the former case $\Lambda_{qn} > 0$ and, hence, modes localized in the x-direction appear only at $g_0 > 0$. Their frequencies are detuned into the stop gap. Thus by using conventional terminology they can be called *donor* modes. Physical interpretation of this phenomenon is quite transparent. The modulation results in the local increasing of the average optical density and the respective region becomes a lock for the electromagnetic field.

In the case where $\Lambda_{qn} < 0$ (i.e., in the vicinity of the bottom of the stop gap) the discrete spectrum appears when $g_0 < 0$. In this case, $\Omega_{qn}^2 > \omega_{qn}^2$ which means the frequency is detuned to the stop gap and is placed near the bottom. These are *acceptor* modes. In this situation, the region with effectively lower optical density becomes a lock for the field. Waveguides having similar property (i.e., guiding waves in a region of a refraction index lower than the index of the bounded media) were originally described by Yeh et al. [2]. However, the layered structure of [2] had strong local destruction of periodicity, while here the smoothly modulated structure is considered.

A number of modes of the effective waveguide is determined by the condition $(2m+1)^2 < 1+4g_0\alpha_{qn}$. They condense toward the stop gap edge. In a particular case when a relation between the parameters is as follows: $\alpha_{qn}g_0 = 2+g_0$, there exists only one mode in the waveguide. Its detuning is $\Delta_{qn} = -\Lambda_{qn}$ and the profile envelope is described by $A_{qn} = a_{qn}\operatorname{sech}(L_p/D)$, a_{qn} characterizing the field amplitude.

5. Discussion and Conclusion

The general form of the basic equation (19) can be examined numerically. Meanwhile, for the analytical predictions the possibility of the expansions (20) and (23) is crucial. This requires additional justification, since for some structures the group velocity dispersion $\omega''(0)$ and $\omega''(\pi/L)$ can be arbitrarily large. Respective estimation can be easily done in the case when $\epsilon(x) = \epsilon' + \epsilon''\cos(2\kappa_0 x)$ with $\epsilon'' \ll \epsilon'$, ϵ' and ϵ'' being constants. Then, one can explore the Kogelnik approximation [7] in order to show that at $\nu_q \gg \kappa_0$ the formulas (20) and (23) are valid if $L^2 D^2 \epsilon'\epsilon''\nu_q^4 \gg g_0$. Note, this condition implies the smallness of the frequency detuning $|\Omega_{qn}^2 - \omega_{qn}^2| \ll \omega_{qn}^2$, which also justifies the definitions of the donor and acceptor modes.

The diversity of effects described by the theory developed here is not restricted only to the waveguides. Thus, for example, Eq. (22) [and, respectively (24)] allows one to calculate a shift of the stop gap edge caused by smooth modulation. Indeed, suppose the function $g(x)$ has different limits $g_\pm$ at $x \to \pm\infty$. The solution of Eq. (22) corresponding to the band edge subject to homogeneous modulation is characterized by $\partial A_{qn}/\partial a = 0$. Thus, one immediately finds the shifts of the gap edges at $\pm\infty$ are given by $\Delta_{qn}^{(\pm)} = -\alpha_{qn}g_\pm(1 + g_\pm)^{-1}$. Thus, if $g_\pm < 0$ (> 0) the gap is shifted up (down).

Equations (22) and (24) are simplified in the case when the modulation is small enough. In this case, assuming smallness of the detuning, $\Delta_{qn} \ll \alpha_{qn}$, Eq. (22) reduces to the conventional stationary Schrödinger equation, in which Δ_{qn} plays a part of the normalized energy and $\alpha_{qn}G(a)$ is a potential. Then, one can apply numerous results of the quantum mechanics [6]. An interesting example arises in the case of the periodic $g(x)$, i.e., when the whole structure is doubly periodic (with periods L and D). Then, spectrum Δ_{qn} consists of allowed and forbidden bands (using the solid state terminology, they can be referred to as *minibands*). Minibands appear at $\Delta_{qn} > 0$. Thus, a doubly periodic structure possesses a miniband structure. Such a spectrum has already been observed in numerical simulations [8]. The theory developed here allows one to obtain quantitative estimates for the miniband spectrum.

By analogy with the ideas of [8] the above theory allows interesting insight into the phenomenon discovered in [9], called gap solitons. Indeed, let us consider a structure with the refractive index depending on the field intensity. Then, a pulse localized in space originates modulation of the dielectric permittivity. The modulation region becomes a lock for the field with the frequency corresponding to donor or acceptor levels (depending on whether the nonlinearity is focusing or defocusing). These levels can be associated with the detuned frequency of the gap soliton. Moreover, as follows from the above example, the shapes of the modulation and squared envelope coincide (and this exactly corresponds to the solitonic situation).

To conclude, the theory of wave propagation in periodic structures with smoothly varying parameters has been developed. It is based on the Wannier function expansion. It is shown that by modulating parameters, one can create waveguides with controlled properties. Such waveguides allow wave propagation in the direction orthogonal to the direction of periodicity.

References

1. P. Yeh and A. Yariv, Opt. Commun. **19**, 427 (1976).
2. P. Yeh, A. Yariv, and C. S. Hong J. Opt. Soc. Am. **67**, 423 (1977).
3. E. Yablonovitch, T. J. Gmitter, R. D. Meade, A. M. Rappe, K. D. Brommer, and J. D. Joannopoulos, Phys. Rev. Lett. **67**, 3380 (1991); R. D. Meade, K. D. Brommer, A. M. Rappe, and J. D. Joannopoulos, Phys. Rev. B **44**, 1372 (1991); E. Yablonovitch, J. Opt. Soc. Am. B **10**, 283 (1993); D. R. Smith, R. Dalichaouch, N. Kroll, S. Schultz, M. L. McCall, and P. M. Platzman, J. Opt. Soc. Am. B **10**, 314 (1993).
4. K. M. Leung, J. Opt. Soc. Am. B **10**, 303 (1993).
5. See e.g., A. I. Anselm, *Introduction to the Theory of Semiconductors*, (Nauka, Moscow, 1978).
6. See e.g., L. D. Landau and E. M. Lifshitz, *Quantum Mechanics*, (Pergamon Press, New York, 1977).
7. H. Kogelnik, Bell Syst. Tech. J. **48**, 2909 (1969).
8. V. M. Agranovich, S. A. Kiselev, and D. L. Mills, Phys. Rev. B. **44**, 10917 (1991).
9. W. Chen and D. L. Mills, Phys. Rev. Lett. **58**, 160 (1987).

PHOTONIC BAND STRUCTURE CALCULATION OF SYSTEM POSSESSING KERR NONLINEARITY

PHUC TRAN
Code 474400D
Research and Technology Division
Naval Air Warfare Center Weapons Dision
China Lake, California 93555 USA

1. Introduction

The Finite Difference Time Domain (FDTD) technique dates back to 1966 when it was first developed by Yee [1]. Since then it has been widely used to calculate the radar cross section of objects as well as normal modes of wave guides. Chan *et al.* [2] recently applied this technique to calculate the band structure of photonic crystal with excellent results. The motivation for their use of this technique is the fact this technique scales linearly with system size. Systems having random defects, which destroy any periodicity, are often studied using the "super cell" method where the system is assumed to be periodic with a very large period. In such a case, it is important the technique for band structure calculation scales favorably with system size. The plane wave expansion technique, for example, is impractical since it scales as N^3 where N is the system size. In this paper I show how the FDTD technique can be extended to calculate the band structure of nonlinear photonic crystals, in particular, those which possess Kerr nonlinearity.

2. Non-linear FDTD

The non-linear FDTD technique is similar to the linear FDTD technique in that it propagates in time the two coupled time dependent Maxwell's equations,

$$\frac{1}{c}\frac{\partial \mathbf{B}}{\partial t} = - \nabla \times \mathbf{E} \tag{1a}$$

$$\frac{1}{c}\frac{\partial \mathbf{D}}{\partial t} = \nabla \times \mathbf{B} . \tag{1b}$$

In Eq. (1) c is the speed of light, and it is assumed the system is non magnetic. The difference between the linear and non-linear FDTD is, of course, the relationship between the displacement field **D** and the electric field **E**. Here I will consider the case of a Kerr non-linear medium where this relationship is given by,

555

C. M. Soukoulis (ed.), Photonic Band Gap Materials, 555–562.
© *1996 Kluwer Academic Publishers. Printed in the Netherlands.*

$$D(r) = \left[\varepsilon(r) + \chi(r)|E(r)|^2\right]E(r) . \tag{2}$$

In Eq. (2) ε is the dielectric constant, and χ is the third order nonlinearity. As can be seen from Eq. (1) where only the B and D field at the next time step can be obtained from the time propagation, the E field at the next time step must be determined before the time propagation can continue. For the linear FDTD this is quite simple as $E = D/\varepsilon$. For the non-linear Kerr medium, the E field can be written in a similar form,

$$E(r) = \frac{D(r)}{\varepsilon(r) + \chi(r)A(r)} , \tag{3}$$

where $A = |E|^2$. The denominator of Eq. (3) can be thought of as an effective dielectric constant. Equation (3) can be used to obtain E if A is known. By taking the modulus square of Eq. (2), a cubic equation for A is obtained,

$$|D|^2 = \left[|\varepsilon|^2 + 2\operatorname{Re}(\varepsilon^*\chi)A + |\chi|^2 A^2\right]A . \tag{4}$$

The dependence on r in Eq. (4) is suppressed for convenience. The solution to a cubic equation is well-known and will not be given here as it can be found in any mathematical handbook. There are three roots to Eq. (4). The task of choosing the correct root is simplified by the fact that A, which is the modulus square of the E field, must be real and non negative. The solution to Eq. (4) can be divided into 2 cases, ε and χ have the same or opposite sign. Figure (1) shows the plot of the RHS of Eq. (4) as a function of A for the two cases. When they have the same sign, there is only one real non-negative root in the region of interest (the upper right quadrant), so the choice for A is obvious. Difficulty arises when they have opposite signs since there can be as many as three real non-negative roots.

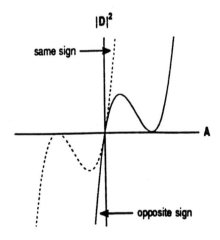

Figure 1. Plot of the RHS of Eq. (4) as a function of A for the case when ε and χ have the same sign (dash curve) and opposite sign (solid curve).

In this case, there is no physical reason, to the author's knowledge, that would allow one root to be chosen over the others. However, one can use information from the previous time step to choose the correct root. The reason for this is that if the time step is small enough, the value for A should vary smoothly. To be more precise, Eq. (4) can be differentiated with respect to time to obtain,

$$\frac{\partial |D|^2}{\partial t} = \left[|\varepsilon|^2 + 4 \operatorname{Re}(\varepsilon^* \chi) A + 3|\chi|^2 A^2 \right] \frac{\partial A}{\partial t} . \qquad (5)$$

From Eq. (5) the value of A at the next time step can be obtained. However, rather than using this value, which may or may not satisfy Eq. (4), Eq. (5) is used to select the correct root from Eq. (4). The root of eq. (4) is chosen such that it minimizes the difference between the value obtained using the Eqs. (4) and (5). In the limit that the time step goes to 0, both equations must give the same value for A. For the remainder of this paper, I will only consider the case of one non-negative real root to illustrate the method.

3. Implementation

The band structure is calculated by starting with some arbitrary initial condition and propagating it in time using Eq. (1). By so doing, one is in effect selecting the normal modes of the system. Note however, the initial field must have non-zero overlap with the eigenmodes of the system in order for them to be projected out in the time propagation. The calculation is carried out by putting the system on a grid of $(N_x x N_y x N_z)$ points and using FFT to calculate the curl in Eq. (1). The reasons for using FFT, at the cost of more computational time compared to the Finite Difference approximation, are better accuracy and reduced complexity in the programming. The programming is simpler because there is no need of a staggered grid (the electric and magnetic fields are evaluated at different grid points in the Finite Difference approximation to get good accuracy) and the periodic boundary conditions are automatically satisfied. Equation (1) implemented with FFT becomes

$$\mathbf{B}(\mathbf{k}, t + \Delta t) = \mathbf{B}(\mathbf{k}, t) - (ic\Delta t)\mathbf{k} \times \mathbf{E}(\mathbf{k}, t + \Delta t / 2) \qquad (6a)$$

$$\mathbf{D}(\mathbf{k}, t + \Delta t / 2) = \mathbf{D}(\mathbf{k}, t - \Delta t / 2) + (ic\Delta t)\mathbf{k} \times \mathbf{B}(\mathbf{k}, t) . \qquad (6b)$$

Note the fields are staggered in time so the error in the time derivative is of order $O(\Delta t^2)$. The electric field is obtained from the displacement field as discussed in section 2, and the time propagation is repeated as long as necessary. This is determined by the desired frequency resolution $\Delta \omega$.

At each time step $\mathbf{B}(\mathbf{k}, t)$ is saved. When the time integration is finished, $\mathbf{B}(\mathbf{k}, t)$ is Fourier analyzed to get $\mathbf{B}(\mathbf{k}, \omega)$. By looking for peaks in the plot of $|\mathbf{B}(\mathbf{k}, \omega)|^2$ versus ω, I obtained the band structure $\omega(\mathbf{k})$. To ensure no mode is missed due to the finite resolution in ω, $\mathbf{B}(\mathbf{k}, t)$ is multiplied with a gaussian in time, which has a width of $1/2\Delta\omega$, to broaden the peaks. Because the periodic boundary condition is satisfied automatically, I can integrate in time as long as necessary to get the desired resolution in ω without worrying about edge effects. The resolution in **k** is determined by the size of the real space grid and is limited by computer memory.

4. Example

As a check of the method, a linear system consisting of a dielectric medium ($\varepsilon = 5.0$) with circular holes drilled through is studied. The holes have a radius of $0.5a$ with a being the lattice constant and form a square lattice. The size of the grid used in the calculation is ($L_x=10a$, $L_y=10a$, $L_z=1a$). Since the hole is infinitely long, the length along the cylinder axis (z-axis) can be chosen arbitraily. The grid has 150x150x5 points. For the initial conditions, the electric and displacement field is set to zero and the magnetic field at each grid point **r** is given a random number between 0 and 1. The same band structure was obtained using the initial condition given by Eq. (7) below. Integration with a time step of $cdt = 0.02a$ was used. A time step of $0.01a$ was also used to check that the results have converged, and the results showed no change. The total integration time is $cT = 400a$ for a frequency resolution of $\omega a/2\pi c = 0.0025$. In Fig. (2) shows the band structure for modes with **E** along the z axis (cylinder axis), as obtained by the FDTD technique, versus that of the plane wave method [3].

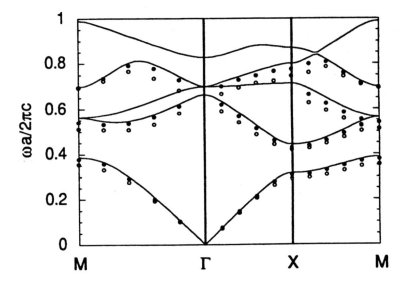

Figure 2. The band structure as calculated by the plane wave expansion method (solid line), the linear FDTD method (solid circle), and the non-linear FDTD (open circle).

Only the four lowest bands from the FDTD method are plotted. A disadvantage of the present method is the initial wave field must have nonzero overlap with the eigenmodes of the system in order to identify the eigenmodes. This problem is clearly seen in Fig. (2) where in the ΓX and ΓM direction the third band is not seen by the FDTD method. Robertson *et al.* [4] also did not see this band in their microwave transmission experiment because this band has a special symmetry [4] not present in the initial wave field. Figure (3) shows the spectral plot, $|B_x(\omega)|^2 + |B_y(\omega)|^2$ as a function of ω at fixed **k**, along the ΓM and ΓX direction from which the modes shown in Fig. (2) are identified. Note the spectrum is very clean (almost no noise) in the linear regime such that extremely weak peaks can still be identified if the scale is magnified.

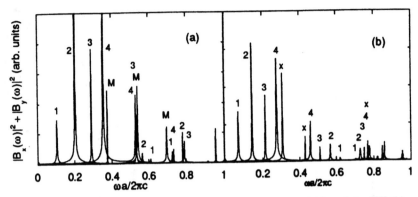

Figure 3. The spectral plot for various points along the ΓM (a) and the ΓX (b) direction. The number besides each peak identifies the point along each direction beginning from the Γ point. For example, number 1 means it is the first dot on either the left or right side of the Γ point in Fig. (2). The relative position of the numbers is meant to indicate the relative position of the peaks, since some peaks are not quite visible on the plot.

Figure (2) shows the band structure for the same system, but with $\chi = 0.005$. The following intial conditions were used

$$\mathbf{B(r)} = \mathbf{D(r)} = \frac{s}{\sqrt{L_x L_y L_z}}\left[\sum_{K \le K_{max}} e^{i\mathbf{K \cdot R}}\right](\hat{x} + \hat{y} + \hat{z}) , \qquad (7)$$

where $\mathbf{R} = (x,y)$, $s = 10$ is a scaling factor, and $K_{max} = 2\pi N_x/8L_x$. The initial **E** field is obtained from the **D** field through Eq. (4). Note the **D** field is assigned, and the **E** field is derive from it rather than the other way. If the order were reversed, the **D** field would have components with a wave vector larger than K_{max}, because of the nonlinear operation to obtain **D** from **E**, causing possible aliasing effects. The quantity $2\pi N_x/L_x$ is the largest wave vector that the FFT can resolve in the x direction,. Therefore, the restriction on the sum over **K** is to prevent any possible aliasing effects. The scaling factor s is a parameter to control the field intensity at fixed χ. The integration time is $cT = 2000a$ for a resolution of $\omega a/2\pi c = 0.0005$.

The non-linear band structure has shifted to a lower energy as compared to the linear case. This is consistent with the variational principle discussed by Joannopolous in another article in this proceeding. To lower its energy, the field will tend to concentrate in the region of high dielctric constant as much as possible. Hence increasing the dielectric contrast will lower the energy of the band. From the denominator of Eq. (3) it is evident that the nonlinearity causes an increase in the effective dielectric constant in the medium. For a nonlinear system, a spectral analysis is strictly incorrect as the linear

560

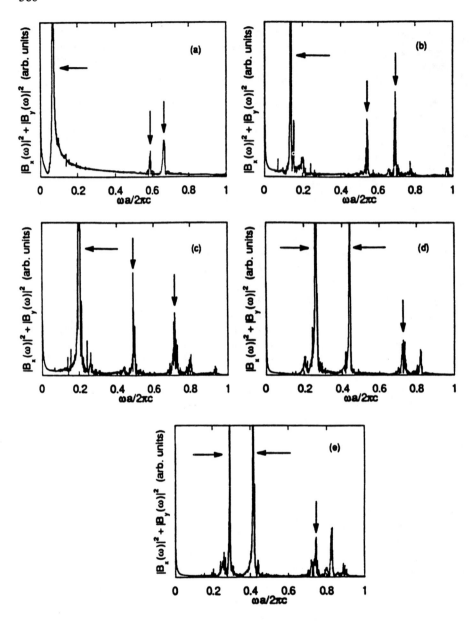

Figure 4. Spectral plot for points along the ΓX direction. Panel (e) corresponds to the X point.

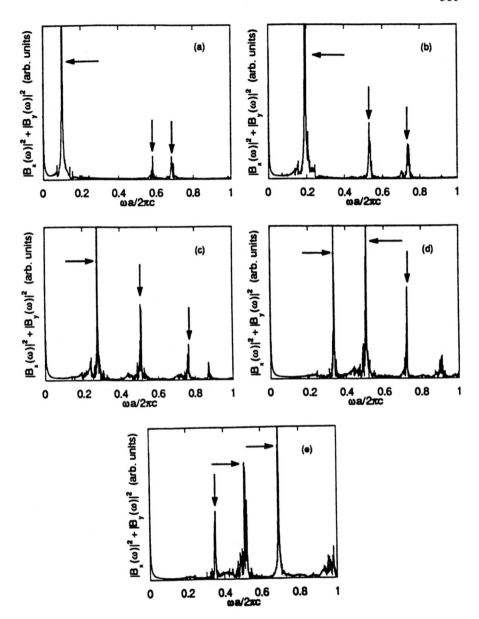

Figure 5. Same as Fig. (4) except along the ΓM direction. Panel (e) would correspond to the M point.

combination of solutions is not a solution. Therefore, each peak in the spectral plot cannot be taken as individual mode. However, in this case the spectral plots are very similar to that of the linear case which indicates that the nonlinearity has not drastically affect the linear solution other than shifting its energy. I believe in this case it is still valid to say that each peak is a mode of the system. Figures (4)-(5) show, for completeness, the spectral plots along the ΓX and the ΓM direction. For each plot the arrows indicate modes that obviously have evolved from the linear case. Again I only focus on the lowest four bands. The spectral plot is quite noisy. A possible explanation for this is that the nonlinearity caused a coupling among the linear mode with different wave vectors and angular frequencies and the nonlinear solution therefore will not have a well-defined wave vector and angular frequency. The question regarding the existence of non-linear mode is left for a more careful study.

5. Conclusion

I described the non-linear FDTD technique for photonic band structure calculation. The new idea presented here is an analytical formula that allow the inversion to obtain the electric field from the displacement field to be done easily. The non-linear band structure calculation through spectral analysis as described above suffers from a problem regarding the initial condition. Since the nonlinearity is dependent upon the field itself, the band structure obtained from the above calculation obviously will depend upon the choice for the initial conditions. It will be difficult to connect theoretical calculation with realizable experiment. For this purpose, it will be better if a transmission calculation is carried out where the initial field is the incident wave so it can be well characterized. This can be done easily by rearranging the geometry to that of a transmission experiment, and the field propagation can be performed the same way as described in this paper.

6. References

1. Yee, K. S. (1966) Numerical Solution of initial boundary value problems involving Maxwell's equations in isotropic media, IEEE Trans. Antennas Propagat AP-14, 302-307.
2. Chan, C. T., Datta, S., Yu, Q. L., Sigalas, M., Ho, K. M., and Soukoulis, C. M. (1993) New structures and algorithms for photonic band gaps, Physica A 211, 411-419.
3. Plihal, M., Shambrook, A., Maradudin, A. A., and Sheng, P. (1991) Two-dimensional photonic band structures, Optics Comm. 80, 199-204.
4. Robertson, W. M., Arjavalingam, G., Meade, R. D., Brommer,K. D., Rappe, A. M., and Joannopoulos, J. D. (1992) Measurement of photonic band structure in a two-dimensional periodic dieectric array, Phys. Rev. Lett 68, 2023-2026.

LOCALIZATION OF LIGHT :
THEORY OF PHOTONIC BAND GAP MATERIALS

SAJEEV JOHN
Department of Physics
University of Toronto
Toronto, Ontario
Canada, M5S 1A7

1. Introduction

There are at least two technological revolutions in the twentieth century which owe their existence to atomic and condensed matter physics. The first was the semiconductor revolution which resulted in the emergence of the electronics industry. The second was a direct consequence of the invention and applications of the laser. A third revolution, may yet emerge from the discovery of "high temperature" superconductivity. In each of these developments, the fundamental elementary particles or "actors" are electrons and photons. The properties of semiconductors, lasers, and superconductors are fundamentally a consequence of the dynamical and cooperative behaviour of electrons and photons in carefully controlled environments. Electrons are strongly interacting fermions which are readily localized by the positive charge of an atomic nucleus. The semiconductor crystal facilitates a highly controlled delocalization of the electronic wavefunction. This selective flow of charge enables a semiconductor device to perform its function. Photons are weakly interacting bosons which propagate readily. The laser facilitates the emergence of a cooperative or coherent state of a large collection of photons. In a superconductor, electrons in a metal form bosonic, Cooper pairs which condense into a cooperative state analogous to the coherent state of photons in a laser field. The emergence of a macroscopic, superconducting, order parameter field is analogous to the appearance of a macroscopic, classical, electric field in the laser. It is evident from these illustrations, that the quantum behaviour electrons and photons, and

C. M. Soukoulis (ed.), Photonic Band Gap Materials, 563–665.

the analogies which one may draw between them provide the fundament for many modern technolgies. Complex phenomena of great practical importance occur from placing electrons and photons in novel environments, which occur either naturally, or are engineered through careful materials fabrication.

Photon localization and photonic band gap (PBG) materials are a new branch of pure and applied science which embraces this philosophy and carries it to a new realm. It asks the question: "Can photons be made to "act" like electrons and if so, what are the consequences?". The search for answers to this question now represents a growing field, involving the disciplines of condensed matter physics, quantum optics, chemistry and engineering. Localization is a "trait" usually associated with massive particles such as the electron. The semiconductor owes its function to the fact that it enables the electron to propagate *coherently*. Propagation is the "trait" of photons. The PBG material exhibits novel functions by enabling the photon to localize *coherently*. The importance of *coherent* localization of photons has long since been implicit in the field of cavity quantum electrodynamics. High Q optical cavities have numerous applications in laser physics and have provided very striking illustrations of the quantum properties of light and photon-atom interactions. They are the pregenitors of true photon localization. In a high Q optical cavity, however, the localization is only partial. The Fabry-Perot resonator, for instance, provides one-dimensional localization of light. This localization is imperfect and is limited by the reflectivity of the resonator walls. More importantly, spontaneous emission of light from atoms can escape the cavity in either of the two non-localizing dimensions. A similar difficulty arises in micro-disk or micro-sphere optical cavities. These systems sustain very high Q factor "whispering gallery modes". Although the structure of these resonances is three-dimensional, they are degenerate in frequency with the continuum modes of the vacuum and spontaneous emission of light at the cavity mode frequency is by no means limited to the high Q mode.

Electromagnetism is the fundamental mediator of interactions in condensed matter and atomic physics. Photonic bandgap materials constitute a fundamentally new class of dielectric materials in which this basic interaction is controllably altered, through the process of localization, and in some cases completely removed over certain frequency and length scales. This leads to a host of new physical phenomena. Unlike ordinary high Q micro-cavity resonators, localized states in a PBG may extend over many optical wavelengths. The ability to control spontaneous emission, while at the same time preserving propagative effects over many wavelengths, make localized states in PBG distinct from their counterparts in conventional cavity quantum electrodynamics. In addition to capturing light emitted by sponteneous emission, these localized states facilitate coherent energy transport and cooperative effects on a scale much larger than the optical wavelength. The ability to tailor the radiative properties of atoms and molecules by means of the structural characteristics of the dielectric host also has applications in photochemistry and catalysis of chemical reactions.

Photonic bandgap materials are the photonic analogues of semiconductors in the electronics industry. Rather than a periodic array of atoms which scatters and modifies the energy-momentum relation of electrons, photonic crystals consist of periodically modulated dielectrics with periodicity on the scale of the wavelength of light. As in the case of semiconductors, the existence of a photonic bandgap does not require long-range periodic order in the dielectric microstructure. It is well known that semiconductors such as amorphous silicon and amorphous germanium have *larger* electronic bandgaps than their crystalline counterparts. The order in these systems is referred to as short-range-order. Here, crystalline order is present on the scale of a few interatomic spacings, and is disrupted by the presence of "topological" defects. The amorphous structure may be regarded as a dense random collection of topological defects. Semiconductors with this structure have a large pseudo-gap containing band tail of strongly localized electronic states. The ability to emulate these amorphous semiconductors with corresponding dielectric microstructures may lead to very inexpensive and large scale photonic bandgap materials. This constitutes an important and unexplored regime of mesoscopic physics with new technological applications.

1.1. HISTORY

The most widely studied example of localization is that of an electron in a disordered solid. Since an electron can have both positive and negative energy states, it is almost obvious that *some* of its quantum states will be localized (bound): deep negative energy potential fluctuations can trap electrons. On the other hand, states of sufficiently high positive energy will have a propagating nature. What is less obvious is the fact that if the disorder is made suitably strong (in three dimensions) that *all* of the electronic states in the energy band of a solid will become localized. This was first pointed out by P.W. Anderson in 1958. This effect now bears his name[7].

In the case of classical waves such as those propagating in a disordered elastic medium, entirely different questions arise. It was shown by S. John and M. Stephen[8, 9] in 1983 that in three-dimensions *some but not all* of the normal modes of vibrations would be localized. This remains true irrespective of the strength of the disorder, and this is due to the phenomenon of Rayleigh scattering. At very low frequencies, the effective scattering strength decreases as the square of the frequency. It follows that low frequency normal modes of an elastic medium are always extended. Phonons in a solid, however, have an upper frequency cutoff associated with the inverse of the lattice constant. This vanishing of the phonon density of states at some large but finite frequency, in the presence of disorder, causes some states to be localized in the vicinity of the cutoff. However, a mobility edge separating high frequency localized states from low frequency extended states always occurs for arbitrarily strong disorder.

There is a more fundamental question that distinguishes electromagnetic waves from either the example of electronic, Anderson localization or phonon localization. This is the question whether *any* of the states of a photon in a dielectric medium (with everywhere real positive refractive index) can be localized in the true sense of a three-dimensional bound state. This possibility was first proposed by S. John in 1984[10]. This was followed by an important commentary by Anderson in 1985[11]. Striking experimental evidence for this possibility came from the work of M. van Albada and A. Lagendijk[12] and P.E. Wolf and G. Maret[13] on coherent backscattering of light. One of the central consequences of photon localization arises from the fact that a localized state behaves as an ideal optical micro-cavity whose quality factor is limited only by bulk absorption in the dielectric material. In a localization regime there is in fact a continuous spectrum of such ideal microcavities, one for each frequency within the localized spectrum. Remarkably, these microcavities while forming a (singular) continuous spectrum are not optically connected with each other and can behave essentially as isolated optical states.

In the second section of this article we focus on the *existence* of localized states of light. Despite the enormous literature on "Anderson localization", the fact that even a single localized state of light could be formed in a real positive dielectric material (3-d) was not experimentally demonstrated until 1991. We begin with a pedagogical introduction to scattering mechanisms and scattering regimes for photons. These differ considerably from those of an electron in a random potential. The criterion for localization of light accordingly differs from that of an electron. The two, however, may be related by a generalized form of the conventional Ioffe-Regel criterion.

The simplest expression of the generalized localization criterion appears in dielectric materials exhibiting a *complete photonic bandgap*. The concept of a photonic band gap in three-dimensions was introduced independently by S. John and E. Yablonovitch in 1987. My own motivation for suggesting the photonic crystal was to strengthen my hypothesis of 1984 that photon localization may occur in strongly scattering dielectrics. The realization of a photonic band gap provides a highly systematic approach to controllably and reproducibly create strongly localized states of light.

As a systematic approach to localization, I published a theory of the electromagnetic bandgap starting from the vector wave equation for the electric field amplitude. Yablonovitch was concerned with dielectric micro-cavities in which spontaneous emission of light from atomic and semiconductor lasers could be strongly controlled. This could clearly be achieved by a drastic modification of the available photon density of states. He published a paper on the possibility of a scalar wave photonic bandgap. The independent submission of Yablonovitch was brought to my attention by the editor of Physical Review Letters. [14, 15]. The fundamental differences between the scalar bandgap and the electromagnetic bandgap are in the nature of the phase space available for propagation as well as the minimum dielectric contrast required to create a complete three-dimensional photonic gap. In section 2, we review the underlying physics of photonic bandgap formation as introduced in my paper of 1987. This provides an existence proof for strong localization of light in lossless dielectrics.

1.2. OVERVIEW

In the perfect photonic crystal, there is a complete absence of linear wave propagation in *any* direction over the frequency interval of the gap. The difference between an optical micro-cavity and a strongly localized state in a photonic bandgap is that the former is actually a scattering resonance whereas the latter is a true bound state which has a quality factor limited only by bulk absorption effects. In the photonic bandgap propagative effects can take place on length scales longer than the wavelength of light, even though the states are eventually localized. This itself leads to new physical phenomena not apparent in micro-cavity resonators of size comparable to the wavelength. Although *linear* wave propagation is forbidden in the bandgap frequency interval, *nonlinear* wave propagation in the form of optical solitary waves can occur if the dielectric has either a positive or negative Kerr coefficient. This type of nonlinear classical electrodynamics is described in detail in section 3.

In section 4, we describe the underlying physics of photon localization from the standpoint of multiple scattering theory and disordered systems. We discuss how wave propagation crosses over to energy diffusion on the scale of the transport mean free path ℓ^*. On length scales longer than ℓ^*, wave interference corrections to diffusion associated with time-reversal symmetry play an important role. The most significant of these is the phenomenon of coherent backscattering. Incorporating the coherent backscattering correction to diffusion in the strong scattering limit leads to the scaling theory of localization. This is discussed in the context of optical experiments which probe the photon mobility edge. The effects of absorption on the photon mobility edge as described in my paper of 1984[10] are reviewed. The localization theory developed in this section is adapted to the case of a disordered photonic bandgap material in which the complete gap is replaced with a pseudogap of strongly localized modes. The quality-factor of a given localized mode within the continuous singular spectrum of the localization pseudogap is evaluated in the presence of aborption, as well as Raman and Brillouin scattering. It is shown that when the localization length is shorter than the length scale on which inelastic scattering of photons takes place, the Q-factor of states within the pseudogap is comparable to that of an isolated defect mode in a photonic crystal.

In section 5, we describe the quantum electrodynamics of atoms and molecules placed within a photonic bandgap material. We begin by discussing model Hamiltonians for the photonic bandgap which facilitate the evaluation of atom-radiation interactions. If the atom or molecule has a first excited state to ground state transition which lies within the forbidden frequency gap of the photonic crystal, an entirely new type of localized state emerges. We refer to this as a *photon-atom bound state*. In this state the photon emitted by the atom is allowed to tunnel many lattice constants away from the atom but is eventually Bragg reflected by the dielectric medium back to the atom. The result is an eigenstate of the atom plus radiation field Hamiltonian which is a linear superposition of an orbiting photon and an excited atom. This dressed atomic state exhibits an anomalous Lamb shift associated with the repulsion of the atomic level from the photonic band edges. If the bare atomic level lies sufficiently close to one of the band edges, we demonstrate that this repulsion gives rise to a highly magnified analog of vacuum-Rabi-splitting. The atomic level splits into a doublet and the magnitude of this splitting is on the order of 10^{-6}-10^{-7} of the atomic transition frequency. One member of this doublet is pulled into the photonic bandgap whereas the other member is pushed into the allowed electromagnetic continuum and behaves as a resonance state. Spontaneous emission from this bound, dressed state exhibits an oscillatory behaviour and the atomic population inversion for a single atom has a novel fractional steady state limit. We discuss the lifetime of the photon-atom bound state due to its interaction with other degrees of freedom in the dielectric material.

In section 6, we point out the emergence of resonance dipole-dipole interactions (RDDI) between atoms placed within the photonic bandgap. In the absence of spontaneous emission, this interaction is the dominant mechanism for excitation transfer and energy transport in the photonic crystal. One of the most significant properties of the PBG is that it protects photons and atoms from radiative effects in the external environment. This opens the door to a new strong coupling regime for quantum optics in which the coupling of atoms to localized modes as well as the coupling of atoms to others atoms by RDDI, is large compared to decay processes that dissipate energy from the system. Under these circumstances, novel forms of cooperative behaviour emerge. One such example is superradiance near a photonic bandedge. We show that N-atom collective spontaneous emission leads to a novel collective state in the steady state limit in which spontaneous symmetry breaking takes place and the atomic dipoles order into a state of macroscopic polarization. This in turn leads to lasing without a cavity mode: Light which is emitted near the band edge remains partially localized in the vicinity of the emitting atoms in the form of a coherent state. For a three-dimensional PBG, the collective emission takes place at an enhanced rate proportional to N^2 (rather than N as is the case of superradiance in ordinary vacuum). Accordingly, the peak intensity for light emission scales as N^3 (rather than N^2 for vacuum superradiance). A more unusual form of cooperative behaviour is predicted to occur if N atoms are placed inside a PBG in which the atomic transition lies deep within the gap. In this case, the coupling to band edge modes is weak and the dominant interaction is the RDDI between pairs of atoms. Deep within the PBG, RDDI takes the form of a random exchange interaction between two-level atoms, represented mathematically by quantum spins. We consider a model in which these two-level atoms are coupled to an isolated localized cavity mode deep within the PBG. If this cavity mode is pumped with coherent light, the dynamical evolution of the system is to the optical analog of a quantum spin glass: The atoms acquire a random spontaneous polarization in the steady state limit. The phases and orientations of these dipoles, however, are random. These leads to a state with zero net (macroscopic) polarization. In equilibrium, the photons in the cavity mode condense into a Bose-glass state.

2. The Existence of Photon Localization

In the case of electron localization, nature provides a variety of readily available materials. Localization is more often the rule than the exception. This is easily seen from the Schrödinger equation for an electron with effective mass m^*:

$$\left[\frac{-\hbar^2}{2m^*} \nabla^2 + V(x) \right] \psi(x) = E\psi(x) \ . \tag{2.1}$$

A random potential $V(x)$ can trap electrons in deep local potential fluctuations when the energy E is sufficiently negative. As the energy increases, however, the probability of finding another nearby potential fluctuation into which the trapped electron can tunnel increases. To quantify these ideas, consider $V(x)$ to have a root mean square amplitude V_{rms} and a length scale a on which random fluctuations take place. The correlation length to the disorder a defines an energy scale $\epsilon_a \equiv \hbar^2/(2m^*a^2)$. For example in an amorphous semiconductor a is the interatomic spacing, ϵ_a plays a role analogous to the conduction band width of the semiconductor, and the zero of energy corresponds to the conduction band edge of the corresponding crystal. In the weak disorder limit $(V_{rms} \ll \epsilon_a)$, a transition takes place as the electron energy is increased to about $-V_{rms}^2/\epsilon_a$ in which successive tunnelling events allow the electron to traverse the entire solid by a slow diffusive process and thereby conduct electricity. This transition energy has been termed a mobility edge by N.F. Mott[16]. At higher and higher energies the scale on which multiple scattering takes place grows larger and larger than the electron's de Broglie wavelength and the electron traverses the solid with relative ease. If, on the other hand, the disorder is made strong $V_{rms} \gg \epsilon_a$, the mobility edge moves into the conduction band continuum $(E > 0)$ and eventually the entire band succumbs to localization. This effect is referred to as the Anderson transition. Since disorder is a nearly universal feature of real materials, electron localization is an important ingredient in determining electrical, optical and other properties of solids.

In the case of monochromatic electromagnetic waves of frequency ω propagating in an inhomogeneous but nondissipative dielectric medium, the classical wave equation for the electric field amplitude $\vec{E}$ may be written in a form resembling the Schrödinger equation:

$$-\nabla^2\vec{E} + \vec{\nabla}(\vec{\nabla} \cdot \vec{E}) - \frac{\omega^2}{c^2}\epsilon_{\text{fluct}}(x)\vec{E} = \epsilon_0\frac{\omega^2}{c^2}\vec{E} \ . \tag{2.2}$$

This follows directly from Maxwell's equations. Here, I have separated the total dielectric constant $\epsilon(x) = \epsilon_0 + \epsilon_{\text{fluct}}(x)$ into its average value ϵ_0 and a spatially fluctuating part $\epsilon_{\text{fluct}}(x)$. The latter plays a role analogous to the random potential $V(x)$ in the Schrödinger equation, and scatters the electromagnetic wave. For the case of a lossless material in which the dielectric constant $\epsilon(x)$ is everywhere real and positive, several important comments concerning the Schrödinger equation-Maxwell equation analogy are in order. First of all, the quantity $\epsilon_0\omega^2/c^2$ which plays a role analogous to an energy eigenvalue is always positive. This precludes the possibility of elementary bound states of light in deep negative potential wells. It must also be noted that the laser frequency ω multiplies the scattering potential $\epsilon_{\text{fluct}}(x)$. Unlike an electronic system, where localization was enhanced by lowering the electron energy, lowering the photon energy instead leads to a complete disappearance of the scattering mechanism itself! In the opposite, high frequency limit, geometric ray optics becomes valid and interference corrections to optical transport become less and less effective. In both limits

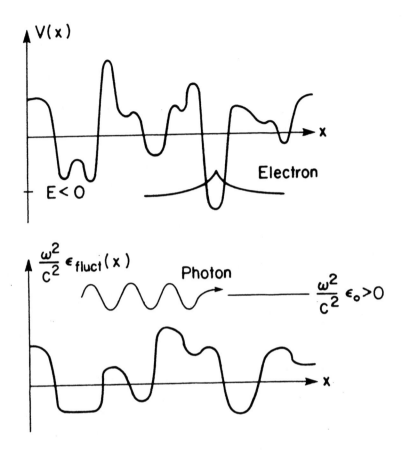

Figure 1. In the case of photon localization $\frac{\omega^2}{c^2}\epsilon_{\text{fluct}}(x)$ plays the role of the scattering potential. Unlike the picture for electrons, the eigenvalue $\frac{\omega^2}{c^2}\epsilon_0$ is always higher than the highest of the potential barriers in the case of a real, positive dielectric constant.

the normal modes of the electromagnetic field are extended, not localized. Finally the condition that $\epsilon_0 + \epsilon_{\text{fluct}} > 0$ everywhere, translates into the requirement that the energy eigenvalue is always greater than the effective potential $|\frac{\omega^2}{c^2}\epsilon_{\text{fluct}}(x)|$. Unlike the familiar picture of electronic localization (figure 1), what we seek in the case of light is an *intermediate frequency window of localization* within the positive energy continuum and *which lies at an energy higher than the highest of the potential barriers!* It is for this simple reason that ordinary dielectrics appearing in nature do not easily exhibit photon localization.

It should be pointed out here, that in the above analogy we have emphasized similarities more than actual differences. The vector nature of the electromagnetic wave equation (2.2) makes it even *more* difficult to localize light than the above analogy would suggest. The polarization density term $\vec{\nabla} \cdot \vec{E}$ has no analog in the electronic case. Even in the absence of scattering, this term leads to a difference between the *microscopic* dielectric constant ϵ_0 and the average *macroscopic* dielectric constant which is measured on scales comparable to the electromagnetic wavelength λ. This latter subtlety, however does not alter the subsequent discussion of scattering and localization provided that $\vec{E}$ in equation (2.2) is interpreted as the coarse-grained electric field which is the spatial average of the true microscopic electric field in which polarization effects on scales smaller than λ have been incorporated.

2.1. INDEPENDENT SCATTERERS AND MICROSCOPIC RESONANCES

In the case of photons, the underlying physics of the high and low frequency limits can be made more precise by considering scattering from a single dielectric sphere. Consider a plane wave of wavelength λ impinging on a small dielectric sphere of radius $a \ll \lambda$ of dielectric constant ϵ_a embedded in a uniform background dielectric ϵ_b in $d = 3$ spatial dimensions. The scattered intensity I_{scatt} at a distance R from the sphere can be a function of only the incident intensity I_0, the dielectric constants ϵ_a and ϵ_b and the lengths R, λ and a. In particular I_{scatt} must be proportional to the square of the induced dipole moment of the sphere which scales as the square of its volume $\sim (a^d)^2$ and by conservation of energy must fall off as $1/R^{d-1}$ with distance from the scattering center:

$$I_{\text{scatt}} = f_1(\lambda, \epsilon_a, \epsilon_b) \frac{a^{2d}}{R^{d-1}} I_0 \ . \tag{2.3}$$

Here f_1 is some yet to be determined function. Since the ratio I_{scatt}/I_0 is dimensionless, it follows that $f_1(\lambda, \epsilon_a, \epsilon_b) = f_2(\epsilon_a, \epsilon_b)/\lambda^{d+1}$ where f_2 is another dimensionless function of the dielectric constants. The vanishing of the scattering cross section for long wavelengths as $\lambda^{-(d+1)}$ is a fundamental result. It is the same reasoning which explains why the sky is blue. This weak scattering is the primary reason that electromagnetic modes are extended in most naturally occurring three dimensional systems. Although I have derived this result for single scattering, it remains true for a dense random collection of scatterers. This behavior is evident in the elastic, photon, mean free path ℓ^* which is proportional to λ^{d+1} for long wavelengths. This generalization of Rayleigh scattering to d spatial dimensions is particularly useful when describing anisotropic dielectric scattering systems. For example a layered random medium in which scattering is confined to directions perpendicular to the layers would be described by setting $d = 1$. Alternatively, a collection of randomly spaced uniaxial rods[17] in which scattering is confined to the plane perpendicular to the axis of the rods would be described by setting $d = 2$. A consequence of the scaling theory of localization (which applies to electrons in disordered solids as well as electromagnetic waves in disordered dielectrics) is that in one and two dimensions, all states are localized but with localization lengths ξ_{loc}, which diverge due to Rayleigh scattering in the low frequency limit. In particular $\xi_{loc} \sim \ell$ in one dimension and $\xi_{loc} \sim \ell \exp(\frac{\omega}{c}\ell)$ in two dimensions[5, 9]. The divergence is apparent when we substitute the Rayleigh behaviour $\ell^* \sim \lambda^{d+1}$ into these formulae.

It is likewise instructive to consider the opposite limit in which the wavelength of light is *small* compared to the scale of the scattering structures. For scattering from a single sphere it is well known[18] that for $\lambda \ll a$ the cross section saturates at a value $2\pi a^2$. The factor of two includes the rays which are weakly diffracted out of the forward direction near the surface of the sphere. This is the result of geometric optics. For a dense random collection of scatterers it is useful to introduce the notion of a *correlation length a*. On scales shorter than a, the dielectric constant does not vary appreciably except for the occasional interface where the physics of refraction and diffraction apply. The essential point is that the elastic mean free path never becomes smaller than the correlation length. This classical elastic mean free path ℓ^* plays a central role in the physics of localization. Wave interference effects lead to large spatial fluctuations in the light intensity within the disordered medium. However, if $\ell^* \gg \lambda$, these fluctuations tend to average out to give a physical picture of essentially noninterfering, multiple scattering paths for electromagnetic transport. When $\ell^* \to \lambda/2\pi$ interference of multiple scattering paths drastically modifies the average transport properties and a transition from extended to localized normal modes takes place. If one adopts the most naïve version of the Ioffe-Regel condition $2\pi\ell^*/\lambda \simeq 1$ for localization[19], with λ being the vacuum wavelength, it follows that extended states are expected at both high and low frequencies. However as depicted in Figure 2, for strong scattering, there arises the distinct possibility of localization within a narrow frequency window when the quantity $\lambda/2\pi \simeq a$. It is this intermediate frequency regime which we wish to analyze in greater detail.

We will refer to the aforementioned criterion for localization as the *free-photon Ioffe-Regel condition*. This particular result is based on perturbation theory about free photon states which undergo multiple scattering from point-like objects and a disorder average is performed over all possible positions of the scatterers. In effect, statistical weight is evenly distributed over all possible configurations of the scatterers and the scatterers have no

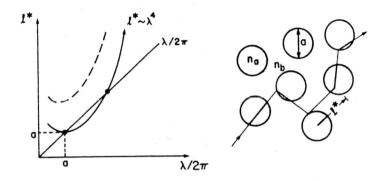

Figure 2. The classical transport mean free path ℓ^* plays the role of the step length in the photon's random walk through the scattering medium. If the scattering microstructures have a single characteristic length scale a, then $\lambda \gg a$ and $\lambda \ll a$ are weak scattering regimes of Rayleigh scattering and classical geometric optics respectively. For low refractive index contrast n_a/n_b all states are extended (upper curve). For $n_a/n_b \gtrsim 2, \ell^* \sim \lambda/2\pi$ when $\lambda/2\pi \simeq a$ suggesting an intermediate frequency window of localization (lower curve).

spatial correlations on average.

The first correction to this picture is to associate some nontrivial structure to the individual scatterers. The theory of Mie resonances[20, 21] for scattering from dielectric spheres immediately tells us that this can have profound consequences on the elastic mean free path. For instance for spheres of radius a, of dielectric constant ϵ_a embedded in a background dielectric ϵ_b, and for a ratio $\epsilon_a/\epsilon_b \simeq 4$, the first Mie resonance which occurs at a frequency given by $\omega/c(2a) \simeq 1$. On this resonance the scattering cross section $\sigma \simeq 6\pi a^2$. Using this result for a relatively dilute collection of spheres of number density n, the classical elastic mean free path is given by $\ell^* \sim \frac{1}{n\sigma} = \frac{2a}{9f}$. Here I have introduced the volume filling fraction f of the spheres. Extrapolating this dilute scattering result to higher density, it is apparent that for a filling fraction $f \simeq 1/9$, the free-photon Ioffe Regel condition is satisfied on resonance. It is tempting to increase the density of scatterers so as to further decrease the mean free path. However, the fact that the cross section on resonance is 6 times the geometrical cross section indicates that a given sphere disturbs the wavefield over distances considerably larger than the actual sphere radius. The existence of the resonance requires that the "spheres of influence" of the scatterers do not overlap. Indeed for higher densities the spheres become *optically connected* in this sense and the mean free path increases rather than decreasing. From the single scattering or microscopic resonance point of view, the free photon criterion for localization is a very delicate one to achieve. This conclusion has been borne out in the initial experiments by A.Z. Genack and M. Drake[22,23] on light scattering from randomly distributed dielectric spheres.

The fundamental problem with microscopic scattering resonances is illustrated even more dramatically in the case of atomic resonance fluorescence. This illustration will be of some value to the discussion of quantum electrodynamics in section 4. We discuss a simple classical model for atomic polarizability and scattering cross section. From this discussion it will be apparent that the free-photon Ioffe-Regel criterion can barely be achieved even for resonant scattering in an atomic vapor.

Consider a charged, simple harmonic oscillator with natural frequency ω_0, charge e and mass m. When radiation of frequency ω and electric field amplitude E_0 impinges on this classical atom, the energy radiated per second is given by the Larmor formula[18]:

$$
\begin{aligned}
\text{Power radiated} &= \tfrac{2}{3}c\left(\tfrac{\omega}{c}\right)^4 |p|^2 \\
&= \sigma\tfrac{c}{8\pi}|E_0|^2
\end{aligned}
\tag{2.4}
$$

where p is the amplitude of the induced atomic dipole moment oscillation and

$$
\sigma = \frac{2c}{3}\left(\frac{\omega}{c}\right)^4 \frac{8\pi}{c}|\alpha|^2
\tag{2.5}
$$

is the scattering cross section. Here, the atomic polarizability α relates the incident electric field to the induced dipole moment by the relation $p = \alpha E_0$. The quantity $c|E_0|^2/8\pi$ may be identified as the incident flux. It follows that the ratio of power radiated to incident flux, denoted by σ, is in fact the scattering cross-section. For a classical oscillator with radiative damping, the polarizability is given by[18]

$$
\alpha = \frac{e^2}{m}\left[\omega_0^2 - \omega^2 - \frac{2i}{3}\frac{e^2}{m}\left(\frac{\omega}{c}\right)^3\right]^{-1}
\tag{2.6}
$$

It follows that the maximum scattering cross-section obtained *on resonance* is given by $\sigma_{\max} = 12\pi/k^2 \simeq \lambda^2$. Here, $k \equiv \omega/c$. If the average spacing between atoms in the vapor is given by a, we may regard the scatterers as independent provided $a > a_{\min}$ with $a_{\min}$ defined by the condition $\pi a_{\min}^2 = \sigma_{\max}$. This is the critical atomic separation at which the optical field of one atom affects the scattering cross-section of a neighbouring atom. When $a < a_{\min}$, the scatterers are optically connected and single atom scattering cross-section becomes damped by the presence of the nearby atom. It follows that the maximum scattering in the atomic vapor occurs when the density of atoms is given by $n = a_{\min}^{-3}$. Using the classical formula[24] for the transport mean free path $\ell^* = (\sqrt{2}n\sigma)^{-1}$ it follows that the shortest possible mean free path is given by $\ell^*_{\min} = \sqrt{6}/(\pi k)$. The free-photon Ioffe-Regel criterion for localization is barely achieved in this case. However, any small change in density n from the optimum value will destroy the localization effect.

2.2. COHERENT SCATTERERS AND MACROSCOPIC RESONANCES

The resonant scattering mechanisms which we have described so far are well known in the field of optics. These resonances occur from local scattering from a single object, such as a dielectric particle or an atom, which is optically isolated from other such scatterers. The approach based on independent, uncorrelated scatterers overlooks a very important possibility. In a sense it is the fundamental theorem of solid state physics: certain geometrical arrangements of identical scatterers can give rise to large scale or macroscopic resonances. The most familiar example is the Bragg scattering of an electron in a perfectly periodic crystal[25]. Such an effect is not given the required statistical weight by a disorder average which improperly averages over all positions of the scatterers.

To illustrate the concept of a macroscopic resonance, we begin by classifying the geometrical arrangements of identical (or nearly identical) scatterers relevant to optical transport. To the reader from an optics background, this is a classification of the different effects of arranging a large collection of microcavity resonators in different geometries. Each individual resonator has a high scattering cross-section for photons which impinge on it. Photons *within* a dielectric resonator will accordingly have a long dwell time before leaking out. The product of the optical frequency and the leakage time is the quality (Q) factor of the microscopic resonator. This Q-factor is inversely proportional to the frequency width of the microscopic scattering resonance. The new physics associated with a *collection* of such resonators is that coherent interference effects between the different resonators can lead to a stronger, *macroscopic*, resonance. In fact, for suitable arrangements of resonators, a localized state of light can be created, which in the absence of absorption or inelastic scattering would have an infinite quality factor. The scale of the localized state of the electromagnetic field is, however, large compared to the optical wavelength and spans many individual micro-cavities.

Consider a collection of N identical dielectric scatterers placed at positions $\vec{R}_i$ where $i = 1, \ldots, N$. The fluctuating part of the dielectric constant may be written as

$$\epsilon_{\text{fluct}}(\vec{x}) = \sum_{i=1}^{N} u(\vec{x} - \vec{R}_i) \ . \tag{2.7}$$

Here, $u(\vec{x})$ is a function which is nonzero only within some finite range of the origin. This range corresponds to the radius of the particles in the case of spheres. When N is very large it is not practical to keep track of the precise position of each scatterer but to describe their positions in a statistical sense. It is useful to regard the collection of scatterers as being divided into a set of smaller subsystems. The precise configuration of scatterers may vary from one subsystem to another as a result of disorder. Each subsystem is made up of a very large number of scatterers and the collection of subsystems is regarded as a statistical ensemble. We consider the Fourier transform of the fluctuating dielectric constant $\epsilon_{\text{fluct}}(\vec{q})$ within each given member of the ensemble and evaluate the average of $|\epsilon_{\text{fluct}}(\vec{q})|^2$ over the entire ensemble.

$$\langle |\epsilon_{\text{fluct}}(\vec{q})|^2 \rangle_{\text{ensemble}} = \frac{1}{N} S(q) |u(\vec{q})|^2 . \qquad (2.8)$$

As indicated, this factorizes into a *structure factor* describing the average positions and statistical correlations between the scatterers:

$$S(q) \equiv \frac{1}{N} \sum_{i,j=1}^{N} \langle e^{-i\vec{q} \cdot (\vec{R}_i - \vec{R}_j)} \rangle_{\text{ensemble}} \qquad (2.9)$$

and a *form factor* $|u(\vec{q})|^2$ describing the nature of individual scatterers. We describe below four distinct classes of geometrical arrangements of the dielectric scatterers.

If the positions $\vec{R}_i$ and $\vec{R}_j$ of the ith and jth scatterers are statistically uncorrelated, then only the $i = j$ survives the summation of equation (2.9). Provided that $q \neq 0$ it follows that $S(q) = 1$, a constant independent of $\vec{q}$. This is the underlying assumption in the previous section, where the free-photon Ioffe-Regel condition was introduced. Quite the opposite situation arises if the dielectric scatterers form a periodic three-dimensional array. For a perfectly crystalline array, this structure factor consists instead of sharp delta-function peaks at values of $\vec{q}$ which correspond to the reciprocal lattice vectors of the crystal (see Figure 3). The only difference between this and the x-ray diffraction pattern from a solid is the difference in length scale. In the present case, the separation between the dielectric scatterers is comparable to the wavelength of light rather than the wavelength of x-rays.

If the photonic crystal described above is weakly disordered, an intermediate form of the static structure factor emerges. The first case we describe is one in which the position of the scatterers is given by $\vec{R}_i = \vec{R}_i^0 + \vec{u}_i$. Here $\{\vec{R}_i^0\}$ are the crystalline positions of the dielectric resonators and $\vec{u}_i$ are some small random deviations which satisfy a gaussian probability distribution: $P(\vec{u}_i) = \exp[-(\vec{u}_i)^2/(2u_{rms}^2)]$. In this case the statistical average

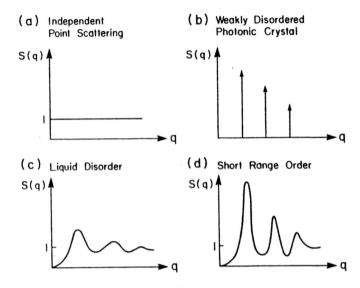

(a) Independent Point Scattering

(b) Weakly Disordered Photonic Crystal

(c) Liquid Disorder

(d) Short Range Order

Figure 3. The static structure factor as a function of wave vector q. For white noise disorder (a) the free-photon Ioffe-Regel criterion applies. For a periodic or weakly disordered photonic crystal, $S(q)$ exhibits Bragg peaks (b). For high refractive index scatterers, a photon localization threshold is barely attainable in the case of liquid-like disorder (c). A strong localization pseudogap is possible when $S(q)$ exhibits correlations analogous to atoms in an amorphous semiconductor (d).

in (2.9) reduces to a Debye-Waller factor[25]:

$$\langle e^{-i\vec{q}\cdot(\vec{u}_i - \vec{u}_j)}\rangle_{\text{ensemble}} = e^{-q^2 u_{rms}^2} . \tag{2.10}$$

The delta function Bragg peaks in $S(q)$ persist, but their intensity is diminished by the Debye-Waller factor.

A more dramatic change in the structure factor occurs if the dielectric scatterers interact with each other but are allowed to be anywhere within the same volume V rather than being tied to lattice sites. In this case a liquid-like structure factor emerges in which the statistical average over the variable $\vec{R} \equiv \vec{R}_i - \vec{R}_j$ is described in terms of a pair distribution function $g(\vec{R})$:

$$\langle e^{-i\vec{q}\cdot(\vec{R}_i - \vec{R}_j)}\rangle_{\text{ensemble}} = \int \frac{d^3\vec{R}}{V} g(\vec{R}) e^{-i\vec{q}\cdot\vec{R}} . \tag{2.11}$$

The function $g(R)$ has peaks at the average nearest neighbour, next nearest neighbour etc... separations and the resulting structure factor is depicted in Figure 3c.

The transport of electromagnetic energy through the sample volume as well as the photon density of states is sensitively dependent on the precise form of $S(q)$. This is particularly evident in the final illustration which is intermediate between the liquid structure factor, Figure (3c), and the weak disorder (Figure (3b)) case. This is an arrangement of scatterers analogous to the positions of atoms in amorphous silicon. Instead of long range crystalline order or liquid like disorder, the dielectric resonators are arranged in a manner in which strong crystalline correlations persist over a few nearest neighbour distances. The array of scatterers is said to have *short-range-order*. In this case, there is a strong remnant of Bragg scattering. The peaks in $S(q)$ are sharper than in the liquid-disorder case, but do not become delta functions (see Figure 3d). The scatterers no longer have any memory of their crystalline positions but have strong crystalline correlations on short length scales.

Dramatic changes in the photon density of states take place as the structure factor evolves from that shown in Figure (3c) to Figure (3d). To illustrate this, we present a result on the *electronic* density of states in the liquid state of silicon and compare it with the density of states for solid, amorphous silicon. This is shown in Figure 4. The results were obtained by R. Carr and M. Parrinello[26] with a computer simulation based on molecular dynamics techniques. It is clear from this computer generated data that changes in the short-range atomic correlations can lead to dramatic changes in the density of states at certain energies. For the case of photons in a dielectric medium, a similar pseudogap in the photon density of states occurs[15]. This is the precursor to photonic bandgap formation. As we will show this leads to a dramatic modification of the free-photon (Ioffe-Regel) criterion for localization. The nature of this modification is most transparent in the case of a weakly disordered *photonic crystal*.

Consider a fluctuating dielectric constant $\epsilon(x) - \epsilon_0 \equiv \epsilon_{fluct}(x) = \epsilon_1(x) + V(x)$ where $\epsilon_1(x) = \epsilon_1 \sum_{\vec{G}} e^{i\vec{G}\cdot\vec{x}}$ is a perfectly periodic Bravais superlattice and $V(x)$ is a small perturbation arising from disorder. Here $\vec{G}$ runs over the appropriate reciprocal lattice and its value for the dominant Fourier component $U_{\vec{G}}$ is chosen so that the Bragg condition $\vec{k} \cdot \hat{G} = (1/2)G$ may be satisfied for a photon of wavevector $\vec{k}$. Such a structure is attainable, albeit in a low dielectric contrast regime, with charged polystyrene balls in aqueous suspension. These exhibit charge induced fcc and bcc superlattice arrangements as well as a number of disordered phases[27]. Setting $V(x) = 0$ for the time being, the effect of the periodic modulation of the photon spectrum may be estimated within a nearly-free-photon approximation. This is the photonic analogue of the nearly free electron approximation well known in solid state physics[23]. Unlike scalar electrons, there is a

580

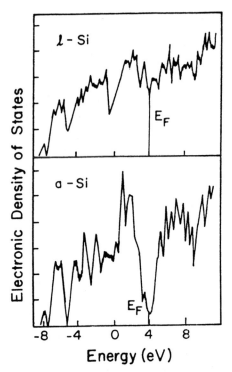

Figure 4. Electronic density of states for liquid (ℓ) and amorphous (a) silicon. Short-range-order in the solid phase leads to the appearance of a sizeable depression near the Fermi energy E_F and dramatic changes in transport properties. This picture is based on a Monte Carlo simulation performed by R. Car and M. Parrinello.

degeneracy for photons between two possible optical polarization states. If the electric field vector is perpendicular to the plane defined by the vectors $\vec{k}$ and $\vec{k}$-$\vec{G}$ (optical S-wave) the resulting photon dispersion is the same as for scalar wave scattering. The nearly-free-photon approximation consists of regarding equation (2.2) as an effective Schrödinger equation in which the free-photon-Hamiltonian is given by $H_0 \equiv -\nabla^2$ and the Bragg scattering is a weak perturbation given by $H_{\text{pert}} = -\frac{\omega^2}{c^2}\epsilon_1 U_G \cos \vec{G} \cdot \vec{x}$. The unperturbed wavefunctions are chosen to be transverse plane waves and so the term $\vec{\nabla}(\vec{\nabla} \cdot \vec{E})$ has no effect. The transverse plane wave of wave-vector $\vec{k}$ is degenerate with that of wave-vector $\vec{k} - \vec{G}$ when $\vec{k}$ lies on a Bragg plane (see Figure 5). Accordingly we apply the rules of degenerate perturbation theory[28] to determine the effective eigenvalue, $\frac{\omega^2}{c^2}\epsilon_0$, of equation (2.2). This yields the determinant condition:

$$\begin{vmatrix} k^2 - \frac{\omega^2}{c^2}\epsilon_0 & -U_G\epsilon_1\frac{\omega^2}{c^2} \\ -U_G\epsilon_1\frac{\omega^2}{c^2} & (\vec{k} - \vec{G})^2 - \frac{\omega^2}{c^2}\epsilon_0 \end{vmatrix} = 0 \ . \tag{2.12}$$

If, on the other hand, the polarization vector lies in the plane of Bragg

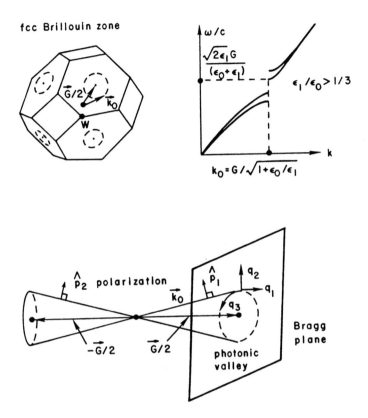

Figure 5. In the case of coherent scattering from a periodic array of dielectric spheres, the phase space available for optical propagation is drastically modified from the vacuum. The dispersion relation exhibits a gap any given direction of the wavevector $\vec{k}$. The inner and outer dispersion branches describe p- and s-polarized Bragg scattering respectively. For a large refractive index contrast, this gap persists over the entire surface of the fcc Brillouin zone giving rise to a gap in the total photon density of states. The phase space for optical propagation near this gap is restricted to a narrow set of symmetry related cones in the nearly free photon approximation.

scattering the scattering amplitude is diminished by a factor of $\cos\theta$, where θ is the angle between $\vec{k}$ and $\vec{k}$-$\vec{G}$ (optical p-wave). The associated photon dispersion relations are depicted in Figure 5, with the gap for the optical p-wave being diminished by precisely $|\cos\theta|$. The band edge frequencies for p-polarized light are obtained from equation (2.12) by replacing U_G by $U_{\vec{G}}|\cos\theta_{\vec{k}}|$. When $\vec{k}$ lies on the Bragg plane, the cosine identity yields $\cos\theta_k = 1 - G^2/(2k^2)$.

It is clear from the above discussion that the effect of the macroscopic Bragg resonance is to create a gap of width $\delta_{\vec{k}}$ in the photon dispersion relation $\omega_{\vec{k}}$ for any given direction in $\vec{k}$-space when $\vec{k}$ crosses a Bragg plane. For a complete photonic bandgap to occur, it is necessary that the gaps $S_{\vec{k}} = (\omega_{\vec{k}} - \Delta_{\vec{k}}/2, \omega_{\vec{k}} + \Delta_{\vec{k}}/2)$ have a nontrivial intersection as $\vec{k}$ spans the surface of the first Brillouin zone of the crystal. We consider for the sake of argument an fcc crystal. In the nearly-free-photon approximation, a weak scattering and a strong scattering regime can be identified. Setting $U_{\vec{G}} = 1$, then for $\epsilon_1/\epsilon_0 \leq 1/3$ (weak scattering) the bottom of the upper band occurs at the center $\vec{k} = \vec{G}/2$ of the Bragg plane. This is the point on the fcc Brillouin zone surface where the magnitude $|\vec{k}|$ is minimum. In a scalar wave theory, as $\vec{k}$ deviates from $\vec{G}/2$ it is natural to expect that this corresponds to higher frequency states. However, for electromagnetic waves this is not necessarily the case. For the p-polarized light, as $\vec{k}$ deviates from $\vec{G}/2$, the factor $|\cos\theta_k|$ diminishes from its maximum value of unity. This means that although $|\vec{k}|$ increases, the magnitude of the gap $\Delta_{\vec{k}}$ decreases. For $\epsilon_1/\epsilon_0 > 1/3$ (strong scattering) the latter effect dominates. The result is that the bottom of the upper band occurs on a circle centered about $\vec{k} = \vec{G}/2$. The magnitude of $\vec{k}$ on this circle is given by $k_0 \equiv G/\sqrt{1 + \epsilon_0/\epsilon_1}$. As it turns out, strong scattering is required for complete photonic bandgap formation. In many practical cases, the radius of this circle becomes so large that the bottom of the upper band occurs at the corner point of the zone boundary labelled as W. It goes without saying that the top of the lower band is dominated by p-polarized light at the W-point. The purpose of using an fcc crystal is also apparent from this discussion. The fcc Brillouin zone boundary exhibits the smallest variation in the magnitude of $\vec{k}$. As a result the *variation* with $\vec{k}$ in the gap interval $S_{\vec{k}}$ is smaller than for other lattices such as bcc and simple cubic. This allows a nontrivial intersection of the intervals $S_{\vec{k}}$ as $\vec{k}$ varies. For the fcc lattice this occurs for a smaller dielectric constant (scattering strength) than the other Bravais lattices.

The existence or near existence of a gap in the photon density of states is of paramount importance in determining transport properties and especially localization [2]. Such a possibility was completely overlooked in the derivation of the free-photon Ioffe-Regel condition which assumed an essentially free-photon density of states. In the vicinity of a band edge the character of propagating states is modified. To a good approximation the electric field amplitude of the propagating wave is a linear superposition of the free photon with wavevector $\vec{k}$ and its Bragg reflected partner at $\vec{k}$-$\vec{G}$. As ω moves into the allowed band, this standing wave is modulated by an envelope function whose wavelength is given by $2\pi/q$ where q is the magnitude of the deviation of $\vec{k}$ from the Bragg plane. Under these circumstances the wavelength which must enter the localization criterion is that of the envelope. In the presence of even very weak disorder, the criterion $2\pi\ell/\lambda_{\text{envelope}} \sim 1$ is automatically satisfied as the photon frequency approaches the band edge frequency. In fact near a bandedge ω_c, $\lambda_{\text{envelope}} \sim |\omega - \omega_c|^{-\frac{1}{2}}$.

In the presence of a complete photonic band gap, the phase space available for photon propagation is restricted to a set of narrow symmetry related cones in $\vec{k}$-space analogous to the pockets of electrons near a conduction band edge well known in semiconductor physics[25]. The perturbative introduction of randomness in the position of the dielectric scatterers leads to a mixing of all nearly degenerate photon branches. In complete analogy with semiconductors, the band gap is replaced by a pseudogap consisting of localized states (Figure 6). Localization is caused here by the severe restriction of the phase space available for propagation. Photon localization arises here not because of a high degree of uncontrolled disorder, but rather as a result of a subtle interplay between order and disorder. The true criterion for localization in fact depends strongly on the underlying static structure factor of the medium. What we have discussed in detail are the two extreme limits of a structureless random medium for which the criterion $2\pi\ell/\lambda \simeq 1$ applies, and of a medium with nearly sharp Bragg peaks and a band gap for which $2\pi\ell/\lambda_{\text{envelope}} \simeq 1$ yields localization. It follows that a continuous crossover occurs between these conditions as the structure factor of a high dielectric material evolves from one limit to the other.

In the strong scattering regime required to produce a significant depression of the photon density of states, important corrections to the nearly free photon picture of band structure emerge. This was evident in initial experimental studies by Yablonovitch [29] who observed a photonic pseudogap in the microwave regime. For a dielectric material with refractive index 3.5 containing an fcc lattice of spherical air cavities, a nearly complete photonic band gap is observed when the solid volume fraction $f \simeq .15$. For a frequency range spanning about 6% of the gap center frequency, propagating electromagnetic modes are absent in all but a few directions. When the solid volume fraction is either increased or decreased from .15, the magnitude of the gap drops sharply. The existence of a well defined optimum value of the microscopic scatterer density, illustrates a very fundamental principle regarding photonic bandgap formation which is not evident in the nearly-free-photon approximation. This is the fact that a true photonic band gap is the result of a synergetic interplay between the *microscopic* and *macroscopic* resonance mechanisms. This is distinct from the case of electronic bandstructure. In a semiconductor, a band of propagating electronic states arises from the overlap of localized atomic orbitals on individual lattice sites. The starting point consists of localized states and the end-

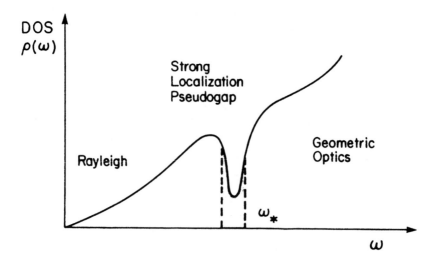

Figure 6. For a disordered lattice of coherent dielectric scatterers the photon density of states exhibits low frequency Rayleigh scattering and high frequency classical ray optics regimes. The photonic band gap however is replaced by a pseudogap of strongly localized states in analogy with pseudogaps in amorphous semiconductors. The existence of a localization window is highly sensitive to the static structure factor of the dielectric material.

product is the band of extended states. In the case of light in a periodic dielectric material, there are no localized states for individual, microscopic scatterers. Photons cannot be bound in any true sense to a single dielectric sphere or any other dielectric microcavity. Instead of bound states, scattering resonances occur when the wavelength of light becomes comparable to the size of the scatterer. Extended states are clearly the starting point and the end-product is localization. The creation of this end-product is guided by the formation of a photonic band gap. In three dimensions, this calls for highly restrictive conditions on the dielectric microstructure. One of these conditions is that the density of scatterers be chosen such that the microscopic scattering resonance of a single unit cell of the structure occur at the same frequency as the macroscopic (Bragg) resonance of the periodic array. This principle may be illustrated (see Figure 7) by a simple example of one-dimensional wave propagation through a periodic array of square wells of width a and spaced by a distance L. Suppose the refractive index is n inside each well and is unity outside. Then the Bragg scattering condition is given by $\lambda = 2L$ where λ is the vacuum wavelength of light. The analog of a Mie resonance in one dimension is a maximum in the reflection coefficient from a single well and this occurs when a quarter wavelength fits into the well: $\lambda/(4n) = a$. Combining these two conditions yields the optimum volume filling fraction $f \equiv a/L = 1/(2n)$. In analogy to the formation of an electronic band, the photonic band gap is the

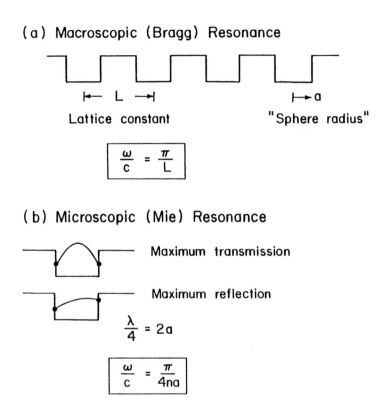

(a) Macroscopic (Bragg) Resonance

$\vdash$ L $\dashv$
Lattice constant

$\vdash$ a
"Sphere radius"

$$\frac{\omega}{c} = \frac{\pi}{L}$$

(b) Microscopic (Mie) Resonance

Maximum transmission

Maximum reflection

$$\frac{\lambda}{4} = 2a$$

$$\frac{\omega}{c} = \frac{\pi}{4na}$$

Figure 7. The photonic bandgap arises from a synergetic interplay between macroscopic and microscopic resonances. This effect is maximized when the lattice constant L and the sphere radius a are chosen in such a way that the two resonance conditions coincide. Maximum scattering from a single well occurs when one-quarter wavelength "fits" into the well.

direct result of the coalescence of scattering resonances of individual microcavities. The generic form of the magnitude of the photonic bandgap with volume filling fraction f of dielectric material is shown in Figure 8. For the case of scalar waves propagating through an fcc lattice of dielectric spheres, there is a threshold refractive index ratio between the spheres and the background of 1.5 before a three-dimensional gap is formed. In the case of electromagnetic waves, it has been shown by detailed bandstructure calculations[30, 31] that the gap associated with the lowest order ($\ell = 1$) Mie resonance of the dielectric spheres does not lead to a complete photonic bandgap but rather a pseudogap. This is because the gap $\Delta_{\vec{k}}$ associated

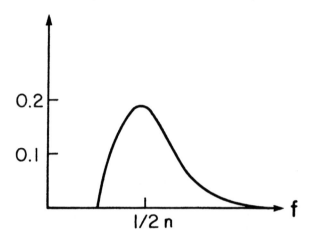

Photonic Band Gap $\Delta\omega/\omega_0$

Figure 8. Magnitude of the photonic band gap $\Delta\omega$ in a periodic dielectric relative to gap center frequency ω_0 as a function of volume filling fraction f of solid material. The maximum gap occurs at $1/(2n)$ where n = refractive index of solid, due to synergetic interplay between microscopic and macroscopic resonances.

with p-polarized Bragg scattering vanishes when $\vec{k}$ approaches the W point of the fcc Brillouin zone whenever the scatterer in the unit cell of the lattice has spherical symmetry. This symmetry can be removed by using nonspherical scatterers or by placing more than one microscopic resonator within the unit cell. Using a diamond lattice of air spheres (2 spheres per unit cell) in a high index background, Ho, Chan and Soukoulis[32] obtained the first theoretical demonstration of a complete photonic bandgap in $d = 3$ for electromagnetic waves. The threshold refractive index ratio in this structure is approximately 2.0. Subsequently, Yablonovitch and co-workers [31] successfully demonstrated a complete gap for air cylinders drilled in a solid host with a gap to center frequency ratio $\Delta\omega/\omega_0$ of 20 percent.

Since the initial work on periodic dielectrics in three dimensions, considerable progress has been made in identifying and fabricating microstructures with a *complete* photonic bandgap. In addition to the synergy between microscopic and macroscopic resonances, *connectivity* of *both* the dielectric and void regions of the material appears to be an important contributing factor to the creation of a complete gap. In the case of electromagnetic waves, it turns out that a *network topology* is more effective than a *cermet topology* in leading to a complete photonic bandgap[34, 35]. In the former

Topology of Dielectric Microstructure

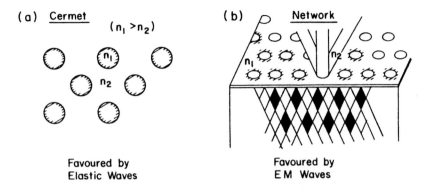

Figure 9. In addition to resonance mechanisms, photonic bandgap formation is facilitated by the connectivity of the underlying dielectric. The cermet topology which is favoured by elastic waves consists of high index n_1 (low velocity) inclusions in a connected low index n_2 (high velocity) background. The network topology favoured by electromagnetic waves consists of two interpenetrating, connected components.

case *both* the high refractive index (low propagation velocity) component as well as the low refractive index (high propagation velocity) component form a continuous, connected network through the material. The cermet topology consists of disconnected high index inclusions in a connected low index (high propagation velocity) background. These two topologies are illustrated in Figure 9. This preference for a *network topology* is associated with the detailed structure of electromagnetic wave equation and the nature of the boundary conditions of the electric and magnetic fields at the dielectric interfaces. By *preference*, we mean that for a given refractive index contrast, the peak in $\Delta\omega/\omega_0$ vs. volume filling fraction (Figure 8) is higher. On the other hand, Economou[36] has shown that for elastic waves (with bulk and shear moduli) and for scalar waves which satisfy an equation of the form

$$\nabla^2\phi = -\frac{\omega^2}{c^2}\epsilon(x)\phi \; , \qquad (2.13)$$

the cermet topology is preferred.

We summarize the conclusions of the previous discussion with a schematic picture (Figure 10) of the overall photon density of states in the various regimes of interest. In the case of independent point scatterers, the density of states (a) has the same form as in the case of free photons. In the case of a periodic dielectric microstructure satisfying the conditions described above, the density of states (b) exhibits a gap between the frequencies labelled by ω_v and ω_c. In analogy with the electronic bandstructure of semiconductors, a deviation from perfect periodicity induces distinct features in the optical spectrum. A defect in the periodic order could arise from a substitutional "impurity" such as a single microscopic scatterer which is different from all of the others in the array either in structure or in optical characteristics. Another possible defect is a deviation of one of the scatterers from its lattice position. In general, this will lead to the creation of one or more localized states within the photonic bandgap. A more precise statement concerning the frequency of this localized state is given in the next section on spectral limit theorems.

From the standpoint of optical device physics, an isolated localized state represents a very high quality (Q) factor optical "cavity". The size of the "cavity" is not the size of the substitutional impurity but rather the localization length of the state. This can typically span many wavelengths. It introduces the possibility of propagative effects taking place within the cavity. In the absence of dissipative or inelastic effects, the localized state is a perfect bound state with an infinite Q-factor. In a real dielectric material, there is a finite absorption length. For a fiber-optic quality material, this absorption length is on the scale of kilometers. This translates into an absorption time $\tau_{abs} \gtrsim 10^{-5}$ sec. For an optical frequency $\omega \sim 10^{15}$ sec^{-1}, the localized state will have a quality factor $Q \equiv \omega \tau_{abs} \gtrsim 10^{10}$. What is much less obvious is the fact that these high Q factors are a property not only of an isolated localized state, but also of localized states in the singular continuous spectrum of a disordered medium. As discussed earlier, this type of strong localization pseudogap replaces the photonic bandgap in disordered systems with short-range-order. We demonstrate this persistence of high Q values in the disordered state in section 4 of this article. Before doing this, it is useful to review in general terms, the nature of localized states created by specific forms of disorder.

2.3. SPECTRAL LIMIT THEOREMS

In general, a defect in an otherwise periodic structure can lead to the appearance of a localized state in the gap or a resonance in the extended state continuum. The spectral limit theorems allow one to predict which of these two possibilities occurs without the need to solve the eigenvalue problem explicitly. The essential point behind the theorem is that the effect of a single defect can be predicted from knowing the effect on the overall bandstructure if such a defect were placed in *each* unit cell of the periodic system. This reduces the impurity problem, in which translational symmetry is ex-

Photon Density of States

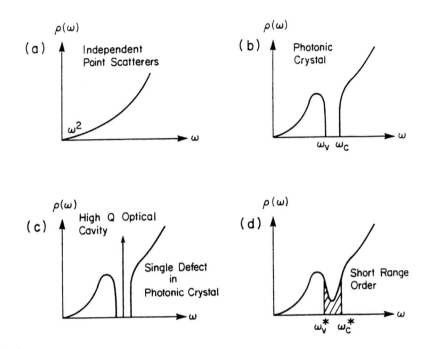

Figure 10. Summary of the photon density of states for four different scattering geometries. For independent point scatterers (a), all states are extended. For a strongly scattering, periodic array (b), a complete photonic bandgap is possible. When a single dielectric defect is created in this photonic crystal, a localized state appears within gap (c). Further disorder in the photonic crystal gives rise to a strong localization pseudogap (d).

plicitly broken, to a simpler problem in which transitional symmetry has been restored. The theorem is most useful when the impurity perturbation is relatively weak. In the case of the photonic bandgap we will show that removing a *small* amount of high dielectric constant material from one unit cell leads to occurrence of a localized state just above the top of the lower band. On the other hand, adding a *small* amount of high dielectric material from a single unit cell causes a single localized state to split off from the upper band edge.

We begin by discussing the spectral limit theorem[37] (due to Hadamard and Gershgorin) in the context of an electronic tight-binding model and then apply it to the photonic band gap. Consider the Hamiltonian for an electron which can reside on a set of sites $\{n\}$ with energy ϵ_n and labelled by state vectors $|n\rangle$. The electron is allowed to hop from site n to site n' with an amplitude of $t_{n,n'}$:

$$\mathcal{H} = \sum_n \epsilon_n |n\rangle\langle n| + \sum_{n,n'} t_{n,n'}(|n\rangle\langle n'| + |n'\rangle\langle n|) \ . \tag{2.14}$$

Let η be an eigenvalue of $\mathcal{H}$ and let $\psi(n)$ denote the wavefunction amplitude at site n. It follows from the associated tight-binding Schrödinger equation, $H\psi = \eta\psi$, that

$$|\eta - \epsilon_n| \le \sum_{n' \ne n} |t_{n,n'}| \frac{|\psi(n')|}{|\psi(n)|} \ . \tag{2.15}$$

In a physically admissable state, the wavefunction amplitude is bounded. It takes on some maximum value at a particular site. Let us label this site by the integer m. With this definition, it follows that $|\psi(n')|/|\psi(m)| \le 1$ for all n'. This results in the inequality

$$|\eta - \epsilon_m| \le \sum_{n' \ne m} |t_{m,n'}| \equiv B_m \ . \tag{2.16}$$

In other words, the eigenvalue η lies within a *"disc"* of radius B_m centered at the energy ϵ_m. Since this is true for at least one of the possible values of n, namely m, it follows that η *is contained in the union of "discs" centered on ϵ_n with radii B_n.*

To illustrate the use of the theorem, we begin with the problem of electrons in a *periodic* one-dimensional chain with nearest neighbour hopping. Here $\epsilon_n = 0$, $t_{n,m} = t(\delta_{n,m+1} + \delta_{n,m-1})$ and $B_n = 2t$ for all n. The theorem correctly reproduces the correct band spectrum for the electron: $|\eta| \le 2t$. Next we consider the effect of a *structural* defect on the electronic spectrum. Suppose, for instance, that the bond connecting site m to site $m+1$ is made *stronger* than all other bonds. If $t_{m,m+1} > t$, then the bounds on the eigenvalue spectrum η extend beyond the band spectrum of the crystal. In fact a detailed eigenvalue calculation reveals that localized states split off both the top and the bottom of the band. On the other hand, if the bond connecting site m to site $m+1$ is made *weaker* than all other bonds ($t_{m,m+1} < t$) the spectral bounds on the crystal remain the same. In fact, if we were to make *all* the bonds weak, the spectral bounds would narrow. It follows that localized states do not split from the band edges in this case. A detailed calculation reveals that the effect of a *single* weak bond is to create resonance states *within* the band.

The same considerations may now be applied to the photonic bandgap. We consider the effects of a *small perturbation* which either adds or removes a small amount of dielectric material from *one* unit cell. The effect of adding an infinitesimal amount of dielectric material to *each* unit cell is to lower the center frequency of the photonic bandgap. This follows from the fact that the *long wavelength* effective speed of light is decreased. Accordingly, the effect of a small addition of dielectric material to a single unit cell, is to split a localized mode from the upper band. Using the terminology of semi-conductor physics, this is the analogy of creating a localized "donor" state below the "conduction band" edge. Similarly by *removing* a *small* amount of dielectric material from one unit cell leads to the splitting of a localized "acceptor" state from the "valence band" edge. It should be borne in mind that these arguments are useful primarily in cases where the perturbation is small compared to the size of the photonic bandgap itself. For example, it is possible that by removing more and more dielectric material that a "valence" band edge "acceptor" state can be moved further and further away from the "valence" band edge and actually approach the "conduction" band edge. It should also be borne in mind that the terminology of "acceptor" and "donor" states as well as "valence" and "conduction" bands do not have the significance that they have in semiconductor physics. Electrons in a semiconductor are fermions. As a result of the Pauli exclusion principle, conduction takes place in the upper band. For photons, there is no limitation to the number of particles in any given mode. Wave propagation occurs in both the "conduction" and "valence" bands.

To summarize this section, we have demonstrated that *the localization of light can occur within the positive energy continuum as the result of an interplay between order and disorder in the dielectric microstructure.* The photonic band gap provides a very useful guide for the choice of material parameters for the realization of this effect. The formation of a photonic band gap involves three basic considerations (i) the *vector nature* of the electromagnetic field, (ii) the synergy between *microscopic* and *macroscopic* resonances and finally (iii) the *network topology* of the dielectric microstructure. Given a periodic, photonic band gap material, localized states of light may be engineered within the photonic band gap. These states may act as ultra-high Q optical cavities with important device applications.

3. Classical Electrodynamics in a Photonic Band Gap

The existence of the photonic bandgap and localized states is a direct consequence of Maxwell's equations. The direct use of Maxwell's equation in describing the electrodynamics of PBG materials, however, is very cumbersome. It is useful to introduce a simplified classical electrodynamics which retains only the essential degree of freedom in describing band gap related phenomena. The PBG in some respects is the bosonic analogue of the Dirac gap for fermions. Accordingly, we derive an effective Dirac equation for electromagnetic waves in a PBG material. This leads to some interesting new phenomena such as the occurence of nonlinear, gap solitary waves. These solitary waves act in many respects like massive particles. They are self-localized state of light which can be stationary or move with any velocity up to the speed of light. As the soliton approaches relativistic speeds, it undergoes Lorentz contraction much like an ordinary massive particle.

It is instructive to consider the nature of wave propagation in the allowed bands near one of the photonic band edges. Let $\vec{k}_0$ be a point on the crystalline Brillouin zone corresponding to a photonic band edge. Then, the dominant contributions to wave propagation at frequencies near this band edge (in the nearly-free-photon approximation) are plane waves at wavevectors $\vec{k}_0$, $\vec{k}_0 - \vec{G}$, and all other symmetry related Bragg partners (in the case of a high symmetry point). When the frequency lies precisely on the band edge, these counter propagating component waves form a standing wave. The group velocity for electromagnetic transport vanishes. When the frequency is moved slightly away from the band edge, the corrections to this standing wave picture may be incorporated in what is known as a *slowly varying envelope approximation* (SVEA). In this approximation the electric field amplitude is expanded in the form:

$$\vec{E}(\vec{r}, t) \simeq e^{-i\omega_0 t} \left[\vec{E}_1(\vec{r}, t) e^{i\vec{k}_0 \cdot \vec{r}} + \vec{E}_2(\vec{r}, t) e^{i(\vec{k}_0 - \vec{G}) \cdot \vec{r}} + \ldots \right] . \qquad (3.1)$$

Here $\vec{E}_1(\vec{r}, t)$, $\vec{E}_2(\vec{r}, t)$ etc... are assumed to have only very slow spatial and temporal variations compared to the wavevector k_0 and the frequency ω_0. For example, a propagating state may be constructed by setting both of the slowly varying envelope functions $\vec{E}_1$ and $\vec{E}_2$ equal to $\hat{x} e^{i\vec{q} \cdot \vec{r}}$ where the wavelength of the envelope function $\lambda' \equiv 2\pi/q \gg$ the vacuum wavelength $\lambda_0 \equiv 2\pi/k_0$. This is illustrated in Figure 11. The essential idea is that the spectral content of the function $\vec{E}(\vec{r}, t)$ is highly peaked in the vicinity of the Bragg scattering points $\vec{k}_0$, $\vec{k}_0 - \vec{G}$, etc... and that width of these peaks $\Delta k/k_0 \ll 1$. With this assumption (valid for near bandgap phenomena), the expansion for $\vec{E}(r, t)$ may be substituted into the wave equation (2.2) and an effective euqation for the slowly varying envelope functions may be derived. We will discuss, below, how this procedure may be used to describe *nonlinear* wave propagation at high light intensity *within* the photonic band gap itself. For the moment, however, we will focus on the effects of static disorder. As discussed earlier, the analogy with semiconductor physics suggests that in the presence of weak disorder, the photonic bandgap is replaced by a pseudogap of localized states. This may seem surprising from the standpoint of independent scatterers and the free-photon Ioffe-Regel criterion described earlier. In the case of the photonic bandgap the criterion that $2\pi\ell^*/\lambda \simeq 1$, where ℓ^* is the transport mean free path

Near Band Edge Behaviour

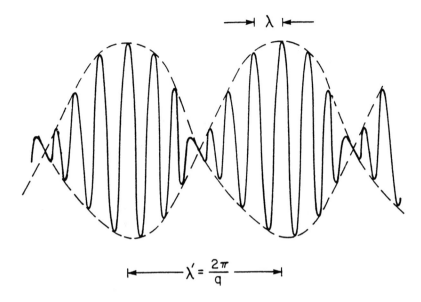

Figure 11. Electromagnetic waves propagation near a photonic band edge takes the form of standing wave modulated by an envelope function with wavelength $\lambda' \gg \lambda$, the vacuum wavelength. As the frequency ω approaches the band edge, λ' approaches infinity.

and λ is the vacuum photon wavelength, is no longer applicable. Instead, the relevant criterion is that $2\pi\ell^*/\lambda' \simeq 1$ where λ' is the wavelength of the modulating envelope function $(\vec{E}_1, \vec{E}_2)$. Since λ' diverges as the frequency approaches the band edge, the latter criterion can be satisfied for very weak disorder (relative to the periodic arrangement) and very long mean free paths. In fact a continuous crossover from one localization criterion to the second one takes place as the static structure factor of the dielectric material $S(\vec{q})$ varies from being independent of $\vec{q}$ to one which has sharp peaks associated with short-range-order.

3.1. LOWER DIMENSIONAL SYSTEMS

Most of the discussion so far has been on three-dimensional localization and three-dimensional photonic bandgaps. A number of important applications arise, nevertheless, in lower dimensional systems. In the case of optical propagation through a dielectric material, lower dimensionality may be achieved in a bulk three-dimensional material with translational symmetry in one or

Lower Dimensional Materials

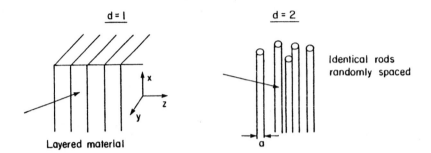

Figure 12. One-dimensional light scattering may be realized in a layered dielectric. Each layer is uniform but the dielectric constant varies from layer to layer. Two-dimensional scattering may be realized in a random collection of uniform, uniaxial dielectric cylinders.

more directions. This is illustrated in Figure 12. The layered material has translational symmetry in the x- and y-directions. Accordingly it behaves as like a one-dimensional system for propagation in the z-direction. Similarly a collection of randomly spaced dielectric rods has a translational symmetry in the x-direction. It behaves as a two-dimensional scattering medium for optical propagation in the y- and z-directions. This dimensional reduction can be quantified by analysis of the wave equation (2.2).

For the layered system, we may consider a propagating solution of the form $\vec{E}(x, y, z) = \hat{y}\psi(z)e^{ik_x x}$. This yields the equation:

$$\frac{-\partial^2}{\partial z^2}\psi(z) - \frac{\omega^2}{c^2}\epsilon_{\text{fluct}}(z)\psi = \left(\frac{\omega^2}{c^2}\epsilon_0 - k_x^2\right)\psi(z) \ . \tag{3.2}$$

Not only is this a one-dimensional wave equation, but the effective "eigenvalue", $\frac{\omega^2}{c^2}\epsilon_0 - k_x^2$, can be varied through the parameter k_x by simply changing the angle of incidence of the incoming light. Lower dimensional systems may not only be easier to fabricate, but they provide additional adjustable parameters such as this, for the study of localization phenomena.

3.2. NONLINEAR OPTICAL SOLITONS IN A 1-D STOP GAP

The photonic bandgap is a frequency regime in which linear electromagnetic effects are absent. This however does not exclude the possibility of *nonlinear* wave propagation. For a large-scale photonic bandgap material, the propagation of high intensity, nonlinear solitary waves may provide a practical way of coupling large amounts of optical energy into, and out of, the otherwise impenetrable photonic bandgap. A stationary solitary wave may be regarded as a self-localized state. In a perfectly periodic material, the high light intensity itself creates a localized dielectric defect through the nonlinear Kerr coefficient. Unlike the localized state induced by static disorder, the localized dielectric defect is free to move with light intensity field. The result is a solitary wave that can move through the bulk photonic bandgap material with any velocity ranging from zero up to the average speed of light in the medium. This macroscopic pulse of optical energy behaves very much like a massive (relativistic) particle whose mass is proportional to the magnitude of the photonic bandgap $\Delta\omega$.

Kerr nonlinearities can arise within the solid dielectric fraction of the photonic bandgap material. This solid fraction would have to be relatively lossless while at the same time exhibiting a large Kerr coefficient. Typical *nonresonant* nonlinearities in semiconductors give rise to a third order nonlinear susceptibility $\chi^{(3)} \sim 10^{-15}$-10^{-17} (cm/V)^2. On the other hand, for certain GaAs-AlAs structures, it has been shown[38] that the optical absorption edge falls off very sharply below the *electronic* bandgap whereas the nonlinear susceptibility[39] remains relatively high. In these systems $\chi^{(3)} \sim 10^{-10}$ (cm/V)^2. For a photonic bandgap with $\Delta\omega/\omega_0 \simeq 0.1$, solitary waves could be created with a peak intensity of 1 kW/cm^2 and electric field strengths no larger than 10^3-10^4 V/cm. I now proceed to demonstrate analytically how solitary wave solutions may arise.

The underlying nonlinear wave equation is the same as (2.2) except with $\epsilon_{\mathrm{fluct}}(x)$ replaced by $\epsilon_{\mathrm{periodic}}(\vec{x}) + \kappa|\vec{E}|^2$. Here, $\epsilon_{\mathrm{periodic}}$ is the periodic dielectric modulation of the photonic bandgap material, $\kappa \equiv 12\pi\chi^{(3)}$ is the nonlinear Kerr coefficient and $\vec{E}$ is the electric field amplitude. Since we are interested primarily in the frequency regime of the photonic bandgap, it is useful to expand the electric field amplitude in the slowly varying envelope function approximation. We define a multicomponent field $\Psi^* = (E_1^*, E_2^*, \ldots E_n^*)$ consisting of the slowly varying fields and derive an equation for this envelope field. The one-dimensional case of optical propagation through a periodic, layered medium serves as a valuable paradigm for higher dimensional systems. We begin with this example. In this case the periodic part of the dielectric constant may be modelled as a sinusoidal modulation, $\epsilon_{\mathrm{periodic}}(\vec{r}) = \epsilon_1 \cos(\vec{G} \cdot \vec{r})$, where $\vec{G} = G\hat{x}$ is the reciprocal lattice vector. We use the (SVEA) expansion $E(x) = E_1(x)e^{ik_0 x} + E_2(x)e^{i(k_0-G)x}$ with $k_0 = G/2$. This may be substituted into the nonlinear wave equation

$$\left[-\nabla^2 - \frac{\omega^2}{c^2}\epsilon_{\mathrm{periodic}}(\vec{r}) - \frac{\omega^2}{c^2}\kappa|E|^2 \right] E = \epsilon_0 \frac{\omega^2}{c^2} E \qquad (3.3)$$

and terms containing the factors e^{ik_0x} and $e^{i(k_0-G)x}$ are grouped together. As a consequence of the linear independence of these functions, the coefficients of these factors can be set to zero independently. These lead to a pair of coupled equations for E_1 and E_2. If we further make use of the slowly varying nature of these envelope functions (by neglecting second order spatial derivatives) we arrive at an "effective nonlinear Dirac equation" which has the following structure

$$\left[i\sigma_z\frac{\partial}{\partial x} + \epsilon_1\sigma_x + \kappa(\Psi, \Psi)\right]\Psi = -(\omega^2 - 1)\Psi \ . \tag{3.4}$$

Here, σ_x and σ_z are the usual 2×2 anticommuting Pauli spin matrices. We have also rescaled the length x by a factor of G^{-1} and the frequency ω by a factor of $\omega_0 = ck_0$, the bandgap center frequency, so that equation (3.4) is in fact dimensionless. It is straightforward to prove that for a stationary solution, we require that $E_1 = E_2^* = \varepsilon e^{i\phi}$, where ε and ϕ are the amplitude and phase respectively of the solitary wave. Substituting this parameterization into equation (3.4) yields the pair of coupled nonlinear differential equations:

$$\frac{\partial \varepsilon}{\partial x} = \epsilon_1 \varepsilon \sin 2\phi \tag{3.5a}$$

$$\frac{\partial \phi}{\partial x} = (\omega^2 - 1) + \epsilon_1 \cos 2\phi + \kappa\varepsilon^2 \ . \tag{3.5b}$$

This is known as the double-Sine-Gordon equation. The exact solution is discussed in detail by D. Mills and S. Trullinger[40]. In order to understand intuitively the nature of this solution, it is useful to make a mechanical analogy for equation (3.5b). Suppose that the variable ϕ represents the position of a particle in a one-dimensional potential $V(\phi) \equiv -(\omega^2 - 1)\phi - (\epsilon_1/2)\sin 2\phi$ and that $x(\equiv t)$ is a time variable. Then equation (3.5b) may be rewritten as

$$m\ddot{\phi} + \dot{\phi} = -\frac{\partial}{\partial\phi}V(\phi) + F_{\text{ext}}(t) \ . \tag{3.6}$$

Here the dot represents a time derivative and $F_{\text{ext}}(t) \equiv \kappa\varepsilon^2$ represents an external force which acts on the particle over and above the potential $V(\phi)$. In order to recapture equation (3.5b) it is necessary to consider the case $(m \to 0)$ in which the "inertia term" is negligible compared to the "viscous damping term" $\dot{\phi}$. We consider for simplicity the case of a solitary wave with average frequency at the middle $(\omega^2 = 1)$ of the photonic bandgap. In this case, the potential $V(\phi)$ is sketched in Figure 13a. The solitary wave solution corresponds to a situation when the particles start at $t = -\infty$ at a local minimum of $V(\phi)$ namely $\phi = \pi/4$. The external force acts so as to move the particle to the unstable equilibrium point at $\phi = -\pi/4$. The

(a) Mechanical Analogy

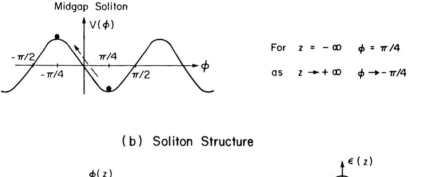

Midgap Soliton

For $z = -\infty$ $\phi = \pi/4$

as $z \to +\infty$ $\phi \to -\pi/4$

(b) Soliton Structure

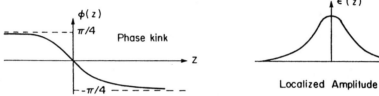

Phase kink

Localized Amplitude

Figure 13. (a). The nonlinear wave equation for the $d = 1$ optical gap soliton is equivalent to an overdamped mechanical oscillator. The soliton arises when the particle at the potential minimum at $\phi = \pi/4$ is subjected to slowly varying external force and asympototically approaches the unstable maximum at $\phi = -\pi/4$. (b) Gap solitons are characterized by a phase kink in the direction of Bragg scattering and an exponentially decreasing amplitude function $\epsilon(z)$.

external force is zero at $t = -\infty$, reaches a maximum at $t = 0$ and then goes back to zero at $t = +\infty$. It is clear that this behaviour is consistent with equation (3.5a). For a finite energy solitary wave, we require that $\varepsilon(x)$ vanish exponentially as $x \to \pm\infty$. This is possible provided $\sin 2\phi \to \mp 1$ in these limits. The motion of the "particle" from $\phi = +\pi/4$ to $\phi = -\pi/4$ clearly satisfies this requirement. The structure of the resulting self-localized state is sketched in Fig. 13b. The distinguishing characteristic of the solitary wave solution is the *self-phase-modulation* or *phase-kink* which accompanies the exponentially decaying amplitude function $\varepsilon(x)$.

Recently, Neset Akozbek and I have generalized this 1-dimensional solitary wave solution to higher dimensional photonic bandgap systems [43]. While an exact solution is no longer possible in general, we have developed a variational method which predicts a variety of different solitary wave solutions in higher dimensions. The variational method reproduces to a high degree of accuracy the exact solution of the one-dimensional double-sine-Gordon equation. It suggests that photonic bandgap materials in higher dimensions may have a variety of interesting bistable switching properties which go beyond the simple, well-known characteristic of one-dimensional dielectrics.

3.3. NONLINEAR OPTICAL SOLITARY WAVES IN A PHOTONIC BAND GAP

In higher dimensional photonic band gap materials, the underlying nonlinear wave equation has not been solved exactly. We present, here, an approximate solution to the nonlinear equation using a new variational method. Our variational method reproduces to a high degree of accuracy the exact soliton solutions in one-dimension. It furthermore makes a number of new predictions concerning the existence and properties of solitary waves in higher dimensions. This includes the occurence of a variety of finite energy solitary waves associated with different symmetry points of the crystalline Brillouin zone. The one-dimensional gap soliton solution with phase and amplitude modulation in the direction of a single reciprocal lattice vector of the crystal is extended in the transverse directions and has an infinite energy in a higher dimensional system. However, if the Kerr coefficient is positive, we show that localized finite energy solitary waves are in fact possible with phase modulation in the direction of a single Bragg scattering vector. Unlike the one-dimensional case, finite energy solutions which pick out a specific reciprocal lattice vector and thus break the symmetry of the crystal, do not occur for negative Kerr coefficient. On the other hand, we find a new class of solitary waves of higher symmetry which occur for both positive and negative Kerr coefficient. These new self-localized states involve Bragg scattering by a complete basis set of reciprocal lattice vectors and exhibit phase modulation in each of these directions.

Another distinguishing property of solitary waves in a photonic band gap is the scaling of the total electromagnetic energy contained in the state with the frequency of the underlying radiation. For one dimensional solitons the total energy vanishes as the frequency approaches a band edge where the effective mass approximation is valid. For a three dimensional solitary wave, a simple scaling argument suggests that the total energy is minimum near the center of the photonic band gap but in fact diverges as the frequency approaches this band edge. These threshold, gap center, solitary waves may prove valuable for coupling energy from external sources into an otherwise inpenetrable photonic band gap material on scales much longer than the tunnelling length.

For the purpose of illustration, we descibe in detail solitary wave solutions in a two dimensional periodic nonlinear dielectric with the point group symmetry of a square lattice. The essential new physics that distinguishes the square lattice from 1-d periodic structures is the existence of multiple symmetry points in the crystalline Brillouin zone which determine the photonic band edges. Each of these symmetry points plays an important role in determining the nature of nonlinear waves within the gap and accordingly a slowly varying envelope function expansion of the true electric field amplitude must be performed about each point. The linear part of the dielectric constant is taken to have the form $\epsilon(x, y) = \epsilon_o + \Delta\epsilon[\cos Gx + \cos Gy]$ where the reciprocal lattice vector has a magnitude $G = 2\pi/a$. It is assumed that the electric field $\vec{E}$ and polarization $\vec{P}$ are perpendicular to the direction in which they vary. Denoting the z- component of the complex electric field amplitude of frequency ω by E, the nonlinear wave equation is

$$[\nabla^2 + k^2\epsilon(x, y) + 12\pi\chi^{(3)}k^2|E|^2]E = 0 \qquad (3.7)$$

where $k = \omega/c$, ω is the optical frequency, c is vacuum speed of light and $\chi^{(3)}$ is the nonlinear susceptibility.

The slowly varying envelope approximation to this equation consists of retaining only those Fourier components of $E(x, y)$ which lie near the crystalline band edges. These occur at the symmetry points X and M of the square lattice Brillouin zone (see Figure 14, inset). We consider first the X symmetry point at $\vec{k}_o = (\frac{\pi}{a})\hat{x}$ and expand $E(x, y) = E_1(x, y)e^{ik_o x} + E_2(x, y)e^{-ik_o x}$ where E_1 and E_2 are slowly varying envelope functions of spectral width $\Delta k \ll k_o$. Substituting this expansion into (3.7), collecting terms of order $e^{ik_o x}$ and $e^{-ik_o x}$, respectively, and keeping only leading terms in the derivative expansion, we obtain a coupled set of nonlinear differential equations for E_1 and E_2. It is straightforward to show [40] that static solutions in which there is no net transport of electromagnetic energy satisfy the condition $|E_1| = |E_2|$ in which case the coupled equations can be conveniently expressed in terms of a two component spinor field:

$$[i\sigma_z\partial_x + \partial_y^2 + (\omega^2 - 1)/4 + \omega^2\beta\sigma_x + \omega^2\lambda(\Psi^\dagger\Psi)]\Psi = 0 \qquad (3.8)$$

where $\Psi^\dagger = (E_1^*, E_2^*)$, x and y are dimensionless coordinate variables measured in units of G^{-1}, $\partial_x = \frac{\partial}{\partial x}$, $\partial_y = \frac{\partial}{\partial y}$, ω is a dimensionless frequency measured in units of the characteristic frequency $\omega_o = c\pi/(a\sqrt{\epsilon_o})$, $\beta = \Delta\epsilon/(8\epsilon_o)$ and $\lambda = 9\pi\chi^{(3)}/(2\epsilon_o)$. σ_x and σ_z are the 2×2 Pauli spin matrices which satisfy an anticommuting algebra. In the absence of the term $\partial_y^2\Psi$, this is precisely the 1-d soliton equation [40]. The second derivative term here is essential, however, to render the solitary wave solution finite in the y-direction.

The underlying bandstructure in the envelope approximation (3.8) is obtained by setting $\lambda = 0$. At $\vec{k} = \frac{\pi}{a}$ there is a frequency gap bounded

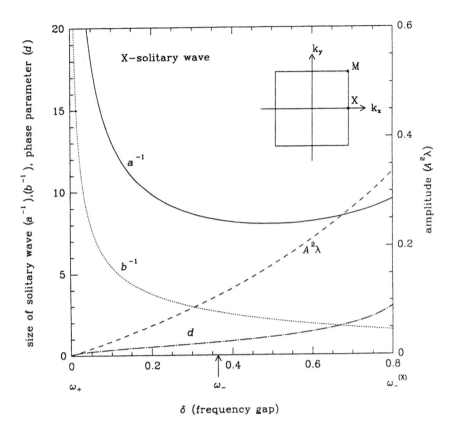

Figure 14. The size parameters a^{-1} (solid-line), b^{-1} (dot-line), phase parameter d (dot-dash-line) [left-scale] and amplitude ($A^2\lambda$) (dash-line) [right-scale] of the broken symmetric X-solitary wave are shown as a function of the band gap frequency $\delta = (\omega_+^2 - \omega^2)$. The absolute band gap is marked on the graph as ω_+ and ω_-. These calculations are done for $\Delta\epsilon_o/\epsilon_o = 0.7$ and $\lambda > 0$.

by the band edge frequencies $\omega_\pm^{(X)} = (1 \mp (\Delta\epsilon/2\epsilon_o))^{-1/2}$. In the overall bandstructure of the crystal the frequency $\omega_+ = \omega_+^{(X)}$ plays the role of the upper band edge of the photonic band gap. As a consequence of the anisotropic dispersion, this model has an indirect gap with the lower band edge ω_- occuring at the corner of the Brillouin zone (M point).

Before discussing the precise nature of solitary wave solutions of (3.8) we describe the possibility of qualitatively different solitary wave states associated with the other symmetry point in the Brillouin zone. Equation (4.2) clearly breaks the symmetry of the square lattice. By interchanging x

and y in equation (4.2) we may obtain solitary waves that are rotated by 90°. To find nonlinear solutions that preserve the symmetry of the crystal it is necessary to perform an envelope expansion about each of the four equivalent X points. $E(x,y) = E_1 e^{ik_o x} + E_2 e^{-ik_o x} + E_3 e^{ik_o y} + E_4 e^{-ik_o y}$ Again assuming a static self-localized state with the magnitude of the slowly varying envelope to be the same, the effective equation for the four component spinor field $\Psi^\dagger = (E_1^*, E_2^*, E_3^*, E_4^*)$ is

$$[i(\gamma_1 \partial_x + \gamma_2 \partial_y) + (\omega^2 - 1)/4 + \omega^2 \beta \gamma_3 + \omega^2 \lambda'(\Psi^\dagger \Psi)]\Psi = 0 \qquad (3.9)$$

where $\lambda' = 3/2\lambda$ and γ_1, γ_2 and γ_3 are 4×4 matrices defined as

$$\gamma_1 = \begin{pmatrix} \sigma_z & 0 \\ 0 & 0 \end{pmatrix} \quad \gamma_2 = \begin{pmatrix} 0 & 0 \\ 0 & \sigma_z \end{pmatrix} \quad \gamma_3 = \begin{pmatrix} \sigma_x & 0 \\ 0 & \sigma_x \end{pmatrix}$$

Unlike equation (3.8) which requires a second derivative to yield a localized solution, we will show that symmetric localized solutions of equation (3.9) are possible even though we have retained only first order derivatives in both the x and y directions.

A distinct class of solitary wave solutions is associated with the fourfold degenerate M point. Since there are two reciprocal lattice vectors which connect a given M point to symmetry related partners, it is necessary to include all four of the corresponding envelope funtions in any expansion of the electric field amplitude. This yields

$$[i(\gamma_1' \partial_x + \gamma_2' \partial_y) + (\omega^2 - 2)/4 + \omega^2 \beta \gamma_3' + \omega^2 \lambda'(\Psi^\dagger \Psi)]\Psi = 0 \qquad (3.10)$$

for the four component envelope field Ψ, where the new 4×4 matrices γ_1' and γ_2' are defined as

$$\gamma_1' = \begin{pmatrix} \sigma_z & 0 \\ 0 & \sigma_z \end{pmatrix} \quad \gamma_2' = \begin{pmatrix} \sigma_z & 0 \\ 0 & -\sigma_z \end{pmatrix} \quad \gamma_3' = \begin{pmatrix} 0 & \sigma_x + I \\ \sigma_x + I & 0 \end{pmatrix}$$

The underlying photonic bandstructure near the M point is determined by setting the nonlinear coefficient $\lambda' = 0$. A spectral gap occurs between the two frequencies $\omega_{\mp}^{(M)} = \sqrt{2}(1 \pm \Delta\epsilon/\epsilon_o)^{-1/2}$.

In order to have a complete photonic band gap, the gaps at the X and M points should overlap. This leads to the requirement that $\Delta\epsilon/\epsilon_o \gtrsim 0.5$. In this case, the true photonic band gap is indirect and is bounded by the lower band edge $\omega_- = \omega_-^{(M)}$ and the upper band edge $\omega_+ = \omega_+^{(X)}$. 2-d photonic band calculations confirm that the absolute band gap frequency is determined by the X and M symmetry points [41].

Equations (3.8), (3.9) and (3.10) describe the properties of three distinct types of elementary solitary waves within the two-dimensional photonic band gap. Unlike their counterpart for the one-dimensional stop gap in which a simple analytic soliton solution exist [40] we seek approximate analytic solutions within the gap. Nevertheless, certain features of the one-dimensional solution remain evident in higher dimensions. The first is the exponential decay of the wave amplitude. Physically we may regard the high light intensity at the center of the solitary wave as creating an effective dielectric defect in the otherwise periodic structure leading to a localized state in the gap. Another essential feature of the solitary wave is the existence of a kink in the phase of electric field near the center of the solution. Both of these features play a vital role in the variational trial function we introduce to approximately solve the non- linear envelope equations. Equation (3.8) may be regarded as a stationary point of the functional

$$F_1 = \int d^2\vec{r}[|\partial_y \Psi|^2 - \Psi^\dagger(i\sigma_z\partial_x + (\omega^2 - 1)/4 + \omega^2\beta\sigma_x + (\omega^2/2)\lambda(\Psi^\dagger\Psi))\Psi]$$

$$(3.11)$$

with respect to arbitrary functions Ψ. The condition $\delta F_1/\delta\Psi^\dagger = 0$ may be implemented approximately by introducing a trial solution $\Psi^\dagger(x, y) = \varepsilon(\vec{r})(e^{-i\Phi(\vec{r})}, e^{i\Phi(\vec{r})})$ with $\varepsilon(\vec{r}) = A\text{sech}(ax)\text{sech}(by)$ and $\Phi(\vec{r}) = c + \arctan[d\tanh(ax)]$ describing a localized state with a kink in the x-direction. The functional F_1 can be evaluated analytically and minimized with respect to the varitional parameters A, a, b, c and d. The results are shown in Figure 15 as a function of the dimensionless detuning frequency $\delta = (\omega_+^2 - \omega^2)$. The phase shift $c = (n + 1/2)\pi$. For this symmetry breaking solitary wave, a nontrival solution exist for $\lambda > 0$. Within the two envelope function approximation this solution persists for all frequencies within the X-gap: $\omega_-^{(X)} < \omega < \omega_+^{(X)}$. However, for frequencies $\omega_-^{(X)} < \omega < \omega_-^{(M)}$ which lie outside of the true photonic band gap, the solution is unstable to decay by interaction with the continuum modes. As shown in Figure 15, within the photonic band gap region, the length scales a^{-1}, b^{-1} of the solitary wave diverge as $\delta^{-1/2}$ and the amplitude $A \sim \delta^{1/2}$ near the upper band edge. A similar behaviour for a^{-1} and A occurs $\omega_-^{(X)}$ but this is preceded by the occurence of the indirect lower band edge $\omega_- = \omega_-^{(M)}$. A more elaborate trial wavefunction is in fact needed to exhibit the divergence of the localization length at the lower direct band edge for $\lambda > 0$. We refer to the above solution as the "X-solitary wave". The total energy of the solitary wave is defined as $U = \int d^d\vec{r}(\Psi^\dagger\Psi)$ for each frequency in the gap. In general, the total electromagnetic energy of the solitary wave scales as the square of the amplitude A and the volume a^{-d}. Near the band edge (small δ) it follows that the energy obeys the scaling behaviour

$$U \sim [\lambda k_o{}^d\delta^{1-d/2}]^{-1}.$$

$$(3.12)$$

$d = 2$ is a marginal dimension in which U is some finite number throughout the gap. For $d = 1$ $U \to 0$ as the band edge is approached and there is no

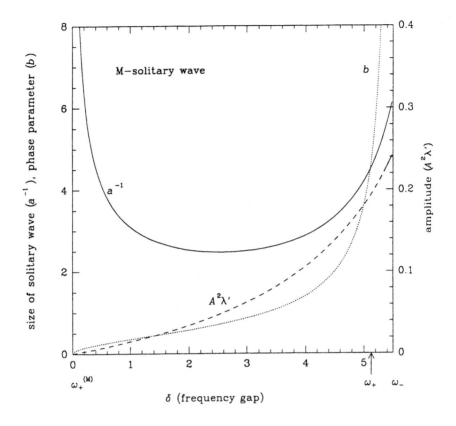

Figure 15. The size parameter a^{-1} (solid-line), phase parameter b (dot-line) [left-scale] and amplitude $(A^2\lambda')$ (dash-line) [right-scale] of the M-solitary wave are shown as function of the band gap frequency $\delta = (\omega_+^{2(M)} - \omega^2)$ for $\lambda' > 0$ and $\Delta\epsilon/\epsilon_o = 0.7$. The absolute band gap is marked on the graph with ω_+ and ω_-.

threshold for creating a soliton. In $d = 3$, however $U \to \infty$ near a band edge and in fact the threshold energy for creating a solitary wave is realized near the gap center, as far from the band edges as possible.

The variational method we have outlined may be used to obtain an "M-solitary wave" as well as a "symmetric X-solitary wave". From equation (3.10) it is apparent that the M solitary wave can be regarded as an extremum of the functional

$$F_2 = \int d^2\vec{r}\,\Psi^\dagger[i(\gamma_1'\partial_x+\gamma_2'\partial_y)+(\omega^2-2)/4+\omega^2\beta\gamma_3'+(\omega^2/2)\lambda'(\Psi^\dagger\Psi)]\Psi \quad (3.13)$$

A solution can be found for all frequencies in the gap for both positive and negative λ' by inserting the varitional function $\Psi^\dagger = \varepsilon(r)(e^{-i\Phi_1}, e^{i\Phi_1}, e^{-i\Phi_2}, e^{i\Phi_2})$ where $\varepsilon(\vec{r}) = A\mathrm{sech}(ax)\mathrm{sech}(ay)$, $\Phi_1 = c_1+\arctan[b\tanh(ax)] + \arctan[b\tanh(ay)]$ and $\Phi_2 = c_2+\arctan[b\tanh(ax)] - \arctan[b\tanh(ay)]$. Minimizing F_2 with respect to the varitional parameters A, a, b and c yields a finite energy solitary wave. The results are given in Fig. 14b for $\lambda' > 0$. Here $c_1 = 2\pi n$ and $c_2 = (2n + 1)\pi$. The case for $\lambda' < 0$ is easily found by mirror reflecting the curves about $\delta = 2.75$ in Figure 15.

We mention finally that a "symmetric X-solitary wave" solution occurs by following essentially the same procedure as above by casting equation (3.9) into a variational form. The trial solution takes a similar form to the symmetric "M-solitary wave". The results are qualitatively similar to those depicted in Fig. 15 so we do not repeat them here. It is interesting to note, nevertheless, that unlike the broken symmetry X-solitary wave, a symmetric solution of finite energy exist for either sign of the nonlinear Kerr coefficient.

The variational method described above provides both a useful qualitative picture as well as a valuable starting point for more quantitative numerical solution of equations (3.8), (3.9) and (3.10). Neset Akozbek has carried out an iterative, finite element numerical solution to these equations. The converged, numerical solutions for the X-solitary wave at midgap (with $\chi^{(3)} > 0$) and the M solitary wave near the lower band (with $\chi^{(3)} < 0$) are depicted in Figure 16.

In summary, we have derived a generalization of the one-dimensional optical gap soliton states to higher dimensional photonic band gap materials. The variational ansatz reproduces very accurately the exact 1-d gap soliton state for most of the band gap. The only significant deviations from the exact solution occur near the lower direct band edge for $\lambda > 0$ and the upper direct band edge for $\lambda < 0$. In our 2-d square lattice model with an indirect gap these regions are not relevant to the true photonic band gap. The 1-d soliton state exhibit a relativistic dynamics when time dependence is included in its governing equation [42]. This follows directly from the anticommutation algebra of the Pauli matrices for the 1-d problem. In higher dimensions this relativistic dynamics, will be lost, since unlike a true Dirac equation, the γ-matrices appearing in equation (3.9) and (3.10) fail to anticommute.

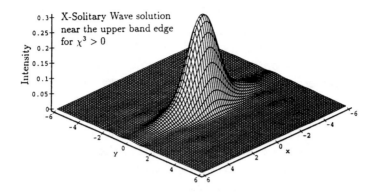

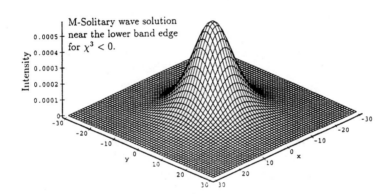

Figure 16. Numerical Solution of 2-D solitary Waves in PBG using finite element expansion. Depicted are (a) the symmetry breaking X-solitary wave at midgap for $\chi^{(3)} > 0$ and (b) the symmetric M-solitary wave near the lower bandedge for $\chi^{(3)} < 0$. The lengths x and y are measured in units of the lattice constant and the intensity units are arbitrary.

4. Scattering and Localization in Disordered Dielectrics

Photonic band gaps and localization are not exclusively the property of photonic crystals. The large electronic pseudo-gaps in amorphous silicon and amorphous germanian provide ample evidence of this fact. PBG's based on *disordered* dielectric microstructures may open the door to very large scale and very inexpensive materials.

Disorder is a nearly ubiquitous feature of real materials. In this section, we present a brief overview of the underlying physics of the photon localization transition from the standpoint of multiple scattering theory. In addition, we show that a disordered dielectric medium with a strong localization pseudogap may perform many of the same functions as a perfectly periodic photonic bandgap material. Localized states have the important property of being optically disconnected even when they are very close in frequency or part of a localized continuum with a finite density of states. We begin by describing the approach to localization. This is the physics of energy diffusion and the wave interference corrections to diffusion. The interference correction responsible for localization is known as coherent backscattering. Since this is the precursor to localization, it is sometimes also referred to as "weak localization". As the name implies, there is an enhanced coherent amplitude for light propagating in a disordered medium with wave-vector $\vec{k}$ to be scattered into the direction $-\vec{k}$. When the scattering is sufficiently strong to produce a noticeable depression in the photon density of states in a given frequency interval, this leads to the localization pseudogap. As discussed in section 2, both the *microscopic* and *macroscopic* resonance scattering mechanisms must be present. The localization frequency window is separated from extended states at both high and low frequencies by a pair of mobility edges ω_c^* and ω_v^*. These are the analogs of photonic band edges in the periodic dielectric. We review the scaling theory of localization [44] and its predictions for optical transport in the vicinity of such mobility edges. Unlike electrons which satisfy a number conservation law, photons can be absorbed. Accordingly we discuss the effects of absorption on the mobility edge scaling by the introduction of an inelastic length scale (ℓ_{inel}). Near the mobility edge, the localization length is divergent. As a result, the inelastic length is shorter than the localization length when ω is sufficiently close to either ω_c^* or ω_v^*. In this case, the inelastic scattering disrupts the coherent interference required for fully developed localization. On the other hand, within a localization pseudogap region, the localization length, ξ_{loc}, may become much shorter than the inelastic length scale. In this case, localized states become relatively immune to dissipative effects and retain their integrity. We calculate the Q-factor of a strongly localized state in the pseudogap regime and show that this can be comparable to that of an *isolated* localized state in an otherwise periodic dielectric.

A disordered dielectric medium can be characterized optically by a set of length scales which define different regions in which different physical processes come into play. Consider for instance a microstructure consisting of dielectric spheres which scatter light weakly. Any given sphere tends to scatter light in a higly anisotropic manner [20, 21]. Although scattering takes place, much of the light is scattered into a direction very close to the forward direction. The length scale on which a small deflection in the photon trajectory takes place is referred to as the scattering length ℓ. For a dilute collection of spheres, the length ℓ is roughly the spacing between spheres. After many scattering events, the photon's direction of propagation bears practically no memory of its initial direction. This defines a new length scale $\ell^* \gg \ell$ which we refer to as the *transport mean free path*. On length scales $L \gg \ell^*$, the transport of optical energy may be regarded as diffusive in nature. In the weak scattering system described above $\ell^* \gg \ell \gg \lambda$. Here λ is the optical wavelength. In the long wavelength limit $(\omega \rightarrow 0)$, the transport mean free path exhibits Rayleigh scattering. For a d-dimensional system, this means that $\ell^*(\omega) \sim \omega^{-(d+1)}$.

Given the fact that on some length scale ℓ^*, the transport of energy is diffusive, it is easy to see that two-dimensions plays the role of a critical dimension for localization effects. On length scales $L \gg \ell^*$ the photon trajectory may be regarded as a random walk. The proability density that a random walker, which is released at time $t = 0$ at the origin $\vec{r} = 0$, will be found at a later time t at a position $\vec{r}$ is given by

$$P(\vec{r}, t) = (4\pi D_0 t)^{-d/2} \exp(-r^2/4D_0 t) .$$

Here, $D_0 = \frac{1}{3} c\ell^*$ is the classical diffusion coefficient and d is the dimension of space. This expression is valid for length scales $\gg \ell^*$ and time scales $t \gg \tau_0$. Here, $\tau_0 \equiv \ell^*/c$ is a transport mean free time. Clearly the probability density that the random walker will return to the origin after a time t is given by $P(0, t) \sim t^{-d/2}$. Moreover, this density, when integrated over time, exhibits strikingly different behaviour for $d > 2$ and $d \leq 2$. In the former case the integrated density

$$\int_{\tau_0}^{T} dt P(0, t)$$

remains finite as $T \rightarrow \infty$, whereas in the latter case, this this quantity is unbounded. That is to say that in dimensions less than or equal to two, the random walker will periodically return to some vicinity of the origin with probability one. In three dimensions, the phase space for propagation is sufficiently large that the *necessity* of returning to the origin no longer exists. In the case of photons, we must replace probabilities with *probability amplitudes*. The *probability amplitudes* to return to the origin interfere constructively (as I will demonstrate shortly) leading to localization of the photon. *In both one and two dimensions all states are localized.* (This, however, does not preclude the possibility that the localization length may be much longer than the sample size. In this case the distinction between extended and localized states is academic.) In three dimensions some reduction of the free-photon phase space is required before localization can occur. This is the requirement of strong scattering. While the diffusion picture takes into some account wave interference effects on length scales short compared to the transport mean free path, the interference of different diffusion paths is completely ignored. The interference of diffusion paths in the backscattering channel leads to a modified picture of the ensemble-averaged optical transport. The ensemble here consists of different realizations of the disordered dielectric. This is the content of the scaling theory of localization. In this picture, the ensemble averaged diffusion coefficient is no longer a local quantity but depends on the entire macroscopic scale L of the illuminated sample. The effective diffusion coefficient is a coarse-grained diffusion coefficient $D(L)$ defined with respect to the scale L of coarse-graining. Extended states are distinguished from localized states by the scaling behaviour of $D(L)$ at some fixed frequency ω. If the states of frequency ω are extended, $D(L)$ remains nonzero as $L \to \infty$. For localized states $D(L) \sim e^{-L/\xi_{loc}}$ for $L \gg \xi_{loc}$, where ξ_{loc} is the localization length.

A detailed derivation of the diffusion approximation starting from the underlying wave equation may be found in the article by Vollhardt and Wolfle [45] for the case of electrons in a Gaussian random white noise potential. A derivation of diffusive and non-diffusive propagation modes for classical waves in a correlated random medium (with nontrivial structure) may be found in John and Stephen [8]. A detailed description of the crossover from ballistic propagation on length scales $L < \ell^*$ to diffusion on length scales $L > \ell^*$ is given by MacKintosh and John[46]. Here it is shown that multiple-light scattering can be used as a spectroscopic probe of disordered media with nontrivial dielectric structure. In the present article, we will take these facts as given and rather focus on the corrections to diffusion due to wave interference which take place on length scales $L \gg \ell^*$. These include the phenomena of coherent backscattering and photon mobility edges.

4.1. PHOTON DIFFUSION AND COHERENT BACKSCATTERING

As discussed above, in a disordered dielectric, a photon propagates by means of a random walk process in which the length of each random step is given by the classical transport mean free path ℓ^*. On length scales long compared to ℓ^* it is convenient to regard this as diffusion of light in which the diffusion coefficient is given by $D = \frac{1}{3}c\ell^*$. Here c is some effective speed of light in the dielectric medium. Unlike a classical random walker, light is a wave and this diffusion process must be described by an amplitude rather than a probability. That is to say, the interference of all possible classical diffusion paths must be considered in evaluating the transport of

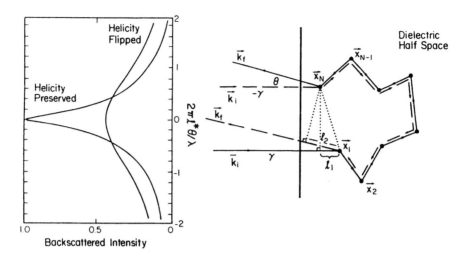

Figure 17. Coherent backscattering is the precursor to localization. For the light incident on a disordered dielectric, the scattering path (γ) and the time reversed path ($-\gamma$) interfere coherently for small angles θ. The result is a pair of peaks in the backscattered intensity I over and above the diffuse background intensity of reflected light labelled as $I = 0$. The larger peak describes backscattering in which the incident photon helicity is preserved whereas the smaller peak describes photons whose final helicity has been flipped during the multiple scattering trajectory.

electromagnetic energy. In the case of optical waves propagating through a disordered dielectric medium, this interference effect has been vividly demonstrated by a beautiful series of experiments initiated by Albada and Lagendijk [12], and Maret and Wolf [13]. This is the phenomenon of coherent backscattering. In these experiments, incident laser light of frequency ω enters a disordered dielectric half space or slab and the angular dependence of the backscattered intensity is measured. For circularly polarized incident light the intensity of the backscattering peak for the helicity preserving channel is a factor of two larger than the incoherent background intensity. Coherent backscattering into the reversed helicity channel however yields a considerably reduced backscattering intensity. The angular width of the peak in either case is roughly $\delta\theta \sim \lambda/(2\pi\ell)$.

As shown in Figure 17, one possible process is that in which incident light with wavevector $\vec{k}_i = \vec{k}_0$ is scattered at points $\vec{x}_1, \vec{x}_2, \ldots \vec{x}_N$ into intermediate (virtual) states with wavevectors $\vec{k}_1, \vec{k}_2, \ldots \vec{k}_{N-1}$ and finally into the state $\vec{k}_N = \vec{k}_f$ which is detected. For scalar waves undergoing an identical set of wavevector transfers, the scattering amplitudes at the points $\vec{x}, \ldots \vec{x}_N$ are the same for the path γ and time reversed path $-\gamma$ (dashed line). To understand the interference of these two paths in greater detail we utilize an argument given by Bergmann [66] in the context of electron localization. For this purpose, we consider backscattering into the direction $-\vec{k} + \vec{q}$, where $\vec{q}$ is a small deviation. Consider the path shown in Figure 18 with wave-vector transfers $\mathbf{g}_j = \mathbf{k}_j - \mathbf{k}_{j-1}(j = 1, 2, \ldots, N)$. By time-reversing the sequence of transfers, it is apparent that the corresponding intermediate states $\mathbf{k}_1', \mathbf{k}_2', \ldots, \mathbf{k}_{N-1}'$ no longer lie on the energy shell for nonzero $\mathbf{q}$. This is allowed since these are virtual states with a lifetime $\tau = l/c$ and the energy shell is accordingly smeared by an amount $\hbar/\tau$. For small $\mathbf{q}$, the corresponding intermediate states differ in energy by an amount $E_{N-j} - E_j' \simeq c\hbar\mathbf{q}\cdot\hat{k}_{N-j}(J = 1, \ldots, N-1)$ where k_{N-j} are unit vectors in the direction of propagation of the intermediate plane wave states. The resulting phase difference $\Delta\phi = \hat{k}_{N-j}\cdot\mathbf{q}l$. Since the direction of intermediate states is random, it follows that after an N step random walk the accumulated root mean square phase difference $\Delta\phi_{\text{rms}} \simeq \sqrt{N/3}(ql)$. Therefore only steps with $N \lesssim 3/(lq)^2$ contribute to the backscattering peak at the angle defined by $\mathbf{q}$. This accounts for the rapid decrease of coherent intensity for large $\mathbf{q}$.

At larger angles few paths contribute and the backscattered intensity decreases rapidly. A detailed derivation of the backscattered lineshape has been given by Akkermans, Wolf and Maynard [47] for scalar waves. This has been extended to electromagnetic waves by Stephen and Cwilich [48].

As shown by F. Mackintosh and S. John[49], these results are most transparent in the helicity representation. In addition to time reversal invariance, there is parity of the right and left hand circular polarization states. These symmetries may be broken by the Faraday effect and by natural optical activity respectively. Detailed lineshapes have been calculated and are shown in Figure 17. Depicted is the excess relative intensity with respect to the incoherent background. For instance an intensity level of 1.0 corresponds to a precise doubling of the light intensity in a particular direction relative Neto the diffuse background level which is labeled with excess intensity 0.0. The angle Θ (measured in radians) is the angle which the wavevector of the backscattered light makes with respect to the vector $-\vec{k}_i$, where $\vec{k}_i$ is the incident direction.

The arguments presented thus far for coherent backscattering enhancement have relied heavily on the existence of the free photon phase space or spherical energy shell. The occurrence of a photonic pseudogap or any other deviation of the photon density of states from that of free photons will accordingly manifest itself in the backscattering lineshape. This is most easily seen in the case of a weakly disordered photonic bandgap crystal. Suppose

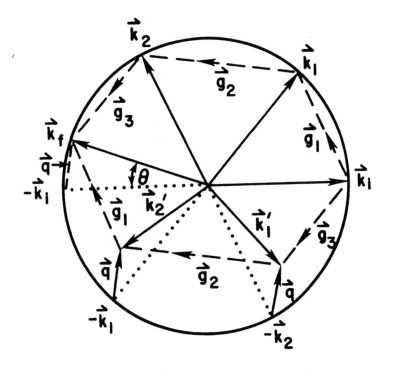

Figure 18. A typical path drawn in momentum space. Equality of scattering amplitudes requires that the wavevector transfers $\mathbf{q}_j$ be the same but in the reversed order for the time-reversed path. Corresponding states differ in energy by an amount $E_{N-j} - E_j' \simeq c\hbar\mathbf{q} \cdot \hat{k}_j$, which leads to a phase difference $(\Delta\phi)_{\mathrm{rms}} \simeq \sqrt{N/3ql}$.

that a backscattering experiment is performed on such a sample with radiation of wavelength short compared to the underlying lattice constant of the crystal. Bloch's theorem ensures that $\omega_{\vec{k}} \simeq \omega_{\vec{k}+\vec{G}}$ in the case of weak disorder whenever $\vec{G}$ is a reciprocal lattice vector of the original crystal. It follows that the previous argument for loss of coherence for backscattering into the direction $-\vec{k}_i + \vec{q}$ (here $\vec{k}_i$ is the incident wave vector) is no longer applicable. In particular, whenever $\vec{q}$ lies in the vicinity of a reciprocal lattice vector, coherent enhancement persists, resulting in a satellite peak. The result, after ensemble averaging over the weak disorder, should be a ring of coherent intensity at an angle $\theta \simeq G/k$ over and above the diffuse intensity that arises from the ensemble average of the random speckle pattern. The occurrence of satellite structure in the backscattering profile has been pointed out by E. Gorodnichev et al [50].

The incorporation of coherent backscattering into the theory of diffusion of wave energy leads to a simple renormalization group picture of transport. When wave interference occurs, the spread of wave energy is not diffusive at all. The photon no longer performs a classical random walk. This presents a very complicated situation. Fortunately, there is a way of applying the concept of classical diffusion here provided we make one major concession in our classical way of thinking: the diffusion coefficient is no longer a local quantity determined by a classical mean free path and a speed of propagation, but depends on the macroscopic coherence properties of the entire illuminated sample. In a random medium it is reasonable to expect that scatterers which are very far apart do not, *on average*, cause large interference corrections to the classical diffusion picture. (The word average here is very important. Changes in distant scatterers *can* give rise to significant fluctuations about the average.) It follows that there exists a coherence length $\xi_{\mathrm{coh}} \gtrsim \ell^*$ which represents a scale on which we must very carefully incorporate interference effects in order to determine the effective diffusion coefficient at any point within the coherence volume. In other words, the possible amplitudes for a photon to diffuse from point A to point B within a coherence volume ξ_{coh}^d interfere significantly with each other on average. Depending on the distance between the point A at which the photon is injected into the medium and the point B at which it is detected, the effective diffusion coefficient of the photon is strongly renormalized by wave interference. Another example is that of a finite size sample of linear size L. By changing the scale of the sample, the number of diffusion paths which can interfere changes, giving rise to an effective diffusion coefficient $D(L)$ at any point within the sample which depends on the macroscopic scale L of the sample. In the vicinity of a mobility edge, on length scales L in the range $\ell^* < L < \xi_{\mathrm{coh}}$ the spread of energy is subdiffusive in nature as a result of coherent backscattering. This gives a significant wave interference correction to classical diffusion. In this range of length scales, the spread of wave energy may be interpreted in terms of a scale dependent diffusion coefficient which behaves roughly as $D(L) \simeq \frac{c\ell^*}{3}(\frac{\ell^*}{L})$. On length scales long compared to ξ_{coh}, the photon resumes its diffusive motion except with a lower or renormalized value $\frac{c\ell^*}{3}(\frac{\ell^*}{\xi_{\mathrm{coh}}})$ of the diffusion coefficient.

4.2. THE PHOTON MOBILITY EDGE

In order to understand the detailed nature of scale-dependent diffusion arising from coherent backscattering, it is useful to consider an electronic analogy. Consider an ideal gas of noninteracting electrons in a disordered metal. In the presence of a uniform applied electric field, conduction takes place by means of a diffusion process. Diffusion is the result of random scattering from impurity potentials. The electronic current density $\vec{J}$ is related to the electric field $\vec{E}$ by a constitutive equation $\vec{J} = \sigma \vec{E}$ where σ is the d.c. electronic conductivity. This is nothing other than Ohm's law. The conductivity at zero temperature in turn is given by the Einstein relation, $\sigma = e^2 \rho(E_f) D_0$. Here e is the electronic charge, $\rho(E_f)$ is the electronic density of states at the Fermi energy E_f and D_0 is the electronic diffusion coefficient. Consider a cubical sample of this metal with side length L as depicted in Figure 19.

Electrodes are placed on opposite faces of the cube and a voltage V is applied. This leads to a current flow I given by Ohm's law $I = \Sigma V$ where the proportionality constant Σ is the total *conductance* of the sample. The *conductance* is related to the *conductivity* by a simple geometrical argument. Σ increases linearly with the cross-sectional area of the sample L^{d-1} and decreases linearly with the length of the conduction path. Putting these factors together yields $\Sigma = \sigma L^{d-2}$, for a d-dimensional system. This simple relation embodies the physics of diffusive transport and Ohm's law. The corresponding logarithmic derivative $\beta \equiv \frac{d\ln\Sigma}{d\ell L} = d - 2$ is a characteristic of extended electronic states at the Fermi energy. The function $\beta(\Sigma)$ may be regarded as a scaling function (in the renormalization group sense) describing how the logarithm of conductance, and hence the diffusion coefficient scales with the size L of the sample. Any deviation from the Ohmic behaviour is an indication of nontrivial wave interference effects and scale-dependent diffusion. To illustrate this, we consider the opposite limit in which electronic states are localized at the Fermi energy.

If electronic wavefunctions at the Fermi energy are exponentially localized with a localization length ξ_{loc}, we expect that the ensemble averaged conductance $\Sigma \sim \exp(-L/\xi_{\text{loc}})$. This is independent of dimensionality. Accordingly, the scaling function $\beta(\Sigma) \sim -L/\xi_{\text{loc}} = \ell n\Sigma$. To summarize, the scaling function goes to a constant $(d-2)$ asymptotically in the high conductance $(\Sigma \to \infty)$ regime and decreases logarithmically in the low conductance $(\Sigma \to 0)$ regime. By drawing a smooth curve between these two asymptotes, we arrive at the picture shown in Figure 19. The arrows on each curve depict how the conductance scales as the sample size L is made larger and larger. In both one and two dimensions, the function $\beta(\Sigma)$ is everywhere negative. Scaling is toward zero conductance on long length scales. This may be interpreted to mean that all states are exponentially localized on a sufficiently large length scale for $d = 1$ and 2. For $d = 3$, the function $\beta(\Sigma)$ passes through zero at a critical conductance Σ_c. This point describes an electronic mobility edge. Depending on the position of the Fermi level, the bare conductance Σ_0 (defined as the conductance on length scales only slightly larger than the transport mean free path ℓ^*) will be above or below the critical value Σ_c. If $\Sigma_0 < \Sigma_c$, scaling is toward zero

614

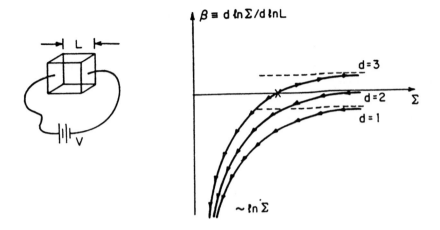

Figure 19. Scaling behaviour of the conductance Σ with length scale L of the sample. In $d = 1$ and 2, $\beta < 0$ so that scaling is toward the low conductance regime and all states are localized. In $d = 3$, a mobility edge separates regions in which scaling is toward large values of Σ from regions where scaling is toward smaller Σ.

conductance and $\Sigma(L) \to 0$ as $L \to \infty$. States corresponding to this energy are localized. If $\Sigma_0 > \Sigma_c$, then $\Sigma(L) \to \sigma L$ as $L \to \infty$ and states corresponding to this energy are extended. This physical picture constitutes the "scaling theory of localization" introduced by Abrahams, Anderson, Licciardello and Ramakrishnan [44] based on the ideas of David Thouless [51, 52].

In the case of photons propagating in a three-dimensional disordered dielectric medium, we may define the analog of a "bare conductance". This is given by $\Sigma_0 \equiv \rho(\omega) D_0 \ell^*$. For scattering from uncorrelated point scatterers (white noise) the photon density of states $\rho(\omega)$ is proportional to $\frac{1}{c}(\frac{\omega}{c})^2$ and the bare diffusion coefficient D_0 is proportional to $c\ell^*$. The word "bare", here, refers to a sampling volume which is on the order of $(\ell^*)^3$. It is clear from the definition that the "bare" conductance for photons is a dimensionless quantity: $\Sigma_0 \sim (\frac{\omega}{c}\ell^*)^2$. The Ioffe-Regel criterion is simply the statement that localization takes place when this bare conductance reaches a critical value $\Sigma_c \simeq 1$, by a suitable choice of frequency and scattering parameters. The condition $\Sigma_c = 1$ reduces to the Ioffe-Regel criterion whenever the *phase space* available for propagation is that of free photons. However, when the scattering microstructures cause a change in the density of states, significant modifications take place in the Ioffe-Regel condition. To see this, it is useful to interpret the factor $4\pi(\frac{\omega}{c})^2$ as representing the *phase space* available for propagation of a photon of frequency ω. The condition $\Sigma_0 \simeq 1$ then reduces to the statement

$$(\text{phase space}) \times (\ell^*)^2 \simeq 4\pi \ . \tag{4.1}$$

In the event that the phase space is reduced by some form of resonant scattering, localization may occur even if ℓ^* is considerably longer than the vaccum photon wavelength λ. This has been graphically illustrated in section 2 for the photonic bandgap. A detailed derivation of the scaling function $\beta(\Sigma)$ depicted in Figure 19 follows from the application of perturbation theory to evaluate corrections to diffusion due to wave interference [45]. In the discussion of coherent backscattering the width of the coherent backscattering peak was proportional to $\frac{1}{(k\ell^*)}$ where $k \equiv \frac{\omega}{c}$. This dimensionless parameter, gives the magnitude of corrections to "Ohmic behaviour". Since $\Sigma \sim (k\ell^*)^2$ in the highly conducting (diffusive) regime, we may likewise regard perturbative corrections to diffusion to be proportional to $1/\Sigma$. A detailed calculation in fact verifies that this is the case [45]. Perturbation theory for the interference of diffusion modes in fact takes the form:

$$\frac{d\ln\Sigma}{d\ln L} \equiv \beta(\Sigma) = (d-2) - \frac{\text{const}}{\Sigma} - \ldots \tag{4.2}$$

For $d = 3$, the constant in this equation is $\Sigma_c \simeq 1$. The nature of diffusion near the photon mobility edge may be obtained by retaining only the leading correction in the above perturbation series. The solution of the differential equation (4.2) may be formally expressed as:

$$\int_{\Sigma_0}^{\Sigma} \frac{d\Sigma}{\Sigma - \Sigma_c} = \int_{\ell^*}^{L} d(\ln L) \qquad (d = 3) \ .$$

The underlying physics of this equation is apparent from the limits of integration. The integration itself, describes wave interference corrections to diffusion which take place on length scales from ℓ^* up to the sampling volume L. On the length scale ℓ^* the conductance is given by its "bare" value $\Sigma_0 \equiv \rho D_0 \ell^*$ with $D_0 \equiv \frac{1}{3}c\ell^*$. On the length scale L, the conductance is given by $\Sigma \equiv \rho D L$ with $D \equiv D(L)$ representing the renormalized, scale-dependent, diffusion coefficient.

To further understand the consequences of the scaling theory of localization, it is useful to focus attention on the free-photon Ioffe-Regel condition. In this case, we assume little or no correlation between scatterers and critical conductance is given by $\Sigma_c \equiv \rho D_c k^{-1}$ where $k \equiv \omega/c$ and $D_c \equiv \frac{1}{3}ck^{-1}$. Explicit evaluation of the preceding integral gives the scale dependent diffusion coefficient:

$$\frac{\Sigma - \Sigma_c}{\Sigma_0 - \Sigma_c} = L/\ell^* \ .$$

Using the above definitions of Σ_0 and Σ_c gives the scale dependent diffusion coefficient:

$$D(L) = D_0 \left[\frac{\ell^*}{\xi_{\text{coh}}} + \frac{1}{(k\ell^*)^2} \frac{\ell^*}{L} \right] \tag{4.3}$$

where the coherence length ξ_{coh} is defined by

$$\frac{\ell^*}{\xi_{\text{coh}}} \equiv 1 - \frac{1}{(k\ell^*)^2} .$$

The behaviour of $D(L)$ near a photon mobility edge is most striking. In this case, $k\ell^* \simeq 1$. As $\omega \to \omega_c^*$, (the mobility edge frequency) from the extended state regime, $k\ell^* \to 1$. It follows that the coherence length is divergent:

$$\frac{\xi_{\text{coh}}}{\ell^*} \propto \frac{\omega_c^*}{|\omega - \omega_c^*|} .$$

For an infinite sample ($L = \infty$) this suggests that $D/D_0 \propto |\omega - \omega_c^*|/\omega_c^*$. A measurement of the diffusion coefficient on length scales $L \gg \ell^*$ will exhibit this linear vanishing with frequency. For frequencies slightly removed from the critical frequency ω_c^*, the coherence length is finite and the character of the "renormalized random walk" taken by individual photons is manifest. If the sampling length L is less than ξ_{coh}, then the second term in equation (4.3) is dominant. It follows that the photon transport on this length scale is strongly scale dependent $D/D_0 \approx \ell^*/L$ and is in fact subdiffusive. The average time of flight of such a photon across a *slab* of thickness L is given by the standard diffusion formula $\tau = L^2/D$. Unlike the familiar classical diffusion, in which the average displacement, R, of the photon scales as the square root of time, scale dependent diffusion yields $R \sim t^{1/3}$. This critical slowing down of optical transport is one of the hallmarks of incipient localization. If on the other hand, the sampling length $L \gg \xi_{\text{coh}}$, it is the first term in equation (4.3) which dominates and the photon is said to resume its diffusive motion on average except with smaller, renormalized value of the diffusion coefficient $D \simeq D_0(\ell^*/\xi_{\text{coh}})$. The coherence length ξ_{coh} describes both the extent and the magnitude of this renormalization. This is depicted in Figure 20a. Some experimental evidence for this behaviour has been reported by A.Z. Genack and N.Garcia [53, 54]. However, this has been criticized by B.A. van Tiggelen et. al [55] who have pointed out that renormalization of the diffusion coefficient may arise purely from *microscopic* resonances. These lead to the long "dwell times" in individual scatterers. This results in an anomalously small energy velocity v_E which enters the *bare* diffusion coefficient $D_0 = \frac{1}{3}v_E\ell^*$. Genack and collaborators have also reported *scale-dependent* diffusion as predicted in equation (4.3). This latter observation provides more direct evidence for mobility edge behaviour.

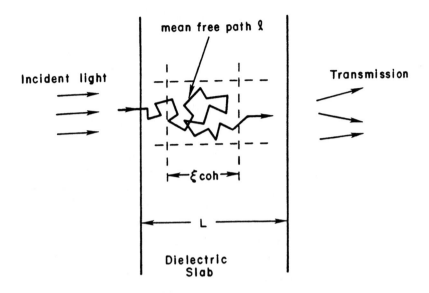

Figure 20. a. A simple physical picture describes optical transport near a photon mobility edge. On scales short compared to the coherence length ξ_{coh}, the spread of electromagnetic energy is subdiffusive in nature due to coherent backscattering. On scales longer than ξ_{coh}, the photon resumes its diffusive behaviour on average except with a renormalized diffusion coefficient $D \simeq c\ell^*/3(\ell^*/\xi_{coh})$.

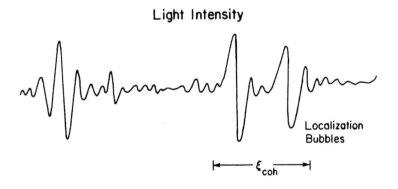

Figure 20. b. The electromagnetic field amplitude exhibits localization bubbles on length scales ranging from ℓ^* up to ξ_{coh}.

From the standpoint of classical optics, an alternative picture of mobility edge behaviour is quite useful. Incipient localization is somewhat reminiscent of what takes place in a cavity resonator. An optical cavity is essentially an imperfect localizer of light. In the absence of absorption, the Q-factor of the cavity is inversely proportional to the degree of imperfection. One may regard light within the cavity as being in a localization bubble rather than a truly localized state. There is high electric field intensity within the cavity but a small amount of leakage out of the cavity. Near a mobility edge, the coherence length describes the length scale of *effective* cavity resonators created randomly within the disordered medium. That is to say, there is a random distribution of cavity resonances on length scales ranging from ℓ^* up to ξ_{coh}. Within these "localization bubbles" the light intensity is high and the transport of photons is subdiffusive. Nevertheless there may be leakage into another nearby localization bubble. Unlike the microcavity resonators which have a scale of the optical wavelength λ, the effective cavity resonators in the disordered medium (near criticality) have a scale $\xi_{coh} \gg \ell^* \approx \lambda/2\pi$. On length scales longer than ξ_{coh} the photon resumes its diffusive motion on average through the process of leakage from one bubble to the next. This is depicted in Figure 20b. At the critical frequency ω_c^*, $\xi_{coh} \to \infty$ and localization bubbles on *all* length scales are present. If the frequency ω passes through ω_c^* and enters the localization regime, the *localization bubbles* become *localized states*. In the absence of inelastic effects and absorption, a localized state acts as a perfect cavity mode. Unlike traditional microcavity resonators, propagative effects may take place within the localization volume. In the localized frequency regime, the coherence length ξ_{coh} is now replaced by the localization length ξ_{loc}. This also diverges as the frequency approaches ω_c^* from the localized side: $\xi_{loc}/\ell^* \propto \omega_c^*/|\omega - \omega_c^*|$. The localized frequency regime has truly unique electromagnetic characteristics especially in regard to the radiative properties of atoms and molecules. In order to substantiate these characteristics, it is necessary to first discuss how localization effects appear in real dielectric materials with inelastic scattering effects and absorption of photons. Remarkably, many of the essential features which we have described, survive this added complication.

4.3. QUALITY FACTOR OF A LOCALIZED STATES IN A PHOTONIC PSEUDO-GAP

Localization of light is a wave interference effect. Since inelastic scattering of light and absorption disrupt this interference effect, it is necessary to examine the interplay between these incoherent effects and localization. As it turns out, there are two distinct regimes of importance. The first is that of a highly disordered medium in which localization is in its incipient stage. That is to say, the localization length ξ_{loc} (or alternatively ξ_{coh}) is very large compared to the inelastic scattering length ℓ_{inel}. In this case, the interference effects leading to complete localization are disrupted on the scale of ℓ_{inel}. The macroscopic diffusion coefficient exhibits scale dependence up to the length ℓ_{inel}, after which it retains a constant residual value given roughly by $D = D_0(\ell^*/\ell_{inel})$. The second regime is that of *strong* localization in which the localization length $\xi_{loc} \ll \ell_{inel}$. It turns out, in this case, that localized states remain relatively immune to inelastic effects and that Q factor remains high. As a consequence, a strong localization pseudogap may be as effective as a complete photonic bandgap for the design of optical devices such as low-threshold microlasers. This is the regime of Mott-variable-range-hopping of photons (in analogy to the hopping conduction of localized electrons [56] in a semiconductor).

In the mobility edge regime, the effect of weak dissipation is to round-off the singularities in transport behaviour associated with the critical point ω_c^*. By weak dissipation I mean that the inelastic mean free path or typical distance between absorption events is large compared to ℓ^*, but nevertheless smaller than the sample size L. This may be introduced by means of a small imaginary part ϵ_2 to the dielectric constant $\epsilon(x) = \epsilon_0 + \epsilon_{fluct}(x) + i\epsilon_2$. The optical absorption coefficient α is defined as the decay constant for the intensity from a source of intensity $I_0 : I = I_0 e^{-\alpha x}$. This absorption coefficient describes the average absorption on scales long compared to the transport mean free path ℓ^*. It should not be confused with the scattering length of the incident beam which may in fact be much shorter than ℓ^*. Within classical diffusion theory, the transport of radiation of frequency ω is given by the steady state equation:

$$D_0 \nabla^2 I - I/\tau_{inel} = \frac{\partial I}{\partial t} = 0 \ . \tag{4.4}$$

Here, the inelastic mean free time $\tau_{inel} = 1/(\epsilon_2 \omega)$. The solution of this equation for a uniform incident intensity I_0 yields $\alpha = (D_0 \tau_{inel})^{-1/2} = (\omega \epsilon_2/D_0)^{1/2}$. This simple picture neglects any interaction of diffusion modes giving rise to localization. The effects of coherent wave interference may be incorporated into this picture by means of the scaling theory of localization. For an infinite medium ($L = \infty$) the diffusion coefficient vanishes as the coherence length diverges. If $\ell_{inel} > \xi_{coh}$, it follows that the absorption coefficient increases in the same manner that the diffusion coefficient $D(\omega) \sim |\omega - \omega_*|$ decreases as the mobility edge frequency ω_* is approached from the extended state side: $\alpha(\omega) \sim \sqrt{\epsilon_2/|\omega - \omega_*|}$. On the other hand, if the coherence length exceeds the inelastic length, then ℓ_{inel} acts as a long distance cuttoff for coherent wave interference. In this case, there is a residual diffusivity given by $D(\omega^*) \simeq \frac{c\ell}{3}(\frac{\ell}{\ell_{inel}})$. Since $\ell_{inel} = \sqrt{D\tau_{inel}}$ and

620

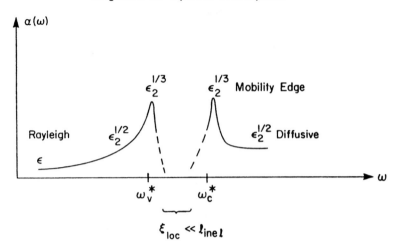

Figure 21. In the ballistic propagation regime ($\omega \to 0$), the optical absorption coefficient scales linearly with the imaginary part of the dielectric constant ϵ_2. In the multiple scattering, diffusive regime $\alpha \propto \sqrt{\epsilon_2}$. At the mobility edges, ω_v^* and ω_c^*, the absorption coefficient exhibits critical behaviour $\alpha \propto \epsilon_2^{1/3}$. Finally, in the localization pseudogap region, $\xi_{\text{loc}} \ll \ell_{\text{inel}}$, optical transport occurs by Mott variable-range-hopping. These localized states may act as high Q optical cavities.

$\tau_{\text{inel}} \sim 1/(\epsilon_2 \omega)$, it follows that the residual diffusivity $D(\omega^*) \sim \epsilon_2^{\frac{1}{3}}$. Substituting the value of the residual diffusivity into the expression for α reveals that the absorption coefficient exhibits an anomalous scaling behaviour with ϵ_2: $\alpha(\omega^*) \sim \epsilon_2^{\frac{1}{3}}$. The physical origin of the critical exponent $\frac{1}{3}$ is the critical slowing down of the photon as it approaches localization. This leads to a greater probability of absorption. The behaviour of the absorption coefficient as a function of frequency for a dielectric material with a localization pseudogap is depicted in Figure 21. The mobility edge frequencies are denoted by ω_v^* and ω_c^*.

The second regime of importance is frequency regime $\omega_c^* < \omega < \omega_c^*$ in which electromagnetic modes are strongly localized. Here, photons no longer propagate elastically and the absorption coefficient described above is no longer meaningful. In this regime, the localization length ξ_{loc} may become considerably shorter than ℓ_{inel}. In this case, the structure of individual localized states remain intact but they have a finite Q-factor. An additional complication arises, however, in the presence of inelastic scattering. When there is a finite density of localized states, scattering by *phonons* or other degrees of freedom in the dielectric host may result in localized photons hopping from one localized state to another. While this effect may appear at first sight to be deleterious to the design of optical devices such as a single-mode light emitting diode, we suggest (following the arguments of Nevill Mott) that the Q-factor of individual localized states can, under suitable circumstances remain nearly as high as for the single mode case.

The high Q-factor of localized states of light in a disordered dielectric is a consequence of a fundamental mathematical property of the singular continuous spectrum. The essential property is that if two localized states are very close together in frequency, then they are (statistically speaking) very far apart spatially and have exponentially vanishing overlap. If two states of nearly identical frequency were spatially overlapping, it would be possible to form a new, lower energy state by constructing a linear superposition of the two. Consider a frequency regime in which $\xi_{loc} \ll \ell_{inel}$. Here, ℓ_{inel} includes absorption and aother inelastic processes such as Brillouin and Raman scattering. The lifetime of a photon in a localized state of frequency ω is then limited by the possibility that inelastic scattering may take place into a nearby state of frequency $\omega + \Delta\omega$. The *decay rate* of the photon is given by a product of three factors. The first is frequency with which the photon is likely to encounter an inelastic scatterer. This is typically a phonon frequency $v_{phonon} \sim 10^{12}/\text{sec}$. The second is the overlap of the initial localized state with the final localized state into which scattering takes place. If the states are separated by a distance R, the overlap factor is given by $\exp(-2R/\xi_{loc})$. Finally, the photon must acquire an energy $\hbar\Delta\omega$ to reach its final state. If we neglect *spontaneous* Raman and Brillouin scattering (for which the cross sections are exceedingly small) and assume that scattering is stimulated by thermal effects, this leads to a Boltzmann factor $\exp(-\hbar\Delta\omega/k_BT)$. Combining these factors yields for the inverse lifetime:

$$\frac{1}{\tau_{\text{lifetime}}} = \nu_{phonon} \exp(-2R/\xi_{loc} - \hbar\Delta\omega/k_BT) \ . \tag{4.5}$$

As mentioned earlier, the spatial separation R and the frequency difference $\Delta\omega$ are statistically correlated variables. In particular, the average number of states within a radius R and frequency $\Delta\omega$ of the initial state is given by $\frac{4}{3}\pi R^3 \rho(\omega)\Delta\omega$. For hopping to take place, we require that there be at least *one* such state. Using the constraint that $\Delta\omega = [\frac{4}{3}\pi R^3 \rho(\omega)]^{-1}$, the decay rate may be expressed as $\frac{1}{\tau_{\text{lifetime}}} = \nu_{phonon}e^{-F(R)}$, where

$$F(R) = 2R/\xi_{loc} + \frac{3\hbar}{4\pi R^3 \rho(\omega)(k_BT)} \ .$$

The decay of the localized photon is then dominated by transfer to a state for which $F(R)$ is minimized. Clearly $R_{\min}^4 = 9\hbar\xi_{loc}/(8\pi\rho(\omega)k_BT)$ and

$F(R_{\min}) = 8R_{\min}/3\xi_{\mathrm{loc}}$. The Q-factor of a localized state is given by:

$$Q \simeq (\omega/\nu_{\mathrm{phonon}})e^{F(R_{\min})} \ . \tag{4.6}$$

Using a typical optical frequency of $10^{15}/\mathrm{sec}$ and a density of states which is suppressed from its vacuum value by a factor S $(\rho(\omega_0) \simeq \frac{4\pi}{c}(\frac{\omega_0}{c})^2/S)$ yields

$$Q \simeq 10^3 \exp\left[\left(\frac{\hbar\omega S}{k_B T}\right)^{\frac{1}{4}} \left(\frac{\lambda}{2\pi\xi_{\mathrm{loc}}}\right)^{\frac{3}{4}}\right] \ .$$

At room temperature, $\frac{\hbar\omega}{k_B T} \simeq 10^2$. If the localization length is a few wavelengths, then a density of states suppression $S \sim 10^7$ will yield a Q-factor of 10^{10}. This can be achieved with less suppression by going to lower temperatures.

To summarize this section, we have described in detail the wave interference effects giving rise to localization in a disordered dielectric. This is the phenomenon of coherent backscattering, a direct consequence of the time-reversal symmetry of Maxwell's equations. The critical behaviour associated with a photon mobility edge is essentially the same as for electrons in a random potential. Light, on the other hand, offers the unique possibility of direct experimental measurement of this effect with high degree of temporal, frequency and angular resolution. Unlike electronic systems, there is no complication of electron-electron interactions. Furthermore, light localization is observable at room temperature. The effect of optical absorption and inelastic scattering is to round off the mobility edge critical behaviour. In the case of a strong localization pseudogap, dissipative effects do not destroy the integrity of localized states. For a sufficiently strong suppression of the density of states and/or sufficiently low temperature, the Q-factors of localized states in dissipative systems may be comparable to those in perfect photonic crystals.

4.4. PAINT-ON LASERS

The interplay between optical localization and optical absorption, suggests that an analogous synergy may occur in a strongly scattering *gain* medium. Rather than a positive absorption coefficient, this medium is characterized by a *negative*, saturable absorption coefficient. In some recent experiments initiated by Lawandy and coworkers laser activity has been reported in multiple-light-scattering media containing dye molecules. [62, 63, 64] The specific system considered consists of Rhodamine 640 dye molecules (in methanol) containing a cooloidal suspension of Titanium Oxide (TiO$_2$) microspheres. When this *paint* is optically pumped with light at 532 nm (in the absorption band of R640), an isotropic laser-like emission is observed at 620 nm and 650 nm. The threshold pump intensity for this amplified laser-like emission depends very strongly on the concentration of the TiO_2 scatterers. The threshold is characterized by a dramatic increase in the emission intensity as well as a collapse of the emission line-width. In particular, the threshold decreases considerably as the transport mean free path ℓ^* in the medium is lowered through the addition of TiO_2 particles. This suggests that multiple-light-scattering plays a crucial role in the amplification process and the medium has been given the name, "Laser-Paint". This remarkable observation also provides a compelling starting point for the investigation of disordered dielectric microstructures as novel sources of coherent light emission. Weak scattering of light has, traditionally, been considered detrimental to laser action since such scattering removes photons from the lasing mode of a conventional cavity. On the other hand, if stronger, multiple, scattering occurs, these photons may return to the amplification region and the amplified mode itself may consist of a multiple scattering path. Under suitable circumstances, the detrimental effects of diffuse scattering and other losses may be offset by the long path length of photons within the gain region giving rise to amplified laser-like emission. While the coherence properties of light emitted by "Laser Paint" have not yet been established, it is plausible that a continuous crossover takes place in these emission properties as the photon transport mean free path, ℓ^*, is decreased. This is the crossover from amplified apontaneous emission well known in pure dye systems with no scatterers, to true laser-like coherent emission as ℓ^* approaches the emission wavelength, λ_0, with a high density of scatterers. At the threshold for photon localization, the diffusion modes of light interfere strongly and are converted into a continuous spectrum of localized cavity modes in the extended medium, in which the emitted light may self-organize into a coherent state. This scenario suggests that some striking effects may be observed if the current experiments on "Laser Paints" are repeated in more strongly scattering dielectrics. A theoretical model describing current experiments (in the classical diffusion regime of multiple-light scattering) has recently been introduced by S. John and G. Pang [65].

The intriguing properties of "Paint-On Lasers" are only a fore-taste of the collective response of atoms placed in a PBG material. In the case of Laser-Paint, the dye molecules interact with a density of electromagnetic modes which is only weakly perturbed from that of ordinary vacuum. In addition, the dye molecules are subject to frequent collisions with solvent molecules leading to rapid dephasing of any optically induced molecular dipole moment. Under these conditions, the system is suitably described by a set of laser rate equations. If, on the other hand, the vacuum density of states is strongly perturbed and dipole-dephasing effects are slow, long-time memory effects occur and a complete quantum mechanical description of the photon-atom interaction is required. The laser rate equations must be replaced by the more fundamental Heisenberg equations of motion for the relevant quantum mechanical operators. This is the subject of the next two sections of this article.

5. Quantum Electrodynamics in a Photonic Bandgap

When an atom or molecule is placed within a dielectric material exhibiting photon localization, the usual laws governing absorption and emission of light from the atom must be re-examined. This is most easily seen in the strong localization limit obtained for a dielectric exhibiting a complete photonic band gap. For a single excited atom with a transition energy $\hbar\omega_0$ to the ground state, which lies within the band gap, there is no true spontaneous emission of light. A photon which is emitted by the atom finds itself within the classically forbidden energy gap of the dielectric. If the nearest bandedge occurs at frequency ω_c, this photon will tunnel a distance $\xi_{\text{loc}} \simeq c/|\omega_0 - \omega_c|$ before being Bragg reflected back to the emitting atom. The result is a coupled eigenstate of the electronic degrees of freedom of the atom and the electromagnetic modes of the dielectric. This photon-atom bound state [57, 58] is the optical analog of an electron-impurity level bound state in the gap of a semiconductor. The atomic polarizability, which is normally limited by the vacuum natural linewidth of the transition, can in the absence of spontaneous emission grow sufficiently large near resonance to produce a localized electromagnetic mode from the nearby propagating band states of the dielectric. The fundamental weakness of the vacuum photon-atom interaction, as expressed by the fine structure constant $\alpha \equiv \frac{1}{137}$, is completely offset by this nearly unrestricted resonance. The alteration of the quantum electrodynamic vacuum by the dielectric host also appears in the spectroscopy of atomic levels. The ordinary Lamb shift of atomic levels is dominated by the emission and reabsorption of high energy virtual photons. Within a photonic band gap, this self dressing is instead dominated by the real, bound photon. In general, this will lead to some anomalous Lamb shift. If this level lies near a photonic band edge, a more striking effect is predicted to occur. In this case the atom is resonantly coupled to photons of vanishing group velocity. The resultant self-dressing of the atom is sufficiently strong to split the atomic level into a doublet. The atomic level is essentially repelled by its electromagnetic coupling to the photonic band edge. One member of the doublet is pulled into the gap and retains a photon bound state, whereas the other member is pushed into the continuum and exhibits resonance fluorescence. In the nearly free photon approximation to electromagnetic band structure, the splitting of a hydrogenic $2p\frac{1}{2}$ level is predicted to be as large as $10^{-7} - 10^{-6}\hbar\omega_0$. This is the analog of the much weaker vacuum Rabi splitting well known for atoms in micro-cavities[59]. In this section, we derive from first principles, the occurrence of these effects using a simple model Hamiltonian for the photonic bandgap.

Additional new phenomena are expected when a collection of impurity atoms is placed into the dielectric. A single excited atom can transfer its bound photon to a neighboring atom by a resonance dipole-dipole interaction (RDDI). For a band gap to center frequency ratio $\Delta\omega/\omega_0 = .05$, the photon tunnelling distance ξ_{loc} is on the scale of $10L$ and the lattice constant L of the dielectric is itself on the scale of the photon wavelength. For impurity atoms spaced by a distance $R = 10\text{Å}-1000\text{Å}$, the dipole-dipole interaction is essentially the same as that in vacuum. The matrix element M describing the hopping of a bound photon from one atom to another is given roughly by $M \sim \mu^2/R^3$ where the atomic dipole $\mu \simeq ea_0$ is given by the product of the electronic charge and the atomic Bohr radius a_0. This can be approximately related to the transition energy $\hbar\omega_0 \sim e^2/a_0$ by writing M as $(e^2/a_0)(a_0/R)^3$. For the case of a finite density of impurity atoms separated by $R\gtrsim 10\text{Å}$, it follows that photonic hopping conduction will occur through a narrow photonic impurity band of width $\sim (\hbar\omega_0)(a_0/R)^3$ within the larger band gap.

The occurrence of photonic impurity band leads to new effects nonlinear optics and laser physics. The strong coupling of light to matter suggests enhanced nonlinear effects which are highly sensitive to the impurity atom spacing. For example when neighbouring impurity atoms A and B are both excited, second harmonic generation may occur by the transfer of the bound state from atom A to atom B. Since atom B is already excited the transfered photon creates a virtual state which may then be emitted as a single photon of energy $2\hbar\omega_0$ outside of the photon band gap. The transfer can take place by dipole (μ) emission from atom A followed by a quadrupole (Q) virtual absorption by atom B. This process has an amplitude $\mu Q/R^4$. The resulting virtual excitation on atom B has odd parity and may then decay by a dipole emission process. The rate of spontaneous second harmonic generation is given by the square of the corresponding amplitude and depends sensitively $\sim (a_0/R)^8$ on the impurity atom spacing. Another significant question is that of laser activity within the impurity band when many photons are present. The impurity band defines a novel quantum many-body system in which the processes of spontaneous and stimulated emission of light are completely confined to and mediated by photonic hopping conduction between atoms.

For N atoms with a radiative transition *near* a *photonic band edge*, Tran Quang and I have shown that the rate of collective spontaneous emission depends sensitively on the nature of the band edge singularity [60, 61]. In particular, the decay *rate* scales as N^ϕ where $\phi = 2/3, 1$ and 2 for one, two and three dimensional band edge singularities respectively. For a three dimensional photonic band edge, this suggests a very rapid rate of superradiance and an anomalously high peak intensity (which scales as N^3 rather than N^2 for vacuum). We have also shown that when the transition frequency lies *on* the band edge frequency ω_c, superradiant emission gives rise to spontaneous symmetry breaking in the long time steady state limit. Unlike vacuum superradiance, in which all of the radiated photons propagate away, in a photonic crystal a finite fraction remain *localized*. This occurs even in the absence of a dielectric defect mode and is a direct consequence of the photon-atom bound state. Consequently, the atomic system acquires a macroscopic polarization in the long time limit. This is analogous to lasing without a cavity mode!

5.1. THEORY OF THE PHOTON-ATOM BOUND STATE

Localization of light leads fundamentally to a novel strong coupling between radiation and matter. To describe this we utilize a quantum description of the radiation field Hamiltonian :

$$H_{rad} = \sum_\lambda \hbar\omega_\lambda a_\lambda^+ a_\lambda \qquad (5.1)$$

where λ is a composite index describing both polarization and wavevector. The vector potential has an expansion in terms of the photon creation a_λ^+ and annihilation operators a_λ of the form

$$\vec{A}(\vec{r}) = \sum_\lambda \sqrt{\frac{\hbar}{2\epsilon_o \omega_\lambda}} [a_\lambda \vec{u}_\lambda(\vec{r}) + a_\lambda^+ \vec{u}_\lambda^*(\vec{r})] \qquad (5.2)$$

where $\vec{u}_\lambda(\vec{r})$ is mode function and the Coulomb constant $\epsilon_o = 8.85 \times 10^{-12}$ $Coul^2 - m^2/N$. The strong coupling effects of interest are most transparent in the case of a periodic dielectric structure with a complete photonic bandgap. A simple model for the photon dispersion relation ω_λ follows from considering a periodic one dimensional system of square wells of refractive index n, width a and lattice constant L. Writing the total dielectric constant as $\epsilon(x) = \epsilon_{avg} + \epsilon_{fluct}(x)$, the classical scalar wave equation for this system is

$$[-\nabla^2 - \frac{\omega^2}{c^2}\epsilon_{fluct}(x)]\phi(x) = \frac{\omega^2}{c^2}\epsilon_{avg}\phi(x) \qquad (5.3a)$$

where $\epsilon_{avg} = 1$,

$$\epsilon_{fluct}(x) = \sum_{m=-\infty}^{\infty} V(x - mL) \qquad (5.3b)$$

and

$$V(x) = \begin{cases} n^2 - 1 & |x| < a \\ 0 & \text{otherwise} \end{cases}$$

The potential wells in this model correspond in the actual material to high dielectric spheres or cylinders in vacuum or alternatively pore regions in a high dielectric background. The detailed derivation of the dispersion ω_k may be found elsewhere [58]. However the salient features of the band structure may be deduced by a simple physical argument. In nearly all observed and calculated band structure, there is a well defined optimum volume filling fraction of high dielectric material for the creation of a complete photonic bandgap. As discussed in section 2, this arises from a synergetic interplay of the microscopic resonance of a single scatterer with the macroscopic Bragg scattering resonance of the entire lattice. In the model (equation 5.3), a microscopic scattering resonance occurs when a quarter wavelength fits into the well diameter : $\frac{2\pi c}{\omega(4n)} = 2a$. At this frequency there is a maximum in the reflection coefficient from a single well. On the other hand, a Bragg resonance occurs when $\frac{\omega}{c} = \pi/L$. If we require that these two resonances coincide in frequency, this leads to the condition on the one dimensional volume filling fraction $f \equiv 2a/L = 1/(2n)$. In a real three dimensional system, the size of the gap diminishes rapidly as one deviates from this optimum volume. One notable feature of this estimate which is borne out experimentally is that for high refractive index materials $n \gtrsim 3.0$, the solid fraction of the microstructure is very small compared to the empty pore space. This is of considerable utility for atomic spectroscopy experiments in which the dielectric microstructure acts as a high quality cavity which radically alters the atom's radiative properties. The definition of the three dimensional model Hamiltonian is completed by symmetrizing the dispersion relation ω_k over all directions in $\vec{k}$ space and adopting a mode expansion (5.2) consisting of transverse plane waves outside the gap region:

$$\vec{u}_{\vec{k}\sigma}(\vec{r}) = \frac{1}{\sqrt{\Omega}} e^{i\vec{k}\cdot\vec{r}} \hat{e}_{\vec{k},\sigma} \; ; \quad \hat{e}_{\vec{k}\sigma} \cdot \vec{k} = 0 \; \sigma = 1,2$$

Here Ω is the sample volume and $\hat{e}_{\vec{k}\sigma}$ are the two transverse unit vectors. The resulting density of states is depicted in Fig. 22.

The total Hamiltonian for an impurity atom interacting with the electromagnetic modes of the dielectric may be written as $H = H_{atom} + H_{int} + H_{rad} + H_{ct}$, where

$$H_{atom} = \frac{\vec{p}^2}{2m} + v(\vec{r}) \; ; \quad v(r) = \frac{e^2}{4\pi\epsilon_o r} \tag{5.4a}$$

$$H_{int} = \frac{e}{m}\vec{p}\cdot\vec{A}(\vec{r}) + \frac{e^2}{2m}A^2 \tag{5.4b}$$

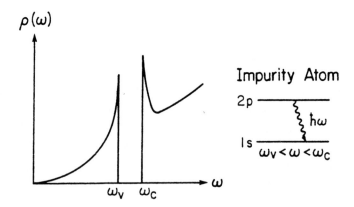

$\rho(\omega)$

Impurity Atom

2p ——————

$\hbar\omega$

1s ——————

$\omega_v < \omega < \omega_c$

ω_v ω_c ω

Figure 22. Photon density of states for the isotropic model of the dispersion relation ω_k. Band edges occur at ω_v and ω_c. An impurity atom with transition frequency ω in the gap region will exhibit a photon-atom bound state.

and

$$H_{ct} = \frac{\delta m}{m} \frac{p^2}{2m} \; . \qquad (5.4c)$$

Here $\vec{p}$ is the momentum of an electron of charge e and observable mass m in a hydrogenic atom. As in standard discussions of quantum electrodynamics [67], the renormalized, observable mass is given by $m = m_o + \delta m$. Here m_0 is the bare electron mass and δm is a correction due to radiative coupling which we will consider to leading order in the fine structure constant $\alpha \equiv \frac{e^2}{4\pi\epsilon_o\hbar c} = 1/137$. The mass renormalization counterterm H_{ct} is accordingly introduced to accommodate the use of m rather than m_0 in H_{atom}. When the atomic $2p \to 1s$ transition lies in the photonic bandgap as depicted in Figure 22, ordinary spontaneous emission of light from the atom is absent. However, the photon can tunnel into the classically forbidden gap. This suggests the occurrence of a photon-atom bound state in analogy to a deep level localized electronic state in the gap of a semiconductor.

A mathematical description of this effect is obtained by introducing a variational trial wavefunction which spans the single photon sector of atom plus electromagnetic Hilbert space:

$$|\psi\rangle = \sum_{n=1}^{\infty} \phi_n |n\rangle + \sum_{n=0}^{\infty} \sum_{\lambda} \psi_{\lambda}^{(n)} |\lambda; n\rangle \qquad (5.5)$$

Here, the state vector $|n\rangle$ describes the atom in its nth excited state with energy E_n and no photons present, whereas the state vector $|\lambda; n\rangle$ describes an atom excited to it's n^{th} level and a single photon in mode λ. The variational amplitudes ϕ_n and $\psi_\lambda^{(n)}$ are chosen to minimize the quantity $\frac{\langle\psi|H|\psi\rangle}{\langle\psi|\psi\rangle}$. In evaluating the matrix elements, the term $\frac{e^2}{2m}A^2$ merely causes a change in the zero of energy and may be neglected. The nontrivial matrix elements are :

$$\langle\lambda; n|H_{int}|\psi\rangle = \frac{e}{m}\sum_{n'}\sqrt{\frac{\hbar}{2\epsilon_o\omega_k}}\vec{u}_\lambda^*(\vec{r})\cdot\vec{p}_{nn'}\phi_{n'}$$

where

$$\vec{p}_{n,n'} \equiv \langle n|\vec{p}|n'\rangle$$

and

$$\langle n|H_{int}|\phi\rangle = \frac{e}{m}\sum_{n'}\sqrt{\frac{\hbar}{2\epsilon_o\omega_k}}\vec{u}_\lambda(\vec{r})\cdot\vec{p}_{nn'}\psi_\lambda^{n'}$$

Also for the mass renormalization counterterm

$$\langle\lambda; n|H_{ct}|\psi\rangle = \frac{\delta m}{m}\sum_{n'}k_{nn'}\psi_\lambda^{n'}$$

and

$$\langle n|H_{ct}|\psi = \frac{\delta m}{m}\sum_{n'}k_{nn'}\phi_n$$

where

$$k_{nn'} \equiv \langle n|p^2/2m|n'\rangle \ .$$

The variational procedure leads to projected Schrodinger equation $H|\psi\rangle = E|\psi\rangle$ which in turn leads to a set of coupled linear equations for the amplitudes ϕ_n and $\psi_\lambda^{(n)}$:

$$E_n\phi_n + \frac{e}{m}\sum_{\lambda,n'}\sqrt{\frac{\hbar}{2\epsilon_o\omega_k}}\vec{u}_\lambda(\vec{r})\cdot\vec{p}_{nn'}\psi_\lambda^{(n')} + \frac{\delta m}{m}\sum_{n'}k_{nn'}\phi_{n'} = E\phi_n \quad (5.6a)$$

$$(E_n + \hbar\omega_k)\psi_\lambda^{(n)} + \frac{e}{m}\sum_{n'}\sqrt{\frac{\hbar}{2\epsilon_o\omega_k}}\vec{u}_\lambda^*\cdot\vec{p}_{nn'}\phi_{n'} + \frac{\delta m}{m}\sum_n k_{nn'}\psi_\lambda^{(n)} = E\Psi_\lambda^{(n)}$$

$$(5.6b)$$

This procedure is completely equivalent to the Brillouin-Wigner perturbation theory commonly used in solid state physics [73].

Before proceeding further it is instructive to physically interpret the nature of the expected solutions to equations (5.6a) and (5.6b). For an

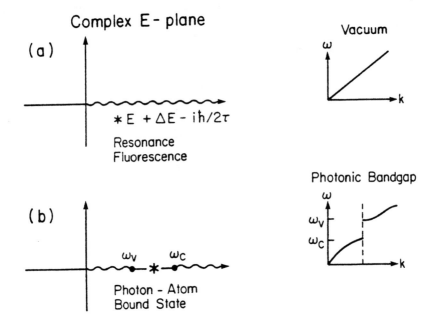

Figure 23. Energy eigenvalue solutions to the projected Schrodinger equation (5.6a) and (5.6b). (a) For photons in vacuum ($\omega_k = ck$), there is a continuum of positive energy scattering states (wiggly line) and a single complex solution on the second Riemann sheet describing resonance scattering. E_1 is the bare atomic level, ΔE is the Lamb shift and $1/\tau$ is the rate of spontaneous emission. (b) In a dielectric with a photonic bandgap the complex solution in (a) migrates to the real axis and describes a photon-atom bound state.

atom in vacuum the dispersion relation is simple $\omega = ck$ where c is the vacuum speed of light. The solutions are depicted in Figure 23. In the complex energy E-plane there is a positive continuum of real energy solutions corresponding to the scattering states of the photon from the atom. This is depicted as a branch cut. In addition there is a single complex solution which appears on the second Riemann sheet of the complex energy plane at $E = E_1 + \Delta E - i\hbar/2\tau$ associated with the phenomenon of resonance fluorescence. Here ΔE describes the Lamb shift of the bare atomic level E_1 and $1/\tau$ is the rate of the spontaneous emission of light from this level. When the same calculation is performed for the atom in a photonic band gap the complex solution migrates to the real energy axis and describes a photon-atom bound state.

The projected Shrodinger equation (5.6) is solved by keeping only the dominant amplitudes $\psi_\lambda^{(0)}$ and ϕ_1. All other amplitudes are perturbatively small in the interaction strength $\alpha \equiv e^2/(4\pi\epsilon_o\hbar c) = 1/137$. Actually, $|\psi_\lambda^{(0)}|^2$ is also of order α, but it can nevertheless become comparable to $|\phi_1|^2$ when the atomic transition frequency ω is near a band edge such as ω_c or ω_v. By substituting, $\psi_\lambda^{(n)}$, in equation (5.6b) into (5.6a) and then neglecting all ϕ_n other than ϕ_1 we arrive at the following eigenvalue equation for the complex energy E:

$$E - E_1 \simeq \frac{\hbar e^2}{2\epsilon_o m^2} \sum_{\lambda,n} \frac{p_{1n}^i p_{n1}^j}{E - E_n - \hbar\omega_\lambda} \frac{1}{\omega_\lambda} u_\lambda^i(r) u_\lambda^{j*}(r) + \frac{\delta m}{m} k_{11} \qquad (5.7)$$

Here we have consistently kept only the leading order contribution in α to mass renormalization.

As in ordinary vacuum quantum electrodynamics, it is apparent that the electromagnetic mode summation over λ in equation (5.7) is linearly divergent in the ultraviolet $\omega_\lambda \to \infty$ limit. This linearly divergent piece may be separated out by means of the identity:

$$\frac{1}{E - E_n - \hbar\omega_\lambda} = -\frac{1}{\hbar\omega_\lambda} + \frac{E - E_n}{\hbar\omega_\lambda(E - E_n - \hbar\omega_\lambda)}$$

The linearly divergent piece then precisely cancels the mass renormalization counterterm (see Chapter 7 of Reference [67]). We regulate the remaining logarithmically divergent term using the cuttoff prescription introduced by Bethe [74]. In this prescription all divergent photon wavevector integrals are cuttoff at the Compton wavevector k_c defined by $\hbar c k_c = mc^2$. Photons of energy greater than the electron rest mass probe the relativistic nature of the electron wavepacket and the Schrodinger description of the electron must be replaced by the Dirac model. Since the wavevector integrals are multiplied by the coupling constant e^2, this cuttoff procedures is equivalent to associating a wavevector dependence to the observable electronic charge : At wavevectors larger than the Compton wavevector the physical vacuum acts as a polarizable dielectric of virtual electron-position pairs which screen out the effective electron charge. The eigenvalue equation after these cancellations becomes

$$z \equiv \frac{E - E_1}{E_1} \simeq \frac{2\alpha}{3\pi} \sum_n \frac{|\vec{p}_{1n}|^2/m}{mc^2} g(E - E_n) \qquad (5.8a)$$

where

$$g(E) \equiv \int_0^{mc^2} \frac{d(\hbar ck)k^2}{(\omega_k/c)^2} \frac{E/E_1}{E - \hbar\omega_k} \qquad (5.8b)$$

Here we have converted the mode sum over transverse plane waves into an integral, made use of the completeness relation

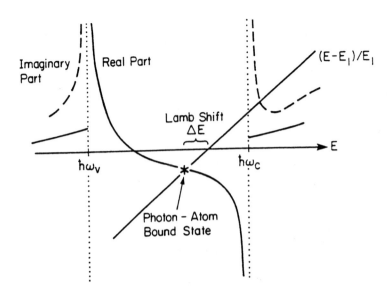

Imaginary Part

Real Part

$(E-E_1)/E_1$

Lamb Shift
ΔE

$\hbar\omega_v$

$\hbar\omega_c$

E

Photon – Atom
Bound State

Figure 24. Graphical solution of eigenvalue equation (5.8a). The energy eigenvalue solution corresponds to a intersection of the straight line $(E - E_1)/E_1$ with the curved solid line representing the real part of the right hand side equation (5.8a). Also depicted (dashed curved line) is the imaginary part of right hand side of equation (5.8a).

$$\sum_{\sigma=1}^{2} e_{k\sigma}^i e_{k\sigma}^j = (\delta_{ij} - \hat{k}_i \hat{k}_j)$$

and performed the angular integral

$$\int d\Omega_k (\delta_{ij} - \hat{k}_i \hat{k}_j) = \frac{8\pi}{3} \delta_{ij} .$$

As discussed earlier, the real and imaginary parts of the eigensolution E describe the Lamb shift and lifetime due to single photon spontaneous emission respectively. The magnitude of these effects is made transparent from the dimensionless form of equation (5.8). For a hydrogen atom the strength of the relative electromagnetic perturbation on the bare atomic level E_1 is given by $z_0 \equiv \frac{2\alpha}{3\pi} \frac{|\vec{p}_{10}|^2/m}{mc^2} \sim 10^{-7} - 10^{-8}$. Anomalies, however, can occur if the wavevector integral in (5.8) has a singularity. This can occur if the atomic level is near a photonic bandedge. As we will see shortly this gives rise to anomalous Lamb shifts and atomic level splittings.

The standard perturbative results for the Lamb shift and radiative life-time of an atom in vacuum ($\omega = ck$) may be reproduced from equation

(5.8) by setting $E = E_1$ everywhere on the right hand side. After some elementary manipulation this gives $\Delta E = \frac{4\alpha^2\hbar^3}{3m^2c}|\psi_1(0)|^2 ln|\frac{mc^2}{E-E_1}|$ where $\psi_1(\vec{r})$ is the electronic wavefunction and for the $2S_{1/2}$ hydrogen level $|\bar{E} - E_{2s}| = 16.64$ Rydbergs. Also $1/\tau = \frac{1}{3\pi\epsilon_0\hbar}(\frac{\omega}{c})^3|\vec{\mu}_{10}|^2$ where $\vec{\mu}_{10} \equiv \langle 1|e\vec{r}|0\rangle$ is the dipole moment matrix element. In the photonic bandgap a graphical solution of equation (5.8) is depicted in Figure 24. Near a photonic bandedge, the dominant contribution to the sum over atomic levels comes from the $n = 0$ term. The singular part of the function $g(E)$ can be isolated by expanding the photon dispersion relation near ω_c: $\hbar\omega_k \simeq \hbar\omega_c + A(k-k_0)^2 + \dots$ In this case, for $E \simeq E_1 \simeq \hbar\omega_c$

$$
\begin{aligned}
g(E) &\simeq \int_0^{mc^2} \frac{d(\hbar ck)}{E - \hbar\omega_c - A(k-k_0)^2} \\
&\simeq -\frac{\hbar\pi c}{\sqrt{A(\hbar\omega_c - E)}}
\end{aligned}
\tag{5.9}
$$

Analytically continuing this expression to the complex E-plane places a branch cut in the multivalued function $g(E)$ for $E > \hbar\omega_c$. A similar analysis near $\hbar\omega_v$ gives a squareroot singularity with opposite sign in the real part of the photon Green's function as $E \to \hbar\omega_v$ from above. A simple dimensional analysis suggests that the coefficient $A \sim \hbar\omega_c/k_0^2$, the constant of proportionality being determined refractive index of the scatterers. Using this fact, and setting $E_1 = \hbar\omega_c$ yields the approximate form $g(E) = -i\pi/\sqrt{z}$, where $z \equiv (E - E_1)/E_1$. The eigenvalue equation then simplifies to

$$
z \simeq -\frac{i\pi z_o}{\sqrt{z}}
\tag{5.10}
$$

On the first Riemann sheet we may write $z = |z_0|^{2/3}e^{i\theta}$ where $0 < \theta < 2\pi$. The second Riemann sheet corresponds to solutions for which $\theta < 0$. Clearly equation (5.10) has two solutions (i) $z = -|z_0|^{2/3}$ which corresponds to a photon-atom bound state in the photonic bandgap and (ii) $z = |z_0|^{2/3}e^{-i\pi/3}$ which is a complex solution on the second Riemann sheet. The existence of two solutions associated with a single atomic level E_1 placed near the bandedge suggests that the resonant coupling of the atom to photons of vanishing group velocity is sufficiently strong to split the atomic level into a doublet. This corresponds an atom strongly dressed by its own radiation field and is analogous to the Mollow [75] splitting of an atomic level dressed by a strong external radiation field. The vacuum Rabi line splitting caused by the dielectric cavity in our model Hamiltonian approach has a magnitude $|z_0|^{2/3} \sim 10^{-5}$ times the transition frequence ω. This anomalously large value is an artifact of the strong bandedge singularity in the photon density of states caused by our isotropic dispersion relation ω_k. In a more realistic anisotropic model [58], this splitting is approximately $10^{-6} - 10^{-7}$ times the transition frequency. This is comparable to the ordinary Lamb splitting of the $2s_{1/2}$ and $2p_{1/2}$ levels of hydrogen.

5.2. LIFETIME OF THE PHOTON-ATOM BOUND STATE

In the previous discussion, the interaction of atoms with single photons was considered. However, the electromagnetic coupling allows the possibility of multi-photon processes. In addition, the impurity atom may interact with other atoms in the dielectric host materials. This gives rise to *phonon* absorption and emission effects. All of these additional interactions determine the lifetime of the photon-atom bound state, which in their absence would be infinite. As it turns out, the dominant factor that limits the lifetime of the photon-atom bound state are non-electromagnetic relaxation events caused either by multiple phonon sidebands and/or collisions of the impurity atoms with surfaces of the solid dielectric. This latter effect may be very deleterious to the *single* impurity atom, and very special experimental conditions may be required to preserve the integrity of the bound state. For example, an impurity atom may be laser-cooled in the void regions of a photonic bandgap material with a network topology. An optical potential well created by interfering laser beams can hold the impurity atom in the void. Another possibility is the use of organic molecules such as pentacene rather than single atoms. These molecules have been shown to exhibit very narrow spectral linewidths when embedded in certain solid hosts [76]. If, on the other hand, there is a finite density of impurity atoms or molecules within the photonic bandgap, we argue that these non-radiative decay mechanisms are less effective in capturing the bound excitation. This is a direct consequence of the spread of the (bound) photon wavefunction due to resonance dipole-dipole interaction and hopping among many impurities. We begin with a discussion of electromagnetic decay channels and then follow with a description of non-radiative relaxation mechanisms.

In the absence of interatomic collisions and non-radiative relaxation, the electromagnetic decay rate of the excited atom due to two-photon spontaneous emission is on the scale of several days [58]. Consider for instance two-photon decay of the excited $2p_{1/2}$ level. A dipole emission from this odd parity state will create a virtual state of even parity which must then decay by a quadrupole transition to the ground state. Unlike the two photon decay of the even parity $2s_{1/2}$ level which proceeds by a pair of dipole emissions and occurs in $1/7$ sec, the lifetime of the odd parity state is considerably longer. Real dielectrics on the other hand have a finite absorption length l_{abs} for the orbiting photon. In present day fiber optic quality materials $l_{abs} \gg 1$ kilometer. The lifetime of the bound state is then given by $\tau = \frac{l_{abs}}{c_{eff} f_p}$ where $c_{eff} \equiv c/n \sim 10^8$m/sec is the effective speed of light in the dielectric and f_p is the fraction of time which the eigenstate is likely to be found as an orbiting photon as opposed to an excited atom. From equation (5.6)

$$f_p \simeq \sum_\lambda |\psi_\lambda^{(0)}|^2 / |\phi_1|^2 \simeq z_0 \int_0^{mc^2} \frac{d(\hbar c k)\hbar c k}{(E - \hbar\omega_k)^2} \left(\frac{ck}{\omega_k}\right) \qquad (5.11)$$

The order of magnitude of f_p for a midgap state is given by the dimensionless parameter $z_0 \sim 10^{-7}$. This yields $\tau \sim 1$ minute per km of absorption length in the dielectric. For a bandgap to center frequency ratio $\Delta\omega/\omega_0 \simeq .05$, the photon tunnelling distance is on the scale of ten optical wavelengths. As the impurity level approaches the bandedge, the photon localization length ξ_{loc} grows larger and eventually diverges near $\omega_c : \xi_{loc} \sim c/\sqrt{\omega_c|\omega_c - \omega|}$. In this case the fraction f_p can be of order unity and the lifetime is made accordingly shorter [58].

This rather extraordinary lifetime, however, depends on very specialized conditions. For this to be realized an impurity atom would have to be laser cooled or otherwise optically trapped within a void region of the photonic bandgap material. If on the other hand, the atom is in thermal contact with the dielectric host, decay of the bound state may occur by phonon scattering, phonon emission and phonon absorption side bands [77]. However, the final radiated photon must have an energy outside of the photonic bandgap. If the bound state is in the center of the gap $\Delta\omega$, it follows that these phonons must either provide or remove an energy of order $\hbar\Delta\omega/2$. The damping rate by thermal phonons must therefore be reduced by a factor of $e^{-\hbar\Delta\omega/(2k_B T)}$ relative to the damping rate in the absence of an electromagnetic bandgap. Even at low temperatures, quantum fluctuations can cause phonon emission sidebands. For a large bandgap, however, this will require a multiphonon process leading to a very weak decay. For example if $\hbar\omega_0 = 2eV$ and $\Delta\omega/\omega_0 = .3$ decay by phonon emission will require on the order of ten optical phonons. For an atom in one of the pore regions of the dielectric the same considerations apply to atom-wall collisions and interatomic collisions. For an atomic vapour at temperature small compared to the gap energy, the atomic collision process will preserve the integrity of the photon bound state.

The effect of multiple phonon emission relaxation of the photon-atom bound state at *zero temperature* may be examined using a simple model of the *optical phonon* density of states in the solid. Consider a *phonon* density of states which is Lorentzian: $N_{\text{phonon}}(\nu) = \Gamma_0/((\nu - \nu_0)^2 + \Gamma_0^2)$. Here ν_0 is the average optical phonon frequency and Γ_0 is a measure of the optical phonon dispersion. It is straightforward to show [57, 58] that probability of phonon assisted decay of an excited impurity atom embedded in a solid host is governed by a dimensionless coupling parameter

$$S \equiv \sum_{\substack{\vec{k} \\ \text{phonons}}} \frac{|V_{\vec{k}}|^2 |\Delta\rho_{\vec{k}}|^2}{(\hbar\nu_{\vec{k}})^2} \ . \qquad (5.12a)$$

Here, $V_{\vec{k}}$ is the coupling strength of the excited electron to phonons of wavevector $\vec{k}$. The detailed form of this electron-phonon interaction may be found in standard texts in solid state physics [77]. The second factor in this dimensionless coupling is the electronic charge density fluctuation:

$$\Delta\rho_k = \int d^3\vec{r}\, e^{i\vec{k}\cdot\vec{r}} [|\phi_m(r)|^2 - |\phi_n(r)|^2] \ . \qquad (5.12b)$$

Here $\phi_n(r)$ is the electronic wave function in the atomic ground state and $\phi_m(\vec{r})$ is the electronic wavefunction in the excited state. Quantum fluctuations in the local *phonon* field may *drive* such charge density oscillations in the atom. This leads to electromagnetic radiation accompanied by phonon emission. It is apparent that *inter*shell atomic transitions involve large charge density oscillations and have a higher probability of phonon assisted decay than *intra*shell atomic transitions for which the atomic charge density redistribution is less severe. Finally $\hbar\nu_{\vec{k}}$ is the detailed phonon spectrum which we have simplified by means of the Lorentzian model density of states. A detailed calculation [58] with this model reveals that the *decay rate* due to *all* multiple phonon emission sidebands is given by $\Gamma \equiv \Gamma_0 S$. If the *photonic* band gap is sufficiently large to exclude the *one-phonon* sideband, then the rate is given by $\Gamma = \Gamma_0 S^2$. If the photonic bandgap $\hbar\Delta\omega$ is larger than *two* optical phonon energies, then both one- and two-phonon emission processes are forbidden and $\Gamma = \Gamma_0 S^3$ etc... Assuming that the dimensionless coupling strength $S \ll 1$, it is clear that the larger the photonic bandgap, the slower the rate of phonon assisted decay.

The stability of the photon-atom bound state may be enhanced by another mechanism which we discuss in greater detail in section 6. Suppose that there is a large collection $N_a \gg 1$ of identical impurity atoms. As described at the beginning of this section, a photon which is bound to any given atom can "hop" to other impurity atoms by means of the resonance dipole-dipole interaction. This leads to a spreading of the photon "wavefunction" over N_a atoms and a consequent decrease in the amplitude of local charge density oscillations in any given atom. Denoting the atomic positions by $\vec{R}_i$, $i = 1, \ldots, N_a$, the charge density fluctuation in real space becomes $\frac{1}{N_a} \sum_{i=1}^{N_a} \Delta\rho(\vec{r} - \vec{R}_i)$ rather than simply $\Delta\rho(\vec{r})$ for a single impurity atom at the origin. Replacing $\Delta\rho_{\vec{k}}$ by this *distributed* charge density fluctuation and then averaging over the atomic positions $\vec{R}_i$ yields the modified dimensionless coupling

$$S' = \frac{1}{N_a} \sum_{\substack{\vec{k} \\ \text{phonons}}} \frac{|V_k|^2 |\Delta\rho_{\vec{k}}|^2}{(\hbar\nu_{\vec{k}})^2} S_{\text{imp}}(\vec{k}) \ . \qquad (5.13a)$$

Here, $S_{\text{imp}}(\vec{k})$ is the ensemble averaged, impurity atom, structure factor:

$$S_{\text{imp}}(\vec{k}) \equiv 1 + \frac{1}{N_a} \sum_{i \neq j} \langle e^{i\vec{k}\cdot(\vec{R}_i - \vec{R}_j)} \rangle_{\text{ensemble}} \ . \qquad (5.13b)$$

The important point to note here is that the new coupling S' has been diminished by a factor of order N_a from the single impurity coupling S. Resonant energy transfer among identical atoms, therefore, serves to suppress phonon mediated decay of the atomic excitation.

5.3. SPONTANEOUS EMISSION DYNAMICS NEAR A PHOTONIC BAND EDGE

It is instructive to consider the proeprties of photon-atom bound state as revealed in the dynamical behaviour a single excited atom near a photonic band edge. This is facilitated by retaining only the $n = 0$ and $n = 1$ terms in equation (5.5). That is to say, we treat the atoms as a quantum two-level system.

Instead of a simple exponential decay as it is in the vacuum, spontaneous emission displays an oscillatory behavior near a photonic band edge. A photon-atom bound dressed state occurs even when the atomic resonant frequency lies outside (near) the gap. This bound dressed-state leads to a novel fractional steady-state of the single atom population in the excited state. For a three-level atom, this fractionalized state can be probed experimentally. In particular, we derive the spectral splitting and subnatural linewidth of spontaneous emission into the third level. We also investigate spontaneous emission of an excited two-level atom in the presence of $N - 1$ unexcited atoms. In particular we find that the collective time scale factor for emission is equal to N^ϕ, where $\phi = 2/3$ for an isotropic bandgap and $\phi = 1$ or 2 for anisotropic 2-d and 3-d bandedges. This novel feature is distinct from both cavity QED and the free space cases, where the collective scale factors are equal to $N^{1/2}$ and N, respectively.

We begin by investigating a two-level atom coupled to the radiation field in a three-dimensional periodic dielectric. The atom has excited state $|2\rangle$, ground state $|1\rangle$, and resonant transition frequency ω_{21}. The Hamiltonian of the system in the interaction picture takes the form

$$H = \sum_\lambda \hbar\Delta_\lambda a_\lambda^\dagger a_\lambda + i\hbar \sum_\lambda g_\lambda(a_\lambda^\dagger \sigma_{12} - \sigma_{21} a_\lambda), \qquad (5.14a)$$

where $\sigma_{ij} = |i\rangle\langle j|$ $(i,j = 1,2)$ are the atomic operators; a_λ and $a_\lambda^\dagger$ are the radiation field annihilation and creation operators; $\Delta_\lambda = \omega_\lambda - \omega_{21}$ is a detuning of the radiation mode frequency ω_λ from the atomic resonant frequency ω_{21} and g_λ is the atomic field coupling constant

$$g_\lambda = \frac{\omega_{21} d_{21}}{\hbar} \left(\frac{\hbar}{2\epsilon_0 \omega_\lambda V}\right)^{\frac{1}{2}} \vec{e}_\lambda \cdot \vec{u}_d. \tag{5.14b}$$

Here d_{21} and $\vec{u}_d$ are the absolute value and unit vector of the atomic dipole moment, V is the sample volume, $\vec{e}_\lambda \equiv \vec{e}_{\vec{k},\sigma}$ are the two transverse (polarization) unit vectors, and ϵ_0 is the Coulomb constant.

Assume the atom is initially on the excited state $|2\rangle$ and the field is in the vacuum state. The wave function of the system then has the form

$$|\psi(t)\rangle = b_2(t)|2, \{0\}\rangle + \sum_\lambda b_{1,\lambda}(t)|1, \{\lambda\}\rangle e^{-i\Delta_\lambda t}. \tag{5.15}$$

The state vector $|2, \{0\}\rangle$ describes the atom in its excited state $|2\rangle$ and no photons present, whereas the state vector $|1, \{\lambda\}\rangle$ describes the atom in its ground state $|1\rangle$ and a single photon in mode $\{\lambda\}$. The time-dependent Schrödinger equation projected on the one-photon sector of the Hilbert space takes the form:

$$\frac{d}{dt} b_2(t) = -\sum_\lambda g_\lambda b_{1,\lambda}(t) e^{-i\Delta_\lambda t}, \tag{5.16a}$$

$$\frac{d}{dt} b_{1,\lambda}(t) = g_\lambda b_2(t) e^{i\Delta_\lambda t}. \tag{5.16b}$$

The formal solution of Eq. (5.16b) is

$$b_{1,\lambda}(t) = g_\lambda \int_0^t b_2(t') e^{i\Delta_\lambda t'} dt'. \tag{5.17}$$

Substituting Eq. (5.17) into Eq. (5.16a) we have

$$\frac{d}{dt} b_2(t) = -\sum_\lambda g_\lambda^2 \int_0^t b_2(t') e^{-i\Delta_\lambda(t-t')} dt'. \tag{5.18}$$

The Laplace transform

$$\tilde{b}_2(s) = \int_0^\infty e^{-st} b_2(t) dt,$$

can be found from Eq. (5.18) as

$$\tilde{b}_2(s) = \left[s + \sum_\lambda g_\lambda^2 \cdot \frac{1}{s + i(\omega_\lambda - \omega_a)}\right]^{-1}. \tag{5.19}$$

Converting the mode sum over transverse plane waves into an integral and performing the angular integral we obtain

$$\tilde{b}_2(s) = \left(s + \frac{\omega_{21}^2 d_{21}^2}{6\pi^2\epsilon_0\hbar} \int_0^\Lambda \frac{k^2 dk}{\omega_k[s + i(\omega_k - \omega_{21})]}\right)^{-1}. \qquad (5.20)$$

and Λ is the Compton cutoff in the photon wave vector since photons of energy higher than the electron rest mass mc^2 probe the relativistic structure of the electron wave packet.

For a broad band of the density of states, such as in vacuum, one can use the Wigner-Weisskopf approximation. That is, only the pole contribution of s ($s \to 0^+$) in the integral of Eq. (5.20) is retained:

$$\lim_{s \to 0^+} \frac{1}{s + i(\omega_k - \omega_a)} = -iP\frac{1}{\omega - \omega_a} + \pi\delta(\omega_k - \omega_a).$$

Substituting Eq. (5.21) into Eq. (5.20) yields

$$\tilde{b}_2(s) = [s + i\delta_{21} + \frac{1}{2}\gamma_{21}]^{-1}, \qquad (5.21)$$

where δ_{21} and γ_{21} are the usual Lamb shift and spontaneous emission rate respectively. Clearly, the spontaneous emission decay is purely exponential.

The Wigner-Weisskopf perturbation theory, however, is inadequate when the density of electromagnetic modes changes rapidly in the vicinity of the atomic transition frequency ω_{21}. In this case we must perform an exact integration in Eq. (5.20). For the purpose of discussion we consider a simple model Hamiltonian for electromagnetic waves in a three-dimensional periodic dielectric described by equation (5.3a) and (5.3b).

By symmetrizing ω_k, given in by this model, to all directions in $\vec{k}$ space, we produce photonic bandgaps at the spheres $|\vec{k}| = m\pi/L$ with $m = 1, 2, 3...$ Near the band gap edges the density of states becomes singular, the atom-field interaction becomes strong and we can expect new dynamical features of spontaneous emission decay. For $k \cong k_0 \equiv \frac{\pi}{L}$ the dispersion relation near the bandedge ω_c can be approximated by

$$\omega_k = \omega_c + A(k - k_0)^2, \qquad (5.22)$$

where $A \cong \omega_c/k_0^2$.

Using Eq. (5.22), we evaluate Eq. (5.20) as

$$\tilde{b}_2(s) = \frac{(s - i\delta)^{\frac{1}{2}}}{s(s - i\delta)^{\frac{1}{2}} - (i\beta)^{\frac{3}{2}}}. \qquad (5.23)$$

Here

$$\beta^{\frac{3}{2}} = \frac{\omega_{21}^{\frac{7}{2}} d^2}{6\pi\epsilon_0 \hbar c^3},$$

and $\delta = \omega_{21} - \omega_c$. The amplitude $b_2(t)$ is given by the inverse Laplace transform

$$b_2(t) = \frac{1}{2\pi i} \int_{\epsilon-i\infty}^{\epsilon+i\infty} e^{st} \tilde{b}_2(s) ds,$$

where the real number ϵ is chosen so that $s = \epsilon$ lies to the right of all the singularities (poles and branch point) of function $\tilde{b}_2(s)$. The probability that the atom remains excited is given by $P(t) \equiv |b_2(t)|^2$. The results is plotted if Figure 25 for various choices of the atomic detuning frequency δ.

A photon which is emitted by the atom exhibits tunneling on a length scale given by the localization length before being Bragg reflected back to the emitting atom. The photon-atom bound state *inside* the band gap has been predicted in Ref. [57]. We emphasize here that the photon-atom bound dressed state is present even when the resonant atomic frequency ω_{21} lies *outside* the band gap. This bound dressed state leads to the novel fractionalized steady-state atomic population in the excited state $P_s = \lim_{t\to\infty} |b_2(t)|^2$ (Fig. 25b). Clearly, the non-zero steady-state atomic population in the excited state is present even when the resonant atomic frequency ω_{21} lies *outside* the bandgap, where the density of states is not equal zero. The atomic level splitting, oscillatory behavior and fractionalized steady-state atomic population in the excited state are all direct consequences of strong interaction between the atom and its own localized radiation when the atomic resonant frequency lies near the edge of a perfect photonic bandgap. These properties are strongly dependent on the detuning of the resonant atomic frequency ω_{21} from the bandedge frequency ω_c. Physically, the atom exchanges energy back and forth with its own radiation, backscattered after tunnelling a localized distance. This in turn is a result of the vacuum Rabi splitting of the atomic level by the photonic band edge. One level of the doublet is a localized state within the photonic bandgap, whereas the other level is a resonance in the extended state continuum. The frequency of oscillations is directly determined by the magnitude of the atomic level splitting. It is distinct from the well known Jaynes-Cummings oscillations which arise from the interaction of the atom with an isolated cavity or dielectric mode [68, 69]. In our model, no defect mode is present.

The excited-state population density can be measured via absorption of a probe beam at different decay times [70]. Alternatively, the nature of the fractionalized excited state can be probed by spontaneous emission

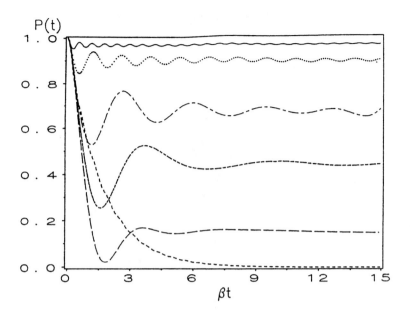

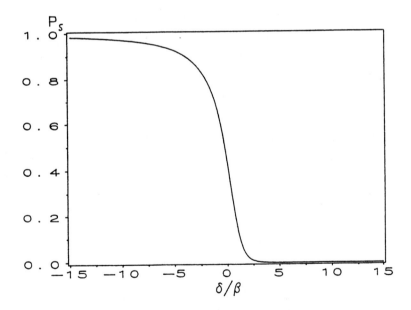

Figure 25. (a). Atomic population on the excited state, $P(t) = |b_2(t)|^2$ as a function of βt and for various values of detuning from photonic bandedge $\delta = -10\beta$ (solid curve), $\delta = -4\beta$ (dotted curve), $\delta = -\beta$ (long-short-dashed curve), $\delta = 0$ (long-short-short-dashed curve), $\delta = \beta$ (long-dashed curve), and $\delta = 10\beta$ (short-dashed curve). (b). Steady-state atomic population $P_s = lim_{t\to\infty} |b_2(t)|^2$ as a function of δ/β.

from the excited state into the third level $|3\rangle$ of a λ-configuration (Fig. 26). Assume that the transition frequency ω_{23} lies far from the gap, so that we can use the Wigner-Weisskopf approximation (5.24) for spontaneous emission $|2\rangle \rightarrow |3\rangle$. The Laplace transform $\tilde{b}_2(s)$, Eq. (2.8), can be found in this case as

$$\tilde{b}_2(s) = \frac{(s - i\delta)^{\frac{1}{2}}}{s(s - i\delta)^{\frac{1}{2}} + (i\delta_{23} + \frac{1}{2}\gamma_{23})(s - i\delta)^{\frac{1}{2}} - (i\beta)^{\frac{3}{2}}}, \qquad (5.24)$$

where δ_{23} and γ_{23} are the Lamb shift and spontaneous emission decay of the transition $|2\rangle \rightarrow |3\rangle$.

The spectrum of spontaneous emission $|2\rangle \rightarrow |3\rangle$ also exhibits interesting properties. This spectrum is given by

$$S(\omega_\lambda) \sim |\tilde{b}_2(-i(\omega_\lambda - \omega_{23}))|^2, \qquad (5.25)$$

where $\tilde{b}_2(s)$ is given in Eq. (5.24). In Fig. 26 we plot spectrum $S(\omega_\lambda)$ for different values of $\delta = \omega_{21} - \omega_c$. Clearly, the spectrum $S(\omega_\lambda)$ splits into a doublet. This splitting is analogous to the Autler-Townes splitting [71]. In our case, however, there is no external field and splitting is caused entirely by strong interaction between the atom and its own radiation field. The linewidth of the left sideband can be much smaller than γ_{21} (solid curve in Fig. 26), the natural linewidth of the spontaneous transition $|2\rangle \rightarrow |3\rangle$.

5.4. COLLECTIVE TIME SCALE FACTORS

We now generalize the previous discussion to spontaneous emission of an excited two-level atom in a presence of $N - 1$ unexcited atoms in a PBG. We limit our studies to the Dicke model in a perfect photonic bandgap. The Hamiltonian (5.14a) for the multi-atom case is given by

$$H = \sum_\lambda \hbar \Delta_\lambda a_\lambda^\dagger a_\lambda + i\hbar \sum_\lambda g_\lambda (a_\lambda^\dagger J_{12} - J_{21} a_\lambda), \qquad (5.26a)$$

where

$$J_{ij} = \sum_{k=1}^{N} \sigma_{ij}^{(k)}, (i, j = 1, 2). \qquad (5.26b)$$

Assume that the atomic system is initially in the symmetrical superradiant state [72] $|J, M = 1 - J\rangle$ with only one atom being in the excited state and the field is in the vacuum state. Here $|J, M\rangle$ states are the normalized eigenstates of operators $J_3 = \frac{1}{2}(J_{22} - J_{11})$ and $J^2 = \frac{1}{2}(J_{21}J_{12} + J_{12}J_{21}) + J_3^2$.

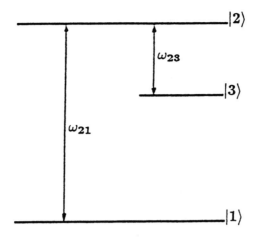

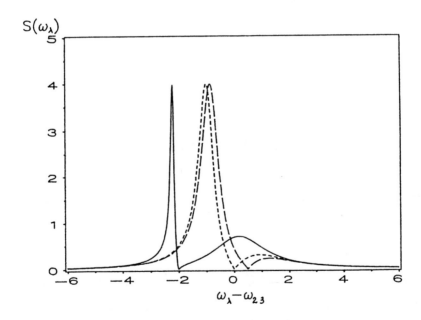

Figure 26. a. Three-level atom of a λ-configuration. The transition frequency ω_{21} lies near the bandedge while ω_{23} is assumed to be far from the gap. b. Autler-Townes spectrum $S(\omega_\lambda)$ of spontaneous emission $|2\rangle \rightarrow |3\rangle$ (in a system of the resonant frequency splitting $\beta = 1$) for $\delta_{23} = 0, \gamma_{23} = 1$ and for various values of detuning from the bandedge $\delta = -0.5$ (long-dashed curve), $\delta = 0.$ (short-dashed curve) and $\delta = 2$ (solid curve).

The wave function of the system then has the form

$$|\psi(t)\rangle_N = b_{2N}|J, M = 1 - J, \{0\}\rangle + \sum_\lambda b_{1N,\lambda}(t)|J, M = -J, \{\lambda\}\rangle e^{-i\Delta_\lambda t}$$

$$(5.27)$$

The state vector $|J, M = 1 - J, 0\rangle$ describes the atomic system in the state $|J, M = 1 - J\rangle$ and no photons present, whereas the state vector $|J, M = -J, \lambda\rangle$ describes the all atoms are in their ground state and a single photon in mode $\{\lambda\}$. It is straightforward to verify that the time-dependent Schrödinger equation yields a solution analogous to (5.20) except that d_{12}^2 is replaced by Nd_{12}^2. In free space with a continuous broad band density of states we can again use the Wigner-Weisskopf approximation (5.21). The atomic population in the excited state is given by:

$$|b_{2N}(t)|^2 = e^{-N\gamma_{21}t}.$$

That is, the collective time scale factor is proportional to N.

In the case when the atomic resonant frequency lies near the edge of a PBG, we used the dispersion relation (5.22) and the Laplace transform $\tilde{b}_{2N}(s)$ is found as

$$\tilde{b}_{2N}(s) = \frac{(s - i\delta)^{\frac{1}{2}}}{s(s - i\delta)^{\frac{1}{2}} - (i\beta_N)^{\frac{3}{2}}},$$

$$(5.28)$$

where

$$\beta_N^{\frac{3}{2}} = N\beta^{\frac{3}{2}}.$$

Clearly $b_{2N}(t)$ has the same form as $b_2(t)$ except with the factor β of a single atom case replaced by $N^{2/3}\beta$. This means that in the presence of N-1 unexcited atoms, the resonant frequency splitting increases by the factor of $N^{2/3}$. Clearly the rate of spontaneous emission is also increased by a factor of $N^{2/3}$.

The collective time scale factor $N^{2/3}$ found above made use of the isotropic model of a PBG described by Eq. (5.22). The exponent of N, however depends sensitively on the dimension of the phase space occupied by band-edge photons of vanishing group velocity and the resulting bandedge singularity in the overall photon density of states. In an isotropic bandedge, we have overestimated this phase space using the entire sphere $|\vec{k}| = \pi/L$. For a real dielectric crystal in three dimensions with an allowed point-group symmetry, the band edge is associated with a specific point $\vec{k} = \vec{k}_0$ rather than the entire sphere $|\vec{k}| = |\vec{k}_0|$

$$\omega_{\vec{k}} \cong \omega_c + A(\vec{k} - \vec{k}_0)^2.$$

$$(5.29)$$

The dispersion relation (5.29) leads to a photonic density of states $\rho(\omega)$ at a bandedge ω_c which behaves as $(\omega - \omega_c)^{\frac{d}{2}-1}$ for $\omega \geq \omega_c$. Here d is the dimensionality of space. We discuss, finally the influence of the anisotropic dispersion relation on the collective scale factor. The Laplace transform $\tilde{b}_{2N}(s)$ in this case can be found as

$$\tilde{b}_{2N}(s) = [s + \frac{N\alpha}{(2\pi)^d} \int \frac{d^d\vec{k}}{\omega_{\vec{k}}[s + i(\omega_{\vec{k}} - \omega_{21})]}]^{-1}, \qquad (5.30)$$

where

$$\alpha \simeq \frac{\omega_{21}^2 d_{21}^2}{2\epsilon_0 \hbar},$$

and $d = 1, 2, 3$ is the band-edge dimension. Using the dispersion relation (5.29) and changing variables of integration to $\vec{q} = \vec{k} - \vec{k}_0$ we can write Eq. (5.30) in the form

$$\tilde{b}_{2N}(s) = [s - \frac{iN\alpha}{(2\pi)^d} \int \frac{d^d\vec{q}}{(\omega_c + Aq^2)[Aq^2 - is]}]^{-1}. \qquad (5.31)$$

For a simplicity we assume $\omega_c = \omega_{21}$, i.e. the atomic resonance frequency lies at the edge of a PBG. The integral in Eq. (5.31) can be evaluated by contour integration and $\tilde{b}_{2N}(s)$ is given by

$$\tilde{b}_{2N}(s) = \begin{cases} [s - i^{3/2}\frac{N\alpha_1}{\sqrt{s}}]^{-1}, & \text{if } d = 1; \\ [s - iN\alpha_2 \ln(\frac{i\omega_c}{s})]^{-1}, & \text{if } d = 2; \\ [s - iN\alpha_3 + \sqrt{i}N\alpha_4\sqrt{s}]^{-1}, & \text{if } d = 3, \end{cases} \qquad (5.32)$$

where $\alpha_1 \ldots \alpha_4$ are constants.

Dynamical properties of spontaneous emission can be studied in detail from the inverse Laplace transform of $\tilde{b}_{2N}(s)$ given in Eq. (5.32). The most important distinction of anisotropic band edges from isotropic band gaps is in the collective scale factor, which we can evaluate directly from Eq. (5.32). For 1-d band edges, $\tilde{b}_{2N}(s)$ has a pole when $s^{\frac{2}{3}} \sim N$, that is the collective scale factor is equal $N^{\frac{2}{3}}$, the same as for the isotropic bandgap. For 2-d band-edges, $\tilde{b}_{2N}(s)$ has a pole $s \sim N$ with a weak logarithmic correction, i.e. the collective scale factor is approximately equal to N as for free space. In the 3-d bandedge case, $\tilde{b}_{2N}(s)$ has a pole $s \sim N^2$. As a result the collective scale factor becomes N^2. The last case may lead to the intensity of superrradiance being proportional to N^3, that is much more intense than Dicke superradiance. We now proceed to discuss the interesting consequences of the PBG on collective spontaneous emission.

6. Collective Phenomena and Quantum Coherence in a Photonic Band Gap

One of the most remarkable properties of PBG systems is that they provide a highly protected environment in which new forms of macroscopic quantum coherence involving photons and atoms may emerge. In general, the random interaction of a quantum system with a heat bath leads to decoherence of the quantum mechanical wave- function: The phase of the wavefunction is disrupted through random and irregular interactions and quantum mechanical time evolution of the system is much different than it would be if the system were perfectly isolated. This is especially evident in a many-body system consisting of a large collection of atoms occupying a region large compared to a cubic wavelength. Macroscopic quantum coherence requires not only protection from random electromagnetic interaction but also the possibility of coherent energy transport between the atoms comprising the many-body system. These are precisely the properties which the photonic bandgap exhibits.

6.1. SUPERRANDIANCE NEAR A PHOTONIC BAND EDGE

Although photonic bandgaps are analogous to electronic bandgaps in semiconductors, there are many intriguing aspects of photons which are not shared by electronic systems. Among these are laser action and superradiance. These are related to the bosonic nature of light through which many photons can occupy the same mode. In this section we derive theoretically the nature of collective spontaneous emission of N two-level atoms whose resonance frequency lies at the edge of an isotropic or anisotropic 3-d photonic bandgap. As discussed in Section 5 the collective decay rate is proportional to $N^{2/3}$ and N^2 for isotropic and anisotropic 3-d bandgaps, respectively. The corresponding peak intensity is proportional to $N^{5/3}$ and N^3, respectively. Furthermore, if atomic population inversion exists at the outset, then a fraction of the superradiant emission remains localized in the vicinity of the atoms leading to a steady state in which the atomic system acquires a macroscopic polarization. A nonzero atomic population in the excited state remains in the long time, steady state limit. This novel form of spontaneous symmetry breaking is the analog of lasing without a cavity mode! The collective emission near the photonic bandedge is accompanied by self-induced oscillations, a simple illustration of the "ringing" regime in superradiance. In addition to being a fundamental phenomenon, localization of superradiance may play an important role in low threshold microlasers based on photonic bandgap engineering. It suggests that a light emitting diode operating near a photonic band edge will exhibit very high modulation speed and coherence properties without recourse to external mirrors or even a true cavity mode.

We consider a Dicke model of N identical two-level atoms coupled to the radiation field in a three-dimensional periodic dielectric. The atoms have excited state $|2\rangle$, ground state $|1\rangle$, and resonant transition frequency ω_{21}. The Hamiltonian of the system is precisely that of equation (5.26).

Assume that the radiation field is initially in the vacuum state. When many atoms are initially excited, the wavefunction approach of section 5 is no longer useful. Instead, we must consider the Heisenberg equations of motion for the atomic operators. The equations of motion for $\langle J_{12}(t)\rangle$ and $\langle J_3(t)\rangle = \langle J_{22}(t)\rangle - \langle J_{11}(t)\rangle$ are:

$$\frac{d}{dt}\langle J_{12}(t)\rangle = \int_0^t G(t-t')\langle J_3(t)J_{12}(t')\rangle dt', \qquad (6.1a)$$

$$\frac{d}{dt}\langle J_3(t)\rangle = -2\int_0^t G(t-t')\langle J_{21}(t)J_{12}(t')\rangle dt' + c.c. \qquad (6.1b)$$

Here $G(t-t') = \sum_\lambda g_\lambda^2 e^{-i\Delta_\lambda(t-t')}$ is the delay Green's function, and $\langle A \rangle$ indicates the expectation value of the system operator A. The Green's function $G(t-t')$ depends strongly on the dispersion relation and density of states of the medium. For the purpose of discussion we consider two simple models of a PBG for electromagnetic waves in a three-dimensional periodic dielectric. In model I, we assume the dispersion is isotropic with respect to the wavevector $\vec{k}$. The simplest model dispersion relation which exhibits an isotropic PBG while retaining the correct behavior in the limit of very low and very high frequencies is

$$\omega_{\vec{k}}/c = sgn(k-k_0)\sqrt{(k-k_0)^2 + \gamma^2} + \sqrt{k_0^2 + \gamma^2}. \qquad (6.2)$$

Here $k \equiv |\vec{k}|$ and k_0 and γ are parameters related to the dielectric microstructure. The two-valued nature of the squareroot function is made explicit by the presence of the function $sgn(k-k_0) = +1$ for $k > k_0$ and -1 for $k < k_0$. The squareroot function has branch point singularities at $k = k_0 \pm i\gamma$. The presence of the sign function indicates that the branch cut should be placed along the line connecting these two branch points. Physically, this corresponds to placing an isotropic photonic bandgap of width $\Delta\omega/c = 2\gamma$ centered about the frequency $\omega_0/c = \sqrt{k_0^2 + \gamma^2}$. Also $\omega_0/c \to k(k_0/\sqrt{k_0^2 + \gamma^2})$ as $k \to 0$ and $\omega_k/c \simeq k + (\sqrt{k_0^2 + \gamma^2} - k_0)$ for $k \gg k_0$. Near the band edge $\omega_c/c = \sqrt{k_0^2 + \gamma^2} + \gamma$, the photon density of states is singular. For $k \simeq k_0$, we may simplify the dispersion relation by the effective mass approximation $\omega_{\vec{k}} \simeq \omega_c + A(k-k_0)^2$, where $A = 1/(2\gamma)$.

648

The singular density of states is an artifact of the isotropic model. In the anisotropic model II, which we describe later, the density of states in fact vanishes at ω_c. While both models exhibit localized superradiance and spontaneous symmetry breaking, the collective time scale factors for superradiant emission are qualitatively different for the two cases.

The delay Green's function $G(t - t')$ can be written for the isotropic PBG of model I as

$$G(t - t') = \frac{\omega_{21}^2 d_{21}^2}{6\pi^2 \epsilon_0 \hbar} \int_0^\Lambda \frac{k^2}{\omega_k} e^{-i(\omega_k - \omega_{21})(t - t')} dk. \tag{6.3}$$

Here, we have converted the mode sum over transverse plane wave into an integral and performed the angular integral. $\Lambda = mc/\hbar$ is the cutoff in the photon wave vector. Photons of energy higher than the electron rest mass mc^2 probe the relativistic structure of the electron wave packet. Using the effective mass isotropic dispersion relation, integration of Eq. (6.3) using the stationary phase method yields $G(t - t') = \beta^{\frac{3}{2}} e^{-i\frac{\pi}{4}} / (\sqrt{\pi(t - t')})$, where $\beta^{\frac{3}{2}} = \omega_{21}^{7/2} d_{21}^2 / (6\pi\epsilon_0\hbar c^3)$. For simplicity we assume $\omega_{21} = \omega_c$, i.e. the atomic resonance frequency lies at the band edge frequency ω_c.

To discuss the possibility of spontaneous symmetry breaking during the process of superradiant emission, we introduce a very small external perturbation which endows the atomic system with an infinitesimal polarization. This is analogous to the addition of a small magnetic field, h, in describing the thermodynamic phases of a collection of N magnetic moments. A ferromagnetic phase transition at zero field is described by taking the limit $h \to 0$ only after the thermodynamic limit $N \to \infty$. Accordingly, we find that an infinitesimal initial polarization of the atomic dipoles gives rise to a macroscopic polarization in the steady state limit $t \to \infty$. Assume that initially the atomic system is in the state

$$|\psi_N\rangle = \prod_{k=1}^N (\sqrt{r}|1\rangle + \sqrt{1 - r}|2\rangle)_k, \tag{6.4}$$

where $r \ll 1$, i.e. atoms are mostly populated in the excited state $|2\rangle$ and the atomic coherence is infinitesimal. Such a state can be created by interaction of atoms with an external pulse. Qualitatively similar results to the ones we present occur for various values of the initial atomic inversion per atom $\langle J_3(0)\rangle/N$ and for an infinitesimal initial polarization $\langle J_{12}(0)\rangle/N$. The system can be considered semiclassical and equations of motion for $x(t) \equiv \langle J_{12}(t)\rangle/N$ and $y(t) \equiv \langle J_3(t)\rangle/N$ may be obtained from (6.1) by factorizing the quantum expectation value of the operator products:

$$\frac{dx}{dt} = Ny(t) \int_0^t G(t - t') x(t') dt', \tag{6.5a}$$

$$\frac{dy}{dt} = -2Nx^*(t) \int_0^t G(t-t')x(t')dt' + c.c., \qquad (6.5b)$$

It is easy to verify, using the isotropic, effective mass solution to the Green's function (6.3), that x and y are functions of the dimensionless, scaled, time variable $\beta N^{2/3}t$. The factor $\beta N^{2/3}$ is analogous to a bandwidth parameter in solid state physics. As a result of the band edge mediated interaction between atoms, the effective Rabi splitting is enhanced and the spectrum is broadened by a factor of $N^{2/3}$.

We have solved system of Eqs. (6.5a-b) exactly using numerical methods. In Fig. 27 we plot the atomic population inversion $\langle J_3(t)\rangle/N$ (solid curve) and atomic dipole moment $D(t)/N = |\langle J_{12}(t)\rangle|/N$ (dashed curve) as a function of $\beta N^{2/3}t$. In Fig. 27, we plot the phase $\mu(t)$ of the atomic polarization $\langle J_{12}(t)\rangle$ for the same initial condition (6.4). Clearly, the collective spontaneous emission at the edge of a PBG displays striking distinctions from the free space case: (i) In the steady-state limit the population inversion $\langle J_3(t)\rangle/N$ is not equal -1. This follows from the fact that the single atomic population inversion in the excited state $|2\rangle$ remains nonzero. This signifies localization of superradiant emission in the vicinity of the atoms. (ii) The atomic polarization evolves from its infitesimal initial value to a steady-state macroscopic value. This is distinct from the free space superradiance where the atomic steady-state polarization is equal to zero. This spontaneous symmetry breaking in the atomic polarization field is analogous to lasing without a cavity mode. It suggests the possibility of observing macroscopic quantum coherent superpositions of states. (iii) The evolution of $\langle J_3(t)\rangle$, $D(t)$ and $\mu(t)$ displays collective self-induced oscillation instead of a simple decay as it is in free space. These oscillations are analogous to the collective Rabi oscillations of N Rydberg's atoms in a resonant high-Q cavity [...]. In addition to amplitude oscillations, the phase of the macroscopic polarization rotates in the steady-state limit with a frequency proportional to the magnitude of vacuum Rabi splitting. (iv) The collective time scale factor for the isotropic PBG is proportional to $N^{2/3}$ rather than N as it is in free space. That is, the collective decay rate of superradiance is proportional to $N^{2/3}$ and the peak superradiance intensity, which is proportional to $-\frac{d}{dt}\langle J_3(t)\rangle$, is proportional to $N^{5/3}$ rather than N^2.

Our numerical results reveal qualitatively similar behavior for different initial conditions. In particular, macroscopic polarization emerges for *any* initial state for which $\langle J_3(0)\rangle/N > 0$ and the steady state limit $\langle J_{12}(\infty)\rangle/N$ is independent of the initial (infinitesimal) value $\langle J_{12}(0)\rangle/N$. The delay time required for superradiant emission, however decreases noticeably as $\langle J_{12}(0)\rangle/N$ was varied from 10^{-4} to 10^{-3}. The magnitude of the macroscopic steady state polarization $\langle J_{12}(\infty)\rangle/N$ decreases monotonically from 0.42 to 0.15 as the initial inversion $\langle J_3(0)\rangle/N$ was decreased from 0.95 to

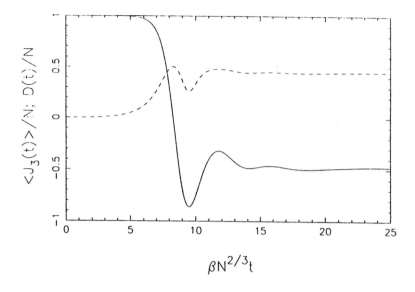

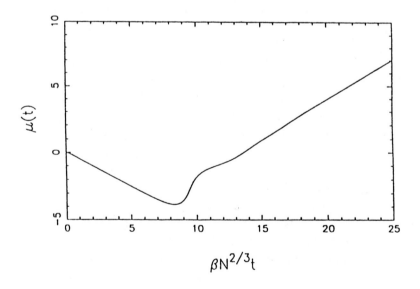

Figure 27. a. Atomic inversion $\langle J_3(t)\rangle/N$ (solid curve) and amplitude of the atomic polarization $D(t)/N = |\langle J_{12}(t)\rangle|/N$ (dashed curve) in an isotropic PBG as a function of the scaled time $\beta N^{2/3}t$ for initial condition (6.4) with $r = 10^{-6}$. b. Phase angle of the atomic polarization as a function of $\beta N^{2/3}t$ for the same parameter as in (a)

0.3. In the absence of population inversion ($\langle J_3(0)\rangle/N < 0$), we find that macroscopic polarization (in the long time limit) occurs only if the initial state itself has a macroscopic polarization.

The collective time scale factor of $N^{2/3}$ was found above using the isotropic PBG (model I). This exponent of N, however, depends sensitively on the dimension of the phase space occupied by band-edge photons of vanishing group velocity and the resulting band edge singularity in the overall photon density of states. In an isotropic band edge, we have overestimated this phase space using the entire sphere $|\vec{k}| = k_0$. For a real dielectric crystal in three dimensions with an allowed point-group symmetry, the band edge is associated with a point $\vec{k} = \vec{k}_0$ (or a finite collection of symmetry related points) rather than the entire sphere $|\vec{k}| = |\vec{k}_0|$. In model II, we choose the "effective mass" dispersion relation to be of the form (5.29). Using this anisotropic dispersion relation the Green's function in Eq. (6.1) and its integral for the case of $\omega_c t \gg 1$ become

$$G(t - t') \cong -\frac{1}{\sqrt{2}}\beta_3^{1/2}e^{i\frac{\pi}{4}}/(t - t')^{3/2}, \qquad (6.6a)$$

$$\int_0^t G(t - t')dt' = \sqrt{2}\beta_3^{1/2}e^{i\frac{\pi}{4}}/\sqrt{t} - i\sqrt{2\pi}\omega_c^{1/2}\beta_3^{1/2}, \qquad (6.6b)$$

where $\beta_3^{1/2} = \omega_{21}^2 d_{21}^2/(8\sqrt{2}\hbar\epsilon_0\pi^{3/2}A^{3/2}\omega_c)$. It is straightforward to verify, using Eqs. (6.5a-b) and the Green's function (6.6a), that x and y are now functions of a new dimensionless time variable $\beta_3 N^2 t$.

Model II also exhibits both localization of superradiance and spontaneous symmetry breaking. The resulting pictures are qualitatively similar to those shown in Fig. 27 provided that the horizontal axis is replaced with $\beta_3 N^2 t$. The details may be found in Reference [61].

In conclusion, we have demonstrated that localization and macroscopic coherence in superradiant emission occur near a photonic band edge even in the absence of a dielectric defect mode or other cavity mode. This suggests the possibility that ordinary light emission in a perfectly periodic dielectric may exhibit coherence properties with infinitesimal threshold. For the case of a physical three-dimensional, anisotropic gap, the superradiant emission occurs much faster and with higher peak intensity than conventional superradiance. These results are based on a simple model of point superradiance in which N atoms are confined to a region smaller than the wavelength of light. The spontaneous atomic polarization in the steady state is analogous to the emergence of a "superfluid" order parameter for photons.

6.2. RESONANCE DIPOLE-DIPOLE INTERACTION

Based on current projections, photonic band gap materials with bandgap to midgap ratios $\Delta\omega/\omega_0$ in the twenty to thirty percent range can be fabricated. If ω_0 lies int he visible spectrum, the total bandgap $\hbar\Delta\omega$ can be on the scale of 0.5eV. If a collection of atoms with a resonant transition at the gap center ω_0 is placed within the PBG material, spontaneous emission into bandedge electromagnetic modes is not energy conserving. The dominant interaction is the direct atom-atom coupling by the emission and re-absorption of energy non-conserving, virtual photons.

We describe in detail the nature of hopping conduction of a bound photon in a photonic band gap containing many identical impurity atoms. As it turns out the only significant modification of the resonance dipole-dipole interaction (RDDI) caused by the bandgap occurs in the wave-zone. When the separation, R, between the two impurity atoms is greater than the wavelength λ of resonant radiation, the RDDI exhibits phase shifts relative to its vacuum value and is eventually cutoff exponentially on scales $R \gtrsim \xi_{\text{loc}} \gtrsim \lambda$. Here ξ_{loc} is the localization length or tunneling distance associated with photons of a given frequency within the photonic bandgap. In the near zone, $R \ll \lambda$, the hopping process is dominated by high energy virtual photons rather than a, real, tunneling photon. This results in essentially the vacuum hopping rate. That is to say, on length scales short compared to λ, photons are unaware of the existence of a photonic bandgap!

This can be made more precise by considering second order perturbation theory in the atom-radiation field interaction. Let A and B be identical atoms with nondegenerate ground and excited states $|E_0\rangle$ and $|E_n\rangle$ connected by an electric dipole transition. Assume that, initially, atom A is in its ground state and atom B is in the excited state. Because of spontaneous emission, atom B can emit a photon and make a downward transition to the ground state. Atom A can absorb this photon and go to the excited state. This resonance energy transfer between the pair states $|E_n^A, E_0^B\rangle$ and $|E_0^A, E_n^B\rangle$ is what we wish to quantify.

As in the analysis of the single photon-atom bound state, we assume that the electromagnetic mode spectrum consists of transverse plane waves in the allowed continuum and that the dispersion relation is isotropic and has gaps. The atomic transition frequency $\omega_0 \equiv E_{n0}/\hbar$ is chosen to lie within the lowest gap. In the multipolar formalism [67], the atom-radiation interaction Hamiltonian is given by the dot product of the atom dipole moment operator $\vec{\mu}$ with the transverse part of the electric displacement field $\vec{d}^\perp$ evaluated at the atomic position:

$$H_{\text{int}} = -\frac{1}{\epsilon_0}\mu(A)\cdot\vec{d}^\perp(\vec{R}_A) - \frac{1}{\epsilon_0}\mu(B)\cdot\vec{d}^\perp(\vec{R}_B) \ . \qquad (6.7)$$

The matrix element is given by

$$
\begin{aligned}
M = \ & \sum_{\vec{k},\lambda} \frac{\hbar\omega_k}{2\epsilon_0 V} e_i^{(\lambda)*}(\vec{k}) e_j^\lambda(\vec{k}) \\
& \times \left(\mu_i^{on}(A)\mu_j^{no}(B)\frac{e^{i\vec{k}\cdot\vec{R}}}{E_{no}-\hbar\omega_k} \right. \\
& + \left. \mu_j^{on}(A)\mu_i^{no}(B)\frac{e^{-i\vec{k}\cdot\vec{R}}}{-E_{no}-\hbar\omega_k} \right) ,
\end{aligned}
\tag{6.8}
$$

where R is the distance between atoms A and B. Here $e_i^\lambda(\vec{k})$ is the ith component of the field polarization vector associated mode $(\lambda,\vec{k})$. $\lambda = 1,2$ describes the two transverse polarization channels. Converting the $\vec{k}$-summation to an integral, using the fact that $\sum_\lambda e_i^\lambda(\vec{k})e_j^\lambda(\vec{k}) = \delta_{ij} - \hat{k}_i\hat{k}_j$ and the isotropic nature of $\omega_{\vec{k}}$, the matrix element can be simplified to

$$
M = \frac{1}{2\pi^2\epsilon_0}\mu_i^{on}(A)\mu_j^{no}(B)\int_0^\infty \frac{k^2\omega_k^2 \tau_{ij}(kR)}{\omega_0^2 - \omega_k^2}dk ,
\tag{6.9a}
$$

where

$$
\begin{aligned}
\tau_{ij}(kR) = \ & (\delta_{ij} - \hat{R}_i\hat{R}_j)\frac{\sin kR}{kR} \\
& + (\delta_{ij} - 3\hat{R}_i\hat{R}_j)\left(\frac{\cos(kR)}{k^2R^2} - \frac{\sin(kR)}{k^3R^3}\right) .
\end{aligned}
\tag{6.9b}
$$

For the dispersion relation $\omega_{\vec{k}}$, the integrand above has many branch cuts singularities in the complex k-plane. Unlike the evaluation of RDDI in vacuum, it is not straightforward to apply the residue theorem. Instead, we calculate the integral numerically as a function of distance R and transition frequency ω_0. Numerical results [58] show that RDDI can be either inhibited or enhanced at intermediate distances $R \lesssim L$. Here, L is the lattice constant of the periodic dielectric. At longer distances ($R \geq L$), the matrix element oscillates with period of $2L$ and decays exponentially. Here L is the lattice constant of the periodic dielectric. A typical RDDI exhibits the asymptotic behaviour

$$
M \sim \begin{cases} \frac{A}{R^n}\cos\left(\frac{\pi R}{L} + \phi\right)e^{-R/\xi_{loc}}, & R \gg \lambda \\ \frac{A'}{R^{n'}}, & R \ll \lambda, \end{cases}
\tag{6.10}
$$

for some amplitudes A, A', exponents n, n', and phase shift ϕ. The nontrivial matrix structure of the function $\tau_{ij}(kR)$ arises from the occurrence of three distinct atomic dipole configurations. For a 2p to 1s atomic transition there are three possible relative orientations of the induced atomic dipoles. There are two orthogonal configurations in which both atomic dipoles are parallel to each other but perpendicular to the line joining the two atoms. These are referred to as π-states. The classical dipolar energy for the π-states takes the form $+1/R^3$ for $R \ll \lambda$. In the third configuration, the atomic dipoles are parallel to each other *and* parallel to the line joining the atoms. In this case the classical energy takes the form $-2/R^3$. This is referred to as the Σ-state. A detailed numerical analysis of RDDI in a photonic bandgap may be found in reference [58]. We summarize the main conclusions of this analysis here. For the Σ-state, the exponent n describing wave zone behaviour is equal to 2. At the center of the band gap, the phase shift is approximately $\phi = \pi/2$ whereas near the band edge the phase shift goes to zero. In the near zone $(R \ll \lambda)$ there is no phase shift and the exponent $n' = 3$ as in vacuum. For the π-states the exponent n is unity describing the tunnelling of a real photon. The phase shift is zero at the center of the bandgap but approaches $\pi/2$ near the photonic band edge. In the near-zone, however $n' = 3$ as before. *For both the π- and Σ-states, for $R \lesssim \cdot 1L$, there is virtually no difference between vacuum RDDI and RDDI in a photonic bandgap.*

6.3. QUANTUM OPTICAL SPIN-GLASS STATE OF IMPURITY ATOMS

One important fact about the short range component of RDDI is that on average, it favours neither alignment nor anti-alignment of the atomic dipoles. That is to say, if a statistical average were performed over all possible impurity atom positions, the energetic tendency for Σ-states to align would be cancelled by the equal and opposite tendency of the π-states (of which there are twice as many) to anti-align. RDDI does not favour the formation of any *macroscopic* polarization. This quite different from the situation where a collection of atoms (confined to a volume λ^3) interacts resonantly with an isolated cavity mode.

Consider for instance a collection of N two-level atoms confined to a volume V with a resonant transition frequency ω_0. The quantum dynamics of the two-level atoms is described by the set of 2×2 Pauli spin operators [78]. σ_z describes the atomic inversion, σ^+ and σ^- describe atomic excitation and de-excitation respectively. The coupling of these atoms to a single radiation mode of energy $\hbar\nu$, described by the photon creation operator a^+, is given by the Dicke model Hamiltonian [81]:

$$\mathcal{H}_{Dicke} = \sum_{j=1}^{N} \frac{\hbar\omega_0}{2}\sigma_j^z + \hbar\nu a^+ a + \frac{g}{\sqrt{V}}\sum_{j=1}^{N}(\sigma_j^+ a + \sigma_j^- a^+) \ . \qquad (6.11)$$

Here g is the radiation-atom coupling constant and $V \equiv \xi_{loc}^3$ is the volume of the localized cavity mode. The equilibrium statistical mechanics of this model was studied by Tavis and Cummings [82] in 1968 and later by Hepp and Lieb [83]. They showed that for sufficiently strong coupling $g^2(\frac{N}{V}) > (\hbar\nu)$ there is a low temperature phase transition to a coherent state of the radiation field: $\langle a \rangle_T > 0$. This is accompanied by a macroscopic atomic polarization $\langle \sigma^+ \rangle_T \neq 0$. Here, $\langle \ \rangle_T$ denotes an equilibrium thermal average. If the system is optically pumped, the energy scale $\hbar\nu$ must be replaced by $\hbar\nu - \mu$, where μ is an effective chemical potential for photons introduced by the pump. In the Dicke model of point superradiance ($\xi_{\text{loc}} \sim \lambda$), the resonant cavity mode serves a mediator of an effective atom-atom interaction. For atomic separations $R \ll \lambda$, this is a *ferroelectric* interaction. The energetic advantage of *parallel* induced atomic dipole moments becomes more pronounced at low temperature than the entropic advantage of disorder. This leads to an equilibrium state in which spontaneously broken symmetry of both the macroscopic atomic polarization and the electric field amplitude [84]. In a periodic dielectric with a photonic bandgap and a single localized defect mode, this model Hamiltonian may be realized experimentally. This leads to low threshold laser activity. The lasing efficiency factor β which measures the ratio of photons emitted *into the lasing mode* to the total number of photons emitted spontaneously by the impurity atoms is close to unity. If the two-level atoms are replaced by electron-hole pairs injected by a driving current into a semiconductor, the result is a light emitting diode (LED) which exhibits laser coherence properties. This can be realized if the backbone of the photonic crystal is doped semiconductor which can be electrically pumped.

For a collection of impurity atoms within a photonic bandgap, alternate forms of quantum coherence can arise from RDDI among atoms. By the nature of RDDI, there is no preference on average for the atomic dipoles to align parallel to each other as opposed to antiparallel. "Average" in this context takes on two meanings. As in the Dicke model for superradiance there is a thermal average $\langle \ \rangle_T$. In the case of RDDI, the relative atomic *positions* determine the magnitude *and sign* of the effective interaction. The configurational average over the random atomic positions is denoted by a square bracket [ ]$_c$. The model Hamiltonian for interacting two-level atoms in a perfect photonic crystal takes the form

$$\mathcal{H} = \mathcal{H}_{Dicke} + \sum_{i \neq j} J_{ij}(\sigma_i^+ \sigma_j^- + \sigma_i^- \sigma_j^+) \tag{6.12}$$

Here, J_{ij} denotes the random RDDI between atoms i and j. For simplicity we may take this to be a Gaussian random variable with mean value $\overline{J_{ij}} = 0$ and variance J. The probability distribution for J_{ij} is given by $P(J_{ij}) \propto e^{-J_{ij}^2/2J^2}$. The configuration average [ ]$_c$ denotes an averaging of thermodynamic variables with respect to this probability distribution. Equationn (6.12) is a *scalar* model for RDDI which mimicks the true vector RDDI (6.9) for the case of an atomic transition deep within a large PBG. In this case, the localization length ξ_{loc} is comparable to the optical wavelength λ. The true RDDI involves a *transverse* component proportional to $(\delta_{ij} - \hat{R}_i\hat{R}_j)$ and a *traceless* component proportional to $(\delta_{ij} - 3\hat{R}_i\hat{R}_j)$. In the near- zone, $R \lesssim \lambda$, the traceless component (which diverges as $1/R^3$ for small R) is dominant. The long-ranged transverse component, which is or- dinary vacuum would dominate the wave-zone, is exponentially cuff off by ξ_{loc} in the PBG. We model the *traceless* component of the vector RDDI by a gaussian random scalar RDDI in which $[J_{ij}]_c = 0$. Near a photonic band edge, the localization length $\xi_{\text{loc}} \to \infty$. This leads to the re-emergence of the transverse component of RDDI. The transverse component is *not* trace- less and would be mimicked by a nonzero average $[J_{ij}]_c$ in the scalar model. This in turn would favour a ferro-electric alignment of the atomic dipoles as implied in our earlier discussion of superrandiance near the photonic band edge. Models similar to (6.12) have been widely studied in condensed matter physics. As in the Dicke model, spontaneous symmetry breaking can occur in which $\langle\sigma^+\rangle_T > 0$. However, in the absence of a resonant cav- ity mode, the effective interaction between atoms is random in sign. As a result, the quantity $[\langle\sigma^+\rangle_T]_c = 0$. Each atom may acquire a spontaneous polarization, but the phases and orientations of the dipoles are random. On the other hand, the quantity $[|\langle\sigma^+\rangle_T|^2]_c > 0$. This latter quantity is the order parameter introduced by Edwards and Anderson protect[85] to distinguish a "spin-glass" state [86, 87] from a ferromagnetic state for a collection of magnetic moments with "frustrated" interactions. A possible thermodynamic state of excited impurity atoms in a photonic bandgap is the optical analog of a *quantum* spin-glass.

In quantum optics, a system of photons and atoms can be prepared in a variety of initial states by means of optical pumping. These initial states are generally far from thermodynamic equilibrium and it is of con- siderable interest to determine the non-equilibrium dynamical evolution of the system to its long time, steady state. The previous discussion of spin-glass order parameters and the equilibrium thermodynamics of quan- tum spin systems in solid state physics motivates this search for an optical glass state. We demonstrate, below, that initial states in a PBG in which there is either atomic population inversion or in which the localized cav- ity mode is pumped with coherent light, dynamically evolve into a novel steady state which is the optical analogue of a Bose-glass. This state of the localized mode radiation field is intermediate between incoherent light arising from a thermal source and coherent light arising from conventional laser emission. Corresponding to this novel, collective, optical state, the impurity atoms acquire a steady state polarization (dipole moment). Un- like the case of superradiance near a photonic band edge discussed earlier, the nature of this spontaneous symmetry breaking is entirely different. As a result of the coherent (but random) RDDI between atoms, the phase of the frozen atomic polarization varies randomly from atom to atom resulting in a collective steady state which is the optical analog of a quantum spin-1/2 dipolar glass. We consider a model in which there are a large number of atoms per cubic wavelength and the typical interatomic spacing $R_{ij} \ll \lambda$. On the other hand, the typical energy scale of RDDI is given by $(ea_0)^2/R_{ij}^3 \sim \hbar\omega_a(a_0/R_{ij})^3$ where a_0 is the atomic Bohr radius, e is the electronic charge and $\hbar\omega_a$ is the atomic transition energy (which we assume lies near the centre, $\hbar\omega_0$, of the PBG). We assume, therefore, that the atomic density is sufficiently low that $J_{ij} \ll \hbar\Delta\omega$, where $\Delta\omega$ is the overall size of the PBG. In this case, the resulting *photonic impurity band* has a width much smaller than the width of the PBG. Coherent energy transfer is possible between atoms within a cubic wavelength, and this transport process is protected from the outside environment by the PBG.

We consider a collection of N two-level atoms within a PBG interacting with a single resonant, localized dielectric, defect mode. Far inside a PBG, $(|\omega_a - \omega_0| \ll \Delta\omega)$ where spontaneous emission is nearly absent, the interaction between atoms and the electromagnetic field may be described (in the interaction picture) by the model Hamiltonian:

$$\mathcal{H} = \sum_{i=1}^{N} \frac{\hbar\delta_i}{2}\sigma_i^z + g\sum_{i=1}^{N}(\sigma_i^\dagger a + a^\dagger\sigma_i) + \sum_{i\neq j}^{N} J_{ij}\sigma_i^\dagger\sigma_j . \qquad (6.13)$$

Here, $\sigma_i^\dagger$ and σ_j describe atomic excitation and de-excitation of ith atom respectively; σ^z describes the atomic inversion; $J_{ij} = J_{ji}$ denotes the RDDI between atoms i and j; a and $a^\dagger$ are the annihilation and the creation operators for photons in the resonant dielectric defect mode; and $\delta_i = \omega_i - \omega_a$ is the atomic frequency shift (from its average value ω_a) caused by the random, static field in the photonic crystal (inhomogeneous atomic line broadening). Here the magnitude of coupling constant g is related to the volume, ξ_{loc}^3, of the defect mode, where ξ_{loc} is the localization length. In particular [78], $g = \hbar(\omega_a\mu\backslash\hbar c)(2\pi\hbar c^2\backslash\omega_a\xi_{loc}^3)^{1/2}$. Here, ω_a is the atomic resonant frequency and $\mu \sim ea_0$ is the atomic dipole moment.

In Eq. (6.13) we further assume that the N atoms are confined within a region small compared to ξ_{loc}. This allows us to neglect the exponential decay of the mode amplitude functions.

We note that in the absence of a localized defect mode $(g = 0)$ and for $\delta_i = 0$, the Hamiltonian (6.13) is the quantum spin-1/2 version fo the Sherrington-Kirkpatrick (SK) model of spin glasses [79]. An important difference here is that the atomic transition energy $\hbar\omega_a$ is much larger than a typical energy k_BT of the thermal background. Consequently we consider this problem under non equilibrium boundary conditions with optical

658

pumping, rather than under thermal equilibrium boundary conditions as it has been done previously in spin-glass theory.

In analogy to spin-glass theory, we characterize the atomic system in terms of the order parameters $m = \frac{1}{N}\Sigma_{i=1}^{N}[\langle\sigma_i\rangle]_c, q = \frac{1}{N}\Sigma_{i=1}^{N}[\langle\sigma_i^\dagger\rangle\langle\sigma_i\rangle]_c$. Here $\langle\ldots\rangle$ denotes quantum expectation value and $[\ldots]_c$ denotes the configuration average over the random atomic positions. Here m is the global polarization density fo the atomic system and q is the so-called Edwards-Anderson order parameter, describing local, spontaneous, atomic polarization. For the optical system, $m \neq 0, q \neq 0$ corresponds to a superrandiant (ferromagnetic) state, $m = 0, q = 0$ corresponds to an incoherent (paramagnetic) state, and $m = 0, q \neq 0$ corresponds to the intermediate (spin-glass) state. As discussed in [80], m and q, defined above, also characterize the local order parameters $m_i = [\langle\sigma_i\rangle]_c$ and $q_i = [\langle\sigma_i^\dagger\rangle\langle\sigma_i\rangle]_c$.

In the case of an initial state with high excitation density, *spontaneous* symmetry breaking takes place in the course of dynamical evolution. In particular, if the initial state has population inversion (and infinitesimal initial polarization), we find that the steady-state limit, a macroscopic value of the glass order parameter can build up from an infinitesimal seed. This follows from solving the Heisenberg equations of motion resulting from the Hamiltonian (6.13):

$$\frac{d}{dt}\langle\sigma_i\rangle = -i\delta_i\langle\sigma_i\rangle + ig\langle\sigma_i^z a\rangle + i\langle\sigma_i^z \sum_{j(\neq i)} J_{ij}\sigma_j\rangle, \qquad (6.14a)$$

$$\frac{d}{dt}\langle\sigma_i^\dagger\rangle = -i\delta_i\langle\sigma^\dagger\rangle + ig\langle a^\dagger\sigma_i^z\rangle - i\langle(\sum_{j(\neq i)} J_{ij}\sigma_j^\dagger)\sigma_i^z\rangle, \qquad (6.14b)$$

$$\frac{d}{dt}\langle\sigma_i^z\rangle = 2ig\langle\sigma_i^\dagger a\rangle - 2i\langle\sigma_i^\dagger \sum_{j\neq i} J_{ij}\sigma_j\rangle + c.c., \qquad (6.14c)$$

$$\frac{d}{dt}\langle a\rangle = -ig\sum_i\langle\sigma_j\rangle. \qquad (6.14d)$$

This equations may be rendered tractable by assuming that each subsystem behaves in a stochastically uncorrelated way with respect to the other. This is equivalent ot a mean-field approximation which ignores quantum correlation effects in the system. With this assumptions, the expectation values of operator products in Eq. (6.14) can be factorized. This leads to a closed set of differential equations which can be integrated for each set of Gaussian random numbers J_{ij} and δ_i. We then take a configurational average over a large number ($\sim 2.10^3$) of sets of the random number J_{ij} and sets of random number δ_i.

For illustration purposes we set the variances of these random numbers J and δ respectively, to be equal to g. The glass order parameter q and macroscopic atomic polarization m are evaluated in this process.

In Fig. 28 we plot the macroscopic polarizatio m (solid curves) and the Edwards-Anderson order parameter q (dashed curves) as function of gt for the initial condition in which atoms are mostly populated in the excited state $|2\rangle$ and the initial atomic coherence $\langle \sigma^\dagger(0)\rangle$ is infinitesimal. Such a state can be created by interaction of atoms with an external pulse. The defect mode is assumed to be initially in the vacuum state. Clearly, at the outset, the photon hopping conduction between atoms is minimal because atoms are mostly in the excited state and there is no "hole" for photon hopping conduction to take place. At the outset, the dominant process is superradiance. That is, excited atoms emit photons into the defect mode. As a result, the macroscopic polarization $m(t)$ as well as $q(t)$ are built up. This is followed by photon hopping conduction between atoms which leads to decay of the macroscopic polarization to zero in the steady state limit. In Fig. 28b we plot $|m_c| = |[\langle a\rangle]_c|$ (solid curve) and $q_c^{1/2} = [\langle a^\dagger\rangle\langle a\rangle]_c^{1/2}$ (dashed curve) as a function of gt for the same parameters. Clearly, photons in the defect mode tend to the new state with $m_c = 0$ and $q_c \neq 0$. We note here that this new state is distinct from the coherent state with $m_c \neq 0, q_c \neq 0$ and from the incoherent state with zero off-diagonal density matrix elements in which $m_c = q_c = 0$. Accompanying the spin glass state of the impurity atoms is a Bose-glass state of the electromagnetic field.

The optical spin-glass and "Bose-glass" states can be also obtained in the case when all atoms are initially in the ground state and the photons occupying the defect mode initially in a coherent state (Fig. 29a,b). In this case, we also find that if the mean number of photons in the defect mode is much larger than the number of atoms or when the coupling constant $g \gg J$, the macroscopic atomic polarization persists for a much longer time before giving way to the glassy state.

In conclusion, we have shown that a system of impurity atoms inside a PBG can evolve into a new collective state, the optical analog of a quantum spin-glass. Photons in the accompanying resonant dielectric mode evolve into a steady-state Bose-glass. Since the range of the RDDI photon hopping condition is much longer than the typical atomic separation, the system is the quantum analog of the classical Sherrington-Kirkpatrick model. The latter system has been widely studied as a paradigm for neutral networks. Programmable, classical spin-glasses have applications in computing and optimization problem. This requires that the interaction parameters J_{ij} can be controllably altered by external input. For the case of atoms in a photonic bandgap these RDDI matrix elements are determined by the atomic positions which may be controlled through laser coooling. Cooling

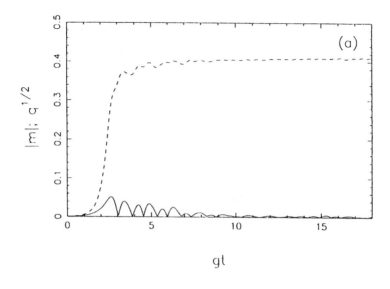

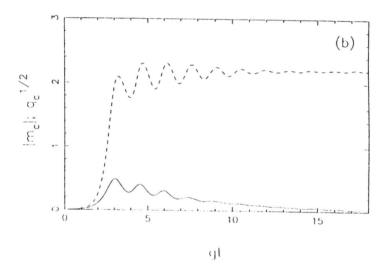

Figure 28. Macroscopic atomic polarization $|m|$ (solid curve), and spin-glass order parameter $q^{1/2}$ (dashed curve) (Fig. 28a), coherent state amplitude $|m_c| = |[\langle a \rangle]_c|$ (solid curve), and Bose-glass order parameter $q_c^{1/2} = [\langle a^\dagger \rangle \langle a \rangle]_c^{1/2}$ (dashed curve) (Fig. 28b) as function of scale time gt for $N = 10, J = \delta = g$. All atoms are initially in a coherent superpositions of states (mostly excited) with $\langle \sigma^\dagger(0) \rangle = 10^{-3}$ and the defect mode is initially in the vacuum state.

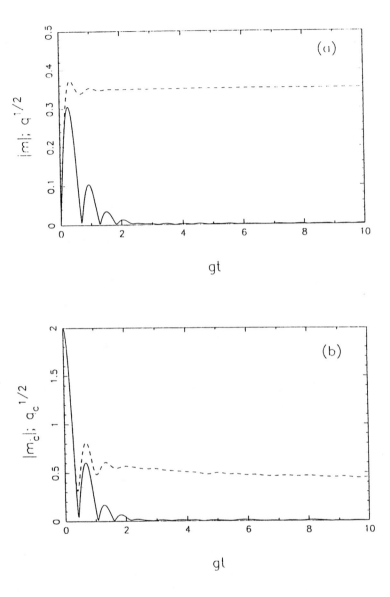

Figure 29. Macroscopic atomic polarization $|m|$ (solid curve), and spin-glass order parameter $q^{1/2}$ (dashed curve) (Fig. 29a), coherent state amplitude $|m_c| = |[\langle a \rangle]_c|$ (solid curve), and Bose-glass order parameter $q_c^{1/2} = [\langle a^\dagger \rangle \langle a \rangle]_c^{1/2}$ (dashed curve) (Fig. 29b) as a function of scale time gt for $N = 20, J = \delta = g$. All atoms are initially in the ground state and the defect mode is initially in the coherent state with $\langle a \rangle = 2$.

662

may be achieved with laser beam that couple to a third atomic level lying outside of the PBG. In this case it is possible that the photonic bandgap may provide an ideal environment for the quantum analogue of a neural network and exhibits its associated quantum computing capabilities [88, 89].

References

1. John S. (1984) Localization of Waves in a Disordered Medium, Ph.D. thesis Harvard University
2. John, S. (May 1991) Localization of Light, *Physics Today*, pg. 32 and cover story.
3. Russell, P. St. John (August 1992) Photonic Bandgaps, *Physics World*, pg. 37.
4. (1993) Soukoulis, C.M. (ed.) *Photonic Bandgaps and Localization* NATO ASI series B in Physics, vol. **308**, Plenum Press .
5. (1990) Ping Sheng (ed.) *Scattering and Localization of Classical Waves in Random Media*, World Scientific Series on Directions in Condensed Matter Physics – vol. **8**, World Scientific Press .
6. (January 1993) Development and Application of Materials Exhibiting Photonic Bandgaps, feature issue of *Journal of the Optical Society of America* **B**.
7. Anderson, P.W. (1958) Absence of Diffusion in Certian Random Lattices, *Phys. Rev.* **109**, 1492.
8. John, S. and Stephen, M.J. (1983) Wave Propagation and Localization in a Long-range Correlated Random Potential, *Physical Review* **B28**, 6358.
9. John, S., Sompolinsky, H. and Stephen, M.J. (1983) Localization in a Disordered Elastic Medium Near Two Dimensions *Physical Review* **27**, 5592.
10. John, S. (1984) Electromagnetic Absorption in a Disordered Medium Near a Photon Mobility Edge, *Physical Review Letters* **53**, 2169.
11. Anderson, P.W. (1985) On the Question of Classical Localization: A Theory of White Paint? *Philosophical Magazine* **B52**, 505,
12. van Albada, M.P. and Lagendijk, A. (1985) Observation of Weak Localization of Light in a Random Medium, *Physical Review Letters* **55**, 2692.
13. Wolf, P.E. and Maret, G. (1985) Weak Localization and Coherent Backscattering of Photons in Disordered Media, *Physical Review Letters* **55**, 2696.
14. Yablonovitch, E. (1987) Inhibited Spontaneous Emission in Solid-State Physics and Electronics, *Physical Review Letters* **58**, 2059.
15. John, S. (1987) Strong Localization of Photons in Certain Dielectric Superlattices, *Physical Review Letters* **58**, 2486.
16. see for instance Mott, N.F. (1974) *Metal-Insulator Transitions*, Taylor and Francis Publishers, London.
17. Freund, I., Rosenbluh, M., Berkovits, R. and Kaveh, M. (1989) Coherent Backscattering of Light in a Quasi-Two-Dimensional System, *Physical Review Letters* **61**, 1214.
18. see for instance Jackson, J.D. (1975) *Classical Electrodynamics*, published by J. Wiley.
19. Ioffe, A.F. and Regel, A.R. (1960) *Prog. Semicond.* **4**, 237.
20. van de Hulst, H.C. (1981) *Light Scattering by Small Particles*, Dover Publications Inc.
21. Bohren, C.F. and Huffman, D.R. (1983) *Absorption and Scattering of Light by Small Particles*, J. Wiley Interscience Publication.
22. Genack, A.Z. (1987) Optical Transmission in Disordered Media, *Physical Review Letters* **58**, 2043.
23. Drake, M. and Genack, A.Z. (1989) Observation of Nonclassical Optical Diffusion, *Physical Review Letters* **63**, 259.
24. (1976) *Statistical Mechanics*, section 16.3, D. McQuarrie, Harper and Row

663

Publishers.
25. Ashcroft, N. and Mermin, D. (1976) *Solid State Physics*, Holt, Rinehart and Winston Publishers.
26. Car, R. and Parrinello, M. (1988) Structural, Dynamical, and Electronic Properties of Amorphous Silicon: An Ab Initio Molecular-Dynamics Study, *Physical Review Letters* **60**.
27. see for instance Chaikin, P. *et. al.* (1987) Colloidal Crystals in S. Safran and N.A. Clark (eds.), *Physics of Complex and Supermolecular Fluids*, Wiley-Interscience Publications.
28. see for instance (1968) *Elementary Quantum Mechanics*, D. Saxon, Holden-Day Publishing.
29. Yablonovitch, E. and Gmitter, T.J. (1989) Photonic Band Structure: The Face-Centered-Cubic Case, *Physical Review Letters* **63**, 1950.
30. Leung, K.M. and Lui, Y.F. (1990) Full Vector Wave Calculation of Photonic Band Structures in fcc Dielectric Media, *Physical Reivew Letters* **65**, 2646.
31. Zhang, Z. and Satpathy, S. (1990) Electromagnetic Wave Propagation in Periodic Structures: Bloch Wave Solution to Maxwell's Equations, *Physical Review Letters* **65**, 2650.
32. Ho, K.M., Chan, C.T. and Soukoulis, C.M. (1990) Existence of Photonic Bandgap in Periodic Dielectric Structures, *Physical Review Letters* **65**, 3152.
33. Yablonovitch, E., Gmitter, T.J. and Leung, K.M. (1991) Photonic Band Strucutre: the fcc case employing nonspherical atoms, *Physical Review Letters* **67**, 2295.
34. Meade, R.D., Rappe, A.M., Brommer, K.D. and Joannopoulos, J.D. (1991) Electromagnetic Bloch Waves at the Surface of a Photonic Crystal, *Physical Review* **B44**, 10961.
35. Economou, E.N. and Sigalas, M.M. (1993) Spectral Gaps for Classical Waves in Periodic Structures, *Physical Review* **B48**, 13434.
36. Sigalas, M.M., Economou, E.N. and Kafesaki, M. (in press) Spectral Gaps for Electromagnetic and Scalar Waves: Possible explanation for certain differences, *Physical Review B*.
37. for a review see Ziman, J. (1979) *Models of Disorder*, Section 9.6, Cambridge University Press.
38. He, J. and Cada, M. (1991) *IEEE J. Quantum Electron*, **27**, 1182.
39. For a general discussion of nonlinear optical susceptibilities see Butcher, P.N. and Cotter, D. (1990) *The Elements of Nonlinear Optics*, Cambridge University Press.
40. Mills, D.L. and Trullinger, S.E. (1987) Gap Solitons in Nonlinear Periodic Structures, *Physical Review* **B36**, 947.
41. Philal, M. and Maradudin, A.A. (1991) *Phys. Rev. B* **44**, 8565; Meade, R.D., Brommer, K.D., Rappe, A.M. and Joannopoulos, J.D. (1992) *Appl. Phys. Lett.* **61**, 495. ; Villeneuve P.R. and Piché, M. (1992) *Phys. Rev. B* **46**, 4973.
42. Aceves, A. B., Wabnitz, S. (1989) *Physics Letters A*, **141**, 37.
43. John S. and Akozbek, N. (1993) Nonlinear Optical Solitary Waves in a Photonic Band Gap, *Physical Review Letters* **71**, 1168.
44. Abrahams, E., Anderson, P.W., Licciardello, D.C., Ramakrishnan, T.V. (1979) Scaling Theory of Localization: Absence of Quantum Diffusion in Two Dimensions, *Physical Review Letters* **42**, 673.
45. Vollhardt, D. and Wölfle, P. (1980) Diagrammatic, self-consistent treatment of the Anderson localizaiton problem, *Physical Review* **B22**, 4666.
46. MacKintosh, F.C. and John, S. (1989) Diffusing-wave spectroscopy and multiple scattering of light in correlated random media, *Physical Review* **B40**, 2383.
47. Akkermans, E., Wolf, P.E. and Maynard, R. (1986) Coherent Backscattering of Light by Disordered Media: Analysis of the Peak Line Shape, *Physical Review Letters* **56**, 1471.
48. Stephen, M.J. and Cwilich, G. (1986) Rayleigh Scattering and Weak Localization: Effects of Polarization, *Physical Review B* **34**, 7564.

49. MacKintosh, F.C. and John, S. (1988) Coherent Backscattering of Light in the Presence of Time-reversal Non-invariant and Parity Violating Media, *Phys. Rev.* **B37**, 1884.

50. Gorodnichev, E.E., Dudarev, S.L., Rogozkin, D.B. and Ryazanov, M.I. (1989) Weak Localization of Waves in Incoherent Scattering in Crystals, *Soviet Physics JETP* **69**, 1017.

51. Licciardello, D.C. and Thouless, D.J. (1975) Constancy of Minimum Metallic Conductivity in Two Dimensions, *Physical Review Letters* **35**, 1475.

52. Edwards, J.T. and Thouless, D.J. (1972) Numerical studies of localization in disordered systems, *J. Phys.* **C5**, 807.

53. Garcia, N. and Genack, A.Z. (1991) Anomalous Photon Diffusion at the Threshold of the Anderson Localization Transition, *Physical Review Letters* **66**, 1850.

54. Genack, A.Z. and Garcia, N. (1991) Observation of Photon Localization in a Three-Dimensional Disordered System, *Physical Review Letters* **66**, 2064.

55. van Tiggelen, B.A., Lagendijk, A., van Albada, M.P. and Tip, A. (1992) Speed of Light in Random Media, *Physical Review* **B45**, 12, 233.

56. see for instance Mott, N.F. and Davis, E.A. (1979) *Electronic Processes in Non-crystalline Materials*, Clarendon Press, Oxford.

57. John, S. and Wang, J. (1990) Quantum Electrodynamics near a Photonic Bandgap: Photon Bound States and Dressed Atoms, *Physical Review Letters*, **64**, 2418.

58. John, S. and Wang, J. (1991) Quantum Optics of Localized Light, *Physical Review* **B43**, 12772.

59. Raizen, M.G., Thompson, R.J., Brecha, R.J., Kimble, H.J. and Carmichael, H.J. (1989), Normal-Mode Splitting and Linewidth Averaging for Two-State Atoms in an Optical Cavity, *Physical Review Letters* **63**, 240.

60. John, S. and Tran Quang (1994) Spontaneous Emission near the Edge of a Photonic Band Gap, *Physical Review* **A50** 1976.

61. John, S. and Tran Quang, (1995) Localization of Superradiance near a Photonic Band Edge, *Physical Review Letters* **74**, 3419.

62. Lawandy, N.M. *et.al.* (1994) *Nature* **368**, 436; Balachandran, R.M. and Lawandy, N.M. (1995) *Optics Letters* **20**, 1271.

63. Sha, W.L., Liu, C.H. and Alfano, R.R. (1994) *Optics Letters* **19**, 1922.

64. Zhang, D. *et.al.* (1995) *Optics Communications* **118**, 462.

65. John, S. and Pang, G., Theory of Paint-On Lasers, to be published.

66. Bergmann, Gerd (1984) Weak localization in Thin Films: a time of flight experiment with conduction electrons, *Physics Reports* **107**, 1.

67. Craig, D.P. and Thirunamachandran, T. (1984) *Molecular Quantum Electrodynamics*, Academic Press.

68. Jaynes, E.T. and Cummings, F.W. (1963) *Proc. I.E.E.E.* **51**, 89.

69. Sanchez-Mondragon, J.J., Narozhny, N.B. and Eberly, J.H. (1983) *Phys. Rev. Lett.* **51**, 550; Nabiev, R.F., Yeh, P. and Sanchez-Mondragon, J.J. (1993) *Phys. Rev.* **A47**, 3380. .

70. Martorell, J. and Lawandy, N.M. (1990) *Phys. Rev. Lett.* **65**, 1877.

71. Autler, S.H. and Townes, C. H. (1955) *Phys. Rev.* **100**, 703.

72. Gross, M. and Haroche, S. (1982) Superrandiance: An Essay on the Theory of Collective Spontaneous Emission, *Physics Reports* **93**, No. 5, 301; Haroche, S. and Raimond, J.M. (1985) *Adv. At. Mol. Phys.* **30**, 347.

73. Ziman, J.M. (1969) *Elements of Advanced Quantum Theory*, Cambridge University Press.

74. Bethe, H.A. (1947) The Electromagnetic Shift of Energy Levels, *Physical Review* **72**, 339.

75. Mollow, B.R. (1969) Power Spectrum of Light Scattered by Two-Level Systems, *Physical Review* **188**, 1969.

76. Basche, Th. and Moerner, W.E. (January 23, 1992) Optical Modification of a Single Impurity Molecule in a Solid, *Nature* volume **355**, 355.

77. A detailed discussion of exciton lineshapes in solids due to phonon interaction may be found in Nakajima, S., Toyozawa, Y. and Abe, R. (1980) *The Physics of Elementary Excitations*, Springer-Verlag Publishers, Berlin.

78. see for instance Allen, L. and Eberly, J.H. (1987) *Optical Resonance and Two-Level Atoms*, Dover Publications Inc., New York.

79. Sherrington, D. and Kirkpatrick, S. (1975) *Phys. Rev. Lett.* **35**, 1793.

80. Mezard, M., Parisi, G. and Virasoro, M. *Spin-Glass Theory and Beyond*, World Scientific Press, 986; Parisi, G. (1983) *Phys. Rev. Lett.* **50**, 1946 .

81. Dicke, R.H. (1954) Coherence in Spontaneous Radiation Processes, *Physical Review* **93**, 99.

82. Tavis, M. and Cummings, F.W. (1968) Exact Solution for an N-Molecule-Radiation-Field Hamiltonian, *Physical Review* **170**, 379.

83. Hepp, K. and Lieb, E.H. (1973) On the Superradiant Phase Transition for Molecules in a Quantized Radiation Field: the Dicke Maser Model, *Annals of Physics*, 360.

84. for an elementary derivation using coherent states see Wang, Y.K. and Hioe, F.T. (1973) Phase Transition in the Dicke Model of Superradiance, *Physical Review* **A7**, 831.

85. Edwards, S.F. and Anderson, P.W. (1975) *Journal of Physics* **F5**, 965.

86. Chowdhury, D., Spin Glasses and Other Frustrated Systems in *Princeton Series in Physics*, Princeton University Press.

87. Binder, K. (1986) Spin Glasses; Experimental Facts, Theoretical Concepts and Open Questions *Reviews of Modern Physics* **58**, 801.

88. Bennet, C. (1995) Quantum Information and Computation, *Physics Today* October issue, 24.

89. Lloyd, S. (1995) Quantum Mechanical Computers, *Scientific American*, October issue, 140

ENERGY TRANSPORT VELOCITY IN RANDOM MEDIA

K. BUSCH[1,2] AND C. M. SOUKOULIS,[1]
[1] *Ames Laboratory and Department of Physics and Astronomy,*
Iowa State University, Ames, IA 50011
[2] *Institut für Theorie der Kondensierten Materie*
Universität Karlsruhe, 76128, Karlsruhe, Germany

Abstract. We present a new method for efficient, accurate calculations of the transport properties of random media based on the principle that the wave energy density should be uniform when averaged over length scales larger than the size of the scatterers. This new scheme captures the effects of the resonant scattering of the individual scatterer exactly, as well as the multiple scattering in a mean-field sense. It has been successfully applied to both "scalar" and "vector" classical wave calculations. Results for the energy transport velocity give pronounced dips for low concentration of scatterers while as the concentration increases, the dips are smeared out in excellent agreement with experiment. This new approach is of general use and can be easily extended to treat different types of wave propagation in random media.

1. Introduction

In recent years, there has been a growing interest in studies of the propagation of classical waves in random media[1]. While some of the features associated with weak localization, such as enhanced coherent backscattering, have been detected in light scattering experiments[1], the localization of classical waves in random systems has not been established beyond doubt. Recent experimental results[2] for the diffusion coefficient D and the transport mean free path ℓ_t demonstrated that in a disordered medium, low values of the diffusion constant $D = v_E \ell_t / 3$ were caused by extremely small values of the energy transport velocity v_E and not by the small values

C. M. Soukoulis (ed.), Photonic Band Gap Materials, 667–678.

of ℓ_t, which signifies strong localization. To explain this low value of the transport velocity, a theory was developed by van Albada et al. [2], in the low-concentration limit, of the Bethe-Salpeter equation. They argued that their approach gives the correct transport velocity observed experimentally, which is the energy transport velocity v_E and not the phase velocity v_p. v_p is approximately equal to the velocity of light, c, divided by an appropriate index of refraction. v_E is always less than v_p, especially close to the Mie resonances. The renormalization of the diffusion coefficient near resonances in random media has been extensively studied [3-7] after its introduction by the pioneer work of van Albada et al. [2]. It is by now well understood that to *lowest order in density* of the dielectric scatterers, the strong decrease in the transport velocity is due to the Mie resonances. Near resonances, a lot of energy is temporarily stored inside the dielectric scatterer or equivalently the wave spends a lot of time (dwell time) inside the dielectric scatterer.

Experimental results[9] for alumina spheres have shown that as the volume fraction of the scatterers f increases towards close packing (f≃0.60), there is no structure in the diffusion coefficient versus frequency. This clearly suggests there is no structure in the transport velocity. It is, therefore, inappropriate to calculate the transport velocity using the v_E of van Albada et al. [2] in this high f-regime, since their theory for v_E is a low-concentration theory. But, if we calculate[10, 11] v_E according to Ref. 2 for this high f = 0.60, a strong structure in v_E is obtained in disagreement with the experimental results. An extension of the well-known coherent-potential-approximation (CPA) was recently developed[10] and obtained a CPA phase velocity for f = 0.60, which is qualitatively consistent with experiment, in not showing any structure as a function of the frequency. Not surprisingly, the newly developed[10] coated CPA for low f gives a CPA phase velocity which reduces to the regular phase velocity which is higher than the velocity of light near Mie resonances. This is an undesirable feature of the CPA, which had to be fixed. Thus, for small f, the theory of van Albada et al.[2], seems to give the correct transport velocity v_E, while for large f, it is the coated CPA approach[10-12] which seems to give transport velocities consistent with experiment[9].

It is, therefore, very important to develop a theoretical scheme that will give results for the transport properties of random media for all frequencies and scatterer concentrations. It is well known that wave propagation in random media is understood in the high- and low-frequency limit. In the high-frequency limit, wave propagation can be reduced to geometric optics, while in the low-frequency limit, where the wave cannot resolve the disorder, CPA approaches are used. The intermediate frequency regime, where the wavelength of the wave is comparable to the scatterer's size is the most difficult to treat theoretically, but the most interesting physically.

In the present paper, we present a new approach in calculating the transport properties of random media that takes into account the multi-scattering interactions in a mean-field sense. The main new physical idea is that in a random medium, *the energy density should be uniform when averaged over the correlation length of the microstructure.* This approach has been applied to *both scalar* and *vector* classical wave propagation in random media with many successes. For both the scalar and vector cases, we obtain results for the energy transport velocity that give pronounced dips in v_E for low f, while as f increases the dips are smeared out, as expected and in agreement with experiment. In addition, this new energy density – CPA gives the correct long wavelength limit for the effective dielectric constant for both the scalar and vector case. The energy density for the vector case is calculated exactly, where both the electric and magnetic field contributions are taken into account. For the vector case, analytical as well as numerical results of this approach, give that the long-wavelength dielectric constant is given by the Maxwell-Garnett formula. The formalism that has been developed in this letter can be easily extended to treat different types of wave propagation in disordered systems and is therefore of general use.

2. Review of Different CPAs

We consider a composite medium of two lossless materials, with dielectric constants ϵ_1 and ϵ_2. Our composite medium is assumed to consist of spheres with diameter d = 2R and dielectric constant ϵ_1 randomly placed within the host material with dielectric constant $\epsilon_2=1$. The random medium is characterized also by f, the volume fraction occupied by the spheres. We consider first the propagation of classical waves in a random medium described by the wave equation for the scalar field Ψ, $(\nabla^2 + \omega^2\epsilon(r)/c^2)\ \Psi=0$, where $\epsilon(r)$ is a random variable. Different papers,[1, 13] have been devoted to the derivation of a Boltzmann equation for scalar classical waves, starting with the Bethe-Salpeter equation. If the disorder is weak, the diffusion coefficient D can be calculated from the Boltzmann equation and is given by the familiar expression $v_p\ell_t/3$, where v_p is the phase velocity and ℓ_t is the transport mean free path. By correctly handling the Ward identities, van Albada et al.[2, 7, 8] reported that in low density limit, D = $v_E\ell_t/3$, where v_E is the energy transport velocity and is much lower than v_p, close to the Mie resonances of the isolated dielectric scatterer. The main theoretical result of Refs. 2, and 7 is

$$v_E = \frac{c^2}{v_p(1+\delta)} \tag{1}$$

The quantity δ is given by

$$\delta = n \int d^3r \, |\Psi_k^+(r)|^2 \, [\epsilon(r) - 1] \; , \tag{2}$$

with n the number density of the scatterers, and $\Psi_k^+(r)$ is the one-scatterer eigenfunction with incident wave vector k for a single dielectric scatterer. Physically, δ can be larger when the incident wave frequency coincides with an internal resonance of the scatterer. When this happens, $|\Psi_k^+(r)|^2$ has a large magnitude inside the scatterer, leading to a large δ, and therefore, to small values of v_E. Another interpretation of resonant scattering is the incident wave can get trapped inside the scatterer, bouncing around many times before emerging again. This tells us in an intuitive way that wave diffusion slows down when resonant scattering is present. A more elegant and most convenient for numerical purposes representation of δ for scalar waves was given[14] by van Tiggelen, where v_E was written with respect to the Van de Hulst scattering coefficients of the scalar dielectric sphere, (see Eq. (3.87) of Ref. 14). The Amsterdam group[2, 7, 8, 14] extended their scalar results for the renormalization of v_E to the vector case, by simply replacing the scalar single-scatterer t-matrix with the vector t-matrix. This is an oversimplified approximation of the real vector problem. The polarization of the EM waves must be taken into account on a fully vector calculation in deriving the Boltzmann equation, starting with the Bethe-Salpeter equation. This is still the outstanding problem of the field. If indeed one makes this approximation, a v_E is obtained (see Eqs. 28 and 29 of Ref. 2) that is much lower than v_p and it has pronounced dips close to the Mie resonances of the isolated dielectric scatterer. Unfortunately, as in the scalar case v_E has pronounced dips not only for small f but also for high f in disagreement with experiment.[9] In addition, the long wavelength limit of v_E and therefore of the dielectric constant $\bar{\epsilon}$ is given by $\bar{\epsilon} = 1 + 3f(\epsilon_1 - \epsilon_2)/(\epsilon_1 + 2\epsilon_2)$ and not by the Maxwell-Garnett theory result, which is the "correct" result for the vector case.

Sheng and his colleagues[12] have re-formulated the well-known CPA to a new scheme that numerically calculates the maxima of the spectral function to identify the modes and the quasimodes of the disordered system. This scheme also gives $v_E > c$ at low f, near the Mie resonances. Soukoulis et al. [10] have extended the approach of van Albada et al. [2] for calculating v_E the following way. The coated CPA is used to calculate a frequency-dependent effective dielectric function $\bar{\epsilon}$; then the energy velocity expressions of van Albada et al. [2] were used to calculate v_E with this $\bar{\epsilon}$ as the outside medium. The energy transport velocity calculated this way gives correct results[10, 11] for most of the cases and extrapolates smoothly between low and high concentrations. However, it is not a very clean theoretical approach, since it combines the low-concentration theory of van Albada et al. [2], with the coated CPA approach [10].

3. Energy Density CPA

It is well know that the CPA scheme is designed to give a description of the wave propagation characteristics when the wavelength is large when compared to the size of the inhomogeneities. To calculate the effective macroscopic properties of the random system, the scattering unit, which can be either a coated sphere[10, 12], or a sphere[1], is embedded in an effective medium with an effective dielectric constant $\bar{\epsilon}$. The quantity $\bar{\epsilon}$ is self-consistently determined by demanding that the average forward-scattering amplitude $f(0)$ is equal to zero. Once the quantity $\bar{\epsilon}$ is determined, an effective propagation constant $q = \bar{\epsilon}^{1/2}\omega/c$ is defined, and, therefore, one can immediately find[10, 11] the mean free path $\ell = 0.5/Im(q)$, the renormalized wave vector $k = Re(q)$, and other effective macroscopic properties of the random system [1,10-13]. The validity of this approach is verified by comparison with experimental results, as well as with numerical calculations. As we mentioned before, all the CPA schemes based on $< f(0) >= 0$ for low f give a CPA phase velocity which is higher than c near Mie resonances. This is an undesirable feature of the CPA, which be fixed. Here, we present a new approach for calculating the transport properties, based on the physical idea that in a random medium the energy density should be uniform when averaged over length scales larger than the size of the inhomogeneities. This choice captures both the effect of Mie resonances and interparticle interactions. Consider, for example, a random medium composed of a dispersion of spheres as shown in Fig. 1. The basic structural unit may be regarded as a coated sphere, as represented by the dashed lines in Fig. 1a. The radius of the coated sphere $R_c = R/f^{1/3}$, where R is the radius of the solid sphere. Let $k_e = \sqrt{\bar{\epsilon}}\,\omega/c$ characterize the effective medium, which has an average dielectric constant $\bar{\epsilon}$ to be determined self-consistently. The self-consistent condition for the determination of k_e or $\bar{\epsilon}$ is the energy content of the coated sphere embedded in the effective medium (see Fig. 1b) is equal to the energy content of a sphere of radius R_c with $\bar{\epsilon}$ dielectric constant (see Fig. 1c) i.e.,

$$\int_0^{R_c} d^3r \; \rho_E^{(1)}(\vec{r}) = \int_0^{R_c} d^3r \; \rho_E^{(2)}(\vec{r}), \qquad (3)$$

which is also shown schematically in Fig. 1. $\rho_E^{(1)}(r)$ and $\rho_E^{(2)}(r)$ are the energy densities for the configurations shown in Figs. 1a and 1b, respectively. For scalar waves the energy density is

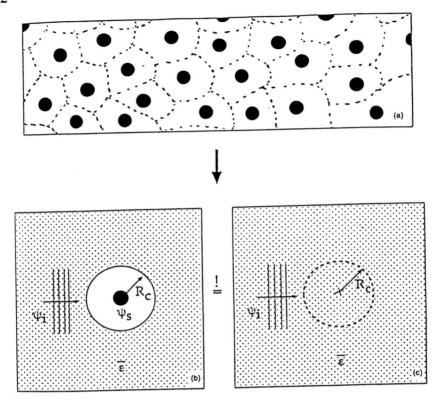

Figure 1. (a) In a random medium composed of dielectric spheres, the basic scattering unit may be regarded as a coated sphere, as represented by the dashed lines. To calculate the effective dielectric constant $\bar{\epsilon}$, a coated sphere of radius $R_c = R/f^{1/3}$ is embedded in a uniform medium. The self-consistent condition for the determination of $\bar{\epsilon}$ is the energy of a coated sphere (b) is equal to the energy of a sphere with radius R_c and dielectric constant $\bar{\epsilon}$ (c).

$$\rho_E(\vec{r}) = \frac{1}{2}(\omega^2 \epsilon(\vec{r}) |\Psi(\vec{r})|^2 / c^2 + |\vec{\nabla}\Psi(\vec{r})|^2), \qquad (4)$$

whereas the energy density of vector waves is given by

$$\rho_E(\vec{r}) = \frac{1}{2}(\epsilon(\vec{r}) |\vec{E}(\vec{r})|^2 + \mu |\vec{H}(\vec{r})|^2), \qquad (5)$$

where μ is the magnetic permeability taken equal to one and $\Psi(r), \vec{E}(r)$, and $\vec{H}(r)$ are the scattering wave function and the scattered electric and magnetic fields for a plane wave incident on a coated sphere, respectively. In Eq. (3), $\bar{\epsilon}$ is the parameter to be determined. It should be noted that the energy density, and, therefore, $\Psi(r)$ being a scattering wave function, implicitly depends on $\bar{\epsilon}$. We, therefore, have Eq. (3) as the self-consistent condition for our energy-density CPA. We find this condition is easily satisfied with a few numbers of iterations (of the order of 10) for all frequencies and filling ratios. This was not the case for all the previous CPAs, which were based on $< f(0) >= 0$. Here the solution of the corresponding self-consistent equation can disappear or jump abruptly or even get multiple solutions. This new scheme can easily follow the unique solution of the self-consistent equation. We feel the integration over all angles in Eq. (3) is responsible for the well behaved solution. It should be noted that $\Psi(r), \vec{E}(r)$, and $\vec{H}(r)$, being scattering wave functions, implicitly depend on $\bar{\epsilon}$. When Eq. (3) is satisfied, we have the energy transport velocity

$$v_E = \frac{c}{\sqrt{\bar{\epsilon}}} \sqrt{1 - Re\Sigma/k_e^2} \qquad (6)$$

where the self-energy $\Sigma = 4\pi n f(0)$, is calculated with the embedding medium characterized by $\bar{\epsilon}$. Eqs. (3), (4), and (5), together define a mean-field approach to the calculation of the transport velocity. The multiple-scattering interactions are implicitly taken into account by the requirement of energy density homogeneity.

We have systematically calculated the energy content of the sphere and coated sphere for the scalar[15] and vector[16] cases. It is very remarkable that the energy stored in a dielectric sphere or coated sphere for both the scalar and vector case are given by these relatively simple forms, which are very convenient for numerical calculations. In Fig. 2, we present the results obtained for the frequency dependence of energy transport velocity for scalar waves and for three different filling ratios. We presented the frequency as d/λ_i, where d is the diameter of the dielectric sphere and $\lambda_i = 2\pi c/\omega\sqrt{\epsilon_1}$ is the wavelength inside the sphere. This choice is based on the fact that the Mie resonances appear in the total scattering cross section

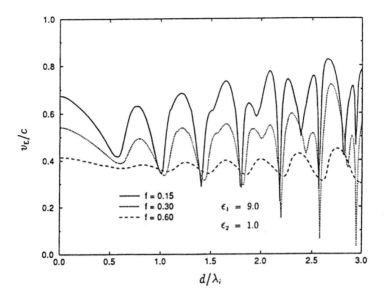

Figure 2. The energy transport velocity v_E for the scalar waves calculated by the energy density – CPA versus d/λ_i for alumina spheres with dielectric constant 9.0 for different values of filling ratios.

from the isolated sphere and in the limit $\epsilon_1/\epsilon_2 \to \infty$, when $d/\lambda_i = (n+1)/2$ with $n = 1, 2, 3 \ldots$ for the vector, and $n = 0, 1, 2, 3 \ldots$ for the scalar case. Notice v_E exhibits pronounced dips as a function of frequency only at low scatterer concentrations, in agreement with the results of the Amsterdam group, [2, 7, 8] and experiments[9]. At higher scatterer concentrations, the variation with frequency is reduced, as expected. Remember the low density approximation of the Amsterdam group gives negative values of v_E for f = 0.60. For the scalar case, we also calculated analytically the long wavelength limit of $\bar{\epsilon}$, and indeed we find that $\bar{\epsilon} = f\epsilon_1 + (1-f)\epsilon_2$, as expected. In Eq. (3), if one approximates the energy density $\rho_E^{(2)}(r)$ of the effective medium, by $\rho_E^{(1)}(r)$, one obtains the following self-consistent equation[13, 14] $\int_0^{R_c} d^3 r \rho_E^{(1)}(r)[\epsilon(r) - \bar{\epsilon}] = 0$. This equation is exactly equal to Eq. (1), provided that the background dielectric constant is $\bar{\epsilon}$ and not unity.

This new energy-density CPA scheme can be easily applied to the vector case. The energy density for the vector case is given in Eq. (4b), and contains the contribution from both the electric and magnetic fields. Up to now, all the low density limit theories, as well as the CPA were developed by only transferring all the scalar wave formalism to the vector case,

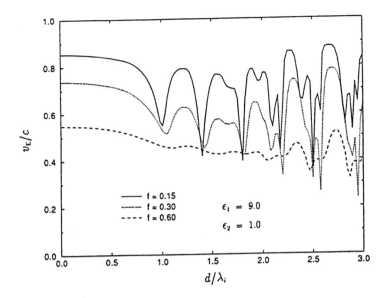

Figure 3. The energy transport velocity v_E for vector waves calculated by the energy density-CPA versus d/λ_i for alumina spheres with dielectric constant 9.0 for different values of filling ratios.

without taking the vector character of the wave function into account. The vector character was only used in calculating the t-matrix or the forward-scattering amplitude for the dielectric scatterer. In Fig. 3, we present the results for the frequency dependence of v_E for the vector case, for three values of f. Notice that for f = 0.15, v_E has pronounced dips close to the Mie resonances, but these dips become weaker or disappear altogether as f increases. These results capture the correct physics, by exhibiting dips near resonant frequencies for low f and no structure for high f, in agreement with experiment. As we have mentioned before, the v_E obtained by the Amsterdam group [2, 7, 8] exhibits strong dips even at f = 0.60 (see Fig. 11 of Ref. 10 for a comparison with Fig. 3 of this paper), because of its low density limit. We have also analytically calculated the long wavelength limit results for $\bar{\epsilon}$ and found that the energy density – CPA yields that $\bar{\epsilon}$ is given by

$$\bar{\epsilon} = \epsilon_2 \left(1 + \frac{3f\alpha}{1 - f\alpha} \right) \quad \text{where} \quad \alpha = \frac{\epsilon_1 - \epsilon_2}{\epsilon_1 + 2\epsilon_2}, \tag{7}$$

which is the Maxwell-Garnett theory result. Finally, we want to mention that within the energy density – CPA, we can also calculate[13] the scattering mean path ℓ_s from the following expression

$$\ell_s = \frac{1}{\sqrt{2}Im\Sigma} \left[(k_e^2 - Re\Sigma) + \sqrt{(k_e^2 - Re\Sigma)^2 + (Im\Sigma)^2} \right]^{1/2} \tag{8}$$

Preliminary results for ℓ_s agree with the weak scattering results for low f and low frequency. At higher f, there is structure in the frequency dependence of ℓ_s, near the Mie resonances, which becomes weaker as f increases.

4. Conclusions

In summary, we have presented a new scheme for calculating the transport properties of random media. This new scheme captures the effects of Mie resonances, always present in cases of finite scatterers, and the interparticle interactions. It is proposed that the choice of an effective medium is based on the principle that the wave energy density should be uniform when averaged over length scales larger than the size of inhomogeneities. This energy density – CPA has been applied to scalar and vector waves. Results for the transport velocity have shown to display dips at scatterer resonance frequencies for low concentrations the variation with frequency is reduced. This new scheme can be easily applied to different types of classical waves.

5. Acknowledgments

We want to thank Ping Sheng for discussing an energy homogenization idea as applied to the scalar wave in Chapter 8 of his book (Ref. 13), and E. N. Economou, and P. Woelfe for useful discussions. Ames Laboratory is operated for the U. S. Department of Energy by Iowa State University under contract No. W-7405-ENG-82. This work was supported by the Director of Energy Research, Office of Basic Energy Sciences and NATO Grant No. CRG 940647.

References

1. For a review, see *Scattering and Localization of Classical Waves in Random Media*, Ed. by Ping Sheng (World Scientific, Singapore, 1990); *Photonic Band Gaps and Localization*, Ed. by C. M. Soukoulis (Plenum, New York, 1993).
2. M. P. van Albada, B. A. van Tiggelen, A. Lagendijk, and A. Tip, Phys. Rev. Lett. **66**, 3132 (1991); Phys. Rev. B **45**, 12233 (1992).
3. E. Kogan and M. Kaveh, Phys. Rev. B **46**, 10636 (1992).
4. G. Cwilich and Y. Fu, Phys. Rev. B **46**, 12015 (1992).
5. J. Kroha, C. M. Soukoulis, and P. Wolfle, Phys. Rev. B **47**, 9208 (1992).
6. Yu. N. Barabanenkov and V. Ozrin, Phys. Rev. Lett. **69**, 1364 (1992).
7. B. A. van Tiggelen, A. Lagendijk, and A. Tip, Phys. Rev. Lett. **71**, 1284 (1993).
8. B. A. van Tiggelen and A. Lagendijk, Europhys. Lett. **23**, 311 (1993).

9. N. Garcia, A. Z. Genack, and A. A. Lisyansky, Phys. Rev. B **46**, 14475 (1992); A. A. Lisyansky et al., in *Photonic Band Gaps and Localization*, Ed. by C. M. Soukoulis (Plenum, New York, 1993), p. 171.

10. C. M. Soukoulis, S. Datta, and E. N. Economou, Phys. Rev. B **49**, 3800 (1994).

11. K. Busch, C. M. Soukoulis, and E. N. Economou, Phys. Rev. B **50**, 93 (1994).

12. X. Jing, P. Sheng, and M. Zhou, Phys. Rev. A **46**, 6513 (1992); Physica A **207**, 37 (1994).

13. Ping Sheng, *Introduction to Wave Scattering, Localization and Mesoscopic Phenomena*, (Academic, New York, 1995), Chapters 3, 4, and 8.

14. B. A. van Tiggelen, Ph.D. Thesis, University of Amsterdam (1992).

15. The energy content of a coated sphere in the scalar case is given by $E = E_s + E_c$ where

$$E_s = \frac{1}{2} \frac{\omega^2}{c^2} \epsilon_1 \sum_{l=0}^{\infty} \frac{1}{k_1^3} \int_0^{k_1 R_i} \rho^2 \, d\rho * |c_l|^2 \, W_l^{(s)}(j_l, j_l; \rho)$$

$$E_c = \frac{1}{2} \frac{\omega^2}{c^2} \epsilon_2 \sum_{l=0}^{\infty} \frac{1}{k_2^3} \int_{k_2 R_i}^{k_2 R_c} \rho^2 \, d\rho * \left[|c_l|^2 \, \phi_l^2 \, W_l^{(s)}(j_l, j_l; \rho) + \right.$$
$$\left. |c_l|^2 \, \zeta_l^2 \, W_l^{(s)}(n_l, n_l; \rho) + 2 \, |c_l|^2 \, \phi_l \, \zeta_l \, W_l^{(s)}(j_l, n_l; \rho) \right]$$

$$W_l^{(s)}(z_l, \overline{z}_l; \rho) = (2l+1) z_l(\rho) \overline{z}_l(\rho) + l \, z_{l-1}(\rho) \overline{z}_{l-1}(\rho) + (l+1) z_{l+1}(\rho) \overline{z}_{l+1}(\rho)$$

$$\phi_l = (k_2 R)^{1/2} (j_l(k_1 R) n_l'(k_2 R) - (k_1/k_2) j_l'(k_1 R) n_l(k_2 R))$$
$$\zeta_l = (k_2 R)^{1/2} ((k_1/k_2) j_l(k_2 R) j_l'(k_1 R) - j_l'(k_2 R) j_l(k_1 R))$$

where $k_i = \epsilon_i^{1/2} \omega/c$ and $i = 1, 2$. j_l and n_l denote the spherical Bessel functions of first and second kind, respectively. The c_l are the scattering coefficients for the field inside the core.

16. The energy content of a coated sphere in the vector case is given by $E = E_s + E_c$ where

$$E_s = \frac{1}{2} \frac{\omega^2}{c^2} \epsilon_1 \sum_{l=1}^{\infty} \frac{1}{k_1^3} \int_0^{k_1 R_i} \rho^2 \, d\rho * (|c_l|^2 + |d_l|^2) \, W_l^{(v)}(j_l, j_l)$$

$$E_c = \frac{1}{2} \frac{\omega^2}{c^2} \epsilon_2 \sum_{l=1}^{\infty} \frac{1}{k_2^3} \int_{k_2 R_i}^{k_2 R_c} \rho^2 \, d\rho *$$
$$\left[(|c_l|^2 \, \phi_l^2 + |d_l|^2 \, \gamma_l^2) \, W_l^{(v)}(j_l, j_l) + \right.$$
$$(|c_l|^2 \, \zeta_l^2 + |d_l|^2 \, \eta_l^2) \, W_l^{(v)}(n_l, n_l) +$$
$$\left. 2 \, (|c_l|^2 \, \phi_l \, \zeta_l + |d_l|^2 \, \gamma_l \, \eta_l) \, W_l^{(v)}(j_l, n_l) \right]$$

$$W_l^{(v)}(z_l, \overline{z}_l) = (2l+1) z_l(\rho) \overline{z}_l(\rho) + (l+1) z_{l-1}(\rho) \overline{z}_{l-1}(\rho) + l \, z_{l+1}(\rho) \overline{z}_{l+1}(\rho)$$

$$\phi_l = \psi_l(k_1 R) \chi_l'(k_2 R) - (k_2/k_1) \psi_l'(k_1 R) \chi_l(k_2 R)$$
$$\zeta_l = \psi_l(k_2 R) \psi_l'(k_1 R) - (k_2/k_1) \psi_l'(k_2 R) \psi_l(k_1 R)$$
$$\gamma_l = (k_2/k_1) \chi_l(k_2 R) \psi_l'(k_1 R) - \chi_l'(k_2 R) \psi_l(k_1 R)$$
$$\eta_l = (k_2/k_1) \psi_l(k_2 R) \psi_l'(k_1 R) - \psi_l'(k_2 R) \psi_l(k_1 R)$$

where $k_i = \epsilon_i^{1/2}\omega/c$ and $i = 1, 2$. ψ and χ denote the Ricatti-Bessel functions of the first and second kind, respectively. The c_l and d_l are the scattering coefficients for the field inside the core.

PHOTONIC BAND STRUCTURES OF ATOMIC LATTICES

RUDOLF SPRIK[1], AD LAGENDIJK[1,2]
[1] van der Waals-Zeeman Instituut, Universiteit van Amsterdam,
Valckenierstraat 65-67, 1018 XE Amsterdam, The Netherlands
[2] FOM Institute for Atomic and Molecular Physics, 1098 SJ
Amsterdam, The Netherlands

AND

BART A. VAN TIGGELEN[3]
Laboratoire de Physique Numérique des Systèmes Complexes,
Université Joseph Fourier/CNRS, 38042 Grenoble Cedex 9,
France

Abstract. Light scattering from resonant two-level atoms on three-dimensional lattices can be described by a classical energy conserving t-matrix. The optical band structure of lattices filled with such atoms are calculated exactly in scalar approximation and displays, e.g., the formation of polariton gaps. The Einstein coefficient for spontaneous emission in a system with an inhomogeneous dielectric constant is shown to be proportional to part of the total density of states. This part is calculated for the dipolar lattice model.

1. Introduction

The first steps towards three-dimensional lattices of laser trapped atoms have been successfully taken with the use of laser cooling techniques [1]. The propagation of light with wavelengths near the optical resonances in the atoms is dominated by multiple scattering from occupied unit cells and will lead to the formation of well-defined optical band structures when all cells are filled.

The band formation for the propagating light is similar to photonic band structures in periodic three-dimensional dielectrics ('photonic band gap materials') [2]. The main difference is the strong resonant character of

C. M. Soukoulis (ed.), Photonic Band Gap Materials, 679–690.

the scatterers near an optical resonance. Furthermore, in the limit of weak light fields and if recoil effects are ignored, the propagation of the light is coherent ([3] p. 414) and without dissipation. Some of these aspects are also encountered in photonic crystals with dielectric and metallic components [4], although dissipation is usually present in these systems.

2. Elastic t-matrix for Dipole Scatterers

A proper description of the photonic band structure of the filled trapped atom lattice may be based on the t-matrix of an energy conserving, resonant dipole oscillator, both in the scalar approximation to the Maxwell equations and the full vector form [5]. This t-matrix is the exact classical representation for a dipole transition in a two-level atom and describes all the multiple elastic scattering events near one center. The t-matrix of the individual scatterers is, in general, complex and depends explicitly on frequency ω. It obeys a detailed balance between the energy of the incoming wave and all the scattered waves (i.e., fulfills the *optical theorem*). The exact representation of the t-matrix of such dipole scatterers with a resonance has been developed earlier. The energy dependent t-matrix for optical point scatterers fulfilling the optical theorem can be derived by using semi-classical arguments[6] or from a more fundamental approach [5]. In scalar approximation the t-matrix reduces to:

$$t(\omega) = \frac{-4\pi\omega^2}{\alpha^{-1} - \beta\omega^2 - i\omega^3} \tag{1}$$

$$= -\frac{4\pi}{\omega}\sin(\eta)\exp(i\eta),$$

with

$$\tan(\eta) = \frac{1}{\beta}\frac{\omega^3}{\omega_0^2 - \omega^2}, \tag{2}$$

α the polarizability of the scatterer and β the width of the resonance. In this form, the t-matrix has a resonance near $\omega_0 = \sqrt{1/\alpha\beta}$. It is different from the conventional t-matrix of an elastic scatterer in the Schrödinger equation [7]. In particular the ω^3 energy dependence in the numerator reflects the Rayleigh limit for scattering of light at low energies. The t-matrix of the point scatterer explicitly depends on the energy but not on the wave vectors of the incoming and outgoing waves.

3. Optical Bandstructure of Dipolar Lattices

With the use of formalisms developed for solid state band calculations (see e.g., [2, 7]) to exploit the symmetry of the lattice, the photonic band structure of the optical atomic lattice can be calculated by diagonalizing the secular matrix for plane wave Bloch states in the crystal [8]. The coupling terms between the plane waves are essentially determined by the phase shift of the t-matrix in (1) and are equal for all the plane waves. In the context of bandstructure calculations, the standing wave solutions of an infinitely extended system are relevant. Such standing wave solutions do not exist for systems with absorption. Therefore, it is essential the t-matrix fulfills the optical theorem [9].

In the following we will limit our calculations to the scalar approximation. Many of the features discussed for the scalar case, are also relevant for the full vector approximation[12].

In scalar approximation the secular equation reduces to:

$$\det \left| \left[(\vec{k} - \vec{g})^2 - \omega^2 \right] \delta_{(\vec{k}-\vec{g})(\vec{k}-\vec{g}\,')} + \Gamma(\omega, \vec{k} - \vec{g}, \vec{k} - \vec{g}\,') \right| = 0, \tag{3}$$

with

$$\begin{aligned}
\Gamma(\omega, \vec{k} - \vec{g}, \vec{k} - \vec{g}\,') &= -\frac{4\pi}{\omega\Omega} \tan(\eta(\omega)) \\
&= -\frac{4\pi}{\beta\Omega} \frac{\omega^2}{\omega_0^2 - \omega^2}. \tag{4}
\end{aligned}$$

Here $\vec{g}$, $\vec{g}\,'$ are reciprocal lattice vectors, $\vec{k}$ the Bloch wave vectors, and Ω the volume of the unit cell.

Since all the off-diagonal elements in the matrix are equal, the secular equation may be simplified considerably. In general, given a set of constants $\{a_1, ..., a_N\}$, the determinant of the matrix of order N with elements $D_{ij} = \delta_{ij}a_i + \Gamma$, $i = 1, .., N$ $j = 1, .., N$ is given by:

$$|D_{ij}| = \left\{ \Gamma \sum_{i=1}^{N} \left[\frac{1}{a_i} \right] + 1 \right\} \left\{ \prod_{i=1}^{N} a_i \right\}. \tag{5}$$

Choosing $a_i = (\vec{k} - \vec{g}_i)^2 - \omega^2$, Γ given by (4) and with the use of (5), the secular equation simplifies to:

$$\left\{ \sum_{i=1}^{N} \frac{1}{(\vec{k} - \vec{g}_i)^2 - \omega^2} \right\} \Gamma(\omega) + 1 = 0. \tag{6}$$

This is essentially the same result as the first term in a Green's function approach to the bandstructure calculation exploiting the isotropic character of the pointlike scatterer and the exact representation of the phase shift terms associated with the elastic t-matrix [9].

Equation (6) can be reproduced by regularizing the full lattice sum instead of the individual scatterers by introducing a cut-off g_c in the summation. Splitting the sum over all $\vec{g}$ in a part with $|\vec{g}| < g_c$ as in (6) and a part with $|\vec{g}| > g_c$ enables regularization. In particular for $|\vec{g}|$ much larger than $|\vec{k}|$,

$$
\sum_{\vec{g}} \frac{1}{(\vec{k} - \vec{g}_i)^2 - \omega^2} = \sum_{|\vec{g}| < g_c} \left\{ \frac{1}{(\vec{k} - \vec{g}_i)^2 - \omega^2} \right\} + \sum_{|\vec{g}| > g_c} \left\{ \frac{1}{(\vec{k} - \vec{g}_i)^2 - \omega^2} \right\}
$$

$$
\approx \sum_{|\vec{g}| < g_c} \frac{1}{(\vec{k} - \vec{g}_i)^2 - \omega^2} + \sum_{\vec{g}_i} \frac{1}{\vec{g}_i^2} - \sum_{|\vec{g}| < g_c} \frac{1}{\vec{g}^2}. \tag{7}
$$

The sum for $\sum_{\vec{g}_i} 1/\vec{g}_i^2$ diverges in 3D and the regularization procedure replaces the *frequency independent* sum $\sum_i 1/|\vec{g}_i|^2$ by $\beta/(4\pi) > 0$. This establishes the resonant expression of the 'optical potential' $\Gamma(\omega)$ in the dispersion laws described by Eq.(3) and Eq.(5) (in 1D no regularization is necessary and we recover the Kronig-Penney model).

We solved the secular Eq.(3) numerically for an fcc lattice of resonant dipoles with a resonance frequency ω_0. Two characteristic cases for ω_0 with respect to the typical frequency of the first Brillouin zone (ω_{BZ}) are discussed here:

1. $\omega_0 < \omega_{BZ}$ (Fig. 1.a)

   The dispersive effects of the scatterers cause a distortion of the band structure and the formation of gaps near the Brillouin zone. Whether such a splitting occurs depends on the symmetry properties of the particular point in the Brillouin zone. Furthermore, near the resonance frequency ω_0 of the two-level system, another genuine band gap develops. The width of the gap depends on the coupling strength and signifies essentially a polariton-type propagation in the crystal [10].

2. $\omega_0 \approx \omega_{BZ}$ (Fig. 2.a)

   When ω_0 is tuned close to ω_{BZ}, combinations of distortion and polariton gap occur. The polariton gap disappears due to the anisotropic nature of the bandstructure near the Brillouin zone.

4. Einstein A Coefficient in a Inhomogeneous Dielectric Medium

The spontaneous emission properties of an atom inside a photonic band gap material is an interesting problem both for applications and for a more fundamental understanding of the quantum properties of an inhomogeneous dielectric system.

It is well known that a 'density of states' of the radiation field appears in the expression for the Einstein coefficient [3]. It turns out that in dielectric systems with a spatially varying dielectric constant, it is a subtle question which density of states is appearing in the Einstein coefficient. For this purpose we will first discuss the proper definition of the density of states. We would like to use the eigenmodes of the scalar wave equation as described by the Helmholtz equation:

$$\left[-\nabla^2 - \omega_j^2 \varepsilon\right] |\psi_j\rangle = 0, \tag{8}$$

where $\langle \vec{r}|\varepsilon|\vec{r}\rangle = \varepsilon(\vec{r})$ is the scalar equivalent of the 'dielectric constant'. Unfortunately, Eq. (8) is not a conventional eigenvalue equation. Let us try to manipulate this scalar wave equation into a proper eigenvalue equation. Dividing by ε is tempting, and gives

$$\left[-\varepsilon^{-1}\nabla^2 - \omega_j^2\right] |\psi_j\rangle = 0 . \tag{9}$$

This looks like a conventional eigenvalue equation but it is not, as the operator $\varepsilon^{-1}\nabla^2$ is not hermitian. To cope with this complication, we symmetrize and find,

$$\left[-\varepsilon^{-1/2}\nabla^2\varepsilon^{-1/2} - \omega_j^2\right] \left|\varepsilon^{1/2}\psi_j\right\rangle = 0 . \tag{10}$$

This is a conventional eigenvalue equation and can be written compactly as

$$\mathcal{L}|\Lambda_j\rangle = \omega_j^2 |\Lambda_j\rangle , \text{where} \left|\varepsilon^{1/2}\psi_j\right\rangle \equiv |\Lambda_j\rangle . \tag{11}$$

The hermitian operator $\mathcal{L}$ is defined as

$$\mathcal{L} \equiv -\varepsilon^{-1/2}\nabla^2\varepsilon^{-1/2} . \tag{12}$$

The set, $\{|\Lambda_j\rangle\}$, can be normalized according to: $\langle\Lambda_j|\Lambda_k\rangle = \delta_{jk}$. The secular Eq.(3) given earlier is essentially the matrix of $\mathcal{L}$ with respect to the basis $|\varepsilon^{1/2}\vec{p}\rangle$.

As $\mathcal{L}$ is Hermitian its eigenvalues are real and we can count the density of states in the usual way. The total density of states (TDOS) with eigenvalue ω^2 between ω^2 and $\omega^2 + d\omega^2$ is

$$N_T(\omega^2) = \sum_j \delta(\omega^2 - \omega_j^2), \tag{13}$$

which can be written as a trace according to,

$$N_T(\omega^2) = \text{Tr}\left[\delta(\omega^2 - \mathcal{L})\right] . \tag{14}$$

It is often more convenient to have the density of states as a function of ω rather than ω^2,

$$N_T(\omega) = 2\omega N(\omega^2) . \tag{15}$$

It is now also possible to decompose $N_T(\omega)$ into local contributions (LT-DOS) according to

$$N_T(\omega, \vec{r}) = 2\omega \langle \vec{r} | \, \delta(\omega^2 - \mathcal{L}) \, | \vec{r} \rangle. \tag{16}$$

We have obtained a well-defined density of states (16) that is found by diagonalizing an Hermitian operator and by counting eigenvalues. This genuine density of states $N_T(\omega, \vec{r})$ is, for instance, important when discussing the transport of radiative energy in the dielectric [11]. However, $N_T(\omega, \vec{r})$ is not the local density that is featuring in the Einstein coefficient. To make this point clear, it is worth while to manipulate the delta-function in Eq. (14),

$$\delta(\omega^2 + \varepsilon^{-1/2}\nabla^2\varepsilon^{-1/2}) = \varepsilon^{1/2}\delta(\varepsilon\omega^2 + \nabla^2)\varepsilon^{1/2}. \tag{17}$$

Here the general property of the delta function: $\delta(ax) = (1/a)\delta(x)$, is used. The new delta function on the rhs. contains the wave operator of the wave equation, $\varepsilon\omega^2 + \nabla^2$. By the cyclic permutation property of a trace, the square root can be eliminated to give:

$$N_T(\omega) = 2\omega \mathrm{Tr}\left[\varepsilon^{1/2}\delta(\varepsilon\omega^2 + \nabla^2)\varepsilon^{1/2}\right] = 2\omega \mathrm{Tr}\left[\varepsilon\delta(\varepsilon\omega^2 + \nabla^2)\right]. \tag{18}$$

We will now derive the correct expression for the Einstein A coefficient inside a spatial dependent dielectric structure. The emission rate τ is a function of the position $\vec{R}$ of the impurity atom that is incorporated in the dielectric, and obtained by applying Fermi's golden rule [3]:

$$\tau(\vec{R}) = \frac{2\pi}{\hbar} \sum_f \left| < f|\vec{\mu} \cdot \vec{E}_{\mathrm{op}}(\vec{R})|i > \right|^2 \delta(E_f - E_i). \tag{19}$$

Here the initial state $|i\rangle = |\sigma\rangle_i \otimes |\psi\rangle_i$ and final state $|f\rangle = |\sigma\rangle_f \otimes |\psi\rangle_f$ describe the state of the atom $|\sigma\rangle$ *and* the light field in scalar approximation $|\psi\rangle$. The electric field is represented by the quantum-mechanical operator $\vec{E}_{\mathrm{op}}$.

The set $\{|\Lambda_j\rangle\}$ can serve as a basis to quantize the radiation field:

$$\vec{E}_{\mathrm{op}}(\vec{r}) = \sum_j \left\{ \sqrt{\frac{\hbar\omega_j}{2\varepsilon(\vec{r})}} \, ia_j^\dagger \, \Lambda_j(\vec{r}) \, \exp(i\omega_j t) + \mathrm{h.c.} \right\}, \tag{20}$$

with $a_j^\dagger$ the creation operator of the field mode.

With this definition, the Hamiltonian of the radiation be can written as $\mathcal{H} = \sum_j \hbar\omega_j a_j^\dagger a_j$. Substitution of (20) in the field part of the Einstein coefficient (19) gives:

$$\tau(\vec{R}) = \mathbf{A}\, 2\omega \sum_{j} \varepsilon(\vec{R})^{-1} \Lambda_j(\vec{R}) \delta(\omega^2 - \omega_j^2) \Lambda_j(\vec{R})$$

$$= \mathbf{A}\, 2\omega\varepsilon(\vec{R})^{-1} \langle \vec{R} | Tr(\omega^2 - \mathcal{L}) | \vec{R} \rangle \equiv \mathbf{A}\, 2\omega N_R(\omega, \vec{R}), \quad (21)$$

where $\mathbf{A}$ is proportional with the dipole transition moment of the two level atom.

This demonstrates that the Einstein coefficient is proportional to what we call the local density of states of the radiation,

$$N_R(\omega, \vec{r}) \equiv \varepsilon(\vec{r})^{-1} N_T(\omega, \vec{r}), \quad (22)$$

and *not* the total DOS is featuring as given in (18). Writing $\varepsilon(\vec{r}) \equiv 1 + \delta\varepsilon(\vec{r})$ we find

$$N_T(\omega, \vec{r}) = N_R(\omega, \vec{r}) + \delta\varepsilon(\vec{r}) N_R(\omega, \vec{r}) \equiv N_R(\omega, \vec{r}) + N_M(\omega, \vec{r}). \quad (23)$$

Recapitulating, the effect of the inhomogeneous dielectric constant $\varepsilon(\vec{r})$ is twofold. Firstly, the emission rate depends on the position. This was allready found in the measurements of the emission lifetime of atoms in a homogeneous dielectric medium near an interface[13, 14] and in micro-cavities[15]. For the dipole lattice this means that the emission rate may depend considerably on the position of the radiating dipole in the unit cell [16, 17, 18]. Secondly, the emission rate is not proportional to the local density of states based on the mode density given in Eq.(18), but to the local density of states for radiation given in Eq.(23), (LRDOS). Only part of the modes are actually contributing to the emission rate.

5. Emission Rates in a Lattice of Resonant Dipoles

We will illustrate numerically the dependence of the emission rate on the position in the unit cell and the difference between the total density of states (18) and the radiation density of states (23) using the earlier mentioned model of resonant dipoles on a lattice. For the dipolar lattice the contribution to the total RDOS and the local MDOS can be separated explicitly [12]:

$$N_R^{dip}(\omega) = \sum_{\vec{k}} \frac{F(\omega, \vec{k})}{F(\omega, \vec{k}) + 1} \delta(\omega^2 - \omega_{\vec{k}}^2)$$

$$N_M^{dip}(\omega) = \sum_{\vec{k}} \frac{1}{F(\omega, \vec{k}) + 1} \delta(\omega^2 - \omega_{\vec{k}}^2), \quad (24)$$

686

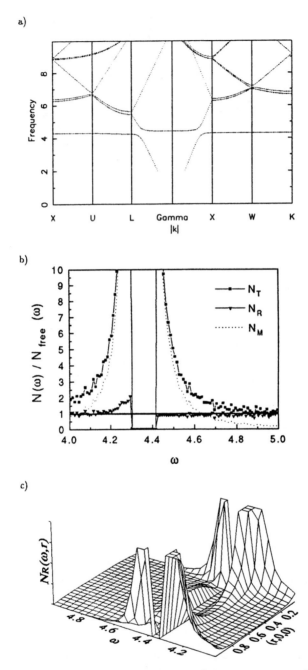

Figure 1. Photonic properties of an fcc lattice of resonant atoms with a resonance frequency at $\omega_0\, a = 5 < \omega_{BZ}$ and $\beta\Omega/a^2 = 25$ a) The photonic bandstructure. b) The total density of states $N_T(\omega)$ and the total density of states for the radiation $N_R(\omega)$ and matter $N_M(\omega)$ scaled to the density of states in vacuum $N_{free}(\omega)$. c) The local density of states for the radiating field $N_R(\omega, \vec{r})$ on a traject moving from $\vec{r} = (0, 0, 0)$ to $\vec{r} = (a, 0, 0)$ with $a = 1$ the fcc latticeconstant.

a)

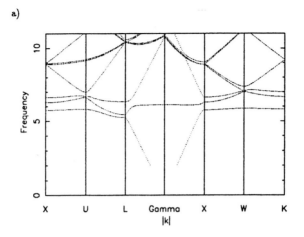

b)

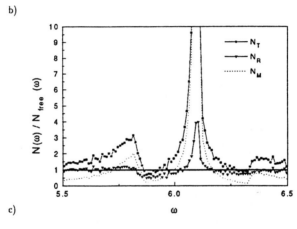

c)

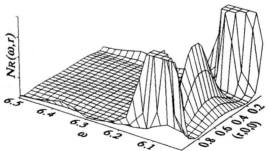

Figure 2. Same as Fig.1, with $\omega_0 a = 7.0 \approx \omega_{BZ}$ and $\beta\Omega/a^2 = 25$.

with

$$F^{-1}(\omega, \vec{k}) = 4\pi\alpha\omega^4 \sum_{\vec{g}} \left[(\vec{k} - \vec{g})^2 - \omega^2\right]^{-2}. \tag{25}$$

The separation of the local properties of radiation and matter in Eq. (23) can be performed explicitly for a lattice of dipoles. Due to the pointlike character of the dipole:

$$N_M(\omega, \vec{r}) = \sum_i \delta(\vec{r} - \vec{r}_i) N_M^{dip}(\vec{r}), \tag{26}$$

with $\vec{r}_i$ the position of an atom on the lattice.
For positions $\vec{r} \neq \vec{r}_i$, $N_M(\omega, \vec{r}) = 0$ and the radiative part is:

$$N_R^{dip}(\omega, \vec{r}) = \sum_{\vec{k}} |\psi_{\vec{k}}(\vec{r})|^2 \delta(\omega^2 - \omega_{\vec{k}}^2), \tag{27}$$

with $\psi_{\vec{k}}(\vec{r})$ the eigenfunctions obtained by diagonalizing the secular equation.

We calculated the density of states for the two cases discussed previously.

1. ω_0 well within the first BZ (Fig. 1.b,c)
   The polariton gap in the band structure is associated with a suppressed DOS (Fig. 1.b). The LRDOS is plotted in Fig. 1.c and shows clearly a spatial and energy dependence. As expected, the DOS vanishes in the polariton gap. Above and below the gap the LRDOS has a distinct structure and is well below the values on the atomic positions. In particular, for energies above the gap, at $\vec{r} \approx (0.1a, 0, 0)$, the LRDOS is almost completely suppressed for an excited atom located there. This would correspond to an infinite emission lifetime and could lead to a bound light state as discussed by John *et al.* [19].

2. ω_0 near the first BZ (Fig. 2.b,c)
   Although the polariton gap is destroyed by the influence of the Brillouin zone, there is still a considerable structure in the total DOS (Fig. 2.b) and in the LRDOS (Fig. 2.c). Similar suppression of the LRDOS as seen in case 1 survives.

6. Conclusion

In this work we described the properties of atomic lattices by applying an elastic t-matrix formalism. The polariton character of the resonance results in a gap in the frequency spectrum. Near the Brillouin zone of the crystalline structure the influence of the anisotropy destroyes the polariton gap. The emission properties inside the crystal are drastically modified at positions in between atomic sites and is proportional to the local density of states of the radiation, which is only a part of the total density of states. This should also hold for photonic band gap materials in general. Tayloring the positions of the active centers inside the unit cell relaxes the need for a full band gap to efficiently suppress spontaneous emission. We only presented calculations in scalar approximation. Extension of the results to the vector equations is under way maintaining many of the features in the present scalar approximation.

7. Acknowledgments

This research has been supported by the Stichting voor Fundamenteel Onderzoek der Materie (FOM), which is financially supported by the Nederlandse Organisatie voor Wetenschappelijk Onderzoek (NWO). BvT acknowledges a grant from the French ministery for Foreign Affairs.

References

1. P. Verkerk, B. Lounis, C. Salomon, C. Cohen-Tannoudji, J.-Y. Courtois, G. Grynberg, Phys. Rev. Lett. **68**, 3861 (1992). P.S. Jessen, C. Gerz, P.D. Lett, W.D. Phillips, S.L. Rolston, R.J.C. Spreeuw, C.I. Westbrook, Phys. Rev. Lett. **69**, 49 (1992). G. Grynberg, B. Lounis, P. Verkerk, J.-Y. Courtois, C. Salomon, Phys. Rev. Lett. **70**, 2249 (1993). A. Hemmerich, C. Zimmerman, T.W. Hänsch, Europhysics Lett. **22**, 89 (1993).
2. See e.g. NATO Workshop, 'Photonic Band Gaps and Localization' edited by C.M. Soukoulis, Plenum Press, New York, (1993).
3. P. Meystre, M. Sargent III, 'Elements of Quantum Optics' 2nd edition, Springer (1991).
4. J.B. Pendry, J. Mod. Opt. **41**, 209 (1994).
5. Th. M. Nieuwenhuizen, A. Lagendijk, B.A. van Tiggelen, Phys. Lett. A **169**, 191 (1992). B. van Tiggelen, 'Multiple Scattering and Localization of Light', Ph.D. Thesis University of Amsterdam, (september 1992).
6. R. Loudon, 'The Quantum Theory of Light', Clarendon, Oxford (1973), first edition.
7. A. Gonis, 'Green Functions for Ordered and Disordered Systems', Studies in Mathematical Physics vol.4, editors E. van Groesen, E.M. de Jager, North Holland (1992).
8. R. Sprik, A. Lagendijk, 'Coupled Dipole Approximation of Photonic Band Structures in Colloidal Crystals' contribution QWA6, European Quantum Electronics Conference (1994).
9. J.M. Ziman, Proc. Phys. Soc. **86**, 337 (1965).
10. N.W. Ashcroft and N.D. Mermin, 'Solid State Physics', Saunders (1976).
11. M.P. van Albada, B.A. van Tiggelen, A. Lagendijk, A. Tip, Phys. Rev. Lett. **66**, 3132 (1991).
12. Publication with details in preparation.
13. E. Yablonovitch, T.J. Gmitter, R. Bhat, Phys. Rev. Lett. **61**, 2546 (1988).
14. E. Snoeks, A. Lagendijk, A. Polman, Phys. Rev. Lett. **74**, 2459 (1995).
15. F. de Martini, M. Marrocco, P. Mataloni, D. Murra, R. Loudon, J. Opt. Soc. Am. B **10**, 360 (1993).

690

16. G. Kurizki and A.Z. Genack, Phys. Rev. Lett. **61**, 2269 (1988); G. Kurizki, Phys. Rev. A **42**, 2915 (1990).
17. J.P. Dowling, C.M. Bowden, Phys. Rev. A **46**, 612 (1992).
18. T. Suzuki and P.K.L. Yu, J. Opt. Soc. Am. B **12**, 570 (1995).
19. S. John, Phys. Rev. Lett. **63**, 2169 (1984); **58**, 2486 (1987). S. John, and J. Wang, Phys. Rev. B **43**, 12 772 (1991). S. John, and T. Quang, Phys. Rev. A **50**, 1764 (1994). S. John, T. Quang, Phys. Rev. Lett. **74**, 3419 (1995).

INTERACTION OF DIFFERENT SCATTERING MECHANISMS IN A ONE-DIMENSIONAL RANDOM PHOTONIC LATTICE

S.A. BULGAKOV[1,2] AND M. NIETO-VESPERINAS,[1]

[1]*Instituto de Ciencia de Materiales, Sede B, C.S.I.C.,*
Departamento de Física de la Materia Condensada,
Universidad Autónoma de Madrid, Facultad de Ciencias,
C-III, Cantoblanco, E-28049, Madrid, Spain
[2]*Radioastronomy Institute Ukrainian Academy of Sciences,*
4 Krasnoznamennaya str., 310002 Kharkov, Ukraine

Abstract.
    A 1D random photonic lattice has been studied both by the plane-wave method and by the transfer matrix approach for arbitrary contrast in its dielectric permittivity. The role and competition of different scattering mechanisms such as Bragg diffraction, one scatterer resonances, and Bragg remnants have been investigated. It is shown that strong localization results not only from Bragg diffraction and disorder in the scatterers distribution, but also Bragg remnants produce localization for a medium with small contrast. Based on the properties of the random lattice in the light frequency band, highly sensitive sensor could be devised.

1. Introduction

In recent years, there is an increased interest in studies of photonic band gap (PBG) structures due to experiments with photonic materials [1] in one-dimensional configurations, which permit detailed investigations of the scattering process resulting in wave transmission [2, 3], and also because of their important application in layered media devices [4].

    There are three different scattering mechanisms appearing in the propagation of a wave through a stratified medium. They are Bragg diffraction, one scatterer resonances, and the effect of Bragg remnants associated to

C. M. Soukoulis (ed.), Photonic Band Gap Materials, 691–702.

the periodical properties of the random superlattice [5-8]. All three are important and must be studied both for regular and random lattices.

On the other hand, until now there is no detailed account of PBG for higher frequency bands. However, such an analysis is important both from a practical viewpoint and for the understanding of the fundamentals.

The aim of the present paper is to study 1-D photonic lattices under the consideration of PBG, light localization, and competition of the scattering mechanisms quoted above. A random structure is necessary to study the role of each of these three mechanisms separately as well as their contributions together. We pay special attention to the relationship between the characteristics of light localization and scattering by both regular and random photonic structures [9].

2. Model and Mathematical Description

We address the problem of electromagnetic wave propagation through a 1-D structure represented by a sequence of dielectric slabs with a stochastic refractive index, Poisson distributed, built by three different ways as follows. The first way is a random removal of layers from an initial regular distribution. Let us assume a sequence of l unit cells. Each cell of length d contains one slab with dielectric permittivity ε_{back}, one slab with permittivity $\varepsilon = \varepsilon_{slab}$ of length a. $\varepsilon_{back} < \varepsilon_{slab}$. We substitute some layers with ε_{slab} by those with ε_{back} in some cells chosen in a random manner. Such a random structure has some analogy with the 2-D lattice studied in the experiment of Ref. [2]. The next type of the randomness is the stochastic variation from slab to slab in the filling factor $\alpha = a/d$. The third type of disorder is the randomly distributed refractive index $\sqrt{\varepsilon_{slab}}$ of the scatterers.

Our study applies to arbitrary differences between the refractive indices of the layers and also to a high degree of disorder in all kinds of the randomness.

a) We study the dispersion relation for the superlattice, whose period (L) contains l unit cells ($L = ld$) with the random distribution of the dielectric constant built as shown above. This allows us to apply the plane-wave method based on Bloch's theorem [10, 11].

We shall consider light or other electromagnetic wave linearly polarized with its electric vector in a plane perpendicular to the Z-axis. In this case, we can use a scalar approach. We shall then represent $E(z,t)$ as one transversal component of the electric vector of this wave. The periodicity in our structure allows us to find a solution in the form $E(z,t) = \exp(iKz - i\omega t)E_K(z)$, $E_K(z) = E_K(z + L)$, K being the Bloch wavevector. We expand the periodic Bloch functions E_K in terms of the reciprocal lattice vectors $G = 2\pi m/L$, where $m = 0, \pm 1, \pm 2, \ldots$, obtaining for each value of G:

$$(G+K)^2 E_G^K - \frac{\omega_K^2}{c^2} \sum_{G'=-\infty}^{\infty} \varepsilon_{G-G'} E_{G'}^K = 0. \tag{1}$$

Here $\varepsilon_{G-G'}$ is the coefficient determined by $\varepsilon_n = \frac{1}{L} \int_0^L \varepsilon(z) \exp(-\frac{i2\pi nz}{L}) dz$, $n = 0, \pm 1, \pm 2, \ldots$, and

$$\varepsilon_n = \frac{a}{dl} \exp\left[-\frac{i\pi n}{l}(2 - \frac{a}{d})\right] S(\frac{\pi n}{l}\frac{a}{d}) \sum_{j=1}^{l} \bar{\varepsilon}(j) \exp\left[-\frac{i2\pi n(j-1)}{l}\right] +$$

$$+ \frac{\varepsilon_{back}}{l}(1 - \frac{a}{d}) \exp\left[-\frac{i\pi n}{l}(1 - \frac{a}{d})\right] S\left[\frac{\pi n}{l}(1 - \frac{a}{d})\right] \sum_{j=1}^{l} \exp\left[-\frac{i2\pi n(j-1)}{l}\right], \tag{2}$$

where $\bar{\varepsilon}(j)$ is a random dielectric permittivity of the j-th slab. $S(x) = \sin(x)/x$.

It should be noted that Eq. (2) is general for both regular and random superlattices. In particular, the case $l = 1$ corresponds to the regular structure ($\bar{\varepsilon}(1) = \varepsilon_{slab}$) and would lead to results already obtained in Ref. [10].

We solve the matrix Eq. (1) and obtain for each value of the Bloch K vector a set of eigenfrequencies ω_K and eigenvectors E_G^K.

b) We now calculate the electric field intensity at the end of the structure containing l cells with two slabs: $\{\varepsilon_{back}\}$ and $\{\bar{\varepsilon}(j)\}$. We use the Transfer Matrix method [4, 5] which allows us to connect the fields at the beginning and at the end of whole structure:

$$\begin{pmatrix} 1 \\ r \end{pmatrix} = \hat{T} \begin{pmatrix} t \\ 0 \end{pmatrix}. \tag{3}$$

In deriving Eq. (3) it is assumed there is an incident plane wave of unit intensity and no wave coming from the right end. In Eq. (3) r and t are the reflection and transmission coefficients, respectively. Also, $\hat{T} = \prod_{j=1}^{l} \bar{T}^j$, where $\bar{T}^j$ is the transfer matrix of the j-th cell, whose elements are well-known functions of the filling factor $\alpha = a/d$ and wavevectors inside each slab of the j-th unit cell, namely, $\kappa_1^j = \frac{\omega d}{c}\sqrt{\varepsilon_{back}}$, $\kappa_2^j = \frac{\omega d}{c}\sqrt{\bar{\varepsilon}(j)}$ (cf. Ref. [4]). We shall use Eq. (3) to study the wave propagation through the random sequence of the slabs having dielectric permittivity being distributed as shown above. We shall also study the effect of localization and the contribution of the different scattering mechanisms.

3. Localization Length and Band Gap Width in the Higher Frequency Bands

To investigate photonic band gaps and the effects of light localization in a random sequence of dielectric slabs having an alternate dielectric constant, we shall proceed with several realizations of the structure.

3.1. REGULAR LATTICE

In order to study the photonic band gap width in a regular lattice, we write the dispersion relation between the Bloch wavevector K and the frequency of the incident wave ω [4]:

$$K(\omega) = \frac{1}{d} \cos^{-1} \left[\frac{1}{2} \left(\bar{T}_{11} + \bar{T}_{22} \right) \right] \qquad (4)$$

In Eq. (4) we take into account that for a regular lattice, $\bar{\varepsilon} = \varepsilon_{slab}$ for any slab j . From Eq. (4) the boundaries of the PBG can be obtained from the solutions of :

$$|\bar{T}_{11} + \bar{T}_{22}| = 2. \qquad (5)$$

We are interested in the band gap width. Let us number each gap, starting from $N = 1$ for the lowest frequency. From Eq. (5) we also calculate the normalized middle gap frequency $F_{md} \equiv (\omega d/c)_{middle}$. It should be noted that the same result can be obtained by solving the matrix Eq. (1), which is sometimes preferable for the sake of accuracy.

The values of the band gap width (BGW) as functions of the filling factor $\alpha = a/d$ are represented in Fig. 1. We have the following remarks:

- Cases with $\alpha = 0$ or 1, correspond to a transparent medium with $\varepsilon = \varepsilon_{back}$ and $\varepsilon = \varepsilon_{slab}$: the BGW is zero for every band (Fig. 1).

- At each band N, F_{md} starts from the value $\pi N/\sqrt{\varepsilon_{back}}$ for $\alpha = 0$ and decreases to the value $\pi N/\sqrt{\varepsilon_{slab}}$. Some aspects of such dependence have been studied in Ref. [10], where the dependence of F_{md} vs ε_{slab} is represented. By choosing the Bloch vector K in the corresponding Brillouin zone, $Kd/2\pi = N/2$, the dispersion relation for a homogeneous medium ($Kc = \omega\sqrt{\varepsilon}$) results in for arbitrary band N:

$$F_{md}(N) \equiv \frac{\omega d}{c} = \frac{\pi N}{\sqrt{\varepsilon}}; \varepsilon = \varepsilon_{back} , \alpha = 0; \; \varepsilon = \varepsilon_{slab} , \alpha = 1. \qquad (6)$$

The decrease of F_{md} is "almost" proportional to the average refractive index:

$$\bar{n} = \alpha\sqrt{\varepsilon_{slab}} + (1 - \alpha)\sqrt{\varepsilon_{back}} \qquad (7)$$

Due to Eqs. (6)-(7), $F_{md}(\alpha)\bar{n}(\alpha)|_{\alpha=0} = F_{md}(\alpha)\bar{n}(\alpha)|_{\alpha=1} \approx F_{md}(\alpha)\bar{n}(\alpha)$. A shift of the F_{md} value from the law Eq. (6) (given by the mean line of

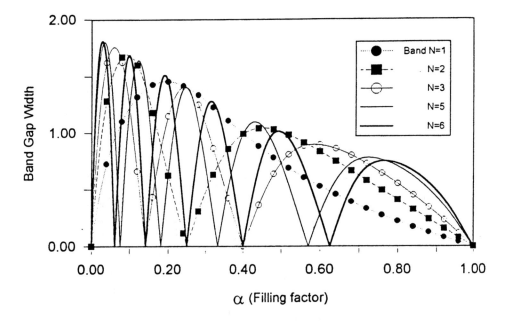

Figure 1. Band gap width as a function of the filling factor for a regular lattice for different frequency bands; $\varepsilon_{back} = 1$, $\varepsilon_{slab} = 9$.

the oscillations) means a larger inhomogeneity and hence an increase of the scattering of the wave by the structure. The amplitudes of these oscillations increase in higher bands.

- As seen from Fig. 1, all curves have a common envelope, which corresponds to the largest absolute value of the BGW in any band and for a given filling factor α. This maximum value depends on the parameters of the scattering structure.

- The number of BGW maxima within one band is equal to the number N of the corresponding band. The number of minima of the BGW (i.e., PBG with zero width) is equal to $N - 1$. This last fact is very important and in the next paragraph we shall give a mathematical and physical justification of such an effect that corresponds to *band gap closing*.

3.2. RESONANCE OF A SINGLE SCATTERER

Let us obtain the point of band gap closing, i.e., the zero BGW value in Fig. 1. As shown in Ref. [12], the boundaries of the PBG (the roots of

Eq. (5)) coincide in a certain band for the frequency:

$$W_c \equiv \frac{\omega_c d}{c} = \frac{k\pi}{\sqrt{\varepsilon_{slab}}} \frac{1}{\alpha} \quad , \tag{8}$$

if the following condition is satisfied:

$$\frac{\sqrt{\varepsilon_{back}}}{\sqrt{\varepsilon_{slab}}} \frac{1-\alpha}{\alpha} = \frac{m}{k}, \qquad k, m \text{ are integers.} \tag{9}$$

This means that for frequency ω_c, PBG is closed. This value of ω_c is independent of the period of the lattice, but depends on parameters of the single scatterer, namely, refractive index $n_{slab} = \sqrt{\varepsilon_{slab}}$ and slab width a.

All values W_c from Eq. (8) can be obtained directly from Eq. (3) if we put $|r|^2 = 0$; $|t|^2 = 1$, which corresponds to transmittance of the total energy, i.e., the medium is transparent for frequency W_c. Such an effect was mentioned as a single scatterer resonance (SSR) in Refs. [7, 6]. The SSR is also known as "antigap" [13]. When $W_c(k)$ coincides with the F_{md} values, which is what actually happens for the conditions involved in Eq. (9), the photonic band gap is closed: the BGW equals zero (see Fig. 1).

From Eq. (6) it follows that the function $F_{md}(\alpha)\bar{n}(\alpha)$ is approximately a constant equal to $F_{md}(\alpha = 0)\bar{n}(\alpha = 0) = N\pi$. Then, from the equality $W_c(\alpha)\bar{n}(\alpha) = F_{md}(0)\bar{n}(0)$ we obtain: $m + k = N$. It can easily be verified that for a given value N, $m = 1, 2, \ldots ; k = 1, 2, \ldots$, the last relation can be satisfied for $(N-1)$ pairs of m and k. This solution is independent on the parameters of the lattice.

3.3. LOCALIZATION LENGTH IN RANDOM LATTICES: GAPS AND ANTIGAPS

In order to study the localization length (LL), we use the well known definition [5, 14, 15]:

$$\ell = -\frac{L}{< \ln(T)} \tag{10}$$

T is the transmittance and $<$ represents an average over several realizations of the structure. In our calculations, we have used a finite structure containing a number of the slabs with randomness as described in Section 2. In order to reach a good average, 70-100 realizations were enough.

We address the main results of this study on the lattice with a randomness built by random "removing" of the slabs. In Fig. 2 the localization length as a function of dimensionless frequency $(W = \omega d/c)$ is represented for different values of α and ε_{back}. It is seen that all resonances connected with particular scales and appearing in a given random realization (like

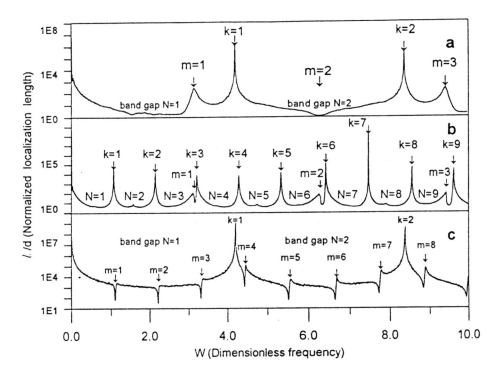

Figure 2. Localization length as a function of frequency W; model of "removed" slabs; $\varepsilon_{slab} = 9$; (a) $\alpha = 0.25, \varepsilon_{back} = 1$; (b) $\alpha = 0.98, \varepsilon_{back} = 1$; (c) $\alpha = 0.25, \varepsilon_{back} = 8$; "k" and "m" stand for the frequencies of SSRs and BRs, respectively.

the defect modes), disappear after averaging. Only SSRs exist as described by Eq. (8). These resonances cannot disappear after averaging, because we preserve the parameters of each scatterer, namely, its width a and $\varepsilon = \varepsilon_{slab}$. Therefore, the existence of SSR is independent on the strength of the disorder. This effect has also been observed in Ref. [8] for weak disorder.

Now let us consider the frequencies

$$W_{BR} = \frac{m\pi}{\sqrt{\varepsilon_{back}}} \tag{11}$$

in Fig. 2. For these frequencies W_{BR}, we observe "peaks," whose positions are independent of the parameter α (compare Figs. 2(a) and 2(b)). In Fig. 2(c) the same dependence as in Fig. 2(a) is represented, but now calculated for $\varepsilon_{back} = 8$. In contrast with Fig. 2(a), we observe deep "slots" of the localization length at localities of the frequencies W_{BR}. They are Bragg remnants (BR) [15] of the corresponding periodic lattice.

The smooth curves between peaks of SSRs in Figs. 2(a)-2(c) can be considered as average levels of the contribution of Bragg diffraction (BDC) to the localization length. Comparing Figs. 2(a) and 2(c) we observe that the BR effect can bring the system either to "transparency" (peaks) or to "reflectivity" (slots) in comparison with the averaged BDC. For $\varepsilon_{back} \rightarrow \varepsilon_{slab}$, the localization of light due to Bragg remnants is stronger than that arising from Bragg diffraction effects. Eliminating the difference between ε_{back} and ε_{slab} brings transparency to the structure in such a way that the average contribution of Bragg diffraction decreases. For the frequencies W_{BR}, we have resonant effects on a scale of the period of the original regular lattice, namely of width d, which differs from the nature of the Bragg diffraction effect. It can be shown that for the weak contrast medium the structure is more reflecting for frequencies W_{BR} than for the frequencies at which the Bragg diffraction mechanism dominates [12].

Concerning media with a large contrast between ε_{slab} and ε_{back}, as seen in Fig. 2(a) that the LL presents peaks at the BR frequencies W_{BR} ($m = 1$ and 3) when W_{BR} does not coincide with a MGF F_{md} (which is what happens at $m = 2$). At $W_{BR} = F_{md}$, there is a significant dip of the LL. Therefore, Bragg diffraction destroys the resonance of the BR mechanism [12]. The particularities of the scattering process by a random lattice having stochasticity in a filling factor distribution are represented in Fig. 3. We study the same dependence of ℓ/d vs W for random lattices and for a different disorder degree, averaged over several realizations of the random structure.

We observe that the disorder eliminates the SSR peaks in the first place, because the variation of the filling factor destroys the parameters of a single scatterer, namely, α. The amplitudes of the SSR peaks decrease as the frequency increases. According to Eq. (8), a small variation of any parameter of a single scatterer, either α or ε_{slab}, leads to the larger shift of the frequency W_c for the larger integer k. Therefore, SSR-peaks corresponding to high frequency will disappear in the first place after averaging over several realizations of the random medium (Fig. 3, curves 1 and 2).

It should be noted that the degree of disorder, which eliminates the SSRs, is not enough to destroy the Bragg remnants effect because the BR frequency is independent of the parameters of the single scatterer. Nevertheless, a further increase in disorder also destroys the BR-peaks in a high frequency band, namely, for $m = 3$ (Fig. 3, curve 3). This is due to an

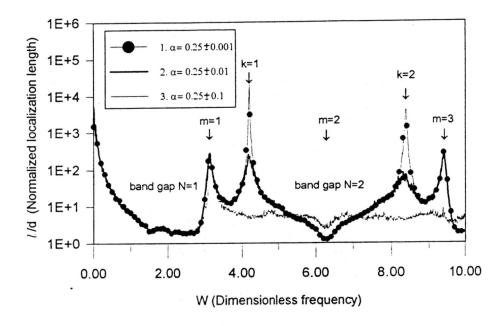

Figure 3. Localization length as a function of frequency W; model of the randomly distributed filling factor; $\varepsilon_{slab} = 9$, $\varepsilon_{back} = 1$; other designations are as in Fig. 2.

overlapping of the BR with the nearest band gaps, namely, the gaps N=2 and N=3 with BR m=3. In contrast with BR frequencies, the position of the band gap (F_{md}) depends on the parameters of a single scatterer as it follows from Section 3.1 and Eq. (7).

We study the wave scattering by a lattice having a randomly distributed refractive index of the scatterers. The randomness is inserted in the dielectric permittivity of each slab as $\varepsilon_{slab} = \varepsilon_d[1 + B(\mathbf{rnd} - 1)]$, $\mathbf{rnd}$ is uniformly distributed between 0 and 1 random number. The parameter B, $0 \leq B \leq 1$, plays the role of the degree of disorder.

In Fig. 4 we represent the localization length as a function of both ε_d and the dimensionless frequency W. Each point of the surface represented in this figure is obtained after averaging over 70 realizations. Case in Fig. 4(a) shows the regular lattice ($B = 0$). The "valleys" correspond to the band gap whose localization length is about $10^{-2}d$, the highest points of the "ridges" correspond to the frequency of the SSR where structure possesses a total transparency. Introducing disorder ($B \neq 0$), we observe an effect analogous to that of Fig. 3. The scattering by the random lattice manifests

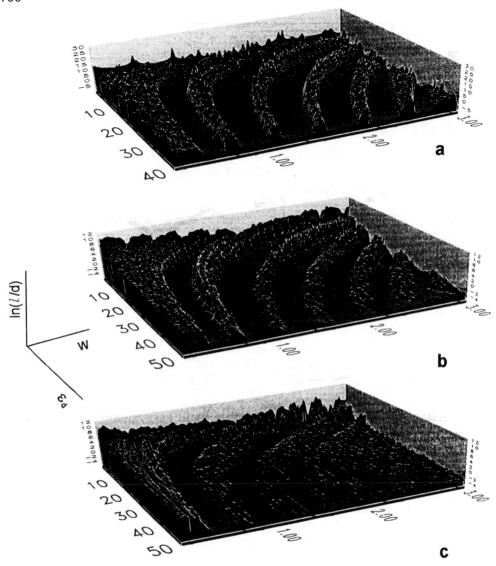

Figure 4. Logarithm of the localization length as a function of both frequency W and ε_d; model of the random distribution of the $\varepsilon_{slab} = \varepsilon_d[1 - B(\mathbf{rnd} - 1)], \alpha = 1, \varepsilon_{back} = 1$; a)$B = 0$; b)$B = 0.2$; c)$B = 0.5$.

the resonant properties at the frequencies: $W_a \equiv \pi k/\bar{n}$, where $\bar{n}$ is the average refractive index of the scatterers. The resonant peaks disappear after averaging for high frequency and large value of ε_d (Fig. 4(b),(c)).

4. Conclusions

We have considered the concurrence of different scattering mechanisms which take place in the 1-D problem concerning the electromagnetic wave propagation through either periodic or random sequences of slabs having different dielectric permittivities.

The wave resonance of one scatterer (analogous to the Mie resonance in 3D [9]) is the strongest scattering mechanism. Its characteristic frequency depends on parameters of the single scatterer only. Such an effect suppresses all other contributions when the wave frequency (at this resonance value) coincides either with the middle gap frequency or with the frequency of the Bragg remnants effect. Bragg diffraction in the middle of the gap destroys the effect of Bragg remnants. Nevertheless, for structures with weak contrast in the permittivities, $(0.8 \leq \varepsilon_{back}/\varepsilon_{slab} < 1)$ this Bragg remnants phenomenon, being a resonant scattering effect can bring reflectivity higher than Bragg diffraction to the scattering process in the sense of producing smaller values of the localization length.

The above mentioned scattering mechanisms exist while the wave is scattered by any kind of disordered medium. The particularities of each scattering picture result from the interplay of these scattering mechanisms, but the influence of each strongly depends on the particular realization of the structure.

We have investigated the higher frequency bands, and the effect of gap closing (or antigap) has been explained as the coincidence of the frequency of the single scatterer resonance and the middle gap frequency. When this happens, the band gap in the 1-D regular lattice for any band number N is closed $N - 1$ times and the band gap width has N maxima as the filling factor varies from 0 to 1. The largest width of the band gap can be exceeded at some frequency band and depends on both the background and slab permittivities. It should be noted that not only disorder and Bragg diffraction are causes of localization, but another scattering mechanism, namely Bragg remnants effect depending on the permittivity contrast, can produce localization length either larger or smaller than those due to Bragg diffraction.

Randomness in the distribution of dielectric slabs leads to the narrow selection of certain frequencies of the single scatterer resonance at which the disordered medium exhibits total transparence. Taking into account that the position of the SSR depends on the parameters of the scatterer

702

and does not depend on the characteristics of the embedded medium, a highly sensitive sensor could be devised on the basis of such an effect. Frequency properties of the random lattice can also serve as the basic physical principles in its applications in the light frequency band. They are control of the periodicity quality and/or disorder degree in the distribution of the layers in a stratified medium; measuring of the averaged characteristics, namely, the filling factor and refractive index of the random sequence of the dielectric slabs.

This work has been supported by the M.E.C. and C.Y.C.I.T. (Spain) and by the European Union.

References

1. E. Yablonovitch, J. Opt. Soc. Am. B **10**, 283 (1993); S. John, Phys.Today May, 32 (1991).
2. R. Dalichaouch J.P. Armstrong, S. Schultz, P.M. Platzman, and S.L. McCall, Nature **354**, 53 (1991).
3. Z. Daozhong, H. Wei, Z. Youlong, Z. Zhaolin, C. Bingying, and Y. Guozhen, Phys. Rev. B **50**, 9810 (1994).
4. D. Yeh, A. Yariv, and C.S. Hong, J. Opt. Soc. Am. **67**, 423 (1977); P. Yeh and A. Yariv, Opt. Comm. **19**, 427 (1976).
5. A.R. McGurn, K.T. Christensen, F.M. Mueller, and A.A. Maradudin, Phys. Rev. B **47**, 13120 (1993).
6. P. Sheng, *Scattering and Localization of Classical Waves in Random Media* (World Scientific, Singapore, 1990).
7. J.B. Pendry, *Symmetry and Transport of Waves in 1D Disordered Systems* (Preprint of Imperial College, London, 1994).
8. A. Kondilis and P. Tzanetakis, J. Opt. Soc. Am. A **11**, 1661 (1994).
9. C.M. Soukoulis, S. Datta and E.N. Economou, Phys. Rev. B **49**, 3800 (1994); E.N. Economou and A.D. Zdetsis, Phys. Rev. B **40**, 1334 (1989).
10. D.R. Smith, R. Dalichaouch, N. Kroll, S. Schultz, S.L. McCall, and P.M. Platzman, J. Opt. Soc. Am. B **10**, 314 (1993).
11. M. Plihal, A. Shambrook, A.A. Maradudin, and P. Sheng, Opt. Comm. **80**, 199 (1991).
12. S.A. Bulgakov and M. Nieto-Vesperinas. Submitted to J. Opt. Soc. Am. A.
13. A. Kondilis and P. Tzanetakis, Phys. Rev. B **46**, 15426 (1992).
14. C.M. Soukoulis, *Photonic Band Gaps and Localization* (Plenum Press, New York , 1993).
15. C. Martijn de Sterke and R.C. McPhedran, Phys. Rev. B **47**, 7780 (1993).

WAVE CONFINEMENT AND LOCALIZATION: DIMENSIONAL CROSSOVER EFFECT

ZHAO-QING ZHANG[1] and PING SHENG[1,2]

[1]Department of Physics, Hong Kong University of Science
and Technology, Clear Water Bay, Kowloon, Hong Kong
[2]Exxon Research and Engineering Co., Rt. 22 East, Clinton,
Annandale, N J 08801, U.S.A.

1. Introduction

One of the most important properties of photonic band-gap (PBG) materials is the confinement of light whose frequency lies in the bandgap [1]. Many of the envisioned applications of PBG materials rely on the degree to which this confinement characteristic is realized. It follows, therefore, that while it is important to examine the special properties of PBG materials in its perfect form, it may also be important to study the wave behavior in imperfect PBG materials, so as to gain the knowledge about PBG material requirements in practical applications. In particular, it is of interest to see what happens to waves when there is a density of imperfections, or impurities, in PBG materials.

The main intent of this paper is to address a question related to the problem posed above, that is, to consider a randomly layered medium in which waves of all frequencies are confined in the direction perpendicular to the layers due to the Anderson mechanism of wave interference. This is an effect which can be as strong as Bragg reflection in terms of producing localized wave states. The confinement is only in one direction, however, and the waves are delocalized in the two other directions for the perfect system. If on top of this randomly-layered medium one adds some weak isotropic randomness, can the waves still be localized as before? Or, will there be some qualitative change in wave propagation/localization characteristics?

The results of our study indicate the answer depends on the direction of wave propagation. For a randomly-layered system, there is generally a critical angle, measured relative to the unit normal to the layers, which separates wave propagation into two angular regions. When a plane wave is launched at an angle smaller than the critical angle, it is found that the small isotropic randomness does not change the basic character of wave localization in the direction perpendicular to the layers. However, for a plane wave launched at an angle larger than the critical angle so it is essentially propagating parallel to the layers, we show the weak isotropic randomness can induce qualitatively different behavior than the pure case. In particular, for a two-dimensional layered system, the resulting finite-size scaling characteristics indicate a spatial dimensionality of 1.5.

2. Model Hamiltonian and Method of Calculations

We consider here a 2D tight-binding Hamiltonian

703

C. M. Soukoulis (ed.), Photonic Band Gap Materials, 703–714.

$$H = \sum_{\alpha} \varepsilon_{\alpha} a_{\alpha}^{+} a_{\alpha} + t \sum_{[\alpha,\beta]} a_{\alpha}^{+} a_{\beta}, \tag{1}$$

where t = 1 is the hopping matrix element, α, β are the site indices of a simple square lattice on the x-z plane, [α, β] indicates α and β are nearest neighbors, and

$$\varepsilon_{\alpha} = \eta_z + \delta_{\alpha} \tag{2}$$

consists of two parts: η_z is a number that is a constant for all x and fixed z, but varies randomly as z varies with a flat distribution

$$P_1(\eta_z) = \begin{cases} 1/W_1 & |\eta_z| \le W_1 / 2, \\ 0 & otherwise. \end{cases} \tag{3}$$

δ_{α} is a random number that varies independently from site to site with a flat distribution

$$P_2(\delta_{\alpha}) = \begin{cases} 1/W_2 & |\delta_{\alpha}| \le W_2 / 2, \\ 0 & otherwise. \end{cases} \tag{4}$$

In other words, η_z represents the random layering and δ_{α} the additive, isotropic randomness. In the limit of $W_2 = 0$, we have a randomly layered system; whereas the isotropic random system is recovered if $W_1 = 0$. Thus, the model considered here is intermediate between the 1D layered system and the 2D isotropic system. It should be mentioned that in 3D the localization phase diagram of this model has been studied previously by us and our collaborators using both diagrammatic and numerical methods [2]. The results of both calculations seem to indicate there exists a critical anisotropic ratio of $W_2/(W_2+W_1)$, denoted as θ_c, below which all the states are localized for arbitrarily small randomness. The Anderson localization transition occurs only above this critical anisotropy. This critical anisotropy is wave energy dependent and has the θ_c value of about 0.2 at the wave energy E = 0. Since 2D is the marginal dimensions for the Anderson transition in an isotropic system, the introduction of layered randomness will only make the system more localized. In the limit of $W_1 \ll W_2$, the system becomes a nearly isotropic 2D system and the localization behavior has all the 2D characteristics. When $W_1 \gg W_2$, on the other hand, wave propagation is expected to behave differently in the directions perpendicular and parallel to the layered planes due to the strong anisotropy. In the limit of $W_2=0$, the wave is localized in the layering direction (z-axis) and the localization effect in the z direction acts as a waveguide for propagation in the x direction. When a small amount of W_2 is present, the localization length in the z direction is expected to increase due to the channel crossings caused by the isotropic scattering. Wave transport in the x direction is more interesting. Due to the overlaps among various guided "channels," isotropic scatterings introduce hoppings between them. As we will show below, this hoppings would result in a new transport behavior intermediate between the randomly layered and the isotropic limits.

In this work, we would like to study the case where the layering randomness is strong, i.e., W_1 large, in which the eigenfunctions will be highly localized with a localization length $\xi_\perp \cong$ 1-2 layers. Even in this case, however, an important point to note is the eigenfunctions can still have significant spatial overlaps; although in the absence of isotropic randomness they are orthogonal to each other in the sense that if initially only a particular eigenfunction is excited, waves will propagate in the x direction as in a waveguide, i.e., confined in the z direction. In the presence of isotropic randomness W_2 > 0, we will study, in a quantitative way, the effects of channel crossings on $\xi_\perp$ and and the transfer of energy among various channels in the parallel direction.

The transport properties of a sample strip of width M and length L can be studied by calculating the transmission coefficients of all the propagating channels. In doing so, we connect both sides of the sample by pure leads of width M. The hard-wall boundary condition was used at the two strip boundaries. Properties of the perfect leads are chosen to be different for the study of wave propagation behavior in two orthogonal propagation directions. For properties in the layering direction (z-axis), the perfect leads have W_1 = W_2 = 0. For propagation in the x direction, the perfect leads have the same layer energies as those of the sample except $W_2 = 0$, and, therefore, these channels are the waveguides described above. With these leads, the channel energies and wave functions can be obtained by diagonalizing the Hamiltonian of any cross sectional layer of the lead. For a given width M of the lead, the wave energy E can be written as

$$E = E_x^{(i)} + 2\cos k_z^{(i)}a \qquad or \qquad E = E_z^{(i)} + 2\cos k_x^{(i)}a, \qquad (5)$$

where a = 1 is the lattice constant. $E_x^{(i)}$ and $E_z^{(i)}$ are the ith channel energies of the leads when the wave is propagating in the z and x directions, respectively. In the x direction propagation, when $|E - E_z^{(i)}| > 2$, $k_x^{(i)}$ is imaginary and the channel i is denoted "evanescent." When $|E - E_z^{(i)}| < 2$ and $k_x^{(i)}$ is real, the channel i is denoted "propagating." This is also true for propagation in the z direction. In this work, we have chosen E = 0. Except for the energy near the band edge, all the transport behaviors found here would also be expected for other values of E. In the case E = 0, most of the channels in the z direction are propagating in nature. Since W_1 is large, there are far less propagating channels in the x direction than in the z direction. With all the channels defined, we can calculate Green's function $G_{ij}^+(0,L)$ by using the recursive Green's function technique [3], where $G_{ij}^+(0,L)$ is the retarded Green's function with the source at one side of the sample of channel i and receiver at the other side of the sample of channel j, and the L is the sample length. The transmission coefficients $|t_{ij}|^2$ is related to Green's function by [3,4]

$$|t_{ij}|^2 = v_i v_j |G_{ij}^+(0,L)|^2, \qquad (6)$$

where v_i is the channel velocity and is given by $\partial E / \partial k^{(i)}$ evaluated at $k^{(i)}$. The localization length for samples of width M is then related to the total transmission by

$$\frac{1}{\xi(M)} = < \frac{-\ln[T(L)]}{2L} > \qquad with \qquad T(L) = \sum_{i,j} |t_{ij}|^2, \qquad (7)$$

where $\langle \bullet \rangle$ denotes the configurational average. Due to the self-averaging property of $1/\xi(M)$, the number of configurations required depends on the length L of the sample, which is always much greater than the width M. Given a configuration of channels, i.e., a set of $E_z^{(i)}$ or $E_x^{(i)}$ in Eq. (5), in the leads, if L is sufficiently large, $\xi(M)$ can be accurately obtained from a single configuration. For propagation in the z direction, there is only one configuration of channels in the leads for any given strip width M because $W_1 = 0$. However, for propagation in the x direction, the channel configuration in the leads also varies due to the randomly layered energies in the leads. In this case, one should also take the average over different layer configurations that also includes the lead.

3. Results and Discussion

In the case of strong anisotropy, e.g., $W_2 = 0.1\ W_1$, we have calculated the localization lengths $\xi_\perp(M)$ and $\xi_\parallel(M)$ for directions perpendicular and parallel to the layers (or z and x), respectively. The results for the cases of $W_1 = 15$ and 10, or $W_2 = 1.5$ and 1.0, are plotted in Fig. 1 from M = 10 to 50. The curves A and B show $\xi_\perp(M)$ for $W_1 = 15$ and 10, respectively. The corresponding $\xi_\parallel(M)$ are given by curves C and D. From the curves A and B, it is found that the value of $\xi_\perp(M)$ starts to saturate around M $\cong$ 30 - 40, i.e., an order of magnitude larger than $\xi_\perp(M)$. Thus, if we plotted the M/$\xi_\perp(M)$ as a function of M, we are able to obtain the bulk localization length in the z direction $\xi_\perp$ by extrapolating the slope to the infinite M limit. The values of $\xi_\perp$ are found to be 1.456 and 2.786 for $W_1 = 15$ and 10 ($W_2 = 0.1\ W_1$), respectively. In fact, the localizing character of the randomly layered media [2,5] is not appreciably altered for waves propagating in the z direction. We have also calculated the localization length $\xi_\perp$ for different values of W_2 ranging from 0 to 3. The result of $\xi_\perp$ is plotted as a function of $\sqrt{W_2}$ in Fig. 2 for the case of $W_1 = 15$. This linear curve shows that the localization in the layering direction increases with the isotropic randomness due to the effect of channel crossings and the increment depends on the square root of the strength of scattering. Because of the strong overlaps among all the channels, the channel crossing is very effective so the relative probability distribution of electrons among all the channels becomes uniform after some scatterings, which is independent of the incident channel.

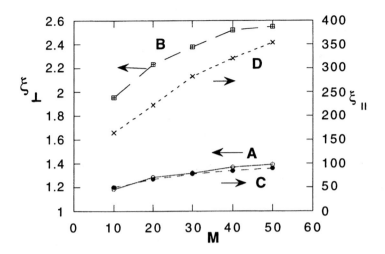

Figure 1. The localization lengths $\xi_\perp(M)$ and $\xi_\parallel(M)$ in the directions perpendicular and parallel to the layers, respectively, as a function of sample width M. Curves A and B show the data of $\xi_\perp(M)$ (using the left scale) for $W_1 = 15$ ($W_2 = 1.5$), and $W_1 = 10$ ($W_2 = 1.0$), respectively. The corresponding curves for the data of $\xi_\parallel(M)$ are C and D (using the right scale).

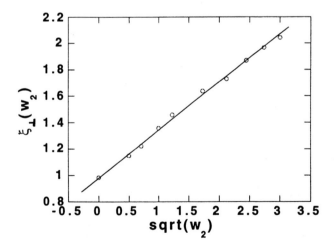

Figure 2. The perpendicular localization length $\xi_\perp$ as a function of the square root of the isotropic scattering strength W_2 for the case of $W_1 = 15$.

The transport along the x direction is more interesting. Due to the strong anisotropy, the width of the propagating channels is localized to a few lattice constants and overlaps among various channels are small. Unlike the propagation in the z direction, it takes many more scatterings before the relative probability distribution of electrons can become stabilized. The length it takes for electrons to reach a stationary state is denoted the equilibration length. The final stationary distribution is highly non-uniform and always exhibits the property there is one channel in which most of the probability amplitude is concentrated. The channel with the final largest relative probability of electrons is called the dominant channel. The formation of the dominant channel is due to the asymmetry of the transmission amplitude t_{ij} between two channels i and j, i.e., $t_{ij} \neq t_{ji}$ [6]. Physically, this is due to the difference in the channel properties, like the channel speed, of the channels i and j as will be shown explicitly later. The localization length in the x direction $\xi_{\parallel}(M)$ is mainly determined by the largest localization length of all the propagating channels in the absence of channel crossings. The data of $\xi_{\parallel}(M)$ are shown as curves C and D in Fig. 1. They are actually the decay length of the dominant channel. By taking various configurations of layer energies, the distribution of $1/\xi_{\parallel}(M)$ is shown in Fig. 3 for the case of $W_1 = 15$, $W_2 = 1.5$ and $M = 40$. Curve A of Fig. 3 shows the distribution of occurrence of $1/\xi_{\parallel}(M)$ from 460 configurations. Curve B is the best Gaussian fit. Since the distribution is normal, the configurational average of $1/\xi_{\parallel}(M)$ can be represented by

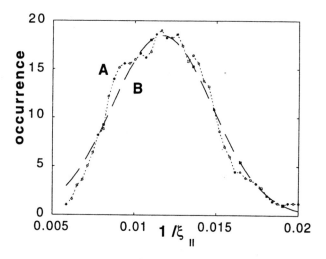

Figure 3. The distribution of the inverse of parallel localization length, $1/\xi_{\parallel}(M)$, for the case of $W_1 = 15$ ($W_2 = 1.5$) and $M = 40$.

its most probable value, from which we obtain curves C and D in Fig. 1. The fact the function $\xi_{\parallel}(M)$ is not saturated even when $M \gg \xi_{\perp}$, which is only a few lattice constants, demonstrates the important mechanism of channel crossings in determining the transport behavior in the x direction. Here, we would prefer to use the word "channel hoppings" instead of "channel crossing" in the x direction, due to the strongly localized nature of the channel. If there were no channel hoppings, we would expect

$$\xi_{\parallel}(M) \propto \xi_{\perp} / W_2^2. \tag{8}$$

This represents the localization length in a 1D strip of width $\xi_{\perp}$. Since $\xi_{\perp}$ is M independent when $M \gg \xi_{\perp}$, we would normally expect no M dependence in $\xi_{\parallel}(M)$ if there were no channel hoppings.

The channel-hopping behavior can be seen explicitly from the transmission coefficients T_{ij} ($=|t_{ij}|^2$) as a function of the sample length L between the various incoming channel i and outgoing channel j. A typical example is shown in Fig. 4. This is the result of a particular sample of width $M = 10$, $W_1 = 15$ and $W_2 = 1.5$. In this case, there are only two spatially well separated propagating channels, a and b. The wave is initially incident from the channel b. As the sample length L is increased, the transmission coefficient T_{bb} (curve B) decreases exponentially due to the localization effect. However, there is also a transfer of energy from channel b to a as the result of channel hoppings. This is evident from the initial rise of T_{ab} (curve A) from zero to a maximum value and the subsequent decay with the same rate as T_{bb}. Albeit some local

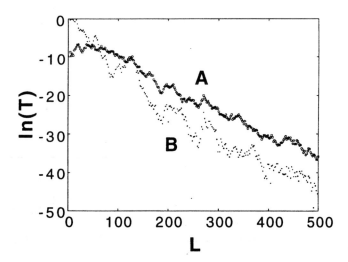

Figure 4. Spatial decay of two channels along the x direction, with wave incident on channel b. Note the two-segment behavior of the b-channel decay. T denotes the amplitude transmission coefficient of $|t_{ba}|^2$ (curve A) and $|t_{bb}|^2$ (curve B). Here $M = 10$, $W_1 = 15$, $W_2 = 1.5$, and the result is from a single configuration calculation.

fluctuations, the ratio of T_{ab}/T_{bb} eventually settles to a constant value after some equilibration length, which is around 200 in this case. Thus, for channel b there is a two-segment behavior in its spatial decay: The initial decay involves not only the localizing effect of the isotropic randomness, but also the transfer of energy to channel a; whereas in the second segment, the behavior of channel b is controlled by channel a. If the wave energy were initially injected into channel a, then only the second segment of b would be evident. In this example, a is clearly seen to be the dominant channel. In fact, for each configuration there is always a dominant channel, which is also the one whose localization length is the largest in the absence of channel hoppings. Thus, to study the behavior of wave propagation parallel to the layers is equivalent to the study of the dominant channel behavior.

What determines the dominant channel behavior? Here a search for dependence on $\xi_\perp$ [as given by Eq. (8)] yields no perceivable correlation. A more relevant parameter is the channel speed $v_x = |\partial E_x / \partial k_x|$. In Fig. 5 the v_x distribution of the dominant channel is plotted. A clear trend for the dominant channel to have a large v_x is seen. This correlation is intuitively plausible by the following arguments. In 1D the localization length has a maximum at the band center and decreases monotonically as the energy moves away from the center [7]. A channel with a larger v_x has the channel energy closer to zero, as can be seen from Eq. (5) that $v_x = 2$ when $E = E_z = 0$. Therefore, a channel with a larger v_x has a larger localization length and should take longer to slow down and be localized than one with a smaller v_x. However, even for $v_x \cong 2$, the maximum value, the amount of wave intensity in the dominant channel can still be relatively small at a given value of x. To study this aspect of the problem quantitatively, we define the occupation ratio of a particular channel, in terms of the transmission coefficients, as

$$\theta_i = \frac{1}{M_c} \left\langle \sum_j \frac{|t_{ij}|^2}{\sum_i |t_{ij}|^2} \right\rangle, \tag{9}$$

where M_c is the number of propagating channels in the leads, and we have averaged over all the incident channel j. The average $\langle \bullet \rangle$ here is taken along the x positions after the equilibration length where the statistics of θ_i becomes stationary. Apart from minor fluctuations, the final answer for the occupation ratio has been tested to be independent from the initial injection channel. In Fig. 6, we plot the distribution of the occupation ratio of the dominant channel for sample configurations with M = 20 and 50. For M = 20, it is seen that while there are some cases where the dominant channel's occupation ratio is in the range of 0.8-0.9, the most probable occupation ratio of the dominant channel is on the order of 0.2-0.3, i.e., (70-80)% the wave energy at a given x is spread to the other channels. As M is increased to 50, the most probable occupation ratio decreases to $\cong 0.1$, with a corresponding decrease in the probability for occupation ratio higher than 0.5.

The key to the understanding of this delocalizing behavior lies in the evanescent channels. In the example shown in Fig. 4, while a and b are the only two propagating channels, there are many other evanescent channels which overlap spatially with a and b and with each other. These channels cannot carry energy in the x direction, but they

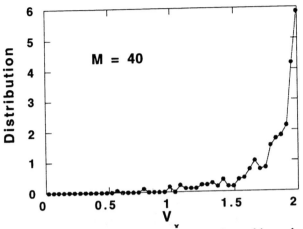

Figure 5. The v_x distribution of the dominant channel in each configuration. Here M = 40, W_1 = 15, W_2 = 1.5, and the result comprises 460 configurations.

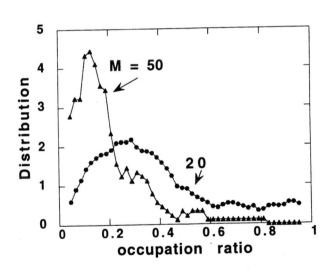

Figure 6. Distributions of the dominant-channel occupation ratio for M = 20 and 50. Here W_1 = 15, W_2 = 1.5, and the distribution for the M = 20 case has 460 configurations, and the distribution for the M = 50 case has 90 configurations.

provide the means of interchannel energy transfer in the presence of isotropic randomness. Wave energy can thus "tunnel" from one channel to the next, resulting in significant delocalization of the wave energy in the z direction. As M increases, the total number of channels increases proportionally, and interchannel tunneling is expected to become progressively more efficient as the average energy spacing between the E_z's decreases, so the overlap between the E_z levels increases.

Because of the existence of the dominant channel, the distribution of wave energy between the propagating (non evanescent) channels is never ergodic. A clean demonstration of this fact is via the calculation of the average number of occupied channels

$$N_{oc}(M_c) = \frac{1}{M_c} \sum_i \frac{[\sum_j |t_{ij}|^2]^2}{\sum_j |t_{ij}|^4}, \tag{10}$$

which is recognized to be the inverse of the average participation ratio. A plot of $N_{oc}(M_c)$ for these different models is shown in Fig. 7. Data for the case under consideration (model A) are denoted by open circles, whereas crosshatched squares and open triangles denote the cases of the 2D isotropic model (model B) and the 2D anisotropic hopping model (model C) [8], respectively. In the anisotropic hopping model, ε_α in Eq. (2) contains only the δ_α term, but the hopping matrix element t is anisotropic. It is seen that models B and C both exhibit linear dependence on the number of propagating channels M_c, as dictated by ergodicity, whereas model A shows an excellent $\sqrt{M_c}$ dependence. The same $\sqrt{M_c}$ dependence is also observed for the most probable value of the $\xi_\parallel(M)$ distribution. If a "truly 1D" behavior is defined as where the wave energy is always confined in each channel so that the resulting dependence on M_c is flat, then Fig. 7 shows our case to be intermediate between 1D and 2D. Of course, the $\sqrt{M_c}$ dependence is not expected to be maintained beyond the limit of M_c = the localization length of the isotropic model with randomness W_2. However, for small W_2, this upper bound can be transcendentally large. The precise origin of the $\sqrt{M_c}$ behavior is under further study.

4. Conclusions

The implication of our findings for the time evolution of a localized pulse is as follows. For a given channel, wave tunneling is analogous to a scattering extinction effect. The energy of a localized pulse is therefore expected to leak out of the channel region defined by $\xi_\perp$ with a time constant determined by the ratio W_2/W_1. In the limit of small W_2/W_1, the decay is expected to be transcendentally slow due to the fact that the energy spreading occurs mostly in the exponential tail region of the intensity distribution. However, while the intensity in the tail region is arguably small, the large value of the cutoff for the $\sqrt{M_c}$ behavior (the isotropic 2D ξ) can mean that a significant fraction of the total intensity will eventually be outside of the initial channel. As an example, for M = 50 an estimated (20-50)% of the total intensity lies outside of the initial channel.

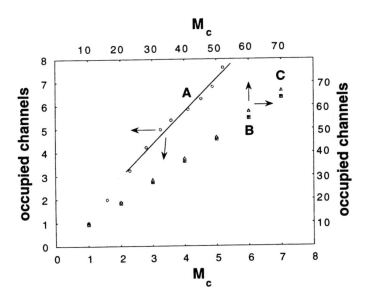

Figure 7. Number of occupied channels plotted as a function of M_C, the number of propagating channels. Whereas the isotropic model [model B ($\oplus$)] and the anisotropic hopping model [model C (Δ)] exhibit linear dependence on M_C (both have $W_1 = 15$, $W_2 = 1.5$, with 1 configuration; the anisotropic model has $t_x = 1$, $t_z = 0.1$), the presence of isotropic randomness in a randomly layered medium [model A (O)] results in a $\sqrt{M_c}$ variation. It should be noted that $M_C = 20$ in the present model means M = 80-100. whereas for the isotropic model M_C corresponds roughly with M. Therefore, the sizes of the actual calculations in all three models are comparable. Here $W_1 = 15$, $W_2 = 1.5$, and 460 configurations are used in averaging for M = 10, 15, 20, 25, and 30. Eighty configurations are used in averaging for M = 70, 75 and 80, and 30 configurations are used for M = 85. The rms deviation of the overall data is about 10%.

ACKNOWLEDGMENTS

One of the authors (P.S.) wishes to thank the Institute for Advanced Study for a visiting appointment in the 1992-93 academic year, during which this work was started and stimulated by numerous discussions with Thomas Spencer. The authors also wish to thank George Papanicolaou for many useful discussions.

714

REFERENCES

1. *Photonic Band Gaps and Localization*, C.M. Soukoulis ed., Plenum, New York (1993).
2. W. Xue, P. Sheng, Q.J. Chu, and Z.Q. Zhang, Phys Rev. Lett. **63**, 2837 (1989); Phys. Rev. B **42**, 4613 (1990).
3. P. A. Lee and D. Fisher. Phys. Rev. Lett. **47**, 882 (1981).
4. D. Fisher and P.A. Lee, Phys. Rev. B **23**, 6851 (1981).
5. V. Freilikher and Yu, Tarasov, IEEE Trans. AP **39**, 197 (1991).
6. The fact that the tunneling need not be symmetric can easily be seen from the fact that $t_{ij} \neq t_{ji}$, in general. This can be demonstrated in the case of two intermediate scatterings by impurity potentials denoted as v_n and v_m:

$$\int \phi_j(z')v_n(z')G_{nm}(z',z)v_m(z)\phi_i(z)dz'\,dz$$
$$\neq \int \phi_i(z')v_n(z')G_{nm}(z',z)v_m(z)\phi_j(z)dz'\,dz,$$

where f_i is the wave function of the ith channel and G denotes the Green's function. Only in the case of single scattering does $t_{ij} \neq t_{ji}$.

7. P. Sheng, *Introduction to Wave Scattering, Localization, and Mesoscopic Phenomena*, Academic Press (1995) p. 258.
8. W. Apel and T.M. Rice, J. Phys. C **16**. L1151 (1983); Q. Li, C.M. Soukoulis, E.N. Economou, and G.S. Grest, Phys. Rev. B **40**, 2825 (1989).

TRANSITION FROM BALLISTIC TO DIFFUSIVE BEHAVIOR FOR MULTIPLY SCATTERED WAVES

ZHAO-QING ZHANG[1] and PING SHENG[1,2]

[1]Department of Physics, Hong Kong University of Science and Technology Clear Water Bay, Kowloon, Hong Kong

[2]Exxon Research and Engineering Co., Rt. 22 East, Clinton, Annandale, N J 08801, U.S.A.

1. Introduction

Wave propagation in random media has been the focus of considerable attention in the past decade [1,2]. Various interesting wave phenomena have been uncovered, and studied, e.g., the coherent backscattering effect[3], short-range and long-range spatial intensity correlations [4,5], and dynamical intensity correlations [6], etc. In case when the sample size is much greater than transport mean free path, great simplification in theoretical treatments has been found in the diffusion approximation. The diffusion approach has been successful in describing most of the phenomena observed, including the coherent backscattering cone and its sample size dependence [7], and the spatial intensity correlations [8,9], etc. It works well even in the case when there are internal reflections in the random media [10,11]. The application to the coherent backscattering of the scalar acoustic waves is also excellent [12]. It should be noted that the diffusive property of light has recently been used to advantage as a new optical imaging technique [13].

The validity of the diffusion approximation results from the multiple scattering of waves in random media. It has been observed that diffusion approximaton fails to describe experimental results when the slab thickness L of the random medium becomes small and comparable to the transport mean free path ℓ. Where the diffusion approximation breaks down depends on the physical quantity we are interested. For instance, by measuring the temporal distribution of the transmitted photon intensity, it has been found that the deviation from the diffusion approximation set when the ratio L/ℓ has the value around 10 [14]. However, by choosing properly the coherent penetration depth and the extrapolation length, the measurement of total photon transmission showed the validity of the diffusion approximation when $L/\ell \geq 4$ [15]. In the same experiment, if the transmitted intensity profiles are compared, the diffusion approximation fails at a larger thickness [15].

It is well-known that as the wave travels through a random medium the coherent part of the wave decays exponentially according to the scattering mean free path ℓ_s due to successive scatterings. In the mean time the scattered portion of the wave builds up gradually and travels diffusely. The purpose of this work is to study the transition between the ballistic and diffusive behaviors. Given a slab geometry and considering a normally incident pulse, we are interested in studying the temporal distribution of the transmitted intensity. Our method is to solve directly by numerical iterations the Bethe-Salpeter equation in the space-frequency domain. The solutions of this equation give the

C. M. Soukoulis (ed.), Photonic Band Gap Materials, 715–722.
© 1996 Kluwer Academic Publishers. Printed in the Netherlands.

field-field correlation functions in the frequency space. In the case of large L/ℓ ratio, this correlation function can be obtained by using the diffusion approximation [4,9]. By taking the Fourier transform of the frequency correlation functions, the temporal profile of transmitted intensity may be obtained.

2. Method of Calculation

Consider a slab of random medium of width L. The wave is normally incident on the medium at the front surface and detected at the back surface. In order to study the temporal behavior of the transmitted intensity, the quantity we are interested in is the following space-frequency correlation function,

$$C_\Omega(\omega;\vec{r},\vec{r}') = < \phi_{\Omega+\omega/2}(\vec{r})\phi^*_{\Omega-\omega/2}(\vec{r}') >, \qquad (1)$$

measured at any position on the back surface $z = L$, i.e., $\vec{r} = (\vec{\rho}, z = L) = \vec{r}'$. Here we have taken $z = 0$ as the front surface. The function $C_\Omega(\omega;\vec{r},\vec{r}')$ satisfies the Bethe-Salpeter equation,

$$C_\Omega(\omega;\vec{r},\vec{r}') = < \varphi_{\Omega^+}(\vec{r}) >< \varphi^*_{\Omega^-}(\vec{r}') > + \qquad (2)$$

$$\int d\vec{r}_1 d\vec{r}_2 d\vec{r}_3 d\vec{r}_4 < G_{\Omega^+}(\vec{r},\vec{r}_1) >< G^*_{\Omega^-}(\vec{r}',\vec{r}_2) > U(\vec{r}_1,\vec{r}_3;\vec{r}_2,\vec{r}_4) C_\Omega(\omega;\vec{r}_3,\vec{r}_4),$$

where we have denoted $\Omega \pm \omega/2$ by $\Omega^\pm$ and the integrations are over the entire slab of width L. The function $< \varphi_\Omega(\vec{r}) >$ is the configurational averaged field of the incident wave and $<G>$ and U are, respectively, the averaged Green's function and the irreducible vertex function. In the weak scattering regime, the averaged Green's function has the approximate form [7],

$$< G_\Omega(\vec{r}_1,\vec{r}_2) > \approx -\frac{\exp(iK|\vec{r}_1 - \vec{r}_2|)}{4\pi|\vec{r}_1 - \vec{r}_2|}; \qquad K = \frac{\Omega}{v} + \frac{i}{2\overline{\ell}}, \qquad (3)$$

where v is the wave speed in the medium and $\overline{\ell}$ is the extinction length of the coherent intensity, which arises from both the elastic scattering and the absorption by the medium. A simple expression for $\overline{\ell}$ is

$$\frac{1}{\overline{\ell}} = \frac{1}{\ell_s} + \frac{1}{\ell_a}. \qquad (4)$$

Here, ℓ_s and ℓ_a are, respectively, the scattering mean free path and the absorption length. The latter is related to the absorption time by $\ell_a = v\tau_a$. We have also assumed that Green's function remains translational invariant inside the slab. In the small ω limit, we can ignore the frequency dependence of $\overline{\ell}$, i.e., $\overline{\ell}(\Omega \pm \omega/2) \approx \overline{\ell}(\Omega)$. By assuming the isotropic scattering, the scattering mean free path ℓ_s is equal to the transport mean free path ℓ and the lowest order contribution to the vertex function becomes [7]

$$U(\vec{r}_1,\vec{r}_3;\vec{r}_2,\vec{r}_4) = \frac{4\pi}{\ell}\delta(\vec{r}_1 - \vec{r}_3)\delta(\vec{r}_1 - \vec{r}_2)\delta(\vec{r}_3 - \vec{r}_4). \quad (5)$$

With this U, Eq. (2) generates a sum of ladder diagrams. In the absence of absorption, it can be shown that the expressions of <G> and U given by Eqs. (3) and (5) satisfy the Ward Identity in the $\omega = 0$ limit [7]. Furthermore, we consider here the correlation function resulting from the plane wave incidence at two different frequencies $\Omega \pm \omega/2$. By substituting Eqs. (3) and (5) into Eq. (2), the equation for $C_\Omega(\omega;\vec{r},\vec{r})$ becomes

$$C_\Omega(\omega;\vec{r},\vec{r}) = \exp(\frac{i\omega z}{v} - \frac{z}{\ell}) +$$

$$\frac{1}{4\pi\ell}\int d\vec{r}_1 \frac{\exp[(\frac{i\omega}{v} - \frac{1}{\ell})(|\vec{r} - \vec{r}_1|)]}{|\vec{r} - \vec{r}_1|^2} C_\Omega(\omega;\vec{r}_1,\vec{r}_1). \quad (6)$$

In this case, the correlation function $C_\Omega(\omega;\vec{r},\vec{r})$ depends only on the z coordinate and is independent of the transverse coordinate $\vec{\rho}$ as can be seen from Eq. (6). Now we define an integrated correlation function

$$B_\Omega(\omega,z) = \int d\vec{\rho} C_\Omega(\omega;\vec{r},\vec{r}). \quad (7)$$

Substituting Eq. (7) into Eq. (6), we obtain

$$B_\Omega(\omega,z) = \exp(\frac{i\omega z}{v} - \frac{z}{\ell}) + \frac{1}{4\pi\ell}\int_0^L dz_1 H(z - z_1)B_\Omega(\omega,z_1), \quad (8)$$

where

$$H(z - z_1) = \int d\vec{\rho} \frac{\exp[(\frac{i\omega}{v} - \frac{1}{\ell})\sqrt{\rho^2 + (z - z_1)^2}]}{\rho^2 + (z - z_1)^2}. \quad (9)$$

It is easy to see from Eqs. (8) and (9) that $B_\Omega(-\omega,L) = B_\Omega^*(\omega,L)$. For a given set of parameters v, L/ℓ and ℓ_a/ℓ, the correlation function at the back surface $B_\Omega(\omega, L)$ can be calculated numerically by solving Eq. (8). It should be noted that the dependence of $B_\Omega(\omega, L)$ on Ω only comes from its dependence on the mean free path $\ell(\Omega)$.

In order to study the temporal distribution of the transmitted intensity, we use a pulse as the source. Again, we assume the width of the pulse is larger than the mean free

718

path so no $\vec{\rho}$ dependence in the transmitted intensity exists. In general, the time dependence of the intensity can be expressed in a convolution form,

$$|\psi(t,\vec{r})|^2 = |\int dt' \, f(t')\phi(t-t',\vec{r})|^2, \tag{10}$$

where f(t) is the profile of the input pulse, and $\phi(t,\vec{r})$ is the Fourier transform of the function $\phi_\Omega(\vec{r})$ given in Eq. (1). In the frequency space, by taking the configurational average of Eq. (10), we have

$$<|\psi(t,\vec{r})|^2> = \int d\Omega [\int d\omega f(\Omega^+) f^*(\Omega^-) < \phi_{\Omega^+}(\vec{r})\phi^*_{\Omega^-}(\vec{r}) > \exp(i\omega t)]. \tag{11}$$

Using Eqs. (1), (7), and (8), the integrated intensity profile on the output surface becomes

$$I(t,L) = \int d\Omega [\int d\omega f(\Omega^+) f^*(\Omega^-) B_\Omega(\omega,L)\exp(i\omega t)]. \tag{12}$$

Now we assume that the source has a power spectrum $|f(\overline{\Omega})|^2$ centered at frequency Ω with a width $\Delta\overline{\Omega}$ much greater than the width of the absolute value of B_Ω (ω,L). In the diffusion approximation, the order of the magnitude estimate of the width of B_Ω (ω,L) is $\Delta\omega \approx D_o / L^2 \approx \ell v / (3L^2)$, where D_o is the diffusion constant. This condition can be realized experimentally by using either short laser or acoustic pulses [14, 16]. With this assumption, the product $f(\Omega^+) f^*(\Omega^-)$ in Eq. (12) can be approximated by $|f(\Omega)|^2$, and the normalized intensity profile becomes

$$\bar{I}(t,L) = \frac{\int d\omega B_\Omega(\omega,L)\exp(i\omega t)}{2\pi B_\Omega(0,L)}. \tag{13}$$

It is worth noting that the integration bounds in Eq. (13) are from $-\infty$ to ∞. Due to the symmetry relation $B_\Omega(-\omega,L) = B_\Omega^*(\omega,L)$, the intensity profile given in Eq. (13) is real.

3. Results and Discussion

In the absence of absorption, we have numerically solved Eq. (8) for $B_\Omega(\omega,L)$ at various ratios of $L/\ell(\Omega) = 2, 3, 4, 6,$ and 8 in the units of v=1. The intensity profiles $\bar{I}[t,L/\ell(\Omega)]$ are plotted in Fig. 1. In obtaining these curves, we have subtracted the coherent source term of the wave, which is the exponential term in Eq. (8), from the final solution of Eq. (8). This source term decays exponentially with the slab width L. The position of the initial rise of each curve in Fig. 1 corresponds to the ballistic travel time, which is equal to $L/v = L$. The sharp rise in intensity in the case of L = 2 and 3 represents the coherent part of the wave. The long tail comes from multiple scatterings.

As the sample becomes thicker and thicker, the coherent part diminishes while the incoherent part picks up and the wave becomes more diffusive. The coherent part disappear when $L/\ell(\Omega) \cong 5$. The transition from ballistic to diffusive is clearly shown. Our results of Fig. 1 are consistent with the experimentally observed transmitted light profiles scattered by latex beads in water [14]. The plot of the peak position vs. sample

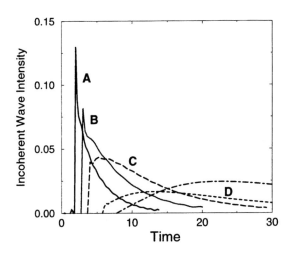

Figure 1. Temporal profile of incoherent wave intensity for various sample thickness $L = 2, 3, 4, 6$, and 8 (curves A, B, C, D, and E, respectively) in units of $\ell = 1$, $v = 1$.

thickness is given in Fig. 2 in log-log plot. It is quite apparent this curve consists of two distinct segments. The slope of the first segment $L/\ell(\Omega) \leq 3$ is one, while the slope of the second segment $L/\ell(\Omega) \geq 3$ is two. This change of slope again demonstrates the crossover from ballistic to diffusive behaviors. The smallness of the dividing ratio $L/\ell(\Omega) \cong 3$ may explain the experimental data of sample size dependence of the total transmitted light intensity reported in Ref. [15], where the deviation of diffusion approximation becomes noticeable only when $L/\ell(\Omega) \leq 4$. As to other quantities, such as spatial [15] or temporal [14] intensity profiles, the deviation from the diffusion approximation occurs at a larger value of $L/\ell(\Omega)$. This becomes clear if we compare the entire frequency correlation function calculated by the present method with the diffusion theory. This is shown in Fig. 3, where the vertical axis is the absolute value of the normalized frequency correlation function $|B_\Omega(\omega,L)/B_\Omega(0,L)|$, and the horizontal axis is the scaled variable $x = \alpha L$ with $\alpha = (3\omega/2)^{0.5}$ in units of $v = 1$ and $\ell = 1$. The solid curve of Fig. 3 is the plot of the diffusion result [4,9]:

$$\left| \frac{B(\omega,L)}{B(0,L)} \right| = \left(\frac{L}{z_0} \right) \left(\frac{\cosh(2z_0\alpha) - \cos(2z_0\alpha)}{\cosh(2L\alpha) - \cos(2L\alpha)} \right)^{0.5}, \tag{14}$$

for $L \geq 5$. In the above equation, we have omitted the subscript Ω in the function B. z_0 is the extrapolation length, which has a value around 0.66-0.71 ℓ in the absence of internal reflections. In Fig. 3 we plot the results of calculations for $L = 2$ (heavy circle),

720

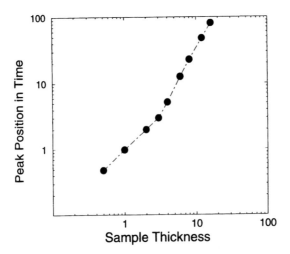

Figure 2. The peak position in the temporal profile of the incoherent wave intensity as a function of sample size in units of $\ell = 1$ and $v = 1$.

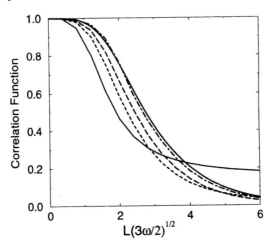

Figure 3. The absolute value of the normalized frequency correlation function as a function of the scaled variable $(3\omega/2)^{0.5}L$ in units of $\ell = 1$ and $v = 1$ for $L = 2$ (heavy circles), 5 (dashed), 10 (long-dashed), and 30 (dot-dashed). The solid curve is the diffusion result for $L \geq 5$.

5 (dashed curve), 10 (long-dashed curve), and 20 (dot-dashed curve). It is apparent the present method converges to the diffusion result in the large L limit. By using QE. (13), we have also computed the corresponding temporal intensity profiles and compared them with those of the diffusion result. It turns out that at $L = 20$ our calculation reproduces the diffusion result for the entire profile, except for some small deviation in the tail region of the initial rise. This is due to the small discrepancy in the correlation function in the large ω region. Some noticeable discrepancy in the temporal profile between the present calculations and the diffusion result arises when $L \leq 10$. This again is consistent with the experimentally observed transmitted light profiles scattered by latex

beads in water [14]. Thus, from the temporal profile of the transmitted intensity, the length required to crossover to the diffusion result is on the order of 10, which is much greater than the crossover region $L/\ell(\Omega) \cong 3\text{-}5$ predicted from the decay of the coherent part of the wave.

We have also used this method to study the frequency correlation function of the transmitted sound through samples comprised of glass beads [16]. Given the experimental values of the phase velocity v =1.3 mm/μsec and absorption time τ_a = 12 μsec for short pulse centered around $\Omega/2\pi$ = 2.5 MHz, we calculated the frequency correlation function for three ratios of L/ℓ = 4.3, 6.7, and 13.4. These results are plotted in curves A, B, and C of Fig. 4, respectively. When experimental data are available, comparison between the calculated and measured frequency correlation function at two thicknesses can determine the mean free path, sound speed and the correlation functions at other thicknesses.

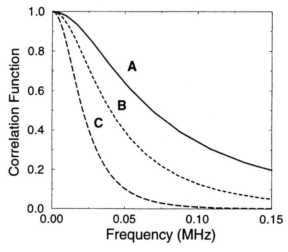

Figure 4. The absolute value of the normalized frequency correlation function as a function of frequency calculated from experimental parameters of v =1.7mm/μsec and τ_a = 12 μsec for L/ℓ = 5.5 (A), 8.3 (B), and 16.7 (C).

Finally, we would like to make two remarks. First, because of the use of point scatterers in the Bethe-Salpeter equation, the transport mean free path ℓ equals the scattering mean free path ℓ_s in our problem. This is unphysical when the size of scatterers becomes comparable to the wavelength, a situation which occurrs in many experiments. The anisotropic scattering caused by the finite size of a scatterer normally gives rise to a larger transport mean free path, as the scattering is favored in the forward direction. Since ℓ is the length where the direction of propagation becomes random, it should provide a lower cutoff length for a diffusive behavior. In large length scale, the energy propagation is still diffusive with a diffusion constant proportional to ℓ. For the dynamical quantities, it is plausible to use ℓ as the unit length in measuring the onset of deviation from diffusion approximation. The experimental data indeed showed the deviation when L/ℓ is about 10 [14]. Since ballistic regime is bounded by a few ℓ_s, we have a larger region of transition if the scattering is anisotropic, i.e., $\ell_s < L < \ell$.

The second point is that the transition from ballistic to diffusive behavior has been studied formally in the past in the context of diffusing wave spectroscopy [17]. By taking into account the finite size of scatterers, it has been found that higher angular momentum eigenstates of a "two-particle" operator become dominant at short length scale rather than the lowest eigenstate of diffusion mode. The importance of these modes is to retain the direction character of energy transport. This is a signature of plane wave propagation in the ballistic regime. However, the quantity interested in this work was the field-field correlation function in time, $\langle \Psi(t)\Psi^*(0)\rangle$, arising from the Brownian motion of scatterers under a static source. This is different from the problem of pulse propagation we studied here. The quantity we are interested in is the time-resolved intensity response from a pulse source, i.e., $\langle \Psi(t)\Psi^*(t)\rangle$. The latter one requires the information of correlation between two different frequencies, while the former one involves only one frequency and the dynamics of the scatterers can be treated adiabatically [17].

REFERENCES

1. P. Sheng, *Introduction to Wave Scattering, Localization, and Mesoscopic Phenomena*, Academic Press, Boston (1995).
2. Photonic Band Gaps and Localization, C. M. Soukoulis, ed., Plenum, New York 1993).
3. M.P. van Albada and A. Lagendijk, Phys. Rev. Lett. **55**, 2692 (1985); P.E. Wolf and G. Maret, Phys. Rev. Lett. **55**, 2696 (1985).
4. J.F. de Boer and M.P. van Albada, and A Lagendijk, Phys. Rev. B **45**, 658 (1992).
5. A. Z. Genack, in *The Scattering and Localization of Classical Waves*, P. Sheng, ed., World Scientific Press, Singapore (1990), P. 207; and also S. Feng in the same volume P. 179.
6. D.J. Pine, D.A. Weitz, P. M. Chaikin, and E. Herbolzheimer, Phys. Rev. Lett. **60**, 1134 (1988).
7. E. Akkermans, P.E. Wolf, and R. Maynard, Phys. Rev. Lett. **56**, 1471 (1986); M.B. van der Mark, M.P. van Albada and A. Lagendijk, Phys. Rev. B**37**, 3575 (1988).
8. B. Shapiro, Phys. Rev. Lett. **57**, 2168 (1986).
9. R.Pnini and B. Shapiro, Phys. Rev. B **39**, 6986 [1989).
10. J.X. Zhu, D.J. Pine, and D.A. Weitz, Phys. Rev. A **44**, 3948 (1991).
11. A.Z. Genack, J.H. Li, N. Garcia, and A.A. Liayansky, in *Photonic Band Gaps and Localization*, C. M. Soukoulis, ed, Plenum Press, New York (1993), p. 23.
12. G. Bayer and T. Niederdrank, Phys. Rev. Lett. **70**, 3884 (1993).
13. *Proceedings of Photon Migration and Imaging in Random Media and Tissue*, B. Chance and R.R. Alfano, ed, Bellingham, Wash. USA : SPIE c1993.
14. K.M. Yoo, F. Liu, and R.R. Alfano, Phys. Rev. Lett. **64**, 2647 (1990); also Opt. Lett. **19**, 740 (1994).
15. J.H. Li, A.A. Lisyansky, T.D. Cheung, D. Livdan, and A. Z. Genack, Europhys. Lett. **22**, 675 (1993).
16. J.H. Page, H.P. Schriemer, A.E. Bailey, and D.A. Weitz, Phya. Rev. E, in press (1995).
17. F. C. Mackintosh and S. John, Phys. Rev. B **40**, 2383 (1989).

AUTHOR INDEX

SUBJECT INDEX